P9-AFP-521

THE INSECTS

An Outline of Entomology

Third Edition

THE INSECTS

AN OUTLINE OF ENTOMOLOGY

P.J. Gullan and P.S. Cranston

Department of Entomology, University of California, Davis, USA

With illustrations by
K. Hansen McInnes

BLACKWELL PUBLISHING
350 Main Street, Malden, MA 02148-5020, USA
9600 Garsington Road, Oxford OX4 2DQ, UK
550 Swanston Street, Carlton, Victoria 3053, Australia

First published 1994 by Chapman & Hall
Second edition published 2000 by Blackwell Science
Third edition published 2005 by Blackwell Publishing Ltd

6 2008

Library of Congress Cataloging-in-Publication Data

Gullan, P.J.
The insects: an outline of entomology/P.J. Gullan & P.S. Cranston; with illustrations by K. Hansen McInnes. – 3rd ed.
p. cm.
Includes bibliographical references and index.
ISBN 978-1-4051-1113-3 (hardback: alk. paper)
1. Insects. I. Cranston, P.S. II. Title.

QL463.G85 2004
595.7–dc22

2004000124

A catalogue record for this title is available from the British Library.

Set in 9/11pt Photina
by Graphicraft Limited, Hong Kong
Printed and bound in Singapore
by Fabulous Printers Pte Ltd

The publisher's policy is to use permanent paper from mills that operate a sustainable forestry policy, and which has been manufactured from pulp processed using acid-free and elementary chlorine-free practices. Furthermore, the publisher ensures that the text paper and cover board used have met acceptable environmental accreditation standards.

For further information on
Blackwell Publishing, visit our website:
http://www.blackwellpublishing.com

CONTENTS

Color plates fall between pp. 14 and 15

LIST OF COLOR PLATES

consumed by the flies and distributed after passing through the insects' guts (P.J. Gullan).

PLATE 4

4.1 A tree trunk and under-branch covered in silk galleries of the webspinner *Antipaluria urichi* (Embiidina: Clothodidae), from Trinidad (refer to Box 9.5) (J.S. Edgerly-Rooks).
4.2 A female webspinner of *Antipaluria urichi* defending the entrance of her gallery from an approaching male, from Trinidad (J.S. Edgerly-Rooks).
4.3 An adult stonefly, *Neoperla edmundsi* (Plecoptera: Perlidae), from Brunei, Borneo (P.J. Gullan).
4.4 A female thynnine wasp of *Zaspilothynnus trilobatus* (Hymenoptera: Tiphiidae) (on the right) compared with the flower of the sexually deceptive orchid *Drakaea glyptodon*, which attracts pollinating male wasps by mimicking the female wasp (see p. 282) (R. Peakall).
4.5 A male thynnine wasp of *Neozeloboria cryptoides* (Hymenoptera: Tiphiidae) attempting to copulate with the sexually deceptive orchid *Chiloglottis trapeziformis* (R. Peakall).
4.6 Pollination of mango flowers by a flesh fly, *Australopierretia australis* (Diptera: Sarcophagidae), in northern Australia (D.L. Anderson).
4.7 The wingless adult female of the whitemarked tussock moth, *Orgyia leucostigma* (Lepidoptera: Lymantriidae), from New Jersey, USA (D.C.F. Rentz).

PLATE 5

5.1 Mealybugs of an undescribed *Planococcus* species (Hemiptera: Pseudococcidae) on an *Acacia* stem attended by ants of a *Polyrhachis* species (Hymenoptera: Formicidae), coastal Western Australia (P.J. Gullan).
5.2 A camouflaged late-instar caterpillar of *Plesanemma fucata* (Lepidoptera: Geometridae) on a eucalypt leaf in eastern Australia (P.J. Gullan).
5.3 A female of the scorpionfly *Panorpa communis* (Mecoptera: Panorpidae) from the UK (P.H. Ward).
5.4 The huge queen termite (approximately 7.5 cm long) of *Odontotermes transvaalensis* (Isoptera: Termitidae: Macrotermitinae) surrounded by her king (mid front), soldiers, and workers, from the Transvaal in South Africa (J.A.L. Watson).
5.5 A parasitic *Varroa* mite (see p. 320) on a pupa of the bee *Apis cerana* (Hymenoptera: Apidae) in a hive from Irian Jaya, New Guinea (D.L. Anderson).
5.6 An adult moth of *Utetheisa ornatrix* (Lepidoptera: Arctiidae) emitting defensive froth containing pyrrolizidine alkaloids that it sequesters as a larva from its food plants, legumes of the genus *Crotalaria* (T. Eisner).
5.7 A snake-mimicking caterpillar of the spicebush swallowtail, *Papilio troilus* (Lepidoptera: Papilionidae), from New Jersey, USA (D.C.F. Rentz).

PLATE 6

6.1 The cryptic adult moths of four species of *Acronicta* (Lepidoptera: Noctuidae): *A. alni*, the alder moth (top left); *A. leporina*, the miller (top right); *A. aceris*, the sycamore (bottom left); and *A. psi*, the grey dagger (bottom right) (D. Carter and R.I. Vane-Wright).
6.2 Aposematic or mechanically protected caterpillars of the same four species of *Acronicta*: *A. alni* (top left); *A. leporina* (top right); *A. aceris* (bottom left); and *A. psi* (bottom right); showing the divergent appearance of the larvae compared with their drab adults (D. Carter and R.I. Vane-Wright).
6.3 A blister beetle, *Lytta polita* (Coleoptera: Meloidae), reflex-bleeding from the knee joints; the hemolymph contains the toxin cantharidin (sections 14.4.3 & 15.2.2) (T. Eisner).
6.4 One of Bates' mimicry complexes from the Amazon Basin involving species from three different lepidopteran families – *Methona confusa confusa* (Nymphalidae: Ithomiinae) (top), *Lycorea ilione ilione* (Nymphalidae: Danainae) (second from top), *Patia orise orise* (Pieridae) (second from bottom), and a day-flying moth of *Gazera heliconioides* (Castniidae) (R.I. Vane-Wright).
6.5 An aposematic beetle of the genus *Lycus* (Coleoptera: Lycidae) on the flower spike of *Cussonia* (Araliaceae) from South Africa (P.J. Gullan).
6.6 A mature cottony-cushion scale, *Icerya purchasi* (Hemiptera: Margarodidae), with a fully formed ovisac, on the stem of a native host plant from Australia (P.J. Gullan).
6.7 Adult male gypsy moth, *Lymantria dispar* (Lepidoptera: Lymantriidae), from New Jersey, USA (D.C.F. Rentz).

LIST OF BOXES

PREFACE TO THE THIRD EDITION

Since writing the earlier editions of this textbook, we have relocated from Canberra, Australia, to Davis, California, where we teach many aspects of entomology to a new cohort of undergraduate and graduate students. We have come to appreciate some differences which may be evident in this edition. We have retained the regional balance of case studies for an international audience. With globalization has come unwanted, perhaps unforeseen, consequences, including the potential worldwide dissemination of pest insects and plants. A modern entomologist must be aware of the global status of pest control efforts. These range from insect pests of specific origin, such as many vectors of disease of humans, animals, and plants, to noxious plants, for which insect natural enemies need to be sought. The quarantine entomologist must know, or have access to, global databases of pests of commerce. Successful strategies in insect conservation, an issue we cover for the first time in this edition, are found worldwide, although often they are biased towards Lepidoptera. Furthermore, all conservationists need to recognize the threats to natural ecosystems posed by introduced insects such as crazy, big-headed, and fire ants. Likewise, systematists studying the evolutionary relationships of insects cannot restrict their studies to a regional subset, but also need a global view.

Perhaps the most publicized entomological event since the previous edition of our text was the "discovery" of a new order of insects – named as Mantophasmatodea – based on specimens from 45-million-year-old amber and from museums, and then found living in Namibia (south-west Africa), and now known to be quite widespread in southern Africa. This finding of the first new order of insects described for many decades exemplifies several aspects of modern entomological research. First, existing collections from which mantophasmatid specimens initially were discovered remain important research resources; second, fossil specimens have significance in evolutionary studies; third, detailed comparative anatomical studies retain a fundamental importance in establishing relationships, even at ordinal level; fourth, molecular phylogenetics usually can provide unambiguous resolution where there is doubt about relationships based on traditional evidence.

The use of molecular data in entomology, notably (but not only) in systematic studies, has grown apace since our last edition. The genome provides a wealth of characters to complement and extend those obtained from traditional sources such as anatomy. Although analysis is not as unproblematic as was initially suggested, clearly we have developed an ever-improving understanding of the internal relationships of the insects as well as their relationships to other invertebrates. For this reason we have introduced a new chapter (Chapter 7) describing methods and results of studies of insect phylogeny, and portraying our current understanding of relationships. Chapter 8, also new, concerns our ideas on insect evolution and biogeography. The use of robust phylogenies to infer past evolutionary events, such as origins of flight, sociality, parasitic and plant-feeding modes of life, and biogeographic history, is one of the most exciting areas in comparative biology.

Another growth area, providing ever-more challenging ideas, is the field of molecular evolutionary development in which broad-scale resemblances (and unexpected differences) in genetic control of developmental processes are being uncovered. Notable studies provide evidence for identity of control for development of gills, wings, and other appendages across phyla. However, details of this field are beyond the scope of this textbook.

We retain the popular idea of presenting some tangential information in boxes, and have introduced seven new boxes: Box 1.1 Collected to extinction?; Box 1.2 Tramp ants and biodiversity; Box 1.3 Sustainable use of mopane worms; Box 4.3 Reception of communication molecules; Box 5.5 Egg-tending fathers – the giant water bugs; Box 7.1 Relationships of the Hexapoda to other Arthropoda; Box 14.2 Backpack bugs – dressed to kill?, plus a taxonomic box (Box 13.3) concerning the Mantophasmatodea (heel walkers).

We have incorporated some other boxes into the text, and lost some. The latter include what appeared to be a very neat example of natural selection in action, the peppered moth *Biston betularia*, whose melanic *carbonaria* form purportedly gained advantage in a sooty industrial landscape through its better crypsis from bird predation. This interpretation has been challenged lately, and we have reinterpreted it in Box 14.1 within an assessment of birds as predators of insects.

Our recent travels have taken us to countries in which insects form an important part of the human diet. In southern Africa we have seen and eaten mopane, and have introduced a box to this text concerning the sustainable utilization of this resource. Although we have tried several of the insect food items that we mention in the opening chapter, and encourage others to do so, we make no claims for tastefulness. We also have visited New Caledonia, where introduced ants are threatening the native fauna. Our concern for the consequences of such worldwide ant invasives, that are particularly serious on islands, is reflected in Box 1.2.

Once again we have benefited from the willingness of colleagues to provide us with up-to-date information and to review our attempts at synthesizing their research. We are grateful to Mike Picker for helping us with Mantophasmatodea and to Lynn Riddiford for assisting with the complex new ideas concerning the evolution of holometabolous development. Matthew Terry and Mike Whiting showed us their unpublished phylogeny of the Polyneoptera, from which we derived part of Fig. 7.2. Bryan Danforth, Doug Emlen, Conrad Labandeira, Walter Leal, Brett Melbourne, Vince Smith, and Phil Ward enlightened us or checked our interpretations of their research speciality, and Chris Reid, as always, helped us with matters coleopterological and linguistic. We were fortunate that our updating of this textbook coincided with the issue of a compendious resource for all entomologists: *Encyclopedia of Insects*, edited by Vince Resh and Ring Cardé for Academic Press. The wide range of contributors assisted our task immensely: we cite their work under one header in the "Further reading" following the appropriate chapters in this book.

We thank all those who have allowed their publications, photographs, and drawings to be used as sources for Karina McInnes' continuing artistic endeavors. Tom Zavortink kindly pointed out several errors in the second edition. Inevitably, some errors of fact and interpretation remain, and we would be grateful to have them pointed out to us.

This edition would not have been possible without the excellent work of Katrina Rainey, who was responsible for editing the text, and the staff at Blackwell Publishing, especially Sarah Shannon, Cee Pike, and Rosie Hayden.

PREFACE TO THE SECOND EDITION

Since writing the first edition of this textbook, we have been pleasantly surprised to find that what we consider interesting in entomology has found a resonance amongst both teachers and students from a variety of countries. When invited to write a second edition we consulted our colleagues for a wish list, and have tried to meet the variety of suggestions made. Foremost we have retained the chapter sequence and internal arrangement of the book to assist those that follow its structure in their lecturing. However, we have added a new final (16th) chapter covering methods in entomology, particularly preparing and conserving a collection. Chapter 1 has been radically reorganized to emphasize the significance of insects, their immense diversity and their patterns of distribution. By popular request, the summary table of diagnostic features of the insect orders has been moved from Chapter 1 to the end pages, for easier reference. We have expanded insect physiology sections with new sections on tolerance of environmental extremes, thermoregulation, control of development and changes to our ideas on vision. Discussion of insect behaviour has been enhanced with more information on insect–plant interactions, migration, diapause, hearing and predator avoidance, "puddling" and sodium gifts. In the ecological area, we have considered functional feeding groups in aquatic insects, and enlarged the section concerning insect–plant interactions. Throughout the text we have incorporated new interpretations and ideas, corrected some errors and added extra terms to the glossary.

The illustrations by Karina McInnes that proved so popular with reviewers of the first edition have been retained and supplemented, especially with some novel chapter vignettes and additional figures for the taxonomic and collection sections. In addition, 41 colour photographs of colourful and cryptic insects going about their lives have been chosen to enhance the text.

The well-received boxes that cover self-contained themes tangential to the flow of the text are retained. With the assistance of our new publishers, we have more clearly delimited the boxes from the text. New boxes in this edition cover two resurging pests (the phylloxera aphid and *Bemisia* whitefly), the origins of the aquatic lifestyle, parasitoid host-detection by hearing, the molecular basis of development, chemically protected eggs, and the genitalia-inflating phalloblaster.

We have resisted some invitations to elaborate on the many physiological and genetic studies using insects – we accept a reductionist view of the world appeals to some, but we believe that it is the integrated whole insect that interacts with its environment and is subject to natural selection. Breakthroughs in entomological understanding will come from comparisons made within an evolutionary framework, not from the technique-driven insertion of genes into insect and/or host.

We acknowledge all those who assisted us with many aspects of the first edition (see Preface for first edition following) and it is with some regret that we admit that such a breadth of expertise is no longer available for consultation in one of our erstwhile research institutions. This is compensated for by the following friends and colleagues who reviewed new sections, provided us with advice, and corrected some of our errors. Entomology is a science in which collaboration remains the norm – long may it continue. We are constantly surprised at the rapidity of freely given advice, even to electronic demands: we hope we haven't abused the rapidity of communication. Thanks to, in alphabetical order: Denis Anderson – varroa mites; Andy Austin – wasps and polydnaviruses; Jeff Bale – cold tolerance; Eldon Ball – segment development; Paul Cooper – physiological updates; Paul De Barro – *Bemisia*; Hugh Dingle – migration; Penny Greenslade – collembola facts; Conrad Labandeira – fossil insects; Lisa Nagy – molecular basis for limb development; Rolf Oberprieler – edible insects; Chris Reid – reviewing

Chapter 1 and coleopteran factoids; Murray Upton – reviewing collecting methods; Lars-Ove Wikars – mycangia information and illustration; Jochen Zeil – vision. Dave Rentz supplied many excellent colour photographs, which we supplemented with some photos by Denis Anderson, Janice Edgerly-Rooks, Tom Eisner, Peter Menzel, Rod Peakall, Dick Vane-Wright, Peter Ward, Phil Ward and the late Tony Watson. Lyn Cook and Ben Gunn provided help with computer graphics. Many people assisted by supplying current names or identifications for particular insects, including from photographs. Special thanks to John Brackenbury, whose photograph of a soldier beetle in preparation for flight (from Brackenbury, 1990) provided the inspiration for the cover centerpiece.

When we needed a break from our respective offices in order to read and write, two Dons, Edward and Bradshaw, provided us with some laboratory space in the Department of Zoology, University of Western Australia, which proved to be rather too close to surf, wineries and wildflower sites – thank you anyway.

It is appropriate to thank Ward Cooper of the late Chapman & Hall for all that he did to make the first edition the success that it was. Finally, and surely not least, we must acknowledge that there would not have been a second edition without the helping hand put out by Blackwell Science, notably Ian Sherman and David Frost, following one of the periodic spasms in scientific publishing when authors (and editors) realize their minor significance in the "commercial" world.

PREFACE AND ACKNOWLEDGMENTS FOR FIRST EDITION

Insects are extremely successful animals and they affect many aspects of our lives, despite their small size. All kinds of natural and modified, terrestrial and aquatic, ecosystems support communities of insects that present a bewildering variety of life-styles, forms and functions. Entomology covers not only the classification, evolutionary relationships and natural history of insects, but also how they interact with each other and the environment. The effects of insects on us, our crops and domestic stock, and how insect activities (both deleterious and beneficial) might be modified or controlled, are amongst the concerns of entomologists.

The recent high profile of biodiversity as a scientific issue is leading to increasing interest in insects because of their astonishingly high diversity. Some calculations suggest that the species richness of insects is so great that, to a near approximation, all organisms can be considered to be insects. Students of biodiversity need to be versed in entomology.

We, the authors, are systematic entomologists teaching and researching insect identification, distribution, evolution and ecology. Our study insects belong to two groups – scale insects and midges – and we make no apologies for using these, our favourite organisms, to illustrate some points in this book.

This book is not an identification guide, but addresses entomological issues of a more general nature. We commence with the significance of insects, their internal and external structure, and how they sense their environment, followed by their modes of reproduction and development. Succeeding chapters are based on major themes in insect biology, namely the ecology of ground-dwelling, aquatic and plant-feeding insects, and the behaviours of sociality, predation and parasitism, and defence. Finally, aspects of medical and veterinary entomology and the management of insect pests are considered.

Those to whom this book is addressed, namely students contemplating entomology as a career, or studying insects as a subsidiary to specialized disciplines such as agricultural science, forestry, medicine or veterinary science, ought to know something about insect systematics – this is the framework for scientific observations. However, we depart from the traditional order-by-order systematic arrangement seen in many entomological textbooks. The systematics of each insect order are presented in a separate section following the ecological–behavioural chapter appropriate to the predominant biology of the order. We have attempted to keep a phylogenetic perspective throughout, and one complete chapter is devoted to insect phylogeny, including examination of the evolution of several key features.

We believe that a picture is worth a thousand words. All illustrations were drawn by Karina Hansen McInnes, who holds an Honours degree in Zoology from the Australian National University, Canberra. We are delighted with her artwork and are grateful for her hours of effort, attention to detail and skill in depicting the essence of the many subjects that are figured in the following pages. Thank you Karina.

This book would still be on the computer without the efforts of John Trueman, who job-shared with Penny in second semester 1992. John delivered invertebrate zoology lectures and ran lab classes while Penny revelled in valuable writing time, free from undergraduate teaching. Aimorn Stewart also assisted Penny by keeping her research activities alive during book preparation and by helping with labelling of figures. Eva Bugledich acted as a library courier and brewed hundreds of cups of coffee.

The following people generously reviewed one or more chapters for us: Andy Austin, Tom Bellas, Keith Binnington, Ian Clark, Geoff Clarke, Paul Cooper, Kendi Davies, Don Edward, Penny Greenslade, Terry Hillman, Dave McCorquodale, Rod Mahon, Dick Norris, Chris Reid, Steve Shattuck, John Trueman and Phil Weinstein. We also enjoyed many discussions on hymenopteran phylogeny and biology with Andy. Tom sorted out our chemistry and Keith gave expert advice on insect cuticle. Paul's broad knowledge of insect physiology was absolutely invaluable. Penny put us straight with springtail facts. Chris' entomological knowledge, especially on beetles, was a constant source of information. Steve patiently answered our endless questions on ants. Numerous other people read and commented on sections of chapters or provided advice or helpful discussion on particular entomological topics. These people included John Balderson, Mary Carver, Lyn Cook, Jane Elek, Adrian Gibbs, Ken Hill, John Lawrence, Chris Lyal, Patrice Morrow, Dave Rentz, Eric Rumbo, Vivienne Turner, John Vranjic and Tony Watson. Mike Crisp assisted with checking on current host-plant names. Sandra McDougall inspired part of Chapter 15. Thank you everyone for your many comments which we have endeavoured to incorporate as far as possible, for your criticisms which we hope we have answered, and for your encouragement.

We benefited from discussions concerning published and unpublished views on insect phylogeny (and fossils), particularly with Jim Carpenter, Mary Carver, Niels Kristensen, Jarmila Kukalová-Peck and John Trueman. Our views are summarized in the phylogenies shown in this book and do not necessarily reflect a consensus of our discussants' views (this was unattainable).

Our writing was assisted by Commonwealth Scientific and Industrial Research Organization (CSIRO) providing somewhere for both of us to work during the many weekdays, nights and weekends during which this book was prepared. In particular, Penny managed to escape from the distractions of her university position by working in CSIRO. Eventually, however, everyone discovered her whereabouts. The Division of Entomology of the CSIRO provided generous support: Carl Davies gave us driving lessons on the machine that produced reductions of the figures, and Sandy Smith advised us on labelling. The Division of Botany and Zoology of the Australian National University also provided assistance in aspects of the book production: Aimorn Stewart prepared the SEMs from which Fig. 4.7 was drawn, and Judy Robson typed the labels for some of the figures.

Chapter 1

THE IMPORTANCE, DIVERSITY, AND CONSERVATION OF INSECTS

Charles Darwin inspecting beetles collected during the voyage of the *Beagle*. (After various sources, especially Huxley & Kettlewell 1965 and Futuyma 1986.)

Curiosity alone concerning the identities and lifestyles of the fellow inhabitants of our planet justifies the study of insects. Some of us have used insects as totems and symbols in spiritual life, and we portray them in art and music. If we consider economic factors, the effects of insects are enormous. Few human societies lack honey, provided by bees (or specialized ants). Insects pollinate our crops. Many insects share our houses, agriculture, and food stores. Others live on us, our domestic pets, or our livestock, and yet more visit to feed on us where they may transmit disease. Clearly, we should understand these pervasive animals.

Although there are millions of kinds of insects, we do not know exactly (or even approximately) how many. This ignorance of how many organisms we share our planet with is remarkable considering that astronomers have listed, mapped, and uniquely identified a comparable diversity of galactic objects. Some estimates, which we discuss in detail below, imply that the species richness of insects is so great that, to a near approximation, all organisms can be considered to be insects. Although dominant on land and in freshwater, few insects are found beyond the tidal limit of oceans.

In this opening chapter, we outline the significance of insects and discuss their diversity and classification and their roles in our economic and wider lives. First, we outline the field of entomology and the role of entomologists, and then introduce the ecological functions of insects. Next, we explore insect diversity, and then discuss how we name and classify this immense diversity. Sections follow in which we consider past and some continuing cultural and economic aspects of insects, their aesthetic and tourism appeal, and their importance as foods for humans and animals. We conclude with a review of the conservation significance of insects.

1.1 WHAT IS ENTOMOLOGY?

Entomology is the study of insects. Entomologists, the people who study insects, observe, collect, rear, and experiment with insects. Research undertaken by entomologists covers the total range of biological disciplines, including evolution, ecology, behavior, anatomy, physiology, biochemistry, and genetics. The unifying feature is that the study organisms are insects. Biologists work with insects for many reasons: ease of culturing in a laboratory, rapid population turnover, and availability of many individuals are important factors. The minimal ethical concerns regarding responsible experimental use of insects, as compared with vertebrates, are a significant consideration.

Modern entomological study commenced in the early 18th century when a combination of rediscovery of the classical literature, the spread of rationalism, and availability of ground-glass optics made the study of insects acceptable for the thoughtful privately wealthy. Although people working with insects hold professional positions, many aspects of the study of insects remain suitable for the hobbyist. Charles Darwin's initial enthusiasm in natural history was as a collector of beetles (as shown in the vignette for this chapter). All his life he continued to study insect evolution and communicate with amateur entomologists throughout the world. Much of our present understanding of worldwide insect diversity derives from studies of non-professionals. Many such contributions come from collectors of attractive insects such as butterflies and beetles, but others with patience and ingenuity continue the tradition of Henri Fabre in observing close-up activities of insects. We can discover much of scientific interest at little expense concerning the natural history of even "well known" insects. The variety of size, structure, and color in insects (see Plates 1.1–1.3, facing p. 14) is striking, whether depicted in drawing, photography, or film.

A popular misperception is that professional entomologists emphasize killing or at least controlling insects, but in fact entomology includes many positive aspects of insects because their benefits to the environment outweigh their harm.

1.2 THE IMPORTANCE OF INSECTS

We should study insects for many reasons. Their ecologies are incredibly variable. Insects may dominate food chains and food webs in both volume and numbers. Feeding specializations of different insect groups include ingestion of detritus, rotting materials, living and dead wood, and fungus (Chapter 9), aquatic filter feeding and grazing (Chapter 10), herbivory (= phytophagy), including sap feeding (Chapter 11), and predation and parasitism (Chapter 13). Insects may live in water, on land, or in soil, during part or all of their lives. Their lifestyles may be solitary, gregarious, subsocial, or highly social (Chapter 12). They may be conspicuous, mimics of other objects, or concealed (Chapter 14), and may be active by day or by night. Insect life cycles (Chapter 6) allow survival under a wide range of conditions, such as extremes of heat and cold, wet and dry, and unpredictable climates.

Insects are essential to the following ecosystem functions:

- nutrient recycling, via leaf-litter and wood degradation, dispersal of fungi, disposal of carrion and dung, and soil turnover;
- plant propagation, including pollination and seed dispersal;
- maintenance of plant community composition and structure, via phytophagy, including seed feeding;
- food for insectivorous vertebrates, such as many birds, mammals, reptiles, and fish;
- maintenance of animal community structure, through transmission of diseases of large animals, and predation and parasitism of smaller ones.

Each insect species is part of a greater assemblage and its loss affects the complexities and abundance of other organisms. Some insects are considered "keystones" because loss of their critical ecological functions could collapse the wider ecosystem. For example, termites convert cellulose in tropical soils (section 9.1), suggesting that they are keystones in tropical soil structuring. In aquatic ecosystems, a comparable service is provided by the guild of mostly larval insects that breaks down and releases the nutrients from wood and leaves derived from the surrounding terrestrial environment.

Insects are associated intimately with our survival, in that certain insects damage our health and that of our domestic animals (Chapter 15) and others adversely affect our agriculture and horticulture (Chapter 16). Certain insects greatly benefit human society, either by providing us with food directly or by contributing to our food or materials that we use. For example, honey bees provide us with honey but are also valuable agricultural pollinators worth an estimated several billion US$ annually in the USA. Estimates of the value of non-honey-bee pollination in the USA could be as much as $5–6 billion per year. The total value of pollination services rendered by all insects globally has been estimated to be in excess of $100 billion annually (2003 valuation). Furthermore, valuable services, such as those provided by predatory beetles and bugs or parasitic wasps that control pests, often go unrecognized, especially by city-dwellers.

Insects contain a vast array of chemical compounds, some of which can be collected, extracted, or synthesized for our use. Chitin, a component of insect cuticle, and its derivatives act as anticoagulants, enhance wound and burn healing, reduce serum cholesterol, serve as non-allergenic drug carriers, provide strong biodegradable plastics, and enhance removal of pollutants from waste water, to mention just a few developing applications. Silk from the cocoons of silkworm moths, *Bombyx mori*, and related species has been used for fabric for centuries, and two endemic South African species may be increasing in local value. The red dye cochineal is obtained commercially from scale insects of *Dactylopius coccus* cultured on *Opuntia* cacti. Another scale insect, the lac insect *Kerria lacca*, is a source of a commercial varnish called shellac. Given this range of insect-produced chemicals, and accepting our ignorance of most insects, there is a high likelihood of finding novel chemicals.

Insects provide more than economic or environmental benefits; characteristics of certain insects make them useful models for understanding general biological processes. For instance, the short generation time, high fecundity, and ease of laboratory rearing and manipulation of the vinegar fly, *Drosophila melanogaster*, have made it a model research organism. Studies of *D. melanogaster* have provided the foundations for our understanding of genetics and cytology, and these flies continue to provide the experimental materials for advances in molecular biology, embryology, and development. Outside the laboratories of geneticists, studies of social insects, notably hymenopterans such as ants and bees, have allowed us to understand the evolution and maintenance of social behaviors such as altruism (section 12.4.1). The field of sociobiology owes its existence to entomologists' studies of social insects. Several theoretical ideas in ecology have derived from the study of insects. For example, our ability to manipulate the food supply (grains) and number of individuals of flour beetles (*Tribolium* spp.) in culture, combined with their short life history (compared to mammals, for example), gave insights into mechanisms regulating populations. Some early holistic concepts in ecology, for example ecosystem and niche, came from scientists studying freshwater systems where insects dominate. Alfred Wallace (depicted in the vignette of Chapter 17), the independent and contemporaneous discoverer with Charles Darwin of the theory of evolution by natural selection, based his ideas on observations of tropical insects. Theories concerning the many forms of mimicry and sexual selection have been derived from observations of insect behavior, which continue to be investigated by entomologists.

Lastly, the sheer numbers of insects means that their impact upon the environment, and hence our lives, is highly significant. Insects are the major component of macroscopic biodiversity and, for this reason alone, we should try to understand them better.

1.3 INSECT BIODIVERSITY

1.3.1 The described taxonomic richness of insects

Probably slightly over one million species of insects have been described, that is, have been recorded in a taxonomic publication as "new" (to science that is), accompanied by description and often with illustrations or some other means of recognizing the particular insect species (section 1.4). Since some insect species have been described as new more than once, due to failure to recognize variation or through ignorance of previous studies, the actual number of described species is uncertain.

The described species of insects are distributed unevenly amongst the higher taxonomic groupings called orders (section 1.4). Five "major" orders stand out for their high species richness, the beetles (Coleoptera), flies (Diptera), wasps, ants, and bees (Hymenoptera), butterflies and moths (Lepidoptera), and the true bugs (Hemiptera). J.B.S. Haldane's jest – that "God" (evolution) shows an inordinate "fondness" for beetles – appears to be confirmed since they comprise almost 40% of described insects (more than 350,000 species). The Hymenoptera have nearly 250,000 described species, with the Diptera and Lepidoptera having between 125,000 and 150,000 species, and Hemiptera approaching 95,000. Of the remaining orders of living insects, none exceed the 20,000 described species of the Orthoptera (grasshoppers, locusts, crickets, and katydids). Most of the "minor" orders have from some hundreds to a few thousands of described species. Although an order may be described as "minor", this does not mean that it is insignificant – the familiar earwig belongs to an order (Dermaptera) with less than 2000 described species and the ubiquitous cockroaches belong to an order (Blattodea) with only 4000 species. Nonetheless, there are only twice as many species described in Aves (birds) as in the "small" order Blattodea.

1.3.2 The estimated taxonomic richness of insects

Surprisingly, the figures given above, which represent the cumulative effort by many insect taxonomists from all parts of the world over some 250 years, appear to represent something less than the true species richness of the insects. Just how far short is the subject of continuing speculation. Given the very high numbers and the patchy distributions of many insects in time and space, it is impossible in our time-scales to inventory (count and document) all species even for a small area. Extrapolations are required to estimate total species richness, which range from some three million to as many as 80 million species. These various calculations either extrapolate ratios for richness in one taxonomic group (or area) to another unrelated group (or area), or use a hierarchical scaling ratio, extrapolated from a subgroup (or subordinate area) to a more inclusive group (or wider area).

Generally, ratios derived from temperate : tropical species numbers for well-known groups such as vertebrates provide rather conservatively low estimates if used to extrapolate from temperate insect taxa to essentially unknown tropical insect faunas. The most controversial estimation, based on hierarchical scaling and providing the highest estimated total species numbers, was an extrapolation from samples from a single tree species to global rainforest insect species richness. Sampling used insecticidal fog to assess the little-known fauna of the upper layers (the canopy) of neotropical rainforest. Much of this estimated increase in species richness was derived from arboreal beetles (Coleoptera), but several other canopy-dwelling groups were much more numerous than believed previously. Key factors in calculating tropical diversity included identification of the number of beetle species found, estimation of the proportion of novel (previously unseen) groups, allocation to feeding groups, estimation of the degree of host-specificity to the surveyed tree species, and the ratio of beetles to other arthropods. Certain assumptions have been tested and found to be suspect: notably, host-plant specificity of herbivorous insects, at least in Papua New Guinean tropical forest, seems very much less than estimated early in this debate.

Estimates of global insect diversity calculated from experts' assessments of the proportion of undescribed versus described species amongst their study insects tend to be comparatively low. Belief in lower numbers of species comes from our general inability to confirm the prediction, which is a logical consequence of the high species-richness estimates, that insect samples ought to contain very high proportions of previously unrecognized and/or undescribed ("novel") taxa. Obviously any expectation of an even spread of novel species is unrealistic, since some groups and regions of the world are poorly known compared to others. However, amongst the minor (less species-rich) orders there is little or no scope for dramatically increased, unrecognized species richness. Very high levels of novelty, if they exist, realistically could only be amongst the Coleoptera, drab-colored Lepidoptera, phytophagous Diptera, and parasitic Hymenoptera.

Some (but not all) recent re-analyses tend towards lower estimates derived from taxonomists' calculations and extrapolations from regional sampling rather than those derived from ecological scaling: a figure of between four and six million species of insects appears realistic.

1.3.3 The location of insect species richness

The regions in which additional undescribed insect species might occur (i.e. up to an order of magnitude greater number of novel species than described) cannot be in the northern hemisphere, where such hidden diversity in the well-studied faunas is unlikely. For example, the British Isles inventory of about 22,500 species of insects is likely to be within 5% of being complete and the 30,000 or so described from Canada must represent over half of the total species. Any hidden diversity is not in the Arctic, with some 3000 species present in the American Arctic, nor in Antarctica, the southern polar mass, which supports a bare handful of insects. Evidently, just as species-richness patterns are uneven across groups, so too is their geographic distribution.

Despite the lack of necessary local species inventories to prove it, tropical species richness appears to be much higher than that of temperate areas. For example, a single tree surveyed in Peru produced 26 genera and 43 species of ants: a tally that equals the total ant diversity from all habitats in Britain. Our inability to be certain about finer details of geographical patterns stems in part from the inverse relationship between the distribution of entomologists interested in biodiversity issues (the temperate northern hemisphere) and the centers of richness of the insects themselves (the tropics and southern hemisphere).

Studies in tropical American rainforests suggest much undescribed novelty in insects comes from the beetles, which provided the basis for the original high richness estimate. Although beetle dominance may be true in places such as the Neotropics, this might be an artifact of the collection and research biases of entomologists. In some well-studied temperate regions such as Britain and Canada, species of true flies (Diptera) appear to outnumber beetles. Studies of canopy insects of the tropical island of Borneo have shown that both Hymenoptera and Diptera can be more species rich at particular sites than the Coleoptera. Comprehensive regional inventories or credible estimates of insect faunal diversity may eventually tell us which order of insects is globally most diverse.

Whether we estimate 30–80 million species or an order of magnitude less, insects constitute at least half of global species diversity (Fig. 1.1). If we consider only life on land, insects comprise an even greater proportion of extant species, since the radiation of insects is a predominantly terrestrial phenomenon. The relative contribution of insects to global diversity will be somewhat lessened if marine diversity, to which insects make a negligible contribution, actually is higher than currently understood.

1.3.4 Some reasons for insect species richness

Whatever the global estimate is, insects surely are remarkably speciose. This high species richness has been attributed to several factors. The small size of insects, a limitation imposed by their method of gas exchange via tracheae, is an important determinant. Many more niches exist in any given environment for small organisms than for large organisms. Thus, a single acacia tree, that provides one meal to a giraffe, may support the complete life cycle of dozens of insect species; a lycaenid butterfly larva chews the leaves, a bug sucks the stem sap, a longicorn beetle bores into the wood, a midge galls the flower buds, a bruchid beetle destroys the seeds, a mealybug sucks the root sap, and several wasp species parasitize each host-specific phytophage. An adjacent acacia of a different species feeds the same giraffe but may have a very different suite of phytophagous insects. The environment can be said to be more fine-grained from an insect perspective compared to that of a mammal or bird.

Small size alone is insufficient to allow exploitation of this environmental heterogeneity, since organisms must be capable of recognizing and responding to environmental differences. Insects have highly organized

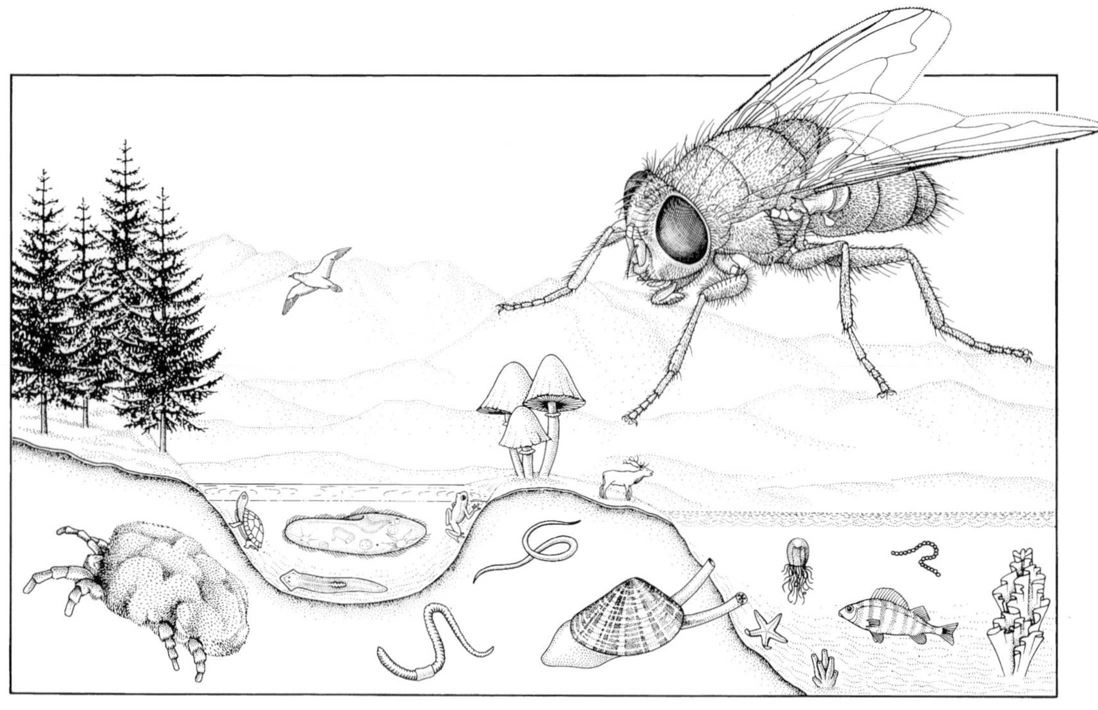

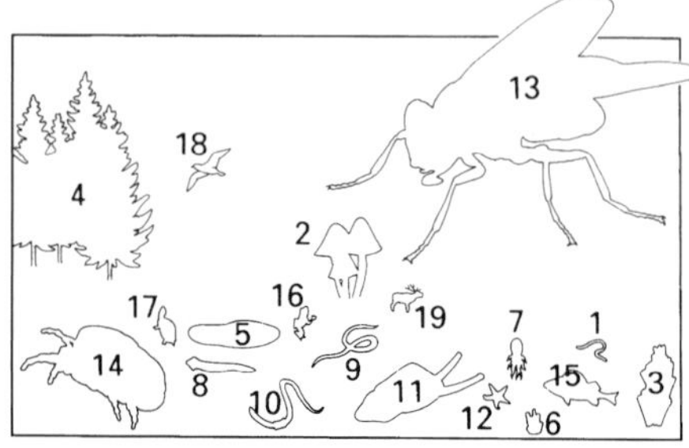

1 Prokaryotes
2 Fungi
3 Algae
4 Plantae (multicellular plants)
5 Protozoa
6 Porifera (sponges)
7 Cnidaria (jellyfish, corals, etc.)
8 Platyhelminthes (flatworms)
9 Nematoda (roundworms)
10 Annelida (earthworms, leeches, etc.)
11 Mollusca (snails, bivalves, octopus, etc.)
12 Echinodermata (starfish, sea urchins, etc.)
13 Insecta
14 Non-insect Arthropoda
15 Pisces (fish)
16 Amphibia (frogs, salamanders, etc.)
17 Reptilia (snakes, lizards, turtles)
18 Aves (birds)
19 Mammalia (mammals)

Fig. 1.1 Speciescape, in which the size of individual organisms is approximately proportional to the number of described species in the higher taxon that it represents. (After Wheeler 1990.)

sensory and neuro-motor systems more comparable to those of vertebrate animals than other invertebrates. However, insects differ from vertebrates both in size and in how they respond to environmental change. Generally, vertebrate animals are longer lived than insects and individuals can adapt to change by some degree of learning. Insects, on the other hand, normally respond to, or cope with, altered conditions (e.g. the application of insecticides to their host plant) by genetic change between generations (e.g. leading to insecticide-resistant insects). High genetic heterogeneity or elasticity within insect species allows persistence in the face of environmental change. Persistence exposes species to processes that promote speciation, predominantly involving phases of range expansion and/or subsequent fragmentation. Stochastic processes (genetic drift) and/or selection pressures provide the genetic alterations that may become fixed in spatially or temporally isolated populations.

Insects possess characteristics that expose them to other potential diversifying influences that enhance their species richness. Interactions between certain groups of insects and other organisms, such as plants in the case of herbivorous insects, or hosts for parasitic insects, may promote the genetic diversification of eater and eaten. These interactions are often called coevolutionary and are discussed in more detail in Chapters 11 and 13. The reciprocal nature of such interactions may speed up evolutionary change in one or both partners or sets of partners, perhaps even leading to major radiations in certain groups. Such a scenario involves increasing specialization of insects at least on plant hosts. Evidence from phylogenetic studies suggests that this has happened – but also that generalists may arise from within a specialist radiation, perhaps after some plant chemical barrier has been overcome. Waves of specialization followed by breakthrough and radiation must have been a major factor in promoting the high species richness of phytophagous insects.

Another explanation for the high species numbers of insects is the role of sexual selection in the diversification of many insects. The propensity of insects to become isolated in small populations (because of the fine scale of their activities) in combination with sexual selection (section 5.3) may lead to rapid alteration in intra-specific communication. When (or if) the isolated population rejoins the larger parental population, altered sexual signaling deters hybridization and the identity of each population (incipient species) is maintained in sympatry. This mechanism is seen to be much more rapid than genetic drift or other forms of selection, and need involve little if any differentiation in terms of ecology or non-sexual morphology and behavior.

Comparisons amongst and between insects and their close relatives suggest reasons for insect diversity. We can ask what are the shared characteristics of the most speciose insect orders, the Coleoptera, Hymenoptera, Diptera, and Lepidoptera? Which features of insects do other arthropods, such as arachnids (spiders, mites, scorpions, and their allies) lack? No simple explanation emerges from such comparisons; probably design features, flexible life-cycle patterns and feeding habits play a part (some of these factors are explored in Chapter 8). In contrast to the most speciose insect groups, arachnids lack winged flight, complete transformation of body form during development (metamorphosis) and dependence on specific food organisms, and are not phytophagous. Exceptionally, mites, the most diverse and abundant of arachnids, have many very specific associations with other living organisms.

High persistence of species or lineages or the numerical abundance of individual species are considered as indicators of insect success. However, insects differ from vertebrates by at least one popular measure of success: body size. Miniaturization is the insect success story: most insects have body lengths of 1–10 mm, with a body length around 0.3 mm of mymarid wasps (parasitic on eggs of insects) being unexceptional. At the other extreme, the greatest wingspan of a living insect belongs to the tropical American owlet moth, *Thysania agrippina* (Noctuidae), with a span of up to 30 cm, although fossils show that some insects were appreciably larger than their extant relatives. For example, an Upper Carboniferous silverfish, *Ramsdelepidion schusteri* (Zygentoma), had a body length of 6 cm compared to a modern maximum of less than 2 cm. The wingspans of many Carboniferous insects exceeded 45 cm, and a Permian dragonfly, *Meganeuropsis americana* (Protodonata), had a wingspan of 71 cm. Notably amongst these large insects, the great size comes predominantly with a narrow, elongate body, although one of the heaviest extant insects, the 16 cm long hercules beetle *Dynastes hercules* (Scarabaeidae), is an exception in having a bulky body.

Barriers to large size include the inability of the tracheal system to diffuse gases across extended distances from active muscles to and from the external environment (Box 3.2). Further elaborations of the tracheal system would jeopardize water balance in a large insect. Most large insects are narrow and have not greatly extended the maximum distance between the external oxygen source and the muscular site of gaseous exchange, compared with smaller insects. A possible explanation for the gigantism of some Palaeozoic insects is considered in section 8.2.1.

In summary, many insect radiations probably depended upon (a) the small size of individuals, combined with (b) short generation time, (c) sensory and neuro-motor sophistication, (d) evolutionary interactions with plants and other organisms, (e) metamorphosis, and (f) mobile winged adults. The substantial time since the origin of each major insect group has allowed many opportunities for lineage diversification (Chapter 8). Present-day species diversity results from either higher rates of speciation (for which there is limited evidence) and/or lower rates of species extinction (higher persistence) than other organisms. The high species richness seen in some (but not all) groups in the tropics may result from the combination of higher rates of species formation with high accumulation in equable climates.

1.4 NAMING AND CLASSIFICATION OF INSECTS

The formal naming of insects follows the rules of nomenclature developed for all animals (plants have a slightly different system). Formal scientific names are required for unambiguous communication between all scientists, no matter what their native language. Vernacular (common) names do not fulfill this need: the same insects even may have different vernacular names amongst peoples that speak the same language. For instance, the British refer to "ladybirds", whereas the same coccinellid beetles are "ladybugs" to many people in the USA. Many insects have no vernacular name, or one common name is given to many species as if only one is involved. These difficulties are addressed by the Linnaean system, which provides every described species with two given names (a binomen). The first is the generic (genus) name, used for a usually broader grouping than the second name, which is the specific (species) name. These latinized names are always used together and are italicized, as in this book. The combination of generic and specific names provides each organism with a unique name. Thus, the name *Aedes aegypti* is recognized by any medical entomologist, anywhere, whatever the local name (and there are many) for this disease-transmitting mosquito. Ideally, all taxa should have such a latinized binomen, but in practice some alternatives may be used prior to naming formally (section 17.3.2).

In scientific publications, the species name often is followed by the name of the original describer of the species and perhaps the year in which the name first was published legally. In this textbook, we do not follow this practice but, in discussion of particular insects, we give the order and family names to which the species belongs. In publications, after the first citation of the combination of generic and species names in the text, it is common practice in subsequent citations to abbreviate the genus to the initial letter only (e.g. *A. aegypti*). However, where this might be ambiguous, such as for the two mosquito genera *Aedes* and *Anopheles*, the initial two letters *Ae.* and *An.* are used, as in Chapter 15.

Table 1.1 Taxonomic categories (obligatory categories are shown in **bold**).

Taxon category	Standard suffix	Example
Order		Hymenoptera
Suborder		Apocrita
Superfamily	-oidea	Apoidea
Family	-idae	Apidae
Subfamily	-inae	Apinae
Tribe	-ini	Apini
Genus		*Apis*
Subgenus		
Species		*A. mellifera*
Subspecies		*A. m. mellifera*

Various taxonomically defined groups, also called **taxa** (singular **taxon**), are recognized amongst the insects. As for all other organisms, the basic biological taxon, lying above the individual and population, is the species, which is both the fundamental nomenclatural unit in taxonomy and, arguably, a unit of evolution. Multi-species studies allow recognition of genera, which are discrete higher groups. In a similar manner, genera can be grouped into tribes, tribes into subfamilies, and subfamilies into families. The families of insects are placed in relatively large but easily recognized groups called orders. This hierarchy of ranks (or categories) thus extends from the species level through a series of "higher" levels of greater and greater inclusivity until all true insects are included in one class, the Insecta. There are standard suffixes for certain ranks in the taxonomic hierarchy, so that the rank of some group names can be recognized by inspection of the ending (Table 1.1).

Depending on the classification system used, some 30 orders of Insecta are recognized. Differences arise principally because there are no fixed rules for deciding the taxonomic ranks referred to above – only general agreement that groups should be monophyletic, comprising all the descendants of a common ancestor (Chapter 7). Orders have been recognized rather arbitrarily in the past two centuries, and the most that can be said is that presently constituted orders contain similar insects differentiated from other insect groups. Over time, a relatively stable classification system has developed but differences of opinion remain as to the boundaries around groups, with "splitters" recognizing a greater number of groups and "lumpers" favoring broader categories. For example, some North American taxonomists group ("lump") the alderflies, dobsonflies, snakeflies, and lacewings into one order, the Neuroptera, whereas others, including ourselves, "split" the group and recognize three separate (but clearly closely related) orders, Megaloptera, Raphidioptera, and a more narrowly defined Neuroptera (Fig. 7.2). The order Hemiptera sometimes was divided into two orders, Homoptera and Heteroptera, but the homopteran grouping is invalid (non-monophyletic) and we advocate a different classification for these bugs shown stylized on our cover and in detail in Fig. 7.5 and Box 11.8.

In this book we recognize 30 orders for which the physical characteristics and biologies of their constituent taxa are described, and their relationships considered (Chapter 7). Amongst these orders, we distinguish "major" orders, based upon the numbers of species being much higher in Coleoptera, Diptera, Lepidoptera, Hymenoptera, and Hemiptera than in the remaining "minor" orders. Minor orders often have quite homogeneous ecologies which can be summarized conveniently in single descriptive/ecological boxes following the appropriate ecologically based chapter (Chapters 9–15). The major orders are summarized ecologically less readily and information may appear in two chapters. A summary of the diagnostic features of all 30 orders and cross references to fuller identificatory and ecological information appears in tabular form in the Appendix.

1.5 INSECTS IN POPULAR CULTURE AND COMMERCE

People have been attracted to the beauty or mystique of certain insects throughout time. We know the importance of scarab beetles to the Egyptians as religious items, but earlier shamanistic cultures elsewhere in the Old World made ornaments that represent scarabs and other beetles including buprestids (jewel beetles). In Old Egypt the scarab, which shapes dung into balls, is identified as a potter; similar insect symbolism extends also further east. Egyptians, and subsequently the Greeks, made ornamental scarabs from many materials including lapis lazuli, basalt, limestone, turquoise, ivory, resins, and even valuable gold and silver. Such adulation may have been the pinnacle that an insect lacking economic importance ever gained in popular and religious culture, although many human societies recognized insects in their ceremonial lives. Cicadas were regarded by the ancient Chinese as symbolizing rebirth or immortality. In Mesopotamian literature the *Poem of Gilgamesh* alludes to odonates (dragonflies/damselflies) as signifying the impossibility of immortality. For the San ("bushmen") of the Kalahari, the praying mantis carries much cultural symbolism, including creation and patience in zen-like waiting. Amongst the personal or clan totems of Aboriginal Australians of the Arrernte language groups are *yarumpa* (honey ants) and *udnirringitta* (witchety grubs). Although these insects are important as food in the arid central Australian environment (see section 1.6.1), they were not to be eaten by clan members belonging to that particular totem.

Totemic and food insects are represented in many Aboriginal artworks in which they are associated with cultural ceremonies and depiction of important locations. Insects have had a place in many societies for their symbolism – such as ants and bees representing hard workers throughout the Middle Ages of Europe, where they even entered heraldry. Crickets, grasshoppers, cicadas, and scarab and lucanid beetles have long been valued as caged pets in Japan. Ancient Mexicans observed butterflies in detail, and lepidopterans were well represented in mythology, including in poem and song. Amber has a long history as jewellery, and the inclusion of insects can enhance the value of the piece.

Urbanized humans have lost much of this contact with insects, excepting those that share our domicile, such as cockroaches, tramp ants, and hearth crickets which generally arouse antipathy. Nonetheless, specialized exhibits of insects notably in butterfly farms and insect zoos are very popular, with millions of people per year visiting such attractions throughout the world. Natural occurrences of certain insects attract ecotourism, including aggregations of overwintering monarch butterflies in coastal central California (see Plate 3.5) and Mexico, the famous glow worm caves of Waitomo, New Zealand, and Costa Rican locations such as Selva Verde representing tropical insect biodiversity.

Although insect ecotourism may be in its infancy, other economic benefits are associated with interest in insects. This is especially so amongst children in

Japan, where native rhinoceros beetles (Scarabaeidae, *Allomyrina dichotoma*) sell for US$3–7 each, and longer-lived common stag beetles for some US$10, and may be purchased from automatic vending machines. Adults collect too with a passion: a 7.5 cm example of the largest Japanese stag beetles (Lucanidae, *Dorcus curvidens*, called *o-kuwagata*) may sell for between 40,000 and 150,000 yen (US$300 and US$1250), depending on whether captive reared or taken from the wild. Largest specimens, even if reared, have fetched several million yen (>US$10,000) at the height of the craze. Such enthusiasm by Japanese collectors can lead to a valuable market for insects from outside Japan. According to official statistics, in 2002 some 680,000 beetles, including over 300,000 each of rhinoceros and stag beetles, were imported, predominantly originating from south and south-east Asia. Enthusiasm for valuable specimens extends outside Coleoptera: Japanese and German tourists are reported to buy rare butterflies in Vietnam for US$1000–2000, which is a huge sum of money for the generally poor local people.

Entomological revenue can enter local communities and assist in natural habitat conservation when tropical species are reared for living butterfly exhibits in the affluent world. An estimated 4000 species of butterflies have been reared in the tropics and exhibited live in butterfly houses in North America, Europe, Malaysia, and Australia. Farming butterflies for export is a successful economic activity in Costa Rica, Kenya, and Papua New Guinea. Eggs or wild-caught larvae are reared on appropriate host plants, grown until pupation, and freighted by air to butterfly farms. Papilionidae, including the well-known swallowtails, graphiums, and birdwings, are most popular, but research into breeding requirements allows an expanded range of potential exhibits to be located, reared, and shipped. In East Africa, the National Museums of Kenya has combined with local people of the Arabuko-Sukoke forest in the Kipepeo Project to export harvested butterflies for live overseas exhibit, thereby providing a cash income for these otherwise impoverished people.

In Asia, particularly in Malaysia, there is interest in rearing, exhibiting, and trading in mantises (Mantodea), including orchid mantises (*Hymenopus* species; see pp. 329 and 358) and stick-insects (Phasmatodea). Hissing cockroaches from Madagascar and burrowing cockroaches from tropical Australia are reared readily in captivity and can be kept as domestic pets as well as being displayed in insect zoos in which handling the exhibits is encouraged.

Questions remain concerning whether wild insect collection, either for personal interest or commercial trade and display, is sustainable. Much butterfly, dragonfly, stick-insect, and beetle trade relies more on collections from the wild than rearing programs, although this is changing as regulations increase and research into rearing techniques continues. In the Kenyan Kipepeo Project, although specimens of preferred lepidopteran species originate from the wild as eggs or early larvae, walk-through visual assessment of adult butterflies in flight suggested that the relative abundance rankings of species was unaffected regardless of many years of selective harvest for export. Furthermore, local appreciation has increased for intact forest as a valuable resource rather than viewing it as "wasted" land to clear for subsistence agriculture. In Japan, although expertise in captive rearing has increased and thus undermined the very high prices paid for certain wild-caught beetles, wild harvesting continues over an ever-increasing region. The possibility of over-collection for trade is discussed in section 1.7, together with other conservation issues.

1.6 INSECTS AS FOOD

1.6.1 Insects as human food: entomophagy

In this section we review the increasingly popular study of insects as human food. Probably 1000 or more species of insects in more than 370 genera and 90 families are or have been used for food somewhere in the world, especially in central and southern Africa, Asia, Australia, and Latin America. Food insects generally feed on either living or dead plant matter, and chemically protected species are avoided. Termites, crickets, grasshoppers, locusts, beetles, ants, bee brood, and moth larvae are frequently consumed insects. Although insects are high in protein, energy, and various vitamins and minerals, and can form 5–10% of the annual animal protein consumed by certain indigenous peoples, western society essentially overlooks entomological cuisine.

Typical "western" repugnance of entomophagy is cultural rather than scientific or rational. After all, other invertebrates such as certain crustaceans and mollusks are favored culinary items. Objections to eating insects cannot be justified on the grounds of taste or food value. Many have a nutty flavor and studies report favorably on the nutritional content of insects,

Table 1.2 Proximate, mineral, and vitamin analyses of four edible Angolan insects (percentages of daily human dietary requirements/100 g of insects consumed). (After Santos Oliviera et al. 1976, as adapted by DeFoliart 1989.)

Nutrient	Requirement per capita (reference person)	*Macrotermes subhyalinus* (Termitidae)	*Imbrasia ertli* (Saturniidae)	*Usta terpsichore* (Saturniidae)	*Rhynchophorus phoenicus* (Curculionidae)
Energy	2850 kcal	21.5%	13.2%	13.0%	19.7%
Protein	37 g	38.4	26.3	76.3	18.1
Calcium	1 g	4.0	5.0	35.5	18.6
Phosphorus	1 g	43.8	54.6	69.5	31.4
Magnesium	400 mg	104.2	57.8	13.5	7.5
Iron	18 mg	41.7	10.6	197.2	72.8
Copper	2 mg	680.0	70.0	120.0	70.0
Zinc	15 mg	–	–	153.3	158.0
Thiamine	1.5 mg	8.7	–	244.7	201.3
Riboflavin	1.7 mg	67.4	–	112.2	131.7
Niacin	20 mg	47.7	–	26.0	38.9

although their amino acid composition needs to be balanced with suitable plant protein. Nutritional values obtained from analyses conducted on samples of four species of insects cooked according to traditional methods in central Angola, Africa are shown in Table 1.2. The insects concerned are: reproductive individuals of a termite, *Macrotermes subhyalinus* (Isoptera: Termitidae), which are de-winged and fried in palm oil; the large caterpillars of two species of moth, *Imbrasia ertli* and *Usta terpsichore* (Lepidoptera: Saturniidae), which are de-gutted and either cooked in water, roasted, or sun-dried; and the larvae of the palm weevil, *Rhynchophorus phoenicis* (Coleoptera: Curculionidae), which are slit open and then fried whole in oil.

Mature larvae of *Rhynchophorus* species have been appreciated by people in tropical areas of Africa, Asia, and the Neotropics for centuries. These fat, legless grubs (Fig. 1.2), often called palmworms, provide one of the richest sources of animal fat, with substantial amounts of riboflavin, thiamine, zinc, and iron (Table 1.2). Primitive cultivation systems, involving the cutting down of palm trees to provide suitable food for the weevils, are known from Brazil, Colombia, Paraguay, and Venezuela. In plantations, however, palmworms are regarded as pests because of the damage they can inflict on coconut and oil palm trees.

In central Africa, the people of southern Zaire (presently Democratic Republic of Congo) eat caterpillars belonging to 20–30 species. The calorific value of these

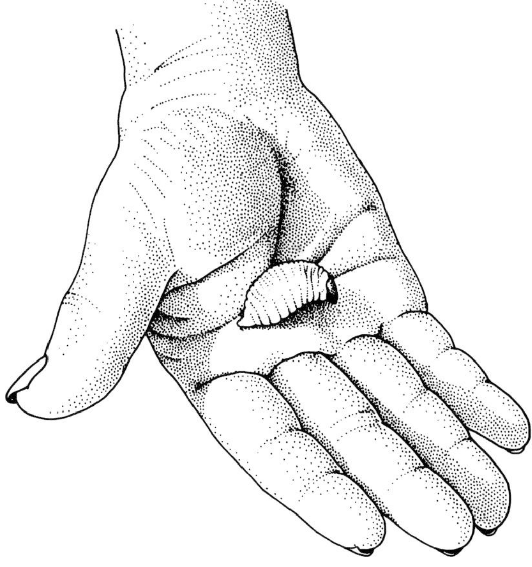

Fig. 1.2 A mature larva of the palm weevil, *Rhynchophorus phoenicis* (Coleoptera: Curculionidae) – a traditional food item in central Angola, Africa. (Larva after Santos Oliveira et al. 1976.)

caterpillars is high, with their protein content ranging from 45 to 80%, and they are a rich source of iron. For instance, caterpillars are the most important source of animal protein in some areas of the Northern Province of Zambia. The edible caterpillars of species of *Imbrasia* (Saturniidae), an emperor moth, locally called mumpa, provide a valuable market. The caterpillars contain 60–70% protein on a dry-matter basis and offset malnutrition caused by protein deficiency. Mumpa are fried fresh or boiled and sun-dried prior to storage. Further south in Africa, *Imbrasia belina* moth (see Plate 1.4) caterpillars (see Plate 1.5), called mopane, mopanie, mophane, or phane, are utilized widely. Caterpillars usually are de-gutted, boiled, sometimes salted, and dried. After processing they contain about 50% protein and 15% fat – approximately twice the values for cooked beef. Concerns that harvest of mopane may be unsustainable and over-exploited are discussed under conservation in Box 1.3.

In the Philippines, June beetles (melolonthine scarabs), weaver ants (*Oecophylla smaragdina*), mole crickets, and locusts are eaten in some regions. Locusts form an important dietary supplement during outbreaks, which apparently have become less common since the widespread use of insecticides. Various species of grasshoppers and locusts were eaten commonly by native tribes in western North America prior to the arrival of Europeans. The number and identity of species used have been poorly documented, but species of *Melanoplus* were consumed. Harvesting involved driving grasshoppers into a pit in the ground by fire or advancing people, or herding them into a bed of coals. Today people in central America, especially Mexico, harvest, sell, cook, and consume grasshoppers.

Australian Aborigines use (or once used) a wide range of insect foods, especially moth larvae. The caterpillars of wood or ghost moths (Cossidae and Hepialidae) (Fig. 1.3) are called witchety grubs from an Aboriginal word "witjuti" for the *Acacia* species (wattles) on the roots and stems of which the grubs feed. Witchety grubs, which are regarded as a delicacy, contain 7–9% protein, 14–38% fat, 7–16% sugars as well as being good sources of iron and calcium. Adults of the bogong moth, *Agrotis infusa* (Noctuidae), formed another important Aboriginal food, once collected in their millions from estivating sites in narrow caves and crevices on mountain summits in south-eastern Australia. Moths cooked in hot ashes provided a rich source of dietary fat.

Aboriginal people living in central and northern Australia eat the contents of the apple-sized galls of *Cystococcus pomiformis* (Hemiptera: Eriococcidae), commonly called bush coconuts or bloodwood apples (see Plate 2.3). These galls occur only on bloodwood

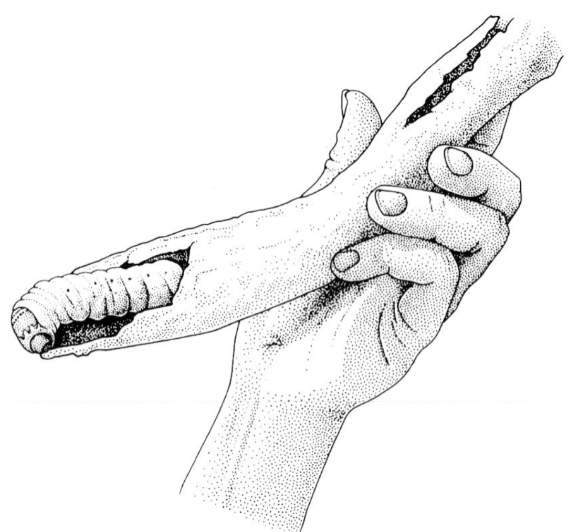

Fig. 1.3 A delicacy of the Australian Aborigines – a witchety (or witjuti) grub, a caterpillar of a wood moth (Lepidoptera: Cossidae) that feeds on the roots and stems of witjuti bushes (certain *Acacia* species). (After Cherikoff & Isaacs 1989.)

eucalypts (*Corymbia* species) and can be very abundant after a favorable growing season. Each mature gall contains a single adult female, up to 4 cm long, which is attached by her mouth area to the base of the inner gall and has her abdomen plugging a hole in the gall apex. The inner wall of the gall is lined with white edible flesh, about 1 cm thick, which serves as the feeding site for the male offspring of the female (see Plate 2.4). Aborigines relish the watery female insect and her nutty-flavored nymphs, then scrape out and consume the white coconut-like flesh of the inner gall.

A favorite source of sugar for Australian Aboriginals living in arid regions comes from species of *Melophorus* and *Camponotus* (Formicidae), popularly known as honeypot ants. Specialized workers (called repletes) store nectar, fed to them by other workers, in their huge distended crops (Fig. 2.4). Repletes serve as food reservoirs for the ant colony and regurgitate part of their crop contents when solicited by another ant. Aborigines dig repletes from their underground nests, an activity most frequently undertaken by women, who may excavate pits to a depth of a meter or more in search of these sweet rewards. Individual nests rarely supply more than 100 g of a honey that is essentially similar in composition to commercial honey. Honeypot ants in the western USA and Mexico belong to a different genus, *Myrmecocystus*. The repletes, a highly valued food, are collected by the rural people of Mexico, a difficult process in the hard soil of the stony ridges where the ants nest.

Perhaps the general western rejection of entomophagy is only an issue of marketing to counter a popular conception that insect food is for the poor and protein-deprived of the developing world. In reality, certain sub-Saharan Africans apparently prefer caterpillars to beef. Ant grubs (so called "ant eggs") and eggs of water boatmen (Corixidae) and backswimmers (Notonectidae) are much sought after in Mexican gastronomy as "caviar". In parts of Asia, a diverse range of insects can be purchased (see Plate 2.1). Traditionally desirable water beetles for human consumption are valuable enough to be farmed in Guangdong. The culinary culmination may be the meat of the giant water bug *Lethocerus indicus* (see Plate 1.6) or the Thai and Laotian *mangda* sauces made with the flavors extracted from the male abdominal glands, for which high prices are paid. Even in the urban USA some insects may yet become popular as a food novelty. The millions of 17-year cicadas that periodically plague cities like Chicago are edible. Newly hatched cicadas, called tenerals, are best for eating because their soft body cuticle means that they can be consumed without first removing the legs and wings. These tasty morsels can be marinated or dipped in batter and then deep-fried, boiled and spiced, roasted and ground, or stir-fried with favorite seasonings.

Large-scale harvest or mass production of insects for human consumption brings some practical and other problems. The small size of most insects presents difficulties in collection or rearing and in processing for sale. The unpredictability of many wild populations needs to be overcome by the development of culture techniques, especially as over-harvesting from the wild could threaten the viability of some insect populations. Another problem is that not all insect species are safe to eat. Warningly colored insects are often distasteful or toxic (Chapter 14) and some people can develop allergies to insect material (section 15.2.3). However, several advantages derive from eating insects. The encouragement of entomophagy in many rural societies, particularly those with a history of insect use, may help diversify peoples' diets. By incorporating mass harvesting of pest insects into control programs, the use of pesticides can be reduced. Furthermore, if carefully regulated, cultivating insects for protein should be less environmentally damaging than cattle ranching, which devastates forests and native grasslands. Insect farming (the rearing of mini-livestock) is compatible with low input, sustainable agriculture and most insects have a high food conversion efficiency compared with conventional meat animals.

1.6.2 Insects as feed for domesticated animals

If you do not relish the prospect of eating insects yourself, then perhaps the concept of insects as a protein source for domesticated animals is more acceptable. The nutritive significance of insects as feed for fish, poultry, pigs, and farm-grown mink certainly is recognized in China, where feeding trials have shown that insect-derived diets can be cost-effective alternatives to more conventional fish meal diets. The insects involved are primarily the pupae of silkworms (*Bombyx mori*) (see Plate 2.1), the larvae and pupae of house flies (*Musca domestica*), and the larvae of mealworms (*Tenebrio molitor*). The same or related insects are being used or investigated elsewhere, particularly as poultry or fish feedstock. Silkworm pupae, a by-product of the silk industry, can be used as a high-protein supplement for chickens. In India, poultry are fed the meal that remains after the oil has been extracted from the pupae. Fly larvae fed to chickens can recycle animal manure and the development of a range of insect recycling systems for converting organic wastes into feed supplements is inevitable, given that most organic substances are fed on by one or more insect species.

Clearly, insects can form part of the nutritional base of people and their domesticated animals. Further research is needed and a database with accurate identifications is required to handle biological information. We must know which species we are dealing with in order to make use of information gathered elsewhere on the same or related insects. Data on the nutritional value, seasonal occurrence, host plants, or other dietary needs, and rearing or collecting methods must be collated for all actual or potential food insects. Opportunities for insect food enterprises are numerous, given the immense diversity of insects.

1.7 INSECT CONSERVATION

Biological conservation typically involves either setting aside large tracts of land for "nature", or addressing and remediating specific processes that threaten large and charismatic vertebrates, such as endangered mammals and birds, or plant species or communities. The concept of conserving habitat for insects, or species thereof, seems of low priority on a threatened planet. Nevertheless, land is reserved and plans exist specifically to conserve certain insects. Such conservation efforts often are associated with human aesthetics, and many (but not all) involve the "charismatic megafauna" of entomology – the butterflies and large, showy beetles. Such charismatic insects can act as "**flagship**" species to enhance wider public awareness and engender financial support for conservation efforts. Single-species conservation, not necessarily of an insect, is argued to preserve many other species by default, in what is known as the "**umbrella effect**". Somewhat complementary to this is advocacy of a habitat-based approach, which increases the number and size of areas to conserve many insects, which are not (and arguably "do not need to be") understood on a species-by-species approach. No doubt efforts to conserve habitats of native fish globally will preserve, as a spin-off, the much more diverse aquatic insect fauna that depends also upon waters being maintained in natural condition. Equally, preservation of old-growth forests to protect tree-hole nesting birds such as owls or parrots also will conserve habitat for wood-mining insects that use timber across a complete range of wood species and states of decomposition. Habitat-based conservationists accept that single-species oriented conservation is important but argue that it may be of limited value for insects because there are so many species. Furthermore, rarity of insect species may be due to populations being localized in just one or a few places, or in contrast, widely dispersed but with low density over a wide area. Clearly, different conservation strategies are required for each case.

Migratory species, such as the monarch butterfly (*Danaus plexippus*), require special conservation. Monarchs from east of the Rockies overwinter in Mexico and migrate northwards as far as Canada throughout the summer (section 6.7). Critical to the conservation of these monarchs is the safeguarding of the overwintering habitat at Sierra Chincua in Mexico. A most significant insect conservation measure implemented in recent years is the decision of the Mexican government to support the Monarch Butterfly Biosphere Reserve established to protect the phenomenon. Although the monarch butterfly is an excellent flagship insect, the preservation of western overwintering populations in coastal California (see Plate 3.5) protects no other native species. The reason for this is that the major resting sites are in groves of large introduced eucalypt trees, especially blue gums, which are faunistically depauperate in their non-native habitat.

A successful example of single-species conservation involves the El Segundo blue, *Euphilotes battoides* ssp. *allyni*, whose principal colony in sand dunes near Los Angeles airport was threatened by urban sprawl and golf course development. Protracted negotiations with many interests resulted in designation of 80 hectares as a reserve, sympathetic management of the golf course "rough" for the larval food plant *Erigonum parvifolium* (buckwheat), and control of alien plants plus limitation on human disturbance. Southern Californian coastal dune systems are seriously endangered habitats, and management of this reserve for the El Segundo blue conserves other threatened species.

Land conservation for butterflies is not an indulgence of affluent southern Californians: the world's largest butterfly, the Queen Alexandra's birdwing (*Ornithoptera alexandrae*), of Papua New Guinea (PNG) is a success story from the developing world. This spectacular species, whose caterpillars feed only on *Aristolochia dielsiana* vines, is limited to a small area of lowland rainforest in northern PNG and has been listed as endangered. Under PNG law, this birdwing species has been protected since 1966, and international commercial trade was banned by listing on Appendix I of the Convention on International Trade in Endangered Species of Wild Fauna and Flora (CITES). Dead specimens in good condition command a high price, which can be more than US$2000. In 1978, the PNG governmental Insect Farming and Trading Agency (IFTA), in Bulolo, Morobe Province, was established to control conservation and exploitation and act as a clearing-house for trade in Queen Alexandra's birdwings and other valuable butterflies. Local cultivators, numbering some 450 village farmers associated with IFTA, "ranch" their butterflies. In contrast to the Kenyan system described in section 1.5, farmers plant appropriate host vines, often on land already cleared for vegetable gardens at the forest edge, thereby providing food plants for a chosen local species of butterfly. Wild adult butterflies emerge from the forest to feed and lay their eggs; hatched larvae feed on the vines until pupation when they are collected and protected in hatching cages. According to species, the purpose for which they

1.1 1.2 1.3 1.4 1.5 1.6

Plate 1

1.1 An atlas moth, *Attacus atlas* (Lepidoptera: Saturniidae), which occurs in southern India and south-east Asia, is one of the largest of all lepidopterans, with a wingspan of about 24 cm and a larger wing area than any other moth (P.J. Gullan).

1.2 A violin beetle, *Mormolyce phyllodes* (Coleoptera: Carabidae), from rainforest in Brunei, Borneo (P.J. Gullan).

1.3 The moon moth, *Argema maenas* (Lepidoptera: Saturniidae), is found in south-east Asia and India; this female, from rainforest in Borneo, has a wingspan of about 15 cm (P.J. Gullan).

1.4 The mopane emperor moth, *Imbrasia belina* (Lepidoptera: Saturniidae), from the Transvaal in South Africa (R. Oberprieler).

1.5 A "worm" or "phane" – the caterpillar of *Imbrasia belina* – feeding on the foliage of *Schotia brachypetala*, from the Transvaal in South Africa (R. Oberprieler).

1.6 A dish of edible water bugs, *Lethocerus indicus* (Hemiptera: Belostomatidae), on sale at a market in Lampang Province, Thailand (R.W. Sites).

[*Facing p.* 14]

2.1 2.2 2.3 2.4 2.5 2.6

Plate 2

2.1 Food insects at a market stall in Lampang Province, Thailand, displaying silk moth pupae (*Bombyx mori*), beetle pupae, adult hydrophiloid beetles, and water bugs, *Lethocerus indicus* (R.W. Sites).

2.2 Adult Richmond birdwing (*Troides richmondia*) butterfly and cast exuvial skin on native pipevine (*Pararistolochia* sp.) host (see p. 15) (D.P.A. Sands).

2.3 A bush coconut or bloodwood apple gall of *Cystococcus pomiformis* (Hemiptera: Eriococcidae), cut open to show the cream-colored adult female and her numerous, tiny nymphal male offspring covering the gall wall (P.J. Gullan).

2.4 Close-up of the second-instar male nymphs of *Cystococcus pomiformis* feeding from the nutritive tissue lining the cavity of the maternal gall (see p. 12) (P.J. Gullan).

2.5 Adult male scale insect of *Melaleucococcus phacelopilus* (Hemiptera: Margarodidae), showing the setiferous antennae and the single pair of wings (P.J. Gullan).

2.6 A tropical butterfly, *Graphium antiphates itamputi* (Lepidoptera: Papilionidae), from Borneo, obtaining salts by imbibing sweat from a training shoe (refer to Box 5.2) (P.J. Gullan).

3.1 3.2 3.3 3.5 3.6 3.7

Plate 3

3.1 A female katydid of an undescribed species of *Austrosalomona* (Orthoptera: Tettigoniidae), from northern Australia, with a large spermatophore attached to her genital opening (refer to Box 5.2) (D.C.F. Rentz).

3.2 Pupa of a Christmas beetle, *Anoplognathus* sp. (Coleoptera: Scarabaeidae), removed from its pupation site in the soil in Canberra, Australia (P.J. Gullan).

3.3 Egg mass of *Tenodera australasiae* (Mantodea: Mantidae) with young mantid nymphs emerging, from Queensland, Australia (refer to Box 13.2) (D.C.F. Rentz).

3.4 Eclosing (molting) adult katydid of an *Elephantodeta* species (Orthoptera: Tettigoniidae), from the Northern Territory, Australia (D.C.F. Rentz).

3.5 Overwintering monarch butterflies, *Danaus plexippus* (Lepidoptera: Nymphalidae), from Mill Valley in California, USA (D.C.F. Rentz).

3.6 A fossilized worker ant of *Pseudomyrmex oryctus* (Hymenoptera: Formicidae) in Dominican amber from the Oligocene or Miocene (P.S. Ward).

3.7 A diversity of flies (Diptera), including calliphorids, are attracted to the odor of this Australian phalloid fungus, *Anthurus archeri*, which produces a foul-smelling slime containing spores that are consumed by the flies and distributed after passing through the insects' guts (P.J. Gullan).

Plate 4

4.1 A tree trunk and under-branch covered in silk galleries of the webspinner *Antipaluria urichi* (Embiidina: Clothodidae), from Trinidad (refer to Box 9.5) (J.S. Edgerly-Rooks).

4.2 A female webspinner of *Antipaluria urichi* defending the entrance of her gallery from an approaching male, from Trinidad (J.S. Edgerly-Rooks).

4.3 An adult stonefly, *Neoperla edmundsi* (Plecoptera: Perlidae), from Brunei, Borneo (P.J. Gullan).

4.4 A female thynnine wasp of *Zaspilothynnus trilobatus* (Hymenoptera: Tiphiidae) (on the right) compared with the flower of the sexually deceptive orchid *Drakaea glyptodon*, which attracts pollinating male wasps by mimicking the female wasp (see p. 282) (R. Peakall).

4.5 A male thynnine wasp of *Neozeloboria cryptoides* (Hymenoptera: Tiphiidae) attempting to copulate with the sexually deceptive orchid *Chiloglottis trapeziformis* (R. Peakall).

4.6 Pollination of mango flowers by a flesh fly, *Australopierretia australis* (Diptera: Sarcophagidae), in northern Australia (D.L. Anderson).

4.7 The wingless adult female of the whitemarked tussock moth, *Orgyia leucostigma* (Lepidoptera: Lymantriidae), from New Jersey, USA (D.C.F. Rentz).

5.1 5.3 5.4 5.5 5.6 5.7

Plate 5

5.1 Mealybugs of an undescribed *Planococcus* species (Hemiptera: Pseudococcidae) on an *Acacia* stem attended by ants of a *Polyrhachis* species (Hymenoptera: Formicidae), coastal Western Australia (P.J. Gullan).

5.2 A camouflaged late-instar caterpillar of *Plesanemma fucata* (Lepidoptera: Geometridae) on a eucalypt leaf in eastern Australia (P.J. Gullan).

5.3 A female of the scorpionfly *Panorpa communis* (Mecoptera: Panorpidae) from the UK (P.H. Ward).

5.4 The huge queen termite (approximately 7.5 cm long) of *Odontotermes transvaalensis* (Isoptera: Termitidae: Macrotermitinae) surrounded by her king (mid front), soldiers, and workers, from the Transvaal in South Africa (J.A.L. Watson).

5.5 A parasitic *Varroa* mite (see p. 320) on a pupa of the bee *Apis cerana* (Hymenoptera: Apidae) in a hive from Irian Jaya, New Guinea (D.L. Anderson).

5.6 An adult moth of *Utetheisa ornatrix* (Lepidoptera: Arctiidae) emitting defensive froth containing pyrrolizidine alkaloids that it sequesters as a larva from its food plants, legumes of the genus *Crotalaria* (T. Eisner).

5.7 A snake-mimicking caterpillar of the spicebush swallowtail, *Papilio troilus* (Lepidoptera: Papilionidae), from New Jersey, USA (D.C.F. Rentz).

6.1
6.2
6.3
6.5
6.6
6.7

Plate 6

6.1 The cryptic adult moths of four species of *Acronicta* (Lepidoptera: Noctuidae): *A. alni*, the alder moth (top left); *A. leporina*, the miller (top right); *A. aceris*, the sycamore (bottom left); and *A. psi*, the grey dagger (bottom right) (D. Carter and R.I. Vane-Wright).

6.2 Aposematic or mechanically protected caterpillars of the same four species of *Acronicta*: *A. alni* (top left); *A. leporina* (top right); *A. aceris* (bottom left); and *A. psi* (bottom right); showing the divergent appearance of the larvae compared with their drab adults (D. Carter and R.I. Vane-Wright).

6.3 A blister beetle, *Lytta polita* (Coleoptera: Meloidae), reflex-bleeding from the knee joints; the hemolymph contains the toxin cantharidin (sections 14.4.3 & 15.2.2) (T. Eisner).

6.4 One of Bates' mimicry complexes from the Amazon Basin involving species from three different lepidopteran families – *Methona confusa confusa* (Nymphalidae: Ithomiinae) (top), *Lycorea ilione ilione* (Nymphalidae: Danainae) (second from top), *Patia orise orise* (Pieridae) (second from bottom), and a day-flying moth of *Gazera heliconioides* (Castniidae) (R.I. Vane-Wright).

6.5 An aposematic beetle of the genus *Lycus* (Coleoptera: Lycidae) on the flower spike of *Cussonia* (Araliaceae) from South Africa (P.J. Gullan).

6.6 A mature cottony-cushion scale, *Icerya purchasi* (Hemiptera: Margarodidae), with a fully formed ovisac, on the stem of a native host plant from Australia (P.J. Gullan).

6.7 Adult male gypsy moth, *Lymantria dispar* (Lepidoptera: Lymantriidae), from New Jersey, USA (D.C.F. Rentz).

are being raised, and conservation legislation, butterflies can be exported live as pupae, or dead as high-quality collector specimens. IFTA, a non-profit organization, sells some $400,000 worth of PNG insects yearly to collectors, scientists, and artists around the world, generating an income for a society that struggles for cash. As in Kenya, local people recognize the importance of maintaining intact forests as the source of the parental wild-flying butterflies of their ranched stock. In this system, the Queen Alexandra's birdwing butterfly has acted as a flagship species for conservation in PNG and the success story attracts external funding for surveys and reserve establishment. In addition, conserving PNG forests for this and related birdwings undoubtedly results in conservation of much diversity under the umbrella effect.

The Kenyan and New Guinean insect conservation efforts have a commercial incentive, providing impoverished people with some recompense for protecting natural environments. Commerce need not be the sole motivation: the aesthetic appeal of having native birdwing butterflies flying wild in local neighborhoods, combined with local education programs in schools and communities, has saved the subtropical Australian Richmond birdwing butterfly (*Troides* or *Ornithoptera richmondia*) (see Plate 2.2). Larval Richmond birdwings eat *Pararistolochia* or *Aristolochia* vines, choosing from three native species to complete their development. However, much coastal rainforest habitat supporting native vines has been lost, and the alien South American *Aristolochia elegans* ("Dutchman's pipe"), introduced as an ornamental plant and escaped from gardens, has been luring females to lay eggs on it as a prospective host. This oviposition mistake is deadly since toxins of this plant kill young caterpillars. The answer to this conservation problem has been an education program to encourage the removal of Dutchman's pipe vines from native vegetation, from sale in nurseries, and from gardens and yards. Replacement with native *Pararistolochia* was encouraged after a massive effort to propagate the vines. Community action throughout the native range of the Richmond birdwing appears to have reversed its decline, without any requirement to designate land as a reserve.

Evidently, butterflies are flagships for invertebrate conservation – they are familiar insects with a non-threatening lifestyle. However, certain orthopterans, including New Zealand wetas, have been afforded protection, and we are aware also of conservation plans for dragonflies and other freshwater insects in the context of conservation and management of aquatic environments, and of plans for firefly (beetle) and glow worm (fungus gnat) habitats. Agencies in certain countries have recognized the importance of retention of fallen dead wood as insect habitat, particularly for long-lived wood-feeding beetles.

Designation of reserves for conservation, seen by some as the answer to threat, rarely is successful without understanding species requirements and responses to management. The butterfly family Lycaenidae (blues, coppers, and hairstreaks) includes perhaps 50% of the butterfly diversity of some 6000 species. Many have relationships with ants (myrmecophily; see section 12.3), some being obliged to pass some or all of their immature development inside ant nests, others are tended on their preferred host plant by ants, yet others are predators on ants and scale insects, while tended by ants. These relationships can be very complex, and may be rather easily disrupted by environmental changes, leading to endangerment of the butterfly. Certainly in western Europe, species of Lycaenidae figure prominently on lists of threatened insect taxa. Notoriously, the decline of the large blue butterfly *Maculinea arion* in England was blamed upon over-collection and certainly some species have been sought after by collectors (but see Box 1.1). Action plans in Europe for the reintroduction of this and related species and appropriate conservation management of other *Maculinea* species have been put in place: these depend vitally upon a species-based approach. Only with understanding of general and specific ecological requirements of conservation targets can appropriate management of habitat be implemented.

Box 1.1 Collected to extinction?

The large blue butterfly (*Maculinea arion*) was reported to be in serious decline in southern England in the late 19th century, a phenomenon ascribed then to poor weather. By the mid-20th century this attractive species was restricted to some 30 colonies in south-western England. Only one or two colonies remained by 1974 and the estimated adult population had declined from about 100,000 in 1950 to 250 in some 20 years. Final extinction of the species in England in 1979 followed two successive hot, dry breeding seasons. Since the butterfly is beautiful and sought by collectors, excessive collecting was presumed to have caused at least the long-term decline that made the species vulnerable to deteriorating climate. This decline occurred even though a reserve was established in the 1930s to exclude both collectors and domestic livestock in an attempt to protect the butterfly and its habitat.

Evidently, habitat had changed through time, including a reduction of wild thyme (*Thymus praecox*), which provides the food for early instars of the large blue's caterpillar. Shrubbier vegetation replaced short-turf grassland because of loss of grazing rabbits (through disease) and exclusion of grazing cattle and sheep from the reserved habitat. Thyme survived, however, but the butterflies continued to decline to extinction in Britain.

A more complex story has been revealed by research associated with reintroduction of the large blue to England from continental Europe. The larva of the large blue butterfly in England and on the European continent is an obligate predator in colonies of red ants belonging to species of *Myrmica*. Larval large blues must enter a *Myrmica* nest, in which they feed on larval ants. Similar predatory behavior, and/or tricking ants into feeding them as if they were the ants' own brood, are features

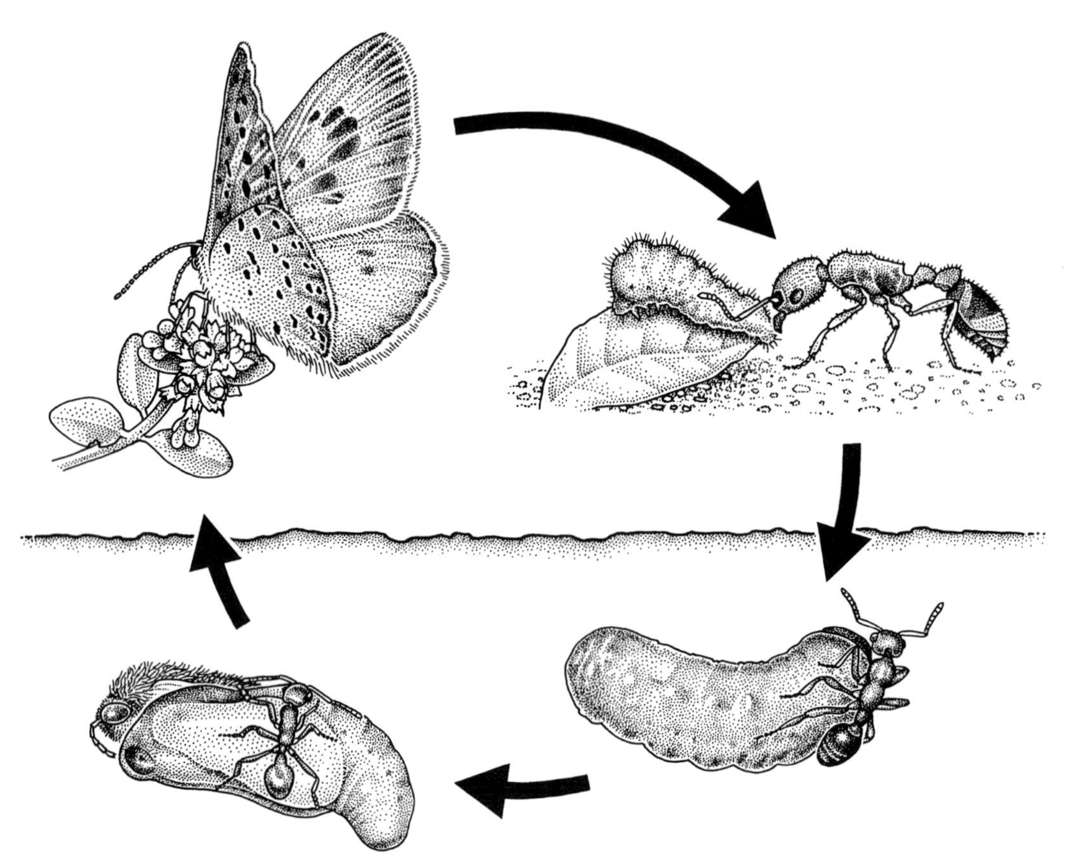

in the natural history of many Lycaenidae (blues and coppers) worldwide (see p. 15). After hatching from an egg laid on the larval food plant, the large blue's caterpillar feeds on thyme flowers until the molt into the final (fourth) larval instar, around August. At dusk, the caterpillar drops to the ground from the natal plant, where it waits inert until a *Myrmica* ant finds it. The worker ant attends the larva for an extended period, perhaps more than an hour, during which it feeds from a sugar gift secreted from the caterpillar's dorsal nectary organ. At some stage the caterpillar becomes turgid and adopts a posture that seems to convince the tending ant that it is dealing with an escaped ant brood, and it is carried into the nest. Until this stage, immature growth has been modest, but in the ant nest the caterpillar becomes predatory on ant brood and grows for 9 months until it pupates in early summer of the following year. The caterpillar requires an average 230 immature ants for successful pupation. The adult butterfly emerges from the pupal cuticle in summer and departs rapidly from the nest before the ants identify it as an intruder.

Adoption and incorporation into the ant colony turns out to be the critical stage in the life history. The complex system involves the "correct" ant, *Myrmica sabuleti*, being present, and this in turn depends on the appropriate microclimate associated with short-turf grassland. Longer grass causes cooler near-soil microclimate favoring other *Myrmica* species, including *M. scabrinodes* that may displace *M. sabuleti*. Although caterpillars associate apparently indiscriminately with any *Myrmica* species, survivorship differs dramatically: with *M. sabuleti* approximately 15% survive, but an unsustainable reduction to <2% survivorship occurs with *M. scabrinodes*. Successful maintenance of large blue populations requires that >50% of the adoption by ants must be by *M. sabuleti.*

Other factors affecting survivorship include the requirements for the ant colony to have no alate (winged) queens and at least 400 well-fed workers to provide enough larvae for the caterpillar's feeding needs, and to lie within 2 m of the host thyme plant. Such nests are associated with newly burnt grasslands, which are rapidly colonized by *M. sabuleti*. Nests should not be so old as to have developed more than the founding queen: the problem here being that the caterpillar becomes imbued with the chemical odors of queen larvae while feeding and, with numerous alate queens in the nest, can be mistaken for a queen and attacked and eaten by nurse ants.

Now that we understand the intricacies of the relationship, we can see that the well-meaning creation of reserves that lacked rabbits and excluded other grazers created vegetational and microhabitat changes that altered the dominance of ant species, to the detriment of the butterfly's complex relationships. Over-collecting is not implicated, although climate change on a broader scale must play a role. Now five populations originating from Sweden have been reintroduced to habitat and conditions appropriate for *M. sabuleti*, thus leading to thriving populations of the large blue butterfly. Interestingly, other rare species of insects in the same habitat have responded positively to this informed management, suggesting an umbrella role for the butterfly species.

Box 1.2 Tramp ants and biodiversity

No ants are native to Hawai'i yet there are more than 40 species on the island – all have been brought from elsewhere within the last century. In fact all social insects (honey bees, yellowjackets, paper wasps, termites, and ants) on Hawai'i arrived with human commerce. Almost 150 species of ants have hitchhiked with us on our global travels and managed to establish themselves outside their native ranges. The invaders of Hawai'i belong to the same suite of ants that have invaded the rest of the world, or seem likely to do so in the near future. From a conservation perspective one particular behavioral subset is very important, the so-called invasive "tramp" ants. They rank amongst the world's most serious pest species, and local, national, and international agencies are concerned with their surveillance and control. The big-headed ant (*Pheidole megacephala*), the long legged or yellow crazy ant (*Anoplolepis longipes*), the Argentine ant (*Linepithema humile*), the "electric" or little fire ant (*Wasmannia auropunctata*), and tropical fire ants (*Solenopsis* species) are considered the most serious of these ant pests.

Invasive ant behavior threatens biodiversity, especially on islands such as Hawai'i, the Galapagos and other Pacific Islands (see section 8.7). Interactions with other insects include the protection and tending of aphids and scale insects for their carbohydrate-rich honeydew secretions. This boosts densities of these insects, which include invasive agricultural pests. Interactions with other arthropods are predominantly negative, resulting in aggressive displacement and/or predation on other species, even other tramp ant species encountered. Initial founding is often associated with unstable environments, including those created by human activity. Tramp ants' tendency to be small and short-lived is compensated by year-round increase and rapid production of new queens. Nestmate queens show no hostility to each other. Colonies reproduce by the mated queen and workers relocating only a short distance from the original nest – a process known as budding. When combined with the absence of intra-specific antagonism between newly founded and natal nests, colony budding ensures the gradual spreading of a "supercolony" across the ground.

Although initial nest foundation is associated with human- or naturally disturbed environments, most invasive tramp species can move into more natural habitats and displace the native biota. Ground-dwelling insects, including many native ants, do not survive the encroachment, and arboreal species may follow into local extinction. Surviving insect communities tend to be skewed towards subterranean species and those with especially thick cuticle such as carabid beetles and cockroaches, which also are chemically defended. Such an impact can be seen from the effects of big-headed ants during the monitoring of rehabilitated sand mining sites, using ants as indicators (section 9.7). Six years into rehabilitation, as seen in the graph (from Majer 1985), ant diversity neared that found in unimpacted control sites, but the arrival of *P. megacephala* dramatically restructured the system, seriously reducing diversity relative to controls. Even large animals can be threatened by ants – land crabs on Christmas Island, horned lizards in southern California, hatchling turtles in south-eastern USA, and ground-nesting birds everywhere. Invasion by Argentine ants of fynbos, a mega-diverse South African plant assemblage, eliminates ants that specialize in carrying and burying large seeds, but not those which carry smaller seeds (see section 11.3.2). Since the vegetation originates by germination after periodic fires, the shortage of buried large seeds is predicted to cause dramatic change to vegetation structure.

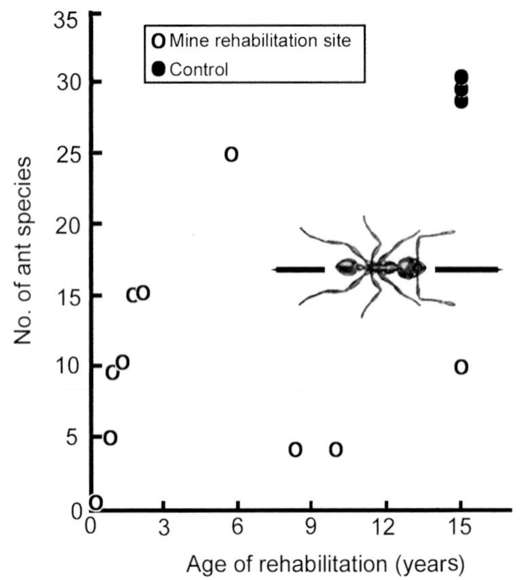

Introduced ants are very difficult to eradicate: all attempts to eliminate fire ants in the USA have failed. We will see if an A$123 million ($US50 million), five-year campaign to rid Australia of *Solenopsis invicta* will prevent it from establishing as an "invasive" species. The first fire ant sites were found around Brisbane in February 2001, and two years later the peri-urban area under surveillance for fire ants extended to some 47,000 ha. Potential economic damage in excess of A$100 billion over the next 30 years is estimated if control fails, with inestimable damage to native biodiversity continent-wide. Although intensive searching and destruction of nests appears to be successful, all must be eradicated to prevent resurgence. Undoubtedly the best strategy for control of invasive ants is quarantine diligence to prevent their entry, and public awareness to detect accidental entry.

Box 1.3 Sustainable use of mopane worms

An important economic insect in Africa is the larva (caterpillar) of emperor moths, especially *Imbrasia belina* (see Plates 1.4 & 1.5, facing p. 14), which is harvested for food across much of southern Africa, including Angola, Namibia, Zimbabwe, Botswana, and Northern Province of South Africa. The distribution coincides with that of mopane (*Colophospermum mopane*), a leguminous tree which is the preferred host plant of the caterpillar and dominates the "mopane woodland" landscape.

Early-instar larvae are gregarious and forage in aggregations of up to 200 individuals: individual trees may be defoliated by large numbers of caterpillars, but regain their foliage if seasonal rains are timely. Throughout their range, and especially during the first larval flush in December, mopane worms are a valued protein source to frequently protein-deprived rural populations. A second cohort may appear some 3–4 months later if conditions for mopane trees are suitable. It is the final-instar larva that is harvested, usually by shaking the tree or by direct collecting from foliage. Preparation is by degutting and drying, and the product may be canned and stored, or transported for sale to a developing gastronomic market in South African towns. Harvesting mopane produces a cash input into rural economies – a calculation in the mid-1990s suggested that a month of harvesting mopane generated the equivalent to the remainder of the year's income to a South African laborer. Not surprisingly, large-scale organized harvesting has entered the scene accompanied by claims of reduction in harvest through unsustainable over-collection. Closure of at least one canning plant was blamed on shortfall of mopane worms.

Decline in the abundance of caterpillars is said to result from both increasing exploitation and reduction in mopane woodlands. In parts of Botswana, heavy commercial harvesting is claimed to have reduced moth numbers. Threats to mopane worm abundance include deforestation of mopane woodland and felling or branch-lopping to enable caterpillars in the canopy to be brought within reach. Inaccessible parts of the tallest trees, where mopane worm density may be highest, undoubtedly act as refuges from harvest and provide the breeding stock for the next season, but mopane trees are felled for their mopane crop. However, since mopane trees dominate huge areas, for example over 80% of the trees in Etosha National Park are mopane, the trees themselves are not endangered.

The problem with blaming the more intensive harvesting for reduction in yield for local people is that the species is patchy in distribution and highly eruptive. The years of reduced mopane harvest seem to be associated with climate-induced drought (the El Niño effect) throughout much of the mopane woodlands. Even in Northern Province of South Africa, long considered to be over-harvested, the resumption of seasonal, drought-breaking rains can induce large mopane worm outbreaks. This is not to deny the importance of research into potential over-harvesting of mopane, but evidently further study and careful data interpretation are needed.

Research already undertaken has provided some fascinating insights. Mopane woodlands are prime elephant habitat, and by all understanding these megaherbivores that uproot and feed on complete mopane trees are keystone species in this system. However, calculations of the impact of mopane worms as herbivores showed that in their six week larval cycle the caterpillars could consume 10 times more mopane leaf material per unit area than could elephants over 12 months. Furthermore, in the same period 3.8 times more fecal matter was produced by mopane worms than by elephants.

Elephants notoriously damage trees, but this benefits certain insects: the heartwood of a damaged tree is exposed as food for termites providing eventually a living but hollow tree. Native bees use the resin that flows from elephant-damaged bark for their nests. Ants nest in these hollow trees and may protect the tree from herbivores, both animal and mopane worm. Elephant populations and mopane worm outbreaks vary in space and time, depending on many interacting biotic and abiotic factors, of which harvest by humans is but one.

FURTHER READING

Berenbaum, M.R. (1995) *Bugs in the System. Insects and their Impact on Human Affairs.* Helix Books, Addison-Wesley, Reading, MA.

Bossart, J.L. & Carlton, C.E. (2002) Insect conservation in America. *American Entomologist* **40**(2), 82–91.

Collins, N.M. & Thomas, J.A. (eds.) (1991) *Conservation of Insects and their Habitats.* Academic Press, London.

DeFoliart, G.R. (ed.) (1988–1995) *The Food Insects Newsletter.* Department of Entomology, University of Wisconsin, Madison, WI. [See Dunkel reference below.]

DeFoliart, G.R. (1989) The human use of insects as food and as animal feed. *Bulletin of the Entomological Society of America* **35**, 22–35.

DeFoliart, G.R. (1995) Edible insects as minilivestock. *Biodiversity and Conservation* **4**, 306–21.

DeFoliart, G.R. (1999) Insects as food; why the western attitude is important. *Annual Review of Entomology* **44**, 21–50.

Dunkel, F.V. (ed.) (1995–present) *The Food Insects Newsletter.* Department of Entomology, Montana State University, Bozeman, MT.

Erwin, T.L. (1982) Tropical forests: their richness in Coleoptera and other arthropod species. *The Coleopterists Bulletin* **36**, 74–5.

Gaston, K.J. (1994) Spatial patterns of species description: how is our knowledge of the global insect fauna growing? *Biological Conservation* **67**, 37–40.

Gaston, K.J. (ed.) (1996) *Biodiversity. A Biology of Numbers and Difference.* Blackwell Science, Oxford.

Gaston, K.J. & Hudson, E. (1994) Regional patterns of diversity and estimates of global insect species richness. *Biodiversity and Conservation* **3**, 493–500.

Hammond, P.M. (1994) Practical approaches to the estimation of the extent of biodiversity in speciose groups. *Philosophical Transactions of the Royal Society, London B* **345**, 119–36.

International Commission of Zoological Nomenclature (1985) *International Code of Zoological Nomenclature,* 3rd edn. International Trust for Zoological Nomenclature, London, in association with British Museum (Natural History) and University of California Press, Berkeley, CA.

May, R.M. (1994) Conceptual aspects of the quantification of the extent of biodiversity. *Philosophical Transactions of the Royal Society, London B* **345**, 13–20.

New, T.R. (1995) *An Introduction to Invertebrate Conservation Biology.* Oxford University Press, Oxford.

Novotny, V., Basset, Y., Miller, S.E., Weiblen, G.D., Bremer, B., Cizek, L. & Drozi, P. (2002) Low host specificity of herbivorous insects in a tropical forest. *Nature* **416**, 841–4.

Price, P.W. (1997) *Insect Ecology,* 3rd edn. John Wiley & Sons, New York.

Roberts, C. (1998) Long-term costs of the mopane worm harvest. *Oryx* **32**(1), 6–8.

Samways, M.J. (1994) *Insect Conservation Biology.* Chapman & Hall, London.

Speight, M.R., Hunter, M.D. & Watt, A.D. (1999) *Ecology of Insects. Concepts and Applications.* Blackwell Science, Oxford.

Stork, N.E. (1988) Insect diversity: facts, fiction and speculation. *Biological Journal of the Linnean Society* **35**, 321–37.

Stork, N.E. (1993) How many species are there? *Biodiversity and Conservation* **2**, 215–32.

Stork, N.E., Adis, J. & Didham, R.K. (eds.) (1997) *Canopy Arthropods.* Chapman & Hall, London.

Tsutsui, N.D. & Suarez, A.V. (2003) The colony structure and population biology of invasive ants. *Conservation Biology* **17**, 48–58.

Vane-Wright, R.I. (1991) Why not eat insects? *Bulletin of Entomological Research* **81**, 1–4.

Wheeler, Q.D. (1990) Insect diversity and cladistic constraints. *Annals of the Entomological Society of America* **83**, 1031–47.

See also articles in "Conservation Special" *Antenna* **25**(1) (2001) and "Arthropod Diversity and Conservation in Southern Africa" *African Entomology* **10**(1) (2002).

Chapter 2

EXTERNAL ANATOMY

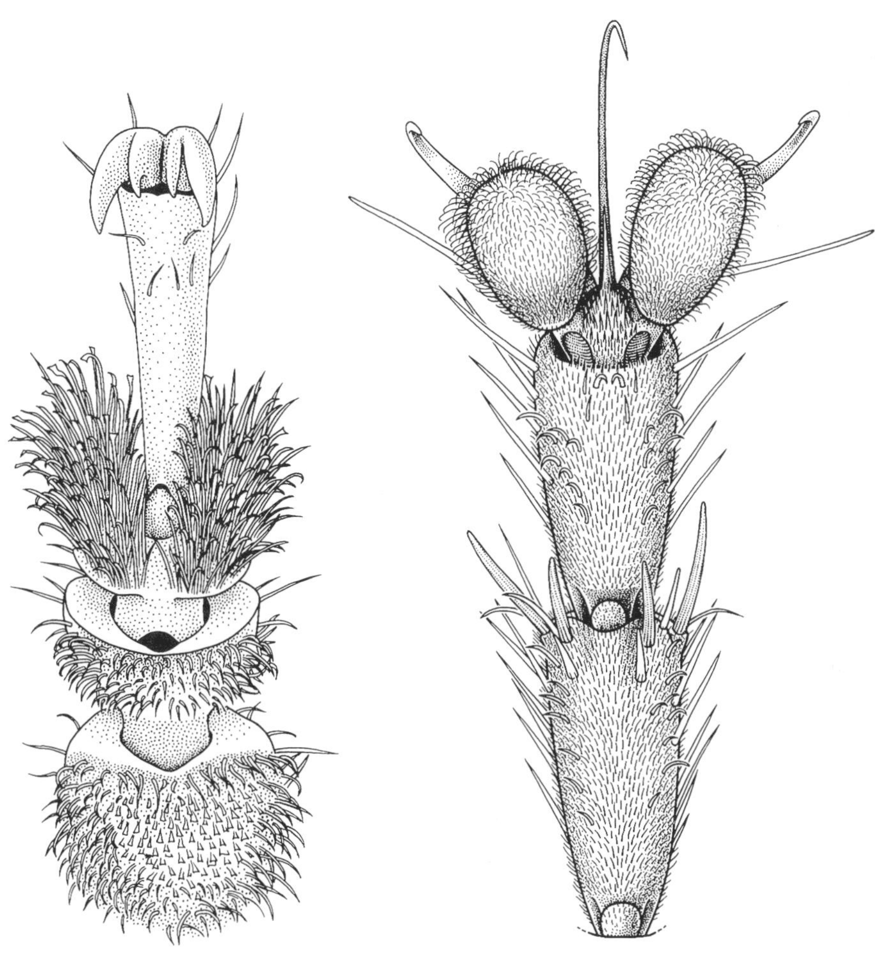

"Feet" of leaf beetle (left) and bush fly (right). (From scanning electron micrographs by C.A.M. Reid & A.C. Stewart.)

Insects are segmented invertebrates that possess the articulated external skeleton (exoskeleton) characteristic of all arthropods. Groups are differentiated by various modifications of the exoskeleton and the appendages – for example, the Hexapoda to which the Insecta belong (section 7.2) is characterized by having six-legged adults. Many anatomical features of the appendages, especially of the mouthparts, legs, wings, and abdominal apex, are important in recognizing the higher groups within the hexapods, including insect orders, families, and genera. Differences between species frequently are indicated by less obvious anatomical differences. Furthermore, the biomechanical analysis of morphology (e.g. studying how insects fly or feed) depends on a thorough knowledge of structural features. Clearly, an understanding of external anatomy is necessary to interpret and appreciate the functions of the various insect designs and to allow identification of insects and their hexapod relatives. In this chapter we describe and discuss the cuticle, body segmentation, and the structure of the head, thorax, and abdomen and their appendages.

First some basic classification and terminology needs to be explained. Adult insects normally have wings (most of the Pterygota), the structure of which may diagnose orders, but there is a group of primitively wingless insects (the "apterygotes") (see section 7.4.1 and Box 9.3 for defining features). Within the Insecta, three major patterns of development can be recognized (section 6.2). Apterygotes (and non-insect hexapods) develop to adulthood with little change in body form (**ametaboly**), except for sexual maturation through development of gonads and genitalia. All other insects either have a gradual change in body form (**hemimetaboly**) with external wing buds getting larger at each molt, or an abrupt change from a wingless immature insect to winged adult stage via a pupal stage (**holometaboly**). Immature stages of hemimetabolous insects are generally called **nymphs**, whereas those of holometabolous insects are referred to as **larvae**.

Anatomical structures of different taxa are **homologous** if they share an evolutionary origin, i.e. if the genetic basis is inherited from an ancestor common to them both. For instance, the wings of all insects are believed to be homologous; this means that wings (but not necessarily flight; see section 8.4) originated once. Homology of structures generally is inferred by comparison of similarity in **ontogeny** (development from egg to adult), composition (size and detailed appearance), and position (on the same segment and same relative location on that segment). The homology of insect wings is demonstrated by similarities in venation and articulation – the wings of all insects can be derived from the same basic pattern or groundplan (as explained in section 2.4.2). Sometimes association with other structures of known homologies is helpful in establishing the homology of a structure of uncertain origin. Another sort of homology, called **serial homology**, refers to corresponding structures on different segments of an individual insect. Thus, the appendages of each body segment are serially homologous, although in living insects those on the head (antennae and mouthparts) are very different in appearance from those on the thorax (walking legs) and abdomen (genitalia and cerci). The way in which molecular developmental studies are confirming these serial homologies is described in Box 6.1.

2.1 THE CUTICLE

The cuticle is a key contributor to the success of the Insecta. This inert layer provides the strong **exoskeleton** of body and limbs, the **apodemes** (internal supports and muscle attachments), and wings, and acts as a barrier between living tissues and the environment. Internally, cuticle lines the tracheal tubes (section 3.5), some gland ducts and the foregut and hindgut of the digestive tract. Cuticle may range from rigid and armor-like, as in most adult beetles, to thin and flexible, as in many larvae. Restriction of water loss is a critical function of cuticle vital to the success of insects on land.

The cuticle is thin but its structure is complex and still the subject of some controversy. A single layer of cells, the **epidermis**, lies beneath and secretes the cuticle, which consists of a thicker **procuticle** overlaid with thin **epicuticle** (Fig. 2.1). The epidermis and cuticle together form an **integument** – the outer covering of the living tissues of an insect.

The epicuticle ranges from 3 μm down to 0.1 μm in thickness, and usually consists of three layers: an **inner epicuticle**, an **outer epicuticle**, and a **superficial layer**. The superficial layer (probably a glycoprotein) in many insects is covered by a lipid or wax layer, sometimes called a free-wax layer, with a variably discrete cement layer external to this. The chemistry of the epicuticle and its outer layers is vital in preventing dehydration, a function derived from water-repelling (hydrophobic) lipids, especially hydrocarbons. These

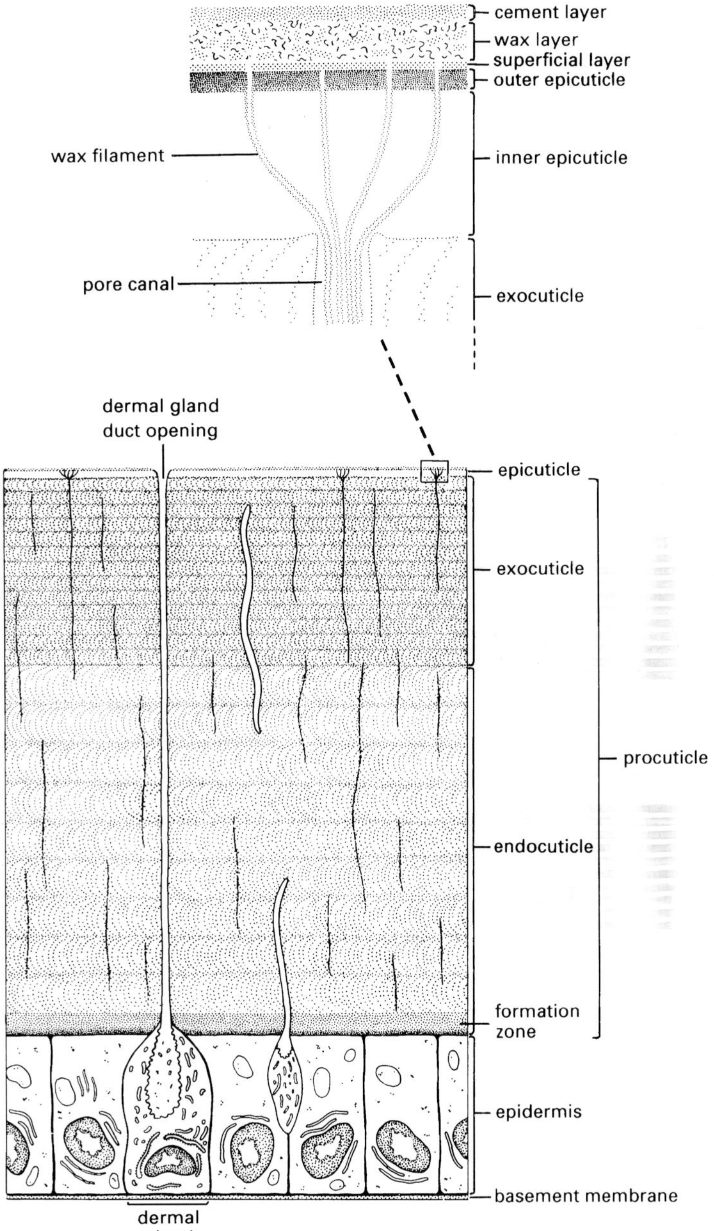

Fig. 2.1 The general structure of insect cuticle; the enlargement above shows details of the epicuticle. (After Hepburn 1985; Hadley 1986; Binnington 1993.)

compounds include free and protein-bound lipids, and the outermost waxy coatings give a bloom to the external surface of some insects. Other cuticular patterns, such as light reflectivity, are produced by various kinds of epicuticular surface microsculpturing, such as close-packed, regular or irregular tubercles, ridges, or tiny hairs. Lipid composition can vary and waxiness can increase seasonally or under dry conditions. Besides being water retentive, surface waxes may deter predation, provide patterns for mimicry or camouflage, repel

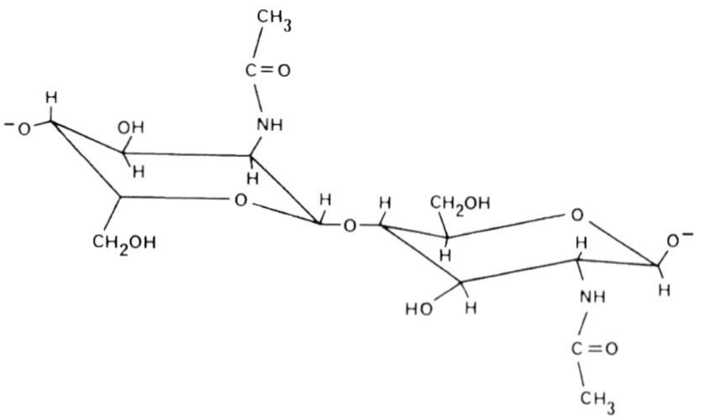

Fig. 2.2 Structure of part of a chitin chain, showing two linked units of *N*-acetyl-D-glucosamine. (After Cohen 1991.)

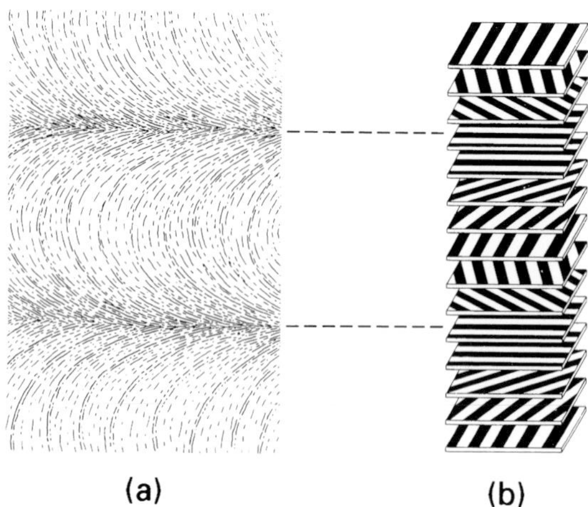

Fig. 2.3 The ultrastructure of cuticle (from a transmission electron micrograph). (a) The arrangement of chitin microfibrils in a helicoidal array produces characteristic (though artifactual) parabolic patterns. (b) Diagram of how the rotation of microfibrils produces a lamellar effect owing to microfibrils being either aligned or non-aligned to the plane of sectioning. (After Filshie 1982.)

excess rainwater, reflect solar and ultraviolet radiation, or give species-specific olfactory cues.

The epicuticle is inextensible and unsupportive. Instead, support is given by the underlying chitinous cuticle known as procuticle when it is first secreted. This differentiates into a thicker **endocuticle** covered by a thinner **exocuticle**, due to **sclerotization** of the latter. The procuticle is from 10 μm to 0.5 mm thick and consists primarily of chitin complexed with protein. This contrasts with the overlying epicuticle which lacks chitin.

Chitin is found as a supporting element in fungal cell walls and arthropod exoskeletons, and is especially important in insect extracellular structures. It is an unbranched polymer of high molecular weight – an amino-sugar polysaccharide predominantly composed of β-(1–4)-linked units of *N*-acetyl-D-glucosamine (Fig. 2.2).

Chitin molecules are grouped into bundles and assembled into flexible microfibrils that are embedded in, and intimately linked to, a protein matrix, giving great tensile strength. The commonest arrangement of chitin microfibrils is in a sheet, in which the microfibrils are in parallel. In the exocuticle, each successive sheet lies in the same plane but may be orientated at a slight angle relative to the previous sheet, such that a thickness of many sheets produces a helicoid arrangement, which in sectioned cuticle appears as alternating light and dark bands (lamellae). Thus the parabolic patterns and lamellar arrangement, visible so clearly in sectioned cuticle, represent an optical artifact resulting from microfibrillar orientation (Fig. 2.3). In the endocuticle, alternate stacked or helicoid arrangements of microfibrillar sheets may occur, often giving rise to thicker lamellae than in the exocuticle. Different arrangements may be laid down during darkness compared with daylight, allowing precise age determination in many adult insects.

Much of the strength of cuticle comes from extensive hydrogen bonding of adjacent chitin chains. Additional stiffening comes from **sclerotization**, an irreversible process that darkens the exocuticle and results in the proteins becoming water-insoluble. Sclerotization may result from linkages of adjacent protein chains by phenolic bridges (quinone tanning), or from controlled dehydration of the chains, or both. Only exocuticle becomes sclerotized. The deposition of pigment in the cuticle, including deposition of melanin, may be associated with quinones, but is additional to sclerotization and not necessarily associated with it.

In contrast to the solid cuticle typical of sclerites and mouthparts such as mandibles, softer, plastic, highly flexible or truly elastic cuticles occur in insects in varying locations and proportions. Where elastic or spring-like movement occurs, such as in wing ligaments or for the jump of a flea, **resilin** – a "rubber-like" protein – is present. The coiled polypeptide chains of this protein function as a mechanical spring under tension or compression, or in bending.

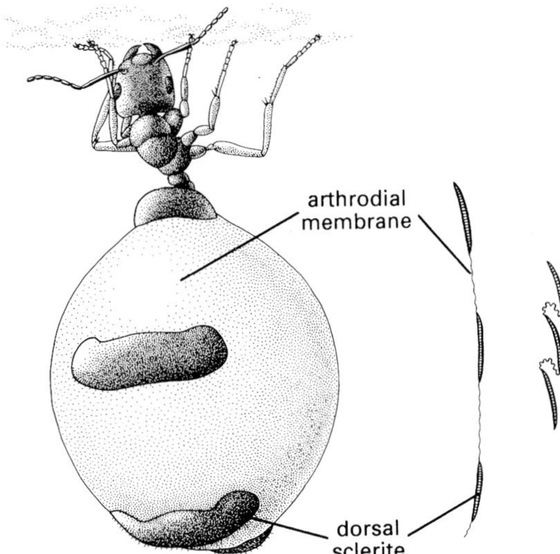

Fig. 2.4 A specialized worker, or replete, of the honeypot ant, *Camponotus inflatus* (Hymenoptera: Formicidae), which holds honey in its distensible abdomen and acts as a food store for the colony. The arthrodial membrane between tergal plates is depicted to the right in its unfolded and folded conditions. (After Hadley 1986; Devitt 1989.)

In soft-bodied larvae and in the membranes between segments, the cuticle must be tough, but also flexible and capable of extension. This "soft" cuticle, sometimes termed **arthrodial membrane**, is evident in gravid females, for example in the ovipositing migratory locust, *Locusta migratoria* (Orthoptera: Acrididae), in which intersegmental membranes may be expanded up to 20-fold for oviposition. Similarly, the gross abdominal dilation of gravid queen bees, termites, and ants is possible through expansion of the unsclerotized cuticle. In these insects, the overlying unstretchable epicuticle expands by unfolding from an originally highly folded state, and some new epicuticle is formed. An extreme example of the distensibility of arthrodial membrane is seen in honeypot ants (Fig. 2.4; see also section 12.2.3). In *Rhodnius* nymphs (Hemiptera: Reduviidae), changes in molecular structure of the cuticle allow actual stretching of the abdominal membrane to occur in response to intake of a large fluid volume during feeding.

Cuticular structural components, waxes, cements, pheromones (Chapter 4), and defensive and other compounds are products of the epidermis, which is a near-continuous, single-celled layer beneath the cuticle. Many of these compounds are secreted to the outside of the insect epicuticle. Numerous fine **pore canals** traverse the procuticle and then branch into numerous finer **wax canals** (containing **wax filaments**) within the epicuticle (enlargement in Fig. 2.1); this system transports lipids (waxes) from the epidermis to the epicuticular surface. The wax canals may also have a structural role within the epicuticle. **Dermal glands**, part of the epidermis, produce cement and/or wax, which is transported via larger ducts to the cuticular surface. Wax-secreting glands are particularly well developed in mealybugs and other scale insects (Fig. 2.5). The epidermis is closely associated with molting – the events and processes leading up to and including **ecdysis** (eclosion), i.e. the shedding of the old cuticle (section 6.3).

Insects are well endowed with cuticular extensions, varying from fine and hair-like to robust and spine-like. Four basic types of protuberance (Fig. 2.6), all with sclerotized cuticle, can be recognized on morphological, functional, and developmental grounds:

1 **spines** are multicellular with undifferentiated epidermal cells;

2 **setae**, also called **hairs**, **macrotrichia**, or **trichoid sensilla**, are multicellular with specialized cells;

3 **acanthae** are unicellular in origin;

4 **microtrichia** are subcellular, with several to many extensions per cell.

Setae sense much of the insect's tactile environment. Large setae may be called bristles or chaetae, with the most modified being **scales**, the flattened setae found on butterflies and moths (Lepidoptera) and sporadically elsewhere. Three separate cells form each seta, one for hair formation (**trichogen** cell), one for socket formation (**tormogen** cell), and one sensory cell (Fig. 4.1).

There is no such cellular differentiation in multicellular spines, unicellular acanthae, and subcellular microtrichia. The functions of these types of protuberances are diverse and sometimes debatable, but their sensory function appears limited. The production of pattern, including color, may be significant for some of the microscopic projections. Spines are immovable, but if they are articulated, then they are called **spurs**. Both spines and spurs may bear unicellular or subcellular processes.

2.1.1 Color production

The diverse colors of insects are produced by the interaction of light with cuticle and/or underlying cells or

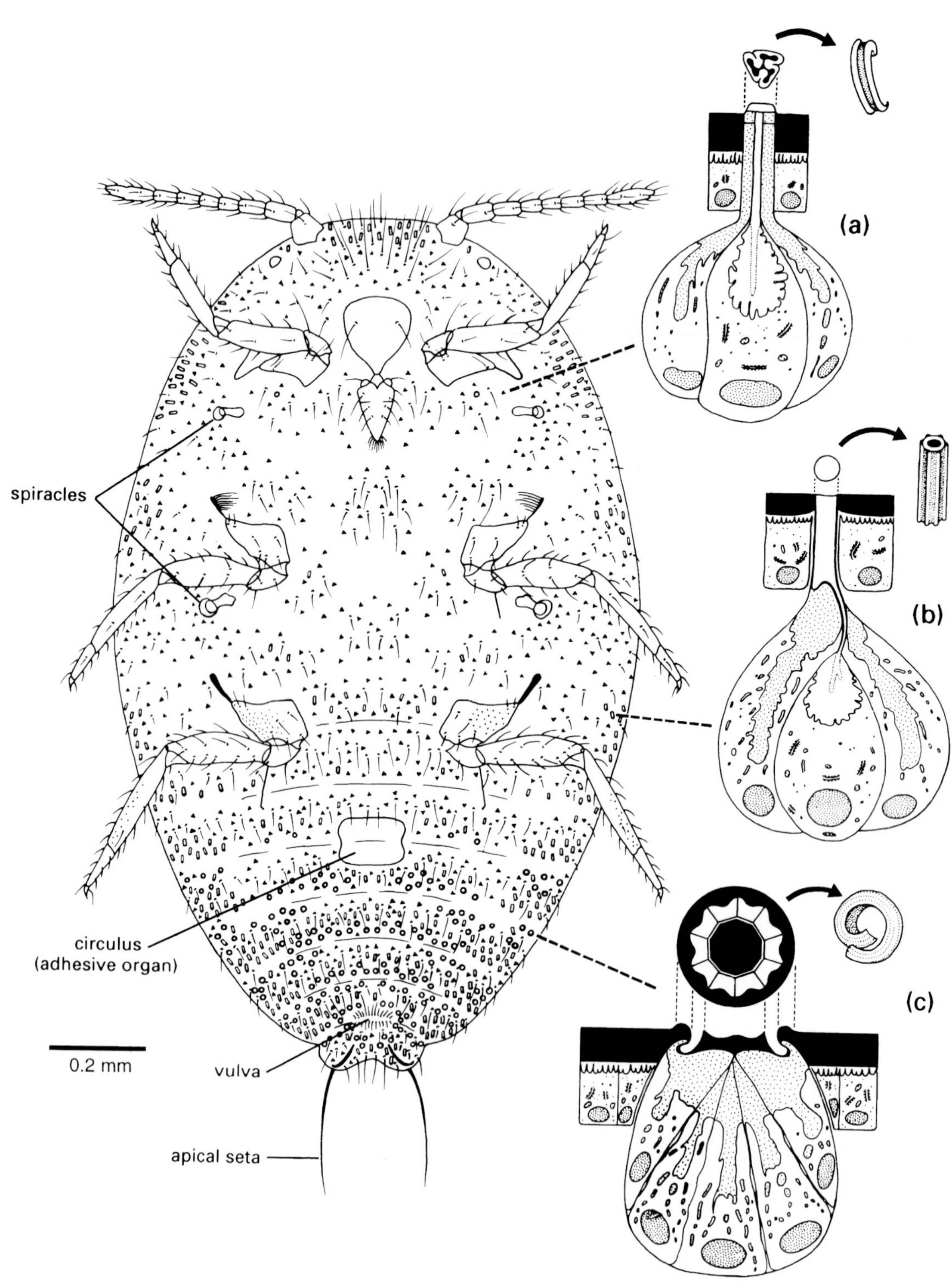
spiracles
circulus
(adhesive organ)
0.2 mm
vulva
apical seta
(a)
(b)
(c)

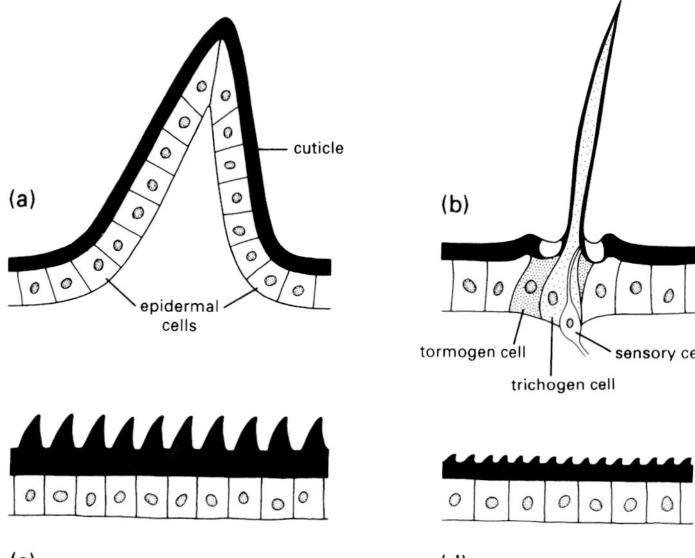

Fig. 2.6 The four basic types of cuticular protuberances: (a) a multicellular spine; (b) a seta, or trichoid sensillum; (c) acanthae; and (d) microtrichia. (After Richards & Richards 1979.)

fluid by two different mechanisms. Physical (structural) colors result from light scattering, interference, and diffraction, whereas pigmentary colors are due to the absorption of visible light by a range of chemicals. Often both mechanisms occur together to produce a color different from either alone.

All physical colors derive from the cuticle and its protuberances. **Interference** colors, such as iridescence and ultraviolet, are produced by refraction from varyingly spaced, close reflective layers produced by microfibrillar orientation within the exocuticle, or, in some beetles, the epicuticle, and by diffraction from regularly textured surfaces such as on many scales. Colors produced by light **scattering** depend on the size of surface irregularities relative to the wavelength of light. Thus, whites are produced by structures larger than the wavelength of light, such that all light is reflected, whereas blues are produced by irregularities that reflect only short wavelengths.

Insect pigments are produced in three ways:

1 by the insect's own metabolism;

2 by sequestering from a plant source;

3 rarely, by microbial endosymbionts.

Pigments may be located in the cuticle, epidermis, hemolymph, or fat body. Cuticular darkening is the most ubiquitous insect color. This may be due to either sclerotization (unrelated to pigmentation) or the exocuticular deposition of melanins, a heterogeneous group of polymers that may give a black, brown, yellow, or red color. Carotenoids, ommochromes, papiliochromes, and pteridines (pterins) mostly produce yellows to reds, flavonoids give yellow, and tetrapyrroles (including breakdown products of porphyrins such as chlorophyll and hemoglobin) create reds, blues, and greens. Quinone pigments occur in scale insects as red and yellow anthraquinones (e.g. carmine from cochineal insects), and in aphids as yellow to red to dark blue–green aphins.

Colors have an array of functions in addition to the obvious roles of color patterns in sexual and defensive display. For example, the ommochromes are the main visual pigments of insect eyes, whereas black melanin, an effective screen for possibly harmful light rays, can

Fig. 2.5 (*opposite*) The cuticular pores and ducts on the venter of an adult female of the citrus mealybug, *Planococcus citri* (Hemiptera: Pseudococcidae). Enlargements depict the ultrastructure of the wax glands and the various wax secretions (arrowed) associated with three types of cuticular structure: (a) a trilocular pore; (b) a tubular duct; and (c) a multilocular pore. Curled filaments of wax from the trilocular pores form a protective body-covering and prevent contamination with their own sugary excreta, or honeydew; long, hollow, and shorter curled filaments from the tubular ducts and multilocular pores, respectively, form the ovisac. (After Foldi 1983; Cox 1987.)

convert light energy into heat, and may act as a sink for free radicals that could otherwise damage cells. The red hemoglobins which are widespread respiratory pigments in vertebrates occur in a few insects, notably in some midge larvae and a few aquatic bugs, in which they have a similar respiratory function.

2.2 SEGMENTATION AND TAGMOSIS

Metameric segmentation, so distinctive in annelids, is visible only in some unsclerotized larvae (Fig. 2.7a). The segmentation seen in the sclerotized adult or nymphal insect is not directly homologous with that of larval insects, as sclerotization extends beyond each primary segment (Fig. 2.7b,c). Each apparent segment represents an area of sclerotization that commences in front of the fold that demarcates the primary segment and extends almost to the rear of that segment, leaving an unsclerotized area of the primary segment, the **conjunctival** or **intersegmental membrane**. This **secondary** segmentation means that the muscles, which are always inserted on the folds, are attached to solid rather than to soft cuticle. The apparent segments of adult insects, such as on the abdomen, are secondary in origin, but we refer to them simply as segments throughout this text.

In adult and nymphal insects, and hexapods in general, one of the most striking external features is the amalgamation of segments into functional units. This process of **tagmosis** has given rise to the familiar **tagmata** (regions) of **head**, **thorax**, and **abdomen**. In this process the 20 original segments have been divided into an embryologically detectable six-segmented head, three-segmented thorax, and 11-segmented abdomen (plus primitively the telson), although varying degrees of fusion mean that the full complement is never visible.

Before discussing the external morphology in more detail, some indication of orientation is required. The bilaterally symmetrical body may be described according to three axes:

1 **longitudinal**, or **anterior** to **posterior**, also termed **cephalic** (head) to **caudal** (tail);
2 **dorsoventral**, or **dorsal** (upper) to **ventral** (lower);
3 **transverse**, or **lateral** (outer) through the longitudinal axis to the opposite lateral (Fig. 2.8).

For appendages, such as legs or wings, **proximal** or **basal** refers to near the body, whereas **distal** or **apical** means distant from the body. In addition, structures are **mesal**, or **medial**, if they are nearer to the **midline** (**median line**), or **lateral** if closer to the body margin, relative to other structures.

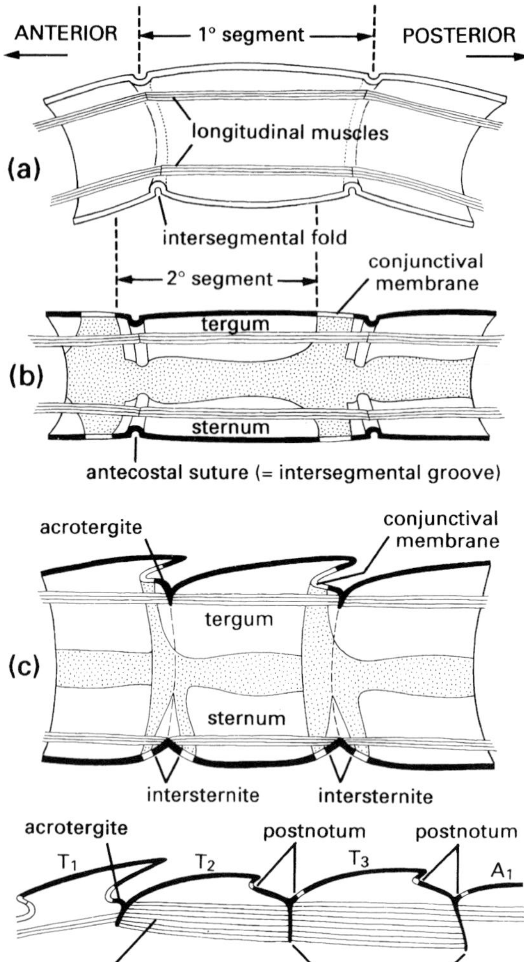

Fig. 2.7 Types of body segmentation. (a) Primary segmentation, as seen in soft-bodied larvae of some insects. (b) Simple secondary segmentation. (c) More derived secondary segmentation. (d) Longitudinal section of dorsum of the thorax of winged insects, in which the acrotergites of the second and third segments have enlarged to become the postnota. (After Snodgrass 1935.)

Four principal regions of the body surface can be recognized: the **dorsum** or upper surface; the **venter** or lower surface; and the two lateral **pleura** (singular:

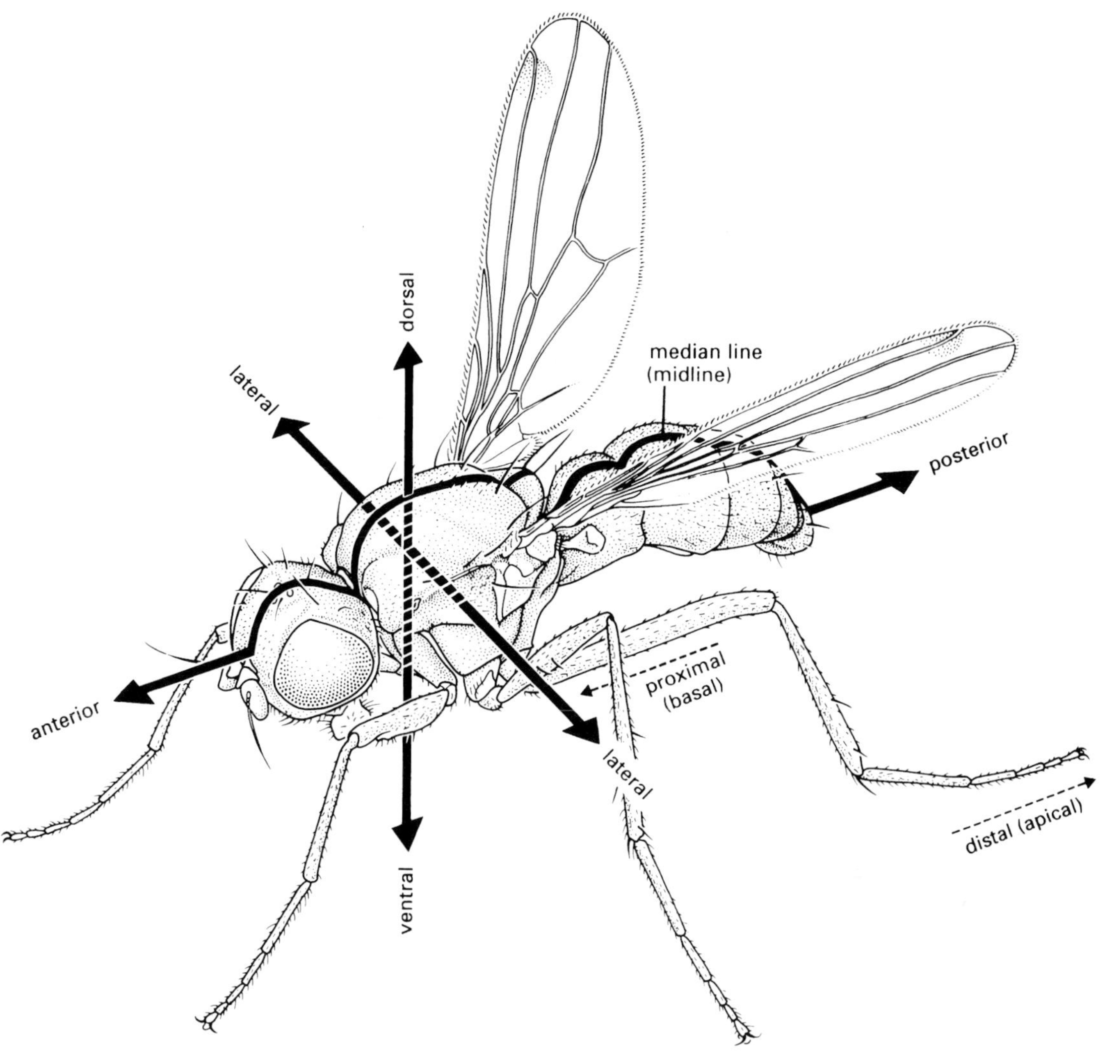

Fig. 2.8 The major body axes and the relationship of parts of the appendages to the body, shown for a sepsid fly. (After McAlpine 1987.)

pleuron), separating the dorsum from the venter and bearing limb bases, if these are present. Sclerotization that takes place in defined areas gives rise to plates called **sclerites**. The major segmental sclerites are the **tergum** (the dorsal plate; plural: **terga**), the **sternum** (the ventral plate; plural: **sterna**), and the pleuron (the side plate). If a sclerite is a subdivision of the tergum, sternum, or pleuron, the diminutive terms **tergite**, **sternite**, and **pleurite** may be applied.

The abdominal pleura are often at least partly membranous, but on the thorax they are sclerotized and usually linked to the tergum and sternum of each segment. This fusion forms a box, which contains the leg muscle insertions and, in winged insects, the flight muscles. With the exception of some larvae, the head sclerites are fused into a rigid capsule. In larvae (but not nymphs) the thorax and abdomen may remain membranous and tagmosis may be less apparent (such as in most wasp larvae and fly maggots) and the terga, sterna, and pleura are rarely distinct.

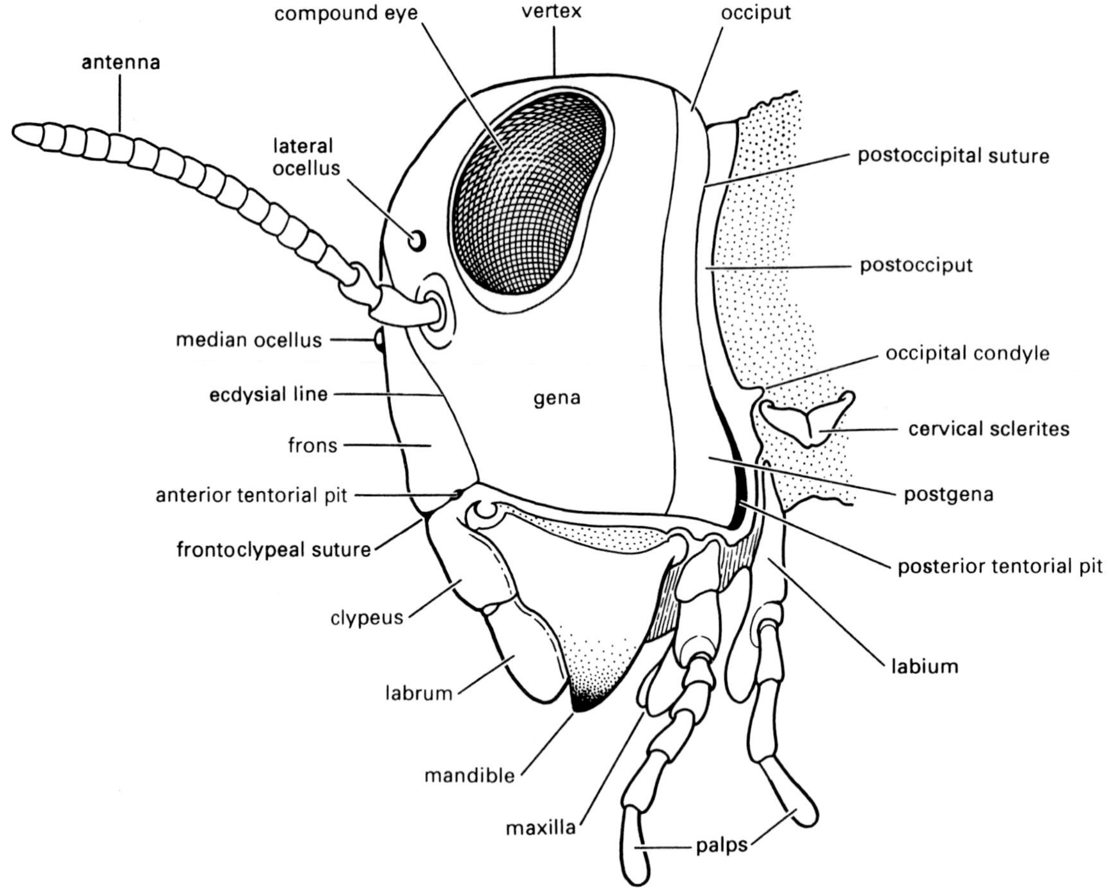

Fig. 2.9 Lateral view of the head of a generalized pterygote insect. (After Snodgrass 1935.)

2.3 THE HEAD

The rigid cranial capsule has two openings, one posteriorly through the **occipital foramen** to the prothorax, the other to the mouthparts. Typically the mouthparts are directed ventrally (**hypognathous**), although sometimes anteriorly (**prognathous**) as in many beetles, or posteriorly (**opisthognathous**) as in, for example, aphids, cicadas, and leafhoppers. Several regions can be recognized on the head (Fig. 2.9): the posterior horseshoe-shaped **posterior cranium** (dorsally the **occiput**) contacts the **vertex** dorsally and the **genae** (singular: **gena**) laterally; the vertex abuts the **frons** anteriorly and more anteriorly lies the **clypeus**, both of which may be fused into a **frontoclypeus**. In adult and nymphal insects, paired **compound eyes** lie more or less dorsolaterally between the vertex and genae, with a pair of sensory **antennae** placed more medially. In many insects, three light-sensitive "simple" eyes, or **ocelli**, are situated on the anterior vertex, typically arranged in a triangle, and many larvae have stemmatal eyes.

The head regions are often somewhat weakly delimited, with some indications of their extent coming from **sutures** (external grooves or lines on the head). Three sorts may be recognized:

1 remnants of original segmentation, generally restricted to the **postoccipital suture**;

2 **ecdysial lines** of weakness where the head capsule of the immature insect splits at molting (section 6.3), including an often prominent inverted "Y", or **epi-**

cranial suture, on the vertex (Fig. 2.10); the frons is delimited by the arms (also called **frontal sutures**) of this "Y";

3 grooves that reflect the underlying internal skeletal ridges, such as the **frontoclypeal** or **epistomal** suture, which often delimits the frons from the more anterior clypeus.

The head endoskeleton consists of several invaginated ridges and arms (**apophyses**, or elongate apodemes), the most important of which are the two pairs of **tentorial arms**, one pair being posterior, the other anterior, sometimes with an additional dorsal component. Some of these arms may be absent or, in pterygotes, fused to form the **tentorium**, an endoskeletal strut. Pits are discernible on the surface of the cranium at the points where the tentorial arms invaginate. These pits and the sutures may provide prominent landmarks on the head but usually they bear little or no association with the segments.

The segmental origin of the head is most clearly demonstrated by the mouthparts (section 2.3.1). From anterior to posterior, there are six fused head segments:

1 labral;
2 antennal, with each antenna equivalent to an entire leg;
3 postantennal, fused with the antennal segment;
4 mandibular;
5 maxillary;
6 labial.

The neck is mainly derived from the first part of the thorax and is not a segment.

2.3.1 Mouthparts

The mouthparts are formed from appendages of all head segments except the second. In omnivorous insects, such as cockroaches, crickets, and earwigs, the mouthparts are of a biting and chewing type (**mandibulate**) and resemble the probable basic design of ancestral pterygote insects more closely than the mouthparts of the majority of modern insects. Extreme modifications of basic mouthpart structure, correlated with feeding specializations, occur in most Lepidoptera, Diptera, Hymenoptera, Hemiptera, and a number of the smaller orders. Here we first discuss basic mandibulate mouthparts, as exemplified by the European earwig, *Forficula auricularia* (Dermaptera: Forficulidae) (Fig. 2.10), and then describe some of the more common modifications associated with more specialized diets.

There are five basic components of the mouthparts:

1 **labrum**, or "upper lip", with a ventral surface called the **epipharynx**;
2 **hypopharynx**, a tongue-like structure;
3 **mandibles**, or jaws;
4 **maxillae** (singular: **maxilla**);
5 **labium**, or "lower lip" (Fig. 2.10).

The labrum forms the roof of the preoral cavity and mouth (Fig. 3.14) and covers the base of the mandibles; it may be formed from fusion of parts of a pair of ancestral appendages. Projecting forwards from the back of the preoral cavity is the hypopharynx, a lobe of probable composite origin; in apterygotes, earwigs, and nymphal mayflies the hypopharynx bears a pair of lateral lobes, the **superlinguae** (singular: **superlingua**) (Fig. 2.10). It divides the cavity into a dorsal food pouch, or **cibarium**, and a ventral **salivarium** into which the salivary duct opens (Fig. 3.14). The mandibles, maxillae, and labium are the paired appendages of segments 4–6 and are highly variable in structure among insect orders; their serial homology with walking legs is more apparent than for the labrum and hypopharynx.

The mandibles cut and crush food and may be used for defense; generally they have an apical cutting edge and the more basal molar area grinds the food. They can be extremely hard (approximately 3 on Moh's scale of mineral hardness, or an indentation hardness of about 30 kg mm^{-2}) and thus many termites and beetles have no physical difficulty in boring through foils made from such common metals as copper, lead, tin, and zinc. Behind the mandibles lie the maxillae, each consisting of a basal part composed of the proximal **cardo** and the more distal **stipes** and, attached to the stipes, two lobes – the mesal **lacinia** and the lateral **galea** – and a lateral, segmented **maxillary palp**, or **palpus** (plural: **palps** or **palpi**). Functionally, the maxillae assist the mandibles in processing food; the pointed and sclerotized lacinae hold and macerate the food, whereas the galeae and palps bear sensory setae (mechanoreceptors) and chemoreceptors which sample items before ingestion. The appendages of the sixth segment of the head are fused with the sternum to form the labium, which is believed to be homologous to the second maxillae of Crustacea. In prognathous insects, such as the earwig, the labium attaches to the ventral surface of the head via a ventromedial sclerotized plate called the **gula** (Fig. 2.10). There are two main parts to the labium: the proximal **postmentum**, closely connected to the posteroventral surface of the

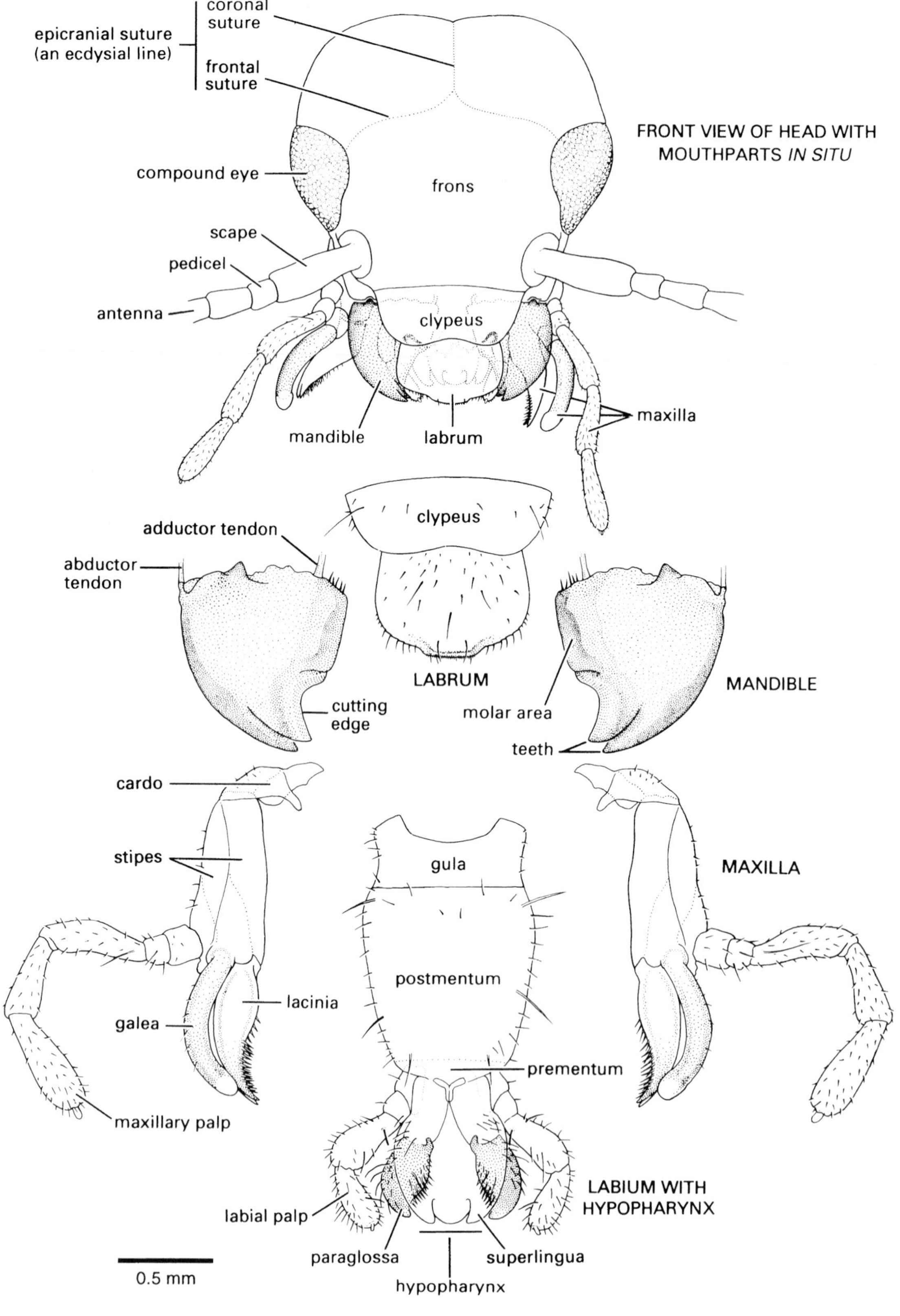
coronal suture
epicranial suture (an ecdysial line)
frontal suture
FRONT VIEW OF HEAD WITH MOUTHPARTS *IN SITU*
compound eye
frons
scape
pedicel
antenna
clypeus
maxilla
mandible
labrum
clypeus
adductor tendon
abductor tendon
LABRUM
MANDIBLE
cutting edge
molar area
teeth
cardo
stipes
gula
MAXILLA
postmentum
lacinia
galea
prementum
maxillary palp
LABIUM WITH HYPOPHARYNX
labial palp
paraglossa
superlingua
hypopharynx
0.5 mm

head and sometimes subdivided into a submentum and mentum; and the free distal **prementum**, typically bearing a pair of **labial palps** lateral to two pairs of lobes, the mesal **glossae** (singular: **glossa**) and the more lateral **paraglossae** (singular: **paraglossa**). The glossae and paraglossae, including sometimes the distal part of the prementum to which they attach, are known collectively as the **ligula**; the lobes may be variously fused or reduced as in *Forficula* (Fig. 2.10), in which the glossae are absent. The prementum with its lobes forms the floor of the preoral cavity (functionally a "lower lip"), whereas the labial palps have a sensory function, similar to that of the maxillary palps.

During insect evolution, an array of different mouthpart types have been derived from the basic design described above. Often feeding structures are characteristic of all members of a genus, family, or order of insects, so that knowledge of mouthparts is useful for both taxonomic classification and identification, and for ecological generalization (see section 10.6). Mouthpart structure is categorized generally according to feeding method, but mandibles and other components may function in defensive combat or even male–male sexual contests, as in the enlarged mandibles on certain male beetles (Lucanidae). Insect mouthparts have diversified in different orders, with feeding methods that include lapping, suctorial feeding, biting, or piercing combined with sucking, and filter feeding, in addition to the basic chewing mode.

The mouthparts of bees are of a chewing and lapping type. Lapping is a mode of feeding in which liquid or semi-liquid food adhering to a protrusible organ, or "tongue", is transferred from substrate to mouth. In the honey bee, *Apis mellifera* (Hymenoptera: Apidae), the elongate and fused labial glossae form a hairy tongue, which is surrounded by the maxillary galeae and the labial palps to form a tubular proboscis containing a food canal (Fig. 2.11). In feeding, the tongue is dipped into the nectar or honey, which adheres to the hairs, and then is retracted so that adhering liquid is carried into the space between the galeae and labial palps. This back-and-forth glossal movement occurs repeatedly. Movement of liquid to the mouth apparently results from the action of the cibarial pump, facilitated by each

Fig. 2.10 (*opposite*) Frontal view of the head and dissected mouthparts of an adult of the European earwig, *Forficula auricularia* (Dermaptera: Forficulidae). Note that the head is prognathous and thus a gular plate, or gula, occurs in the ventral neck region.

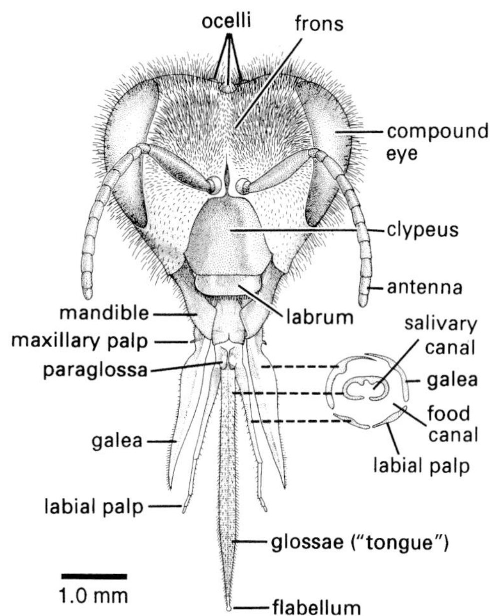

Fig. 2.11 Frontal view of the head of a worker honey bee, *Apis mellifera* (Hymenoptera: Apidae), with transverse section of proboscis showing how the "tongue" (fused labial glossae) is enclosed within the sucking tube formed from the maxillary galae and labial palps. (Inset after Wigglesworth 1964.)

retraction of the tongue pushing liquid up the food canal. The maxillary laciniae and palps are rudimentary and the paraglossae embrace the base of the tongue, directing saliva from the dorsal salivary orifice around into a ventral channel from whence it is transported to the **flabellum**, a small lobe at the glossal tip; saliva may dissolve solid or semi-solid sugar. The sclerotized, spoon-shaped mandibles lie at the base of the proboscis and have a variety of functions, including the manipulation of wax and plant resins for nest construction, the feeding of larvae and the queen, grooming, fighting, and the removal of nest debris including dead bees.

Most adult Lepidoptera and some adult flies obtain their food solely by sucking up liquids using suctorial (**haustellate**) mouthparts that form a proboscis or rostrum (Box 15.5). Pumping of the liquid food is achieved by muscles of the cibarium and/or pharynx. The proboscis of moths and butterflies, formed from the greatly elongated maxillary galeae, is extended (Fig. 2.12a) by increases in hemolymph ("blood") pressure. It is loosely coiled by the inherent elasticity of the cuticle, but tight coiling requires contraction of intrinsic muscles

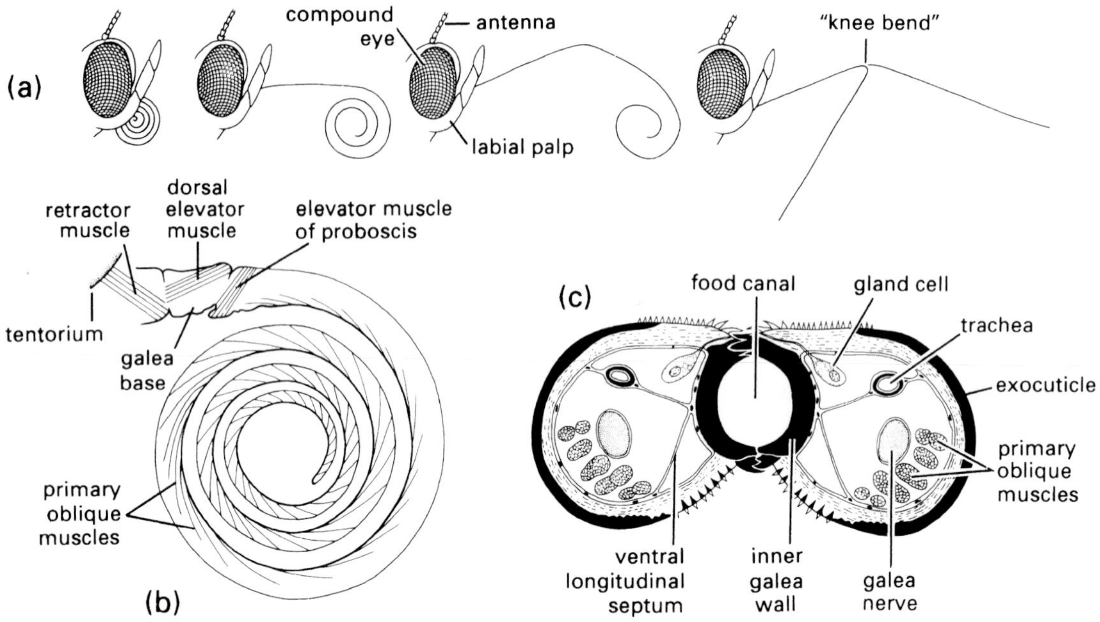

Fig. 2.12 Mouthparts of a cabbage white butterfly, *Pieris* species (Lepidoptera: Pieridae). (a) Positions of the proboscis showing, from left to right, at rest, with proximal region uncoiling, with distal region uncoiling, and fully extended with tip in two of many possible different positions due to flexing at "knee bend". (b) Lateral view of proboscis musculature. (c) Transverse section of the proboscis in the proximal region. (After Eastham & Eassa 1955.)

(Fig. 2.12b). A cross-section of the proboscis (Fig. 2.12c) shows how the food canal, which opens basally into the cibarial pump, is formed by apposition and interlocking of the two galeae. The proboscis of some male hawkmoths (Sphingidae), such as that of *Xanthopan morgani*, can attain great length (Fig. 11.8).

A few moths and many flies combine sucking with piercing or biting. For example, moths that pierce fruit and exceptionally suck blood (species of Noctuidae) have spines and hooks at the tip of their proboscis which are rasped against the skins of either ungulate mammals or fruit. For at least some moths, penetration is effected by the alternate protraction and retraction of the two galeae that slide along each other. Blood-feeding flies have a variety of skin-penetration and feeding mechanisms. In the "lower" flies such as mosquitoes and black flies, and the Tabanidae (horse flies, Brachycera), the labium of the adult fly forms a non-piercing sheath for the other mouthparts, which together contribute to the piercing structure. In contrast, the biting calyptrate dipterans (Brachycera: Calyptratae, e.g. stable flies and tsetse flies) lack mandibles and maxillae and the principal piercing organ is the highly modified labium. Mouthparts of adult Diptera are described in Box 15.5.

Other mouthpart modifications for piercing and sucking are seen in the true bugs (Hemiptera), thrips (Thysanoptera), fleas (Siphonaptera), and sucking lice (Phthiraptera: Anoplura). In each order different mouthpart components form needle-like stylets capable of piercing the plant or animal tissues upon which the insect feeds. Bugs have extremely long, thin paired mandibular and maxillary stylets, which fit together to form a flexible stylet-bundle containing a food canal and a salivary canal (Box 11.8). Thrips have three stylets – paired maxillary stylets (laciniae) plus the left mandibular one (Fig. 2.13). Sucking lice have three stylets – the hypopharyngeal (dorsal), the salivary (median), and the labial (ventral) – lying in a ventral sac of the head and opening at a small eversible proboscis armed with internal teeth that grip the host during blood-feeding (Fig. 2.14). Fleas possess a single stylet derived from the epipharynx, and the laciniae of the maxillae form two long cutting blades that are

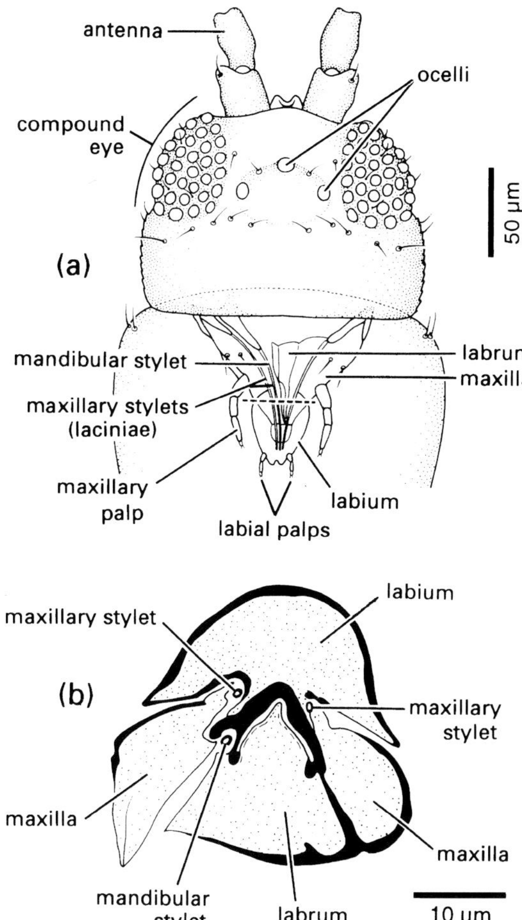

Fig. 2.13 Head and mouthparts of a thrips, *Thrips australis* (Thysanoptera: Thripidae). (a) Dorsal view of head showing mouthparts through prothorax. (b) Transverse section through proboscis. The plane of the transverse section is indicated by the dashed line in (a). (After Matsuda 1965; CSIRO 1970.)

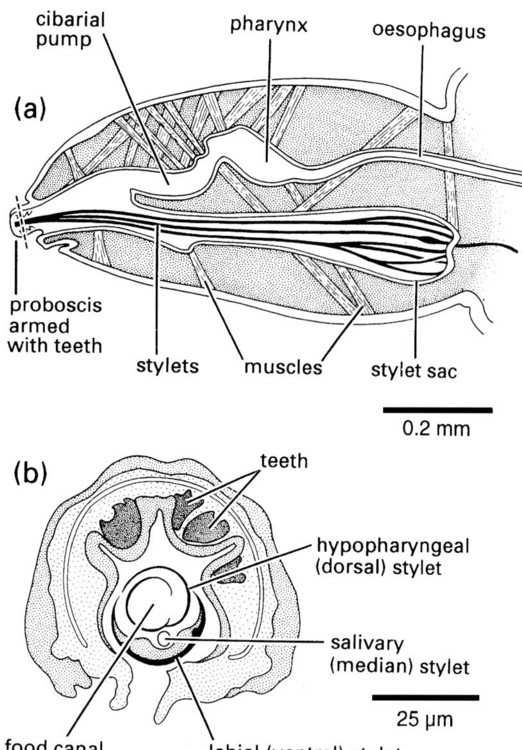

Fig. 2.14 Head and mouthparts of a sucking louse, *Pediculus* (Phthiraptera: Anoplura: Pediculidae). (a) Longitudinal section of head (nervous system omitted). (b) Transverse section through eversible proboscis. The plane of the transverse section is indicated by the dashed line in (a). (After Snodgrass 1935.)

ensheathed by the labial palps (Fig. 2.15). The Hemiptera and the Thysanoptera are sister groups and belong to the same assemblage as the Phthiraptera (Fig. 7.2), but the lice at least had a psocopteroid-like ancestor, presumably with mouthparts of a more generalized, mandibulate type. The Siphonaptera are distant relatives of the other three taxa; thus similarities in mouthpart structure among these orders result largely from parallel or, in the case of fleas, convergent evolution.

Slightly different piercing mouthparts are found in antlions and the predatory larvae of other lacewings (Neuroptera). The stylet-like mandible and maxilla on each side of the head fit together to form a sucking tube (Fig. 13.2c), and in some families (Chrysopidae, Myrmeleontidae, and Osmylidae) there is also a narrow poison channel. Generally, labial palps are present, maxillary palps are absent, and the labrum is reduced. Prey is seized by the pointed mandibles and maxillae, which are inserted into the victim; its body contents are digested extra-orally and sucked up by pumping of the cibarium.

A unique modification of the labium for prey capture occurs in nymphal damselflies and dragonflies (Odonata These predators catch other aquatic organisms by

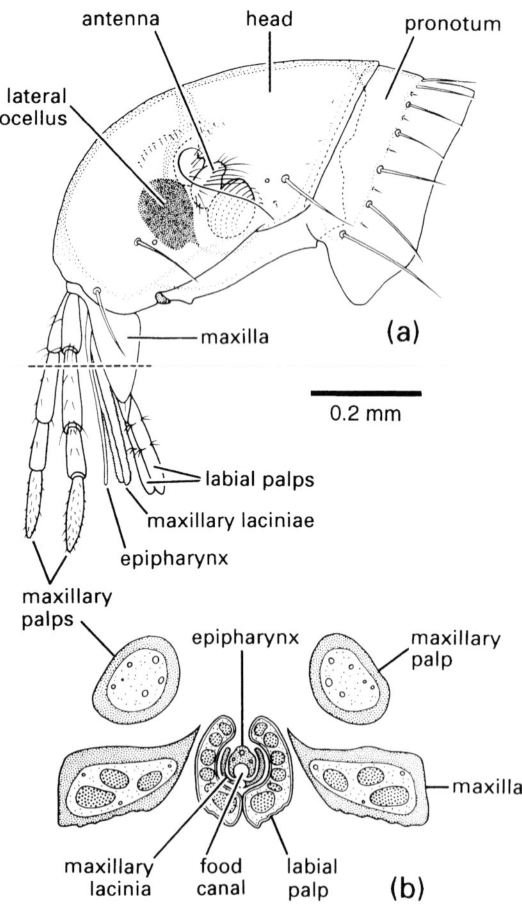

Fig. 2.15 Head and mouthparts of a human flea, *Pulex irritans* (Siphonaptera: Pulicidae): (a) lateral view of head; (b) transverse section through mouthparts. The plane of the transverse section is indicated by the dashed line in (a). (After Snodgrass 1946; Herms & James 1961.)

extending their folded labium (or "mask") rapidly and seizing mobile prey using prehensile apical hooks on modified labial palps (Fig. 13.4). The labium is hinged between the prementum and postmentum and, when folded, covers most of the underside of the head. Labial extension involves the sudden release of energy, produced by increases in blood pressure brought about by the contraction of thoracic and abdominal muscles, and stored elastically in a cuticular click mechanism at the prementum–postmentum joint. As the click mechanism is disengaged, the elevated hydraulic pressure shoots the labium rapidly forwards. Labial retraction then brings the captured prey to the other mouthparts for maceration.

Filter feeding in aquatic insects has been studied best in larval mosquitoes (Diptera: Culicidae), black flies (Diptera: Simuliidae), and net-spinning caddisflies (Trichoptera: many Hydropsychoidea and Philopotamoidea), which obtain their food by filtering particles (including bacteria, microscopic algae, and detritus) from the water in which they live. The mouthparts of the dipteran larvae have an array of setal "brushes" and/or "fans", which generate feeding currents or trap particulate matter and then move it to the mouth. In contrast, the caddisflies spin silk nets that filter particulate matter from flowing water and then use their mouthpart brushes to remove particles from the nets. Thus insect mouthparts are modified for filter feeding chiefly by the elaboration of setae. In mosquito larvae the lateral palatal brushes on the labrum generate the feeding currents (Fig. 2.16); they beat actively, causing particle-rich surface water to flow towards the mouthparts, where setae on the mandibles and maxillae help to move particles into the pharynx, where food boluses form at intervals.

In some adult insects, such as mayflies (Ephemeroptera), some Diptera (warble flies), a few moths (Lepidoptera), and male scale insects (Hemiptera: Coccoidea), mouthparts are greatly reduced and nonfunctional. Atrophied mouthparts correlate with short adult lifespan.

2.3.2 Cephalic sensory structures

The most obvious sensory structures of insects are on the head. Most adults and many nymphs have compound eyes dorsolaterally on head segment 4 and three ocelli on the vertex of the head. The median, or anterior, ocellus lies on segment 1 and is formed from a fused pair; the two lateral ocelli are on segment 3. The only visual structures of larval insects are **stemmata**, or simple eyes, positioned laterally on the head, either singly or in clusters. The structure and functioning of these three types of visual organs are described in detail in section 4.4.

Antennae are mobile, segmented, paired appendages. Primitively, they appear to be eight-segmented in nymphs and adults, but often there are numerous subdivisions, sometimes called **antennomeres**. The entire

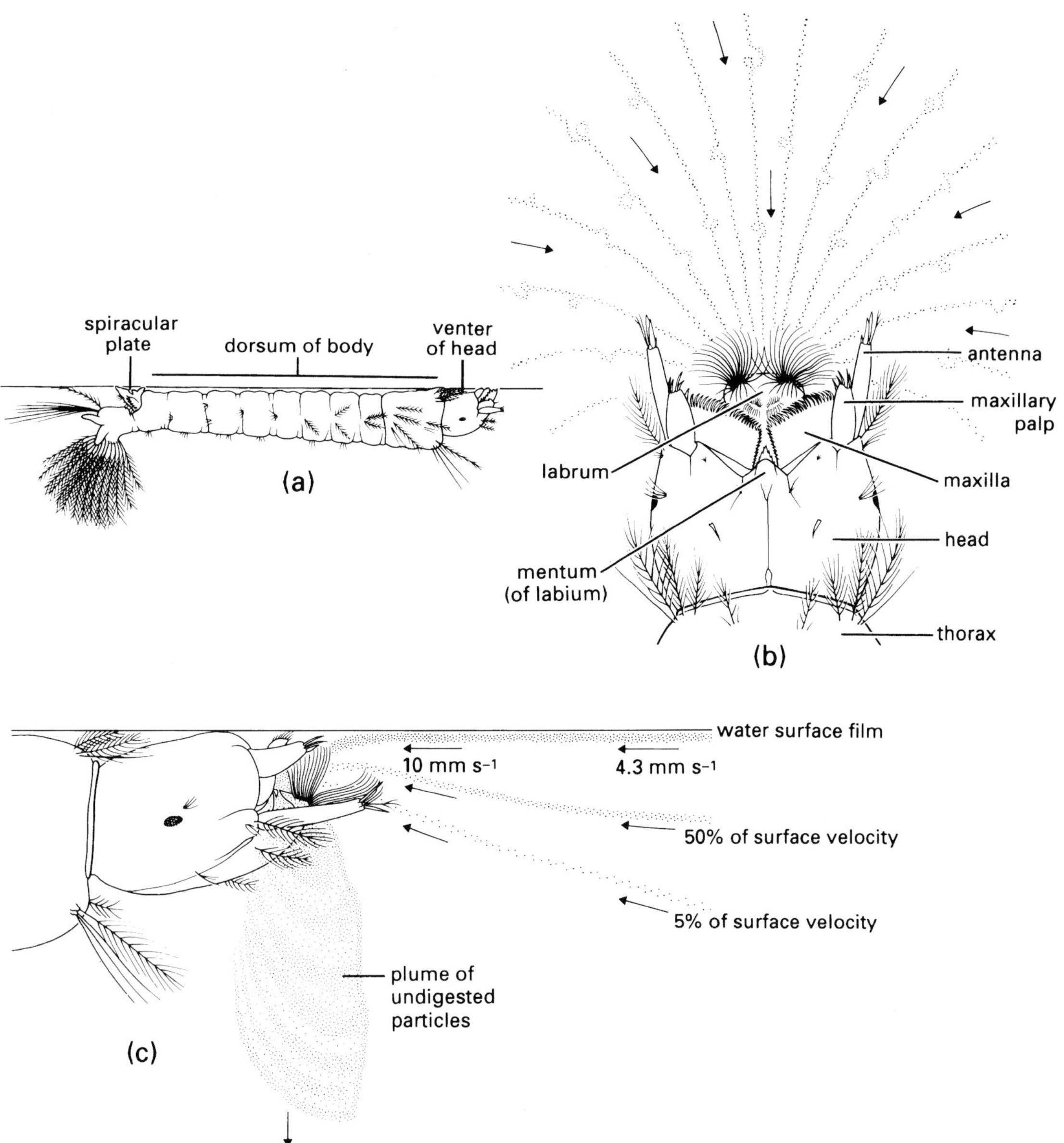

Fig. 2.16 The mouthparts and feeding currents of a mosquito larva of *Anopheles quadrimaculatus* (Diptera: Culicidae). (a) The larva floating just below the water surface, with head rotated through 180° relative to its body (which is dorsum-up so that the spiracular plate near the abdominal apex is in direct contact with the air). (b) Viewed from above showing the venter of the head and the feeding current generated by setal brushes on the labrum (direction of water movement and paths taken by surface particles are indicated by arrows and dotted lines, respectively). (c) Lateral view showing the particle-rich water being drawn into the preoral cavity between the mandibles and maxillae and its downward expulsion as the outward current. ((b,c) After Merritt et al. 1992.)

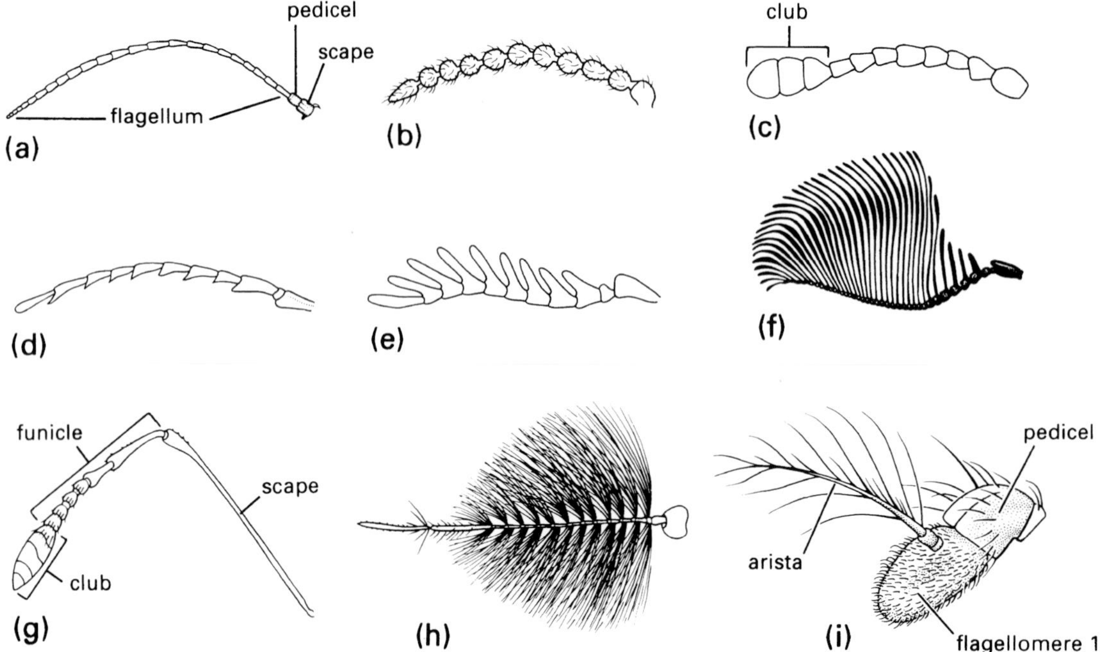

Fig. 2.17 Some types of insect antennae: (a) filiform – linear and slender; (b) moniliform – like a string of beads; (c) clavate or capitate – distinctly clubbed; (d) serrate – saw-like; (e) pectinate – comb-like; (f) flabellate – fan-shaped; (g) geniculate – elbowed; (h) plumose – bearing whorls of setae; and (i) aristate – with enlarged third segment bearing a bristle.

antenna typically has three main divisions (Fig. 2.17a): the first segment, or **scape**, generally is larger than the other segments and is the basal stalk; the second segment, or **pedicel**, nearly always contains a sensory organ known as **Johnston's organ**, which responds to movement of the distal part of the antenna relative to the pedicel; the remainder of the antenna, called the **flagellum**, is often filamentous and multisegmented (with many **flagellomeres**), but may be reduced or variously modified (Fig. 2.17b–i). The antennae are reduced or almost absent in some larval insects.

Numerous sensory organs, or **sensilla** (singular: **sensillum**), in the form of hairs, pegs, pits, or cones, occur on antennae and function as chemoreceptors, mechanoreceptors, thermoreceptors, and hygroreceptors (Chapter 4). Antennae of male insects may be more elaborate than those of the corresponding females, increasing the surface area available for detecting female sex pheromones (section 4.3.2).

The mouthparts, other than the mandibles, are well endowed with chemoreceptors and tactile setae. These sensilla are described in detail in Chapter 4.

2.4 THE THORAX

The thorax is composed of three segments: the first or **prothorax**, the second or **mesothorax**, and the third or **metathorax**. Primitively, and in apterygotes (bristletails and silverfish) and immature insects, these segments are similar in size and structural complexity. In most winged insects the mesothorax and metathorax are enlarged relative to the prothorax and form a **pterothorax**, bearing the wings and associated musculature. Wings occur only on the second and third segments in extant insects although some fossils have prothoracic winglets (Fig. 8.2) and homeotic mutants may develop prothoracic wings or wing buds. Almost all nymphal and adult insects have three pairs of thoracic legs – one pair per segment. Typically the legs are used for walking, although various other functions and associated modifications occur (section 2.4.1). Openings (**spiracles**) of the gas-exchange, or tracheal, system (section 3.5) are present laterally on the second and third thoracic segments at most with one pair per segment. However, a secondary condition in some

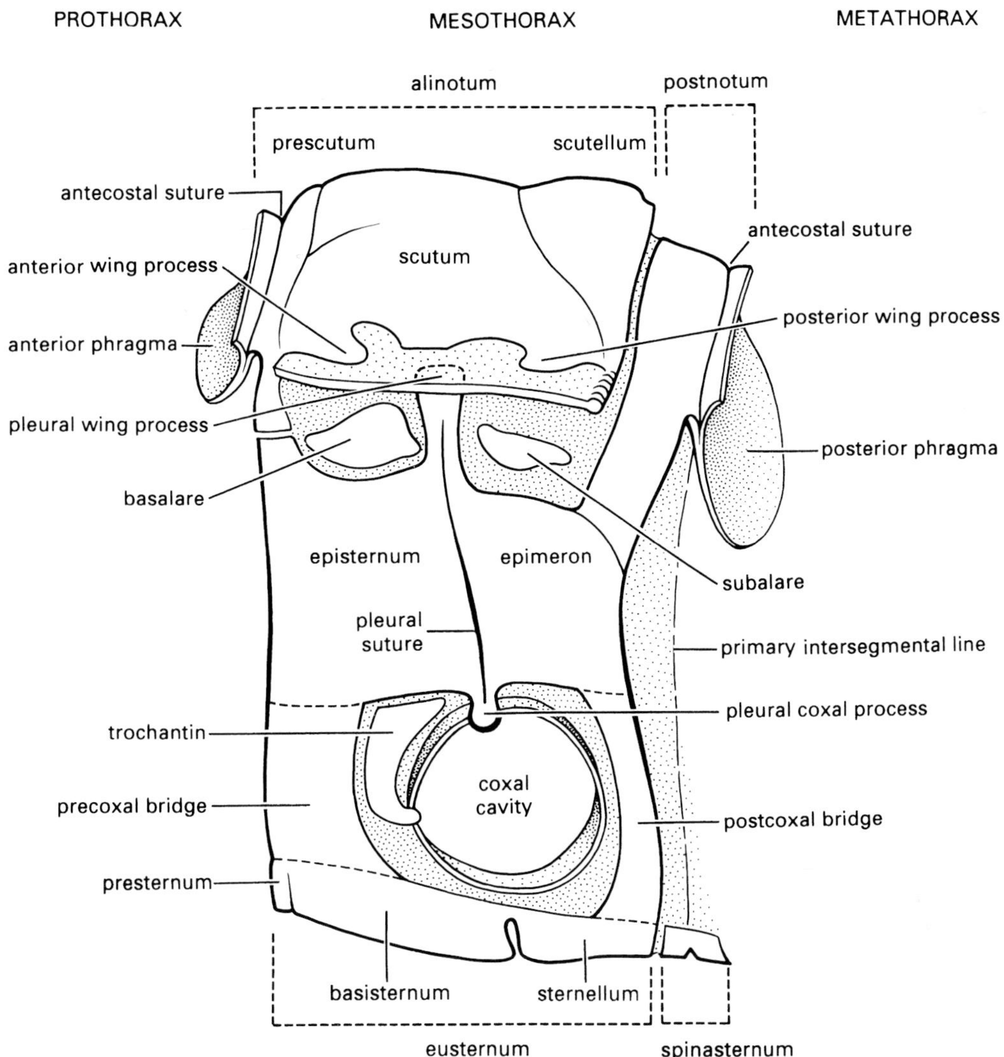

Fig. 2.18 Diagrammatic lateral view of a wing-bearing thoracic segment, showing the typical sclerites and their subdivisions. (After Snodgrass 1935.)

insects is for the mesothoracic spiracles to open on the prothorax.

The tergal plates of the thorax are simple structures in apterygotes and in many immature insects, but are variously modified in winged adults. Thoracic terga are called **nota** (singular: **notum**), to distinguish them from the abdominal terga. The **pronotum** of the prothorax may be simple in structure and small in comparison with the other nota, but in beetles, mantids, many bugs, and some Orthoptera the pronotum is expanded and in cockroaches it forms a shield that covers part of the head and mesothorax. The pterothoracic nota each have two main divisions – the anterior wing-bearing **alinotum** and the posterior phragma-bearing **postnotum** (Fig. 2.18). **Phragmata** (singular: **phragma**) are plate-like apodemes that extend inwards below the **antecostal sutures**, marking the primary intersegmental folds between segments; phragmata provide

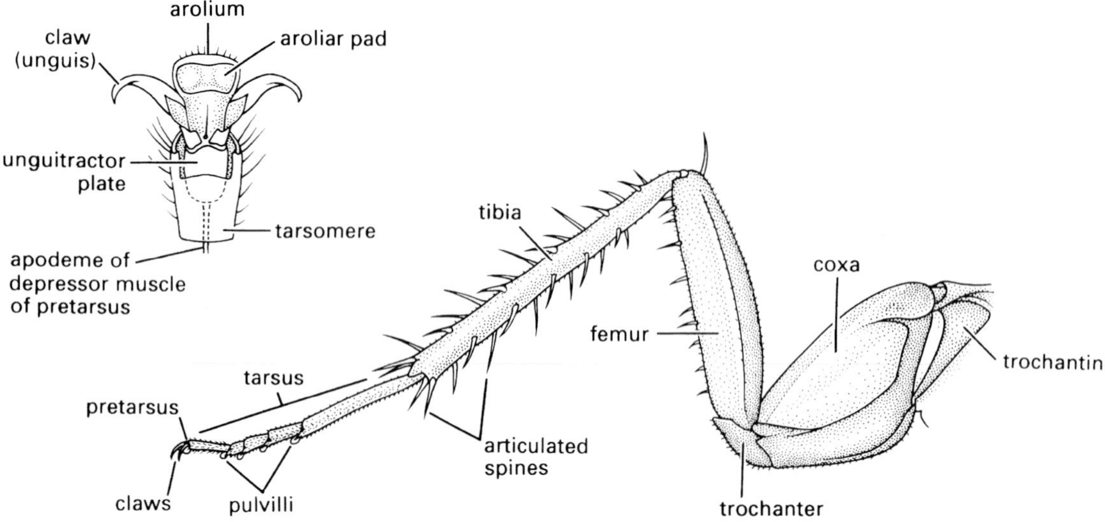

Fig. 2.19 The hind leg of a cockroach, *Periplaneta americana* (Blattodea: Blattidae), with enlargement of ventral surface of pretarsus and last tarsomere. (After Cornwell 1968; enlargement after Snodgrass 1935.)

attachment for the longitudinal flight muscles (Fig. 2.7d). Each alinotum (sometimes confusingly referred to as a "notum") may be traversed by sutures that mark the position of internal strengthening ridges and commonly divide the plate into three areas – the anterior **prescutum**, the **scutum**, and the smaller posterior **scutellum**.

The lateral pleural sclerites are believed to be derived from the subcoxal segment of the ancestral insect leg (Fig. 8.4a). These sclerites may be separate, as in silverfish, or fused into an almost continuous sclerotic area, as in most winged insects. In the pterothorax, the pleuron is divided into two main areas – the anterior **episternum** and the posterior **epimeron** – by an internal **pleural ridge**, which is visible externally as the **pleural suture** (Fig. 2.18); the ridge runs from the **pleural coxal process** (which articulates with the coxa) to the **pleural wing process** (which articulates with the wing), providing reinforcement for these articulation points. The **epipleurites** are small sclerites beneath the wing and consist of the **basalaria** anterior to the pleural wing process and the posterior **subalaria**, but often reduced to just one basalare and one subalare, which are attachment points for some direct flight muscles. The **trochantin** is the small sclerite anterior to the coxa.

The degree of ventral sclerotization on the thorax varies greatly in different insects. Sternal plates, if present, are typically two per segment: the **eusternum** and the following intersegmental sclerite or **intersternite** (Fig. 2.7c), commonly called the **spinasternum** (Fig. 2.18) because it usually has an internal apodeme called the **spina** (except for the metasternum which never has a spinasternum). The eusterna of the prothorax and mesothorax may fuse with the spinasterna of their segment. Each eusternum may be simple or divided into separate sclerites – typically the **presternum**, **basisternum**, and **sternellum**. The eusternum may be fused laterally with one of the pleural sclerites and is then called the **laterosternite**. Fusion of the sternal and pleural plates may form **precoxal** and **postcoxal bridges** (Fig. 2.18).

2.4.1 Legs

In most adult and nymphal insects, segmented **fore**, **mid**, and **hind legs** occur on the prothorax, mesothorax, and metathorax, respectively. Typically, each leg has six segments (Fig. 2.19) and these are, from proximal to distal: **coxa**, **trochanter**, **femur**, **tibia**, **tarsus**, and **pretarsus** (or more correctly **post-tarsus**) with claws. Additional segments – the prefemur, patella, and basitarsus (Fig. 8.4a) – are recognized in some fossil insects and other arthropods, such as arachnids, and one or more of these segments are evident in some

Ephemeroptera and Odonata. Primitively, two further segments lie proximal to the coxa and in extant insects one of these, the epicoxa, is associated with the wing articulation, or tergum, and the other, the subcoxa, with the pleuron (Fig. 8.4a).

The tarsus is subdivided into five or fewer components, giving the impression of segmentation; but, because there is only one tarsal muscle, **tarsomere** is a more appropriate term for each "pseudosegment". The first tarsomere sometimes is called the basitarsus, but should not be confused with the segment called the basitarsus in certain fossil insects. The underside of the tarsomeres may have ventral pads, **pulvilli**, also called **euplantulae**, which assist in adhesion to surfaces. Terminally on the leg, the small pretarsus (enlargement in Fig. 2.19) bears a pair of lateral **claws** (also called **ungues**) and usually a median lobe, the **arolium**. In Diptera there may be a central spine-like or pad-like **empodium** (plural: **empodia**) which is not the same as the arolium, and a pair of lateral pulvilli (as shown for the bush fly, *Musca vetustissima*, depicted on the right side of the vignette of this chapter). These structures allow flies to walk on walls and ceilings. The pretarsus of Hemiptera may bear a variety of structures, some of which appear to be pulvilli, whereas others have been called empodia or arolia, but the homologies are uncertain. In some beetles, such as Coccinellidae, Chrysomelidae, and Curculionidae, the ventral surface of some tarsomeres is clothed with adhesive setae that facilitate climbing. The left side of the vignette for this chapter shows the underside of the tarsus of the leaf beetle *Rhyparida* (Chrysomelidae).

Generally the femur and tibia are the longest leg segments but variations in the lengths and robustness of each segment relate to their functions. For example, walking (**gressorial**) and running (**cursorial**) insects usually have well-developed femora and tibiae on all legs, whereas jumping (**saltatorial**) insects such as grasshoppers have disproportionately developed hind femora and tibiae. In aquatic beetles (Coleoptera) and bugs (Hemiptera), the tibiae and/or tarsi of one or more pairs of legs usually are modified for swimming (**natatorial**) with fringes of long, slender hairs. Many ground-dwelling insects, such as mole crickets (Orthoptera: Gryllotalpidae), nymphal cicadas (Hemiptera: Cicadidae), and scarab beetles (Scarabaeidae), have the tibiae of the fore legs enlarged and modified for digging (**fossorial**) (Fig. 9.2), whereas the fore legs of some predatory insects, such as mantispid lacewings (Neuroptera) and mantids (Mantodea), are specialized for seizing prey (**raptorial**) (Fig. 13.3). The tibia and basal tarsomere of each hind leg of honey bees are modified for the collection and carriage of pollen (Fig. 12.4).

These "typical" thoracic legs are a distinctive feature of insects, whereas abdominal legs are confined to the immature stages of holometabolous insects. There have been conflicting views on whether (i) the legs on the immature thorax of the Holometabola are developmentally identical (serially homologous) to those of the abdomen, and/or (ii) the thoracic legs of the holometabolous immature stages are homologous with those of the adult. Detailed study of musculature and innervation shows similarity of development of thoracic legs throughout all stages of insects with ametaboly (without metamorphosis, as in silverfish) and hemimetaboly (partial metamorphosis and no pupal stage) and in adult Holometabola, having identical innervation through the lateral nerves. Moreover, the oldest known larva (from the Upper Carboniferous) has thoracic and abdominal legs/leglets each with a pair of claws, as in the legs of nymphs and adults. Although larval legs appear similar to those of adults and nymphs, the term **prolegs** is used for the larval leg. Prolegs on the abdomen, especially on caterpillars, usually are lobe-like and each bears an apical circle or band of small sclerotized hooks, or **crochets**. The thoracic prolegs may possess the same number of segments as the adult leg, but the number is more often reduced, apparently through fusion. In other cases, the thoracic prolegs, like those of the abdomen, are unsegmented outgrowths of the body wall, often bearing apical hooks.

2.4.2 Wings

Wings are developed fully only in the adult, or exceptionally in the subimago, the penultimate stage of Ephemeroptera. Typically, functional wings are flap-like cuticular projections supported by tubular, sclerotized veins. The major veins are longitudinal, running from the wing base towards the tip, and are more concentrated at the anterior margin. Additional supporting **cross-veins** are transverse struts, which join the longitudinal veins to give a more complex structure. The major veins usually contain tracheae, blood vessels, and nerve fibers, with the intervening membranous areas comprising the closely appressed dorsal and ventral cuticular surfaces. Generally, the major veins are alternately "convex" and "concave" in relation to the surface plane of the wing, especially near the

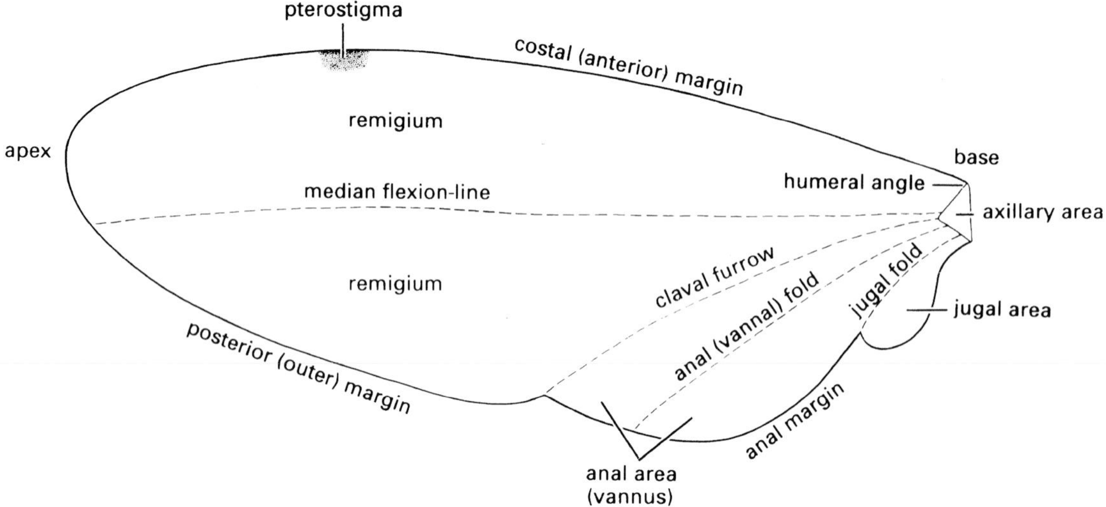

Fig. 2.20 Nomenclature for the main areas, folds, and margins of a generalized insect wing.

wing attachment; this configuration is described by plus (+) and minus (–) signs. Most veins lie in an anterior area of the wing called the **remigium** (Fig. 2.20), which, powered by the thoracic flight muscles, is responsible for most of the movements of flight. The area of wing posterior to the remigium sometimes is called the **clavus**; but more often two areas are recognized: an anterior **anal area** (or **vannus**) and a posterior **jugal area**. Wing areas are delimited and subdivided by **fold-lines**, along which the wing can be folded; and **flexion-lines**, at which the wing flexes during flight. The fundamental distinction between these two types of lines is often blurred, as fold-lines may permit some flexion and vice versa. The **claval furrow** (a flexion-line) and the **jugal fold** (or fold-line) are nearly constant in position in different insect groups, but the **median flexion-line** and the **anal** (or **vannal**) **fold** (or fold-line) form variable and unsatisfactory area boundaries. Wing folding may be very complicated; transverse folding occurs in the hind wings of Coleoptera and Dermaptera, and in some insects the enlarged anal area may be folded like a fan.

The fore and hind wings of insects in many orders are coupled together, which improves the aerodynamic efficiency of flight. The commonest coupling mechanism (seen clearly in Hymenoptera and some Trichoptera) is a row of small hooks, or **hamuli**, along the anterior margin of the hind wing that engages a fold along the posterior margin of the fore wing (**hamulate** coupling). In some other insects (e.g. Mecoptera, Lepidoptera, and some Trichoptera), a jugal lobe of the fore wing overlaps the anterior hind wing (**jugate** coupling), or the margins of the fore and hind wing overlap broadly (**amplexiform** coupling), or one or more hindwing bristles (the **frenulum**) hook under a retaining structure (the **retinaculum**) on the fore wing (**frenate** coupling). The mechanics of flight are described in section 3.1.4 and the evolution of wings is covered in section 8.4.

All winged insects share the same basic wing venation comprising eight veins, named from anterior to posterior of the wing as: **precosta (PC)**, **costa (C)**, **subcosta (Sc)**, **radius (R)**, **media (M)**, **cubitus (Cu)**, **anal (A)**, and **jugal (J)**. Primitively, each vein has an anterior convex (+) **sector** (a branch with all of its subdivisions) and a posterior concave (–) sector. In almost all extant insects, the precosta is fused with the costa and the jugal vein is rarely apparent. The wing nomenclatural system presented in Fig. 2.21 is that of Kukalová-Peck and is based on detailed comparative studies of fossil and living insects. This system can be applied to the venation of all insect orders, although as yet it has not been widely applied because the various schemes devised for each insect order have a long history of use and there is a reluctance to discard familiar systems. Thus in most textbooks, the same vein may be referred to by different names in different insect orders because the structural homologies were not recognized

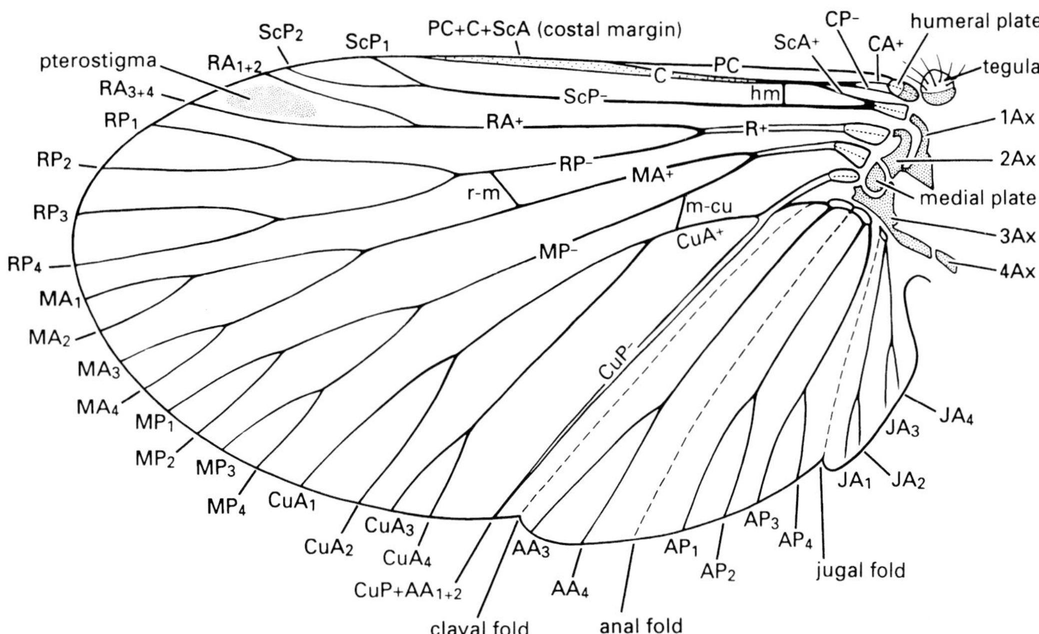

Fig. 2.21 A generalized wing of a neopteran insect (any living winged insect other than Ephemeroptera and Odonata), showing the articulation and the Kukalová-Peck nomenclatural scheme of wing venation. Notation as follows: AA, anal anterior; AP, anal posterior; Ax, axillary sclerite; C, costa; CA, costa anterior; CP, costa posterior; CuA, cubitus anterior; CuP, cubitus posterior; hm, humeral vein; JA, jugal anterior; MA, media anterior; m-cu, cross-vein between medial and cubital areas; MP, media posterior; PC, precosta; R, radius; RA, radius anterior; r-m, cross-vein between radial and median areas; RP, radius posterior; ScA, subcosta anterior; ScP, subcosta posterior. Branches of the anterior and posterior sector of each vein are numbered, e.g. CuA_{1-4}. (After CSIRO 1991.)

correctly in early studies. For example, until 1991, the venational scheme for Coleoptera labeled the radius posterior (RP) as the media (M) and the media posterior (MP) as the cubitus (Cu). Correct interpretation of venational homologies is essential for phylogenetic studies and the establishment of a single, universally applied scheme is essential.

Cells are areas of the wing delimited by veins and may be **open** (extending to the wing margin) or **closed** (surrounded by veins). They are named usually according to the longitudinal veins or vein branches that they lie behind, except that certain cells are known by special names, such as the discal cell in Lepidoptera (Fig. 2.22a) and the triangle in Odonata (Fig. 2.22b). The **pterostigma** is an opaque or pigmented spot anteriorly near the apex of the wing (Figs. 2.20 & 2.22b).

Wing venation patterns are consistent within groups (especially families and orders) but often differ between groups and, together with folds or pleats, provide major features used in insect classification and identification. Relative to the basic scheme outlined above, venation may be greatly reduced by loss or postulated fusion of veins, or increased in complexity by numerous cross-veins or substantial terminal branching. Other features that may be diagnostic of the wings of different insect groups are pigment patterns and colors, hairs, and scales. Scales occur on the wings of Lepidoptera, many Trichoptera, and a few psocids (Psocoptera) and flies. Hairs consist of small microtrichia, either scattered or grouped, and larger macrotrichia, typically on the veins.

Usually two pairs of functional wings lie dorsolaterally as **fore wings** on the mesothorax and as **hind wings** on the metathorax; typically the wings are membranous and transparent. However, from this basic pattern are derived many other conditions, often involving variation in the relative size, shape, and degree of sclerotization of the fore and hind wings. Examples of fore-wing modification include the

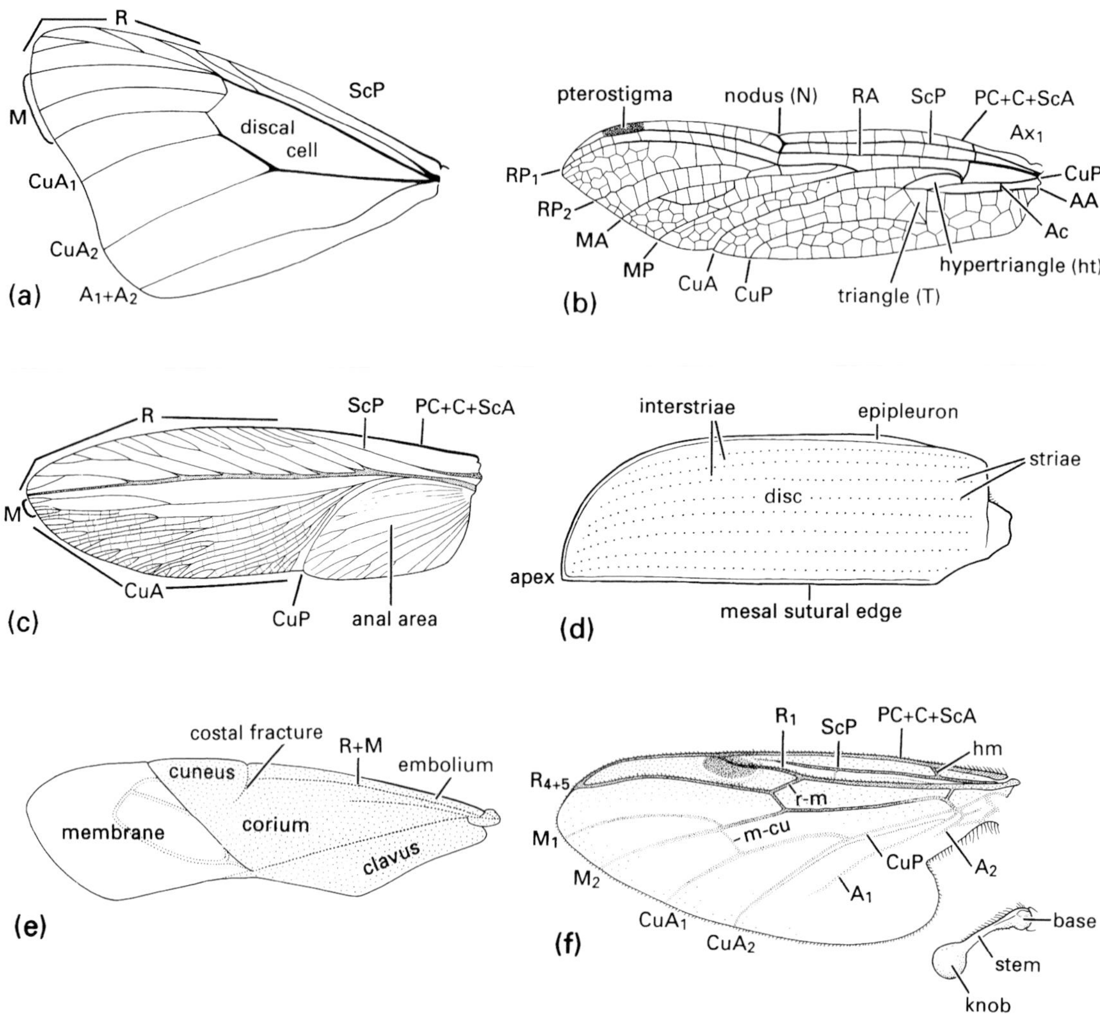

Fig. 2.22 The left wings of a range of insects showing some of the major wing modifications: (a) fore wing of a butterfly of *Danaus* (Lepidoptera: Nymphalidae); (b) fore wing of a dragonfly of *Urothemis* (Odonata: Anisoptera: Libellulidae); (c) fore wing or tegmen of a cockroach of *Periplaneta* (Blattodea: Blattidae); (d) fore wing or elytron of a beetle of *Anomala* (Coleoptera: Scarabaeidae); (e) fore wing or hemelytron of a mirid bug (Hemiptera: Heteroptera: Miridae) showing three wing areas – the membrane, corium, and clavus; (f) fore wing and haltere of a fly of *Bibio* (Diptera: Bibionidae). Nomenclatural scheme of venation consistent with that depicted in Fig. 2.21; that of (b) after J.W.H. Trueman, unpublished. ((a–d) After Youdeowei 1977; (f) after McAlpine 1981.)

thickened, leathery fore wings of Blattodea, Dermaptera, and Orthoptera, which are called **tegmina** (singular: **tegmen**; Fig. 2.22c), the hardened fore wings of Coleoptera that form protective wing cases or **elytra** (singular: **elytron**; Fig. 2.22d & Plate 1.2), and the **hemelytra** (singular: **hemelytron**) of heteropteran Hemiptera with the basal part thickened and the apical part membranous (Fig. 2.22e). Typically, the heteropteran hemelytron is divided into three wing areas: the **membrane**, **corium**, and **clavus**. Sometimes the corium is divided further, with the **embolium** anterior to R + M, and the **cuneus** distal to a **costal fracture**. In Diptera the hind wings are modified as stabilizers (**halteres**) (Fig. 2.22f) and do not function as wings, whereas in male Strepsiptera the fore wings form halteres and the hind wings are used in flight (Box 13.6). In male scale insects (see Plate 2.5, facing p. 14) the fore wings have highly reduced venation and the hind wings form hamulohalteres (different in structure to the halteres) or are lost completely.

Small insects confront different aerodynamic challenges compared with larger insects and their wing area often is expanded to aid wind dispersal. Thrips (Thysanoptera), for example, have very slender wings but have a fringe of long setae or cilia to extend the wing area (Box 11.7). In termites (Isoptera) and ants (Hymenoptera: Formicidae) the winged reproductives, or alates, have large **deciduous** wings that are shed after the nuptial flight. Some insects are wingless, or **apterous**, either primitively as in silverfish (Zygentoma) and bristletails (Archaeognatha), which diverged from other insect lineages prior to the origin of wings, or secondarily as in all lice (Phthiraptera) and fleas (Siphonaptera), which evolved from winged ancestors. Secondary partial wing reduction occurs in a number of short-winged, or **brachypterous**, insects.

In all winged insects (Pterygota), a triangular area at the wing base, the **axillary area** (Fig. 2.20), contains the movable **articular sclerites** via which the wing articulates on the thorax. These sclerites are derived, by reduction and fusion, from a band of articular sclerites in the ancestral wing. Three different types of wing articulation among living Pterygota result from unique patterns of fusion and reduction of the articular sclerites. In Neoptera (all living winged insects except the Ephemeroptera and Odonata), the articular sclerites consist of the **humeral plate**, the **tegula**, and usually three, rarely four, **axillary sclerites** (1Ax, 2Ax, 3Ax, and 4Ax) (Fig. 2.21). The Ephemeroptera and Odonata each has a different configuration of these sclerites compared with the Neoptera (literally meaning "new wing"). Odonate and ephemeropteran adults cannot fold their wings back along the abdomen as can neopterans. In Neoptera, the wing articulates via the articular sclerites with the **anterior** and **posterior wing processes** dorsally, and ventrally with the **pleural wing processes** and two small pleural sclerites (the basalare and subalare) (Fig. 2.18).

2.5 THE ABDOMEN

Primitively, the insect abdomen is 11-segmented although segment 1 may be reduced or incorporated into the thorax (as in many Hymenoptera) and the terminal segments usually are variously modified and/ or diminished (Fig. 2.23a). Generally, at least the first seven abdominal segments of adults (the **pregenital** segments) are similar in structure and lack appendages. However, apterygotes (bristletails and silverfish) and many immature aquatic insects have abdominal appendages. Apterygotes possess a pair of **styles** – rudimentary appendages that are serially homologous with the distal part of the thoracic legs – and, mesally, one or two pairs of **protrusible** (or **exsertile**) **vesicles** on at least some abdominal segments. These vesicles are derived from the coxal and trochanteral **endites** (inner annulated lobes) of the ancestral abdominal appendages (Fig. 8.4b). Aquatic larvae and nymphs may have gills laterally on some to most abdominal segments (Chapter 10). Some of these may be serially homologous with thoracic wings (e.g. the plate gills of mayfly nymphs) or with other leg derivatives. Spiracles typically are present on segments 1–8, but reductions in number occur frequently in association with modifications of the tracheal system (section 3.5), especially in immature insects, and with specializations of the terminal segments in adults.

2.5.1 Terminalia

The anal-genital part of the abdomen, known as the **terminalia**, consists generally of segments 8 or 9 to the abdominal apex. Segments 8 and 9 bear the genitalia; segment 10 is visible as a complete segment in many "lower" insects but always lacks appendages; and the small segment 11 is represented by a dorsal epiproct and pair of ventral paraprocts derived from the sternum (Fig. 2.23b). A pair of appendages, the **cerci**, articulates laterally on segment 11; typically these are annulated and filamentous but have been modified (e.g. the forceps of earwigs) or reduced in different insect orders. An annulated caudal filament, the median **appendix dorsalis**, arises from the tip of the epiproct in apterygotes, most mayflies (Ephemeroptera), and a few fossil insects. A similar structure in nymphal stoneflies (Plecoptera) is of uncertain homology. These terminal abdominal segments have excretory and sensory functions in all insects, but in adults there is an additional reproductive function.

The organs concerned specifically with mating and the deposition of eggs are known collectively as the **external genitalia**, although they may be largely

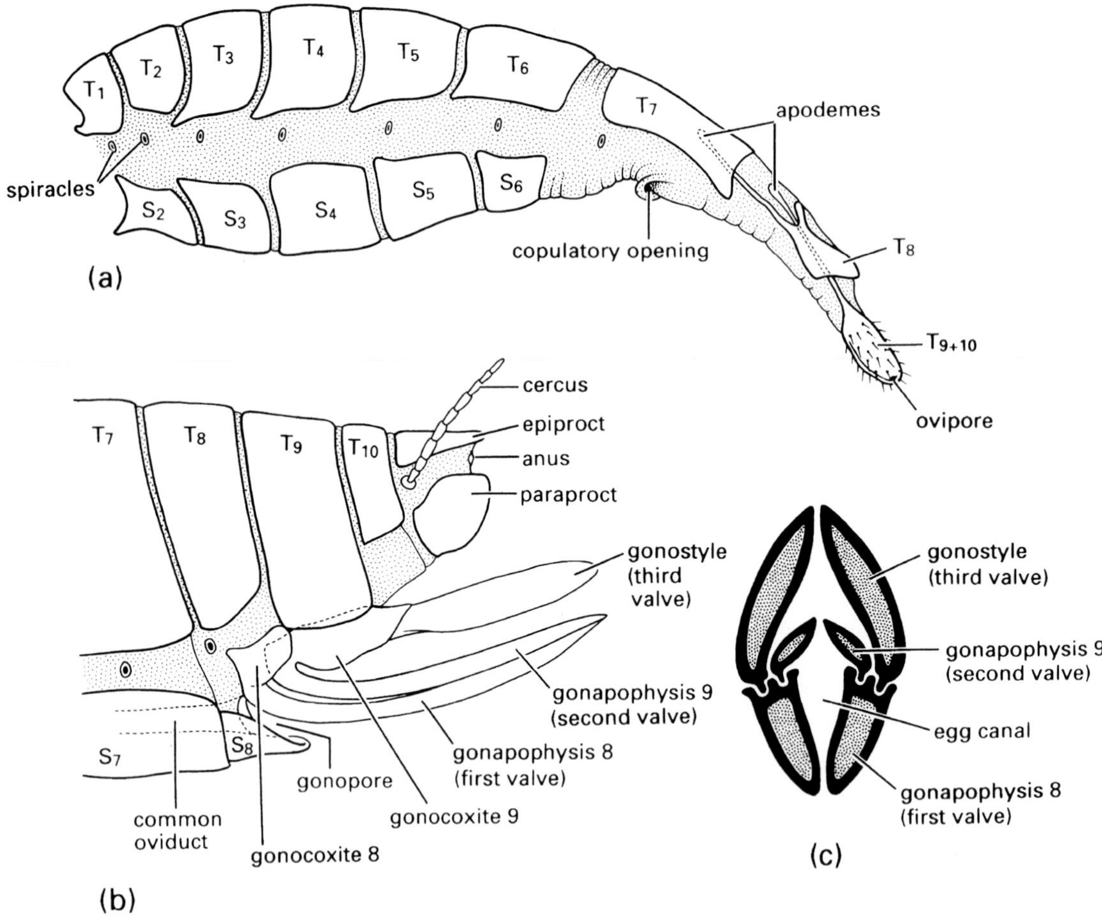

Fig. 2.23 The female abdomen and ovipositor: (a) lateral view of the abdomen of an adult tussock moth (Lepidoptera: Lymantriidae) showing the substitutional ovipositor formed from the extensible terminal segments; (b) lateral view of a generalized orthopteroid ovipositor composed of appendages of segments 8 and 9; (c) transverse section through the ovipositor of a katydid (Orthoptera: Tettigoniidae). T_1–T_{10}, terga of first to tenth segments; S_2–S_8, sterna of second to eighth segments. ((a) After Eidmann 1929; (b) after Snodgrass 1935; (c) after Richards & Davies 1959.)

internal. The components of the external genitalia of insects are very diverse in form and often have considerable taxonomic value, particularly amongst species that appear structurally similar in other respects. The male external genitalia have been used widely to aid in distinguishing species, whereas the female external genitalia may be simpler and less varied. The diversity and species-specificity of genitalic structures are discussed in section 5.5.

The terminalia of adult female insects include internal structures for receiving the male copulatory organ and his spermatozoa (sections 5.4 and 5.6) and external structures used for oviposition (egg-laying; section 5.8). Most female insects have an egg-laying tube, or **ovipositor**; it is absent in Isoptera, Phthiraptera, many Plecoptera, and most Ephemeroptera. Ovipositors take two forms:

1 true, or **appendicular**, formed from appendages of abdominal segments 8 and 9 (Fig. 2.23b);

2 **substitutional**, composed of extensible posterior abdominal segments (Fig. 2.23a).

Substitutional ovipositors include a variable number

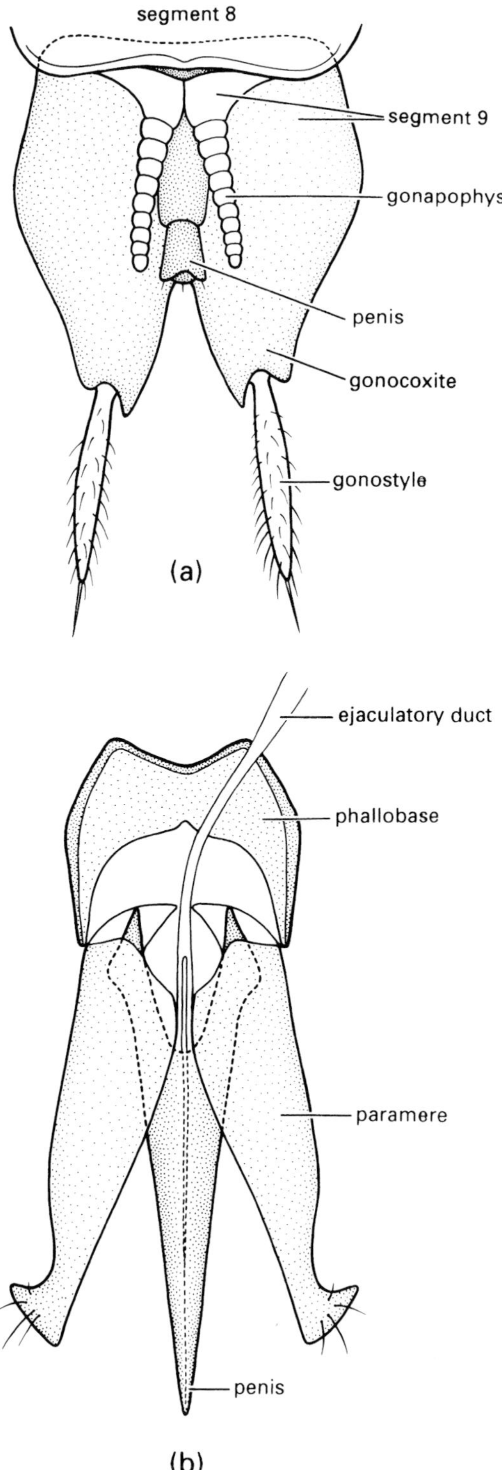

of the terminal segments and clearly have been derived convergently several times, even within some orders. They occur in many insects, including most Lepidoptera, Coleoptera, and Diptera. In these insects, the terminalia are telescopic and can be extended as a slender tube, manipulated by muscles attached to apodemes of the modified terga (Fig. 2.23a) and/or sterna.

Appendicular ovipositors represent the primitive condition for female insects and are present in Archaeognatha, Zygentoma, many Odonata, Orthoptera, some Hemiptera, some Thysanoptera, and Hymenoptera. In some Hymenoptera, the ovipositor is modified as a poison-injecting sting (Fig. 14.11) and the eggs are ejected at the base of the sting. In all other cases, the eggs pass down a canal in the shaft of the ovipositor (section 5.8). The shaft is composed of three pairs of **valves** (Fig. 2.23b,c) supported on two pairs of **valvifers** – the coxae + trochanters, or **gonocoxites**, of segments 8 and 9 (Fig. 2.23b). The gonocoxites of segment 8 have a pair of trochanteral endites (inner lobe from each trochanter), or **gonapophyses**, which form the first valves, whereas the gonocoxites of segment 9 have a pair of gonapophyses (the second valves) plus a pair of **gonostyles** (the third valves) derived from the distal part of the appendages of segment 9 (and homologous with the styles of the apterygotes mentioned above). In each half of the ovipositor, the second valve slides in a tongue-and-groove fashion against the first valve (Fig. 2.23c), whereas the third valve generally forms a sheath for the other valves.

The external genitalia of male insects include an organ for transferring the spermatozoa (either packaged in a **spermatophore**, or free in fluid) to the female and often involve structures that grasp and hold the partner during mating. Numerous terms are applied to the various components in different insect groups and homologies are difficult to establish. Males of Archaeognatha, Zygentoma, and Ephemeroptera have relatively simple genitalia consisting of gonocoxites, gonostyles, and sometimes gonapophyses on segment 9 (and also on segment 8 in Archaeognatha), as in the female, except with a median **penis** (**phallus**) or, if paired or bilobed, **penes**, on segment 9 (Fig. 2.24a). The penis (or penes) is believed to be derived from the

Fig. 2.24 (*left*) Male external genitalia. (a) Abdominal segment 9 of the bristletail *Machilis variabilis* (Archaeognatha: Machilidae). (b) Aedeagus of a click beetle (Coleoptera: Elateridae). ((a) After Snodgrass 1957.)

fused inner lobes (endites) of either the ancestral coxae or trochanters of segment 9. In the orthopteroid orders, the gonocoxites are reduced or absent, although gonostyles may be present (called styles), and there is a median penis with a lobe called a **phallomere** on each side of it. The evolutionary fate of the gonapophyses and the origin of the phallomeres are uncertain. In the "higher" insects – the hemipteroids and the holometabolous orders – the homologies and terminology of the male structures are even more confusing if one tries to compare the terminalia of different orders. The whole copulatory organ of higher insects generally is known as the **aedeagus** (edeagus) and, in addition to insemination, it may clasp and provide sensory stimulation to the female. Typically, there is a median tubular penis (although sometimes the term "aedeagus" is restricted to this lobe), which often has an inner tube, the **endophallus**, that is everted during insemination (Fig. 5.4b). The ejaculatory duct opens at the gonopore, either at the tip of the penis or the endophallus. Lateral to the penis is a pair of lobes or **parameres**, which may have a clasping and/or sensory function. Their origin is uncertain; they may be homologous with the gonocoxites and gonostyles of lower insects, with the phallomeres of orthopteroid insects, or be derived *de novo*, perhaps even from segment 10. This trilobed type of aedeagus is well exemplified in many beetles (Fig. 2.24b), but modifications are too numerous to describe here.

Much variation in male external genitalia correlates with mating position, which is very variable between and sometimes within orders. Mating positions include end-to-end, side-by-side, male below with his dorsum up, male on top with female dorsum up, and even venter-to-venter. In some insects, torsion of the terminal segments may take place post-metamorphosis or just prior to or during copulation, and asymmetries of male clasping structures occur in many insects. Copulation and associated behaviors are discussed in more detail in Chapter 5.

FURTHER READING

Binnington, K. & Retnakaran, A. (eds.) (1991) *Physiology of the Insect Epidermis*. CSIRO Publications, Melbourne.

Chapman, R.F. (1998) *The Insects. Structure and Function*, 4th edn. Cambridge University Press, Cambridge.

Hadley, N.F. (1986) The arthropod cuticle. *Scientific American* **255**(1), 98–106.

Lawrence, J.F., Nielsen, E.S. & Mackerras, I.M. (1991) Skeletal anatomy and key to orders. In: *The Insects of Australia*, 2nd edn. (CSIRO), pp. 3–32. Melbourne University Press, Carlton.

Nichols, S.W. (1989) *The Torre-Bueno Glossary of Entomology*. The New York Entomological Society in co-operation with the American Museum of Natural History, New York.

Resh, V.H. & Cardé, R.T. (eds.) (2003) *Encyclopedia of Insects*. Academic Press, Amsterdam. [Particularly see articles on anatomy; head; thorax; abdomen and genitalia; and mouthparts.]

Richards, A.G. & Richards, P.A. (1979) The cuticular protuberances of insects. *International Journal of Insect Morphology and Embryology* **8**, 143–57.

Smith, J.J.B. (1985) Feeding mechanisms. In: *Comprehensive Insect Physiology, Biochemistry and Pharmacology*, Vol. 4: *Regulation: Digestion, Nutrition, Excretion* (eds. G.A. Kerkut & L.I. Gilbert), pp. 33–85. Pergamon Press, Oxford.

Snodgrass, R.E. (1935) *Principles of Insect Morphology*. McGraw-Hill, New York.

Wootton, R.J. (1992) Functional morphology of insect wings. *Annual Review of Entomology* **37**, 113–40.

Chapter 3

INTERNAL ANATOMY AND PHYSIOLOGY

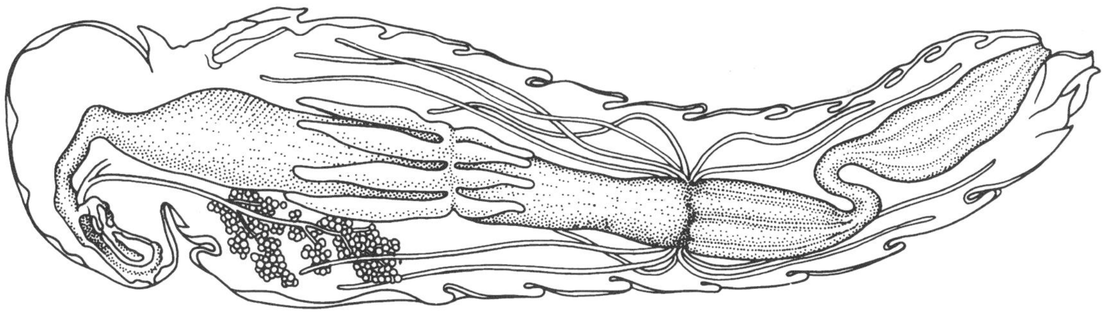

Internal structures of a locust. (After Uvarov 1966.)

What you see if you dissect open the body of an insect is a complex and compact masterpiece of functional design. Figure 3.1 shows the "insides" of two omnivorous insects, a cockroach and a cricket, which have relatively unspecialized digestive and reproductive systems. The digestive system, which includes salivary glands as well as an elongate gut, consists of three main sections. These function in storage, biochemical breakdown, absorption, and excretion. Each gut section has more than one physiological role and this may be reflected in local structural modifications, such as thickening of the gut wall or diverticula (extensions) from the main lumen. The reproductive systems depicted in Fig. 3.1 exemplify the female and male organs of many insects. These may be dominated in males by very visible accessory glands, especially as the testes of many adult insects are degenerate or absent. This is because the spermatozoa are produced in the pupal or penultimate stage and stored. In gravid female insects, the body cavity may be filled with eggs at various stages of development, thereby obscuring other internal organs. Likewise, the internal structures (except the gut) of a well-fed, late-stage caterpillar may be hidden within the mass of fat body tissue.

The insect body cavity, called the **hemocoel** (haemocoel) and filled with fluid **hemolymph** (haemolymph), is lined with endoderm and ectoderm. It is not a true coelom, which is defined as a mesoderm-lined cavity. Hemolymph (so-called because it combines many roles of vertebrate blood (hem/haem) and lymph) bathes all internal organs, delivers nutrients, removes metabolites, and performs immune functions. Unlike vertebrate blood, hemolymph rarely has respiratory pigments and therefore has little or no role in gaseous exchange. In insects this function is performed by the **tracheal system**, a ramification of air-filled tubes (**tracheae**), which sends fine branches throughout the body. Gas entry to and exit from tracheae is controlled by sphincter-like structures called **spiracles** that open through the body wall. Non-gaseous wastes are filtered from the hemolymph by filamentous **Malpighian tubules** (named after their discoverer), which have free ends distributed through the hemocoel. Their contents are emptied into the gut from which, after further modification, wastes are eliminated eventually via the anus.

All motor, sensory, and physiological processes in insects are controlled by the nervous system in conjunction with hormones (chemical messengers). The brain and ventral nerve cord are readily visible in dissected insects, but most endocrine centers, neurosecretion sites, numerous nerve fibers, muscles, and other tissues cannot be seen by the unaided eye.

This chapter describes insect internal structures and their functions. Topics covered are the muscles and locomotion (walking, swimming, and flight), the nervous system and co-ordination, endocrine centers and hormones, the hemolymph and its circulation, the tracheal system and gas exchange, the gut and digestion, the fat body, nutrition and microorganisms, the excretory system and waste disposal, and lastly the reproductive organs and gametogenesis. A full account of insect physiology cannot be provided in one chapter, and we direct readers to Chapman (1998) for a comprehensive treatment, and to relevant chapters in the *Encyclopedia of Insects* (Resh & Cardé 2003).

3.1 MUSCLES AND LOCOMOTION

As stated in section 1.3.4, much of the success of insects relates to their ability to sense, interpret, and move around their environment. Although the origin of flight at least 340 million years ago was a major innovation, terrestrial and aquatic locomotion also is well developed. Power for movement originates from muscles operating against a skeletal system, either the rigid cuticular exoskeleton or, in soft-bodied larvae, a hydrostatic skeleton.

3.1.1 Muscles

Vertebrates and many non-insect invertebrates have **striated** and **smooth** muscles, but insects have only striated muscles, so-called because of overlapping thicker myosin and thinner actin filaments giving a microscopic appearance of cross-banding. Each striated muscle fiber comprises many cells, with a common plasma membrane and **sarcolemma**, or outer sheath. The sarcolemma is invaginated, but not broken, where an oxygen-supplying tracheole (section 3.5, Fig. 3.10b) contacts the muscle fiber. Contractile **myofibrils** run the length of the fiber, arranged in sheets or cylinders. When viewed under high magnification, a myofibril comprises a thin actin filament sandwiched between a pair of thicker myosin filaments. Muscle contraction involves the sliding of filaments past each other, stimulated by nerve impulses. Innervation comes from one to three motor axons per bundle of fibers, each

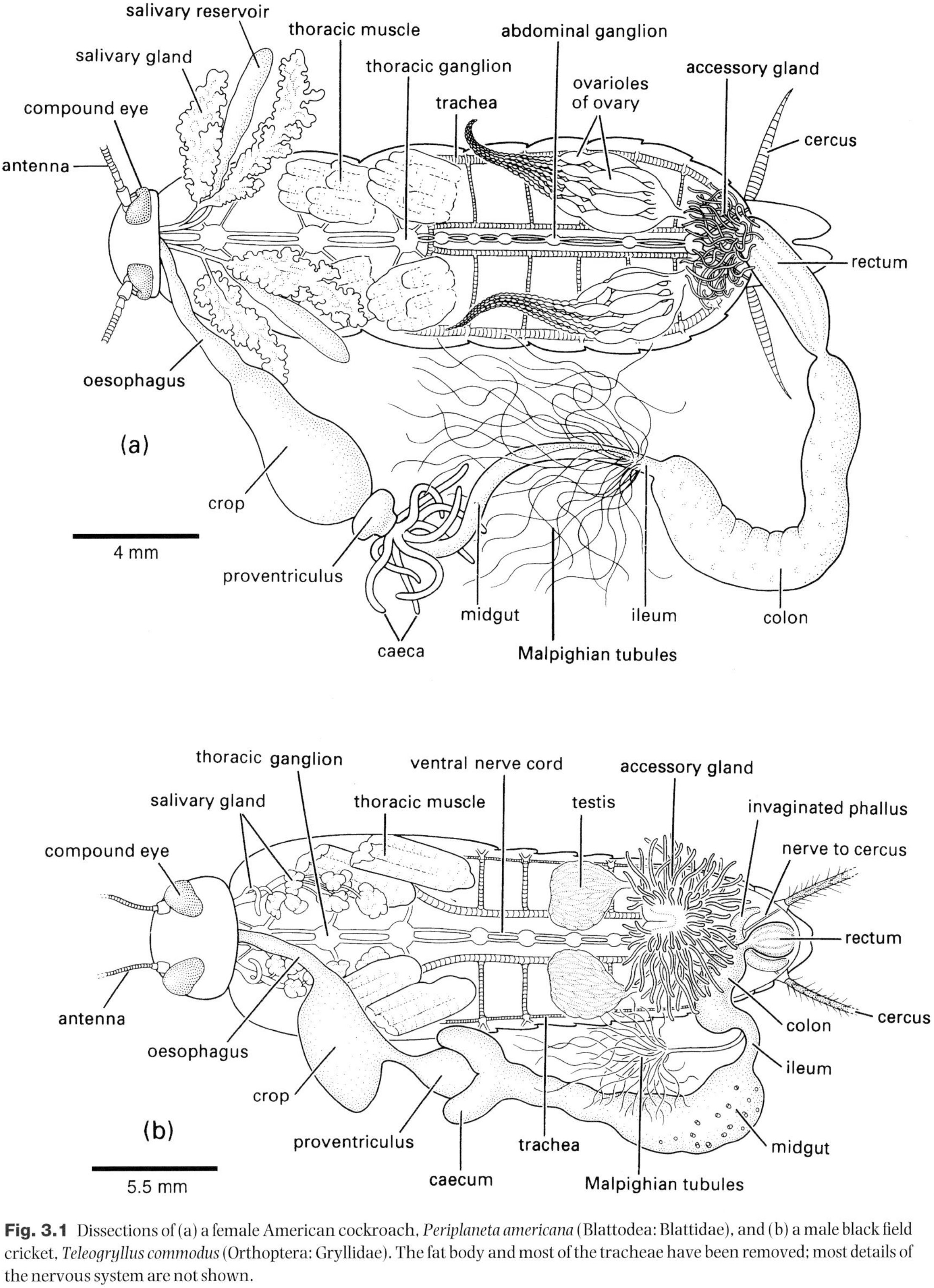

salivary reservoir
thoracic muscle
abdominal ganglion
salivary gland
thoracic ganglion
accessory gland
ovarioles
of ovary
compound eye
trachea
cercus
antenna
rectum
oesophagus
(a)
crop
4 mm
proventriculus
midgut
ileum
colon
caeca
Malpighian tubules
thoracic ganglion
ventral nerve cord
accessory gland
salivary gland
thoracic muscle
testis
invaginated phallus
compound eye
nerve to cercus
rectum
antenna
colon
cercus
oesophagus
ileum
crop
(b)
proventriculus
trachea
midgut
5.5 mm
caecum
Malpighian tubules

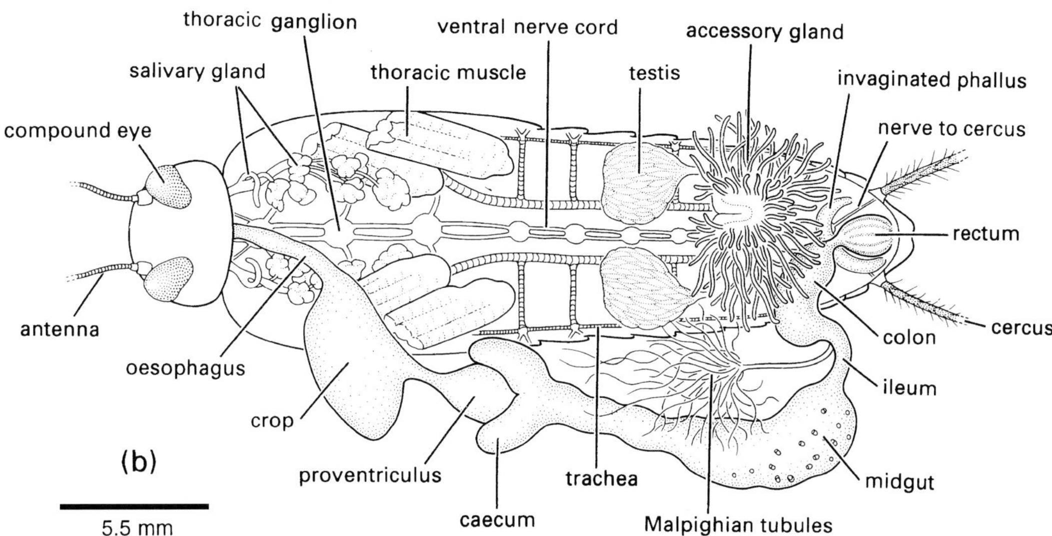

Fig. 3.1 Dissections of (a) a female American cockroach, *Periplaneta americana* (Blattodea: Blattidae), and (b) a male black field cricket, *Teleogryllus commodus* (Orthoptera: Gryllidae). The fat body and most of the tracheae have been removed; most details of the nervous system are not shown.

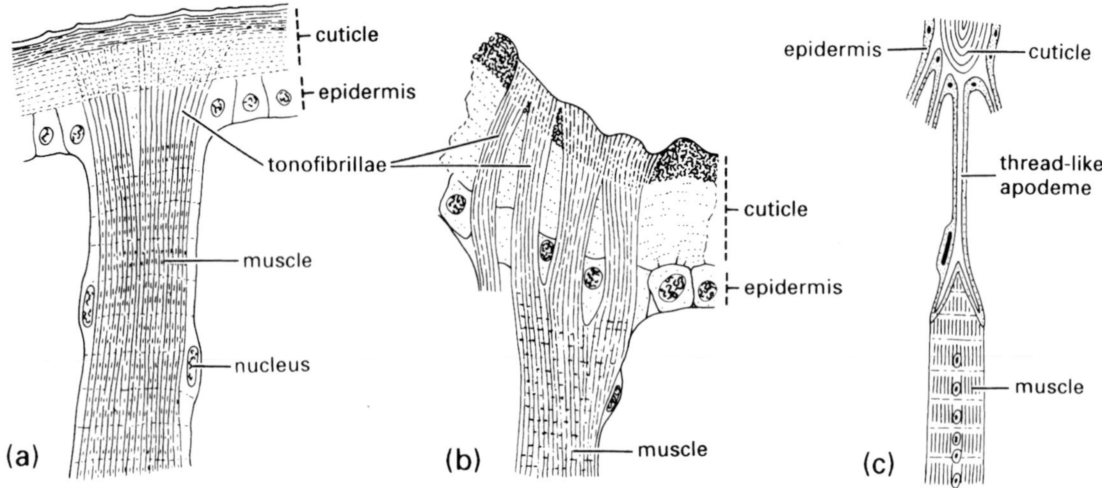

Fig. 3.2 Muscle attachments to body wall: (a) tonofibrillae traversing the epidermis from the muscle to the cuticle; (b) a muscle attachment in an adult beetle of *Chrysobothrus femorata* (Coleoptera: Buprestidae); (c) a multicellular apodeme with a muscle attached to one of its thread-like, cuticular "tendons" or apophyses. (After Snodgrass 1935.)

separately tracheated and referred to as one muscle unit, with several units grouped in a functional muscle.

There are several different muscle types. The most important division is between those that respond synchronously, with a contraction cycle once per impulse, and fibrillar muscles that contract asynchronously, with multiple contractions per impulse. Examples of the latter include some flight muscles (see below) and the tymbal muscle of cicadas (section 4.1.4).

There is no inherent difference in action between muscles of insects and vertebrates, although insects can produce prodigious muscular feats, such as the leap of a flea or the repetitive stridulation of the cicada tympanum. Reduced body size benefits insects because of the relationship between (i) power, which is proportional to muscle cross-section and decreases with reduction in size by the square root, and (ii) the body mass, which decreases with reduction in size by the cube root. Thus the power : mass ratio increases as body size decreases.

3.1.2 Muscle attachments

Vertebrates' muscles work against an internal skeleton, but the muscles of insects must attach to the inner surface of an external skeleton. As musculature is mesodermal and the exoskeleton is of ectodermal origin, fusion must take place. This occurs by the growth of **tonofibrillae**, fine connecting fibrils that link the epidermal end of the muscle to the epidermal layer (Fig. 3.2a,b). At each molt tonofibrillae are discarded along with the cuticle and therefore must be regrown.

At the site of tonofibrillar attachment, the inner cuticle often is strengthened through ridges or **apodemes**, which, when elongated into arms, are termed **apophyses** (Fig. 3.2c). These muscle attachment sites, particularly the long, slender apodemes for individual muscle attachments, often include resilin to give an elasticity that resembles that of vertebrate tendons.

Some insects, including soft-bodied larvae, have mainly thin, flexible cuticle without the rigidity to anchor muscles unless given additional strength. The body contents form a **hydrostatic skeleton**, with turgidity maintained by criss-crossed body wall "turgor" muscles that continuously contract against the incompressible fluid of the hemocoel, giving a strengthened foundation for other muscles. If the larval body wall is perforated, the fluid leaks, the hemocoel becomes compressible and the turgor muscles cause the larva to become flaccid.

3.1.3 Crawling, wriggling, swimming, and walking

Soft-bodied larvae with hydrostatic skeletons move by crawling. Muscular contraction in one part of the body gives equivalent extension in a relaxed part elsewhere on the body. In apodous (legless) larvae, such as dipteran "maggots", waves of contractions and relaxation run from head to tail. Bands of adhesive hooks or tubercles successively grip and detach from the substrate to provide a forward motion, aided in some maggots by use of their mouth hooks to grip the substrate. In water, lateral waves of contraction against the hydrostatic skeleton can give a sinuous, snake-like, swimming motion, with anterior-to-posterior waves giving an undulating motion.

Larvae with thoracic legs and abdominal prolegs, like caterpillars, develop posterior-to-anterior waves of turgor muscle contraction, with as many as three waves visible simultaneously. Locomotor muscles operate in cycles of successive detachment of the thoracic legs, reaching forwards and grasping the substrate. These cycles occur in concert with inflation, deflation, and forward movement of the posterior prolegs.

Insects with hard exoskeletons can contract and relax pairs of agonistic and antagonistic muscles that attach to the cuticle. Compared to crustaceans and myriapods, insects have fewer (six) legs that are located more ventrally and brought close together on the thorax, allowing concentration of locomotor muscles (both flying and walking) into the thorax, and providing more control and greater efficiency. Motion with six legs at low to moderate speed allows continuous contact with the ground by a tripod of fore and hind legs on one side and mid leg of the opposite side thrusting rearwards (**retraction**), whilst each opposite leg is moved forwards (**protraction**) (Fig. 3.3). The center of gravity of the slow-moving insect always lies within this tripod, giving great stability. Motion is imparted through thoracic muscles acting on the leg bases, with transmission via internal leg muscles through the leg to extend or flex the leg. Anchorage to the substrate,

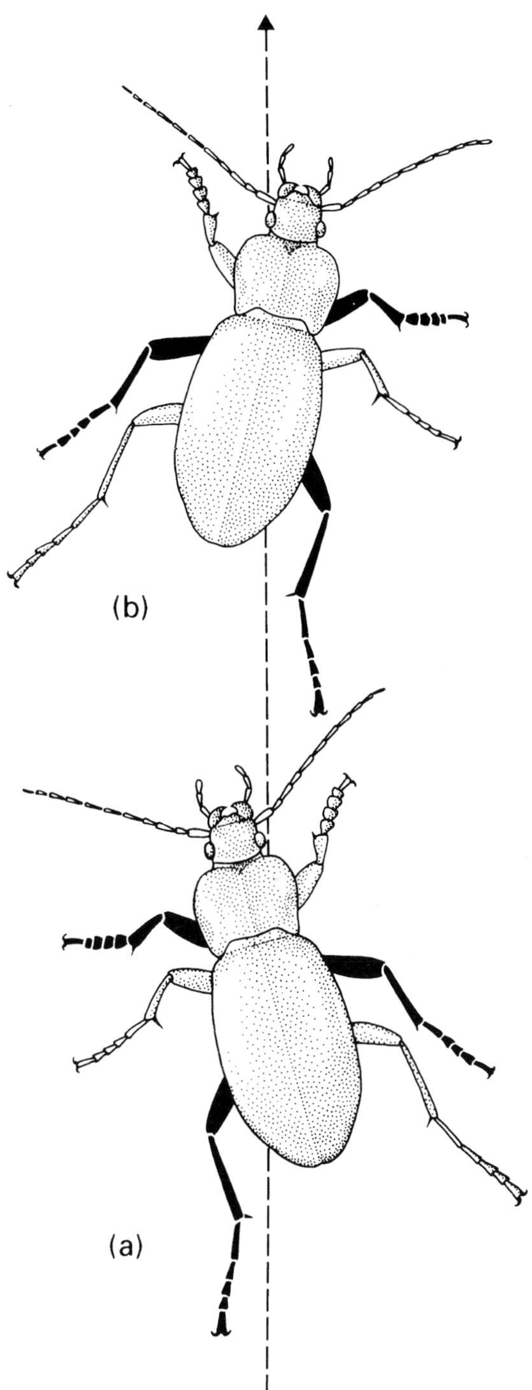

Fig. 3.3 (*right*) A ground beetle (Coleoptera: Carabidae: *Carabus*) walking in the direction of the broken line. The three blackened legs are those in contact with the ground in the two positions illustrated – (a) is followed by (b). (After Wigglesworth 1972.)

needed to provide a lever to propel the body, is through pointed claws and adhesive pads (the arolium or, in flies and some beetles, pulvilli). Claws such as those illustrated in the vignette to Chapter 2 can obtain purchase on the slightest roughness in a surface, and the pads of some insects can adhere to perfectly smooth surfaces through the application of lubricants to the tips of numerous fine hairs and the action of close-range molecular forces between the hairs and the substrate.

When faster motion is required there are several alternatives – increasing the frequency of the leg movement by shortening the retraction period; increasing the stride length; altering the triangulation basis of support to adopt quadrupedy (use of four legs); or even hind-leg bipedality with the other legs held above the substrate. At high speeds even those insects that maintain triangulation are very unstable and may have no legs in contact with the substrate at intervals. This instability at speed seems to cause no difficulty for cockroaches, which when filmed with high-speed video cameras have been shown to maintain speeds of up to 1 m s^{-1} whilst twisting and turning up to 25 times per second. This motion was maintained by sensory information received from one antenna whose tip maintained contact with an experimentally provided wall, even when it had a zig-zagging surface.

Many insects jump, some prodigiously, usually using modified hind legs. In orthopterans, flea beetles (Alticinae), and a range of weevils, an enlarged hind (meta-) femur contains large muscles whose slow contraction produces energy stored by either distortion of the femoro-tibial joint or in some spring-like sclerotization, for example the meta-tibial extension tendon. In fleas, the energy is produced by the trochanter levator muscle raising the femur and is stored by compression of an elastic resilin pad in the coxa. In all these jumpers, release of tension is sudden, resulting in propulsion of the insect into the air – usually in an uncontrolled manner, but fleas can attain their hosts with some control over the leap. It has been suggested that the main benefit for flighted jumpers is to get into the air and allow the wings to be opened without damage from the surrounding substrate.

In swimming, contact with the water is maintained during protraction, so it is necessary for the insect to impart more thrust to the rowing motion than to the recovery stroke to progress. This is achieved by expanding the effective leg area during retraction by extending fringes of hairs and spines (Fig. 10.8). These collapse onto the folded leg during the recovery stroke. We have seen already how some insect larvae swim using contractions against their hydrostatic skeleton. Others, including many nymphs and the larvae of caddisflies, can walk underwater and, particularly in running waters, do not swim routinely.

The surface film of water can support some specialist insects, most of which have hydrofuge (water-repelling) cuticles or hair fringes and some, such as gerrid water-striders (Fig. 5.7), move by rowing with hair-fringed legs.

3.1.4 Flight

The development of flight allowed insects much greater mobility, which helped in food and mate location and gave much improved powers of dispersal. Importantly, flight opened up many new environments for exploitation. Plant microhabitats such as flowers and foliage are more accessible to winged insects than to those without flight.

Fully developed, functional, flying wings occur only in adult insects, although in nymphs the developing wings are visible as wing buds in all but the earliest instars. Usually two pairs of functional wings arise dorsolaterally, as fore wings on the second and hind wings on the third thoracic segment. Some of the many derived variations are described in section 2.4.2.

To fly, the forces of weight (gravity) and drag (air resistance to movement) must be overcome. In gliding flight, in which the wings are held rigidly outstretched, these forces are overcome through the use of passive air movements – known as the relative wind. The insect attains lift by adjusting the angle of the leading edge of the wing when orientated into the wind. As this angle (the attack angle) increases, so lift increases until stalling occurs, i.e. when lift is catastrophically lost. In contrast to aircraft, nearly all of which stall at around 20°, the attack angle of insects can be raised to more than 30°, even as high as 50°, giving great maneuverability. Aerodynamic effects such as enhanced lift and reduced drag can come from wing scales and hairs, which affect the boundary layer across the wing surface.

Most insects glide a little, and dragonflies (Odonata) and some grasshoppers (Orthoptera), notably locusts, glide extensively. However, most winged insects fly by beating their wings. Examination of wing beat is difficult because the frequency of even a large slow-flying butterfly may be five times a second (5 Hz), a bee may beat its wings at 180 Hz, and some midges emit an audible buzz with their wing-beat frequency of greater than 1000 Hz. However, through the use of slowed-down, high-speed cine film, the insect wing beat can be slowed from faster than the eye can see until a single beat can be analyzed. This reveals that a single beat comprises three interlinked movements. First is a cycle of downward, forward motion followed by an upward and backward motion. Second, during the cycle each wing is rotated around its base. The third component occurs as various parts of the wing flex in response to local variations in air pressure. Unlike gliding, in which the relative wind derives from passive air movement, in true flight the relative wind is produced by the moving wings. The flying insect makes constant adjustments, so that during a wing beat, the air ahead of the insect is thrown backwards and downwards, impelling the insect upwards (lift) and forwards (thrust). In climbing, the emergent air is directed more downwards, reducing thrust but increasing lift. In turning, the wing on the inside of the turn is reduced in power by decrease in the amplitude of the beat.

Despite the elegance and intricacy of detail of insect flight, the mechanisms responsible for beating the wings are not excessively complicated. The thorax of the wing-bearing segments can be envisaged as a box with the sides (pleura) and base (sternum) rigidly fused, and the wings connected where the rigid tergum is attached to the pleura by flexible membranes. This membranous attachment and the wing hinge are composed of resilin (section 2.1), which gives crucial elasticity to the thoracic box. Flying insects have one of two kinds of arrangements of muscles powering their flight:

1 **direct flight muscles** connected to the wings;

2 an **indirect system** in which there is no muscle-to-wing connection, but rather muscle action deforms the thoracic box to move the wing.

A few old groups such as Odonata and Blattodea appear to use direct flight muscles to varying degrees, although at least some recovery muscles may be indirect. More advanced insects use indirect muscles for flight, with direct muscles providing wing orientation rather than power production.

Direct flight muscles produce the upward stroke by contraction of muscles attached to the wing base inside the pivotal point (Fig. 3.4a). The downward wing stroke is produced through contraction of muscles that extend from the sternum to the wing base outside the pivot point (Fig. 3.4b). In contrast, indirect flight muscles are attached to the tergum and sternum. Contraction causes the tergum, and with it the very base of the wing, to be pulled down. This movement levers the outer, main part of the wing in an upward stroke (Fig. 3.4c). The down beat is powered by contraction of the second set of muscles, which run from front to back of the thorax, thereby deforming the box and lifting the tergum (Fig. 3.4d). At each stage in the cycle, when the flight muscles relax, energy is conserved because the elasticity of the thorax restores its shape.

Primitively, the four wings may be controlled independently with small variation in timing and rate allowing alteration in direction of flight. However, excessive variation impedes controlled flight and the beat of all wings is usually harmonized, as in butterflies, bugs, and bees, for example, by locking the fore and hind wings together, and also by neural control. For insects with slower wing-beat frequencies (<100 Hz), such as dragonflies, one nerve impulse for each beat can be maintained by **synchronous** muscles. However, in faster-beating wings, which may attain a frequency of 100 to over 1000 Hz, one impulse per beat is impossible and **asynchronous** muscles are required. In these insects, the wing is constructed such that only two wing positions are stable – fully up and fully down. As the wing moves from one extreme to the alternate one, it passes through an intermediate unstable position. As it passes this unstable ("click") point, thoracic elasticity snaps the wing through to the alternate stable position. Insects with this asynchronous mechanism have peculiar fibrillar flight muscles with the property that, on sudden release of muscle tension, as at the click point, the next muscle contraction is induced. Thus muscles can oscillate, contracting at a much higher frequency than the nerve impulses, which need be only periodic to maintain the insect in flight. Harmonization of the wing beat on each side is maintained through the rigidity of the thorax – as the tergum is depressed or relaxed, what happens to one wing must happen identically to the other. However, insects with indirect flight muscles retain direct muscles that are used in making fine adjustments in wing orientation during flight.

Direction and any deviations from course, perhaps caused by air movements, are sensed by insects predominantly through their eyes and antennae. However, the true flies (Diptera) have extremely sophisticated sensory equipment, with their hind wings modified as balancing organs. These halteres, which each comprise a base, stem, and apical knob (Fig. 2.22f), beat in time

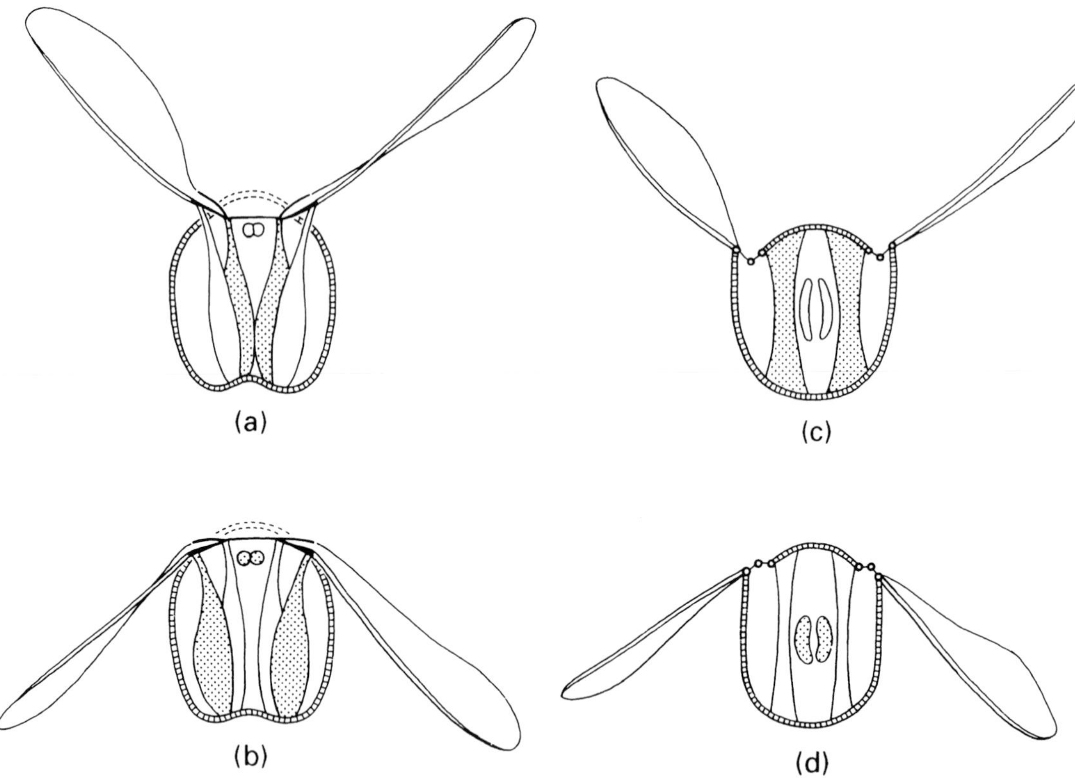

Fig. 3.4 Direct flight mechanisms: thorax during (a) upstroke and (b) downstroke of the wings. Indirect flight mechanisms: thorax during (c) upstroke and (d) downstroke of the wings. Stippled muscles are those contracting in each illustration. (After Blaney 1976.)

but out of phase with the fore wings. The knob, which is heavier than the rest of the organ, tends to keep the halteres beating in one plane. When the fly alters direction, whether voluntarily or otherwise, the haltere is twisted. The stem, which is richly endowed with sensilla, detects this movement, and the fly can respond accordingly.

Initiation of flight, for any reason, may involve the legs springing the insect into the air. Loss of tarsal contact with the ground causes neural firing of the direct flight muscles. In flies, flight activity originates in contraction of a mid-leg muscle, which both propels the leg downwards (and the fly upwards) and simultaneously pulls the tergum downwards to inaugurate flight. The legs are also important when landing because there is no gradual braking by running forwards – all the shock is taken on the outstretched legs, endowed with pads, spines, and claws for adhesion.

3.2 THE NERVOUS SYSTEM AND CO-ORDINATION

The complex nervous system of insects integrates a diverse array of external sensory and internal physiological information and generates some of the behaviors discussed in Chapter 4. In common with other animals, the basic component is the nerve cell, or **neuron** (neurone), composed of a cell body with two projections (fibers) – the **dendrite**, which receives stimuli; and the **axon**, which transmits information, either to another neuron or to an effector organ such as a muscle. Insect neurons release a variety of chemicals at **synapses** to either stimulate or inhibit effector neurons or muscles. In common with vertebrates, particularly important neurotransmitters include acetylcholine and catecholamines such as dopamine. Neurons (Fig. 3.5) are of at least four types:

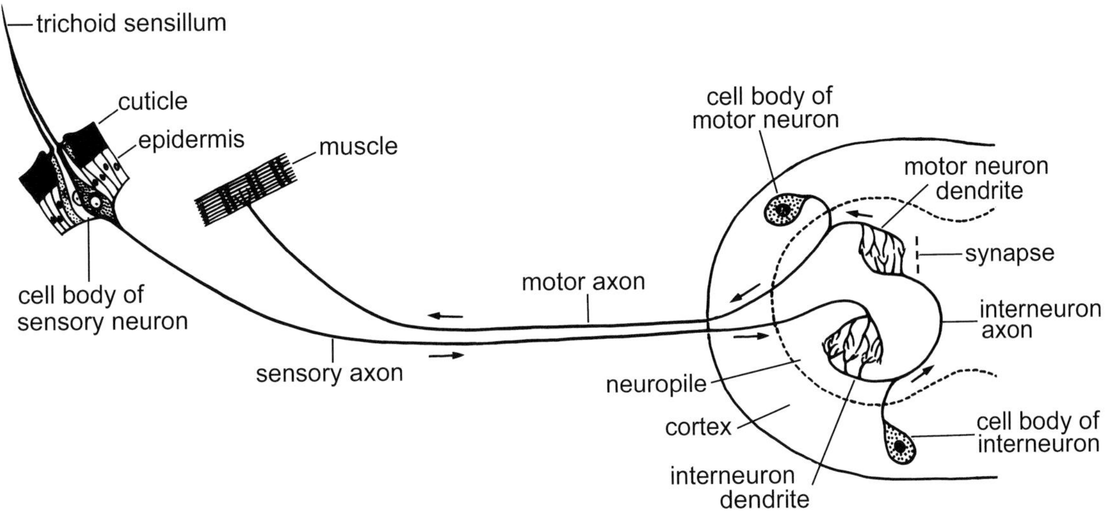

Fig. 3.5 Diagram of a simple reflex mechanism of an insect. The arrows show the paths of nerve impulses along nerve fibers (axons and dendrites). The ganglion, with its outer cortex and inner neuropile, is shown on the right. (After various sources.)

1 sensory neurons receive stimuli from the insect's environment and transmit them to the central nervous system (see below);
2 interneurons (or association neurons) receive information from and transmit it to other neurons;
3 motor neurons receive information from interneurons and transmit it to muscles;
4 neuroendocrine cells (section 3.3.1).

The cell bodies of interneurons and motor neurons are aggregated with the fibers interconnecting all types of nerve cells to form nerve centers called **ganglia**. Simple reflex behavior has been well studied in insects (described further in section 4.5), but insect behavior can be complex, involving integration of neural information within the ganglia.

The **central nervous system** (CNS) (Fig. 3.6) is the principal division of the nervous system and consists of series of ganglia joined by paired longitudinal nerve cords called **connectives**. Primitively there are a pair of ganglia per body segment but usually the two ganglia of each thoracic and abdominal segment are fused into a single structure and the ganglia of all head segments are coalesced to form two ganglionic centers – the **brain** and the **suboesophageal** (subesophageal) **ganglion** (seen in Fig. 3.7). The chain of thoracic and abdominal ganglia found on the floor of the body cavity is called the **ventral nerve cord**. The brain, or the dorsal ganglionic center of the head, is composed of three pairs of fused ganglia (from the first three head segments):
1 protocerebrum, associated with the eyes and thus bearing the optic lobes;
2 deutocerebrum, innervating the antennae;
3 tritocerebrum, concerned with handling the signals that arrive from the body.
Coalesced ganglia of the three mouthpart-bearing segments form the suboesophageal ganglion, with nerves emerging that innervate the mouthparts.

The **visceral** (or **sympathetic**) **nervous system** consists of three subsystems – the **stomodeal** (or **stomatogastric**) (which includes the frontal ganglion); the **ventral visceral**; and the **caudal visceral**. Together the nerves and ganglia of these subsystems innervate the anterior and posterior gut, several endocrine organs (corpora cardiaca and corpora allata), the reproductive organs, and the tracheal system including the spiracles.

The **peripheral nervous system** consists of all of the motor neuron axons that radiate to the muscles from the ganglia of the CNS and stomodeal nervous system plus the sensory neurons of the cuticular sensory structures (the sense organs) that receive mechanical, chemical, thermal, or visual stimuli from an insect's environment. Insect sensory systems are discussed in detail in Chapter 4.

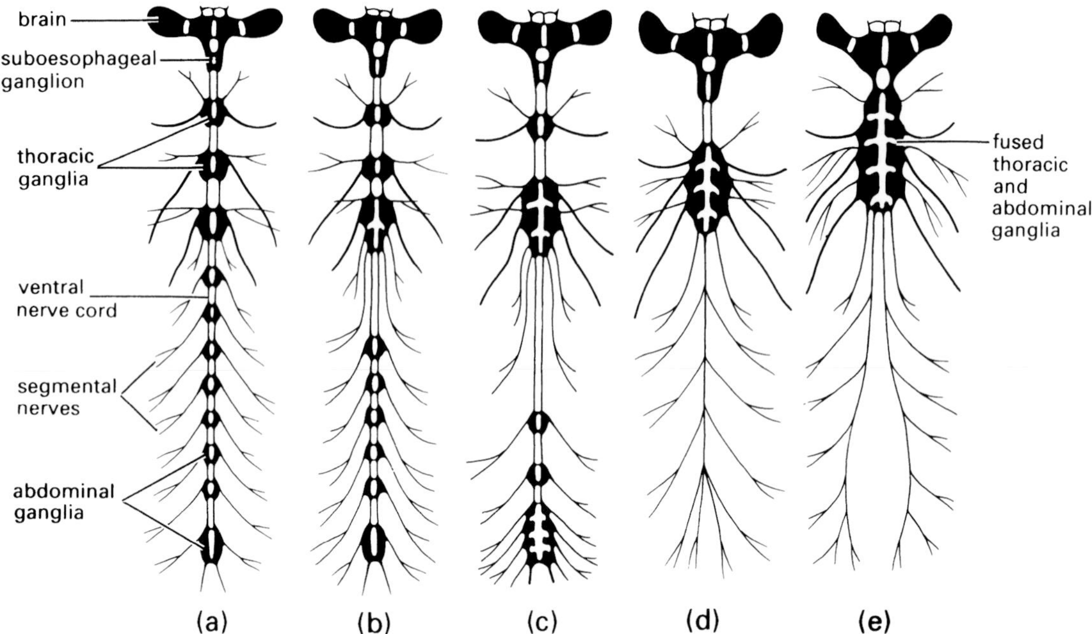

Fig. 3.6 The central nervous system of various insects showing the diversity of arrangement of ganglia in the ventral nerve cord. Varying degrees of fusion of ganglia occur from the least to the most specialized: (a) three separate thoracic and eight abdominal ganglia, as in *Dictyopterus* (Coleoptera: Lycidae) and *Pulex* (Siphonaptera: Pulicidae); (b) three thoracic and six abdominal, as in *Blatta* (Blattodea: Blattidae) and *Chironomus* (Diptera: Chironomidae); (c) two thoracic and considerable abdominal fusion of ganglia, as in *Crabro* and *Eucera* (Hymenoptera: Crabronidae and Anthophoridae); (d) highly fused with one thoracic and no abdominal ganglia, as in *Musca*, *Calliphora*, and *Lucilia* (Diptera: Muscidae and Calliphoridae); (e) extreme fusion with no separate suboesophageal ganglion, as in *Hydrometra* (Hemiptera: Hydrometridae) and *Rhizotrogus* (Scarabaeidae). (After Horridge 1965.)

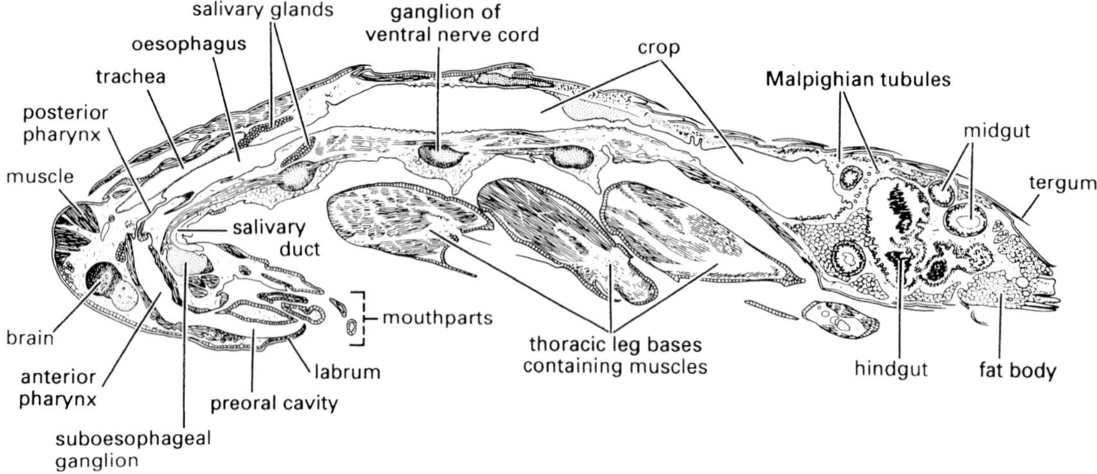

Fig. 3.7 Mediolongitudinal section of an immature cockroach of *Periplaneta americana* (Blattodea: Blattidae) showing internal organs and tissues.

3.3 THE ENDOCRINE SYSTEM AND THE FUNCTION OF HORMONES

Hormones are chemicals produced within an organism's body and transported, generally in body fluids, away from their point of synthesis to sites where they influence a remarkable variety of physiological processes, even though present in extremely small quantities. Insect hormones have been studied in detail in only a handful of species but similar patterns of production and function are likely to apply to all insects. The actions and interrelationships of these chemical messengers are varied and complex but the role of hormones in the molting process is of overwhelming importance and will be discussed more fully in this context in section 6.3. Here we provide a general picture of the endocrine centers and the hormones that they export.

Historically, the implication of hormones in the processes of molting and metamorphosis resulted from simple but elegant experiments. These utilized techniques that removed the influence of the brain (**decapitation**), isolated the hemolymph of different parts of the body (**ligation**), or artificially connected the hemolymph of two or more insects by joining their bodies. Ligation and decapitation of insects enabled researchers to localize the sites of control of developmental and reproductive processes and to show that substances are released that affect tissues at sites distant from the point of release. In addition, critical developmental periods for the action of these controlling substances have been identified. The blood-sucking bug *Rhodnius prolixus* (Hemiptera: Reduviidae) and various moths and flies were the principal experimental insects. More refined technologies allowed microsurgical removal or transplant of various tissues, hemolymph transfusion, hormone extraction and purification, and radioactive labeling of hormone extracts. Today, molecular biological (Box 3.1) and advanced chemical analytical techniques allow hormone isolation, characterization, and manipulation.

3.3.1 Endocrine centers

The hormones of the insect body are produced by neuronal, neuroglandular, or glandular centers (Fig. 3.8). Hormonal production by some organs, such as the ovaries, is secondary to their main function, but several tissues and organs are specialized for an endocrine role.

Neurosecretory cells

Neurosecretory cells (NSC) (also called neuroendocrine cells) are modified neurons found throughout the nervous system (within the CNS, peripheral nervous system, and the stomodeal nervous system), but they occur in major groups in the brain. These cells produce most of the known insect hormones, the notable exceptions being the production by non-neural tissues of ecdysteroids and juvenile hormones. However, the synthesis and release of the latter hormones are regulated by neurohormones from NSC.

Corpora cardiaca

The corpora cardiaca (singular: corpus cardiacum) are a pair of neuroglandular bodies located on either side of the aorta and behind the brain. As well as producing their own neurohormones, they store and release neurohormones, including prothoracicotropic hormone (PTTH, formerly called brain hormone or ecdysiotropin), originating from the NSC of the brain. PTTH stimulates the secretory activity of the prothoracic glands.

Prothoracic glands

The prothoracic glands are diffuse, paired glands generally located in the thorax or the back of the head. In cyclorrhaphous Diptera they are part of the ring gland, which also contains the corpora cardiaca and corpora allata. The prothoracic glands secrete an ecdysteroid, usually ecdysone (sometimes called molting hormone), which, after hydroxylation, elicits the molting process of the epidermis (section 6.3).

Corpora allata

The corpora allata (singular: corpus allatum) are small, discrete, paired glandular bodies derived from the epithelium and located on either side of the foregut. In some insects they fuse to form a single gland. Their function is to secrete **juvenile hormone** (JH), which has regulatory roles in both metamorphosis and reproduction.

3.3.2 Hormones

Three hormones or hormone types are integral to the growth and reproductive functions in insects. These

Box 3.1 Molecular genetic techniques and their application to neuropeptide research*

Molecular biology is essentially a set of techniques for the isolation, analysis, and manipulation of DNA and its RNA and protein products. Molecular genetics is concerned primarily with the nucleic acids, whereas research on the proteins and their constituent amino acids involves chemistry. Thus, genetics and chemistry are integral to molecular biology. Molecular biological tools provide:

- a means of cutting DNA at specific sites using restriction enzymes and of rejoining naked ends of cut fragments with ligase enzymes;
- techniques, such as the polymerase chain reaction (PCR), that produce numerous identical copies by repeated cycles of amplification of a segment of DNA;
- methods for rapid sequencing of nucleotides of DNA or RNA, and amino acids of proteins;
- the ability to synthesize short sequences of DNA or proteins;
- DNA–DNA affinity hybridization to compare the match of the synthesized DNA with the original sequence;
- the ability to search a genome for a specific nucleotide sequence using oligonucleotide probes, which are defined nucleic acid segments that are complementary to the sequence being sought;
- site-directed mutation of specific DNA segments *in vitro*;
- genetic engineering – the isolation and transfer of intact genes into other organisms, with subsequent stable transmission and gene expression;
- cytochemical techniques to identify how, when, and where genes are actually transcribed;
- immunochemical and histochemical techniques to identify how, when, and where a specific gene product functions.

Insect peptide hormones have been difficult to study because of the minute quantities produced by individual insects and their structural complexity and occasional instability. Currently, neuropeptides are the subject of an explosion of studies because of the realization that these proteins play crucial roles in most aspects of insect physiology (see Table 3.1), and the availability of appropriate technologies in chemistry (e.g. gas-phase sequencing of amino acids in proteins) and genetics. Knowledge of neuropeptide amino acid sequences provides a means of using the powerful capabilities of molecular genetics. Nucleotide sequences deduced from primary protein structures allow construction of oligonucleotide probes for searching out peptide genes in other parts of the genome or, more importantly, in other organisms, especially pests. Methods such as PCR and its variants facilitate the production of probes from partial amino acid sequences and trace amounts of DNA. Genetic amplification methods, such as PCR, allow the production of large quantities of DNA and thus allow easier sequencing of genes. Of course, these uses of molecular genetic methods depend on the initial chemical characterization of the neuropeptides. Furthermore, appropriate bioassays are essential for assessing the authenticity of any product of molecular biology. The possible application of neuropeptide research to control of insect pests is discussed in section 16.4.3.

*After Altstein 2003; Hoy 2003.

are the ecdysteroids, the juvenile hormones, and the neurohormones (also called neuropeptides).

Ecdysteroid is a general term applied to any steroid with molt-promoting activity. All ecdysteroids are derived from sterols, such as cholesterol, which insects cannot synthesize *de novo* and must obtain from their diet. Ecdysteroids occur in all insects and form a large group of compounds, of which ecdysone and 20-hydroxyecdysone are the most common members. **Ecdysone** (also called α-ecdysone) is released from the prothoracic glands into the hemolymph and usually is converted to the more active hormone **20-hydroxyecdysone** in several peripheral tissues. The 20-hydroxyecdysone (often referred to as ecdysterone or β-ecdysone in older literature) is the most widespread and physiologically important ecdysteroid in insects. The action of ecdysteroids in eliciting molting has been studied extensively and has the same function in different insects. Ecdysteroids also are produced by the ovary of the adult female insect and may be involved in ovarian maturation (e.g. yolk deposition) or be packaged in the eggs to be metabolized during the formation of embryonic cuticle.

Juvenile hormones form a family of related sesquiterpenoid compounds, so that the symbol JH may denote one or a mixture of hormones, including JH-I, JH-II, JH-III, and JH-0. The occurrence of mixed-JH-producing insects (such as the tobacco hornworm, *Manduca sexta*)

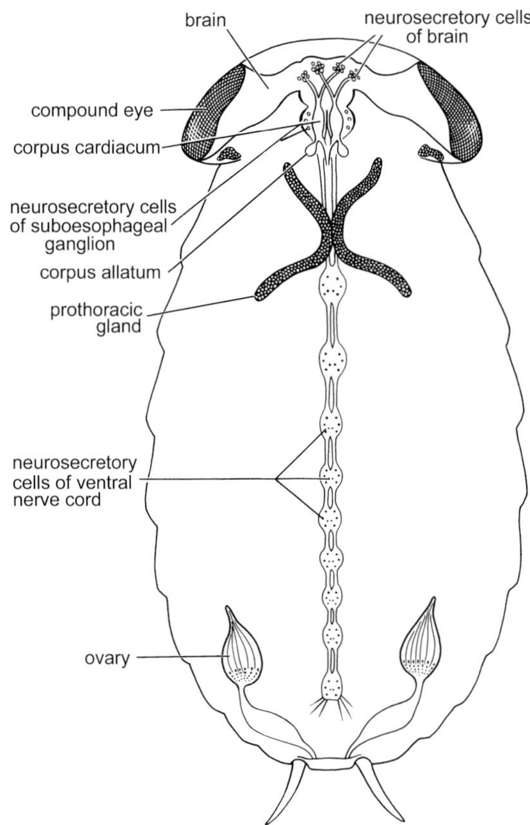

Fig. 3.8 The main endocrine centers in a generalized insect. (After Novak 1975.)

adds to the complexity of unraveling the functions of the homologous JHs. These hormones have two major roles – the control of metamorphosis and regulation of reproductive development. Larval characteristics are maintained and metamorphosis is inhibited by JH; adult development requires a molt in the absence of JH (see section 6.3 for details). Thus JH controls the degree and direction of differentiation at each molt. In the adult female insect, JH stimulates the deposition of yolk in the eggs and affects accessory gland activity and pheromone production (section 5.11).

Neurohormones constitute the third and largest class of insect hormones. They are generally peptides (small proteins) and hence have the alternative name **neuropeptides**. These protein messengers are the master regulators of many aspects of insect development, homeostasis, metabolism, and reproduction, including the secretion of the JHs and ecdysteroids. Nearly 150 neuropeptides have been recognized, and some (perhaps many) exist in multiple forms encoded by the same gene following gene duplication events. From this diversity, Table 3.1 summarizes a representative range of physiological processes reportedly affected by neurohormones in various insects. The diversity and vital co-ordinating roles of these small molecules continue to be revealed thanks to technological developments in peptide molecular chemistry (Box 3.1) allowing characterization and functional interpretation. Structural diversity among peptides of equivalent or related biological activity is a consequence of synthesis from large precursors that are cleaved and modified to form the active peptides. Neuropeptides either reach terminal effector sites directly along nerve axons or via the hemolymph, or indirectly exert control via their action on other endocrine glands (corpora allata and prothoracic glands). Both inhibitory and stimulatory signals are involved in neurohormone regulation. The effectiveness of regulatory neuropeptides depends on stereospecific high-affinity binding sites located in the plasma membrane of the target cells.

Hormones reach their target tissues by transport (even over short distances) by the body fluid or hemolymph. Hormones are often water-soluble but some may be transported bound to proteins in the hemolymph; for example, ecdysteroid-binding proteins and JH-binding proteins are known in a number of insects. These hemolymph-binding proteins may contribute to the regulation of hormone levels by facilitating uptake by target tissues, reducing non-specific binding, or protecting from degradation or excretion.

3.4 THE CIRCULATORY SYSTEM

Hemolymph, the insect body fluid (with properties and functions as described in section 3.4.1), circulates freely around the internal organs. The pattern of flow is regular between compartments and appendages, assisted by muscular contractions of parts of the body, especially the peristaltic contractions of a longitudinal **dorsal vessel**, part of which is sometimes called the heart. Hemolymph does not directly contact the cells because the internal organs and the epidermis are covered in a basement membrane, which may regulate the exchange of materials. This open circulatory system has only a few vessels and compartments to direct hemolymph movement, in contrast to the

Table 3.1 Examples of some important insect physiological processes mediated by neuropeptides. (After Keeley & Hayes 1987; Holman et al. 1990; Gäde et al. 1997; Altstein 2003.)

Neuropeptide	Action
Growth and development	
Allatostatins and allatotropins	Induce/regulate juvenile hormone (JH) production
Bursicon	Controls cuticular sclerotization
Crustacean cardioactive peptide (CCAP)	Switches on ecdysis behavior
Diapause hormone (DH)	Causes dormancy in silkworm eggs
Pre-ecdysis triggering hormone (PETH)	Stimulates pre-ecdysis behavior
Ecdysis triggering hormone (ETH)	Initiates events at ecdysis
Eclosion hormone (EH)	Controls events at ecdysis
JH esterase inducing factor	Stimulates JH degradative enzyme
Prothoracicotropic hormone (PTTH)	Induces ecdysteroid secretion from prothoracic gland
Puparium tanning factor	Accelerates fly puparium tanning
Reproduction	
Antigonadotropin (e.g. oostatic hormone, OH)	Suppresses oocyte development
Ovarian ecdysteroidogenic hormone (OEH = EDNH)	Stimulates ovarian ecdysteroid production
Ovary maturing peptide (OMP)	Stimulates egg development
Oviposition peptides	Stimulate egg deposition
Prothoracicotropic hormone (PTTH)	Affects egg development
Pheromone biosynthesis activating neuropeptide (PBAN)	Regulates pheromone production
Homeostasis	
Metabolic peptides (= AKH/RPCH family)	
Adipokinetic hormone (AKH)	Releases lipid from fat body
Hyperglycemic hormone	Releases carbohydrate from fat body
Hypoglycemic hormone	Enhances carbohydrate uptake
Protein synthesis factors	Enhance fat body protein synthesis
Diuretic and antidiuretic peptides	
Antidiuretic peptide (ADP)	Suppresses water excretion
Diuretic peptide (DP)	Enhances water excretion
Chloride-transport stimulating hormone	Stimulates Cl^- absorption (rectum)
Ion-transport peptide (ITP)	Stimulates Cl^- absorption (ileum)
Myotropic peptides	
Cardiopeptides	Increase heartbeat rate
Kinin family (e.g. leukokinins and myosuppressins)	Regulate gut contraction
Proctolin	Modifies excitation response of some muscles
Chromatotropic peptides	
Melanization and reddish coloration hormone (MRCH)	Induces darkening
Pigment-dispersing hormone (PDH)	Disperses pigment
Corazonin	Darkens pigment

closed network of blood-conducting vessels seen in vertebrates.

3.4.1 Hemolymph

The volume of the hemolymph may be substantial (20–40% of body weight) in soft-bodied larvae, which use the body fluid as a hydrostatic skeleton, but is less than 20% of body weight in most nymphs and adults. Hemolymph is a watery fluid containing ions, molecules, and cells. It is often clear and colorless but may be variously pigmented yellow, green, or blue, or rarely, in the immature stages of a few aquatic and endoparasitic flies, red owing to the presence of hemoglobin. All chemical exchanges between insect tissues are mediated via the hemolymph – hormones are transported, nutrients are distributed from the gut, and wastes are removed to the excretory organs. However, insect hemolymph only rarely contains respiratory pigments and hence has a very low oxygen-carrying capacity. Local changes in hemolymph pressure are important in ventilation of the tracheal system (section 3.5.1), in thermoregulation (section 4.2.2), and at molting to aid splitting of the old and expansion of the new cuticle. The hemolymph serves also as a water reserve, as its main constituent, **plasma**, is an aqueous solution of inorganic ions, lipids, sugars (mainly trehalose), amino acids, proteins, organic acids, and other compounds. High concentrations of amino acids and organic phosphates characterize insect hemolymph, which also is the site of deposition of molecules associated with cold protection (section 6.6.1). Hemolymph proteins include those that act in storage (hexamerins) and those that transport lipids (lipophorin) or complex with iron (ferritin) or juvenile hormone (JH-binding protein).

The blood cells, or **hemocytes** (haemocytes), are of several types (mainly plasmatocytes, granulocytes, and prohemocytes) and all are nucleate. They have four basic functions:

1 phagocytosis – the ingestion of small particles and substances such as metabolites;
2 encapsulation of parasites and other large foreign materials;
3 hemolymph coagulation;
4 storage and distribution of nutrients.

The hemocoel contains two additional types of cells. **Nephrocytes** (sometimes called pericardial cells) generally occur near the dorsal vessel and appear to function as ductless glands by sieving the hemolymph of certain substances and metabolizing them for use or excretion elsewhere. **Oenocytes** may occur in the hemocoel, fat body, or epidermis and, although their functions are unclear in most insects, they appear to have a role in cuticle lipid (hydrocarbon) synthesis and, in some chironomids, they produce hemoglobins.

3.4.2 Circulation

Circulation in insects is maintained mostly by a system of muscular pumps moving hemolymph through compartments separated by fibromuscular septa or membranes. The main pump is the pulsatile dorsal vessel. The anterior part may be called the aorta and the posterior part may be called the heart, but the two terms are inconsistently applied. The dorsal vessel is a simple tube, generally composed of one layer of myocardial cells and with segmentally arranged openings, or **ostia**. The lateral ostia typically permit the one-way flow of hemolymph into the dorsal vessel as a result of valves that prevent backflow. In many insects there also are more ventral ostia that permit hemolymph to flow out of the dorsal vessel, probably to supply adjacent active muscles. There may be up to three pairs of thoracic ostia and nine pairs of abdominal ostia, although there is an evolutionary tendency towards reduction in number of ostia. The dorsal vessel lies in a compartment, the **pericardial sinus**, above a **dorsal diaphragm** (a fibromuscular septum – a separating membrane) formed of connective tissue and segmental pairs of **alary muscles**. The alary muscles support the dorsal vessel but their contractions do not affect heartbeat. Hemolymph enters the pericardial sinus via segmental openings in the diaphragm and/or at the posterior border and then moves into the dorsal vessel via the ostia during a muscular relaxation phase. Waves of contraction, which normally start at the posterior end of the body, pump the hemolymph forwards in the dorsal vessel and out via the aorta into the head. Next, the appendages of the head and thorax are supplied with hemolymph as it circulates posteroventrally and eventually returns to the pericardial sinus and the dorsal vessel. A generalized pattern of hemolymph circulation in the body is shown in Fig. 3.9a; however, in adult insects there also may be a periodic reversal of hemolymph flow in the dorsal vessel (from thorax posteriorly) as part of normal circulatory regulation.

Another important component of the circulation of many insects is the **ventral diaphragm** (Fig. 3.9b) – a

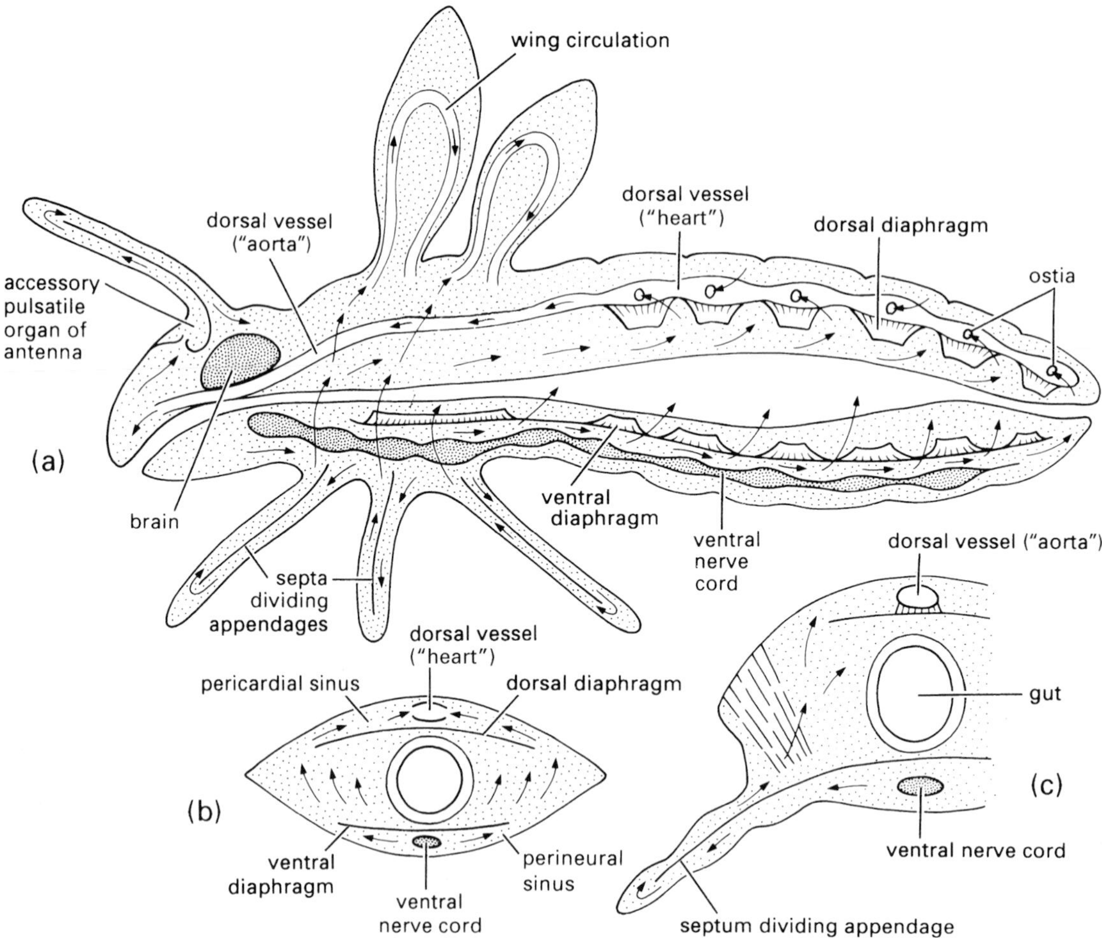

Fig. 3.9 Schematic diagram of a well-developed circulatory system: (a) longitudinal section through body; (b) transverse section of the abdomen; (c) transverse section of the thorax. Arrows indicate directions of hemolymph flow. (After Wigglesworth 1972.)

fibromuscular septum that lies in the floor of the body cavity and is associated with the ventral nerve cord. Circulation of the hemolymph is aided by active peristaltic contractions of the ventral diaphragm, which direct the hemolymph backwards and laterally in the **perineural sinus** below the diaphragm. Hemolymph flow from the thorax to the abdomen also may be dependent, at least partially, on expansion of the abdomen, thus "sucking" hemolymph posteriorly. Hemolymph movements are especially important in insects that use the circulation in thermoregulation (some Odonata, Diptera, Lepidoptera, and Hymenoptera). Another function of the diaphragm may be to facilitate rapid exchange of chemicals between the ventral nerve cord and the hemolymph by either actively moving the hemolymph and/or moving the cord itself.

Hemolymph generally is circulated to appendages unidirectionally by various tubes, septa, valves, and pumps (Fig. 3.9c). The muscular pumps are termed **accessory pulsatile organs** and occur at the base of the antennae, at the base of the wings, and sometimes in the legs. Furthermore, the antennal pulsatile organs may release neurohormones that are carried to the antennal lumen to influence the sensory neurons. Wings have a definite but variable circulation, although it may be apparent only in the young adult. At least in some Lepidoptera, circulation in the wing occurs by the reciprocal movement of hemolymph (in the wing vein sinuses) and air (within the elastic wing tracheae) into and from the wing, brought about by pulsatile organ activity, reversals of heartbeat, and tracheal volume changes.

The insect circulatory system displays an impressive degree of synchronization between the activities of the dorsal vessel, fibromuscular diaphragms, and accessory pumps, mediated by both nervous and neurohormonal regulation. The physiological regulation of many body functions by the neurosecretory system occurs via neurohormones transported in the hemolymph.

3.4.3 Protection and defense by the hemolymph

Hemolymph provides various kinds of protection and defense from (i) physical injury; (ii) the entry of disease organisms, parasites, or other foreign substances; and sometimes (iii) the actions of predators. In some insects the hemolymph contains malodorous or distasteful chemicals, which are deterrent to predators (Chapter 14). Injury to the integument elicits a wound-healing process that involves hemocytes and plasma coagulation. A hemolymph clot is formed to seal the wound and reduce further hemolymph loss and bacterial entry. If disease organisms or particles enter an insect's body, then immune responses are invoked. These include the cellular defense mechanisms of phagocytosis, encapsulation, and nodule formation mediated by the hemocytes, as well as the actions of humoral factors such as enzymes or other proteins (e.g. lysozymes, prophenoloxidase, lectins, and peptides).

The immune system of insects bears little resemblance to the complex immunoglobulin-based vertebrate system. However, insects sublethally infected with bacteria can rapidly develop greatly increased resistance to subsequent infection. Hemocytes are involved in phagocytosing bacteria but, in addition, immunity proteins with antibacterial activity appear in the hemolymph after a primary infection. For example, lytic peptides called cecropins, which disrupt the cell membranes of bacteria and other pathogens, have been isolated from certain moths. Furthermore, some neuropeptides may participate in cell-mediated immune responses by exchanging signals between the neuroendocrine system and the immune system, as well as influencing the behavior of cells involved in immune reactions. The insect immune system is much more complicated than once thought.

3.5 THE TRACHEAL SYSTEM AND GAS EXCHANGE

In common with all aerobic animals, insects must obtain oxygen from their environment and eliminate carbon dioxide respired by their cells. This is **gas exchange**, distinguished from **respiration**, which strictly refers to oxygen-consuming, cellular metabolic processes. In almost all insects, gas exchange occurs by means of internal air-filled tracheae. These tubes branch and ramify through the body (Fig. 3.10). The finest branches contact all internal organs and tissues, and are especially numerous in tissues with high oxygen requirements. Air usually enters the tracheae via spiracular openings that are positioned laterally on the body, primitively with one pair per post-cephalic segment. No extant insect has more than 10 pairs (two thoracic and eight abdominal) (Fig. 3.11a), most have eight or nine, and some have one (Fig. 3.11c), two, or none (Fig. 3.11d–f). Typically, spiracles (Fig. 3.10a) have a chamber, or **atrium**, with an opening-and-closing mechanism, or **valve**, either projecting externally or at the inner end of the atrium. In the latter type, a filter apparatus sometimes protects the outer opening. Each spiracle may be set in a sclerotized cuticular plate called a **peritreme**.

The tracheae are invaginations of the epidermis and thus their lining is continuous with the body cuticle. The characteristic ringed appearance of the tracheae seen in tissue sections (as in Fig. 3.7) is due to the spiral ridges or thickenings of the cuticular lining, the **taenidia**, which allow the tracheae to be flexible but resist compression (analogous to the function of the ringed hose of a vacuum cleaner). The cuticular linings of the tracheae are shed with the rest of the exoskeleton when the insect molts. Usually even the linings of the finest branches of the tracheal system are shed at ecdysis but linings of the fluid-filled blind endings, the **tracheoles**, may or may not be shed. Tracheoles are less than 1 μm in diameter and closely contact the respiring tissues (Fig. 3.10b), sometimes indenting into the cells that they supply. However, the tracheae that supply oxygen to the ovaries of many insects have very few tracheoles, the taenidia are weak or absent, and the tracheal surface is evaginated as tubular spirals projecting into the hemolymph. These aptly named **aeriferous tracheae**

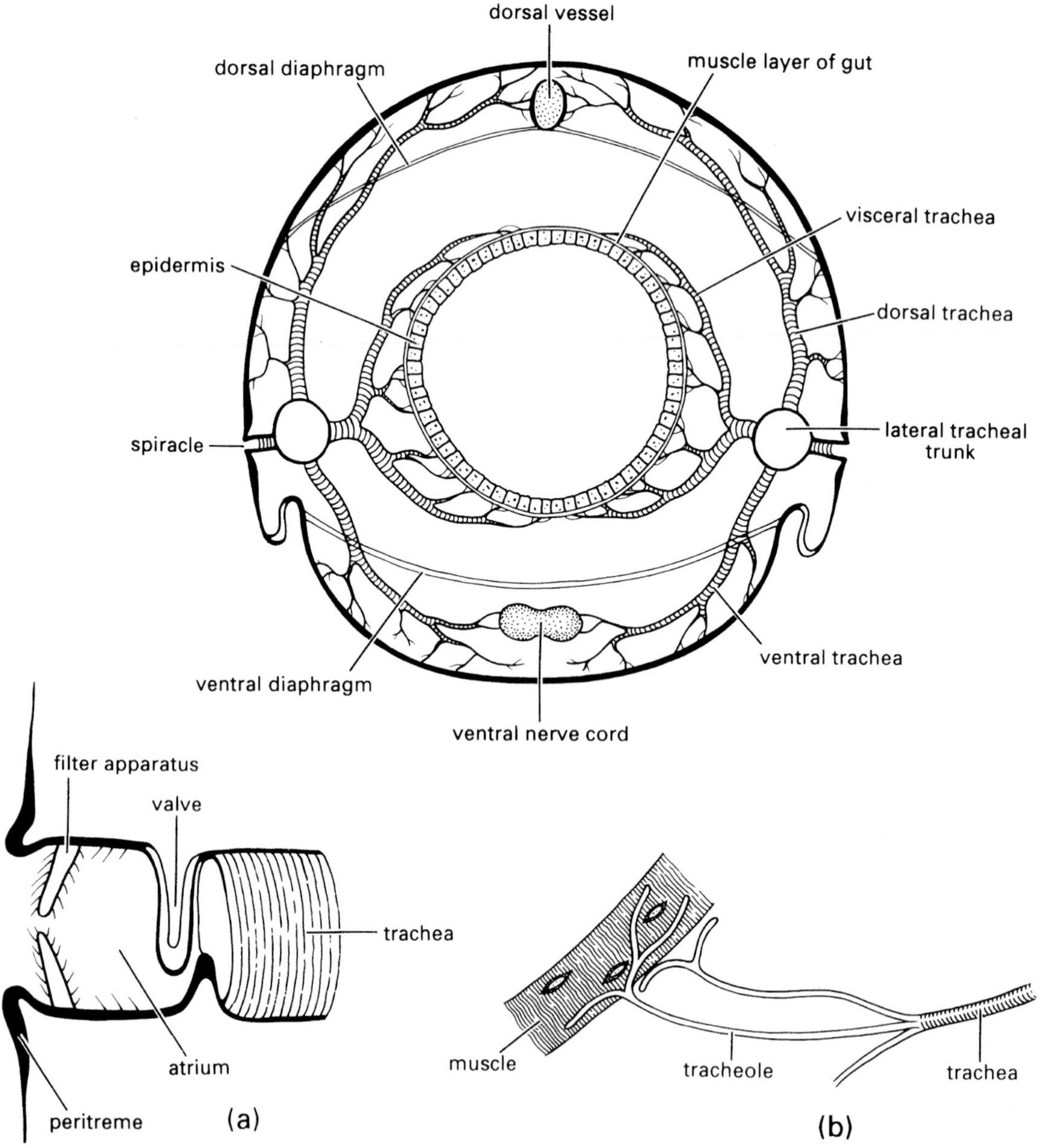

Fig. 3.10 Schematic diagram of a generalized tracheal system seen in a transverse section of the body at the level of a pair of abdominal spiracles. Enlargements show: (a) an atriate spiracle with closing valve at inner end of atrium; (b) tracheoles running to a muscle fiber. (After Snodgrass 1935.)

have a highly permeable surface that allows direct aeration of the surrounding hemolymph from tracheae that may exceed 50 μm in diameter.

In terrestrial and many aquatic insects the tracheae open to the exterior via the spiracles (an **open tracheal system**) (Fig. 3.11a–c). In contrast, in some aquatic and many endoparasitic larvae spiracles are absent (a **closed tracheal system**) and the tracheae divide

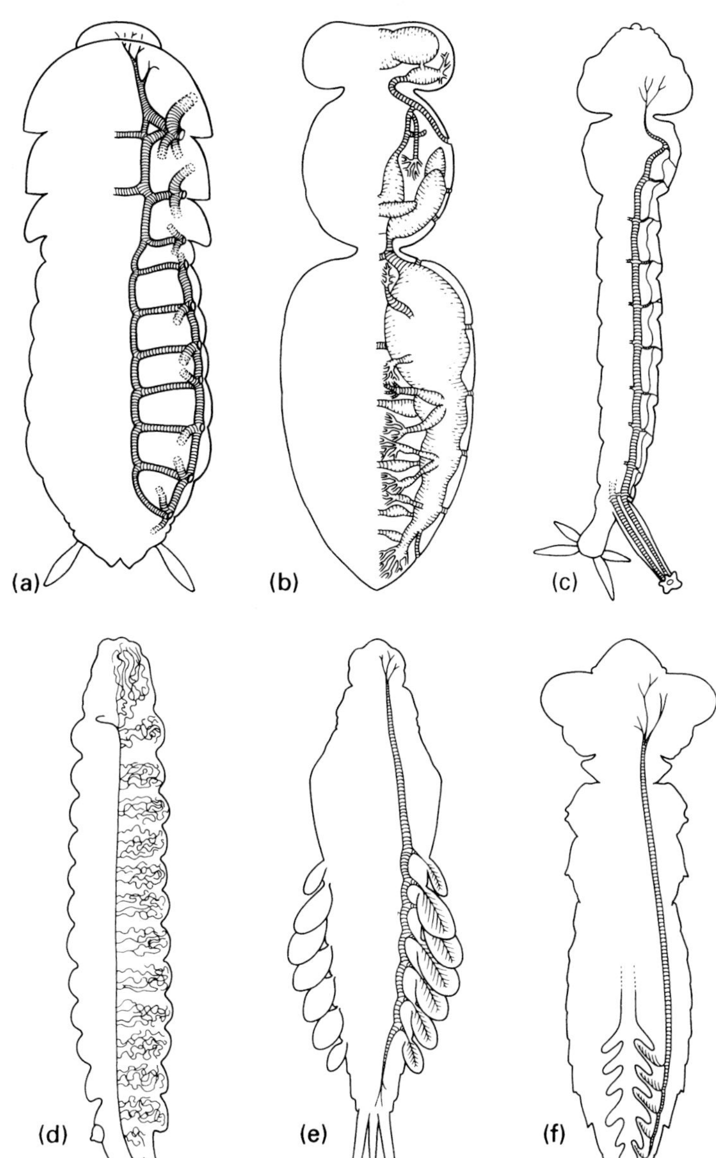

Fig. 3.11 Some basic variations in the open (a–c) and closed (d–f) tracheal systems of insects. (a) Simple tracheae with valved spiracles, as in cockroaches. (b) Tracheae with mechanically ventilated air sacs, as in honey bees. (c) Metapneustic system with only terminal spiracles functional, as in mosquito larvae. (d) Entirely closed tracheal system with cutaneous gas exchange, as in most endoparasitic larvae. (e) Closed tracheal system with abdominal tracheal gills, as in mayfly nymphs. (f) Closed tracheal system with rectal tracheal gills, as in dragonfly nymphs. (After Wigglesworth 1972; details in (a) after Richards & Davies 1977, (b) after Snodgrass 1956, (c) after Snodgrass 1935, (d) after Wigglesworth 1972.)

peripherally to form a network. This covers the general body surface (allowing cutaneous gas exchange) (Fig. 3.11d) or lies within specialized filaments or lamellae (tracheal gills) (Fig. 3.11e,f). Some aquatic insects with an open tracheal system carry **gas gills** with them (e.g. bubbles of air); these may be temporary or permanent (section 10.3.4).

The volume of the tracheal system ranges between 5% and 50% of the body volume depending on species and stage of development. The more active the insect, the more extensive is the tracheal system. In many insects, parts of tracheae are dilated or enlarged to increase the reservoir of air, and in some species the dilations form **air sacs** (Fig. 3.11b), which collapse readily because the taenidia of the cuticular lining are reduced or absent. Sometimes the tracheal volume may decrease within a developmental stage as air sacs are occluded by growing tissues. Air sacs reach their greatest development in very active flying insects, such as bees and cyclorrhaphous Diptera. They may assist flight by increasing buoyancy, but their main function is in ventilation of the tracheal system.

3.5.1 Diffusion and ventilation

Oxygen enters the spiracle, passes through the length of the tracheae to the tracheoles and into the target cells by a combination of ventilation and diffusion along a concentration gradient, from high in the external air to low in the tissue. Whereas the net movement of oxygen molecules in the tracheal system is inward, the net movement of carbon dioxide and (in terrestrial insects) water vapor molecules is outward. Hence gas exchange in most terrestrial insects is a compromise between securing sufficient oxygen and reducing water loss via the spiracles. During periods of inactivity, the spiracles in many insects are kept closed most of the time, opening only periodically. In insects of xeric environments, the spiracles may be small with deep atria or have a mesh of cuticular projections in the orifice.

In insects without air sacs, such as most holometabolous larvae, diffusion appears to be the primary mechanism for the movement of gases in the tracheae and is always the sole mode of gas exchange at the tissues. The efficiency of diffusion is related to the distance of diffusion and perhaps to the diameter of the tracheae (Box 3.2). Recently, rapid cycles of tracheal compression and expansion have been observed in the head and thorax of some insects using X-ray videoing. Movements of the hemolymph and body could not explain these cycles, which appear to be a new mechanism of gas exchange in insects. In addition, large or dilated tracheae may serve as an oxygen reserve when the spiracles are closed. In very active insects, especially large ones, active pumping movements of the thorax and/or abdomen **ventilate** (pump air through) the outer parts of the tracheal system and so the diffusion pathway to the tissues is reduced. Rhythmic thoracic movements and/or dorsoventral flattening or telescoping of the abdomen expels air, via the spiracles, from extensible or some partially compressible tracheae or from air sacs. Co-ordinated opening and closing of the spiracles usually accompanies ventilatory movements and provides the basis for the unidirectional air flow that occurs in the main tracheae of larger insects. Anterior spiracles open during inspiration and posterior ones open during expiration. The presence of air sacs, especially if large or extensive, facilitates ventilation by increasing the volume of tidal air that can be changed as a result of ventilatory movements. If the main tracheal branches are strongly ventilated, diffusion appears sufficient to oxygenate even the most actively respiring tissues, such as flight muscles. However, the design of the gas-exchange system of insects places an upper limit on size because, if oxygen has to diffuse over a considerable distance, the requirements of a very large and active insect either could not be met, even with ventilatory movements and compression and expansion of tracheae, or would result in substantial loss of water through the spiracles. Interestingly, many large insects are long and thin, thereby minimizing the diffusion distance from the spiracle along the trachea to any internal organ.

3.6 THE GUT, DIGESTION, AND NUTRITION

Insects of different groups consume an astonishing variety of foods, including watery xylem sap (e.g. nymphs of spittle bugs and cicadas), vertebrate blood (e.g. bed bugs and female mosquitoes), dry wood (e.g. some termites), bacteria and algae (e.g. black fly and many caddisfly larvae), and the internal tissues of other insects (e.g. endoparasitic wasp larvae). The diverse range of mouthpart types (section 2.3.1) correlates with the diets of different insects, but gut structure and function also reflect the mechanical properties and the nutrient composition of the food eaten. Four major feeding specializations can be identified depending on whether the food is solid or liquid or of plant or animal origin (Fig. 3.12). Some insect species clearly fall into a single category, but others with generalized diets may fall between two or more of them, and most endopterygotes will occupy different categories at different stages of their life (e.g. moths and butterflies switch from solid-plant as larvae to liquid-plant as adults). Gut morphology and physiology relate to these dietary differences in the following ways. Insects that take solid food typically have a wide, straight, short gut with strong musculature and obvious protection from abrasion (especially in the midgut, which has no cuticular lining). These features are most obvious in solid-feeders with rapid throughput of food as in plant-feeding caterpillars. In contrast, insects feeding on blood, sap, or nectar usually have long, narrow, convoluted guts to allow maximal contact with the liquid food; here, protection from

Box 3.2 Tracheal hypertrophy in mealworms at low oxygen concentrations

Resistance to diffusion of gases in insect tracheal systems arises from the spiracular valves when they are partially or fully closed, the tracheae, and the cytoplasm supplied by the tracheoles at the end of the tracheae. Air-filled tracheae will have a much lower resistance per unit length than the watery cytoplasm because oxygen diffuses several orders of magnitude faster in air than in cytoplasm for the same gradient of oxygen partial pressure. Until recently, the tracheal system was believed to provide more than sufficient oxygen (at least in non-flying insects that lack air sacs), with the tracheae offering trivial resistance to the passage of oxygen. Experiments on mealworm larvae, *Tenebrio molitor* (Coleoptera: Tenebrionidae), that were reared in different levels of oxygen (all at the same total gas pressure) showed that the main tracheae that supply oxygen to the tissues in the larvae hypertrophy (increase in size) at lower oxygen levels. The dorsal (D), ventral (V), and visceral (or gut, G) tracheae were affected but not the lateral longitudinal tracheae that interconnect the spiracles (the four tracheal categories are illustrated in an inset on the graph). The dorsal tracheae supply the dorsal vessel and dorsal musculature, the ventral tracheae supply the nerve cord and ventral musculature, whereas the visceral tracheae supply the gut, fat body, and gonads. The graph shows that the cross-sectional areas of the dorsal, ventral, and visceral tracheae were greater when the larvae were reared in 10.5% oxygen (●) than when they were reared in 21% oxygen (as in normal air) (○) (after Loudon 1989). Each point on the graph is for a single larva and is the average of the summed areas of the dorsal, ventral, and visceral tracheae for six pairs of abdominal spiracles. This hypertrophy appears to be inconsistent with the widely accepted hypothesis that tracheae contribute an insignificant resistance to net oxygen movement in insect tracheal systems. Alternatively, hypertrophy may simply increase the amount of air (and thus oxygen) that can be stored in the tracheal system, rather than reduce resistance to air flow. This might be particularly important for mealworms because they normally live in a dry environment and may minimize the opening of their spiracles. Whatever the explanation, the observations suggest that some adjustment can be made to the size of the tracheae in mealworms (and perhaps other insects) to match the requirements of the respiring tissues.

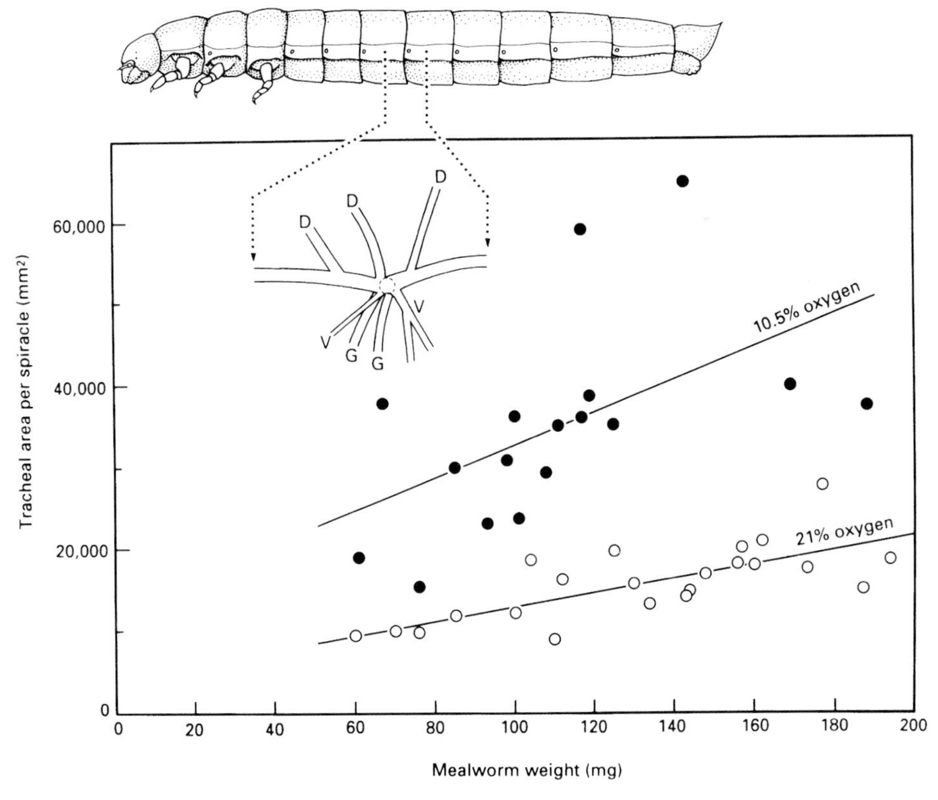

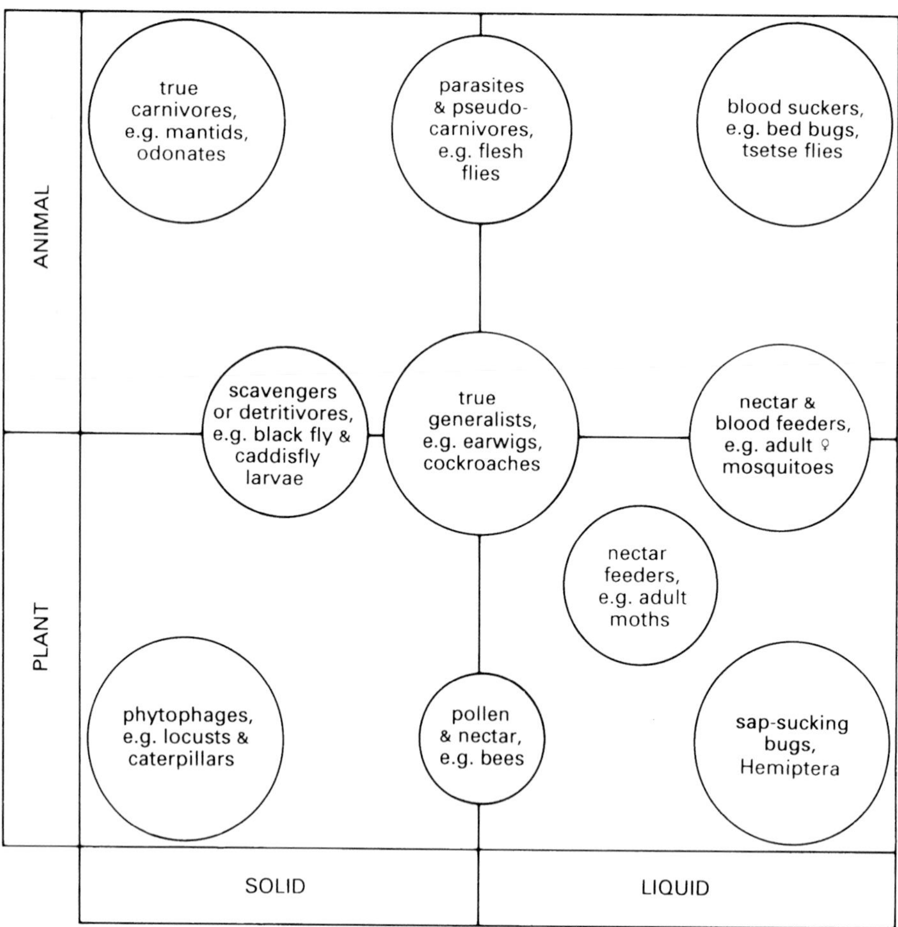

Fig. 3.12 The four major categories of insect feeding specialization. Many insects are typical of one category, but others cross two categories (or more, as in generalist cockroaches). (After Dow 1986.)

abrasion is unnecessary. The most obvious gut specialization of liquid-feeders is a mechanism for removing excess water to concentrate nutrient substances prior to digestion, as seen in hemipterans (Box 3.3). From a nutritional viewpoint, most plant-feeding insects need to process large amounts of food because nutrient levels in leaves and stems are often low. The gut is usually short and without storage areas, as food is available continuously. By comparison, a diet of animal tissue is nutrient-rich and, at least for predators, well balanced. However, the food may be available only intermittently (such as when a predator captures prey or a blood meal is obtained) and the gut normally has large storage capacity.

3.6.1 Structure of the gut

There are three main regions to the insect gut (or alimentary canal), with sphincters (valves) controlling food–fluid movement between regions (Fig. 3.13). The **foregut** (**stomodeum**) is concerned with ingestion, storage, grinding, and transport of food to the next region, the **midgut** (**mesenteron**). Here digestive enzymes are produced and secreted and absorption of the products of digestion occurs. The material remaining in the gut lumen together with urine from the Malpighian tubules then enters the **hindgut** (**proctodeum**), where absorption of water, salts, and other valuable molecules occurs prior to elimination of the

Box 3.3 The filter chamber of Hemiptera

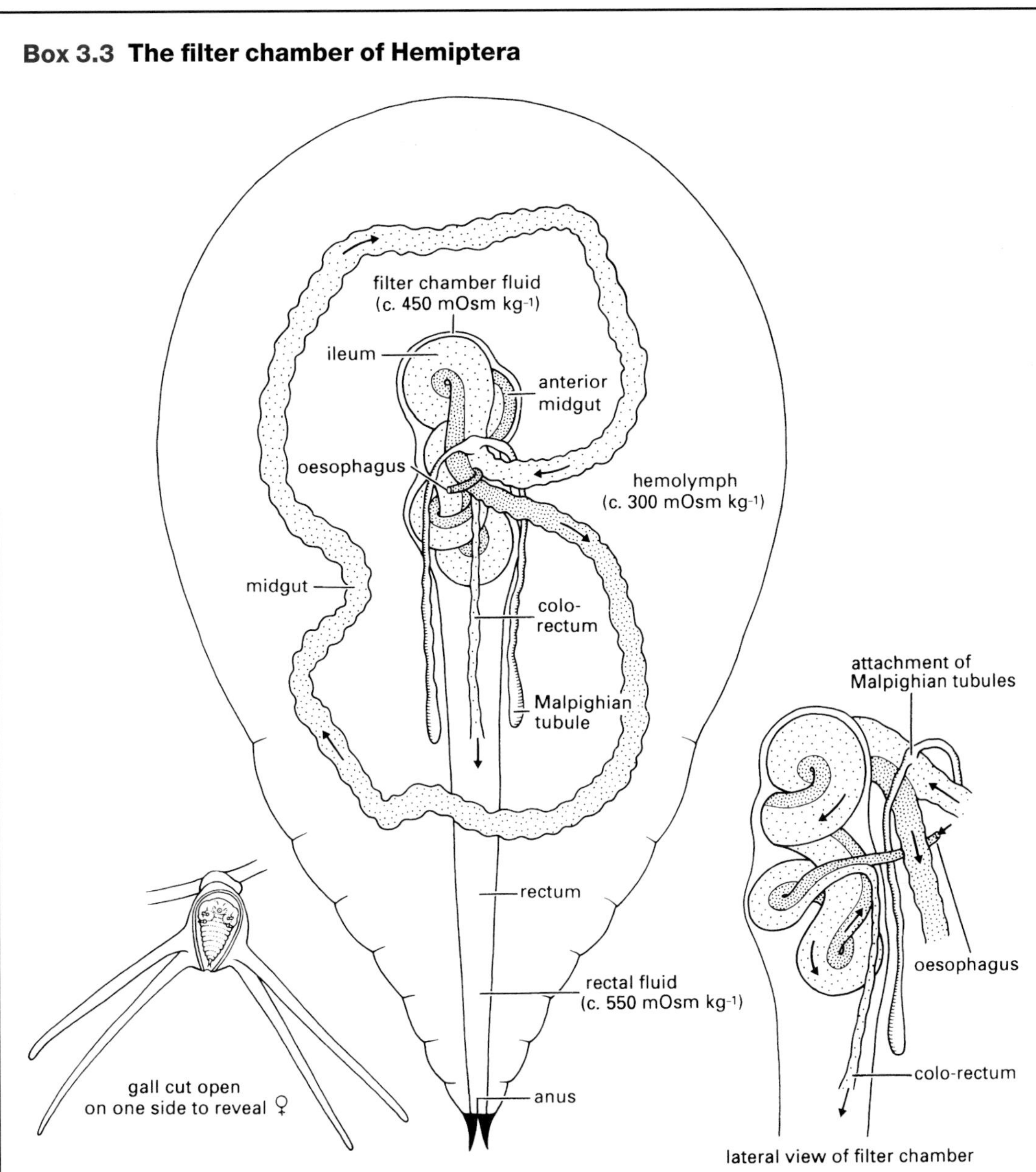

Most Hemiptera have an unusual arrangement of the midgut which is related to their habit of feeding on plant fluids. An anterior and a posterior part of the gut (typically involving the midgut) are in intimate contact to allow concentration of the liquid food. This **filter chamber** allows excess water and relatively small molecules, such as simple sugars, to be passed quickly and directly from the anterior gut to the hindgut, thereby short-circuiting the main absorptive portion of the midgut. Thus, the digestive region is not diluted by water nor congested by superabundant food molecules. Well-developed filter chambers are characteristic of cicadas and spittle bugs, which feed on xylem (sap that is rich in ions, low in organic compounds, and with low osmotic pressure), and leafhoppers and coccoids, which feed on phloem (sap that is rich in nutrients, especially sugars, and with high osmotic pressure). The gut physiology of such sap-suckers has been rather poorly studied because accurate recording of gut fluid composition and osmotic pressure depends on the technically difficult task of taking readings from an intact gut.

Adult female coccoids of gall-inducing *Apiomorpha* species (Eriococcidae) (section 11.2.4) tap the vascular tissue of the gall wall to obtain phloem sap. Some species have a highly developed filter chamber formed from loops of the anterior midgut and anterior hindgut enclosed within the membranous rectum. Depicted here is the gut of an adult female of *A. munita* viewed from the ventral side of the body. The thread-like sucking mouthparts (Fig. 11.4c) in series with the cibarial pump connect to a short oesophagus, which can be seen here in both the main drawing and the enlarged lateral view of the filter chamber. The oesophagus terminates at the anterior midgut, which coils upon itself as three loops of the filter chamber. It emerges ventrally and forms a large midgut loop lying free in the hemolymph. Absorption of nutrients occurs in this free loop. The Malpighian tubules enter the gut at the commencement of the ileum, before the ileum enters the filter chamber where it is closely apposed to the much narrower anterior midgut. Within the irregular spiral of the filter chamber, the fluids in the two tubes move in opposite directions (as indicated by the arrows).

The filter chamber of these coccoids apparently transports sugar (perhaps by active pumps) and water (passively) from the anterior midgut to the ileum and then via the narrow colo-rectum to the rectum, from which it is eliminated as honeydew. In *A. munita*, other than water, the honeydew is mostly sugar (accounting for 80% of the total osmotic pressure of about 550 mOsm kg^{-1}*). Remarkably, the osmotic pressure of the hemolymph (about 300 mOsm kg^{-1}) is much lower than that within the filter chamber (about 450 mOsm kg^{-1}) and rectum. Maintenance of this large osmotic difference may be facilitated by the impermeability of the rectal wall.

*Osmolarity values are from the unpublished data of P.D. Cooper & A.T. Marshall.

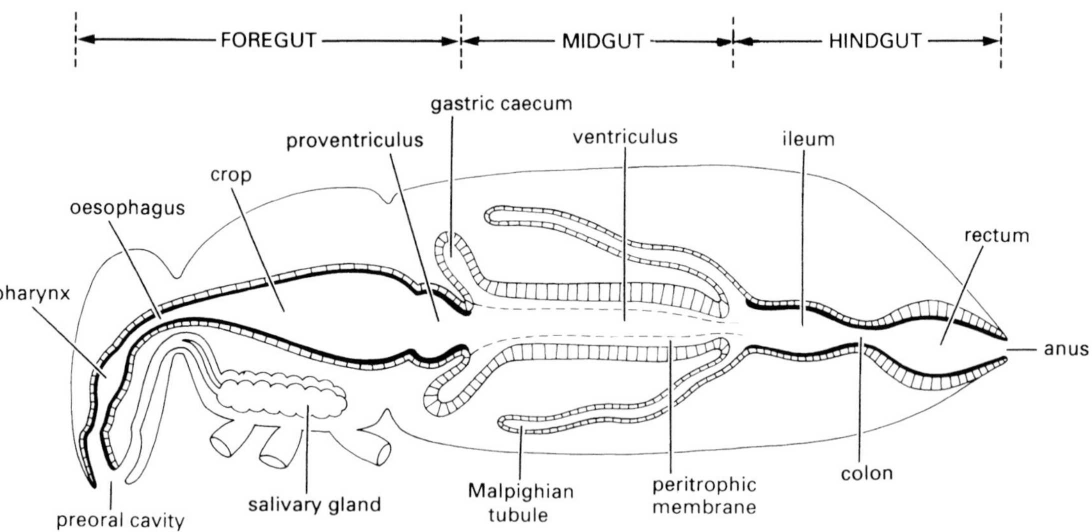

Fig. 3.13 Generalized insect alimentary canal showing division into three regions. The cuticular lining of the foregut and hindgut are indicated by thicker black lines. (After Dow 1986.)

feces through the anus. The gut epithelium is one cell layer thick throughout the length of the canal and rests on a basement membrane surrounded by a variably developed muscle layer. Both the foregut and hindgut have a cuticular lining, whereas the midgut does not.

Each region of the gut displays several local specializations, which are variously developed in different

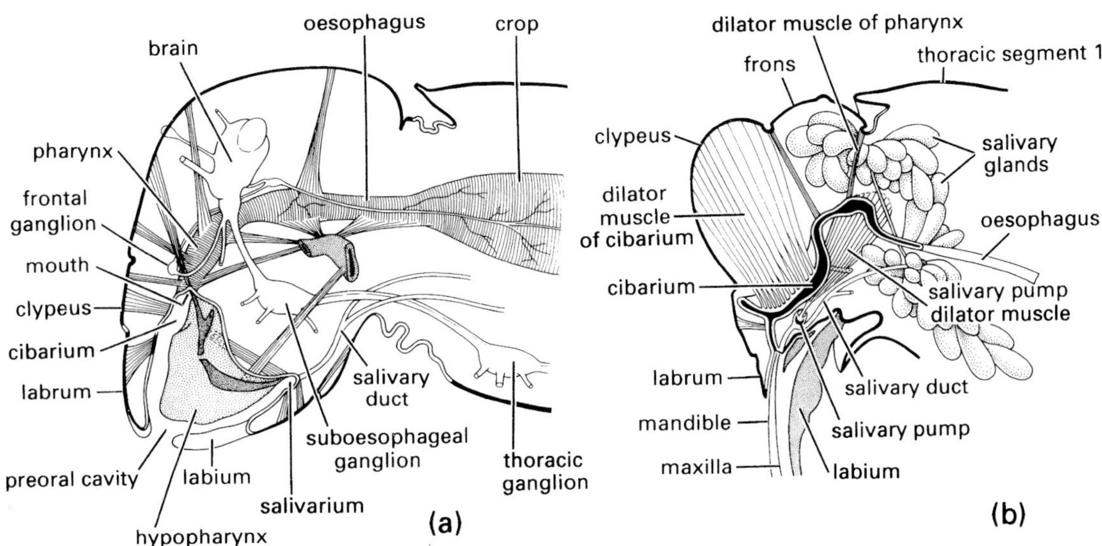

Fig. 3.14 Preoral and anterior foregut morphology in (a) a generalized orthopteroid insect and (b) a xylem-feeding cicada. Musculature of the mouthparts and the (a) pharyngeal or (b) cibarial pump are indicated but not fully labeled. Contraction of the respective dilator muscles causes dilation of the pharynx or cibarium and fluid is drawn into the pump chamber. Relaxation of these muscles results in elastic return of the pharynx or cibarial walls and expels food upwards into the oesophagus. (After Snodgrass 1935.)

insects, depending on diet. Typically the foregut is subdivided into a **pharynx**, an **oesophagus** (esophagus), and a **crop** (food storage area), and in insects that ingest solid food there is often a grinding organ, the **proventriculus** (or gizzard). The proventriculus is especially well developed in orthopteroid insects, such as cockroaches, crickets, and termites, in which the epithelium is folded longitudinally to form ridges on which the cuticle is armed with spines or teeth. At the anterior end of the foregut the mouth opens into a preoral cavity bounded by the bases of the mouthparts and often divided into an upper area, or **cibarium**, and a lower part, or **salivarium** (Fig. 3.14a). The paired **labial** or **salivary glands** vary in size and arrangement from simple elongated tubes to complex branched or lobed structures.

Complicated glands occur in many Hemiptera that produce two types of saliva (see section 3.6.2). In Lepidoptera, the labial glands produce silk, whereas mandibular glands secrete the saliva. Several types of secretory cell may occur in the salivary glands of one insect. The secretions from these cells are transported along cuticular ducts and emptied into the ventral part of the preoral cavity. In insects that store meals in their foregut, the crop may take up the greater portion of the food and often is capable of extreme distension, with a posterior sphincter controlling food retention. The crop may be an enlargement of part of the tubular gut (Fig. 3.7) or a lateral diverticulum.

The generalized midgut has two main areas – the tubular **ventriculus** and blind-ending lateral diverticula called **caeca** (ceca). Most cells of the midgut are structurally similar, being columnar with microvilli (finger-like protrusions) covering the inner surface. The distinction between the almost indiscernible foregut epithelium and the thickened epithelium of the midgut usually is visible in histological sections (Fig. 3.15). The midgut epithelium mostly is separated from the food by a thin sheath called the **peritrophic membrane**, consisting of a network of chitin fibrils in a protein–glycoprotein matrix. These proteins, called peritrophins, may have evolved from gastrointestinal mucus proteins by acquiring the ability to bind chitin. The peritrophic membrane either is delaminated from the whole midgut or produced by cells in the anterior region of the midgut. Exceptionally Hemiptera and Thysanoptera lack a peritrophic membrane, as do just the adults of several other orders.

Typically, the beginning of the hindgut is defined by

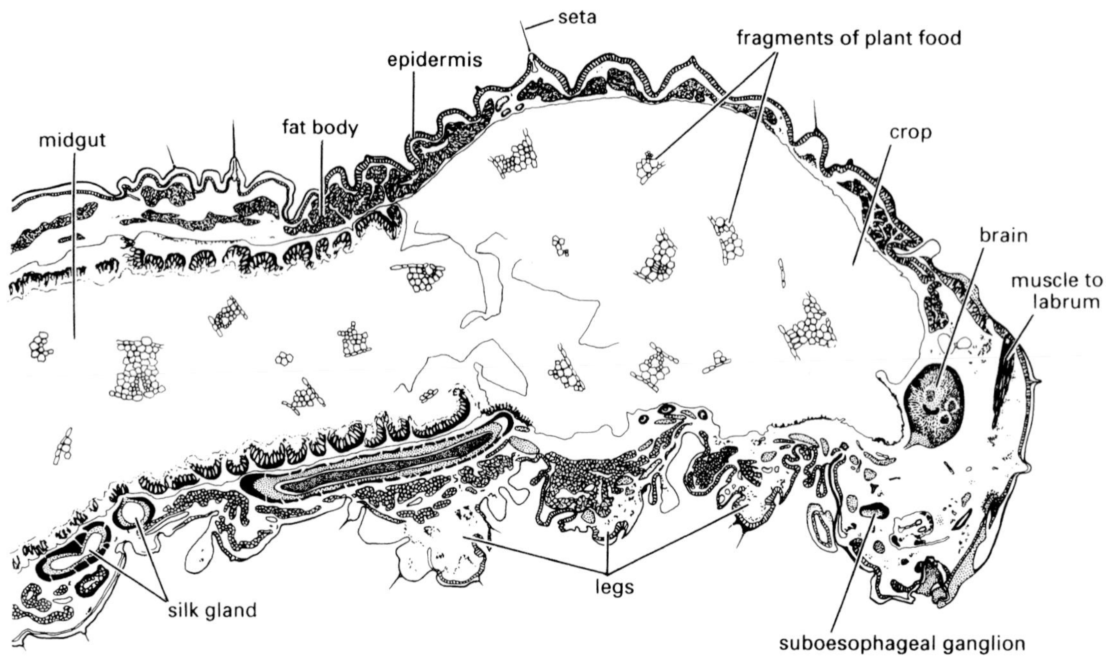

Fig. 3.15 Longitudinal section through the anterior body of a caterpillar of the cabbage white butterfly, *Pieris rapae* (Lepidoptera: Pieridae). Note the thickened epidermal layer lining the midgut.

the entry point of the Malpighian tubules, often into a distinct **pylorus** forming a muscular pyloric sphincter, followed by the ileum, colon, and rectum. The main functions of the hindgut are the absorption of water, salts, and other useful substances from the feces and urine; a detailed discussion of structure and function is presented in section 3.7.1.

3.6.2 Saliva and food ingestion

Salivary secretions dilute the ingested food and adjust its pH and ionic content. The saliva often contains digestive enzymes and, in blood-feeding insects, anticoagulants and thinning agents are present. In insects with extra-intestinal digestion, such as predatory Hemiptera, digestive enzymes are exported into the food and the resulting liquid is ingested. Most Hemiptera produce an alkaline watery saliva that is a vehicle for enzymes (either digestive or lytic), and a proteinaceous solidifying saliva that either forms a complete sheath around the mouthparts (stylets) as they pierce and penetrate the food or just a securing flange at the point of entry (section 11.2.3, Fig. 11.4c). Stylet-sheath feeding is characteristic of phloem- and xylem-feeding Hemiptera, such as aphids, scale insects (coccoids), and spittle bugs, which leave visible tracks formed of exuded solidifying saliva in the plant tissue on which they have fed. The sheath may function to guide the stylets, prevent loss of fluid from damaged cells, and/or absorb necrosis-inducing compounds to reduce defensive reaction by the plant. By comparison, a macerate-and-flush strategy is typical of Heteroptera, such as mirids and coreids. These insects disrupt the tissues of plants or other insects by thrusting of the stylets and/or by addition of salivary enzymes. The macerated and/or partly digested food is "flushed out" with saliva and ingested by sucking.

In fluid-feeding insects, prominent dilator muscles attach to the walls of the pharynx and/or the preoral cavity (**cibarium**) to form a pump (Fig. 3.14b), although most other insects have some sort of pharyngeal pump (Fig. 3.14a) for drinking and air intake to facilitate cuticle expansion during a molt.

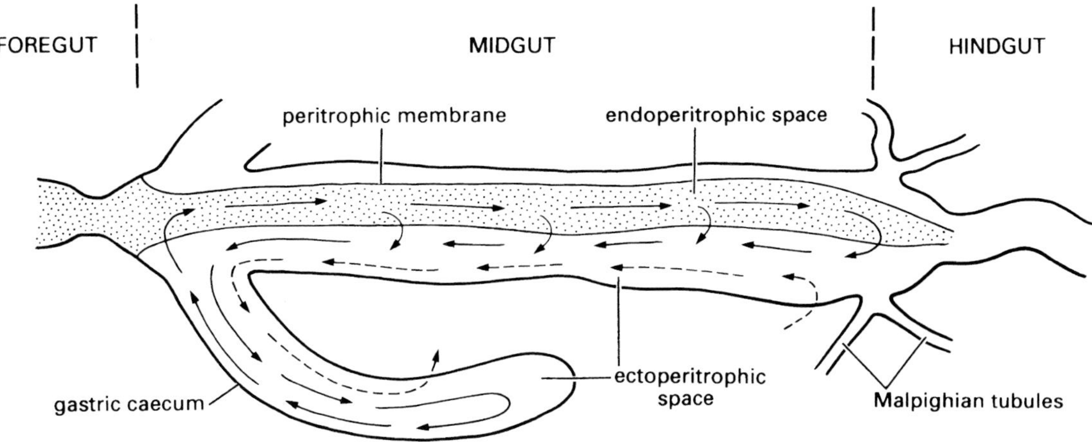

Fig. 3.16 Generalized scheme of the endo–ectoperitrophic circulation of digestive enzymes in the midgut. (After Terra & Ferreira 1981.)

3.6.3 Digestion of food

Most digestion occurs in the midgut, where the epithelial cells produce and secrete digestive enzymes and also absorb the resultant food breakdown products. Insect food consists principally of polymers of carbohydrates and proteins, which are digested by enzymatically breaking these large molecules into small monomers. The midgut pH usually is 6–7.5, although very alkaline values (pH 9–12) occur in many plant-feeding insects that extract hemicelluloses from plant cell walls, and very low pH values occur in many Diptera. High pH may prevent or reduce the binding of dietary tannins to food proteins, thereby increasing the digestibility of ingested plants, although the significance of this mechanism *in vivo* is unclear. In some insects, gut lumenal surfactants (detergents) may have an important role in preventing the formation of tannin–protein complexes, particularly in insects with near neutral gut pH.

In most insects, the midgut epithelium is separated from the food bolus by the peritrophic membrane, which constitutes a very efficient high-flux sieve. It is perforated by pores, which allow passage of small molecules while restricting large molecules, bacteria, and food particles from directly accessing the midgut cells. The peritrophic membrane also may protect herbivorous insects from ingested allelochemicals such as tannins (section 11.2). In some insects, all or most midgut digestion occurs inside the peritrophic membrane in the **endoperitrophic space**. In others, only initial digestion occurs there and smaller food molecules then diffuse out into the **ectoperitrophic space**, where further digestion takes place (Fig. 3.16). A final phase of digestion usually occurs on the surface of the midgut microvilli, where certain enzymes are either trapped in a mucopolysaccharide coating or bound to the cell membrane. Thus the peritrophic membrane forms a permeability barrier and helps to compartmentalize the phases of digestion, in addition to providing mechanical protection of the midgut cells, which was once believed to be its principal function. Fluid containing partially digested food molecules and digestive enzymes is thought to circulate through the midgut in a posterior direction in the endoperitrophic space and forwards in the ectoperitrophic space, as indicated in Fig. 3.16. This endo–ectoperitrophic circulation may facilitate digestion by moving food molecules to sites of final digestion and absorption and/or by conserving digestive enzymes, which are removed from the food bolus before it passes to the hindgut.

3.6.4 The fat body

In many insects, especially the larvae of holometabolous groups, fat body tissue is a conspicuous component of the internal anatomy (Figs. 3.7 & 3.15). Typically, it forms a white or yellow tissue formed of loose sheets, ribbons, or lobes of cells lying in the hemocoel. The structure of this organ is ill-defined and taxonomically variable, but often caterpillars and other larvae have a peripheral layer of fat body beneath the cuticle and a central layer around the gut. The fat body is an organ of multiple metabolic functions, including: the metabolism of carbohydrates, lipids, and nitrogenous compounds; the storage of glycogen, fat, and protein; the synthesis and regulation of blood sugar; and the synthesis of major hemolymph proteins (such as hemoglobins, vitellogenins for yolk formation, and storage proteins). Fat body cells can switch their activities in response to nutritional and hormonal signals to supply the requirements of insect growth, metamorphosis, and reproduction. For example, specific storage proteins are synthesized by the fat body during the final larval instar of holometabolous insects and accumulate in the hemolymph to be used during metamorphosis as a source of amino acids for the synthesis of proteins during pupation. **Calliphorin**, a hemolymph storage protein synthesized in the fat body of larval blow flies (Diptera: Calliphoridae: *Calliphora*), may form about 75% of the hemolymph protein of a late-instar maggot, or about 7 mg; the amount of calliphorin falls to around 3 mg at the time of pupariation and to 0.03 mg after emergence of the adult fly. The production and deposition of proteins specifically for amino acid storage is a feature that insects share with seed plants but not with vertebrates. Humans, for example, excrete any dietary amino acids that are in excess of immediate needs.

The principal cell type found in the fat body is the **trophocyte** (or **adipocyte**), which is responsible for most of the above metabolic and storage functions. Visible differences in the extent of the fat body in different individuals of the same insect species reflect the amount of material stored in the trophocytes; little body fat indicates either active tissue construction or starvation. Two other cell types – urocytes and mycetocytes (also called bacteriocytes) – may occur in the fat body of some insect groups. **Urocytes** temporarily store spherules of urates, including uric acid, one of the nitrogenous wastes of insects. Amongst studied cockroaches, rather than being permanent stores of excreted waste uric acid (storage excretion), urocytes recycle urate nitrogen, perhaps with assistance of mycetocyte bacteria. **Mycetocytes** (**bacteriocytes**) contain symbiotic microorganisms and are scattered through the fat body of cockroaches or contained within special organs, sometimes surrounded by fat body. These bacteria-like symbionts appear important in insect nutrition.

3.6.5 Nutrition and microorganisms

Broadly defined, nutrition concerns the nature and processing of foods needed to meet the requirements for growth and development, involving feeding behavior (Chapter 2) and digestion. Insects often have unusual or restricted diets. Sometimes, although only one or a few foods are eaten, the diet provides a complete range of the chemicals essential to metabolism. In these cases, monophagy is a specialization without nutritional limitations. In others, a restricted diet may require utilization of microorganisms in digesting or supplementing the directly available nutrients. In particular, insects cannot synthesize sterols (required for molting hormone) and carotenoids (used in visual pigments), which must come from the diet or microorganisms.

Insects may harbor extracellular or intracellular microorganisms, referred to as **symbionts** because they are dependent on their insect hosts. These microorganisms contribute to the nutrition of their hosts by functioning in sterol, vitamin, carbohydrate, or amino acid synthesis and/or metabolism. Symbiotic microorganisms may be bacteria or bacteroids, yeasts or other unicellular fungi, or protists. Studies on their function historically were hampered by difficulties in removing them (e.g. with antibiotics, to produce aposymbionts) without harming the host insect, and also in culturing the microorganisms outside the host. The diets of their hosts provided some clues as to the functions of these microorganisms. Insect hosts include many sap-sucking hemipterans (such as aphids, psyllids, whiteflies, scale insects, leafhoppers, and cicadas) and sap- and blood-sucking heteropterans (Hemiptera), lice (Phthiraptera), some wood-feeding insects (such as termites and some longicorn beetles and weevils), many seed- or grain-feeding insects (certain beetles), and some omnivorous insects (such as cockroaches, some termites, and some ants). Predatory insects never seem to contain such symbionts. That microorganisms are required by insects on suboptimal diets has been confirmed by modern studies showing, for example, that critical dietary shortfall in certain essential amino acids in aposymbiotic aphids is compensated for by production by *Buchnera* symbionts. An important role for bacteria is verified in acetogenesis and nitrogen fixation. Although insects were presumed to lack cellulases, they are present at least in termite guts, yet their role in cellulose digestion relative to that of symbionts is unclear.

Extracellular symbionts may be free in the gut lumen or housed in diverticula or pockets of the midgut or hindgut. For example, termite hindguts contain a veritable fermenter comprising many bacteria, fungi, and protists, including flagellates, which assist in the degradation of the otherwise refractory dietary lignocellulose, and in the fixation of atmospheric nitrogen. The process involves generation of methane, and calculations suggest that tropical termites' symbiont-assisted cellulose digestion produces a significant proportion of the world's methane (a greenhouse gas) production.

Transmission of extracellular symbionts from an individual insect to another involves one of two main methods, depending upon where the symbionts are located within the insect. The first mode of transmission, by oral uptake by the offspring, is appropriate for insects with gut symbionts. Microorganisms may be acquired from the anus or the excreta of other individuals or eaten at a specific time, as in some bugs, in which the newly hatched young eat the contents of special symbiont-containing capsules deposited with the eggs.

Intracellular symbionts (**endosymbionts**) may occur in as many as 70% of all insect species. Endosymbionts probably mostly have a mutualistic association with their host insect, but some are best referred to as "guest microbes" because they appear parasitic on their host. Examples of the latter include *Wolbachia* (section 5.10.4), *Spiroplasma*, and microsporidia. Endosymbionts may be housed in the gut epithelium, as in lygaeid bugs and some weevils; however, most insects with intracellular microorganisms house them in symbiont-containing cells called **mycetocytes** or **bacteriocytes**, according to the identity of the symbiont. These cells are in the body cavity, usually associated with the fat body or the gonads, and often in special aggregations of mycetocytes, forming an organ called a **mycetome** or **bacteriome**. In such insects, the symbionts are transferred to the ovary and then to the eggs or embryos prior to oviposition or parturition – a process referred to as **vertical** or **transovarial transmission**. Lacking evidence for lateral transfer (to an unrelated host), this method of transmission found in many Hemiptera and cockroaches indicates a very close association or coevolution of the insects and their microorganisms. Actual evidence of benefits of endosymbionts to hosts is limited, but the provision of the otherwise dietarily scarce essential amino acids to aphids by their bacteriocyte-associated *Buchnera* symbiont is well substantiated. Of interest for further research is the suggestion that aphid biotypes with *Buchnera* bacteriocytes show enhanced ability to transmit certain plant viruses of the genus *Luteovirus* relative to antibiotic-treated, symbiont-free individuals. The relationship between bacteriocyte endosymbionts and their phloem-feeding host insects is a very tight phylogenetic association (cf. *Wolbachia* infections, section 5.10.4), suggesting a very old association with co-diversification.

Some insects that maintain fungi essential to their diet cultivate them external to their body as a means of converting woody substances to an assimilable form. Examples are the fungus gardens of some ants (Formicidae) and termites (Termitidae) (sections 9.5.2 & 9.5.3) and the fungi transmitted by certain timber pests, namely, wood wasps (Hymenoptera: Siricidae) and ambrosia beetles (Coleoptera: Scolytinae).

3.7 THE EXCRETORY SYSTEM AND WASTE DISPOSAL

Excretion – the removal from the body of waste products of metabolism, especially nitrogenous compounds – is essential. It differs from defecation in that excretory wastes have been metabolized in cells of the body rather than simply passing directly from the mouth to the anus (sometimes essentially unchanged chemically). Of course, insect feces, either in liquid form or packaged in pellets and known as **frass**, contain both undigested food and metabolic excretions. Aquatic insects excrete dilute wastes from their anus directly into water, and so their fecal material is flushed away. In comparison, terrestrial insects generally must conserve water. This requires efficient waste disposal in a concentrated or even dry form while simultaneously avoiding the potentially toxic effects of nitrogen. Furthermore, both terrestrial and aquatic insects must conserve ions, such as sodium (Na^+), potassium (K^+), and chloride (Cl^-), that may be limiting in their food or, in aquatic insects, lost into the water by diffusion. Production of insect urine or frass therefore results from two intimately related processes: excretion and **osmoregulation** – the maintenance of a favorable body fluid composition (osmotic and ionic homeostasis). The system responsible for excretion and osmoregulation is referred to loosely as the excretory system, and its activities are performed largely by the Malpighian tubules and hindgut as outlined below. However, in freshwater insects, hemolymph composition must be regulated in response to constant loss of salts (as ions) to the surrounding water, and ionic regulation involves both the typical excretory system and special cells, called **chloride cells**, which usually are associated with the hindgut. Chloride cells are capable of absorbing inorganic ions from very dilute solutions and are best studied in larval dragonflies and damselflies.

3.7.1 The Malpighian tubules and rectum

The main organs of excretion and osmoregulation in insects are the Malpighian tubules acting in concert with the rectum and/or ileum (Fig. 3.17). Malpighian tubules are outgrowths of the alimentary canal and consist of long thin tubules (Fig. 3.1) formed of a single layer of cells surrounding a blind-ending lumen. They range in number from as few as two in most scale insects (coccoids) to over 200 in large locusts. Generally they are free, waving around in the hemolymph, where they filter out solutes. Only aphids lack Malpighian tubules. The vignette for this chapter shows the gut of *Locusta*, but with only a few of the many Malpighian tubules depicted. Similar structures are believed to have arisen convergently in different arthropod groups, such as myriapods and arachnids, in response to the physiological stresses of life on dry land. Traditionally, insect Malpighian tubules are considered to belong to the hindgut and be ectodermal in origin. Their position marks the junction of the midgut and the cuticle-lined hindgut.

The anterior hindgut is called the **ileum**, the generally narrower middle portion is the **colon**, and the expanded posterior section is the **rectum** (Fig. 3.13). In many terrestrial insects the rectum is the only site of water and solute resorption from the excreta, but in other insects, for example the desert locust *Schistocerca gregaria* (Orthoptera: Acrididae), the ileum makes some contribution to osmoregulation. In a few insects, such as the cockroach *Periplaneta americana* (Blattodea:

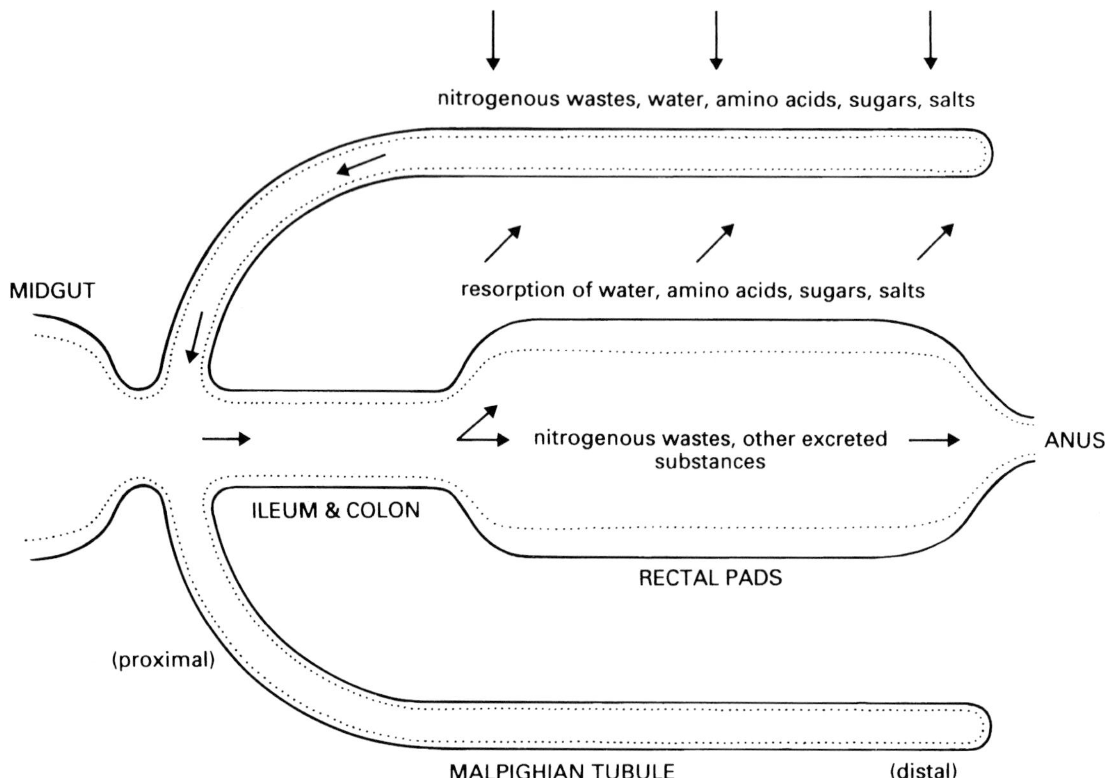

Fig. 3.17 Schematic diagram of a generalized excretory system showing the path of elimination of wastes. (After Daly et al. 1978.)

Box 3.4 Cryptonephric systems*

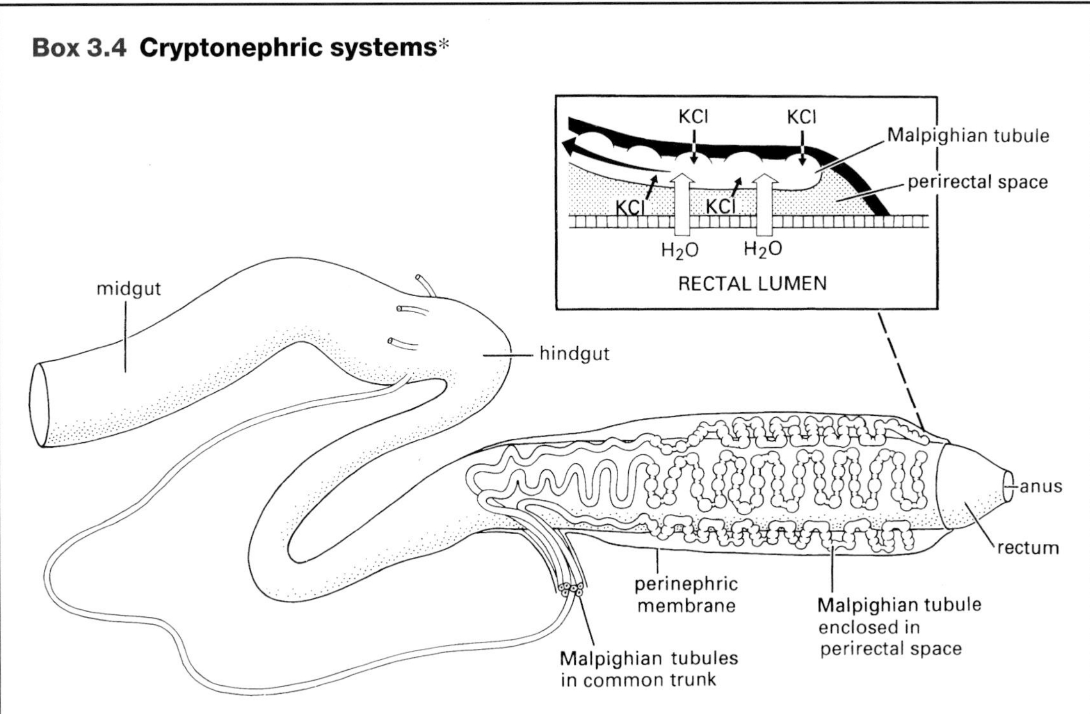

Many larval and adult Coleoptera, larval Lepidoptera, and some larval Symphyta have a modified arrangement of the excretory system that is concerned either with efficient dehydration of feces before their elimination (in beetles) or ionic regulation (in plant-feeding caterpillars). These insects have a **cryptonephric system** in which the distal ends of the Malpighian tubules are held in contact with the rectal wall by the perinephric membrane. Such an arrangement allows some beetles that live on a very dry diet, such as stored grain or dry carcasses, to be extraordinarily efficient in their conservation of water. Water even may be extracted from the humid air in the rectum. In the cryptonephric system of the mealworm, *Tenebrio molitor* (Coleoptera: Tenebrionidae), shown here, ions (principally potassium chloride, KCl) are transported into and concentrated in the six Malpighian tubules, creating an osmotic gradient that draws water from the surrounding perirectal space and the rectal lumen. The tubule fluid is then transported forwards to the free portion of each tubule, from which it is passed to the hemolymph or recycled in the rectum.

*After Grimstone et al. 1968; Bradley 1985.

Blattidae), even the colon may be a potential site of some fluid absorption. The resorptive role of the rectum (and sometimes the anterior hindgut) is indicated by its anatomy. In most insects, specific parts of the rectal epithelium are thickened to form **rectal pads** or papillae composed of aggregations of columnar cells; typically there are six pads arranged longitudinally, but there may be fewer pads or many papillate ones.

The general picture of insect excretory processes outlined here is applicable to most freshwater species and to the adults of many terrestrial species. The Malpighian tubules produce a filtrate (the primary urine) which is isosmotic but ionically dissimilar to the hemolymph, and then the hindgut, especially the rectum, selectively reabsorbs water and certain solutes but eliminates others (Fig. 3.17). Details of Malpighian tubule and rectal structure and of filtration and absorption mechanisms differ between taxa, in relation to both taxonomic position and dietary composition (Box 3.4 gives an example of one type of specialization – cryptonephric systems), but the excretory system of the desert locust *S. gregaria* (Fig. 3.18) exemplifies the

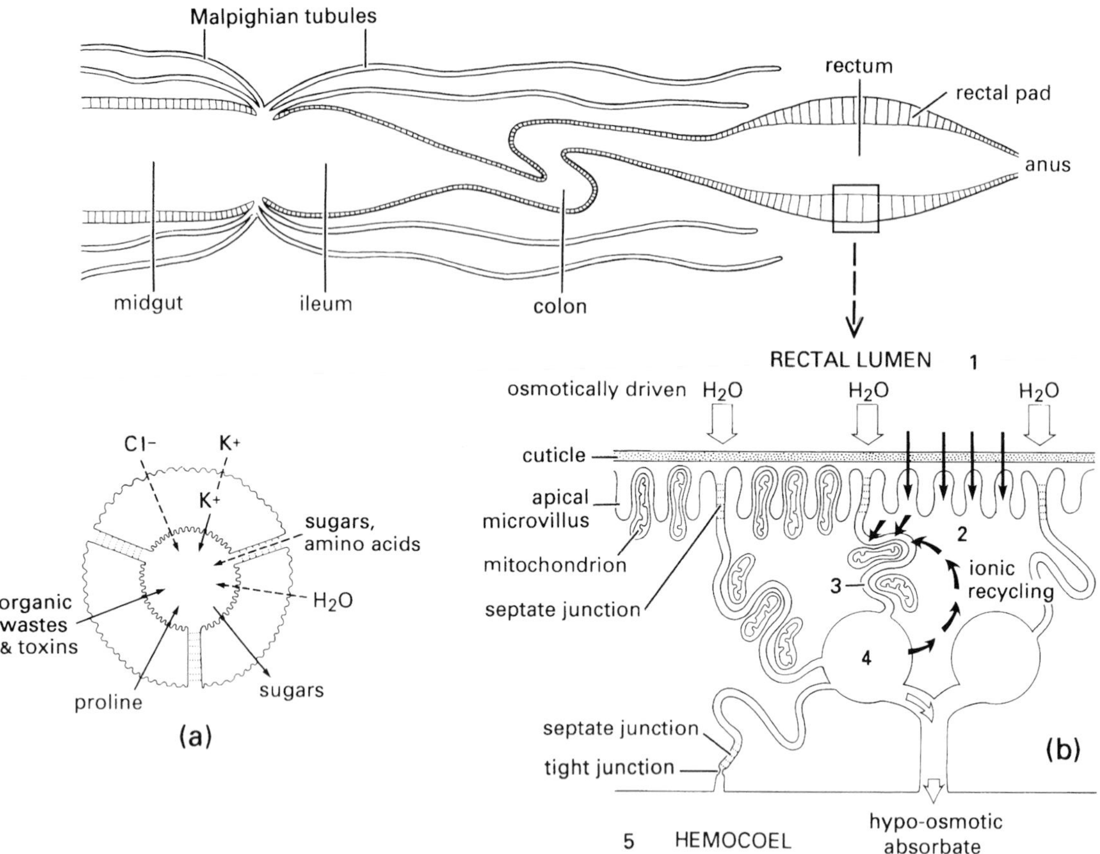

Fig. 3.18 Schematic diagram of the organs in the excretory system of the desert locust *Schistocerca gregaria* (Orthoptera: Acrididae). Only a few of the >100 Malpighian tubules are drawn. (a) Transverse section of one Malpighian tubule showing probable transport of ions, water, and other substances between the surrounding hemolymph and the tubule lumen; active processes are indicated by solid arrows and passive processes by dashed arrows. (b) Diagram illustrating the movements of solutes and water in the rectal pad cells during fluid resorption from the rectal lumen. Pathways of water movement are represented by open arrows and solute movements by black arrows. Ions are actively transported from the rectal lumen (compartment 1) to the adjacent cell cytoplasm (compartment 2) and then to the intercellular spaces (compartment 3). Mitochondria are positioned to provide the energy for this active ion transport. Fluid in the spaces is hyperosmotic (higher ion concentration) to the rectal lumen and draws water by osmosis from the lumen via the septate junctions between the cells. Water thus moves from compartment 1 to 3 to 4 and finally to 5, the hemolymph in the hemocoel. (After Bradley 1985.)

general structure and principles of insect excretion. The Malpighian tubules of the locust produce an isosmotic filtrate of the hemolymph, which is high in K^+, low in Na^+, and has Cl^- as the major anion. The active transport of ions, especially K^+, into the tubule lumen generates an osmotic pressure gradient so that water passively follows (Fig. 3.18a). Sugars and most amino acids also are filtered passively from the hemolymph (probably via junctions between the tubule cells), whereas the amino acid proline (later used as an energy source by the rectal cells) and non-metabolizable and toxic organic compounds are transported actively into the tubule lumen. Sugars, such as sucrose and trehalose, are resorbed from the lumen and returned to the hemolymph. The continuous secretory activity of each Malpighian tubule leads to a flow of primary urine from its lumen towards and into the gut. In the rectum, the urine is modified by removal of solutes and water to

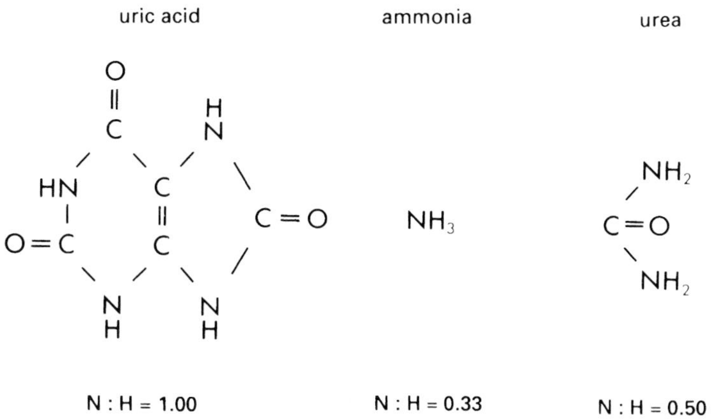

uric acid — N : H = 1.00

ammonia — NH_3 — N : H = 0.33

urea — NH_2–C(=O)–NH_2 — N : H = 0.50

Fig. 3.19 Molecules of the three common nitrogenous excretory products. The high N : H ratio of uric acid relative to both ammonia and urea means that less water is used for uric acid synthesis (as hydrogen atoms are derived ultimately from water).

maintain fluid and ionic homeostasis of the locust's body (Fig. 3.18b). Specialized cells in the rectal pads carry out active recovery of Cl^- under hormonal stimulation. This pumping of Cl^- generates electrical and osmotic gradients that lead to some resorption of other ions, water, amino acids, and acetate.

3.7.2 Nitrogen excretion

Many predatory, blood-feeding and even plant-feeding insects ingest nitrogen, particularly certain amino acids, far in excess of requirements. Most insects excrete nitrogenous metabolic wastes at some or all stages of their life, although some nitrogen is stored in the fat body or as proteins in the hemolymph in some insects. Many aquatic insects and some flesh-feeding flies excrete large amounts of ammonia, whereas in terrestrial insects wastes generally consist of **uric acid** and/or certain of its salts (**urates**), often in combination with urea, pteridines, certain amino acids, and/or relatives of uric acid, such as hypoxanthine, allantoin, and allantoic acid. Amongst these waste compounds, ammonia is relatively toxic and usually must be excreted as a dilute solution, or else rapidly volatilized from the cuticle or feces (as in cockroaches). Urea is less toxic but more soluble, requiring much water for its elimination. Uric acid and urates require less water for synthesis than either ammonia or urea (Fig. 3.19), are non-toxic and, having low solubility in water (at least in acidic conditions), can be excreted essentially dry, without causing osmotic problems. Waste dilution can be achieved easily by aquatic insects, but water conservation is essential for terrestrial insects and uric acid excretion (**uricotelism**) is highly advantageous.

Deposition of urates in specific cells of the fat body (section 3.6.4) was viewed as "excretion" by storage of uric acid. However, it might constitute a metabolic store for recycling by the insect, perhaps with the assistance of symbiotic microorganisms, as in cockroaches that house bacteria in their fat body. These cockroaches, including *P. americana*, do not excrete uric acid in the feces even if fed a high-nitrogen diet but do produce large quantities of internally stored urates.

By-products of feeding and metabolism need not be excreted as waste – for example, the antifeedant defensive compounds of plants may be sequestered directly or may form the biochemical base for synthesis of chemicals used in communication (Chapter 4) including warning and defense. White-pigmented uric acid derivatives color the epidermis of some insects and provide the white in the wing scales of certain butterflies (Lepidoptera: Pieridae).

3.8 REPRODUCTIVE ORGANS

The reproductive organs of insects exhibit an incredible variety of forms, but there is a basic design and function to each component so that even the most aberrant reproductive system can be understood in terms of a generalized plan. Individual components of the reproductive system can vary in shape (e.g. of gonads and accessory glands), position (e.g. of the attachment of accessory glands), and number (e.g. of ovarian or testicular tubes, or sperm storage organs) between

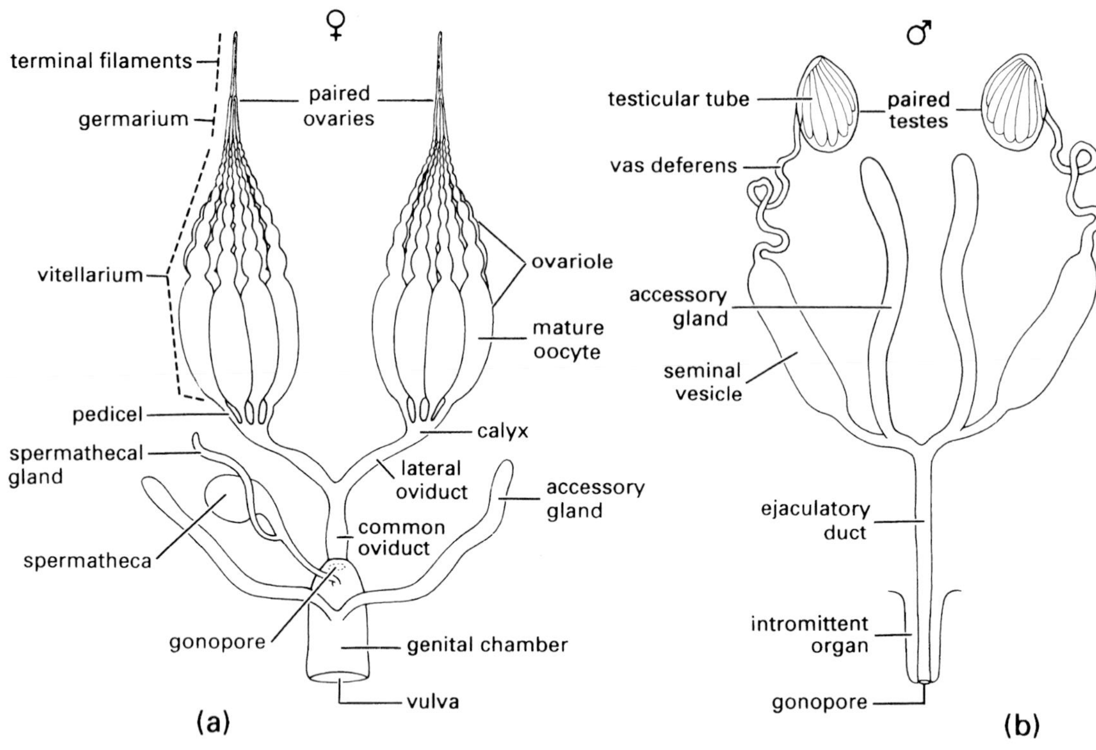

Fig. 3.20 Comparison of generalized (a) female and (b) male reproductive systems. (After Snodgrass 1935.)

different insect groups, and sometimes even between different species in a genus. Knowledge of the homology of the components assists in interpreting structure and function in different insects. Generalized male and female systems are depicted in Fig. 3.20, and a comparison of the corresponding reproductive structures of male and female insects is provided in Table 3.2. Many other aspects of reproduction, including copulation and regulation of physiological processes, are discussed in detail in Chapter 5.

3.8.1 The female system

The main functions of the female reproductive system are egg production, including the provision of a protective coating in many insects, and the storage of the male's spermatozoa until the eggs are ready to be fertilized. Transport of the spermatozoa to the female's storage organ and their subsequent controlled release requires movement of the spermatozoa, which in some species is known to be mediated by muscular contractions of parts of the female reproductive tract.

The basic components of the female system (Fig. 3.20a) are paired **ovaries**, which empty their mature **oocytes** (eggs) via the **calyces** (singular: calyx) into the **lateral oviducts**, which unite to form the **common** (or **median**) **oviduct**. The **gonopore** (opening) of the common oviduct usually is concealed in an inflection of the body wall that typically forms a cavity, the **genital chamber**. This chamber serves as a copulatory pouch during mating and thus often is known as the **bursa copulatrix**. Its external opening is the **vulva**. In many insects the vulva is narrow and the genital chamber becomes an enclosed pouch or tube, referred to as the **vagina**. Two sorts of ectodermal glands open into the genital chamber. The first is the **spermatheca**, which stores spermatozoa until needed for egg fertilization. Typically, the spermatheca is single, generally sac-like with a slender duct and often has a diverticulum that forms a tubular **spermathecal gland**. The gland or glandular cells within the storage

Table 3.2 The corresponding female and male reproductive organs of insects.

Female reproductive organs	Male reproductive organs
Paired ovaries composed of ovarioles (ovarian tubes)	Paired testes composed of follicles (testicular tubes)
Paired oviducts (ducts leading from ovaries)	Paired vasa deferentia (ducts leading from testes)
Egg calyces (if present, reception of eggs)	Seminal vesicles (sperm storage)
Common (median) oviduct and vagina	Median ejaculatory duct
Accessory glands (ectodermal origin: colleterial or cement glands)	Accessory glands (two types): (i) ectodermal origin (ii) mesodermal origin
Bursa copulatrix (copulatory pouch) and spermatheca (sperm storage)	No equivalent
Ovipositor (if present)	Genitalia (if present): aedeagus and associated structures

part of the spermatheca provide nourishment to the contained spermatozoa. The second type of ectodermal gland, known collectively as accessory glands, opens more posteriorly in the genital chamber and has a variety of functions depending on the species (see section 5.8).

Each ovary is composed of a cluster of ovarian or egg tubes, the **ovarioles**, each consisting of a terminal filament, a **germarium** (in which mitosis gives rise to primary oocytes), a **vitellarium** (in which oocytes grow by deposition of yolk in a process known as vitellogenesis; section 5.11.1), and a **pedicel** (or stalk). An ovariole contains a series of developing oocytes, each surrounded by a layer of follicle cells forming an epithelium (the oocyte and its epithelium is termed a follicle); the youngest oocytes occur near the apical germarium and the most mature near the pedicel. Three different types of ovariole are recognized based on the manner in which the oocytes are nourished. A **panoistic ovariole** lacks specialized nutritive cells so that it contains only a string of follicles, with the oocytes obtaining nutrients from the hemolymph via the follicular epithelium. Ovarioles of the other two types contain trophocytes (nurse cells) that contribute to the nutrition of the developing oocytes. In a **telotrophic** (or **acrotrophic**) **ovariole** the trophocytes are confined to the germarium and remain connected to the oocytes by cytoplasmic strands as the oocytes move down the ovariole. In a **polytrophic ovariole** a number of trophocytes are connected to each oocyte and move down the ovariole with it, providing nutrients until depleted; thus individual oocytes alternate with groups of successively smaller trophocytes. Different suborders or orders of insects usually have only one of these three ovariole types.

Accessory glands of the female reproductive tract often are referred to as **colleterial** or **cement glands** because in most insect orders their secretions surround and protect the eggs or cement them to the substrate (section 5.8). In other insects the accessory glands may function as **poison glands** (as in many Hymenoptera) or as **"milk" glands** in the few insects (e.g. tsetse flies, *Glossina* spp.) that exhibit adenotrophic viviparity (section 5.9). Accessory glands of a variety of forms and functions appear to have been derived independently in different orders and even may be non-homologous within an order, as in Coleoptera.

3.8.2 The male system

The main functions of the male reproductive system are the production and storage of spermatozoa and their transport in a viable state to the reproductive tract of the female. Morphologically, the male tract consists of paired **testes** each containing a series of testicular tubes or follicles (in which spermatozoa are produced), which open separately into the mesodermally derived sperm duct or **vas deferens**, which usually expands posteriorly to form a sperm storage organ, or **seminal vesicle** (Fig. 3.20b). Typically, tubular, paired **accessory glands** are formed as diverticula of the vasa deferentia, but sometimes the vasa deferentia themselves are glandular and fulfill the functions of accessory glands (see below). The paired vasa deferentia unite where they lead into the ectodermally derived **ejaculatory**

duct – the tube that transports the semen or the sperm package to the gonopore. In a few insects, particularly certain flies, the accessory glands consist of an enlarged glandular part of the ejaculatory duct.

Thus, the accessory glands of male insects can be classified into two types according to their mesodermal or ectodermal derivation. Almost all are mesodermal in origin and those apparently ectodermal ones have been poorly studied. Furthermore, the mesodermal structures of the male tract frequently differ morphologically from the basic paired sacs or tubes described above. For example, in male cockroaches and many other orthopteroids the seminal vesicles and the numerous accessory gland tubules (Fig. 3.1) are clustered into a single median structure called the **mushroom body**. Secretions of the male accessory glands form the **spermatophore** (the package that surrounds the spermatozoa of many insects), contribute to the seminal fluid which nourishes the spermatozoa during transport to the female, are involved in activation (induction of motility) of the spermatozoa, and may alter female behavior (induce non-receptivity to further males and/or stimulate oviposition; see sections 5.4 & 5.11, and Box 5.4).

FURTHER READING

Barbehenn, R.V. (2001) Roles of peritrophic membranes in protecting herbivorous insects from ingested plant allelochemicals. *Archives of Insect Biochemistry and Physiology* **47**, 86–99.

Bourtzis, K. & Miller, T.A. (eds.) (2003) *Insect Symbiosis*. CRC Press, Boca Raton, FL.

Chapman, R.F. (1998) *The Insects. Structure and Function*, 4th edn. Cambridge University Press, Cambridge.

Davey, K.G. (1985) The male reproductive tract/The female reproductive tract. In: *Comprehensive Insect Physiology, Biochemistry, and Pharmacology*, Vol. 1: *Embryogenesis and Reproduction* (eds. G.A. Kerkut & L.I. Gilbert), pp. 1–14, 15–36. Pergamon Press, Oxford.

Douglas, A.E. (1998) Nutritional interactions in insect–microbial symbioses: aphids and their symbiotic bacteria *Buchnera*. *Annual Review of Entomology* **43**, 17–37.

Dow, J.A.T. (1986) Insect midgut function. *Advances in Insect Physiology* **19**, 187–328.

Gäde, G., Hoffman, K.-H. & Spring, J.H. (1997) Hormonal regulation in insects: facts, gaps, and future directions. *Physiological Reviews* **77**, 963–1032.

Harrison, F.W. & Locke, M. (eds.) (1998) *Microscopic Anatomy of Invertebrates*, Vol. 11B: *Insecta*. Wiley–Liss, New York.

Holman, G.M., Nachman, R.J. & Wright, M.S. (1990) Insect neuropeptides. *Annual Review of Entomology* **35**, 201–17.

Kerkut, G.A. & Gilbert, L.I. (eds.) (1985) *Comprehensive Insect Physiology, Biochemistry, and Pharmacology*, Vol. 3: *Integument, Respiration and Circulation*; Vol. 4: *Regulation: Digestion, Nutrition, Excretion*; Vols. 7 and 8: *Endocrinology I and II*. Pergamon Press, Oxford.

Menn, J.J., Kelly, T.J. & Masler, E.P. (eds.) (1991) *Insect Neuropeptides: Chemistry, Biology, and Action*. American Chemical Society, Washington, DC.

Nijhout, H.F. (1994) *Insect Hormones*. Princeton University Press, Princeton, NJ.

Raabe, M. (1989) *Recent Developments in Insect Neurohormones*. Plenum, New York.

Resh, V.H. & Cardé, R.T. (eds.) (2003) *Encyclopedia of Insects*. Academic Press, Amsterdam. [Particularly see articles on accessory glands; brain and optic lobes; circulatory system; digestion; digestive system; excretion; fat body; flight; hemolymph; immunology; nutrition; respiratory system; salivary glands; tracheal system; walking and jumping; water and ion balance, hormonal control of.]

Terra, W.R. (1990) Evolution of digestive systems of insects. *Annual Review of Entomology* **35**, 181–200.

Terra, W.R. & Ferreira, C. (1994) Insect digestive enzymes: properties, compartmentalization and function. *Comparative Biochemistry and Physiology* **109B**, 1–62.

See also papers and references in *BioScience* (1998) **48**(4), on symbiotic associations involving microorganisms, especially: Moran, N.A. & Telang, A., Bacteriocyte-associated symbionts of insects (pp. 295–304); and Bourtzis, K. & O'Neill, S., *Wolbachia* infections and arthropod reproduction (pp. 287–93).

Chapter 4

SENSORY SYSTEMS AND BEHAVIOR

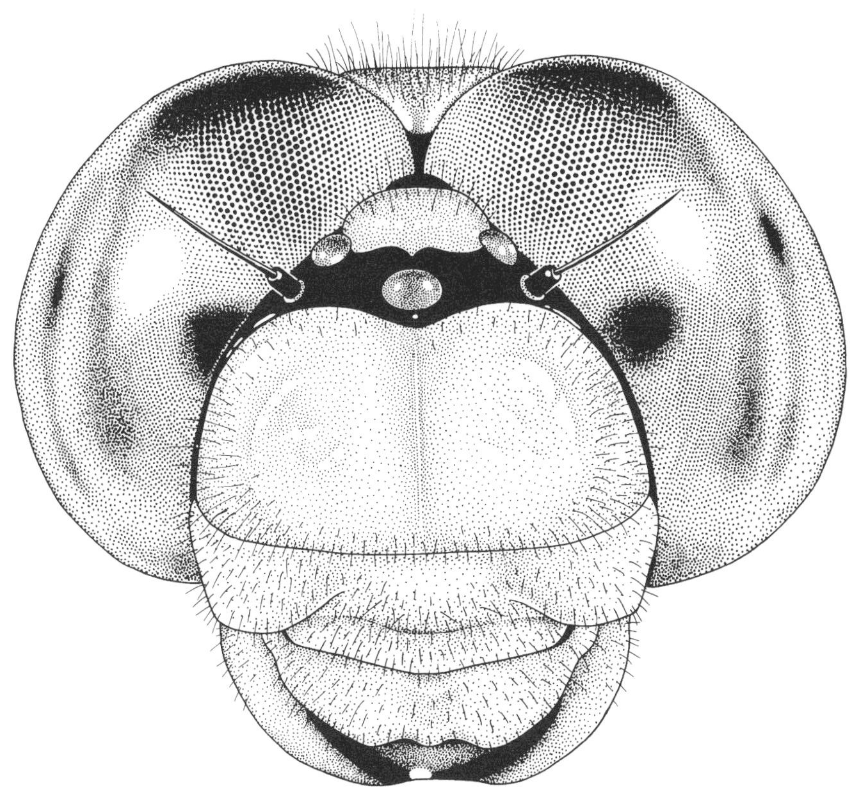

Head of a dragonfly showing enormous compound eyes. (After Blaney 1976.)

In the opening chapter of this book we suggested that the success of insects derives at least in part from their ability to sense and interpret their surroundings and to discriminate on a fine scale. Insects can identify and respond selectively to cues from a heterogeneous environment. They can differentiate between hosts, both plant and animal, and distinguish among many microclimatic factors, such as variations in humidity, temperature, and air flow.

Sensory complexity allows both simple and complex behaviors of insects. For example, to control flight, the aerial environment must be sensed and appropriate responses made. Because much insect activity is nocturnal, orientation and navigation cannot rely solely on the conventional visual cues, and in many night-active species odors and sounds play a major role in communication. The range of sensory information used by insects differs from that of humans. We rely heavily on visual information and although many insects have well-developed vision, most insects make greater use of olfaction and hearing than humans do.

The insect is isolated from its external surroundings by a relatively inflexible, insensitive, and impermeable cuticular barrier. The answer to the enigma of how this armored insect can perceive its immediate environment lies in frequent and abundant cuticular modifications that detect external stimuli. Sensory organs (**sensilla**, singular: **sensillum**) protrude from the cuticle, or sometimes lie within or beneath it. Specialized cells detect stimuli that may be categorized as mechanical, thermal, chemical, and visual. Other cells (the neurons) transmit messages to the central nervous system (section 3.2), where they are integrated. The nervous system instigates and controls appropriate behaviors, such as posture, movement, feeding, and behaviors associated with mating and oviposition.

This chapter surveys sensory systems and presents selected behaviors that are elicited or modified by environmental stimuli. The means of detection and, where relevant, the production of these stimuli are treated in the following sequence: touch, position, sound, temperature, chemicals (with particular emphasis on communication chemicals called pheromones), and light. The chapter concludes with a section that relates some aspects of insect behavior to the preceding discussion on stimuli.

4.1 MECHANICAL STIMULI

The stimuli grouped here are those associated with distortion caused by mechanical movement as a result of the environment itself, the insect in relation to the environment, or internal forces derived from the muscles. The mechanical stimuli sensed include touch, body stretching and stress, position, pressure, gravity, and vibrations, including pressure changes of the air and substrate involved in sound transmission and hearing.

4.1.1 Tactile mechanoreception

The bodies of insects are clothed with cuticular projections. These are called microtrichia if many arise from one cell, or hairs, bristles, setae, or macrotrichia if they are of multicellular origin. Most flexible projections arise from an innervated socket. These are sensilla, termed **trichoid sensilla** (literally hair-like little sense organs), and develop from epidermal cells that switch from cuticle production. Three cells are involved (Fig. 4.1):

1 **trichogen cell**, which grows the conical hair;
2 **tormogen cell**, which grows the socket;
3 **sensory neuron**, or nerve cell, which grows a **dendrite** into the hair and an **axon** that winds inwards to link with other axons to form a nerve connected to the central nervous system.

Fully developed trichoid sensilla fulfill tactile functions. As touch sensilla they respond to the movement of the hair by firing impulses from the dendrite at a frequency related to the extent of the deflection. Touch sensilla are stimulated only during actual movement of the hair. The sensitivity of each hair varies, with some being so sensitive that they respond to vibrations of air particles caused by noise (section 4.1.3).

4.1.2 Position mechanoreception (proprioceptors)

Insects require continuous knowledge of the relative position of their body parts such as limbs or head, and need to detect how the orientation of the body relates to gravity. This information is conveyed by **proprioceptors** (self-perception receptors), of which three types are described here. One type of trichoid sensillum gives a continuous sensory output at a frequency that varies with the position of the hair. Sensilla often form a bed of grouped small hairs, a **hair plate**, at joints or at the neck, in contact with the cuticle of an adjacent body part (Fig. 4.2a). The degree of flexion of the joint gives a variable stimulus to the sensilla, thereby allowing

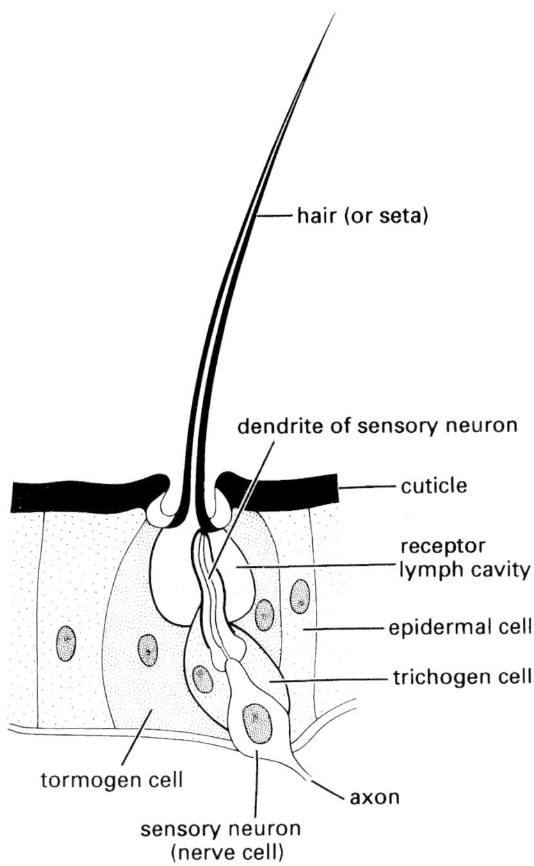

Fig. 4.1 Longitudinal section of a trichoid sensillum showing the arrangement of the three associated cells. (After Chapman 1991.)

monitoring of the relative positions of different parts of the body.

The second type, stretch receptors, comprise internal proprioceptors associated with muscles such as those of the abdominal and gut walls. Alteration of the length of the muscle fiber is detected by multiple-inserted neuron endings, producing variation in the rate of firing of the nerve cell. Stretch receptors monitor body functions such as abdominal or gut distension, or ventilation rate.

The third type are stress detectors on the cuticle via stress receptors called **campaniform sensilla**. Each sensillum comprises a central cap or peg surrounded by a raised circle of cuticle and with a single neuron per sensillum (Fig. 4.2b). These sensilla are located on joints, such as those of legs and wings, and other places liable to distortion. Locations include the haltere (the knob-like modified hind wing of Diptera), at the base of which there are dorsal and ventral groups of campaniform sensilla that respond to distortions created during flight.

4.1.3 Sound reception

Sound is a pressure fluctuation transmitted in a wave form via movement of the air or the substrate, including water. Sound and hearing are terms often applied to the quite limited range of frequencies of airborne vibration that humans perceive with their ears, usually in adults from 20 to 20,000 Hz (1 hertz (Hz) is a frequency of one cycle per second). Such a definition of sound is restrictive, particularly as amongst insects some receive vibrations ranging from as low as 1–2 Hz to ultrasound frequencies perhaps as high as 100 kHz. Specialized emission and reception across this range of frequencies of vibration are considered here. The reception of these frequencies involves a variety of organs, none of which resemble the ears of mammals.

An important role of insect sound is in intraspecific acoustic communication. For example, courtship in most orthopterans is acoustic, with males producing species-specific sounds ("songs") that the predominantly non-singing females detect and upon which they base their choice of mate. Hearing also allows detection of predators, such as insectivorous bats, which use ultrasound in hunting. Probably each species of insect detects sound within one or two relatively narrow ranges of frequencies that relate to these functions.

The insect mechanoreceptive communication system can be viewed as a continuum from substrate vibration reception, grading through the reception of only very near airborne vibration to hearing of even quite distant sound using thin cuticular membranes called **tympani** (singular: **tympanum**; adjective: **tympanal**). Substrate signaling probably appeared first in insect evolution; the sensory organs used to detect substrate vibrations appear to have been co-opted and modified many times in different insect groups to allow reception of airborne sound at considerable distance and a range of frequencies.

Non-tympanal vibration reception

Two types of vibration or sound reception that do not involve tympani (see p. 90) are the detection of substrate-borne signals and the ability to perceive the relatively large translational movements of the

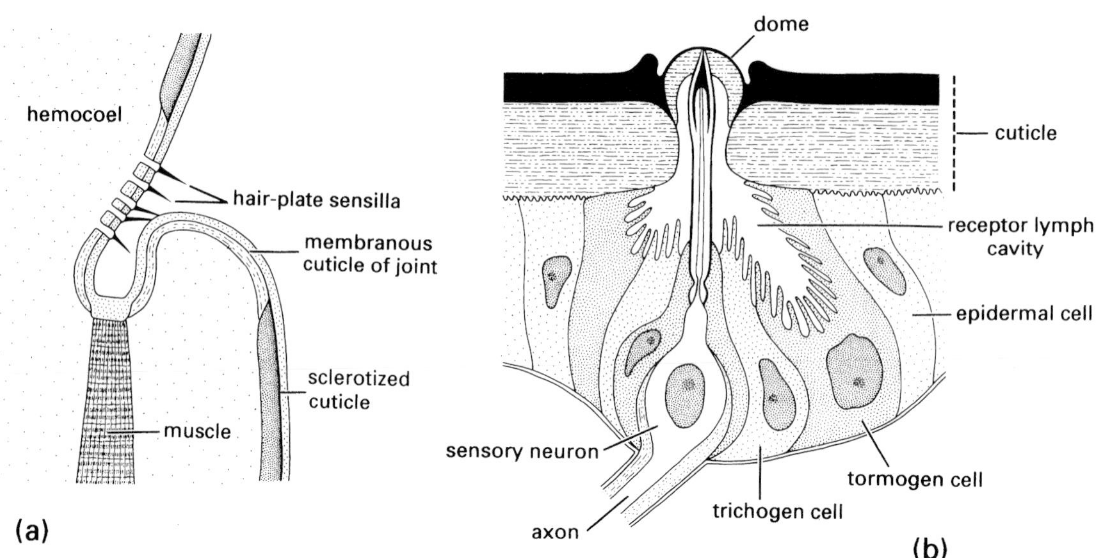

Fig. 4.2 Proprioceptors: (a) sensilla of a hair plate located at a joint, showing how the hairs are stimulated by contacting adjacent cuticle; (b) campaniform sensillum on the haltere of a fly. ((a) After Chapman 1982; (b) after Snodgrass 1935; McIver 1985.)

surrounding medium (air or water) that occur very close to a sound. The latter, referred to as **near-field** sound, is detected by either sensory hairs or specialized sensory organs.

A simple form of sound reception occurs in species that have very sensitive, elongate, trichoid sensilla that respond to vibrations produced by a near-field sound. For example, caterpillars of the noctuid moth *Barathra brassicae* have thoracic hairs about 0.5 mm long that respond optimally to vibrations of 150 Hz. Although in air this system is effective only for locally produced sounds, caterpillars can respond to the vibrations caused by audible approach of parasitic wasps.

The cerci of many insects, especially crickets, are clothed in long, fine trichoid sensilla (filiform setae or hairs) that are sensitive to air currents, which can convey information about the approach of predatory or parasitic insects or a potential mate. The direction of approach of another animal is indicated by which hairs are deflected; the sensory neuron of each hair is tuned to respond to movement in a particular direction. The dynamics (the time-varying pattern) of air movement gives information on the nature of the stimulus (and thus on what type of animal is approaching) and is indicated by the properties of the mechanosensory hairs. The length of each hair determines the response of its sensory neuron to the stimulus: neurons that innervate short hairs are most sensitive to high-intensity, high-frequency stimuli, whereas long hairs are more sensitive to low-intensity, low-frequency stimuli. The responses of many sensory neurons innervating different hairs on the cerci are integrated in the central nervous system to allow the insect to make a behaviorally appropriate response to detected air movement.

For low-frequency sounds in water (a medium more viscous than air), longer distance transmission is possible. Currently, however, rather few aquatic insects have been shown to communicate through underwater sounds. Notable examples are the "drumming" sounds that some aquatic larvae produce to assert territory, and the noises produced by underwater diving hemipterans such as corixids and nepids.

Many insects can detect vibrations transmitted through a substrate at a solid–air or solid–water boundary or along a water–air surface. The perception of substrate vibrations is particularly important for ground-dwelling insects, especially nocturnal species, and social insects living in dark nests. Some insects living on plant surfaces, such as sawflies (Hymenoptera: Pergidae), communicate with each other by tapping the stem. Various plant-feeding bugs (Hemiptera), such as leafhoppers, planthoppers, and pentatomids, produce vibratory signals that are transmitted through the host plant. Water-striders (Hemiptera: Gerridae), which live on the aquatic surface film, send pulsed waves across the water surface to communicate in courtship and aggression. Moreover, they can detect the vibrations produced by the struggles of prey that fall onto the water surface. Whirligig beetles (Gyrinidae; Fig. 10.8) can navigate using a form of echolocation: waves that move on the water surface ahead of them and are reflected from obstacles are sensed by their antennae in time to take evasive action.

The specialized sensory organs that receive vibrations are subcuticular mechanoreceptors called **chordotonal organs**. An organ consists of one to many **scolopidia**, each of which consists of three linearly arranged cells: a sub-tympanal **cap cell** placed on top of a sheath cell (**scolopale cell**), which envelops the end of a nerve cell dendrite (Fig. 4.3). All adult insects and many larvae have a particular chordotonal organ, **Johnston's organ**, lying within the pedicel, the second antennal segment. The primary function is to sense movements of the antennal flagellum relative to the rest of the body, as in detection of flight speed by air movement. Additionally, it functions in hearing in some insects. In male mosquitoes (Culicidae) and midges (Chironomidae), many scolopidia are contained in the swollen pedicel. These scolopidia are attached at one end to the pedicel wall and at the other, sensory end to the base of the third antennal segment. This greatly modified Johnston's organ is the male receptor for the female wing tone (see section 4.1.4), as shown when males are rendered unreceptive to the sound of the female by amputation of the terminal flagellum or arista of the antenna.

Detection of substrate vibration involves the **subgenual organ**, a chordotonal organ located in the proximal tibia of each leg. Subgenual organs are found in most insects except the Coleoptera and Diptera. The organ consists of a semi-circle of many sensory cells lying in the hemocoel, connected at one end to the inner cuticle of the tibia, and at the other to the trachea. There are subgenual organs within all legs: the organs of each pair of legs may respond specifically to substrate-borne sounds of differing frequencies. Vibration reception may involve either direct transfer of low-frequency substrate vibrations to the legs, or there may

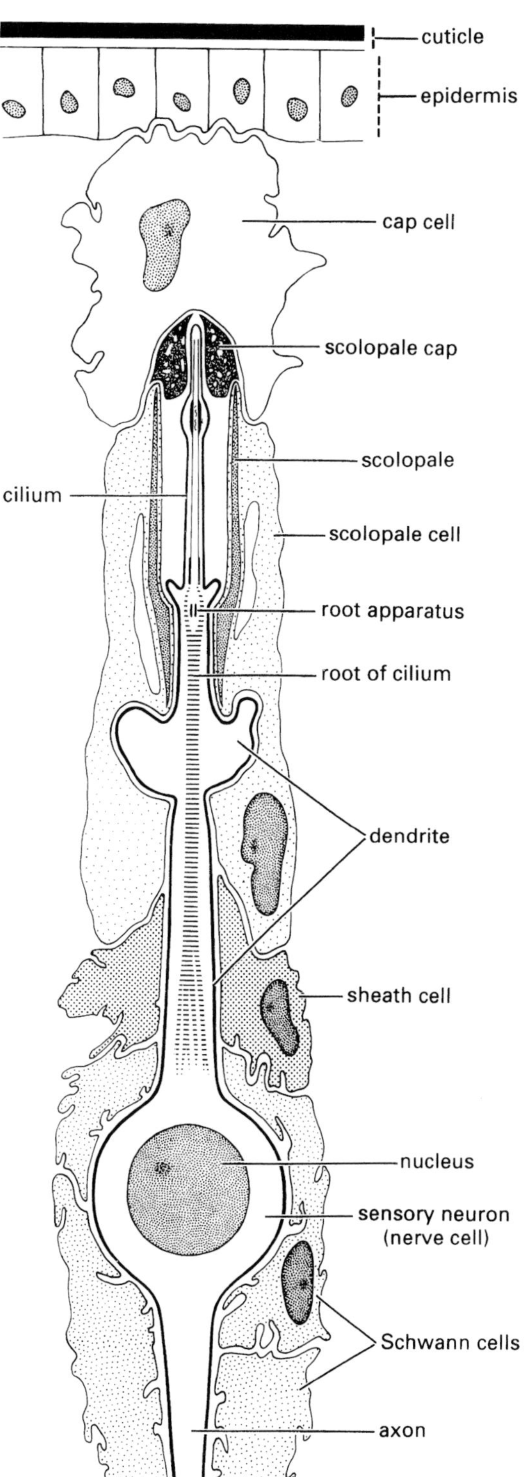

Fig. 4.3 (*right*) Longitudinal section of a scolopidium, the basic unit of a chordotonal organ. (After Gray 1960.)

be more complex amplification and transfer. Airborne vibrations can be detected if they cause vibration of the substrate and hence of the legs.

Tympanal reception

The most elaborate sound reception system in insects involves a specific receptor structure, the **tympanum**. This membrane responds to distant sounds transmitted by airborne vibration. Tympanal membranes are linked to chordotonal organs and are associated with air-filled sacs, such as modifications of the trachea, that enhance sound reception. Tympanal organs are located on the:

- ventral thorax between the metathoracic legs of mantids;
- metathorax of many noctuid moths;
- prothoracic legs of many orthopterans;
- abdomen of other orthopterans, cicadas, and some moths and beetles;
- wing bases of certain moths and lacewings;
- prosternum of some flies (Box 4.1);
- cervical membranes of a few scarab beetles.

The differing location of these organs and their occurrence in distantly related insect groups indicates that tympanal hearing evolved several times in insects. Neuroanatomical studies suggest that all insect tympanal organs evolved from proprioceptors, and the wide distribution of proprioceptors throughout the insect cuticle must account for the variety of positions of tympanal organs.

Tympanal sound reception is particularly well developed in orthopterans, notably in the crickets and katydids. In most of these ensiferan Orthoptera the tympanal organs are on the tibia of each fore leg (Figs. 4.4 & 9.2a). Behind the paired tympanal membranes lies an acoustic trachea that runs from a prothoracic spiracle down each leg to the tympanal organ (Fig. 4.4a).

Crickets and katydids have similar hearing systems. The system in crickets appears to be less specialized because their acoustic tracheae remain connected to the ventilatory spiracles of the prothorax. The acoustic tracheae of katydids form a system completely isolated from the ventilatory tracheae, opening via a separate

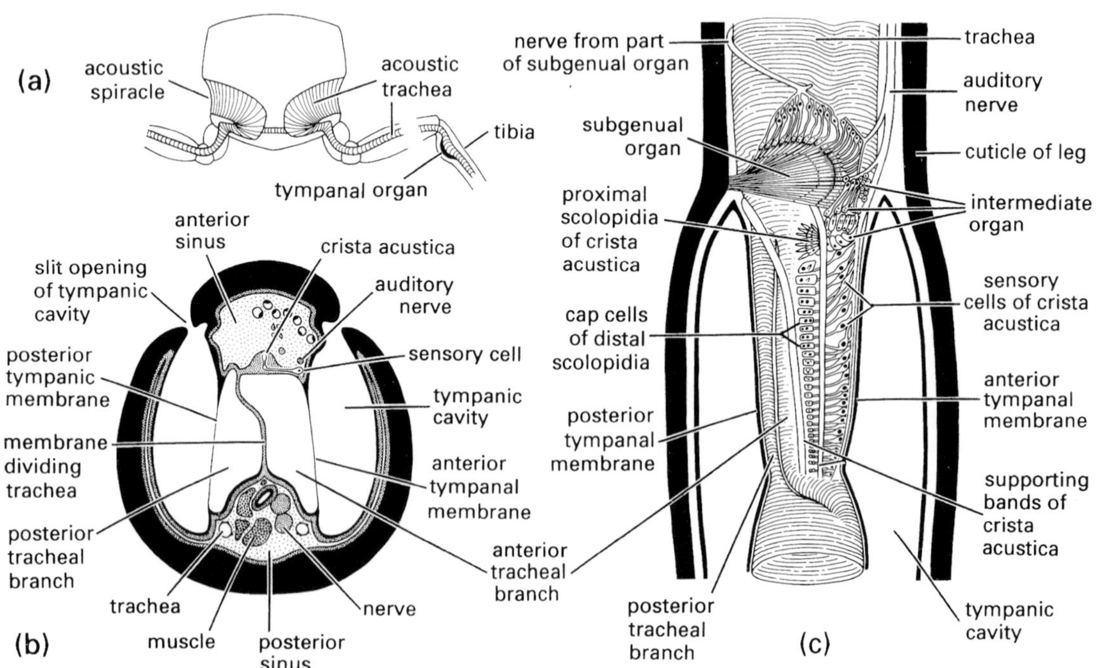

Fig. 4.4 Tympanal organs of a katydid, *Decticus* (Orthoptera: Tettigoniidae): (a) transverse section through the fore legs and prothorax to show the acoustic spiracles and tracheae; (b) transverse section through the base of the fore tibia; (c) longitudinal breakaway view of the fore tibia. (After Schwabe 1906; in Michelsen & Larsen 1985.)

Box 4.1 Aural location of host by a parasitoid fly

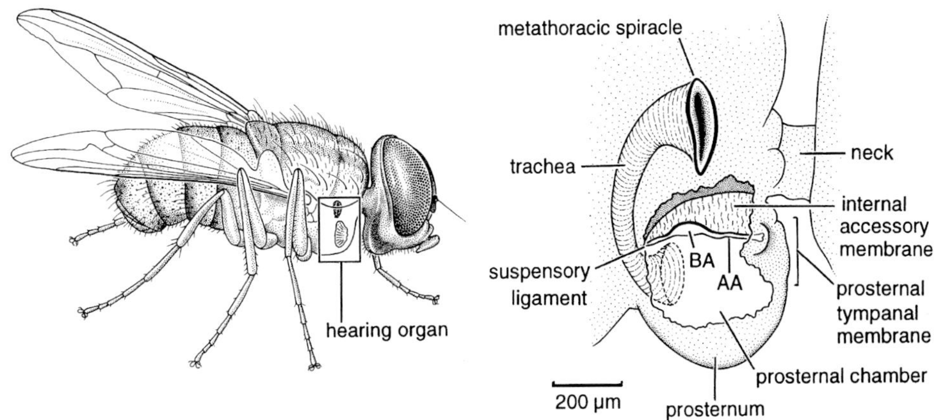

Parasitoid insects track down hosts, upon which their immature development depends, using predominantly chemical and visual cues (section 13.1). Locating a host from afar by orientation towards a sound that is specific for that host is rather unusual behavior. Although close-up low-frequency air movements produced by prospective hosts can be detected, for example by fleas and some blood-feeding flies (section 4.1.3), host location by distant sound is developed best in flies of the tribe Ormiini (Diptera: Tachinidae). The hosts are male crickets, for example of the genus *Gryllus*, and katydids, whose mate-attracting songs (chirps) range in frequency from 2 to 7 kHz. Under the cover of darkness, the female *Ormia* locates the calling host insect, on or near which she deposits first-instar larvae (larviposits). The larvae burrow into the host, in which they develop by eating selected tissues for 7–10 days, after which the third-instar larvae emerge from the dying host and pupariate in the ground.

Location of a calling host is a complex matter compared with simply detecting its presence by hearing the call, as will be understood by anyone who has tried to trace a calling cricket or katydid. Directional hearing is a prerequisite to orientate towards and localize the source of the sound. In most animals with directional hearing, the two receptors ("ears") are separated by a distance greater than the wavelength of the sound, such that the differences (e.g. in intensity and timing) between the sounds received by each "ear" are large enough to be detected and converted by the receptor and nervous system. However, in small animals, such as the house fly-sized ormiine female, with a hearing system spanning less than 1.5 mm, the "ears" are too close together to create interaural differences in intensity and timing. A very different approach to sound detection is required.

As in other hearing insects, the reception system contains a flexible tympanal membrane, an air sac apposed to the tympanum, and a chordotonal organ linked to the tympanum (section 4.1.3). Uniquely amongst hearing insects, the ormiine paired tympanal membranes are located on the prosternum, ventral to the neck (cervix), facing forwards and somewhat obscured by the head (as illustrated here in the side view of a female fly of *Ormia*). On the inner surface of these thin (1 μm) membranes are attached a pair of auditory sense organs, the bulbae acusticae (BA) – chordotonal organs comprising many scolopidia (section 4.1.3). The bulbae are located within an unpartitioned prosternal chamber, which is enlarged by relocation of the anterior musculature and connected to the external environment by tracheae. A sagittal view of this hearing organ is shown above to the right of the fly (after Robert et al. 1994). The structures are sexually dimorphic, with strongest development in the host-seeking female.

What is anatomically unique amongst hearing animals, including all other insects studied, is that there is no separation of the "ears" – the auditory chamber that contains the sensory organs is undivided. Furthermore, the tympani virtually abut, such that the difference in arrival time of sound at each ear is <1 to 2 microseconds. The answer to the physical dilemma is revealed by close examination, which shows that the two tympani actually are joined by a cuticular structure that functions to connect the ears. This mechanical intra-aural coupling involves the connecting cuticle acting as a flexible lever that pivots about a fulcrum and functions to increase the time lag between the nearer-to-noise (ipsilateral) tympanum and the further-from-noise (contralateral) tympanum by about 20-fold. The ipsilateral tympanic membrane is first to be excited to vibrate by incoming sound, slightly before the contralateral one, with the connecting cuticle then commencing to vibrate. In a complex manner involving some damping and cancellation of vibrations, the ipsilateral tympanum produces most vibrations.

This magnification of interaural differences allows very sensitive directionality in sound reception. Such a novel design discovered in ormiine hearing suggests applications in human hearing-aid technology.

pair of acoustic spiracles. In many katydids, the tibial base has two separated longitudinal slits each of which leads into a tympanic chamber (Fig. 4.4b). The acoustic trachea, which lies centrally in the leg, is divided in half at this point by a membrane, such that one half closely connects with the anterior and the other half with the posterior tympanal membrane. The primary route of sound to the tympanal organ is usually from the acoustic spiracle and along the acoustic trachea to the tibia. The change in cross-sectional area from the enlargement of the trachea behind each spiracle (sometimes called a tracheal vesicle) to the tympanal organ in the tibia approximates the function of a horn and amplifies the sound. Although the slits of the tympanic chambers do allow the entry of sound, their exact function is debatable. They may allow directional hearing, because very small differences in the time of arrival of sound waves at the tympanum can be detected by pressure differences across the membrane.

Whatever the major route of sound entry to the tympanal organs, air- and substrate-borne acoustic signals cause the tympanal membranes to vibrate. Vibrations are sensed by three chordotonal organs: the **subgenual organ**, the **intermediate organ**, and the **crista acustica** (Fig. 4.4c). The subgenual organs, which have a form and function like those of non-orthopteroid insects, are present on all legs but the crista acustica and intermediate organs are found only on the fore legs in conjunction with the tympana. This implies that the tibial hearing organ is a serial homologue of the proprioceptor units of the mid and hind legs.

The crista acustica consists of a row of up to 60 scolopidial cells attached to the acoustic trachea and is the main sensory organ for airborne sound in the 5–50 kHz range. The intermediate organ, which consists of 10–20 scolopidial cells, is posterior to the subgenual organ and virtually continuous with the crista acustica. The role of the intermediate organ is uncertain but it may respond to airborne sound of frequencies from 2 to 14 kHz. Each of the three chordotonal organs is innervated separately, but the neuronal connections between the three imply that signals from the different receptors are integrated.

Hearing insects can identify the direction of a point source of sound, but exactly how they do so varies between taxa. Localization of sound directionality clearly depends upon detection of differences in the sound received by one tympanum relative to another, or in some orthopterans by a tympanum within a single leg. Sound reception varies with the orientation of the body relative to the sound source, allowing some precision in locating the source. The unusual means of sound reception and sensitivity of detection of direction of sound source shown by ormiine flies is discussed in Box 4.1.

Night activity is common, as shown by the abundance and diversity of insects attracted to artificial light, especially at the ultraviolet end of the spectrum, and on moonless nights. Night flight allows avoidance of visual-hunting predators, but exposes the insect to specialist nocturnal predators – the insectivorous bats (Microchiroptera). These bats employ a biological sonar system using ultrasonic frequencies that range (according to species) from 20 to 200 kHz for navigating and for detecting and locating prey, predominantly flying insects.

Although bat predation on insects occurs in the darkness of night and high above a human observer, it is evident that a range of insect taxa can detect bat ultrasounds and take appropriate evasive action. The behavioral response to ultrasound, called the acoustic startle response, involves very rapid and co-ordinated muscle contractions. This leads to reactions such as "freezing", unpredictable deviation in flight, or rapid cessation of flight and plummeting towards the ground. Instigation of these reactions, which assist in escape from predation, obviously requires that the insect hears the ultrasound produced by the bat. Physiological experiments show that within a few milliseconds of the emission of such a sound the response takes place, which would precede the detection of the prey by a bat.

To date, insects belonging to five orders have been shown to be able to detect and respond to ultrasound: lacewings (Neuroptera), beetles (Coleoptera), praying mantids (Mantodea), moths (Lepidoptera), and locusts, katydids, and crickets (Orthoptera). Tympanal organs occur in different sites amongst these insects, showing that ultrasound reception has several independent origins amongst these insects. As seen earlier in this chapter (p. 90), the Orthoptera are major acoustic communicators that use sound in intraspecific sexual signaling. Evidently, hearing ability arose early in orthopteran evolution, probably at least some 200 mya, long before bats evolved (perhaps a little before the Eocene (50 mya) from which the oldest fossil comes). Thus, orthopteran ability to hear bat ultrasounds can be seen as an **exaptation** – a morphological–physiological predisposition that has been modified to add sensitivity to ultrasound. The crickets, bush-crickets, and acridid grasshoppers that communicate intraspecifically and also hear ultrasound have sensitivity to high- and low-frequency sound – and perhaps limit their discrimination to only two discrete frequencies. The ultrasound elicits aversion; the other (under suitable conditions) elicits attraction.

In contrast, the tympanal hearing that has arisen independently in several other insects appears to be receptive specifically to ultrasound. The two receptors of a "hearing" noctuoid moth, though differing in threshold, are tuned to the same ultrasonic frequency, and it has been demonstrated experimentally that the moths show behavioral (startle) and physiological (neural) response to bat sonic frequencies. In the parasitic tachinid fly *Ormia* (Box 4.1), in which the female fly locates its orthopteran host by tracking its mating calls, the structure and function of the "ear" is sexually dimorphic. The tympanic area of the female fly is larger, and is sensitive to the 5 kHz frequency of the cricket host and also to the 20–60 kHz ultrasounds made by insectivorous bats, whereas the smaller tympanic area of the male fly responds only to the ultrasound. This suggests that the acoustic response originally was present in both sexes and was used to detect and avoid bats, with sensitivity to cricket calls a later modification in the female sex alone.

At least in these cases, and probably in other groups in which tympanal hearing is limited in taxonomic range and complexity, ultrasound reception appears to have coevolved with the sonic production of the bats that seek to eat them.

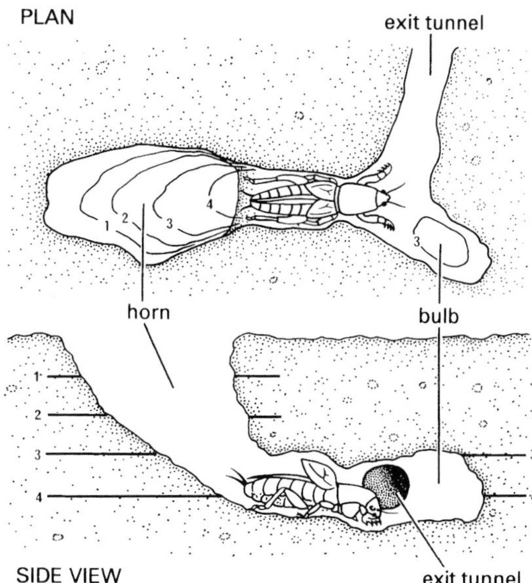

Fig. 4.5 The singing burrow of a mole cricket, *Scapteriscus acletus* (Orthoptera: Gryllotalpidae), in which the singing male sits with his head in the bulb and tegmina raised across the throat of the horn. (After Bennet-Clark 1989.)

4.1.4 Sound production

The commonest method of sound production by insects is by **stridulation**, in which one specialized body part, the **scraper**, is rubbed against another, the **file**. The file is a series of teeth, ridges, or pegs, which vibrate through contact with a ridged or plectrum-like scraper. The file itself makes little noise, and so has to be amplified to generate airborne sound. The horn-shaped burrow of the mole cricket is an excellent sound enhancer (Fig. 4.5). Other insects produce many modifications of the body, particularly of wings and internal air sacs of the tracheal system, to produce amplification and resonance.

Sound production by stridulation occurs in some species of many orders of insects, but the Orthoptera show most elaboration and diversity. All stridulating orthopterans enhance their sounds using the tegmina (the modified fore wings). The file of katydids and crickets is formed from a basal vein of one or both tegmina, and rasps against a scraper on the other wing. Grasshoppers and locusts (Acrididae) rasp a file on the fore femora against a similar scraper on the tegmen.

Many insects lack the body size, power, or sophistication to produce high-frequency airborne sounds, but they can produce and transmit low-frequency sound by vibration of the substrate (such as wood, soil, or a host plant), which is a denser medium. Substrate vibrations are also a by-product of airborne sound production as in acoustic signaling insects, such as some katydids, whose whole body vibrates whilst producing audible airborne stridulatory sounds. Body vibrations, which are transferred through the legs to the substrate (plant or ground), are of low frequencies of 1–5000 Hz. Substrate vibrations can be detected by the female and appear to be used in closer range localization of the calling male, in contrast to the airborne signal used at greater distance.

A second means of sound production involves alternate muscular distortion and relaxation of a specialized area of elastic cuticle, the **tymbal**, to give individual clicks or variably modulated pulses of sound. Tymbal sound production is most audible to the human ear from cicadas, but many other hemipterans and some moths produce sounds from a tymbal. In the cicadas, only the males have these paired tymbals, which are located dorsolaterally, one on each side, on the first abdominal segment. The tymbal membrane is supported by a variable number of ribs. A strong tymbal muscle distorts the membrane and ribs to produce a sound; on relaxation, the elastic tymbal returns to rest. To produce sounds of high frequency, the tymbal muscle contracts asynchronously, with many contractions per nerve impulse (section 3.1.1). A group of chordonotal sensilla is present and a smaller tensor muscle controls the shape of the tymbal, thereby allowing alteration of the acoustic property. The noise of one or more clicks is emitted as the tymbal distorts, and further sounds may be produced during the elastic return on relaxation. The first abdominal segment contains air sacs – modified tracheae – tuned to resonate at or close to the natural frequency of tymbal vibration.

The calls of cicadas generally are in the range of 4–7 kHz, usually of high intensity, carrying as far as 1 km, even in thick forest. Sound is received by both sexes via tympanic membranes that lie ventral to the position of the male tymbal on the first abdominal segment. Cicada calls are species-specific – studies in New Zealand and North America show specificity of duration and cadence of introductory cueing phases inducing timed responses from a prospective mate. Interestingly however, song structures are very homoplasious, with similar songs found in distantly related taxa, but closely related taxa differing markedly in their song.

In other sound-producing hemipterans, both sexes may possess tymbals but because they lack abdominal air sacs, the sound is very damped compared with that of cicadas. The sounds produced by *Nilaparvata lugens* (the brown planthopper; Delphacidae), and probably other non-cicadan hemipterans, are transmitted by vibration of the substrate, and are specifically associated with mating.

Certain moths can hear the ultrasound produced by predatory bats, and moths themselves can produce sound using metepisternal tymbals. The high-frequency clicking sounds that arctiid moths produce can cause bats to veer away from attack, and may have the following (not mutually exclusive) roles:

- interspecific communication between moths;
- interference with bat sonar systems;
- aural mimicry of a bat to delude the predator about the presence of a prey item;
- warning of distastefulness (aposematism; see section 14.4).

The humming or buzzing sound characteristic of swarming mosquitoes, gnats, and midges is a flight tone produced by the frequency of wing beat. This tone, which can be virtually species-specific, differs between the sexes: the male produces a higher tone than the female. The tone also varies with age and ambient temperature for both sexes. Male insects that form nuptial (mating) swarms recognize the swarm site by species-specific environmental markers rather than audible cues (section 5.1); they are insensitive to the wing tone of males of their species. Neither can the male detect the wing tone of immature females – the Johnson's organ in his antenna responds only to the wing tone of physiologically receptive females.

4.2 THERMAL STIMULI

4.2.1 Thermoreception

Insects evidently detect variation in temperature, as seen by their behavior (section 4.2.2), yet the function and location of receptors is poorly known. Most studied insects have antennal sensing of temperature – those with amputated antennae respond differently from insects with intact antennae. Antennal temperature receptors are few in number (presumably ambient temperature is much the same at all points along the antenna), are exposed or concealed in pits, and are associated with humidity receptors in the same sensillum. In the cockroach *Periplaneta americana*, the arolium and pulvilli of the tarsi bear temperature receptors, and thermoreceptors have been found on the legs of certain other insects. Central temperature sensors must exist to detect internal temperature, but the only experimental evidence is from a large moth in which thoracic neural ganglia were found to have a role in instigating temperature-dependent flight muscle activity.

An extreme form of temperature detection is illustrated in jewel beetles (Buprestidae) belonging to the largely Holarctic genus *Melanophila* and also *Merimna atrata* (from Australia). These beetles can detect and orientate towards large-scale forest fires, where they oviposit in still-smoldering pine trunks. Adults of *Melanophila* eat insects killed by fire, and their larvae develop as pioneering colonists boring into fire-killed trees. Detection and orientation in *Melanophila* to distant fires is achieved by detection of infrared radiation (in the wavelength range 3.6–4.1 μm) by pit organs next to the coxal cavities of the mesothoracic legs that are exposed when the beetle is in flight. Within the pits some of the 50–100 small sensillae can respond with heat-induced nanometer-scale deformation, converted to mechanoreceptor signal. The receptor organs in *Merimna* lie on the posterolateral abdomen. These pit organ receptors allow a flying adult buprestid to locate the source of infrared perhaps as far distant as 12 km – a feat of some interest to the US military.

4.2.2 Thermoregulation

Insects are poikilothermic, that is they lack the means to maintain homeothermy – a constant temperature independent of fluctuations in ambient (surrounding) conditions. Although the temperature of an inactive insect tends to track the ambient temperature, many insects can alter their temperature, both upwards and downwards, even if only for a short time. The temperature of an insect can be varied from ambient either behaviorally using external heat (ectothermy) or by physiological mechanisms (endothermy). Endothermy relies on internally generated heat, predominantly from metabolism associated with flight. As some 94% of flight energy is generated as heat (only 6% directed to mechanical force on the wings), flight is not only very energetically demanding but also produces much heat.

Understanding thermoregulation requires some appreciation of the relationship between heat and mass (or volume). The small size of insects in general means any heat generated is rapidly dissipated. In an environment at 10°C a 100 g bumble bee with a body temperature of 40°C experiences a temperature drop of 1°C per second, in the absence of any further heat generation. The larger the body the slower is this heat loss – which is one factor enabling larger organisms to be homeothermic, with the greater mass buffering against heat loss. However, a consequence of the mass–heat relationship is that a small insect can warm up quickly from an external heat source, even one as restricted as a light fleck. Clearly, with insects showing a 500,000-fold variation in mass and 1000-fold variation in metabolic rate, there is scope for a range of variants on thermoregulatory physiologies and behaviors. We review the conventional range of thermoregulatory strategies below, but refer elsewhere to tolerance of extreme temperature (section 6.6.2).

Behavioral thermoregulation (ectothermy)

The extent to which radiant energy (either solar or substrate) influences body temperature is related to the aspect that a diurnal insect adopts. Basking, by which many insects maximize heat uptake, involves both posture and orientation relative to the source of heat. The setae of some "furry" caterpillars, such as gypsy moth larvae (Lymantriidae), serve to insulate the body against convective heat loss while not impairing radiant heat uptake. Wing position and orientation may enhance heat absorption or, alternatively, provide shading from excessive solar radiation. Cooling may include shade-seeking behavior, such as seeking cooler environmental microhabitats or altered orientation on plants. Many desert insects avoid temperature extremes by burrowing. Some insects living in exposed places may avoid excessive heating by "stilting"; that is raising themselves on extended legs to elevate most of the body out of the narrow boundary layer close to the ground. Conduction of heat from the substrate is reduced, and convection is enhanced in the cooler moving air above the boundary layer.

There is a complex (and disputed) relationship between temperature regulation and insect color and surface sculpturing. Amongst some desert beetles (Tenebrionidae), black species become active earlier in the day at lower ambient temperatures than do pale ones, which in turn can remain active longer during hotter times. The application of white paint to black tenebrionid beetles results in substantial body temperature changes: black beetles warm up more rapidly at a given ambient temperature and overheat more quickly compared with white ones, which have greater reflectivity to heat. These physiological differences correlate with certain observed differences in thermal ecology between dark and pale species. Further evidence of the role of color comes from a beclouded cicada (Hemiptera: *Cacama valvata*) in which basking involves directing the dark dorsal surface towards the sun, in contrast to cooling, when the pale ventral surface only is exposed.

For aquatic insects, in which body temperature must follow water temperature, there is little or no ability to regulate body temperature beyond seeking microclimatic differences within a water body.

Physiological thermoregulation (endothermy)

Some insects can be endothermic because the thoracic flight muscles have a very high metabolic rate and produce much heat. The thorax can be maintained at a relatively constant high temperature during flight. Temperature regulation may involve clothing the thorax with insulating scales or hairs, but insulation must be balanced with the need to dissipate any excess heat generated during flight. Some butterflies and locusts alternate heat-producing flight with gliding, which allows cooling, but many insects must fly continuously and cannot glide. Bees and many moths prevent thoracic overheating in flight by increasing the heart rate and circulating hemolymph from the thorax to the poorly insulated abdomen where radiation and convection dissipate heat. At least in some bumble bees (*Bombus*) and carpenter bees (*Xylocopa*) a countercurrent system that normally prevents heat loss is bypassed during flight to augment abdominal heat loss.

The insects that produce elevated temperatures during flight often require a warm thorax before they can take off. When ambient temperatures are low, these insects use the flight muscles to generate heat prior to switching them for use in flight. Mechanisms differ according to whether the flight muscles are synchronous or asynchronous (section 3.1.4). Insects with synchronous flight muscles warm up by contracting antagonistic muscle pairs synchronously and/or synergistic muscles alternately. This activity generally produces some wing vibration, as seen for example in odonates. Asynchronous flight muscles are warmed by operating the flight muscles whilst the wings are uncoupled, or the thoracic box is held rigid by accessory muscles to prevent wing movement. Usually no wing movement is seen, though ventilatory pumping movements of the abdomen may be visible. When the thorax is warm but the insect is sedentary (e.g. whilst feeding), many insects maintain temperature by shivering, which may be prolonged. In contrast, foraging honey bees may cool off during rest, and must then warm up before take-off.

4.3 CHEMICAL STIMULI

In comparison with vertebrates, insects show a more profound use of chemicals in communication, particularly with other individuals of their own species. Insects produce chemicals for many purposes. Their perception in the external environment is through specific chemoreceptors.

4.3.1 Chemoreception

The chemical senses may be divided into **taste**, for detection of aqueous chemicals, and **smell**, for airborne ones – but the distinction is relative. Alternative terms are contact (taste, gustatory) and distant (smell, olfactory) chemoreception. For aquatic insects, all chemicals sensed are in aqueous solution, and strictly all chemoreception should be termed "taste". However, if an aquatic insect has a chemoreceptor that is structurally and functionally equivalent to one in a terrestrial insect that is olfactory, then the aquatic insect is said to "smell" the chemical.

Chemosensors trap chemical molecules, which are transferred to a site for recognition, where they specifically depolarize a membrane and stimulate a nerve impulse. Effective trapping involves localization of the chemoreceptors. Thus, many contact (taste) receptors occur on the mouthparts, such as the labella of higher Diptera (Box 15.5) where salt and sugar receptors occur, and on the ovipositor, to assist with identification of suitable oviposition sites. The antennae, which often are forward-directed and prominent, are first to encounter sensory stimuli and are endowed with many distant chemoreceptors, some contact chemoreceptors, and many mechanoreceptors. The legs, particularly the tarsi which are in contact with the substrate, also have many chemoreceptors. In butterflies, stimulation of the tarsi by sugar solutions evokes an automatic extension of the proboscis. In blow flies, a complex sequence of stereotyped feeding behaviors is induced when a tarsal chemoreceptor is stimulated with sucrose. The proboscis starts to extend and, following sucrose stimulation of the chemoreceptors on the labellum, further proboscis extension occurs and the labellar lobes open. With more sugar stimulus, the source is sucked until stimulation of the mouthparts ceases. When this happens, a predictable pattern of search for further food follows.

Insect chemoreceptors are sensilla with one or more pores (holes). Two classes of sensilla can be defined based on their ultrastructure: **uniporous**, with one pore, and **multiporous**, with several to many pores. Uniporous sensilla range in appearance from hairs to pegs, plates, or simply pores in a cuticular depression, but all have relatively thick walls and a simple permeable pore, which may be apical or central. The hair or peg contains a chamber, which is in basal contact with a dendritic chamber that lies beneath the cuticle. The outer chamber may extrude a viscous liquid, presumed to assist in the entrapment and transfer of chemicals to the dendrites. It is assumed that these uniporous chemoreceptors predominantly detect chemicals by contact, although there is evidence for some olfactory function. Gustatory (contact) neurons are classified best according to their function and thus,

Box 4.2 The electroantennogram

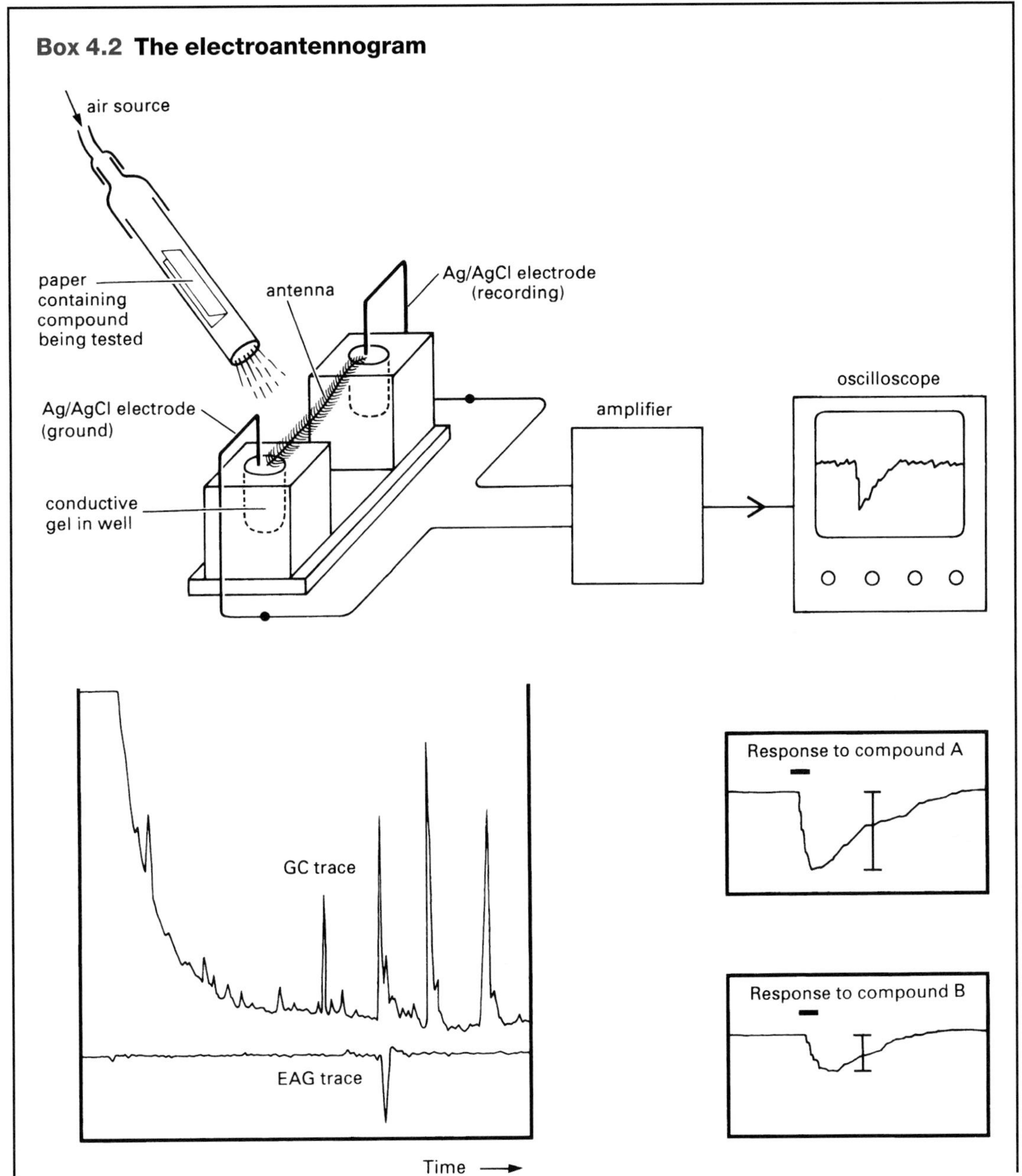

Electrophysiology is the study of the electrical properties of biological material, such as all types of nerve cells, including the peripheral sensory receptors of insects. Insect antennae bear a large number of sensilla and are the major site of olfaction in most insects. Electrical recordings can be made from either individual sensilla on the antenna (single cell recordings) or from the whole antenna (electroantennogram) (as explained by Rumbo 1989). The electroantennogram (EAG) technique measures the total response of insect antennal receptor cells to particular stimuli. Recordings can be made using the antenna either excised, or attached to an isolated head or to the whole insect. In the illustrated example, the effects of a particular biologically active compound (a pheromone) blown across the isolated antenna of a male moth are being assessed. The recording electrode, connected to the apex of the antenna, detects the electrical response, which is amplified and visualized as a trace as in the EAG set-up illustrated in the upper drawing. Antennal receptors are very sensitive and specifically perceive particular odors, such as the sex pheromone of potential conspecific partners or volatile chemicals released by the insect's host. Different compounds usually elicit different EAG responses from the same antenna, as depicted in the two traces on the lower right.

This elegant and simple technique has been used extensively in pheromone identification studies as a quick method of bioassaying compounds for activity. For example, the antennal responses of a male moth to the natural sex pheromone obtained from conspecific female moths are compared with responses to synthetic pheromone components or mixtures. Clean air is blown continuously over the antenna at a constant rate and the samples to be tested are introduced into the air stream, and the EAG response is observed. The same samples can be passed through a gas chromatograph (GC) (which can be interfaced with a mass spectrometer to determine molecular structure of the compounds being tested). Thus, the biological response from the antenna can be related directly to the chemical separation (seen as peaks in the GC trace), as illustrated here in the graph on the lower left (after Struble & Arn 1984).

In addition to lepidopteran species, EAG data have been collected for cockroaches, beetles, flies, bees, and other insects, to measure antennal responses to a range of volatile chemicals affecting host attraction, mating, oviposition, and other behaviors. EAG information is of greatest utility when interpreted in conjunction with behavioral studies.

in relation to feeding, there are cells whose activity in response to chemical stimulation either is to enhance or reduce feeding. These receptors are called phagostimulatory or deterrent.

The major olfactory role comes from multiporous sensilla, which are hair- or peg-like setae, with many round pores or slits in the thin walls, leading into a chamber known as the **pore kettle**. This is richly endowed with pore tubules, which run inwards to meet multibranched dendrites (Box 4.3). Development of an electroantennogram (Box 4.2) allowed revelation of the specificity of chemoreception by the antenna. Used in conjunction with the scanning electron microscope, micro-electrophysiology and modern molecular techniques have extended our understanding of the ability of insects to detect and respond to very weak chemical signals (Box 4.3). Great sensitivity is achieved by spreading very many receptors over as great an area as possible, and allowing the maximum volume of air to flow across the receptors. Thus, the antennae of many male moths are large and frequently the surface area is enlarged by pectinations that form a sieve-like basket (Fig. 4.6). Each antenna of the male silkworm moth (Bombycidae: *Bombyx mori*) has some 17,000 sensilla of different sizes and several ultrastructural morphologies. Sensilla respond specifically to sex-signaling chemicals produced by the female (sex pheromones; see below). As each sensillum has up to 3000 pores, each 10–15 nm in diameter, there are some 45 million pores per moth. Calculations concerning the silkworm moth suggest that just a few molecules could stimulate a nerve impulse above the background rate, and behavioral change may be elicited by less than a hundred molecules.

4.3.2 Semiochemicals: pheromones

Many insect behaviors rely on the sense of smell. Chemical odors, termed **semiochemicals** (from *semion* – signal), are especially important in both interspecific and intraspecific communication. The latter is particularly highly developed in insects, and involves the use of chemicals called **pheromones**. When recognized first in the 1950s, pheromones were defined as: substances that are secreted to the outside by one individual and received by a second individual of the same species in which they release a specific reaction, for example a

Box 4.3 Reception of communication molecules

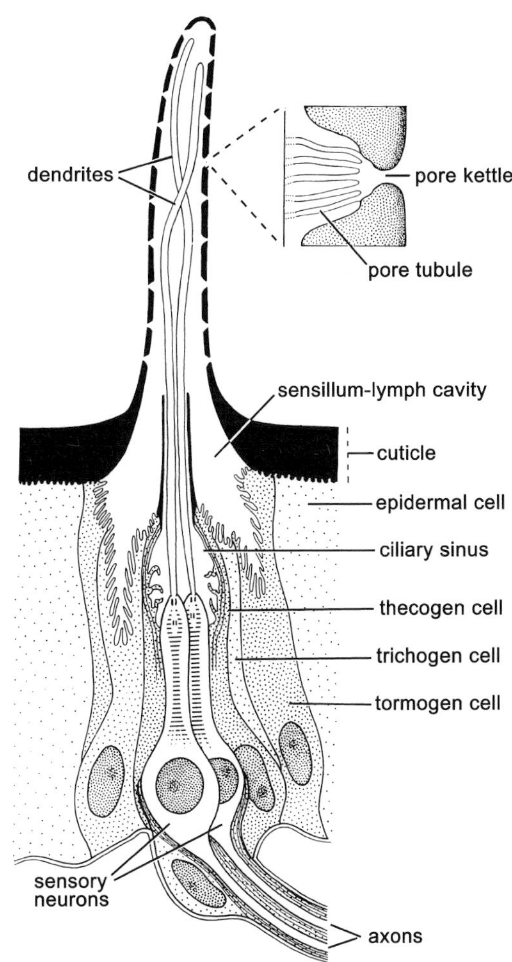

Pheromones, and indeed all signaling chemicals (semiochemicals), must be detectable in the smallest quantities. For example, the moth approaching a pheromone source portrayed in Fig. 4.7, must detect an initially weak signal, and then respond appropriately by orientating towards it, distinguishing abrupt changes in concentration ranging from zero to short-lived concentrated puffs. This involves a physiological ability to monitor continuously and respond to aerial pheromone levels in a process involving extra- and intracellular events.

Ultrastructural studies of *Drosophila melanogaster* and several species of moth allow identification of several types of chemoreceptive (olfactory) sensilla: namely sensilla basiconica, sensilla trichodea, and sensilla coeloconica. These sensillar types are widely distributed across insect taxa and structures but most often are concentrated on the antenna. Each sensilla has from two to multiple subtypes which differ in their sensitivity and tuning to different communication chemicals. The structure of a generalized multiporous olfactory sensillum in the accompanying illustration follows Birch and Haynes (1982) and Zacharuk (1985).

To be detected, first the chemical must arrive at a pore of an olfactory sensillum. In a multiporous sensillum, it enters a pore kettle and contacts and crosses the cuticular lining of a pore tubule. Because pheromones (and other semiochemicals) largely are hydrophobic (lipophilic) compounds they must be made soluble to reach the receptors. This role falls to odorant-binding proteins (OBP) produced in the tormogen and trichogen cells (Fig. 4.1), from which they are secreted into the sensillum-lymph cavity that surrounds the dendrite of the receptor. Specific OBPs bind the semiochemical into a soluble ligand (OBP–pheromone complex) which is protected as it diffuses through the lymph to the dendrite surface. Here, interaction with negatively charged sites transforms the complex, releasing the pheromone to the binding site of the appropriate olfactory receptors located on the dendrite of the neuron, triggering a cascade of neural activity leading to appropriate behavior.

Much research has involved detection of pheromones because of their use in pest management (see section 16.9), but the principles revealed apparently apply to semiochemical reception across a range of organs and taxa. Thus, experiments with the electroantennogram (Box 4.2) using a single sensillum show highly specific responses to particular semiochemicals, and failure to respond even to "trivially" modified compounds. Studied OBPs appear to be one-to-one matched with each semiochemical, but insects apparently respond to more chemical cues than there are OBPs yet revealed. Additionally, olfactory receptors on the dendrite surface seemingly may be less specific, being triggered by a range of unrelated ligands. Furthermore, the model above does not address the frequently observed synergistic effects, in which a cocktail of chemicals provokes a stronger response than any component alone. It remains an open question as to exactly how insects are so spectacularly sensitive to so many specific chemicals, alone or in combination. This is an active research area, with microphysiology and molecular tools providing many new insights.

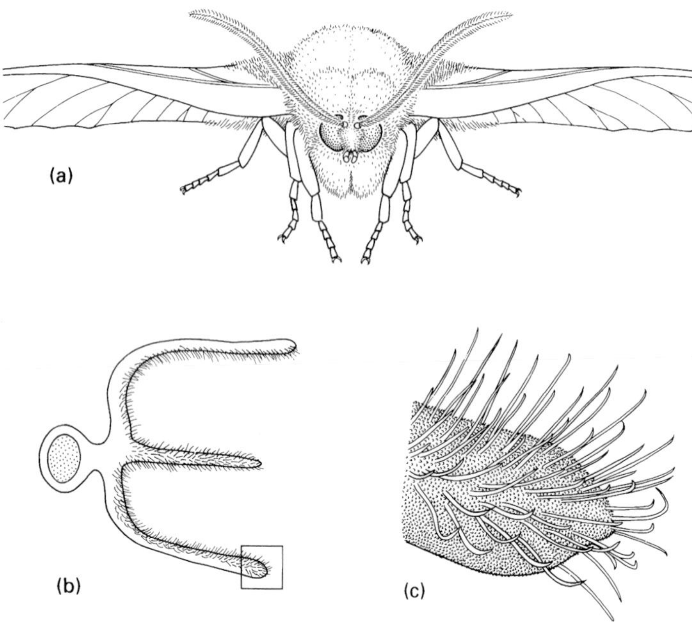

Fig. 4.6 The antennae of a male moth of *Trictena atripalpis* (Lepidoptera: Hepialidae): (a) anterior view of head showing tripectinate antennae of this species; (b) cross-section through the antenna showing the three branches; (c) enlargement of tip of outer branch of one pectination showing olfactory sensilla.

definite behavior or developmental process. This definition remains valid today, despite the discovery of a hidden complexity of pheromone cocktails.

Pheromones are predominantly volatile but sometimes are liquid contact chemicals. All are produced by exocrine glands (those that secrete to the outside of the body) derived from epidermal cells. The scent organs may be located almost anywhere on the body. Thus, sexual scent glands on female Lepidoptera lie in eversible sacs or pouches between the eighth and ninth abdominal segments; the organs are mandibular in the female honey bee, but are located on the swollen hind tibiae of female aphids, and within the midgut and genitalia in cockroaches.

Classification of pheromones by chemical structure reveals that many naturally occurring compounds (such as host odors) and pre-existing metabolites (such as cuticular waxes) have been co-opted by insects to serve in the biochemical synthesis of a wide variety of compounds that function in communication. Chemical classification, although of interest, is of less value for many entomologists than the behaviors that the chemicals elicit. Very many behaviors of insects are governed by chemicals; nevertheless, we can distinguish pheromones that **release** specific behaviors from those that **prime** long-term, irreversible physiological changes. Thus, the stereotyped sexual behavior of a male moth is released by the female-emitted sex pheromone, whereas the crowding pheromone of locusts will prime maturation of gregarious phase individuals (section 6.10.5). Here, further classification of pheromones is based on five categories of behavior associated with sex, aggregation, spacing, trail forming, and alarm.

Sex pheromones

Male and female conspecific insects often communicate with chemical **sex pheromones**. Mate location and courtship may involve chemicals in two stages, with **sex attraction pheromones** acting at a distance, followed by close-up **courtship pheromones** employed prior to mating. The sex pheromones involved in attraction often differ from those used in courtship. Production and release of sex attractant pheromones tends to be restricted to the female, although there are lepidopterans and scorpionflies in which males are the releasers of distance attractants that lure females. The producer releases volatile pheromones that stimulate characteristic behavior in those members of the opposite sex within range of the odorous plume. An aroused recipient raises the antennae, orientates towards the source and walks or flies upwind to the source, often in a zig-zag track (Fig. 4.7) based on ability to respond rapidly to minor changes in pheromone concentration

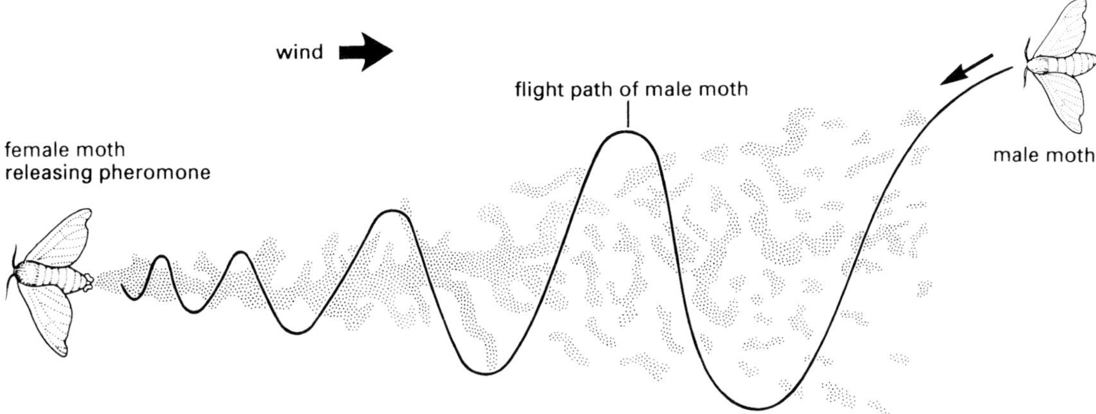

Fig. 4.7 Location of pheromone-emitting female by male moth tacking upwind. The pheromone trail forms a somewhat discontinuous plume because of turbulence, intermittent release, and other factors. (After Haynes & Birch 1985.)

by direction change (Box 4.3). Each successive action appears to depend upon an increase in concentration of this airborne pheromone. As the insect approaches the source, cues such as sound and vision may be involved in close-up courtship behavior.

Courtship (section 5.2), which involves co-ordination of the two sexes, may require close-up chemical stimulation of the partner with a courtship pheromone. This pheromone may be simply a high concentration of the attractant pheromone, but "aphrodisiac" chemicals do exist, as seen in the queen butterfly (Nymphalidae: *Danaus gilippus*). The males of this species, as with several other lepidopterans, have extrusible abdominal hairpencils (brushes), which produce a pheromone that is dusted directly onto the antennae of the female, whilst both are in flight (Fig. 4.8). The effect of this pheromone is to placate a natural escape reaction of the female, who alights, folds her wings and allows copulation. In *D. gilippus*, this male courtship pheromone, a pyrrolixidine alkaloid called danaidone, is essential to successful courtship. However, the butterfly cannot synthesize it without acquiring the chemical precursor by feeding on selected plants as an adult. In the arctiid moth, *Creatonotus gangis*, the precursor of the male courtship pheromone likewise cannot be synthesized by the moth, but is sequestered by the larva in the form of a toxic alkaloid from the host plant. The larva uses the chemical in its defense and at metamorphosis the toxins are transferred to the adult. Both sexes use them as defensive compounds, with the male additionally

Fig. 4.8 A pair of queen butterflies, *Danaus gilippus* (Lepidoptera: Nymphalidae: Danainae), showing aerial "hairpencilling" by the male. The male (above) has splayed hairpencils (at his abdominal apex) and is applying pheromone to the female (below). (After Brower et al. 1965.)

converting them to his pheromone. This he emits from inflatable abdominal tubes, called **coremata**, whose development is regulated by the alkaloid pheromone precursor.

A spectacular example of deceitful sexual signaling occurs in bolas spiders, which do not build a web, but whirl a single thread terminating in a sticky globule towards their moth prey (like gauchos using a bolas to hobble cattle). The spiders lure male moths to within reach of the bolas using synthetic lures of sex-attractant pheromone cocktails. The proportions of the components vary according to the abundance of particular moth species available as prey. Similar principles are applied by humans to control pest insects using lures containing synthetic sex pheromones or other attractants (section 16.9). Certain chemical compounds (e.g. methyl eugenol), that either occur naturally in plants or can be synthesized in the laboratory, are used to lure male fruit flies (Tephritidae) for pest management purposes. These male lures are sometimes called **parapheromones**, probably because the compounds may be used by the flies as a component in the synthesis of their sex pheromones and have been shown to improve mating success, perhaps by enhancing the male's sexual signals.

Sex pheromones once were thought to be unique, species-specific chemicals, but in reality often they are chemical blends. The same chemical (e.g. a particular 14-carbon chain alcohol) may be present in a range of related and unrelated species, but it occurs in a blend of different proportions with several other chemicals. An individual component may elicit only one part of the sex attraction behavior, or a partial or complete mixture may be required. Often the blend produces a greater response than any individual component, a **synergism** that is widespread in insects that produce pheromone mixtures. Chemical structural similarity of pheromones may indicate systematic relationship amongst the producers. However, obvious anomalies arise when identical or very similar pheromones are synthesized from chemicals derived from identical diets by unrelated insects.

Even if individual components are shared by many species, the mixture of pheromones is very often species-specific. It is evident that pheromones, and the stereotyped behaviors that they evoke, are highly significant in maintenance of reproductive isolation between species. The species-specificity of sex pheromones avoids cross-species mating before males and females come into contact.

Aggregation pheromones

The release of an **aggregation pheromone** causes conspecific insects of both sexes to crowd around the source of the pheromone. Aggregation may lead to increased likelihood of mating but, in contrast to many sex pheromones, both sexes may produce and respond to aggregation pheromones. The potential benefits provided by the response include security from predation, maximum utilization of a scarce food resource, overcoming of host resistance, or cohesion of social insects, as well as the chance to mate.

Aggregation pheromones are known in six insect orders, including cockroaches, but their presence and mode of action has been studied in most detail in Coleoptera, particularly in economically damaging species such as stored-grain beetles (from several families) and timber and bark beetles (Curculionidae: Scolytinae). A well-researched example of a complex suite of aggregation pheromones is provided by the Californian western pine beetle, *Dendroctonus brevicomis* (Scolytinae), which attacks ponderosa pine (*Pinus ponderosa*). On arrival at a new tree, colonizing females release the pheromone *exo*-brevicomin augmented by myrcene, a terpene originating from the damaged pine tree. Both sexes of western pine beetle are attracted by this mixture, and newly arrived males then add to the chemical mix by releasing another pheromone, frontalin. The cumulative lure of frontalin, *exo*-brevicomin, and myrcene is synergistic, i.e. greater than any one of these chemicals alone. The aggregation of many pine beetles overwhelms the tree's defensive secretion of resins.

Spacing pheromones

There is a limit to the number of western pine beetles (*D. brevicomis*; see above) that attack a single tree. Cessation is assisted by reduction in the attractant aggregation pheromones, but deterrent chemicals also are produced. After the beetles mate on the tree, both sexes produce "anti-aggregation" pheromones called verbenone and *trans*-verbenone, and males also emit ipsdienol. These deter further beetles from landing close by, encouraging spacing out of new colonists. When the resource is saturated, further arrivals are repelled.

Such semiochemicals, called **spacing**, **epideictic**, or **dispersion pheromones**, may effect appropriate spacing on food resources, as with some phytophagous insects. Several species of tephritid flies lay eggs singly in fruit where the solitary larva is to develop. Spacing occurs because the ovipositing female deposits an oviposition-deterrent pheromone on the fruit on which she has laid an egg, thereby deterring subsequent oviposition. Social insects, which by definition are aggregated, utilize pheromones to regulate many aspects of their behavior, including the spacing between colonies. Spacer pheromones of colony-specific odors may be used to ensure an even dispersal of colonies of conspecifics, as in African weaver ants (Formicidae: *Oecophylla longinoda*).

Trail-marking pheromones

Many social insects use pheromones to mark their trails, particularly to food and the nest. **Trail-marking pheromones** are volatile and short-lived chemicals that evaporate within days unless reinforced (perhaps as a response to a food resource that is longer lasting than usual). Trail pheromones in ants are commonly metabolic waste products excreted by the poison gland. These need not be species-specific for several species share some common chemicals. Dufour's gland secretions of some ant species may be more species-specific chemical mixtures associated with marking of territory and pioneering trails. Ant trails appear to be non-polar, i.e. the direction to nest or food resource cannot be determined by the trail odor.

In contrast to trails laid on the ground, an airborne trail – an odor plume – has directionality because of increasing concentration of the odor towards the source. An insect may rely upon angling the flight path relative to the direction of the wind that brings the odor, resulting in a zig-zag upwind flight towards the source. Each directional shift is produced where the odor diminishes at the edge of the plume (Fig. 4.7).

Alarm pheromones

Nearly two centuries ago it was recognized that workers of honey bees (*Apis mellifera*) were alarmed by a freshly extracted sting. In the intervening years many aggregating insects have been found to produce chemical releasers of alarm behavior – **alarm pheromones** – that characterize most social insects (termites and eusocial hymenopterans). In addition, alarm pheromones are known in several hemipterans, including subsocial treehoppers (Membracidae), aphids (Aphididae), and some other true bugs. Alarm pheromones are volatile, non-persistent compounds that are readily dispersed throughout the aggregation. Alarm is provoked by the presence of a predator, or in many social insects, a threat to the nest. The behavior elicited may be rapid dispersal, such as in hemipterans that drop from the host plant; or escape from an unwinnable conflict with a large predator, as in poorly defended ants living in small colonies. The alarm behavior of many eusocial insects is most familiar to us when disturbance of a nest induces many ants, bees, or wasps to an aggressive defense. Alarm pheromones attract aggressive workers and these recruits attack the cause of the disturbance by biting, stinging, or firing repellent chemicals. Emission of more alarm pheromone mobilizes further defenders. Alarm pheromone may be daubed over an intruder to aid in directing the attack.

Alarm pheromones may have been derived over evolutionary time from chemicals used as general anti-predator devices (allomones; see below), utilizing glands co-opted from many different parts of the body to produce the substances. For example, hymenopterans commonly produce alarm pheromones from mandibular glands and also from poison glands, metapleural glands, the sting shaft, and even the anal area. All these glands also may be production sites for defensive chemicals.

4.3.3 Semiochemicals: kairomones, allomones, and synomones

Communication chemicals (semiochemicals) may function between individuals of the same species (pheromones) or between different species (**allelochemicals**). Interspecific semiochemicals may be grouped according to the benefits they provide to the producer and receiver. Those that benefit the receiver but disadvantage the producer are **kairomones**. **Allomones** benefit the producer by modifying the behavior of the receiver although having a neutral effect on the receiver. **Synomones** benefit both the producer and the receiver. This terminology has to be applied in the context of the specific behavior induced in the recipient, as seen in the examples discussed below. A particular chemical can act as an intraspecific pheromone and may also fulfill all three categories of interspecific communication, depending on circumstances. The use of the same chemical for two or more functions in different contexts is referred to as semiochemical parsimony.

Kairomones

Myrcene, the terpene produced by a ponderosa pine when it is damaged by the western pine beetle (see above), acts as a synergist with aggregation pheromones that act to lure more beetles. Thus, myrcene and other terpenes produced by damaged conifers can be kairomones, disadvantaging the producer by luring damaging timber beetles. A kairomone need not be a product of insect attack: elm bark beetles (Curculionidae: Scolytinae: *Scolytus* spp.) respond to α-cubebene, a product of the Dutch elm disease fungus *Ceratocystis ulmi* that indicates a weakened or dead elm tree (*Ulmus*). Elm beetles themselves inoculate previously healthy elms with the fungus, but pheromone-induced aggregations of beetles form only when the kairomone (fungal α-cubenene) indicates suitability for colonization. Host-plant detection by phytophagous insects also involves reception of plant chemicals, which therefore are acting as kairomones.

Insects produce many communication chemicals, with clear benefits. However, these semiochemicals also may act as kairomones if other insects recognize them. In "hijacking" the chemical messenger for their own use, specialist parasitoids (Chapter 13) use chemicals emitted by the host, or plants attacked by the host, to locate a suitable host for development of its offspring.

Allomones

Allomones are chemicals that benefit the producer but have neutral effects on the recipient. For example, defensive and/or repellent chemicals are allomones that advertise distastefulness and protect the producer from lethal experiment by prospective predators. The effect on a potential predator is considered to be neutral, as it is warned from wasting energy in seeking a distasteful meal.

The worldwide beetle family Lycidae has many distasteful and warning-colored (aposematic) members (see Plate 6.5, facing p. 14), including species of *Metriorrhynchus* that are protected by odorous alkylpyrazine allomones. In Australia, several distantly related beetle families include many mimics that are modeled visually on *Metriorrhynchus*. Some mimics are remarkably convergent in color and distasteful chemicals, and possess nearly identical alkylpyrazines. Others share the allomones but differ in distasteful chemicals, whereas some have the warning chemical but appear to lack distastefulness. Other insect mimicry complexes involve allomones. Mimicry and insect defenses in general are considered further in Chapter 14.

Some defensive allomones can have a dual function as sex pheromones. Examples include chemicals from the defensive glands of various bugs (Heteroptera), grasshoppers (Acrididae), and beetles (Staphylinidae), as well as plant-derived toxins used by some Lepidoptera (section 4.3.2). Many female ants, bees, and wasps have exploited the secretions of the glands associated with their sting – the poison (or venom) gland and Dufour's gland – as male attractants and releasers of male sexual activity.

A novel use of allomones occurs in certain orchids, whose flowers produce similar odors to female sex pheromone of the wasp or bee species that acts as their specific pollinator. Male wasps or bees are deceived by this chemical mimicry and also by the color and shape of the flower (see Plates 4.4 & 4.5), with which they attempt to copulate (section 11.3.1). Thus the orchid's odor acts as an allomone beneficial to the plant by attracting its specific pollinator, whereas the effect on the male insects is near neutral – at most they waste time and effort.

Synomones

The terpenes produced by damaged pines are kairomones for pest beetles, but if identical chemicals are used by beneficial parasitoids to locate and attack the bark beetles, the terpenes are acting as synomones (by benefiting both the producer and the receiver). Thus α-pinene and myrcene produced by damaged pines are kairomones for species of *Dendroctonus* but synomones for pteromalid hymenopterans that parasitize these timber beetles. In like manner, α-cubebene produced by Dutch elm fungus is a synomone for the braconid hymenopteran parasitoids of elm bark beetles (for which it is a kairomone).

An insect parasitoid may respond to host-plant odor directly, like the phytophage it seeks to parasitize, but this means of searching cannot guarantee the parasitoid that the phytophage host is actually present. There is a greater chance of success for the parasitoid if it can identify and respond to the specific plant chemical defenses that the phytophage provokes. If an insect-damaged host plant produced a repellent odor, such as a volatile terpenoid, then the chemical could act as:

• an allomone that deters non-specialist phytophages;
• a kairomone that attracts a specialist phytophage;
• a synomone that lures the parasitoid of the phytophage.

Of course, phytophagous, parasitic, and predatory insects rely on more than odors to locate potential hosts or prey, and visual discrimination is implicated in resource location.

4.4 INSECT VISION

Excepting a few blind subterranean and endoparasitic species, most insects have some sight, and many possess highly developed visual systems. The basic components needed for vision are a lens to focus light onto **photoreceptors** – cells containing light-sensitive molecules – and a nervous system complex enough to process visual information. In insect eyes, the photoreceptive structure is the **rhabdom**, comprising several adjacent **retinula** (or nerve) cells and consisting of close-packed **microvilli** containing visual pigment. Light falling onto the rhabdom changes the configuration of the visual pigment, triggering a change of electrical potential across the cell membrane. This signal is then transmitted via chemical synapses to nerve cells in the brain. Comparison of the visual systems of different kinds of insect eyes involves two main considerations: (i) their resolving power for images, i.e. the amount of fine detail that can be resolved; and (ii) their light sensitivity, i.e. the minimum ambient light level at which the insect can still see. Eyes of different kinds and in different insects vary widely in resolving power and light sensitivity and thus in details of function.

The compound eyes are the most obvious and familiar visual organs of insects but there are three other means by which an insect may perceive light: dermal detection, stemmata, and ocelli. The dragonfly head depicted in the vignette of this chapter is dominated by its huge compound eyes with the three ocelli and paired antennae in the center.

4.4.1 Dermal detection

In insects able to detect light through their body surface, there are sensory receptors below the body cuticle but no optical system with focusing structures. Evidence for this general responsivity to light comes from the persistence of photic responses after covering all visual organs, for example in cockroaches and lepidopteran larvae. Some blind cave insects, with no recognizable visual organs, respond to light, as do decapitated cockroaches. In most cases the sensitive cells and their connection with the central nervous system have yet to be discovered. However, within the brain itself, aphids have light-sensitive cells that detect changes in day length – an environmental cue that controls the mode of reproduction (i.e. either sexual or parthenogenetic). The setting of the biological clock (Box 4.4) relies upon the ability to detect photoperiod.

4.4.2 Stemmata

The only visual organs of larval holometabolous insects are stemmata, sometimes called larval ocelli (Fig. 4.9a). These organs are located on the head, and vary from a single pigmented spot on each side to six or seven larger stemmata, each with numerous photoreceptors and associated nerve cells. In the simplest stemma, a cuticular lens overlies a crystalline body secreted by several cells. Light is focused by the lens onto a single rhabdom. Each stemma points in a different direction so that the insect sees only a few points in space according to the number of stemmata. Some caterpillars increase the field of view and fill in the gaps between the direction of view of adjacent stemmata by scanning movements of the head. Other larvae, such as those of sawflies and tiger beetles, possess more sophisticated stemmata. They consist of a two-layered lens that forms an image on an extended retina composed of many rhabdoms, each receiving light from a different part of the image. In general, stemmata seem designed for high light sensitivity, with resolving power relatively low.

4.4.3 Ocelli

Many adult insects, as well as some nymphs, have dorsal ocelli in addition to compound eyes. These ocelli are unrelated embryologically to the stemmata. Typically, three small ocelli lie in a triangle on top of the head. The cuticle covering an ocellus is transparent and may be curved as a lens. It overlies transparent epidermal cells, so that light passes through to an extended retina made up of many rhabdoms (Fig. 4.9b). Individual groups

Box 4.4 Biological clocks

Seasonal changes in environmental conditions allow insects to adjust their life histories to optimize the use of suitable conditions and minimize the impact of unsuitable ones (e.g. through diapause; section 6.5). Similar physical fluctuations on a daily scale encourage a diurnal (daily) cycle of activity and quiescence. Nocturnal insects are active at night, diurnal ones by day, and crepuscular insect activity occurs at dusk and dawn when light intensities are transitional. The external physical environment, such as light–dark or temperature, controls some daily activity patterns, called exogenous rhythms. However, many other periodic activities are internally driven endogenous rhythms that have a clock-like or calendar-like frequency irrespective of external conditions. Endogenous periodicity is frequently about 24 h (circadian), but lunar and tidal periodicities govern the emergence of adult aquatic midges from large lakes and the marine intertidal zones, respectively. This unlearned, once-in-a-lifetime rhythm which allows synchronization of eclosion demonstrates the innate ability of insects to measure passing time.

Experimentation is required to discriminate between exogenous and endogenous rhythms. This involves observing what happens to rhythmic behavior when external environmental cues are altered, removed, or made invariate. Such experiments show that inception (setting) of endogenous rhythms is found to be day length, with the clock then free-running, without daily reinforcement by the light–dark cycle, often for a considerable period. Thus, if nocturnal cockroaches that become active at dusk are kept at constant temperature in constant light or dark, they will maintain the dusk commencement of their activities at a circadian rhythm of 23–25 h. Rhythmic activities of other insects may require an occasional clock-setting (such as darkness) to prevent the circadian rhythm drifting, either through adaptation to an exogenous rhythm or into arrhythmy.

Biological clocks allow solar orientation – the use of the sun's elevation above the horizon as a compass – provided that there is a means of assessing (and compensating for) the passage of time. Some ants and honey bees use a "light-compass", finding direction from the sun's elevation and using the biological clock to compensate for the sun's movement across the sky. Evidence came from an elegant experiment with honey bees trained to forage in the late afternoon at a feeding table (F) placed 180 m NW of their hive (H), as depicted in the left figure (after Lindauer 1960). Overnight the hive was moved to a new location to prevent use of familiar landmarks in foraging, and a selection of four feeding tables (F_{1-4}) was provided at 180 m, NW, SW, SE, and NE from the hive. In the morning, despite the sun being at a very different angle to that during the afternoon training, 15 of the 19 bees were able to locate the NW table (as depicted in the figure on the right). The honey bee "dance language" that communicates direction and distance of food to other workers (Box 12.1) depends upon the capacity to calculate direction from the sun.

The circadian pacemaker (oscillator) that controls the rhythm is located in the brain; it is not an external photoperiod receptor. Experimental evidence shows that in cockroaches, beetles, and crickets a pacemaker lies in the optic lobes, whereas in some silkworms it lies in the cerebral lobes of the brain. In the well-studied *Drosophila*, a major oscillator site appears to be located between the lateral protocerebellum and the medulla of the optic. However, visualization of the sites of *period* gene activity is not localized, and there is increasing evidence of multiple pacemaker centers located throughout the tissues. Whether they communicate with each other or run independently is not yet clear.

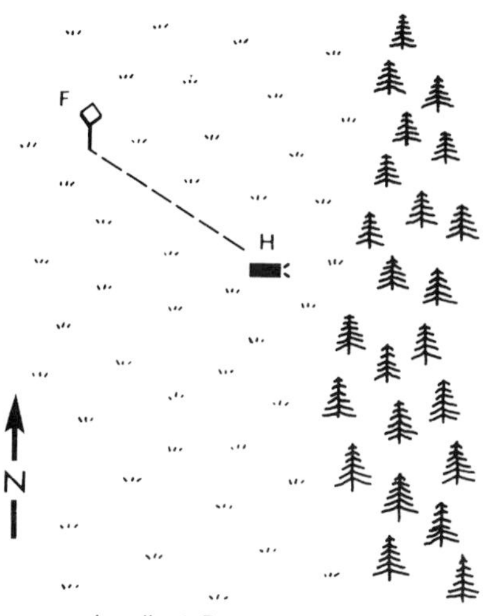

Locality 1, Day 1, Afternoon

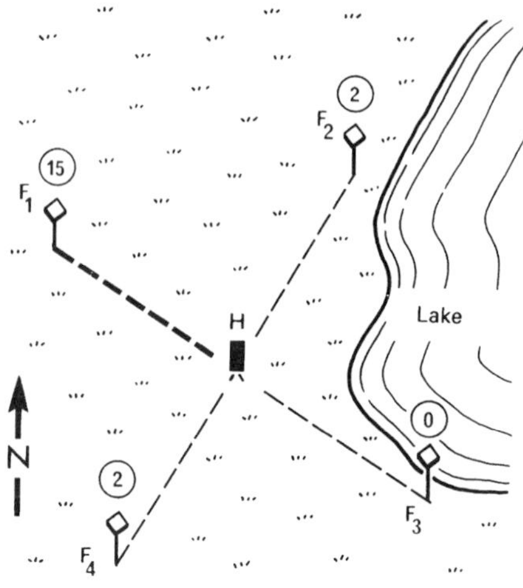

Locality 2, Day 2, Morning

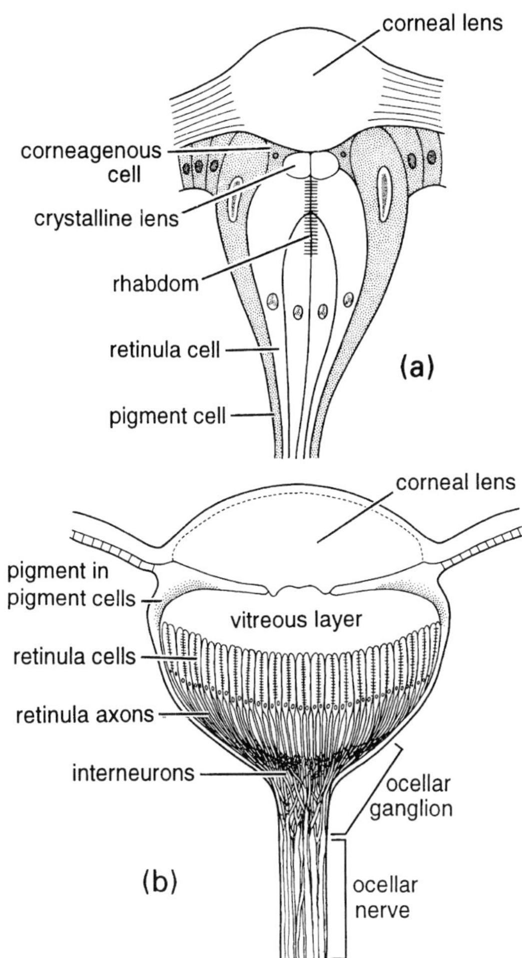

Fig. 4.9 Longitudinal sections through simple eyes: (a) a simple stemma of a lepidopteran larva; (b) a light-adapted median ocellus of a locust. ((a) After Snodgrass 1935; (b) after Wilson 1978.)

of retinula cells that contribute to one rhabdom or the complete retina are surrounded by pigment cells or by a reflective layer. The focal plane of the ocellar lens lies below the rhabdoms so that the retina receives a blurred image. The axons of the ocellar retinula cells converge onto only a few neurons that connect the ocelli to the brain. In the ocellus of the dragonfly *Sympetrum*, some 675 receptor cells converge onto one large neuron, two medium-sized neurons, and a few small ones in the ocellar nerve.

The ocelli thus integrate light over a large visual field, both optically and neurally. They are very sensitive to low light intensities and to subtle changes in light, but they are not designed for high-resolution vision. They appear to function as "horizon detectors" for control of roll and pitch movements in flight and to register cyclical changes in light intensity that correlate with diurnal behavioral rhythms.

4.4.4 Compound eyes

The most sophisticated insect visual organ is the compound eye. Virtually all adult insects and nymphs have a pair of large, prominent compound eyes, which often cover nearly 360 degrees of visual space.

The compound eye is based on repetition of many individual units called **ommatidia** (Fig. 4.10). Each ommatidium resembles a simple stemma: it has a cuticular lens overlying a crystalline cone, which directs and focuses light onto eight (or maybe 6–10) elongate retinula cells (see transverse section in Fig. 4.10). The retinula cells are clustered around the longitudinal axis of each ommatidium and each contributes a rhabdomere to the rhabdom at the center of the ommatidium. Each cluster of retinula cells is surrounded by a ring of light-absorbing pigment cells, which optically isolates an ommatidium from its neighbors.

The corneal lens and crystalline cone of each ommatidium focus light onto the distal tip of the rhabdom from a region about 2–5 degrees across. The field of view of each ommatidium differs from that of its neighbors and together the array of all ommatidia provides the insect with a panoramic image of the world. Thus, the actual image formed by the compound eye is of a series of apposed points of light of different intensities, hence the name **apposition eye**.

The light sensitivity of apposition eyes is limited severely by the small diameter of facet lenses. Crepuscular and nocturnal insects, such as moths and some beetles, overcome this limitation with a modified optical design of compound eyes, called **optical superposition eyes**. In these, ommatidia are not isolated optically from each other by pigment cells. Instead, the retina is separated by a wide clear zone from the corneal facet lenses, and many lenses co-operate to focus light on an individual rhabdom (light from many lenses super-imposes on the retina). The light sensitivity of these eyes is thus greatly enhanced. In some optical superposition eyes screening pigment moves into the

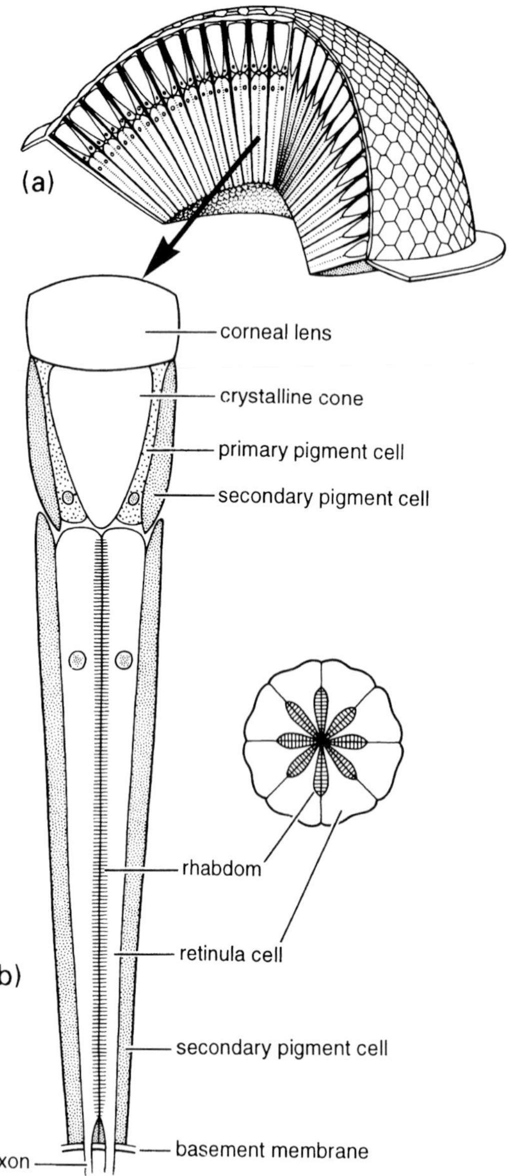

Fig. 4.10 Details of the compound eye: (a) a cutaway view showing the arrangement of the ommatidia and the facets; (b) a single ommatidium with an enlargement of a transverse section. (After CSIRO 1970; Rossel 1989.)

clear zone during light adaptation and by this means the ommatidia become isolated optically as in the apposition eye. At low light levels, the screening pigment moves again towards the outer surface of the eye to open up the clear zone for optical superposition to occur.

Because the light arriving at a rhabdom has passed through many facet lenses, blurring is a problem in optical superposition eyes and resolution generally is not as good as in apposition eyes. However, high light sensitivity is much more important than good resolving power in crepuscular and nocturnal insects whose main concern is to see anything at all. In the eyes of some insects, photon-capture is increased even further by a mirror-like tapetum of small tracheae at the base of the retinula cells; this reflects light that has passed unabsorbed through a rhabdom, allowing it a second pass. Light reflecting from the tapetum produces the bright eye shine seen when an insect with an optical superposition eye is illuminated in the flashlight or car headlight beam at night.

In comparison with a vertebrate eye, the resolving power of insect compound eyes is rather unimpressive. However, for the purpose of flight control, navigation, prey capture, predator avoidance, and mate-finding they obviously do a splendid job. Bees can memorize quite sophisticated shapes and patterns, and flies and odonates hunt down prey insects or mates in extremely fast, aerobatic flight. Insects in general are exquisitely sensitive to image motion, which provides them with useful cues for avoiding obstacles and landing, and for distance judgment. Insects, however, cannot easily use binocular vision for the perception of distance because their eyes are so close together and their resolution is quite poor. A notable exception is the praying mantid, which is the only insect known to make use of binocular disparity to localize prey.

Within one ommatidium, most studied insects possess several classes of retinula cells that differ in their spectral sensitivities; this feature means that each responds best to light of a different wavelength. Variations in the molecular structure of visual pigments are responsible for these differences in spectral sensitivity and are a prerequisite for the color vision of flower visitors such as bees and butterflies. Some insects are pentachromats, with five classes of receptors of differing spectral sensitivities, compared with human di- or trichromats. Most insects can perceive ultraviolet light (which is invisible to us) allowing them to see distinctive alluring flower patterns visible only in the ultraviolet.

Light emanating from the sky and reflected light from water surfaces or shiny leaves is polarized, i.e. it has greater vibration in some planes than in others. Many insects can detect the plane of polarization of light and utilize this in navigation, as a compass or as an indicator of water surfaces. The pattern of polarized skylight changes its position as the sun moves across the sky, so that insects can use small patches of clear sky to infer the position of the sun, even when it is not visible. In like manner, an African dung beetle has been shown to orientate using polarized moonlight in the absence of direct sighting of the moon, perhaps representing a more general ability amongst nocturnal insects. The microvillar organization of the insect rhabdomere makes insect photoreceptors inherently sensitive to the plane of polarization of light, unless precautions are taken to scramble the alignment of microvilli. Insects with well-developed navigational abilities often have a specialized region of retina in the dorsal visual field, the dorsal rim, in which retinula cells are highly sensitive to the plane of polarization of light. Ocelli and stemmata also may be involved in the detection of polarized light.

4.4.5 Light production

The most spectacular visual displays of insects involve light production, or **bioluminescence**. Some insects co-opt symbiotic luminescent bacteria or fungi, but self-luminescence is found in a few Collembola, one hemipteran (the fulgorid lantern bug), a few dipteran fungus gnats, and a diverse group amongst several families of coleopterans. The beetles are members of the Phengodidae, Drilidae, some lesser known families, and notably the Lampyridae, and are commonly given colloquial names including fireflies, glow worms, and lightning bugs. Any or all stages and sexes in the life history may glow, using one to many luminescent organs, which may be located nearly anywhere on the body. Light emitted may be white, yellow, red, or green.

The light-emitting mechanism studied in the lampyrid firefly *Photinus pyralis* may be typical of luminescent Coleoptera. The enzyme **luciferase** oxidizes a substrate, **luciferin**, in the presence of an energy source of adenosine triphosphate (ATP) and oxygen, to produce oxyluciferin, carbon dioxide, and light. Variation in ATP release controls the rate of flashing, and differences in pH may allow variation in the frequency (color) of light emitted.

The principal role of light emission was argued to be in courtship signaling. This involves species-specific variation in the duration, number, and rate of flashes in a pattern, and the frequency of repetition of the pattern (Fig. 4.11). Generally, a mobile male advertises his presence by instigating the signaling with one or more flashes and a sedentary female indicates her location with a flash in response. As with all communication systems, there is scope for abuse, for example that involving luring of prey by a carnivorous female lampyrid of *Photurus* (section 13.1.2). Recent phylogenetic studies have suggested a rather different interpretation of beetle bioluminescence, with it originating only in larvae of a broadly defined lampyrid clade, where it serves as a warning of distastefulness (section 14.4). From this origin in larvae, luminescence appears to have been retained into the adults, serving dual warning and sexual functions. The phylogeny suggests that luminescence then was lost in lampyrid relatives and regained subsequently in the Phengodidae, in which it is present in larvae and adults and fulfills a warning function. In this family it is possible that light is used also in illuminating a landing or courtship site, and perhaps red light serves for nocturnal prey detection.

Bioluminescence is involved in both luring prey and mate-finding in Australian and New Zealand cave-dwelling *Arachnocampa* fungus gnats (Diptera: Mycetophilidae). Their luminescent displays in the dark zone of caves have become tourist attractions in some places. All developmental stages of these flies use a reflector to concentrate light that they produce from modified Malpighian tubules. In the dark zone of a cave, the larval light lures prey, particularly small flies, onto a sticky thread suspended by the larva from the cave ceiling. The flying adult male locates the luminescent female while she is still in the pharate state and waits for the opportunity to mate upon her emergence.

4.5 INSECT BEHAVIOR

Many of the insect behaviors mentioned in this chapter appear very complex, but behaviorists attempt to reduce them to simpler components. Thus, individual **reflexes** (simple responses to simple stimuli) can be

Fig. 4.11 The flash patterns of males of a number of *Photinus* firefly species (Coleoptera: Lampyridae), each of which generates a distinctive pattern of signals in order to elicit a response from their conspecific females. (After Lloyd 1966.)

identified, such as the flight response when the legs lose contact with the ground, and cessation of flight when contact is regained. Some extremely rapid reflex actions, such as the feeding lunge of odonate nymphs, or some "escape reactions" of many insects, depend upon a reflex involving **giant axons** that conduct impulses rapidly from sense organs to the muscles. The integration of multiple reflexes associated with movement of the insect may be divisible into:

- **kinesis** (plural: **kineses**), in which unorientated action varies according to stimulus intensity;
- **taxis** (plural: **taxes**), in which movement is directly towards or away from the stimulus.

Kineses include **akinesis**, unstimulated lack of movement, **orthokinesis**, in which speed depends upon stimulus intensity, and **klinokinesis**, which is a "random walk" with course changes (turns) being made when unfavorable stimuli are perceived and with the frequency of turns depending on the intensity of the stimulus. Increased exposure to unfavorable stimuli leads to increased tolerance (**acclimation**), so that random walking and acclimation will lead the insect to a favorable environment. The male response to the plume of sex attractant (Fig. 4.7) is an example of klinokinesis to a chemical stimulus. Ortho- and klinokineses are effective responses to diffuse stimuli, such as temperature or humidity, but different, more efficient responses are seen when an insect is confronted by less diffuse, gradient or point-source stimuli.

Kineses and taxes can be defined with respect to the type of stimulus eliciting a response. Appropriate prefixes include: **anemo-** for air currents, **astro-** for solar, lunar, or astral (including polarized light), **chemo-** for taste and odor, **geo-** for gravity, **hygro-** for moisture,

phono- for sound, **photo-** for light, **rheo-** for water current, and **thermo-** for temperature. Orientation and movement may be positive or negative with respect to the stimulus source so that, for example, resistance to gravity is termed negative geotaxis, attraction to light is positive phototaxis, repulsion from moisture is negative hygrotaxis.

In **klinotaxic** behavior, an insect moves relative to a gradient (or cline) of stimulus intensity, such as a light source or a sound emission. The strength of the stimulus is compared on each side of the body by moving the receptors from side to side (as in head waving of ants when they follow an odor trail), or by detection of differences in stimulus intensity between the two sides of the body using paired receptors. The tympanal organs detect the direction of the sound source by differences in intensity between the two organs. Orientation with respect to a constant angle of light is termed **menotaxis** and includes the "light-compass" referred to in Box 4.4. Visual fixation of an object, such as prey, is termed **telotaxis**.

Often the relationship between stimulus and behavioral response is complex, as a **threshold** intensity may be required before an action ensues. A particular stimulatory intensity is termed a **releaser** for a particular behavior. Furthermore, complex behavior elicited by a single stimulus may comprise several sequential steps, each of which may have a higher threshold, requiring an increased stimulus. As described in section 4.3.2, a male moth responds to a low-level sex pheromone stimulus by raising the antennae; at higher levels he orientates towards the source; and at an even higher threshold, flight is initiated. Increasing concentration encourages continued flight and a second, higher threshold may be required before courtship ensues. In other behaviors, several different stimuli are involved, such as for courtship through to mating. This sequence can be seen as a long **chain reaction** of stimulus, action, new stimulus, next action, and so on, with each successive behavioral stage depending upon the occurrence of an appropriate new stimulus. An inappropriate stimulus during a chain reaction (such as the presentation of food while courting) is not likely to elicit the usual response.

Most insect behaviors are considered to be **innate**, i.e. they are programmed genetically to arise stereotypically upon first exposure to the appropriate stimulus. However, many behaviors are environmentally and physiologically modified: for example, virgins and mated females respond in very different ways to identical stimuli, and immature insects often respond to different stimuli compared with conspecific adults. Furthermore, experimental evidence shows that learning can modify innate behavior. By experimental teaching (using training and reward), bees and ants can learn to run a maze and butterflies can be induced to alter their favorite flower color. However, study of natural behavior (ethology) is more relevant to understanding the role played in the evolutionary success of insects' behavioral plasticity, including the ability to modify behavior through learning. In pioneering ethological studies, Niko Tinbergen showed that a digger wasp (Crabronidae: *Philanthus triangulum*) can learn the location of its chosen nest site by making a short flight to memorize elements of the local terrain. Adjustment of prominent landscape features around the nest misleads the homing wasp. However, as wasps identify landmark relationships rather than individual features, the confusion may be only temporary. Closely related *Bembix* digger wasps (Sphecidae) learn nest locations through more distant and subtle markers, including the appearance of the horizon, and are not tricked by investigator ethologists moving local small-scale landmarks.

FURTHER READING

Blum, M.S. (1996) Semiochemical parsimony in the Arthropoda. *Annual Review of Entomology* **41**, 353–74.

Chapman, R.F. (1998) *The Insects. Structure and Function*, 4th edn. Cambridge University Press, Cambridge.

Chapman, R.F. (2003) Contact chemoreception in feeding by phytophagous insects. *Annual Review of Entomology* **48**, 455–84.

Dicke, M. (1994) Local and systemic production of volatile herbivore-induced terpenoids: their role in plant–carnivore mutualism. *Journal of Plant Physiology* **143**, 465–72.

Fuller, J.H. & Yack, J.E. (1993) The evolutionary biology of insect hearing. *Trends in Ecology and Evolution* **8**, 248–52.

Heinrich, B. (1993) *The Hot-blooded Insects: Strategies and Mechanisms of Thermoregulation*. Harvard University Press, Cambridge, MA.

Heinrich, B. (1996) *The Thermal Warriors*. Harvard University Press, Cambridge, MA.

Howse, P., Stevens, I. & Jones, O. (1998) *Insect Pheromones and their Use in Pest Management*. Chapman & Hall, London.

Hoy, R.R. & Robert, D. (1996) Tympanal hearing in insects. *Annual Review of Entomology* **41**, 433–50.

Jacobs, G.A. (1995) Detection and analysis of air currents by crickets. *BioScience* **45**, 776–85.

Kerkut, G.A. & Gilbert, L.I. (eds.) (1985) *Comprehensive Insect Physiology, Biochemistry, and Pharmacology*, Vol. 6: *Nervous System: Sensory*; Vol. 9: *Behavior*. Pergamon Press, Oxford.

Landolt, P.J. (1997) Sex attractant and aggregation pheromones of male phytophagous insects. *American Entomologist* **43**, 12–22.

Resh, V.H. & Cardé, R.T. (eds.) (2003) *Encyclopedia of Insects*. Academic Press, Amsterdam. [Particularly see articles on chemoreception; circadian rhythms; eyes and vision; hearing; magnetic sense; mechanoreception; orientation; pheromones; thermoregulation; vibrational communication.]

Wood, D.L. (1982) The role of pheromones, kairomones and allomones in host selection and colonization by bark beetles. *Annual Review of Entomology* **27**, 411–46.

Chapter 5

REPRODUCTION

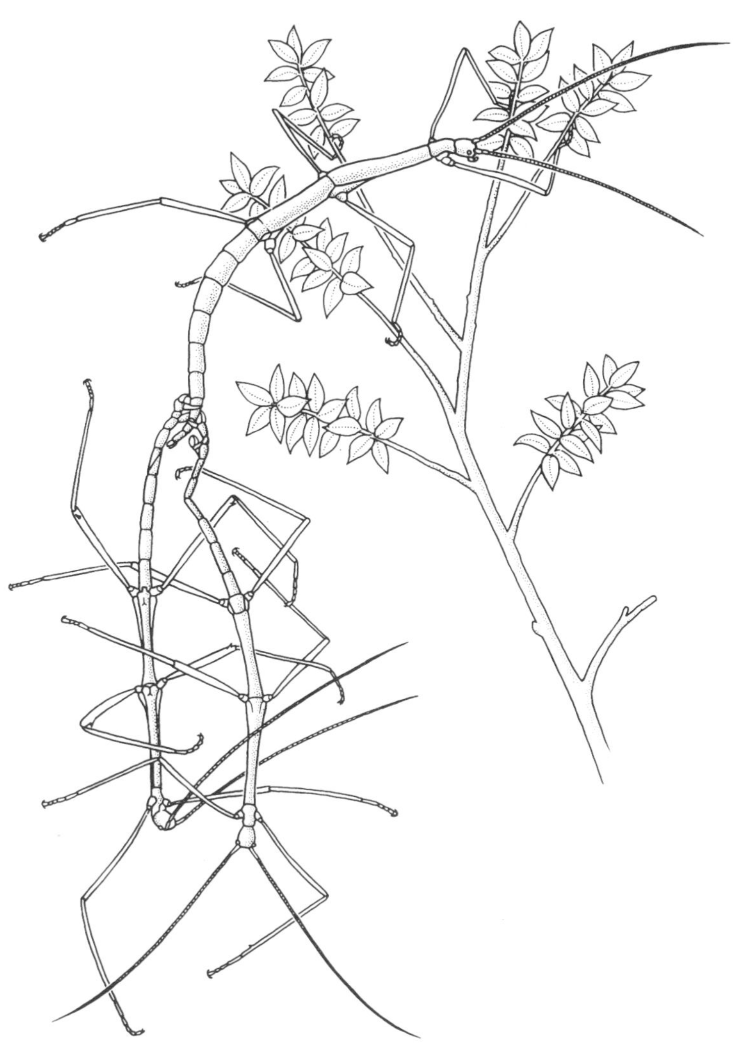

Two male stick-insects fighting over a female. (After Sivinski 1978.)

Most insects are sexual and thus mature males and females must be present at the same time and place for reproduction to take place. As insects are generally short-lived, their life history, behavior, and reproductive condition must be synchronized. This requires finely tuned and complex physiological responses to the external environment. Furthermore, reproduction also depends on monitoring of internal physiological stimuli, and the neuroendocrine system plays a key regulatory role. Mating and egg production in many flies is known to be controlled by a series of hormonal and behavioral changes, yet there is much still to learn about the control and regulation of insect reproduction, particularly if compared with our knowledge of vertebrate reproduction.

These complex regulatory systems are highly successful. For example, look at the rapidity with which pest insect outbreaks occur. A combination of short generation time, high fecundity, and population synchronization to environmental cues allows many insect populations to react extremely rapidly under appropriate environmental conditions, such as a crop monoculture, or release from a controlling predator. In these situations, temporary or obligatory loss of males (**parthenogenesis**) has proved to be another effective means by which some insects rapidly exploit temporarily (or seasonally) abundant resources.

This chapter examines the different mechanisms associated with courtship and mating, avoidance of interspecies mating, ensuring paternity, and determination of sex of offspring. Then we examine the elimination of sex and show some extreme cases in which the adult stage has been dispensed with altogether. These observations relate to theories concerning sexual selection, including those linked to why insects have such remarkable diversity of genitalic structures. The concluding summary of the physiological control of reproduction emphasizes the extreme complexity and sophistication of mating and oviposition in insects.

5.1 BRINGING THE SEXES TOGETHER

Insects often are at their most conspicuous when synchronizing the time and place for mating. The flashing lights of fireflies, the singing of crickets, and cacophony of cicadas are spectacular examples. However, there is a wealth of less ostentatious behavior, of equal significance in bringing the sexes together and signaling readiness to mate to other members of the species. All signals are species-specific, serving to attract members of the opposite sex of the same species, but abuse of these communication systems can take place, as when females of one predatory species of firefly lure males of another species to their death by emulating the flashing signal of that species.

Swarming is a characteristic and perhaps fundamental behavior of insects, as it occurs amongst some insects from ancient lineages, such as mayflies and odonates, and also in many "higher" insects, such as flies and butterflies. Swarming sites are identified by visual markers (Fig. 5.1) and are usually species-specific, although mixed-species swarms have been reported, especially in the tropics or subtropics. Swarms are predominantly of the male sex only, though female-only swarms do occur. Swarms are most evident when many individuals are involved, such as when midge swarms are so dense that they have been mistaken for smoke from burning buildings, but small swarms may be more significant in evolution. A single male insect holding station over a spot is a swarm of one – he awaits the arrival of a receptive female that has responded identically to visual cues that identify the site. The precision of swarm sites allows more effective mate-finding than searching, particularly when individuals are rare or dispersed and at low density. The formation of a swarm allows insects of differing genotypes to meet and outbreed. This is of particular importance if larval development sites are patchy and locally dispersed; inbreeding would occur if adults did not disperse.

In addition to aerial aggregations, some male insects form substrate-based aggregations where they may defend a territory against conspecific males and/or court arriving females. Species in which males hold territories that contain no resources (e.g. oviposition substrates) important to the females and exhibit male–male aggression plus courtship of females are said to have a **lek** mating system. Lek behavior is common in fruit flies of the families Drosophilidae and Tephritidae. Polyphagous fruit flies should be more likely to have a lek mating system than monophagous species because, in the latter, males can expect to encounter females at the particular fruit that serves as the oviposition site.

Insects that form aerial or substrate-based mating aggregations often do so on hilltops, although some swarming insects aggregate above a water surface or use landmarks such as bushes or cattle. Most species probably use visual cues to locate an aggregation site, except that uphill wind currents may guide insects to hilltops.

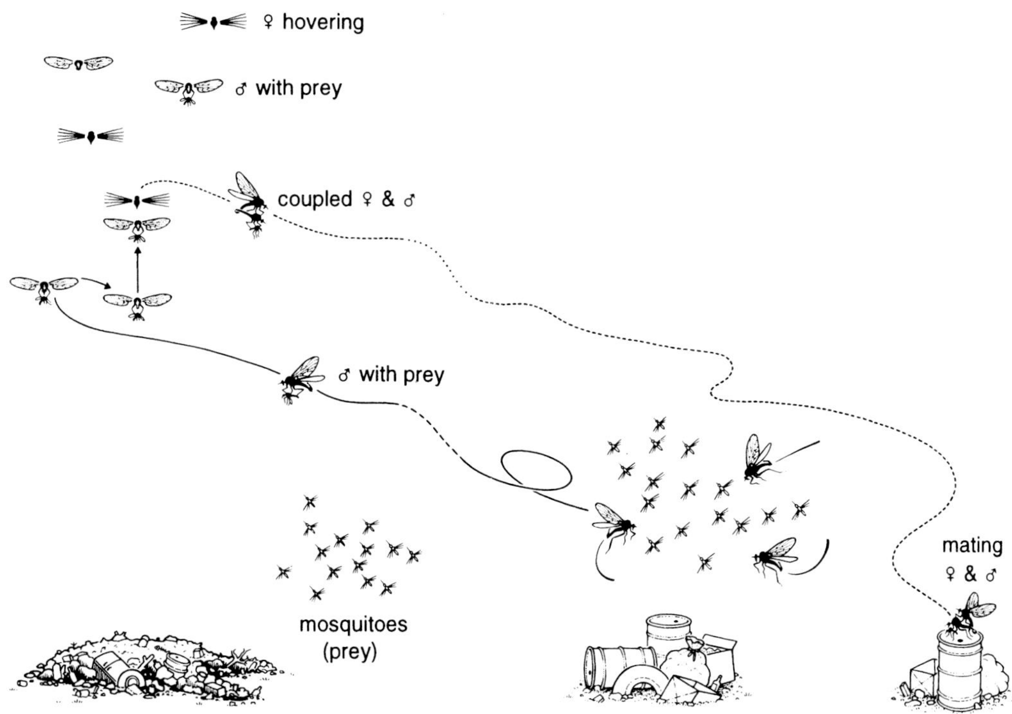

Fig. 5.1 Males of the Arctic fly *Rhamphomyia nigrita* (Diptera: Empididae) hunt for prey in swarms of *Aedes* mosquitoes (lower mid-right of drawing) and carry the prey to a specific visual marker of the swarm site (left of drawing). Swarms of both the empidids and the mosquitoes form near conspicuous landmarks, including refuse heaps or oil drums that are common in parts of the tundra. Within the mating swarm (upper left), a male empidid rises towards a female hovering above, they pair, and the prey is transferred to the female; the mating pair alights (lower far right) and the female feeds as they copulate. Females appear to obtain food only via males and, as individual prey items are small, must mate repeatedly to obtain sufficient nutrients to develop a batch of eggs. (After Downes 1970).

In other insects, the sexes may meet via attraction to a common resource and the meeting site might not be visually located. For species whose larval development medium is discrete, such as rotting fruit, animal dung, or a specific host plant or vertebrate host, where better for the sexes to meet and court? The olfactory receptors by which the female dung fly finds a fresh pile of dung (the larval development site) can be employed by both sexes to facilitate meeting.

Another odoriferous communication involves one or both sexes producing and emitting a **pheromone**, which is a chemical or mixture of chemicals perceptible to another member of the species (section 4.3.2). Substances emitted with the intention of altering the sexual behavior of the recipient are termed **sex pheromones**. Generally, these are produced by the female and announce her presence and sexual availability to conspecific males. Recipient males that detect the odor plume become aroused and orientate from downwind towards the source. More and more insects investigated are found to have species-specific sex pheromones, the diversity and specificity of which are important in maintaining the reproductive isolation of a species.

When the sexes are in proximity, mating in some species takes place with little further ado. For example, when a conspecific female arrives at a swarm of male flies, a nearby male, recognizing her by the particular sound of her wingbeat frequency, immediately copulates with her. However, more elaborate and specialized close-range behaviors, termed courtship, are commonplace.

Box 5.1 Courtship and mating in Mecoptera

Sexual behavior has been well studied in hangingflies (Bittacidae) of the North American *Hylobittacus* (*Bittacus*) *apicalis* and *Bittacus* species and the Australian *Harpobittacus* species, and in the Mexican *Panorpa* scorpionflies (Panorpidae). Adult males hunt for arthropod prey, such as caterpillars, bugs, flies, and katydids. These same food items may be presented to a female as a nuptial offering to be consumed during copulation. Females are attracted by a sex pheromone emitted from one or more eversible vesicles or pouches near the end of the male's abdomen as he hangs in the foliage using prehensile fore tarsi.

Courting and mating in Mecoptera are exemplified by the sexual interactions in *Harpobittacus australis* (Bittacidae). The female closely approaches the "calling" male; he then ends pheromone emission by retracting the abdominal vesicles. Usually the female probes the prey briefly, presumably testing its quality, while the male touches or rubs her abdomen and seeks her genitalia with his own. If the female rejects the nuptial gift, she refuses to copulate. However, if the prey is suitable, the genitalia of the pair couple and the male temporarily withdraws the prey with his hind legs. The female lowers herself until she hangs head downwards, suspended by her terminalia. The male then surrenders the nuptial offering (in the illustration, a caterpillar) to the female, which feeds as copulation proceeds. At this stage the male frequently supports the female by holding either her legs or the prey that she is feeding on. The derivation of the common name "hangingflies" is obvious!

Detailed field observations and manipulative experiments have demonstrated female choice of male partners in species of Bittacidae. Both sexes mate several times per day with different partners. Females discriminate against males that provide small or unsuitable prey either by rejection or by copulating only for a short time, which is insufficient to pass the complete ejaculate. Given an acceptable nuptial gift, the duration of copulation correlates with the size of the offering. Each copulation in field populations of *Ha. australis* lasts from 1 to a maximum of about 17 minutes for prey from 3 to 14 mm long. In the larger *Hy. apicalis*, copulations involving prey of the size of houseflies or larger (19–50 mm^2) last from 20 to 29 minutes, resulting in maximal sperm transfer, increased oviposition, and the induction of a refractory period (female non-receptivity to other males) of several hours. Copulations that last less than 20 minutes reduce or eliminate male fertilization success. (Data after Thornhill 1976; Alcock 1979.)

5.2 COURTSHIP

Although the long-range attraction mechanisms discussed above reduce the number of species present at a prospective mating site, generally there remains an excess of potential partners. Further discrimination among species and conspecific individuals usually takes place. Courtship is the close-range, intersexual behavior that induces sexual receptivity prior to (and often during) mating and acts as a mechanism for species recognition. During courtship, one or both sexes seek to facilitate insemination and fertilization by influencing the other's behavior.

Courtship may include visual displays, predominantly by males, including movements of adorned parts of the body, such as antennae, eyestalks, and "picture" wings, and ritualized movements ("dancing"). Tactile stimulation such as rubbing and stroking often occurs later in courtship, often immediately prior to mating, and may continue during copulation. Antennae, palps, head horns, external genitalia, and legs are used in tactile stimulation.

Insects such as crickets, which use long-range calling, may have different calls for use in close-range courtship. Others, such as fruit flies (*Drosophila*), have no long-distance call and sing (by wing vibration) only in close-up courtship. In some predatory insects, including empidid flies and mecopterans, the male courts a prospective mate by offering a prey item as a nuptial gift (Fig. 5.1; Box 5.1).

If the sequence of display proceeds correctly, courtship grades into mating. Sometimes the sequence need not be completed before copulation commences. On other occasions courtship must be prolonged and repeated. It may be unsuccessful if one sex fails to respond or makes inappropriate responses. Generally, members of different species differ in some elements of their courtships and interspecies matings do not occur. The great specificity and complexity of insect courtship behaviors can be interpreted in terms of mate location, synchronization, and species recognition, and viewed as having evolved as a premating isolating mechanism. Important as this view is, there is equally compelling evidence that courtship is an extension of a wider phenomenon of competitive communication and involves sexual selection.

5.3 SEXUAL SELECTION

Many insects are sexually dimorphic, usually with the male adorned with secondary sexual characteristics, some of which have been noted above in relation to courtship display. In many insect mating systems courtship can be viewed as intraspecific competition for mates, with certain male behaviors inducing female response in ways that can increase the mating success of particular males. Because females differ in their responsiveness to male stimuli, females can be said to choose between mates, and courtship thus is competitive. Female choice might involve no more than selection of the winners of male–male interactions, or may be as subtle as discrimination between the sperm of different males (section 5.7). All elements of communication associated with gaining fertilization of the female, from long-distance sexual calling through to insemination, are seen as competitive courtship between males. By this reasoning, members of a species avoid hybrid matings because of a specific-mate recognition system that evolved under the direction of female choice, rather than as a mechanism to promote species cohesion.

Understanding sexual dimorphism in insects such as staghorn beetles, song in orthopterans and cicadas, and wing color in butterflies and odonates helped Darwin to recognize the operation of sexual selection – the elaboration of features associated with sexual competition rather than directly with survival. Since Darwin's day, studies of sexual selection often have featured insects because of their short generation time, facility of manipulation in the laboratory, and relative ease of observation in the field. For example, dung beetles belonging to the large and diverse genus *Onthophagus* may display elaborate horns that vary in size between individuals and in position on the body between species. Large horns are restricted nearly exclusively to males, with only one species known in which the female has better developed protuberances than conspecific males. Studies show that females preferentially select males with larger horns as mates. Males size each other up and may fight, but there is no lek. Benefits to the female come from long-horned males' better defensive capabilities against intruders seeking to oust the resident from the resource-rich nest site, provisioned with dung, his mate, and their young (Fig. 9.5). However, the system is more complicated, at least in the North American *Onthophagus taurus*. In this dung beetle, male horn size is dimorphic, with insects greater than a certain threshold size having large horns, and those below a certain size having only minimal horns (Fig. 5.2). However, nimble small-horned

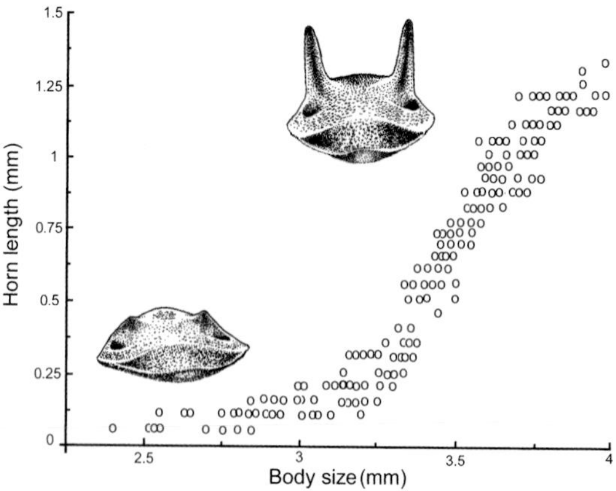

Fig. 5.2 Relationship between length of horn and body size (thorax width) of male scarabs of *Onthophagus taurus*. (After Moczek & Emlen 2000; with beetle heads drawn by S.L. Thrasher.)

males attain some mating success through sneakily circumventing the large-horned but clumsy male defending the tunnel entrance, either by evasion or by digging a side tunnel to access the female.

Darwin could not understand why the size and location of horns varied, but now elegant comparative studies have shown that elaboration of large horns bears a developmental cost. Organs located close to a large horn are diminished in size – evidently resources are reallocated during development so that either eyes, antennae, or wings apparently "pay for" being close to a male's large horn. Regular-sized adjacent organs are developed in females of the same species with smaller horns and male conspecifics with weakly developed horns. Exceptionally, the species with the female having long horns on the head and thorax commensurately has reduced adjacent organs, and a sex reversal in defensive roles is assumed to have taken place. The different locations of the horns appear to be explained by selective sacrifice of adjacent organs according to species behavior. Thus, nocturnal species that require good eyes have their horns located elsewhere than the head; those requiring flight to locate dispersed dung have horns on the head where they interfere with eye or antennal size, but do not compromise the wings. Presumably, the upper limit to horn elaboration either is the burden of ever-increasing deleterious effects on adjacent vital functions, or an upper limit on the volume of new cuticle that can develop sub-epidermally in the pharate pupa within the final-instar larva, under juvenile hormonal control.

Size alone may be important in female choice: in some stick-insects (also called walking sticks) larger males often monopolize females. Males fight over their females by boxing at each other with their legs while grasping the female's abdomen with their claspers (as shown for *Diapheromera veliei* in the vignette for this chapter). Ornaments used in male-to-male combat include the extraordinary "antlers" of *Phytalmia* (Tephritidae) (Fig. 5.3) and the eyestalks of a few other flies (such as Diopsidae), which are used in competition for access to the oviposition site visited by females. Furthermore, in studied species of diopsid (stalk-eyed flies), female mate choice is based on eyestalk length up to a dimension of eye separation that can surpass the body length. Cases such as these provide evidence for two apparently alternative but likely non-exclusive explanations for male adornments: sexy sons or good genes. If the female choice commences arbitrarily for any particular adornment, their selection alone will drive the increased frequency and development of the elaboration in male offspring in ensuing generations (the sexy sons) despite countervailing selection against conventional unfitness. Alternatively, females may choose mates that can demonstrate their fitness by carrying around apparently deleterious elaborations thereby indicating a superior genetic background (good genes). Darwin's interpretation of the enigma of female choice certainly is substantiated, not least by studies of insects.

5.4 COPULATION

The evolution of male external genitalia made it possible for insects to transfer sperm directly from male to female during copulation. All but the most primitive insects were freed from reliance on indirect methods, such as the male depositing a **spermatophore** (sperm packet) for the female to pick up from the substrate, as in Collembola, Diplura, and apterygote insects. In pterygote insects, copulation (sometimes referred to as mating) involves the physical apposition of male and female genitalia, usually followed by insemination – the transfer of sperm via the insertion of part of the male's **aedeagus** (edeagus), the penis, into the reproductive tract of the female. In males of many species the extrusion of the aedeagus during copulation is a two-stage process. The complete aedeagus is extended from

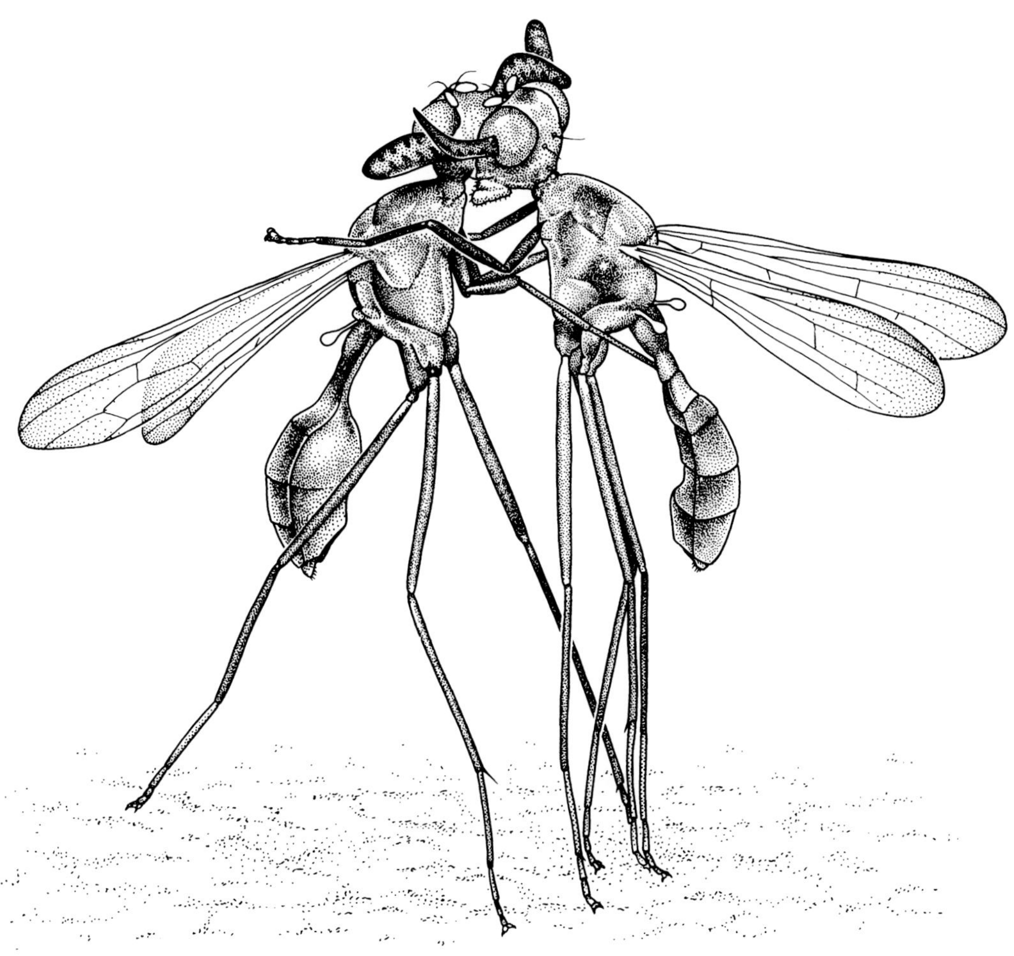

Fig. 5.3 Two males of *Phytalmia mouldsi* (Diptera: Tephritidae) fighting over access to the oviposition site at the larval substrate visited by females. These tropical rainforest flies thus have a resource-defense mating system. (After Dodson 1989, 1997.)

the abdomen, then the intromittent organ is everted or extended to produce an expanded, often elongate structure (variably called the endophallus, flagellum, or vesica) capable of depositing semen deep within the female's reproductive tract (Fig. 5.4). In many insects the male terminalia have specially modified claspers, which lock with specific parts of the female terminalia to maintain the connection of their genitalia during sperm transfer.

This mechanistic definition of copulation ignores the sensory stimulation that is a vital part of the copulatory act in insects, as it is in other animals. In over a third of all insect species surveyed, the male indulges in copulatory courtship – behavior that appears to stimulate the female during mating. The male may stroke, tap, or bite the body or legs of the female, wave antennae, produce sounds, or thrust or vibrate parts of his genitalia.

Sperm are received by the female insect in a copulatory pouch (genital chamber, vagina, or bursa copulatrix) or directly into a spermatheca or its duct (as in *Oncopeltus*; Fig. 5.4). A spermatophore is the means of sperm transfer in most orders of insects; only some Heteroptera, Coleoptera, Diptera, and Hymenoptera deposit unpackaged sperm. Sperm transfer requires lubrication, obtained from the seminal fluids, and, in insects that use a spermatophore, packaging of sperm.

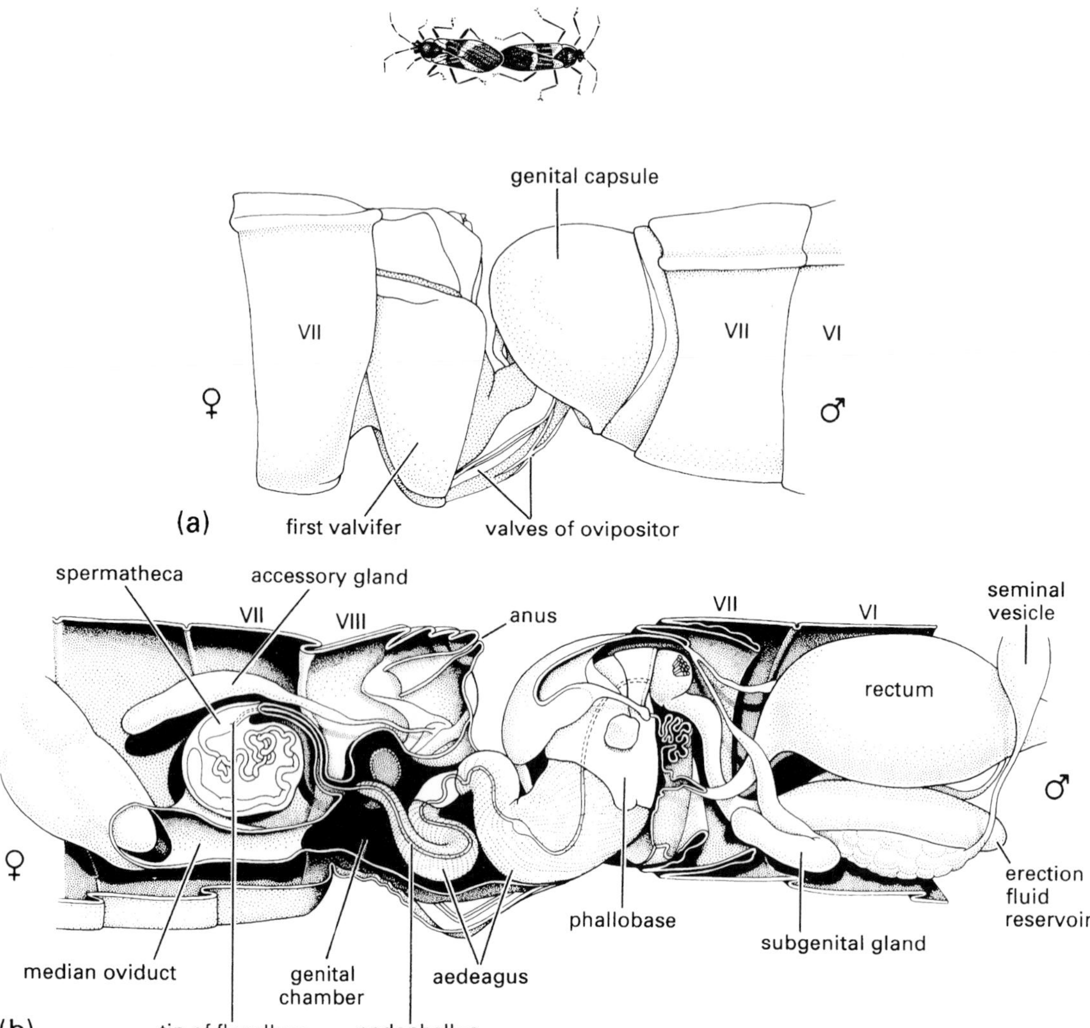

Fig. 5.4 Posterior ends of a pair of copulating milkweed bugs, *Oncopeltus fasciatus* (Hemiptera: Lygaeidae). Mating commences with the pair facing in the same direction, then the male rotates his eighth abdominal segment (90°) and genital capsule (180°), erects the aedeagus and gains entry to the female's genital chamber, before he swings around to face in the opposite direction. The bugs may copulate for several hours, during which they walk around with the female leading and the male walking backwards. (a) Lateral view of the terminal segments, showing the valves of the female's ovipositor in the male genital chamber; (b) longitudinal section showing internal structures of the reproductive system, with the tip of the male's aedeagus in the female's spermatheca. (After Bonhag & Wick 1953.)

Secretions of the male accessory glands serve both of these functions as well as sometimes facilitating the final maturation of sperm, supplying energy for sperm maintenance, regulating female physiology and, in a few species, providing nourishment to the female (Box 5.2). The male accessory secretions may elicit one or two major responses in the female – induction of oviposition (egg-laying) and/or repression of sexual receptivity – by entering the female hemolymph and acting on her nervous and/or endocrine system.

Box 5.2 Nuptial feeding and other "gifts"

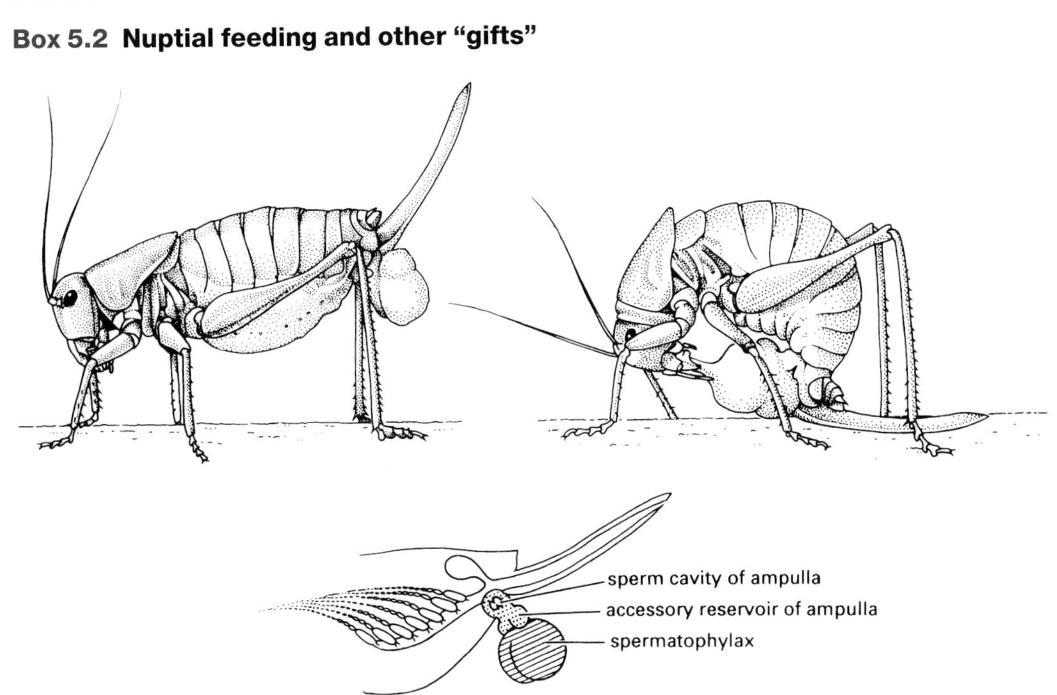

Feeding of the female by the male before, during, or after copulation has evolved independently in several different insect groups. From the female's perspective, feeding takes one of three forms:

1 receipt of nourishment from food collected, captured, or regurgitated by the male (Box 5.1); or

2 obtaining nourishment from a glandular product (including the spermatophore) of the male; or

3 by cannibalization of males during or after copulation.

From the male's perspective, nuptial feeding may represent parental investment (provided that the male can be sure of his paternity), as it may increase the number or survival of the male's offspring indirectly via nutritional benefits to the female. Alternatively, courtship feeding may increase the male's fertilization success by preventing the female from interfering with sperm transfer. These two hypotheses concerning the function of nuptial feeding are not necessarily mutually exclusive; their explanatory value appears to vary between insect groups and may depend, at least partly, on the nutritional status of the female at the time of mating. Studies of mating in Mecoptera, Orthoptera, and Mantodea exemplify the three nuptial feeding types seen in insects, and continuing research on these groups addresses the relative importance of the two main competing hypotheses that seek to explain the selective advantage of such feeding.

In some other insect orders, such as the Lepidoptera and Coleoptera, the female sometimes acquires metabolically essential substances or defensive chemicals from the male during copulation, but oral uptake by the female usually does not occur. The chemicals are transferred by the male with his ejaculate. Such nuptial gifts may function solely as a form of parental investment (as in puddling; see below) but may also be a form of mating effort (Box 14.3).

Puddling and sodium gifts in Lepidoptera

Male butterflies and moths frequently drink at pools of liquid, a behavior known as **puddling**. Anyone who has visited a tropical rainforest will have seen drinking clusters of perhaps hundreds of newly eclosed male butterflies, attracted particularly to urine, feces, and human sweat (see Plate 2.6, facing p. 14). It has long been suggested that puddling – in which copious quantities of liquid are ingested orally and expelled anally – results in uptake of minerals, such as sodium, which are deficient in the larval (caterpillar) folivore diet. The sex bias in puddling occurs because the male uses the sodium obtained by puddling as a nuptial gift for his mate. In the moth *Gluphisia septentrionis* (Notodontidae) the sodium gift amounts to more than half of the puddler's total body sodium and appears to be transferred to the female via his spermatophore (Smedley &

Eisner 1996). The female then apportions much of this sodium to her eggs, which contain several times more sodium than eggs sired by males that have been experimentally prevented from puddling. Such paternal investment in the offspring is of obvious advantage to them in supplying an ion important to body function.

In some other lepidopteran species, such "salted" gifts may function to increase the male's reproductive fitness not only by improving the quality of his offspring but also by increasing the total number of eggs that he can fertilize, assuming that he remates. In the skipper butterfly, *Thymelicus lineola* (Hesperiidae), females usually mate only once and male-donated sodium appears essential for both their fecundity and longevity (Pivnick & McNeil 1987). These skipper males mate many times and can produce spermatophores without access to sodium from puddling but, after their first mating, they father fewer viable eggs compared with remating males that have been allowed to puddle. This raises the question of whether females, which should be selective in the choice of their sole partner, can discriminate between males based on their sodium load. If they can, then sexual selection via female choice also may have selected for male puddling.

In other studies, copulating male lepidopterans have been shown to donate a diversity of nutrients, including zinc, phosphorus, lipids, and amino acids, to their partners. Thus, paternal contribution of chemicals to offspring may be widespread within the Lepidoptera.

Mating in katydids (Orthoptera: Tettigoniidae)

During copulation the males of many species of katydids transfer elaborate spermatophores, which are attached externally to the female's genitalia (see Plate 3.1). Each spermatophore consists of a large, proteinaceous, sperm-free portion, the **spermatophylax**, which is eaten by the female after mating, and a **sperm ampulla**, eaten after the spermatophylax has been consumed and the sperm have been transferred to the female. The illustration (p. 121) shows a recently mated female Mormon cricket, *Anabrus simplex*, with a spermatophore attached to her gonopore; in the illustration on the upper right, the female is consuming the spermatophylax of the spermatophore (after Gwynne 1981). The schematic illustration underneath depicts the posterior of a female Mormon cricket showing the two parts of the spermatophore: the spermatophylax (cross-hatched) and the sperm ampulla (stippled) (after Gwynne 1990). During consumption of the spermatophylax, sperm are transferred from the ampulla along with substances that "turn off" female receptivity to further males. Insemination also stimulates oviposition by the female, thereby increasing the probability that the male supplying the spermatophore will fertilize the eggs.

There are two main hypotheses for the adaptive significance of this form of nuptial feeding. The spermatophylax may serve as a sperm-protection device by preventing the ampulla from being removed until after the complete ejaculate has been transferred. Alternatively, the spermatophylax may be a form of parental investment in which nutrients from the male increase the number or size of the eggs sired by that male. Of course, the spermatophylax may serve both of these purposes, and there is evidence from different species to support each hypothesis. Experimental alteration of the size of the spermatophylax has demonstrated that females take longer to eat larger ones, but in some katydid species the spermatophylax is larger than is needed to allow complete insemination and, in this case, the nutritional bonus to the female benefits the male's offspring. The function of the spermatophylax apparently varies between genera, although phylogenetic analysis suggests that the ancestral condition within the Tettigoniidae was to possess a small spermatophylax that protected the ejaculate.

Cannibalistic mating in mantids (Mantodea)

The sex life of mantids is the subject of some controversy, partly as a consequence of behavioral observations made under unnatural conditions in the laboratory. For example, there are many reports of the male being eaten by the generally larger female before, during, or after mating. Males decapitated by females are even known to copulate more vigorously because of the loss of the suboesophageal ganglion that normally inhibits copulatory movements. Sexual cannibalism has been attributed to food deprivation in confinement but female mantids of at least some species may indeed eat their partners in the wild.

Courtship displays may be complex or absent, depending on species, but generally the female attracts the male via sex pheromones and visual cues. Typically, the male approaches the female cautiously, arresting movement if she turns her head towards him, and then he leaps onto her back from beyond her strike reach. Once mounted, he crouches to elude his partner's grasp. Copulation usually lasts at least half an hour and may continue for several hours, during which sperm are transferred from the male to the female in a spermatophore. After mating, the male retreats hastily. If the male were in no danger of becoming the female's meal, his distinctive behavior in the presence of the female would be inexplicable. Furthermore, suggestions of gains in reproductive fitness of the male via indirect nutritional benefits to his offspring are negated by the obvious unwillingness of the male to participate in the ultimate nuptial sacrifice – his own life!

Whereas there is no evidence yet for an increase in male reproductive success as a result of sexual cannibalism, females that obtain an extra meal by eating their mate may gain a selective advantage, especially if food is limiting. This hypothesis is supported by experiments with captive females of the Asian mantid *Hierodula membranacea* that were fed different quantities of food. The frequency of sexual cannibalism was higher for females of poorer nutritional condition and, among the females on the poorest diet, those that ate their mates produced significantly larger oothecae (egg packages) and hence more offspring. The cannibalized males would be making a parental investment only if their sperm fertilize the eggs that they have nourished. The crucial data on sperm competition in mantids are not available and so currently the advantages of this form of nuptial feeding are attributed entirely to the female.

5.5 DIVERSITY IN GENITALIC MORPHOLOGY

The components of the terminalia of insects are very diverse in structure and frequently exhibit species-specific morphology (Fig. 5.5), even in otherwise similar species. Variations in external features of the male genitalia often allow differentiation of species, whereas external structures in the female usually are simpler and less varied. Conversely, the internal genitalia of female insects often show greater diagnostic variability than the internal structures of the males. However, recent development of techniques to evert the endophallus of the male aedeagus allows increasing demonstration of the species-specific shapes of these male internal structures. In general, external genitalia of both sexes are much more sclerotized than the internal genitalia, although parts of the reproductive tract are lined with cuticle. Increasingly, characteristics of insect internal genitalia and even soft tissues are recognized as allowing species delineation and providing evidence of phylogenetic relationships.

Observations that genitalia frequently are complex and species-specific in form, sometimes appearing to correspond tightly between the sexes, led to formulation of the "lock-and-key" hypothesis as an explanation of this phenomenon. Species-specific male genitalia (the "keys") were believed to fit only the conspecific female genitalia (the "locks"), thus preventing interspecific mating or fertilization. For example, in some katydids interspecific copulations are unsuccessful in transmitting spermatophores because the specific structure of the male claspers (modified cerci) fails to fit the subgenital plate of the "wrong" female. The lock-and-key hypothesis was postulated first in 1844 and has been the subject of controversy ever since. In many (but not all) insects, mechanical exclusion of "incorrect" male genitalia by the female is seen as unlikely for several reasons:

1 morphological correlation between conspecific male and female parts may be poor;

2 interspecific, intergeneric, and even interfamilial hybrids can be induced;

3 amputation experiments have demonstrated that male insects do not need all parts of the genitalia to inseminate conspecific females successfully.

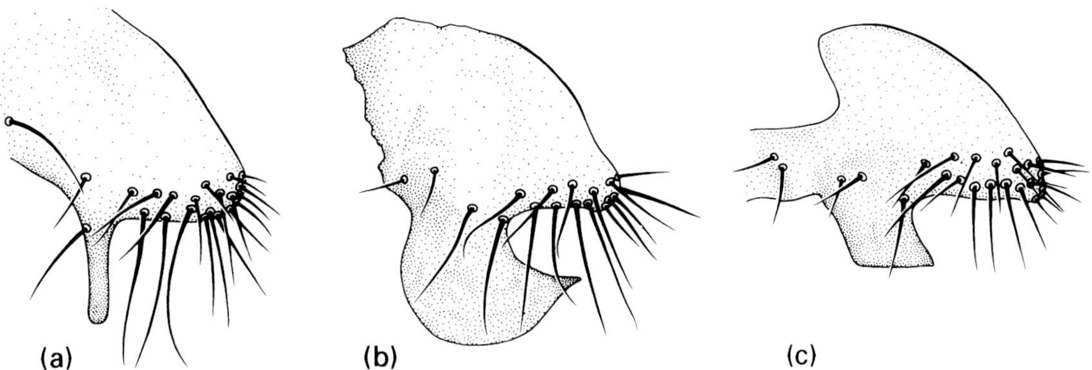

Fig. 5.5 Species-specificity in part of the male genitalia of three sibling species of *Drosophila* (Diptera: Drosophilidae). The epandrial processes of tergite 9 in: (a) *D. mauritiana*; (b) *D. simulans*; (c) *D. melanogaster*. (After Coyne 1983.)

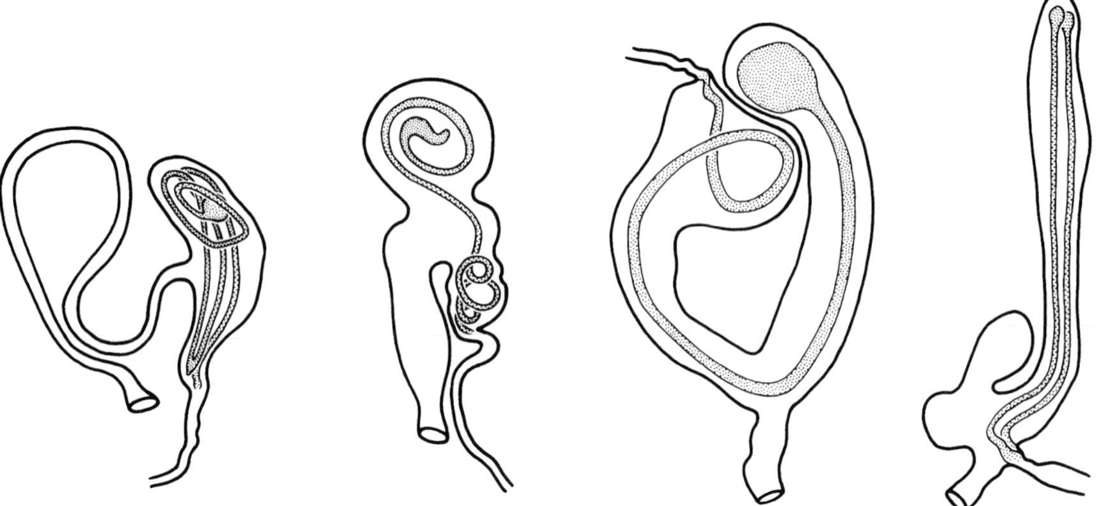

Fig. 5.6 Spermatophores lying within the bursae of the female reproductive tracts of moth species from four different genera (Lepidoptera: Noctuidae). The sperm leave via the narrow end of each spermatophore, which has been deposited so that its opening lies opposite the "seminal duct" leading to the spermatheca (not drawn). The bursa on the far right contains two spermatophores, indicating that the female has remated. (After Williams 1941; Eberhard 1985.)

Some support for the lock-and-key hypothesis comes from studies of certain noctuid moths in which structural correspondence in the internal genitalia of the male and female is thought to indicate their function as a postcopulatory but prezygotic isolating mechanism. Laboratory experiments involving interspecific matings support a lock-and-key function for the internal structures of other noctuid moths. Interspecific copulation can occur, although without a precise fit of the male's vesica (the flexible tube everted from the aedeagus during insemination) into the female's bursa (genital pouch); the sperm may be discharged from the spermatophore to the cavity of the bursa, instead of into the duct that leads to the spermatheca, resulting in fertilization failure. In conspecific pairings, the spermatophore is positioned so that its opening lies opposite that of the duct (Fig. 5.6).

In species of Japanese ground beetle of the genus *Carabus* (subgenus *Ohomopterus*) (Carabidae), the male's copulatory piece (a part of the endophallus) is a precise fit for the vaginal appendix of the conspecific female. During copulation, the male everts his endophallus in the female's vagina and the copulatory piece is inserted into the vaginal appendix. Closely related parapatric species are of similar size and external appearance but their copulatory piece and vaginal appendix are very different in shape. Although hybrids occur in areas of overlap of species, matings between different species of beetles have been observed to result in broken copulatory pieces and ruptured vaginal membranes, as well as reduced fertilization rates compared with conspecific pairings. Thus, the genital lock-and-key appears to select strongly against hybrid matings.

Mechanical reproductive isolation is not the only available explanation of species-specific genital morphology. Five other hypotheses have been advanced: pleiotropy, genitalic recognition, female choice, intersexual conflict, and male–male competition. The first two of these are further attempts to account for reproductive isolation of different species, whereas the last three are concerned with sexual selection, a topic that is addressed in more detail in sections 5.3 and 5.7.

The pleiotropy hypothesis explains genitalic differences between species as chance effects of genes that primarily code for other vital characteristics of the organism. This idea fails to explain why genitalia should be more affected than other parts of the body. Nor can pleiotropy explain genital morphology in groups (such as the Odonata) in which organs other than the primary male genitalia have an intromittent

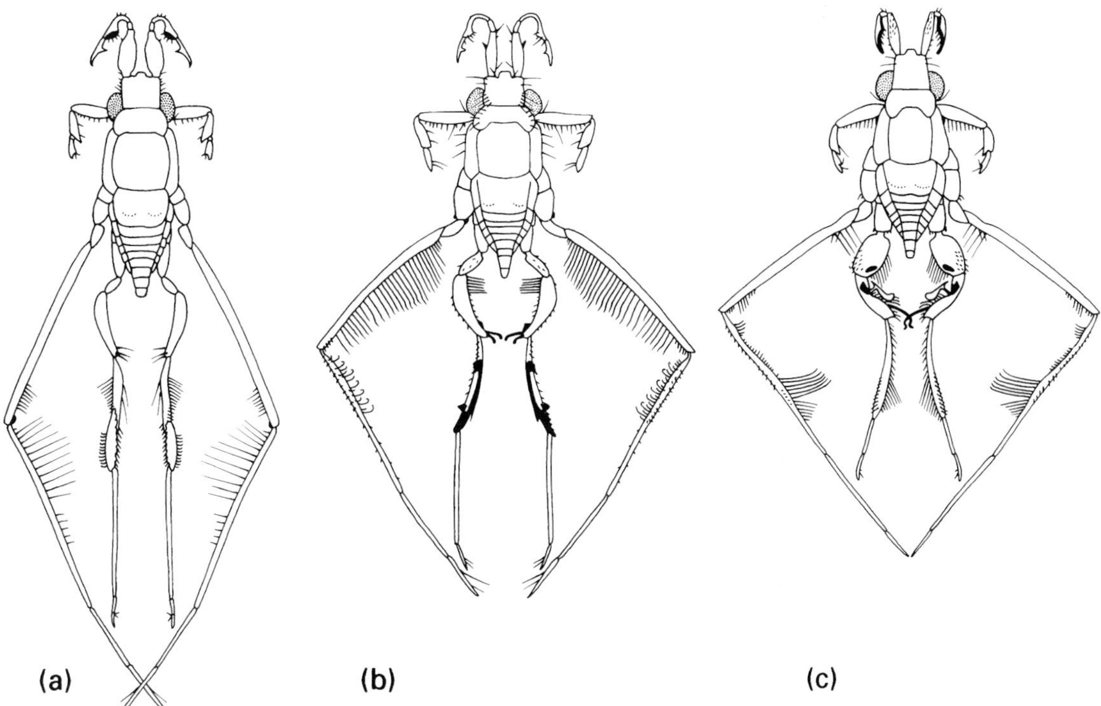

Fig. 5.7 Males of three species of the water-strider genus *Rheumatobates*, showing species-specific antennal and leg modifications (mostly flexible setae). These non-genitalic male structures are specialized for contact with the female during mating, when the male rides on her back. Females of all species have a similar body form. (a) *R. trulliger*; (b) *R. rileyi*; (c) *R. bergrothi*. (After Hungerford 1954.)

function (like those on the anterior abdomen in odonates). Such secondary genitalia consistently become subject to the postulated pleiotropic effects whereas the primary genitalia do not, a result inexplicable by the pleiotropy hypothesis.

The hypothesis of genitalic recognition involves reproductive isolation of species via female sensory discrimination between different males based upon genitalic structures, both internal and external. The female thus responds only to the appropriate genital stimulation of a conspecific male and never to that of any male of another species.

In contrast, the female-choice hypothesis involves female sexual discrimination amongst conspecific males based on qualities that can vary intraspecifically and for which the female shows preference. This idea has nothing to do with the origin of reproductive isolation, although female choice may lead to reproductive isolation or speciation as a by-product. The female-choice hypothesis predicts diverse genitalic morphology in taxa with promiscuous females and uniform genitalia in strictly monogamous taxa. This prediction seems to be fulfilled in some insects. For example, in neotropical butterflies of the genus *Heliconius*, species in which females mate more than once are more likely to have species-specific male genitalia than species in which females mate only once. The greatest reduction in external genitalia (to near absence) occurs in termites, which, as might be predicted, form monogamous pairs.

Variation in genitalic and other body morphology also may result from intersexual conflict over control of fertilization. According to this hypothesis, females evolve barriers to successful fertilization in order to control mate choice, whereas males evolve mechanisms to overcome these barriers. For example, in many species of water-striders (Gerridae) males possess complex genital processes and modified appendages (Fig. 5.7) for grasping females, which in turn exhibit

Box 5.3 Sperm precedence

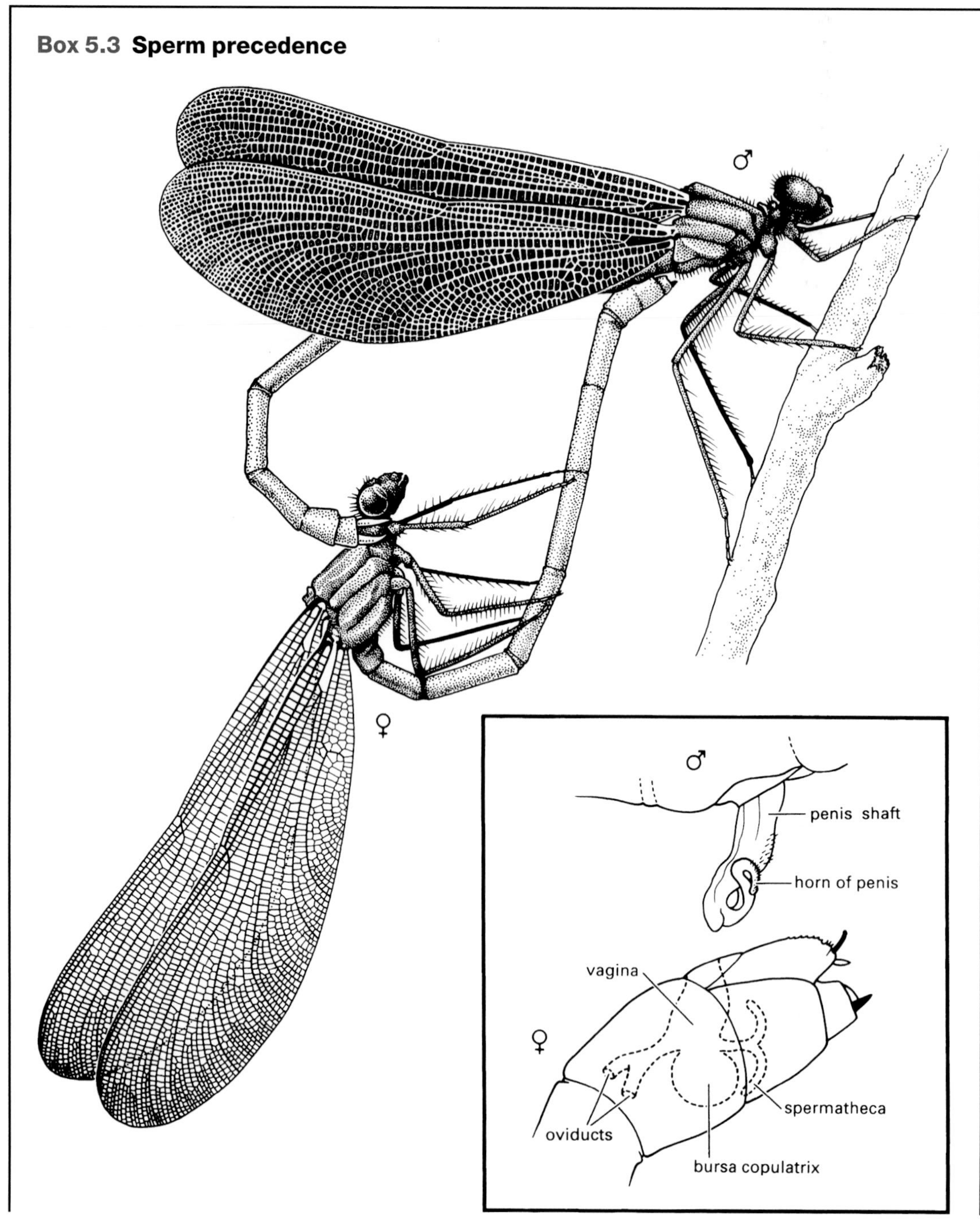

The penis or aedeagus of a male insect may be modified to facilitate placement of his own sperm in a strategic position within the spermatheca of the female or even to remove a rival's sperm. Sperm displacement of the former type, called stratification, involves pushing previously deposited sperm to the back of a spermatheca in systems in which a "last-in-first-out" principle operates (i.e. the most recently deposited sperm are the first to be used when the eggs are fertilized). Last-male sperm precedence occurs in many insect species; in others there is either first-male precedence or no precedence (because of sperm mixing). In some dragonflies, males appear to use inflatable lobes on the penis to reposition rival sperm. Such sperm packing enables the copulating male to place his sperm closest to the oviduct. However, stratification of sperm from separate inseminations may occur in the absence of any deliberate repositioning, by virtue of the tubular design of the storage organs.

A second strategy of sperm displacement is removal, which can be achieved either by direct scooping out of existing sperm prior to depositing an ejaculate or, indirectly, by flushing out a previous ejaculate with a subsequent one. An unusually long penis that could reach well into the spermathecal duct may facilitate flushing of a rival's sperm from the spermatheca. A number of structural and behavioral attributes of male insects can be interpreted as devices to facilitate this form of sperm precedence, but some of the best known examples come from odonates.

Copulation in Odonata involves the female placing the tip of her abdomen against the underside of the anterior abdomen of the male, where his sperm are stored in a reservoir of his secondary genitalia. In some dragonflies and most damselflies, such as the pair of copulating *Calopteryx* damselflies (Calopterygidae) illustrated opposite in the wheel position (after Zanetti 1975), the male spends the greater proportion of the copulation time physically removing the sperm of other males from the female's sperm storage organs (spermathecae and bursa copulatrix). Only at the last minute does he introduce his own. In these species, the male's penis is structurally complex, sometimes with an extensible head used as a scraper and a flange to trap the sperm plus lateral horns or hook-like distal appendages with recurved spines to remove rival sperm (inset to figure; after Waage 1986). A male's ejaculate may be lost if another male mates with the female before she oviposits. Thus, it is not surprising that male odonates guard their mates, which explains why they are so frequently seen as pairs flying in tandem.

behaviors or morphological traits (e.g. abdominal spines) for dislodging males.

Another example is the long spermathecal tube of some female crickets (Gryllinae), fleas (Ceratophyllinae), flies (e.g. Tephritidae), and beetles (e.g. Chrysomelidae), which corresponds to a long spermatophore tube in the male, suggesting an evolutionary contest over control of sperm placement in the spermatheca. In the seed beetle *Callosobruchus maculatus* (Chrysomelidae: Bruchinae) spines on the male's intromittent organ wound the genital tract of the female during copulation either to reduce remating and/or increase female oviposition rate, both of which would increase his fertilization success. The female responds by kicking to dislodge the male, thus shortening copulation time, reducing genital damage and presumably maintaining some control over fertilization. It is also possible that **traumatic insemination** (known in Cimicidae, including bed bugs *Cimex lectularius*), in which the male inseminates the female by piercing her body wall with his aedeagus, evolved as a mechanism for the male to short-circuit the normal insemination pathway controlled by the female. Such examples of apparent intersexual conflict could be viewed as male attempts to circumvent female choice.

Another possibility is that species-specific elaborations of male genitalia may result from interactions between conspecific males vying for inseminations. Selection may act on male genitalic clasping structures to prevent usurpation of the female during copulation or act on the intromittent organ itself to produce structures that can remove or displace the sperm of other males (section 5.7). However, although sperm displacement has been documented in a few insects, this phenomenon is unlikely to be a general explanation of male genitalic diversity because the penis of male insects often cannot reach the sperm storage organ(s) of the female or, if the spermathecal ducts are long and narrow, sperm flushing should be impeded.

Functional generalizations about the species-specific morphology of insect genitalia are controversial because different explanations no doubt apply in different groups. For example, male–male competition (via sperm removal and displacement; see Box 5.3) may be important in accounting for the shape of odonate penes, but appears irrelevant as an explanation in noctuid moths. Female choice, intersexual conflict, and male–male competition may have little selective effect on genitalic structures of insect species in which the female mates with only one male (as in termites). In such species, sexual selection may affect features that determine which male is chosen as a partner, but not how the male's genitalia are shaped. Furthermore, both mechanical and sensory locks-and-keys will be unnecessary if isolating mechanisms, such as courtship behavior or seasonal or ecological differences, are well developed. So we might predict morphological constancy (or a high level of similarity, allowing for some pleiotropy) in genitalic structures among species in a genus that has species-specific precopulatory displays involving non-genital structures followed by a single insemination of each female.

5.6 SPERM STORAGE, FERTILIZATION, AND SEX DETERMINATION

Many female insects store the sperm that they receive from one or more males in their sperm storage organ, or spermatheca. Females of most insect orders have a single spermatheca but some flies are notable in having more, often two or three. Sometimes sperm remain viable in the spermatheca for a considerable time, even three or more years in the case of honey bees. During storage, secretions from the female's spermathecal gland maintain the viability of sperm.

Eggs are fertilized as they pass down the median oviduct and vagina. The sperm enter the egg via one or more **micropyles**, which are narrow canals that pass through the eggshell. The micropyle or micropylar area is orientated towards the opening of the spermatheca during egg passage, facilitating sperm entry. In many insects, the release of sperm from the spermatheca appears to be controlled very precisely in timing and number. In queen honey bees as few as 20 sperm per egg may be released, suggesting extraordinary economy of use.

The fertilized eggs of most insects give rise to both males and females, with the sex dependent upon specific determining mechanisms, which are predominantly genetic. Most insects are **diploid**, i.e. having one set of chromosomes from each parent. The most common mechanism is for sex of the offspring to be determined by the inheritance of sex chromosomes (X-chromosomes; heterochromosomes), which are differentiated from the remaining autosomes. Individuals are thus allocated to sex according to the presence of one (XO) or two (XX) sex chromosomes, but although XX is usually female and XO male, this allocation varies within and between taxonomic groups. Mechanisms involving multiple sex chromosomes also occur and there are related observations of complex fusions between sex chromosomes and autosomes. Frequently we cannot recognize sex chromosomes, particularly as sex is determined by single genes in certain insects, such as some mosquitoes and midges. Additional complications with the determination of sex arise with the interaction of both the internal and external environment on the genome (epigenetic factors). Furthermore, great variation is seen in sex ratios at birth; although the ratio is often one male to one female, there are many deviations ranging from 100% of one sex to 100% of the other.

In **haplodiploidy** (male haploidy) the male sex has only one set of chromosomes. This arises either through his development from an unfertilized egg (containing half of the female chromosome complement following meiosis), called **arrhenotoky** (section 5.10.1), or from a fertilized egg in which the paternal set of chromosomes is inactivated and eliminated, called **paternal genome elimination** (as in many male scale insects). Arrhenotoky is exemplified by honey bees, in which females (queens and workers) develop from fertilized eggs whereas males (drones) come from unfertilized eggs. However, sex is determined in at least some Hymenoptera by a single gene (the complimentary sex-determining locus, characterized recently in honey bees) that is heterozygous in females and hemizygous in (haploid) males. The female controls the sex of offspring through her ability to store sperm and control fertilization of eggs. Evidence points to a precise control of sperm release from storage, but very little is known about this process in most insects. The presence of an egg in the genital chamber may stimulate contractions of the spermathecal walls, leading to sperm release.

5.7 SPERM COMPETITION

Multiple matings are common in many insect species. The occurrence of remating under natural conditions can be determined by observing the mating behavior of individual females or by dissection to establish the amount of ejaculate or the number of spermatophores present in the female's sperm storage organs. Some of the best documentation of remating comes from studies of many Lepidoptera, in which part of each spermatophore persists in the bursa copulatrix of the female throughout her life (Fig. 5.6). These studies show that remating occurs, to some extent, in almost all species of Lepidoptera for which adequate field data are available.

The combination of internal fertilization, sperm storage, multiple mating by females, and the overlap within a female of ejaculates from different males leads to a phenomenon known as **sperm competition**. This occurs within the reproductive tract of the female at the time of oviposition when sperm from two or more males compete to fertilize the eggs. Both physiological and behavioral mechanisms determine the outcome of sperm competition. Thus, events inside the female's reproductive tract, combined with various attributes of mating behavior, determine which sperm will succeed in reaching the eggs. It is important to realize that male reproductive fitness is measured in terms of the number of eggs fertilized or offspring fathered and not simply the number of copulations achieved, although these measures sometimes are correlated. Often there may be a trade-off between the number of copulations that a male can secure and the number of eggs that he will fertilize at each mating. A high copulation frequency is generally associated with low time or energy investment per copulation but also with low certainty of paternity. At the other extreme, males that exhibit substantial parental investment, such as feeding their mates (Boxes 5.1 & 5.2), and other adaptations that more directly increase certainty of paternity, will inseminate fewer females over a given period.

There are two main types of sexually selected adaptations in males that increase certainty of paternity. The first strategy involves mechanisms by which males can ensure that females use their sperm preferentially. Such **sperm precedence** is achieved usually by displacing the ejaculate of males that have mated previously with the female (Box 5.3). The second strategy is to reduce the effectiveness or occurrence of subsequent inseminations by other males. Various mechanisms appear to achieve this result, including mating plugs, use of male-derived secretions that "switch off" female receptivity (Box 5.4), prolonged copulation (Fig. 5.8), guarding of females, and improved structures for gripping the female during copulation to prevent "take-over" by other males. A significant selective advantage would accrue to any male that could both achieve sperm precedence and prevent other males from successfully inseminating the female until his sperm had fertilized at least some of her eggs.

The factors that determine the outcome of sperm competition are not totally under male control. Female choice is a complicating influence, as shown in the above discussions on sexual selection and on morphology of genitalic structures. Female choice of sexual partners may be two-fold. First, there is good evidence that the females of many species choose among potential mating partners. For example, females of many mecopteran species mate selectively with males that provide food of a certain minimum size and quality (Box 5.1). In some insects, such as a few beetles and some moth and katydid species, females have been shown to prefer larger males as mating partners. Second, subsequent to copulation, the female might discriminate between partners as to which sperm will be used. One idea is that variation in the stimuli of the male genitalia induces the female to use one male's sperm in preference to those of another, based upon an "internal courtship". Differential sperm use is possible because females have control over sperm transport to storage, maintenance, and use at oviposition.

5.8 OVIPARITY (EGG-LAYING)

The vast majority of female insects are **oviparous**, i.e. they lay eggs. Generally, ovulation – expulsion of eggs from the ovary into the oviducts – is followed rapidly by fertilization and then oviposition. Ovulation is controlled by hormones released from the brain, whereas oviposition appears to be under both hormonal and neural control. Oviposition, the process of the egg passing from the external genital opening or vulva to the outside of the female (Fig. 5.9), is often associated with behaviors such as digging or probing into an egg-laying site, but often the eggs are simply dropped to the ground or into water. Usually the eggs are deposited on or near the food required by the offspring upon hatching. Care of eggs after laying often is lacking or minimal, but social insects (Chapter 12) have highly developed care, and certain aquatic insects show very unusual paternal care (Box 5.5).

An insect egg within the female's ovary is complete when an oocyte becomes covered with an outer protective coating, the eggshell, formed of the **vitelline membrane** and the **chorion**. The chorion may be composed of any or all of the following layers: wax layer, innermost chorion, endochorion, and exochorion

Box 5.4 Control of mating and oviposition in a blow fly

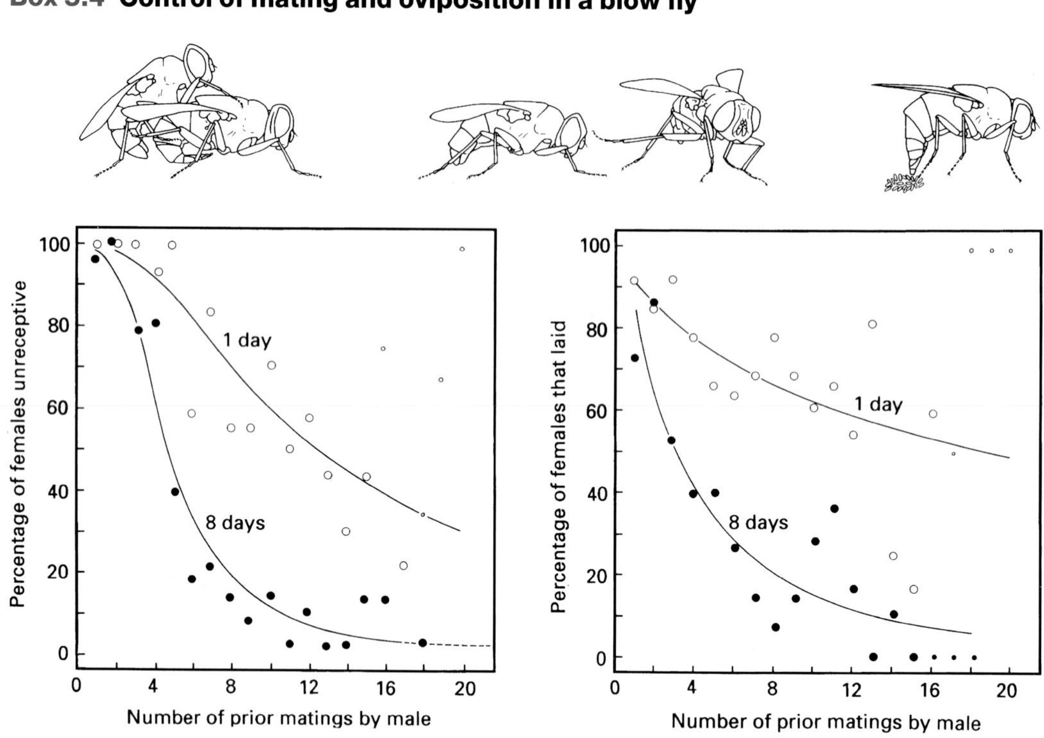

The sheep blow fly, *Lucilia cuprina* (Diptera: Calliphoridae), costs the Australian sheep industry many millions of dollars annually through losses caused by myiases or "strikes". This pestiferous fly may have been introduced to Australia from Africa in the late 19th century. The reproductive behavior of *L. cuprina* has been studied in some detail because of its relevance to a control program for this pest. Ovarian development and reproductive behavior of the adult female are highly stereotyped and readily manipulated via precise feeding of protein. Most females are **anautogenous**, i.e. they require a protein meal in order to develop their eggs, and usually mate after feeding and before their oocytes have reached early vitellogenesis. After their first mating, females reject further mating attempts for several days. The "switch-off" is activated by a peptide produced in the accessory glands of the male and transferred to the female during mating. Mating also stimulates oviposition; virgin females rarely lay eggs, whereas mated females readily do so. The eggs of each fly are laid in a single mass of a few hundred (illustration at top right) and then a new ovarian cycle commences with another batch of synchronously developing oocytes. Females may lay one to four egg masses before remating.

Unreceptive females respond to male mating attempts by curling their abdomen under their body (illustration at top left), by kicking at the males (illustration at top centre), or by actively avoiding them. Receptivity gradually returns to previously mated females, in contrast to their gradually diminishing tendency to lay. If remated, such non-laying females resume laying. Neither the size of the female's sperm store nor the mechanical stimulation of copulation can explain these changes in female behavior. Experimentally, it has been demonstrated that the female mating refractory period and readiness to lay are related to the amount of male accessory gland substance deposited in the female's bursa copulatrix during copulation. If a male repeatedly mates during one day (a multiply-mated male), less gland material is transferred at each successive copulation. Thus, if one male is mated, during one day, to a succession of females, which are later tested at intervals for their receptivity and readiness to lay, then the proportion of females either unreceptive or laying is inversely related to the number of females with which the male had previously mated. The graph on the left shows the percentage of females unreceptive to further mating when tested 1 day (○) or 8 days (●) after having mated with multiply-mated males. The percentage unreceptive values are based on 1–29 tests of different females. The graph on the right shows the percentage of females that laid eggs during 6 h of access to oviposition substrate presented 1 day (○) or 8 days (●) after mating with multiply-mated males. The percentage laid values are based on tests of 1–15 females. These two plots represent data from different groups of 30 males; samples of female flies numbering less than five are represented by smaller symbols. (After Bartell et al. 1969; Barton Browne et al. 1990; Smith et al. 1990.)

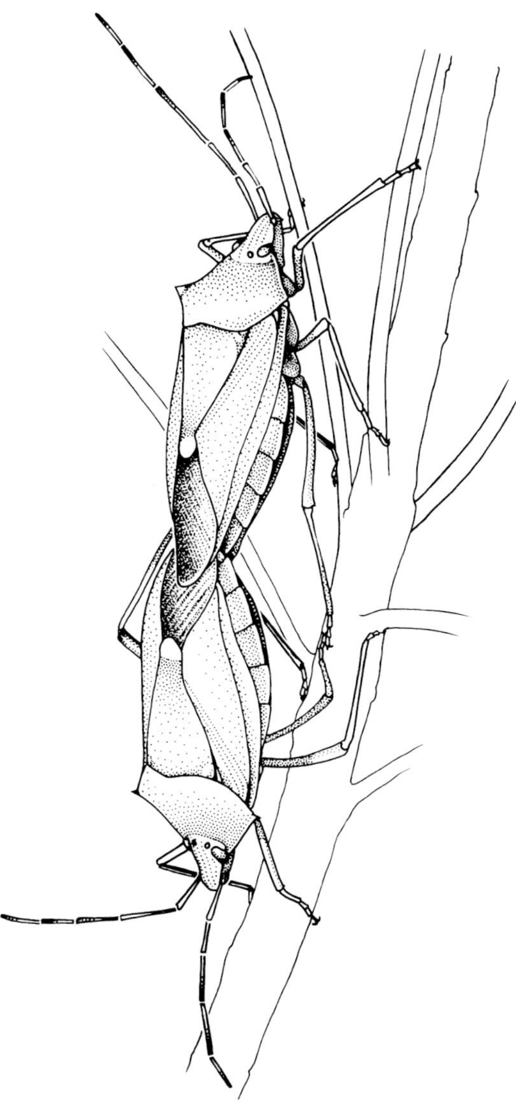

Fig. 5.8 A copulating pair of stink or shield bugs of the genus *Poecilometis* (Hemiptera: Pentatomidae). Many heteropteran bugs engage in prolonged copulation, which prevents other males from inseminating the female until either she becomes non-receptive to further males or she lays the eggs fertilized by the "guarding" male.

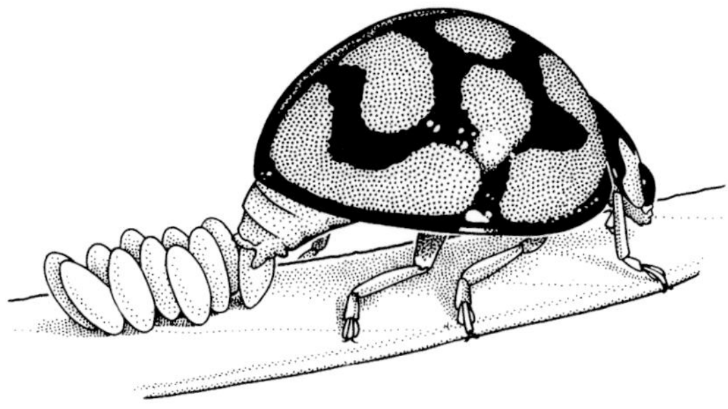

Fig. 5.9 Oviposition by a South African ladybird beetle, *Chilomenes lunulata* (Coleoptera: Coccinellidae). The eggs adhere to the leaf surface because of a sticky secretion applied to each egg. (After Blaney 1976.)

(Fig. 5.10). Ovarian follicle cells produce the eggshell and the surface sculpturing of the chorion usually reflects the outline of these cells. Typically, the eggs are yolk-rich and thus large relative to the size of the adult insect; egg cells range in length from 0.2 mm to about 13 mm. Embryonic development within the egg begins after egg activation (section 6.2.1).

The eggshell has a number of important functions. Its design allows selective entry of the sperm at the time of fertilization (section 5.6). Its elasticity facilitates oviposition, especially for species in which the eggs are compressed during passage down a narrow egg-laying tube, as described below. Its structure and composition afford the embryo protection from deleterious conditions such as unfavorable humidity and temperature, and microbial infection, while also allowing the exchange of oxygen and carbon dioxide between the inside and outside of the egg.

The differences seen in composition and complexity of layering of the eggshell in different insect groups generally are correlated with the environmental conditions encountered at the site of oviposition. In parasitic wasps the eggshell is usually thin and relatively homogeneous, allowing flexibility during passage down the narrow ovipositor, but, because the embryo develops within host tissues where desiccation is not a hazard, the wax layer of the eggshell is absent. In contrast, many insects lay their eggs in dry places and here the problem of avoiding water loss while obtaining oxygen is often acute because of the high surface area to volume ratio of most eggs. The majority of terrestrial eggs have a hydrofuge waxy chorion that contains a meshwork holding a layer of gas in contact with the outside atmosphere via narrow holes, or aeropyles.

The females of many insects (e.g. Zygentoma, many Odonata, Orthoptera, some Hemiptera, some Thysanoptera, and Hymenoptera) have appendages of the eighth and ninth abdominal segments modified to

Box 5.5 Egg-tending fathers – the giant water bugs

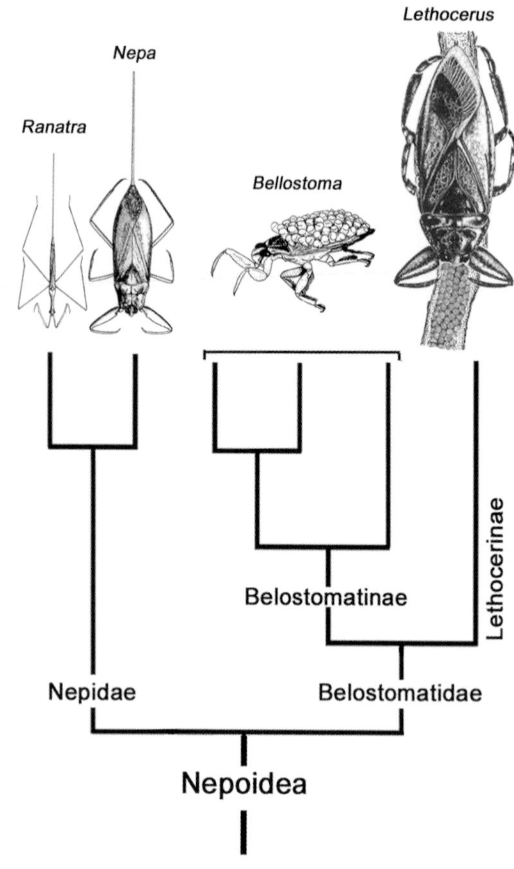

Care of eggs by adult insects is common in those that show sociality (Chapter 12), but tending solely by male insects is very unusual. This behavior is known best in the giant water bugs, the Nepoidea, comprising the families Belostomatidae and Nepidae whose common names – giant water bugs, water scorpions, toe biters – reflect their size and behaviors. These are predators, amongst which the largest species specialize in vertebrate prey such as tadpoles and small fish, which they capture with raptorial forelegs and piercing mouthparts. Evolutionary attainment of the large adult body size necessary for feeding on these large items is inhibited by the fixed number of five nymphal instars in Heteroptera and the limited size increase at each molt (Dyar's rule; section 6.9.1). These phylogenetic (evolutionarily inherited) constraints have been overcome in intriguing ways – by the commencement of development at a large size via oviposition of large eggs, and in one family, with specialized paternal protection of the eggs.

Egg tending in the subfamily Belostomatinae involves the males "back-brooding" – carrying the eggs on their backs, in a behavior shared by over a hundred species in five genera. The male mates repeatedly with a female, perhaps up to a hundred times, thus guaranteeing that the eggs she deposits on his back are his alone, which encourages his subsequent tending behavior. Active male-tending behavior, called "brood-pumping", involves underwater undulating "press-ups" by the anchored male, creating water currents across the eggs. This is an identical, but slowed-down, form of the pumping display used in courtship. Males of other taxa "surface-brood", with the back (and thus eggs) held horizontally at the water surface such that the interstices of the eggs are wet and the apices aerial. This position, which is unique to brooding males, exposes the males to higher levels of predation. A third behavior, "brood-stroking", involves the submerged male sweeping and circulating water over the egg pad. Tending results in >95% successful emergence, in contrast to death of all eggs if removed from the male, whether aerial or submerged.

Members of the Lethocerinae, sister group to the Belostomatinae, show related behaviors that help us to understand the origins of aspects of these paternal egg defenses. Following courtship that involves display pumping as in Belostomatinae, the pair copulate frequently between bouts of laying in which eggs are placed on a stem or other projection above the surface of a pond or lake. After completion of egg-laying, the female leaves the male to attend the eggs and she swims away and plays no further role. The "emergent brooding" male tends the aerial eggs for the few days to a week until they hatch. His roles include periodically submerging himself to absorb and drink water that he regurgitates over the eggs, shielding the eggs, and display posturing against airborne threats. Unattended eggs die from desiccation; those immersed by rising water are abandoned and drown.

Insect eggs have a well-developed chorion that enables gas exchange between the external environment and the developing embryo (see section 5.8). The problem with a large egg relative to a smaller one is that the surface area increase of the sphere is much less than the increase in volume. Because oxygen is scarce in water and diffuses much more slowly than in air (section 10.3) the increased sized egg hits a limit of the ability for oxygen diffusion from water to egg. For such an egg in a terrestrial environment gas exchange is easy, but desiccation through loss of water becomes an issue. Although terrestrial insects use waxes around the chorion to avoid desiccation, the long aquatic history of the Nepoidea means that any such a mechanism has been lost and is unavailable, providing another example of phylogenetic inertia.

In the phylogeny of Nepoidea (shown opposite in reduced form from Smith 1997) a stepwise pattern of acquisition of paternal care can be seen. In the sister family to Belostomatidae, the Nepidae (the water-scorpions), all eggs, including the largest, develop immersed. Gas exchange is facilitated by expansion of the chorion surface area into either a crown or two long horns: the eggs never are brooded. No such chorionic elaboration evolved in Belostomatidae: the requirement by large eggs for oxygen with the need to avoid drowning or desiccation could have been fulfilled by oviposition on a wave-swept rock – although this strategy is unknown in any extant taxa. Two alternatives developed – avoidance of submersion and drowning by egg-laying on emergent structures (Lethocerinae), or, perhaps in the absence of any other suitable substrate, egg-laying onto the back of the attendant mate (Belostomatinae). In Lethocerinae, watering behaviors of the males counter the desiccation problems encountered during emergent brooding of aerial eggs; in Belostomatinae, the pre-existing male courtship pumping behavior is a pre-adaptation for the oxygenating movements of the back-brooding male. Surface-brooding and brood-stroking are seen as more derived male-tending behaviors.

The traits of large eggs and male brooding behavior appeared together, and the traits of large eggs and egg respiratory horns also appeared together, because the first was impossible without the second. Thus, large body size in Nepoidea must have evolved twice. Paternal care and egg respiratory horns are different adaptations that facilitate gas exchange and thus survival of large eggs.

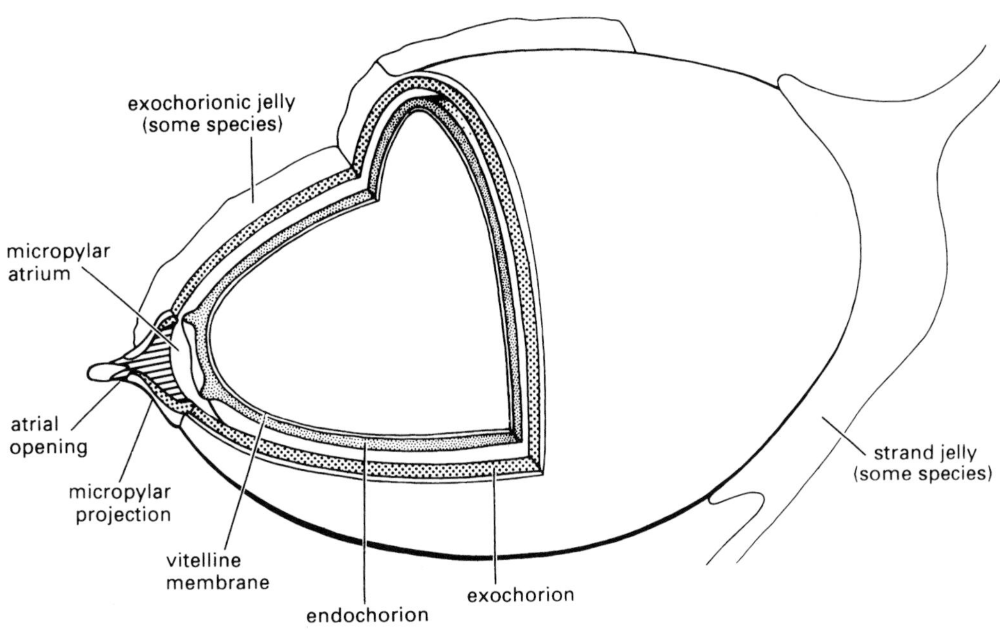

Fig. 5.10 The generalized structure of a libelluloid dragonfly egg (Odonata: Corduliidae, Libellulidae). Libelluloid dragonflies oviposit into freshwater but always exophytically (i.e. outside of plant tissues). The endochorionic and exochorionic layers of the eggshell are separated by a distinct gap in some species. A gelatinous matrix may be present on the exochorion or as connecting strands between eggs. (After Trueman 1991.)

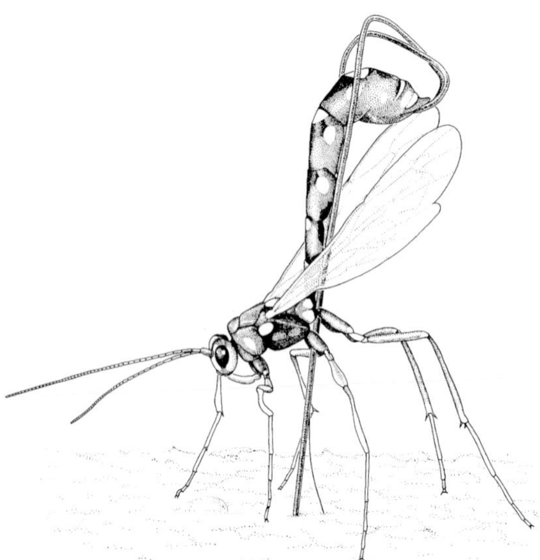

Fig. 5.11 A female of the parasitic wasp *Megarhyssa nortoni* (Hymenoptera: Ichneumonidae) probing a pine log with her very long ovipositor in search of a larva of the sirex wood wasp, *Sirex noctilio* (Hymenoptera: Siricidae). If a larva is located, she stings and paralyses it before laying an egg on it.

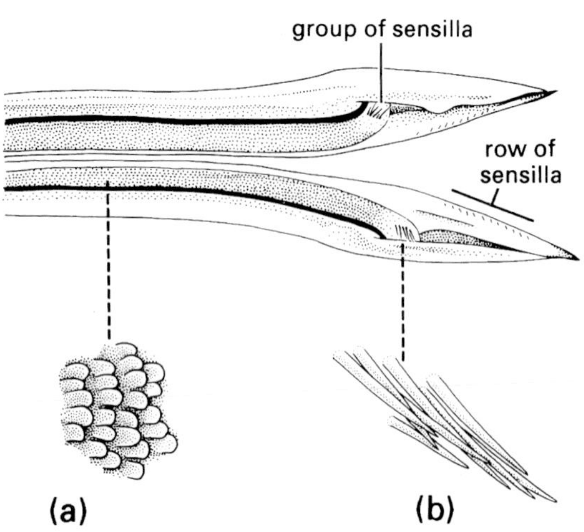

Fig. 5.12 Tip of the ovipositor of a female of the black field cricket, *Teleogryllus commodus* (Orthoptera: Gryllidae), split open to reveal the inside surface of the two halves of the ovipositor. Enlargements show: (a) posteriorly directed ovipositor scales; (b) distal group of sensilla. (After Austin & Browning 1981.)

form an egg-laying organ or ovipositor (section 2.5.1). In other insects (e.g. many Lepidoptera, Coleoptera, and Diptera) it is the posterior segments rather than appendages of the female's abdomen that function as an ovipositor (a "substitutional" ovipositor). Often these segments can be protracted into a telescopic tube in which the opening of the egg passage is close to the distal end. The ovipositor or the modified end of the abdomen enables the insect to insert its eggs into particular sites, such as into crevices, soil, plant tissues, or, in the case of many parasitic species, into an arthropod host. Other insects, such as Isoptera, Phthiraptera, and many Plecoptera, lack an egg-laying organ and eggs are deposited simply on a surface.

In certain Hymenoptera (some wasps, bees, and ants) the ovipositor has lost its egg-laying function and is used as a poison-injecting sting. The stinging Hymenoptera eject the eggs from the opening of the genital chamber at the base of the modified ovipositor. However, in most wasps the eggs pass down the canal of the ovipositor shaft, even if the shaft is very narrow (Fig. 5.11). In some parasitic wasps with very slender ovipositors the eggs are extremely compressed and stretched as they move through the narrow canal of the shaft.

The valves of an insect ovipositor usually are held together by interlocking tongue-and-groove joints, which prevent lateral movement but allow the valves to slide back and forth on one another. Such movement, and sometimes also the presence of serrations on the tip of the ovipositor, is responsible for the piercing action of the ovipositor into an egg-laying site. Movement of eggs down the ovipositor tube is possible because of many posteriorly directed "scales" (microsculpturing) located on the inside surface of the valves. Ovipositor scales vary in shape (from plate-like to spine-like) and in arrangement among insect groups, and are seen best under the scanning electron microscope.

The scales found in the conspicuous ovipositors of crickets and katydids exemplify these variations (Orthoptera: Gryllidae and Tettigoniidae). The ovipositor of the field cricket *Teleogryllus commodus* (Fig. 5.12) possesses overlapping plate-like scales and scattered, short sensilla along the length of the egg canal. These sensilla may provide information on the position of the egg as it moves down the canal, whereas a group of larger sensilla at the apex of each dorsal valve presumably signals that the egg has been expelled. In addition, in *T. commodus* and some other insects, there are scales on the outer surface of the ovipositor tip, which are orientated in the opposite direction to those on the inner surface. These are thought to assist with penetration of the substrate and holding the ovipositor in position during egg-laying.

In addition to the eggshell, many eggs are provided with a proteinaceous secretion or cement which coats and fastens them to a substrate, such as a vertebrate hair in the case of sucking lice, or a plant surface in the case of many beetles (Fig. 5.9). Colleterial glands, accessory glands of the female reproductive tract, produce such secretions. In other insects, groups of thin-shelled eggs are enclosed in an **ootheca**, which protects the developing embryos from desiccation. The colleterial glands produce the tanned, purse-like ootheca of cockroaches (Box 9.8) and the frothy ootheca of mantids (see Plate 3.3, facing p. 14), whereas the foamy ootheca that surrounds locust and other orthopteran eggs in the soil is formed from the accessory glands in conjunction with other parts of the reproductive tract.

5.9 OVOVIVIPARITY AND VIVIPARITY

Most insects are oviparous, with the act of laying involved in initiation of egg development. However, some species are viviparous, with initiation of egg development taking place within the mother. The life cycle is shortened by retention of eggs and even of developing young within the mother. Four main types of viviparity are observed in different insect groups, with many of the specializations prevalent in various higher dipterans.

1 **Ovoviviparity**, in which fertilized eggs containing yolk and enclosed in some form of eggshell are incubated inside the reproductive tract of the female. This occurs in some cockroaches (Blattidae), some aphids and scale insects (Hemiptera), a few beetles (Coleoptera) and thrips (Thysanoptera), and some flies (Muscidae, Calliphoridae, and Tachinidae). The fully developed eggs hatch immediately after being laid or just prior to ejection from the female's reproductive tract.

2 **Pseudoplacental viviparity** occurs when a yolk-deficient egg develops in the genital tract of the female. The mother provides a special placenta-like tissue, through which nutrients are transferred to developing embryos. There is no oral feeding and larvae are laid upon hatching. This form of viviparity occurs in many aphids (Hemiptera), some earwigs (Dermaptera), a few psocids (Psocoptera), and in polyctenid bugs (Hemiptera).

3 **Hemocoelous viviparity** involves embryos developing free in the female's hemolymph, with nutrients taken up by osmosis. This form of internal parasitism occurs only in Strepsiptera, in which the larvae exit through a brood canal (Box 13.6), and in some gall midges (Diptera: Cecidomyiidae), where the larvae may consume the mother (as in pedogenetic development, below).

4 **Adenotrophic viviparity** occurs when a poorly developed larva hatches and feeds orally from accessory ("milk") gland secretions within the "uterus" of the mother's reproductive system. The full-grown larva is deposited and pupariates immediately. The dipteran "pupiparan" families, namely the Glossinidae (tsetse flies), Hippoboscidae (louse or wallaby flies, keds), and Nycteribidae and Streblidae (bat flies), demonstrate adenotrophic viviparity.

5.10 ATYPICAL MODES OF REPRODUCTION

Sexual reproduction (**amphimixis**) with separate male and female individuals (**gonochorism**) is the usual mode of reproduction in insects, and **diplodiploidy**, in which males as well as females are diploid, occurs as the ancestral system in almost all insect orders. However, other modes are not uncommon. Various types of asexual reproduction occur in many insect groups; development from unfertilized eggs is a widespread phenomenon, whereas the production of multiple embryos from a single egg is rare. Some species exhibit alternating sexual and asexual reproduction, depending on season or food availability. A few species possess both male and female reproductive systems in one individual (**hermaphroditism**) but self-fertilization has been established for species in just one genus.

5.10.1 Parthenogenesis, pedogenesis (paedogenesis), and neoteny

Some or a few representatives of virtually every insect order have dispensed with mating, with females producing viable eggs even though unfertilized. In other groups, notably the Hymenoptera, mating occurs but the sperm need not be used in fertilizing all the eggs. Development from unfertilized eggs is called **parthenogenesis**, which in some species may be **obligatory**, but in many others is **facultative**. The female may produce parthenogenetically only female eggs (**thelytokous parthenogenesis**), only male eggs (**arrhenotokous parthenogenesis**), or eggs of both sexes (**amphitokous** or **deuterotokous parthenogenesis**). The largest insect group showing arrhenotoky is the Hymenoptera. Within the Hemiptera, aphids display thelytoky and most whiteflies are arrhenotokous. Certain Diptera and a few Coleoptera are thelytokous, and Thysanoptera display all three types of parthenogenesis. Facultative parthenogenesis, and variation in sex of egg produced, may be a response to fluctuations in environmental conditions, as occurs in aphids that vary the sex of their offspring and mix parthenogenetic and sexual cycles according to season.

Some insects abbreviate their life cycles by loss of the adult stage, or even both adult and pupal stages. In this precocious stage, reproduction is almost exclusively by parthenogenesis. **Larval pedogenesis**, the production of young by the larval insect, has arisen at least three times in the gall midges (Diptera: Cecidomyiidae) and once in the Coleoptera (*Macromalthus debilis*). In some gall midges, in an extreme case of hemocoelous viviparity, the precocially developed eggs hatch internally and the larvae may consume the body of the mother-larva before leaving to feed on the surrounding fungal medium. In the well-studied gall midge *Heteropeza pygmaea*, eggs develop into female larvae, which may metamorphose to female adults or produce more larvae pedogenetically. These larvae, in turn, may be males, females, or a mixture of both sexes. Female larvae may become adult females or repeat the larval pedogenetic cycle, whereas male larvae must develop to adulthood.

In **pupal pedogenesis**, which sporadically occurs in gall midges, embryos are formed in the hemocoel of a pedogenetic mother-pupa, termed a hemipupa as it differs morphologically from the "normal" pupa. This production of live young in pupal pedogenetic insects also destroys the mother-pupa from within, either by larval perforation of the cuticle or by the eating of the mother by the offspring. Pedogenesis appears to have evolved to allow maximum use of locally abundant but ephemeral larval habitats, such as a mushroom fruiting body. When a gravid female detects an oviposition site, eggs are deposited, and the larval population builds up rapidly through pedogenetic development. Adults are developed only in response to conditions adverse to larvae, such as food depletion and overcrowding. Adults may be female only, or males may occur in some species under specific conditions.

In true pedogenetic taxa there are no reproductive adaptations beyond precocious egg development. In contrast, in **neoteny** a non-terminal instar develops reproductive features of the adult, including the ability to locate a mate, copulate, and deposit eggs (or larvae) in a conventional manner. For example, the scale insects (Hemiptera: Coccoidea) appear to have neotenous females. Whereas a molt to the winged adult male follows the final immature instar, development of the reproductive female involves omission of one or more instars relative to the male. In appearance the female is a sedentary nymph-like or larviform instar, resembling a larger version of the previous (second or third) instar in all but the presence of a vulva and developing eggs. Neoteny also occurs in all members of the order Strepsiptera; in these insects female development ceases at the puparium stage. In some other insects (e.g. marine midges; Chironomidae), the adult appears larva-like, but this is evidently not due to neoteny because complete metamorphic development is retained, including a pupal instar. Their larviform appearance therefore results from suppression of adult features, rather than the pedogenetic acquisition of reproductive ability in the larval stage.

5.10.2 Hermaphroditism

Several of the species of *Icerya* (Hemiptera: Margarodidae) that have been studied cytologically are gynomonoecious hermaphrodites, as they are female-like but possess an ovotestis (a gonad that is part testis, part ovary). In these species, occasional males arise from unfertilized eggs and are apparently functional, but normally self-fertilization is assured by production of male gametes prior to female gametes in the body of one individual (protandry of the hermaphrodite). Without doubt, hermaphroditism greatly assists the spread of the pestiferous cottony-cushion scale, *I. purchasi* (Box 16.2), as single nymphs of this and other hermaphroditic *Icerya* species can initiate new infestations if dispersed or accidentally transported to new plants. Furthermore, all iceryine margarodids are arrhenotokous, with unfertilized eggs developing into males and fertilized eggs into females.

5.10.3 Polyembryony

This form of asexual reproduction involves the production of two or more embryos from one egg by subdivision (fission). It is restricted predominantly to parasitic insects; it occurs in at least one strepsipteran and representatives of four wasp families, especially the Encyrtidae. It appears to have arisen independently within each wasp family. In these parasitic wasps, the number of larvae produced from a single egg varies in different genera but is influenced by the size of the host, with from fewer than 10 to several hundred, and in *Copidosoma* (Encyrtidae) up to 3000 embryos, arising from one small, yolkless egg. Nutrition for a large number of developing embryos obviously cannot be supplied by the original egg and is acquired from the host's hemolymph through a specialized enveloping membrane called the **trophamnion**. Typically, the embryos develop into larvae when the host molts to its final instar, and these larvae consume the host insect before pupating and emerging as adult wasps.

5.10.4 Reproductive effects of endosymbionts

Wolbachia, an intracellular bacterium discovered first infecting the ovaries of *Culex pipiens* mosquitoes, causes some inter-populational (intraspecific) matings to produce inviable embryos. Such crosses, in which embryos abort before hatching, could be returned to viability after treatment of the parents with antibiotic – thus implicating the microorganism with the sterility. This phenomenon, termed **cytoplasmic** or **reproductive incompatibility**, now has been demonstrated in a very wide range of invertebrates that host many "strains" of *Wolbachia*. Surveys have suggested that up to 76% of insect species may be infected. *Wolbachia* is transferred vertically (inherited by offspring from the mother via the egg), and causes several different but related effects. Specific effects include the following:

- Cytoplasmic (reproductive) incompatibility, with directionality varying according to whether one, the other, or both sexes of partners are infected, and with which strain. Unidirectional incompatibility typically involves an infected male and uninfected female, with the reciprocal cross (uninfected male with infected female) being compatible (producing viable offspring). Bidirectional incompatibility usually involves both partners being infected with different strains of *Wolbachia* and no viable offspring are produced from any mating.
- Parthenogenesis, or sex ratio bias to the diploid sex (usually female) in insects with haplodiploid genetic systems (sections 5.6, 12.2, & 12.4.1). In the parasitic wasps (*Trichogramma*) studied this involves infected females that produce only fertile female offspring. The mechanism is usually gamete duplication, involving disruption of meiotic chromosomal segregation such that the nucleus of an unfertilized, *Wolbachia*-infected egg contains two sets of identical chromosomes (diploidy), producing a female. Normal sex ratios are restored by treatment of parents with antibiotics, or by development at elevated temperature, to which *Wolbachia* is sensitive.
- Feminization, the conversion of genetic males into functional females, perhaps caused by specific inhibitions of male-determiner genes, thus far only observed in terrestrial isopods and two Lepidoptera species, but perhaps yet to be discovered in other arthropods.

The strategy of *Wolbachia* can be viewed as reproductive parasitism (section 3.6.5), in which the bacterium manipulates its host into producing an imbalance of female offspring (this being the sex responsible for the vertical transmission of the infection), compared with uninfected hosts. Only in a very few cases have infections been shown to benefit the insect host, primarily via enhanced fecundity. Certainly, with evidence derived from phylogenies of *Wolbachia* and their host, *Wolbachia* often has been transferred horizontally between unrelated hosts, and no coevolution is apparent.

Although *Wolbachia* is now the best studied system of a sex-ratio modifying organism, there are other somewhat similar cytoplasm-dwelling organisms, with the most extreme sex-ratio distorters known as male-killers. This phenomenon of male lethality is known across at least five orders of insects, associated with a range of maternally inherited, symbiotic–infectious causative organisms, from bacteria to viruses, and microsporidia. Each acquisition seems to be independent, and others are suspected to exist. Certainly, if parthenogenesis often involves such associations, many such interactions remain to be discovered. Furthermore, much remains to be learned about the effects of insect age, remating frequency, and temperature on the expression and transmission of this bacterium. There is also an intriguing case involving the parasitic wasp *Asobara tabida* (Braconidae) in which the elimination of *Wolbachia* by antibiotics causes the inhibition of egg production rendering the wasps infertile. Such obligatory infection with *Wolbachia* also occurs in filarial nematodes (section 15.5.5).

5.11 PHYSIOLOGICAL CONTROL OF REPRODUCTION

The initiation and termination of some reproductive events often depend on environmental factors, such as temperature, humidity, photoperiod, or availability of food or a suitable egg-laying site. Additionally, these external influences may be modified by internal factors such as nutritional condition and the state of maturation of the oocytes. Copulation also may trigger oocyte development, oviposition, and inhibition of sexual receptivity in the female via enzymes or peptides transferred to the female reproductive tract in male accessory gland secretions (Box 5.4). Fertilization following mating normally triggers embryogenesis via egg activation (Chapter 6). Regulation of reproduction is complex and involves sensory receptors, neuronal transmission, and integration of messages in the brain, as well as chemical messengers (hormones) transported in the hemolymph or via the nerve axons to target tissues or to other endocrine glands. Certain parts of the nervous system, particularly neurosecretory cells in the brain, produce neurohormones or neuropeptides (proteinaceous messengers) and also control the synthesis of two groups of insect hormones – the ecdysteroids and the juvenile hormones (JH). More detailed discussions of the regulation and functions of all of these hormones are provided in Chapters 3 and 6. Neuropeptides, steroid hormones, and JH all play essential roles in the regulation of reproduction, as summarized in Fig. 5.13.

Juvenile hormones and/or ecdysteroids are essential to reproduction, with JH mostly triggering the functioning of organs such as the ovary, accessory glands, and fat body, whereas ecdysteroids influence morphogenesis as well as gonad functions. Neuropeptides play various roles at different stages of reproduction, as they regulate endocrine function (via the corpora allata and prothoracic glands) and also directly influence reproductive events, especially ovulation and oviposition or larviposition.

The role of neuropeptides in control of reproduction (Table 3.1) is an expanding area of research, made possible by new technologies, especially in biochemistry and molecular biology. To date, most studies have concentrated on the Diptera (especially *Drosophila*, mosquitoes, and houseflies), the Lepidoptera (especially the tobacco hornworm, *Manduca sexta*), locusts, and cockroaches.

5.11.1 Vitellogenesis and its regulation

In the ovary, both nurse cells (or trophocytes) and ovarian follicle cells are associated with the oocytes (section 3.8.1). These cells pass nutrients to the growing oocytes. The relatively slow period of oocyte growth is followed by a period of rapid yolk deposition, or **vitellogenesis**, which mostly occurs in the terminal oocyte of each ovariole and leads to the production of fully developed eggs. Vitellogenesis involves the production (mostly by the fat body) of specific female lipoglycoproteins called **vitellogenins**, followed by their passage into the oocyte. Once inside the oocyte these proteins are called **vitellins** and their chemical structure may differ slightly from that of vitellogenins. Lipid bodies – mostly triglycerides from the follicle cells, nurse cells, or fat body – also are deposited in the growing oocyte.

Vitellogenesis has been a favored area of insect hormone research because it is amenable to experimental manipulation with artificially supplied hormones, and analysis is facilitated by the large amounts of vitellogenins produced during egg growth. The regulation of vitellogenesis varies among insect taxa, with JH from the corpora allata, ecdysteroids from the prothoracic glands or the ovary, and brain neurohormones (neuropeptides such as ovarian ecdysteroidogenic hormone, OEH) considered to induce or stimulate vitellogenin synthesis to varying degrees in different insect species (Fig. 5.13).

Inhibition of egg development in ovarian follicles in the previtellogenic stage is mediated by antigonadotropins. This inhibition ensures that only some of the oocytes undergo vitellogenesis in each ovarian cycle. The antigonadotropins responsible for this suppression are peptides termed **oostatic hormones**. In most of the insects studied, oostatic hormones are produced by the ovary or neurosecretory tissue associated with the ovary and, depending on species, may work in one of three ways:

1 inhibit the release or synthesis of OEH (also called egg development neurohormone, EDNH); or

2 affect ovarian development by inhibiting proteolytic enzyme synthesis and blood digestion in the midgut, as in mosquitoes; or

3 inhibit the action of JH on vitellogenic follicle cells thus preventing the ovary from accumulating vitellogenin from the hemolymph, as in the blood-sucking bug *Rhodnius prolixus*.

Originally, it was firmly believed that JH controlled

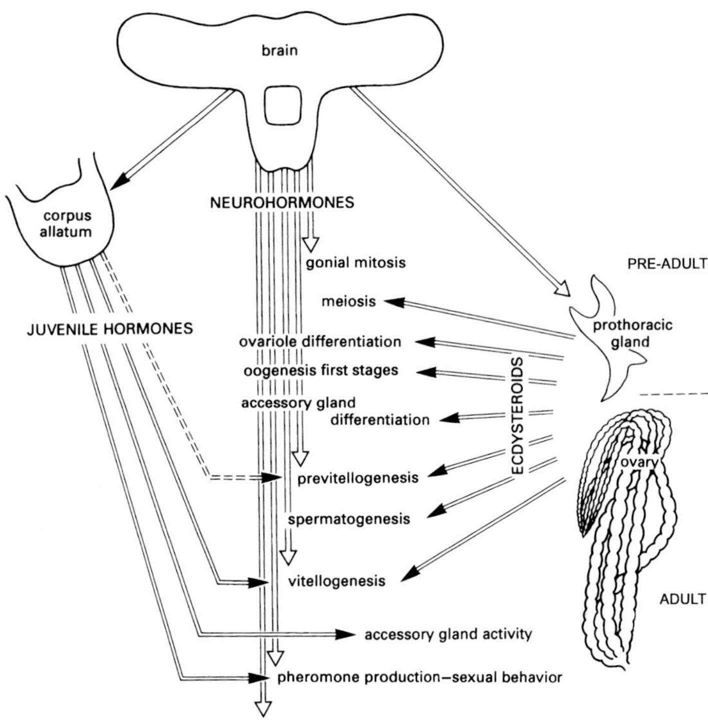

Fig. 5.13 A schematic diagram of the hormonal regulation of reproductive events in insects. The transition from ecdysterone production by the pre-adult prothoracic gland to the adult ovary varies between taxa. (After Raabe 1986.)

vitellogenesis in most insects. Then, in certain insects, the importance of ecdysteroids was discovered. Now we are becoming increasingly aware of the part played by neuropeptides, a group of proteins for which reproductive regulation is but one of an array of functions in the insect body (see Table 3.1).

FURTHER READING

Austin, A.D. & Browning, T.O. (1981) A mechanism for movement of eggs along insect ovipositors. *International Journal of Insect Morphology and Embryology* **10**, 93–108.

Bourtzis, K. & Miller, T.A. (eds.) (2001) *Insect Symbioses*. CRC Press, Boca Raton, FL.

Choe, J.C. & Crespi, B.J. (eds.) (1997) *The Evolution of Mating Systems in Insects and Arachnids*. Cambridge University Press, Cambridge.

Eberhard, W.G. (1985) *Sexual Selection and Animal Genitalia*. Harvard University Press, Cambridge, MA.

Eberhard, W.G. (1994) Evidence for widespread courtship during copulation in 131 species of insects and spiders, and implications for cryptic female choice. *Evolution* **48**, 711–33.

Emlen, D.F.J. (2001) Costs and diversification of exaggerated animal structures. *Science* **291**, 1534–6.

Gwynne, D.T. (2001) *Katydids and Bush-Crickets: Reproductive Behavior and Evolution of the Tettigoniidae*. Comstock Publishing Associates, Ithaca.

Heming, B.-S. (2003) *Insect Development and Evolution*. Cornell University Press, Ithaca, NY.

Judson, O. (2002) *Dr Tatiana's Advice to All Creation. The Definitive Guide to the Evolutionary Biology of Sex*. Metropolitan Books, Henry Holt & Co., New York.

Kerkut, G.A. & Gilbert, L.I. (eds.) (1985) *Comprehensive Insect Physiology, Biochemistry, and Pharmacology*, Vol. 1: *Embryogenesis and Reproduction*. Pergamon Press, Oxford.

Leather, S.R. & Hardie, J. (eds.) (1995) *Insect Reproduction*. CRC Press, Boca Raton, FL.

Mikkola, K. (1992) Evidence for lock-and-key mechanisms in the internal genitalia of the *Apamea* moths (Lepidoptera, Noctuidae). *Systematic Entomology* **17**, 145–53.

Normark, B.B. (2003) The evolution of alternative genetic systems in insects. *Annual Review of Entomology* **48**, 397–423.

O'Neill, S.L., Hoffmann, A.A. & Werren, J.H. (1997) *Influential Passengers. Inherited Microorganisms and Arthropod Reproduction*. Oxford University Press, Oxford.

Raabe, M. (1986) Insect reproduction: regulation of successive steps. *Advances in Insect Physiology* **19**, 29–154.

Resh, V.H. & Cardé, R.T. (eds.) (2003) *Encyclopedia of Insects*. Academic Press, Amsterdam. [Particularly see articles on mating behaviors; parthenogenesis; polyembryony; four chapters on reproduction; sexual selection; *Wolbachia*.]

Ringo, J. (1996) Sexual receptivity in insects. *Annual Review of Entomology* **41**, 473–94.

Shapiro, A.M. & Porter, A.H. (1989) The lock-and-key hypothesis: evolutionary and biosystematic interpretation of insect genitalia. *Annual Review of Entomology* **34**, 231–45.

Simmons, L.W. (2001) *Sperm Competition and its Evolutionary Consequences in the Insects*. Princeton University Press, Princeton, NJ.

Sota, T. & Kubota, K. (1998) Genital lock-and-key as a selective agent against hybridization. *Evolution* **52**, 1507–13.

Weekes, A.R., Reynolds, K.T. & Hoffman, A.A. (2002) *Wolbachia* dynamics and host effects: what has (and has not) been demonstrated? *Trends in Ecology and Evolution* **17**, 257–62.

Vahed, K. (1998) The function of nuptial feeding in insects: a review of empirical studies. *Biological Reviews* **73**, 43–78.

Chapter 6

INSECT DEVELOPMENT AND LIFE HISTORIES

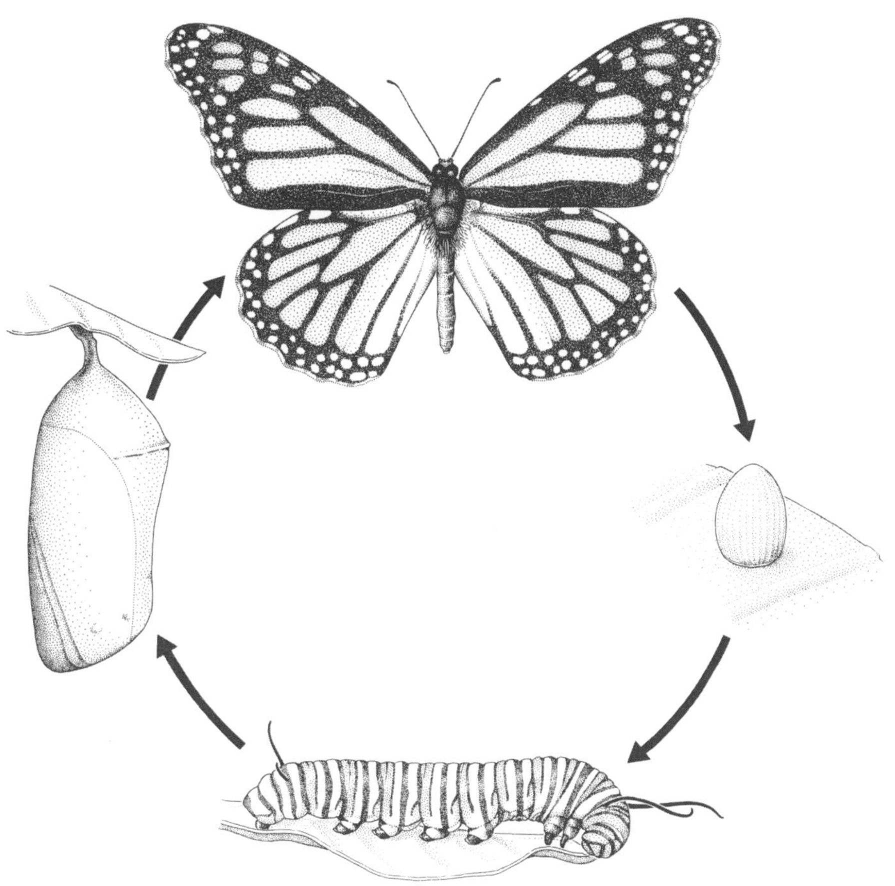

Life cycle of the monarch or wanderer butterfly, *Danaus plexippus*. (After photographs by P.J. Gullan.)

In this chapter we discuss the pattern of growth from egg to adult – the ontogeny – and life histories of insects. The various growth phases from the egg, through immature development, to the emergence of the adult are dealt with. Also, we consider the significance of different kinds of metamorphosis and suggest that complete metamorphosis reduces competition between conspecific juveniles and adults, by providing a clear differentiation between immature and adult stages. Amongst the different aspects of life histories covered are voltinism, resting stages, the coexistence of different forms within a single species, migration, age determination, allometry, and genetic and environmental effects on development. The influence of environmental factors, namely temperature, photoperiod, humidity, toxins, and biotic interactions, upon life-history traits is vital to any applied entomological research. Likewise, knowledge of the process and hormonal regulation of molting is fundamental to insect control.

Insect life-history characteristics are very diverse, and the variability and range of strategies seen in many higher taxa imply that these traits are highly adaptive. For example, diverse environmental factors trigger termination of egg dormancy in different species of *Aedes* although the species in this genus are closely related. However, phylogenetic constraint, such as the restrained instar number of Nepoidea (Box 5.5), undoubtedly plays a role in life-history evolution in insects.

We conclude the chapter by considering how the potential distributions of insects can be modeled, using data on insect growth and development to answer questions in pest entomology, and bioclimatic data associated with current-day distributions to predict past and future patterns.

6.1 GROWTH

Insect growth is discontinuous, at least for the sclerotized cuticular parts of the body, because the rigid cuticle limits expansion. Size increase is by **molting** – periodic formation of new cuticle of greater surface area and shedding of the old cuticle. Thus, for sclerite-bearing body segments and appendages, increases in body dimensions are confined to the postmolt period immediately after molting, before the cuticle stiffens and hardens (section 2.1). Hence, the sclerotized head capsule of a beetle or moth larva increases in dimensions in a saltatory manner (in major increments) during development, whereas the membranous nature of body cuticle allows the larval body to grow more or less continuously.

Studies concerning insect development involve two components of growth. The first, the **molt increment**, is the increment in size occurring between one **instar** (growth stage, or the form of the insect between two successive molts) and the next. Generally, increase in size is measured as the increase in a single dimension (length or width) of some sclerotized body part, rather than a weight increment, which may be misleading because of variability in food or water intake. The second component of growth is the **intermolt period** or interval, better known as the **stadium** or instar duration, which is defined as the time between two successive molts, or more precisely between successive ecdyses (Fig. 6.1 and section 6.3). The magnitude of both molt increments and intermolt periods may be affected by food supply, temperature, larval density, and physical damage (such as loss of appendages) (section 6.10), and may differ between the sexes of a species.

In collembolans, diplurans, and apterygote insects, growth is **indeterminate** – the animals continue to molt until they die. There is no definitive terminal molt in such animals, but they do not continue to increase in size throughout their adult life. In the vast majority of insects, growth is **determinate**, as there is a distinctive instar that marks the cessation of growth and molting. All insects with determinate growth become reproductively mature in this final instar, called the adult or imaginal instar. This reproductively mature individual is called an adult or **imago** (plural: **imagines** or **imagos**). In most insect orders it is fully winged, although secondary wing loss has occurred independently in the adults of a number of groups, such as lice, fleas, and certain parasitic flies, and in the adult females of all scale insects (Hemiptera: Coccoidea). In just one order of insects, the Ephemeroptera or mayflies, a **subimaginal instar** immediately precedes the final or imaginal instar. This **subimago**, although capable of flight, only rarely is reproductive; in the few mayfly groups in which the female mates as a subimago she dies without molting to an imago, so that the subimaginal instar actually is the final growth stage.

In some pterygote taxa the total number of pre-adult growth stages or instars may vary within a species depending on environmental conditions, such as developmental temperature, diet, and larval density. In many other species, the total number of instars (although not necessarily final adult size) is genetically

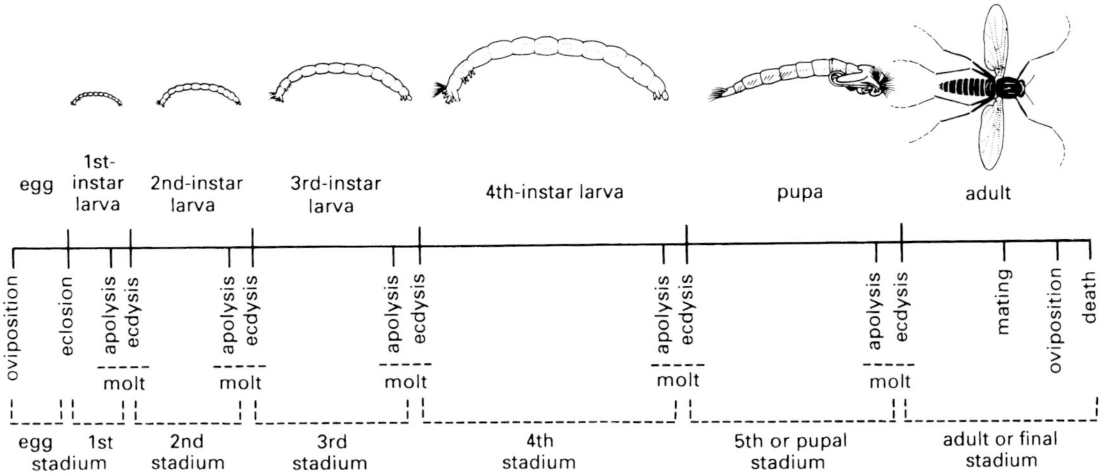

Fig. 6.1 Schematic drawing of the life cycle of a non-biting midge (Diptera: Chironomidae, *Chironomus*) showing the various events and stages of insect development.

determined and constant regardless of environmental conditions.

6.2 LIFE-HISTORY PATTERNS AND PHASES

Growth is an important part of an individual's **ontogeny**, the developmental history of that organism from egg to adult. Equally significant are the changes, both subtle and dramatic, that take place in body form as insects molt and grow larger. Changes in form (morphology) during ontogeny affect both external structures and internal organs, but only the external changes are apparent at each molt. We recognize three broad patterns of developmental morphological change during ontogeny, based on the degree of external alteration that occurs in the postembryonic phases of development.

The primitive developmental pattern, **ametaboly**, is for the hatchling to emerge from the egg in a form essentially resembling a miniature adult, lacking only genitalia. This pattern is retained by the primitively wingless orders, the silverfish (Zygentoma) and bristletails (Archaeognatha) (Box 9.3), whose adults continue to molt after sexual maturity. In contrast, all pterygote insects undergo a more or less marked change in form, a **metamorphosis**, between the immature phase of development and the winged or secondarily wingless (apterous) adult or imaginal phase. These insects can be subdivided according to two broad patterns of development, **hemimetaboly** (partial or incomplete metamorphosis; Fig. 6.2) and **holometaboly** (complete metamorphosis; Fig. 6.3 and the vignette for this chapter, which shows the life cycle of the monarch butterfly).

Developing wings are visible in external sheaths on the dorsal surface of nymphs of hemimetabolous insects except in the youngest immature instars. The term **exopterygote** has been applied to this type of "external" wing growth. In the past, insect orders with hemimetabolous and exopterygote development were grouped into "Hemimetabola" (also called Exopterygota), but this group is recognized now as applying to a grade of organization rather than to a monophyletic phylogenetic unit (Chapter 7). In contrast, pterygote orders displaying holometabolous development share the unique evolutionary innovation of a resting stage or **pupal instar** in which development of the major structural differences between immature (larval) and adult stages is concentrated. The orders that share this unique, derived pattern of development represent a clade called the Endopterygota or **Holometabola**. In the early branching Holometabola, expression of all adult features is retarded until the pupal stage; however, in more derived taxa including *Drosophila*, uniquely adult structures including wings may be present internally in larvae as groups of undifferentiated

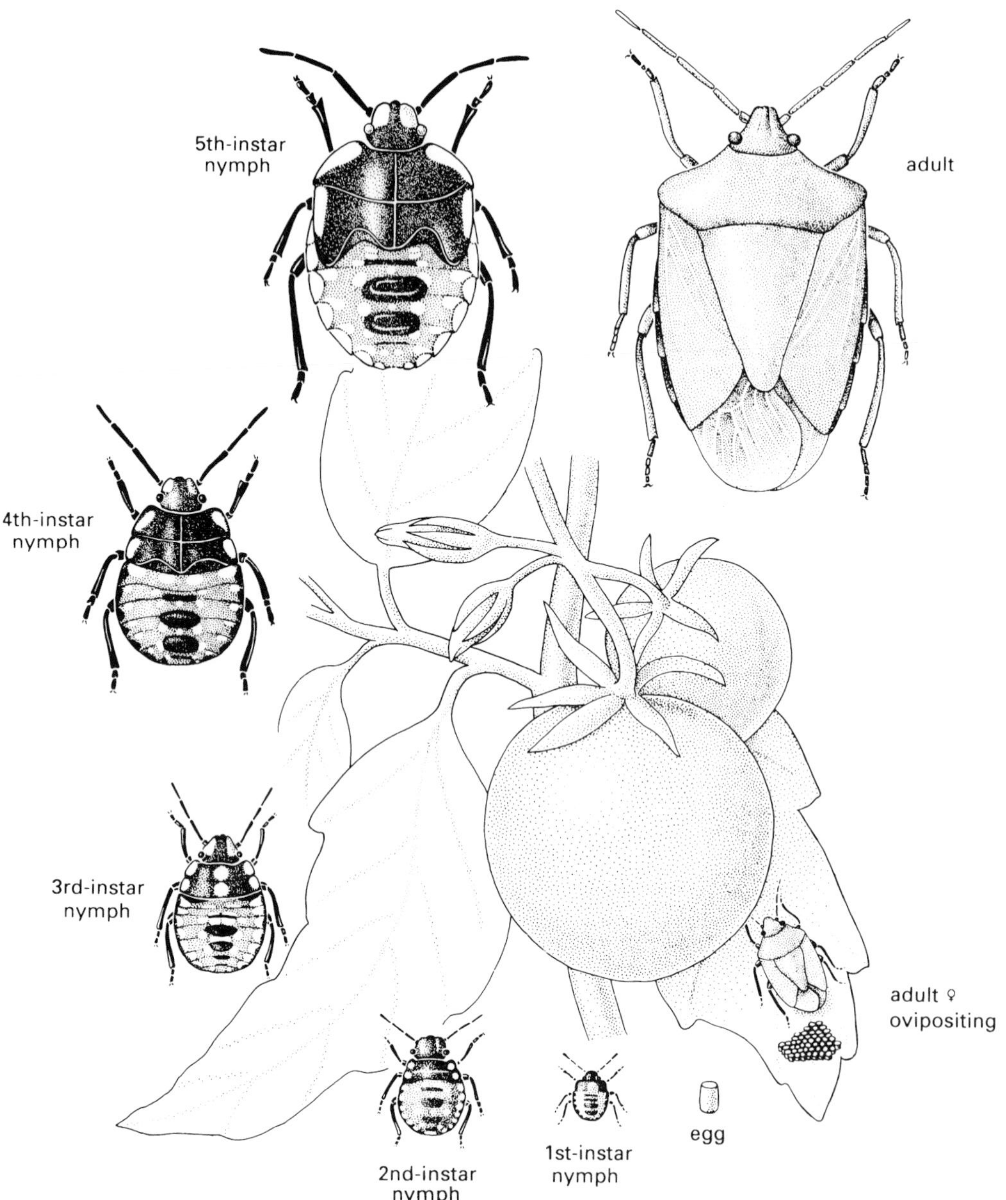

Fig. 6.2 The life cycle of a hemimetabolous insect, the southern green stink bug or green vegetable bug, *Nezara viridula* (Hemiptera: Pentatomidae), showing the eggs, nymphs of the five instars, and the adult bug on a tomato plant. This cosmopolitan and polyphagous bug is an important world pest of food and fiber crops. (After Hely et al. 1982.)

cells called **imaginal discs** (or **buds**) (Fig. 6.4), although they are scarcely visible until the pupal instar. Such wing development is called **endopterygote** because the wings develop from primordia in invaginated pockets of the integument and are everted only at the larval–pupal molt.

The evolution of holometaboly allows the immature and adult stages of an insect to specialize in different

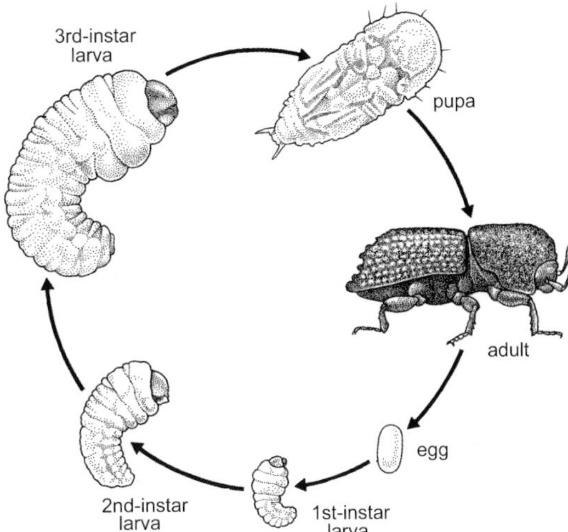

Fig. 6.3 Life cycle of a holometabolous insect, a bark beetle, *Ips grandicollis*, showing the egg, the three larval instars, the pupa, and the adult beetle. (After Johnson & Lyon 1991.)

resources, contributing to the successful radiation of the group (see section 8.5).

6.2.1 Embryonic phase

The egg stage begins as soon as the female deposits the mature egg. For practical reasons, the age of an egg is estimated from the time of its deposition even though the egg existed before oviposition. The beginning of the egg stage, however, need not mark the commencement of an individual insect's ontogeny, which actually begins when embryonic development within the egg is triggered by **activation**. This trigger usually results from fertilization in sexually reproducing insects, but in parthenogenetic species appears to be induced by various events at oviposition, including the entry of oxygen to the egg or mechanical distortion.

Following activation of the insect egg cell, the zygote nucleus subdivides by mitotic division to produce many daughter nuclei, giving rise to a syncytium. These nuclei and their surrounding cytoplasm, called cleavage **energids**, migrate to the egg periphery where the membrane infolds leading to cellularization of the superficial layer to form the one-cell thick blastoderm. This distinctive superficial cleavage during early embryogenesis in insects is the result of the large amount of yolk in the egg. The blastoderm usually gives rise to all the cells of the larval body, whereas the central yolky part of the egg provides the nutrition for the developing embryo and will be used up by the time of **eclosion**, or emergence from the egg.

Regional differentiation of the blastoderm leads to the formation of the germ **anlage** or germ disc (Fig. 6.5a), which is the first sign of the developing embryo, whereas the remainder of the blastoderm becomes a thin membrane, the **serosa**, or embryonic cover. Next, the germ anlage develops an infolding in a process called gastrulation (Fig. 6.5b) and sinks into the yolk, forming a two-layered embryo containing the amniotic cavity (Fig. 6.5c). After gastrulation, the germ anlage becomes the **germ band**, which externally is characterized by segmental organization (commencing in Fig. 6.5d with the formation of the protocephalon). The germ band essentially forms the ventral regions of the future body, which progressively differentiates with the head, body segments, and appendages becoming increasingly well defined (Fig. 6.5e–g). At this time the embryo undergoes movement called **katatrepsis** which brings it into its final position in the egg. Later, near the end of embryogenesis (Fig. 6.5h,i), the edges of the germ band grow over the remaining yolk and fuse mid-dorsally to form the lateral and dorsal parts of the insect – a process called **dorsal closure**.

In the well-studied *Drosophila*, the complete embryo is large, and becomes segmented at the cellularization stage, termed "long germ" (as in all studied Diptera, Coleoptera, and Hymenoptera). In contrast, in "short-germ" insects (phylogenetically earlier branching taxa, including locusts) the embryo derives from only a small region of the blastoderm and the posterior segments are added post-cellularization, during subsequent growth. In the developing long-germ embryo, the syncytial phase is followed by cell membrane intrusion to form the blastoderm phase.

Functional specialization of cells and tissues occurs during the latter period of embryonic development, so that by the time of hatching (Fig. 6.5j) the embryo is a tiny proto-insect crammed into an eggshell. In ametabolous and hemimetabolous insects, this stage may be recognized as a **pronymph** – a special hatching stage (section 8.5). Molecular developmental processes involved in organizing the polarity and differentiation of areas of the body, including segmentation, are reviewed in Box 6.1.

6.2.2 Larval or nymphal phase

Hatching from the egg may be by a pronymph, nymph,

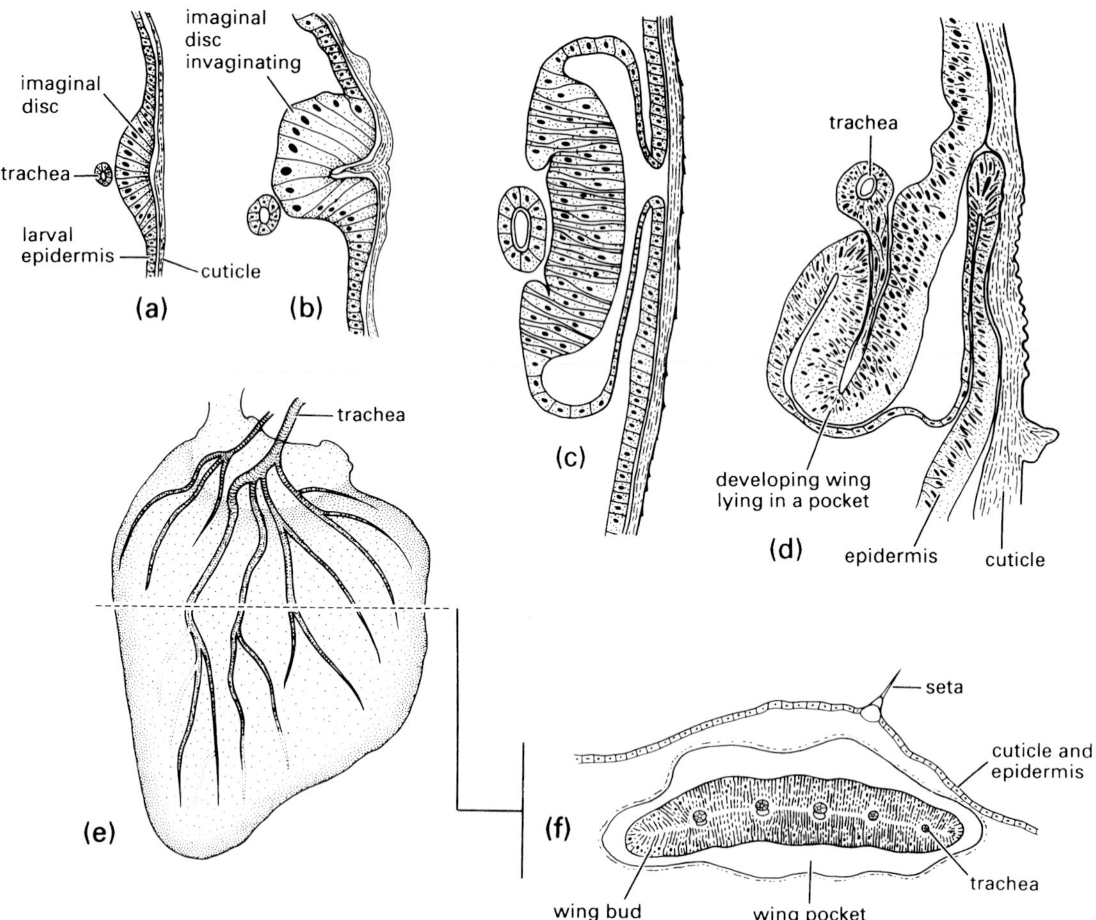

Fig. 6.4 Stages in the development of the wings of the cabbage white or cabbage butterfly, *Pieris rapae* (Lepidoptera: Pieridae). A wing imaginal disc in an (a) first-instar larva, (b) second-instar larva, (c) third-instar larva, and (d) fourth-instar larva; (e) the wing bud as it appears if dissected out of the wing pocket or (f) cut in cross-section in a fifth-instar larva. ((a–e) After Mercer 1900.)

or larva: eclosion conventionally marks the beginning of the first stadium, when the young insect is said to be in its first instar (Fig. 6.1). This stage ends at the first ecdysis when the old cuticle is cast to reveal the insect in its second instar. Third and often subsequent instars generally follow. Thus, the development of the immature insect is characterized by repeated molts separated by periods of feeding, with hemimetabolous insects generally undergoing more molts to reach adulthood than holometabolous insects.

All immature holometabolous insects are called **larvae**. Immature terrestrial insects with hemimetabolous development such as cockroaches (Blattodea), grasshoppers (Orthoptera), mantids (Mantodea), and bugs (Hemiptera) always are called **nymphs**. However, immature individuals of aquatic hemimetabolous insects (Odonata, Ephemeroptera, and Plecoptera), although possessing external wing pads at least in later instars, also are frequently, but incorrectly, referred to as larvae (or sometimes naiads). True larvae look very different from the final adult form in every instar, whereas nymphs more closely approach the adult appearance at each successive molt. Larval diets and lifestyles are very different from those of their adults. In

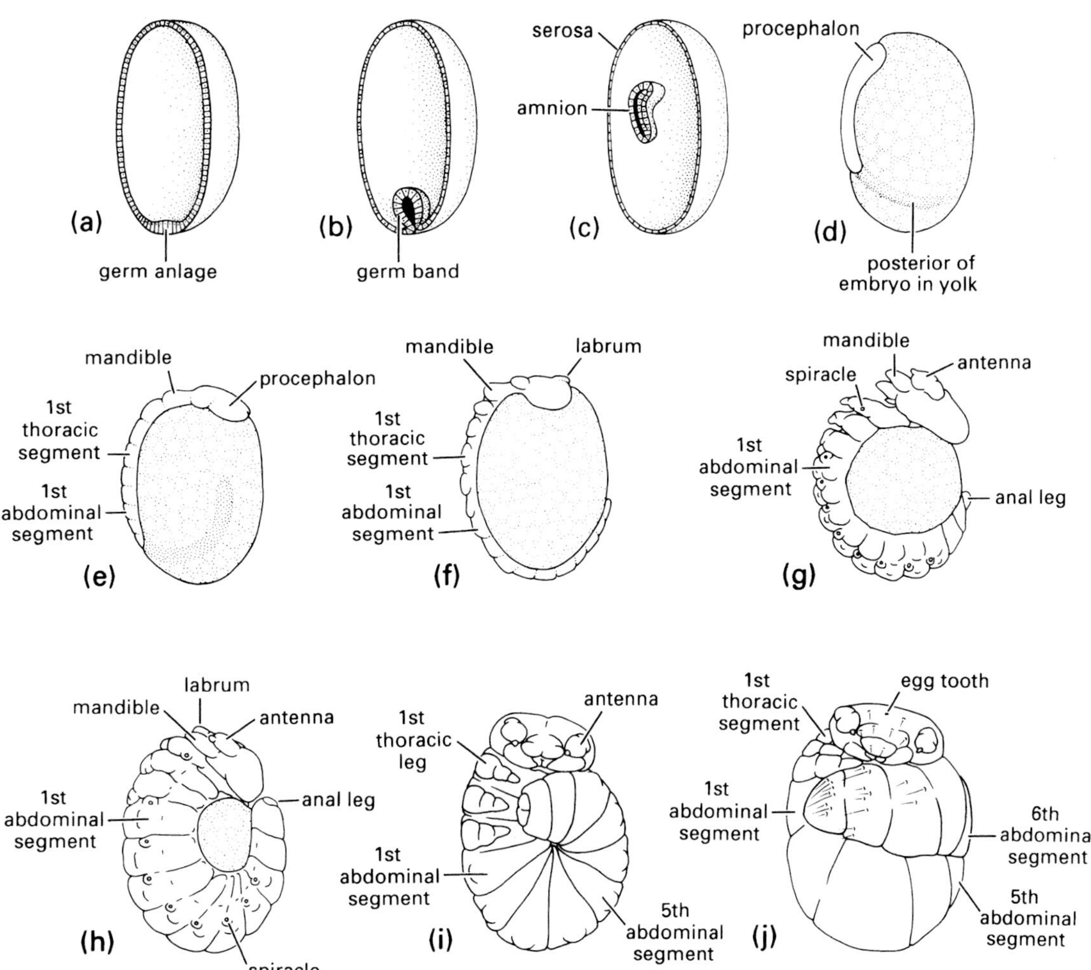

Fig. 6.5 Embryonic development of the scorpionfly *Panorpodes paradoxa* (Mecoptera: Panorpodidae): (a–c) schematic drawings of egg halves from which yolk has been removed to show position of embryo; (d–j) gross morphology of developing embryos at various ages. Age from oviposition: (a) 32 h; (b) 2 days; (c) 7 days; (d) 12 days; (e) 16 days; (f) 19 days; (g) 23 days; (h) 25 days; (i) 25–26 days; (j) full grown at 32 days. (After Suzuki 1985.)

contrast, nymphs often eat the same food and coexist with the adults of their species. Competition thus is rare between larvae and their adults but is likely to be prevalent between nymphs and their adults.

The great variety of endopterygote larvae can be classified into a few functional rather than phylogenetic types. Often the same larval type occurs convergently in unrelated orders. The three commonest forms are the polypod, oligopod, and apod larvae (Fig. 6.6). Lepidopteran caterpillars (Fig. 6.6a,b) are characteristic **polypod larvae** with cylindrical bodies with short thoracic legs and abdominal prolegs (pseudopods). Symphytan Hymenoptera (sawflies; Fig. 6.6c) and most Mecoptera also have polypod larvae. Such larvae are rather inactive and are mostly phytophagous. **Oligopod larvae** (Fig. 6.6d–f) lack abdominal prolegs but have functional thoracic legs and frequently prognathous mouthparts. Many are active predators but others are slow-moving detritivores living in soil or are phytophages. This larval type occurs in at least some members of most orders of insects but not in the Lepidoptera, Mecoptera, Diptera, Siphonaptera, or

Box 6.1 Molecular insights into insect development

The formation of segments in the early embryo of *Drosophila* is understood better than almost any other complex developmental process. Segmentation is controlled by a hierarchy of proteins known as transcription factors, which bind to DNA and act to enhance or repress the production of specific messages. In the absence of a message, the protein for which it codes is not produced; thus ultimately transcription factors act as molecular switches, turning on and off the production of specific proteins. In addition to controlling genes below them in the hierarchy, many transcription factors also act on other genes at the same level, as well as regulating their own concentrations. Mechanisms and processes observed in *Drosophila* have much wider relevance, including to vertebrate development, and information obtained from *Drosophila* has provided the key to cloning many human genes. However, we know *Drosophila* to be a highly derived fly, and it may not be a suitable model from which to derive generalities about insect development.

During oogenesis (section 6.2.1) in *Drosophila,* the anterior–posterior and dorsal–ventral axes are established by localization of maternal messenger RNAs (mRNAs) or proteins at specific positions within the egg. For example, the mRNAs from the *bicoid* (*bcd*) and *nanos* genes become localized at anterior and posterior ends of the egg, respectively. At oviposition, these messages are translated and proteins are produced that establish concentration gradients by diffusion from each end of the egg. These protein gradients differentially activate or inhibit zygotic genes lower in the segmentation hierarchy – as in the upper figure (after Nagy 1998), with zygotic gene hierarchy on the left and representative genes on the right – as a result of their differential thresholds of action. The first class of zygotic genes to be activated is the gap genes, for example *Kruppel* (*Kr*), which divide the embryo into broad, slightly overlapping zones from anterior to posterior. The maternal and gap proteins establish a complex of overlapping protein gradients that provide a chemical framework that controls the periodic (alternate segmental) expression of the pair-rule genes. For example, the pair-rule protein hairy is expressed in seven stripes along the length of the embryo while it is still in the syncytial stage. The pair-rule proteins, in addition to the proteins produced by genes higher in the hierarchy, then act to regulate the segment polarity genes, which are expressed with segmental periodicity and represent the final step in the determination of segmentation. Because there are many members of the various classes of segmentation genes, each row of

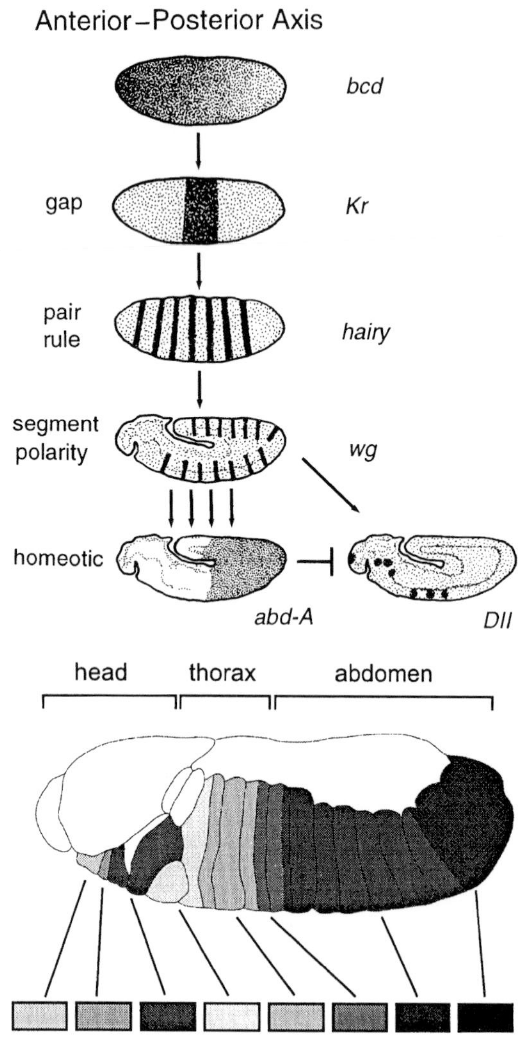

cells in the anterior–posterior axis must contain a unique combination and concentration of the transcription factors that inform cells of their position along the anterior–posterior axis.

Once the segmentation process is complete each developing segment is given its unique identity by the homeotic genes. Although these genes were first discovered in *Drosophila* it has since been established that they are very ancient, and a more or less complete subset of them is found in all multicellular animals. When this was realized it was agreed that this group of genes would be called the Hox genes, although both terms, homeotic and Hox, are still in use for the same group of genes. In many organisms these genes form a single cluster on one chromosome, although in *Drosophila* they are organized into two clusters, an anteriorly expressed Antennapedia complex (Antp-C) and a posteriorly expressed Bithorax complex (Bx-C). The composition of these clusters in *Drosophila* is as follows (from anterior to posterior): (Antp-C) – *labial* (*lab*), *proboscidea* (*pb*), *Deformed* (*Dfd*), *Sex combs reduced* (*Scr*), *Antennapedia* (*Antp*); (Bx-C) – *Ultrabithorax* (*Ubx*), *abdominal-A* (*abd-A*), and *Abdominal-B* (*Abd-B*), as illustrated in the lower figure of a *Drosophila* embryo (after Carroll 1995; Purugganan 1998). The evolutionary conservation of the Hox genes is remarkable for not only are they conserved in their primary structure but they follow the same order on the chromosome, and their temporal order of expression and anterior border of expression along the body correspond to their chromosomal position. In the lower figure the anterior zone of expression of each gene and the zone of strongest expression is shown (for each gene there is a zone of weaker expression posteriorly); as each gene switches on, protein production from the gene anterior to it is repressed.

The zone of expression of a particular Hox gene may be morphologically very different in different organisms so it is evident that Hox gene activities demarcate relative positions but not particular morphological structures. A single Hox gene may regulate directly or indirectly many targets; for example, *Ultrabithorax* regulates some 85–170 genes. These downstream genes may operate at different times and also have multiple effects (pleiotropy); for example, *wingless* in *Drosophila* is involved successively in segmentation (embryo), Malpighian tubule formation (larva), and leg and wing development (larva–pupa).

Boundaries of transcription factor expression are important locations for the development of distinct morphological structures, such as limbs, tracheae, and salivary glands. Studies of the development of legs and wings have revealed something about the processes involved. Limbs arise at the intersection between expression of wingless, engrailed, and decapentaplegic (dpp), a protein that helps to inform cells of their position in the dorsal–ventral axis. Under the influence of the unique mosaic of gradients created by these gene products, limb primordial cells are stimulated to express the gene *distal-less* (*Dll*) required for proximodistal limb growth. As potential limb primordial cells (anlage) are present on all segments, as are limb-inducing protein gradients, prevention of limb growth on inappropriate segments (i.e. the *Drosophila* abdomen) must involve repression of *Dll* expression on such segments. In Lepidoptera, in which larval prolegs typically are found on the third to sixth abdominal segments, homeotic gene expression is fundamentally similar to that of *Drosophila*. In the early lepidopteran embryo *Dll* and *Antp* are expressed in the thorax, as in *Drosophila*, with *abd-A* expression dominant in abdominal segments including 3–6, which are prospective for proleg development. Then a dramatic change occurs, with abd-A protein repressed in the abdominal proleg cell anlagen, followed by activation of *Dll* and up-regulation of *Antp* expression as the anlagen enlarge. Two genes of the Bithorax complex (Bx-C), *Ubx* and *abd-A*, repress *Dll* expression (and hence prevent limb formation) in the abdomen of *Drosophila*. Therefore, expression of prolegs in the caterpillar abdomen results from repression of Bx-C proteins thus derepressing *Dll* and *Antp* and thereby permitting their expression in selected target cells with the result that prolegs develop.

A somewhat similar condition exists with respect to wings, in that the default condition is presence on all thoracic and abdominal segments with Hox gene repression reducing the number from this default condition. In the prothorax, the homeotic gene *Scr* has been shown to repress wing development. Other effects of *Scr* expression in the posterior head, labial segment, and prothorax appear homologous across many insects, including ventral migration and fusion of the labial lobes, specification of labial palps, and development of sex combs on male prothoracic legs. Experimental mutational damage to *Scr* expression leads, amongst other deformities, to appearance of wing primordia from a group of cells located just dorsal to the prothoracic leg base. These mutant prothoracic wing anlagen are situated very close to the site predicted by Kukalová-Peck from paleontological evidence (section 8.4, Fig. 8.4b). Furthermore, the apparent default condition (lack of repression of wing expression) would produce an insect resembling the hypothesized "protopterygote", with winglets present on all segments.

Regarding the variations in wing expression seen in the pterygotes, *Ubx* activity differs in *Drosophila* between the meso- and metathoracic imaginal discs; the anterior produces a wing, the posterior a haltere. *Ubx* is unexpressed in the wing (mesothoracic) imaginal disc but is strongly expressed in the metathoracic disc, where its activity suppresses wing and enhances haltere formation. However, in some studied non-dipterans *Ubx* is expressed as in *Drosophila* – not in the fore-wing but strongly in the hind-wing imaginal disc – despite the elaboration of a complete hind wing as in butterflies or beetles. Thus, very different wing morphologies seem to result from variation in "downstream" response to wing-pattern genes regulated by *Ubx* rather than from homeotic control.

Clearly, much is yet to be learnt concerning the multiplicity of morphological outcomes from the interaction between Hox genes and their downstream interactions with a wide range of genes. It is tempting to relate major variation in Hox pathways with morphological disparities associated with high-level taxonomic rank (e.g. animal classes), more subtle changes in Hox regulation with intermediate taxonomic levels (e.g. orders/suborders), and changes in downstream regulatory/functional genes perhaps with suborder/family rank. Notwithstanding some progress in the case of the Strepsiptera (q.v.), such simplistic relationships between a few well-understood major developmental features and taxonomic radiations may not lead to great insight into insect macroevolution in the immediate future. Estimated phylogenies from other sources of data will be necessary to help interpret the evolutionary significance of homeotic changes for some time to come.

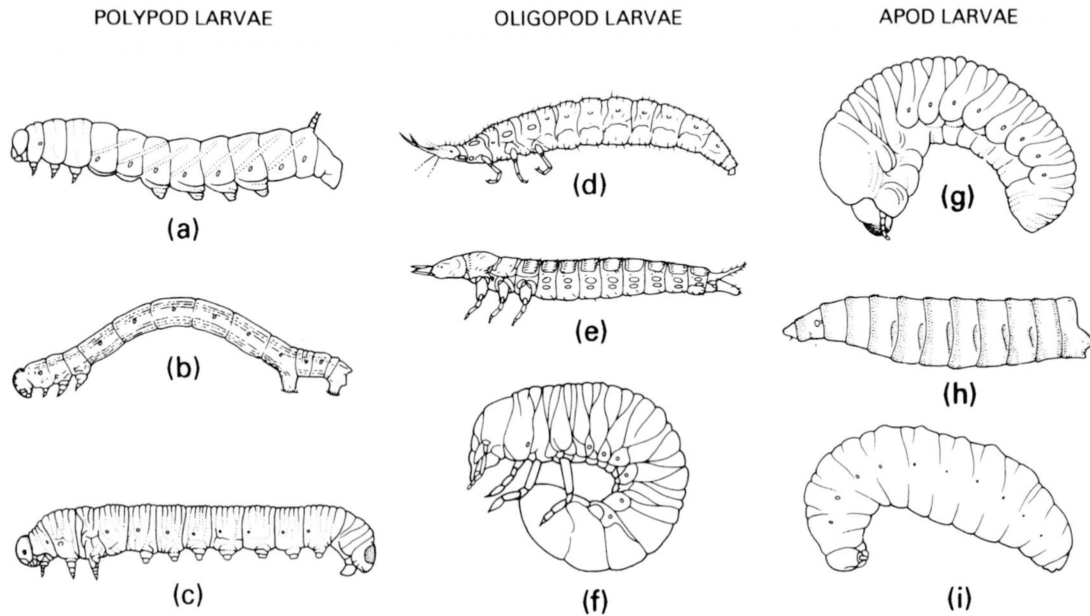

Fig. 6.6 Examples of larval types. Polypod larvae: (a) Lepidoptera: Sphingidae; (b) Lepidoptera: Geometridae; (c) Hymenoptera: Diprionidae. Oligopod larvae: (d) Neuroptera: Osmylidae; (e) Coleoptera: Carabidae; (f) Coleoptera: Scarabaeidae. Apod larvae: (g) Coleoptera: Scolytidae; (h) Diptera: Calliphoridae; (i) Hymenoptera: Vespidae. ((a,e–g) After Chu 1949; (b,c) after Borror et al. 1989; (h) after Ferrar 1987; (i) after CSIRO 1970.)

Strepsiptera. **Apod larvae** (Fig. 6.6g–i) lack true legs and are usually worm-like or maggot-like, living in soil, mud, dung, decaying plant or animal matter, or within the bodies of other organisms as parasitoids (Chapter 13). The Siphonaptera, aculeate Hymenoptera, nematoceran Diptera, and many Coleoptera typically have apod larvae with a well-developed head, whereas in the maggots of higher Diptera the mouth hooks may be the only obvious evidence of the cephalic region. The grub-like apod larvae of some parasitic and gall-inducing wasps and flies are greatly reduced in external structure and are difficult to identify to order level even by a specialist entomologist. Furthermore, the early-instar larvae of some parasitic wasps resemble a naked embryo but change into typical apod larvae in later instars.

A major change in form during the larval phase, such as different larval types in different instars, is called **larval heteromorphosis** (or **hypermetamorphosis**). In the Strepsiptera and certain beetles this involves an active first-instar larva, or **triungulin**, followed by several grub-like, inactive, sometimes legless, later-instar larvae. This developmental phenomenon occurs most commonly in parasitic insects in which a mobile first instar is necessary for host location and entry. Larval heteromorphosis and diverse larval types are typical of many parasitic wasps, as mentioned above.

6.2.3 Metamorphosis

All pterygote insects undergo varying degrees of transformation from the immature to the adult phase of their life history. Some exopterygotes, such as cockroaches, show only slight morphological changes during post-embryonic development, whereas the body is largely reconstructed at metamorphosis in many endopterygotes. Only the Holometabola (= Endopterygota) have a metamorphosis involving a pupal stadium, during which adult structures are elaborated from larval structures. Alterations in body shape, which are the essence of metamorphosis, are brought about by differential growth of various body parts. Organs that will function in the adult but that were undeveloped in the larva grow at a faster rate than the body average. The accelerated growth of wing pads is the most obvious example, but legs, genitalia, gonads, and other internal organs may increase in size and complexity to a considerable extent.

The onset of metamorphosis generally is associated with the attainment of a certain body size, which is thought to program the brain for metamorphosis, resulting in altered hormone levels. Metamorphosis in most studied beetles, however, shows considerable independence from the influence of the brain, especially during the pupal instar. In most insects, a reduction in the amount of circulating juvenile hormone (as a result of reduction of corpora allata activity) is essential to the initiation of metamorphosis. (The physiological events are described in section 6.3.)

The molt into the pupal instar is called **pupation**, or the larval–pupal molt. Many insects survive conditions unfavorable for development in the "resting", non-feeding pupal stage, but often what appears to be a pupa is actually a fully developed adult within the pupal cuticle, referred to as a **pharate** (cloaked) adult. Typically, a protective cell or cocoon surrounds the pupa and then, prior to emergence, the pharate adult; only certain Coleoptera, Diptera, Lepidoptera, and Hymenoptera have unprotected pupae.

Several pupal types (Fig. 6.7) are recognized and these appear to have arisen convergently in different orders. Most pupae are **exarate** (Fig. 6.7a–d) – their appendages (e.g. legs, wings, mouthparts, and antennae) are not closely appressed to the body (see Plate 3.2, facing p. 14); the remaining pupae are **obtect** (Fig. 6.7g–j) – their appendages are cemented to the body and the cuticle is often heavily sclerotized (as in almost all Lepidoptera). Exarate pupae can have articulated mandibles (**decticous**), that the pharate adult uses to cut through the cocoon, or the mandibles can be non-articulated (**adecticous**), in which case the adult usually first sheds the pupal cuticle and then uses its mandibles and legs to escape the cocoon or cell. In some cyclorrhaphous Diptera (the Schizophora) the adecticous exarate pupa is enclosed in a **puparium** (Fig. 6.7e,f) – the sclerotized cuticle of the last larval instar. Escape from the puparium is facilitated by eversion of a membranous sac on the head of the emerging adult, the **ptilinum**. Insects with obtect pupae may lack a cocoon, as in coccinellid beetles and most nematocerous and orthorrhaphous Diptera. If a cocoon is present, as in most Lepidoptera, emergence from the cocoon is either by the pupa using backwardly directed abdominal spines or a projection on the head, or an adult emerges from the pupal cuticle before escaping the cocoon, sometimes helped by a fluid that dissolves the silk.

6.2.4 Imaginal or adult phase

Except for the mayflies, insects do not molt again once the adult phase is reached. The adult, or imaginal, stage has a reproductive role and is often the dispersive stage in insects with relatively sedentary larvae. After the imago emerges from the cuticle of the previous instar (eclosion), it may be reproductively competent almost immediately or there may be a period of maturation in readiness for sperm transfer or oviposition. Depending on species and food availability, there are from one to several reproductive cycles in the adult stadium. The adults of certain species, such as some mayflies, midges, and male scale insects, are very short-lived. These insects have reduced or no mouthparts and fly for only a few hours or at the most a day or two – they simply mate and die. Most adult insects live at least a few weeks, often a few months and sometimes for several years; termite reproductives and queen ants and bees are particularly long-lived.

Adult life begins at eclosion from the pupal cuticle. Metamorphosis, however, may have been complete for some hours, days, or weeks previously and the pharate

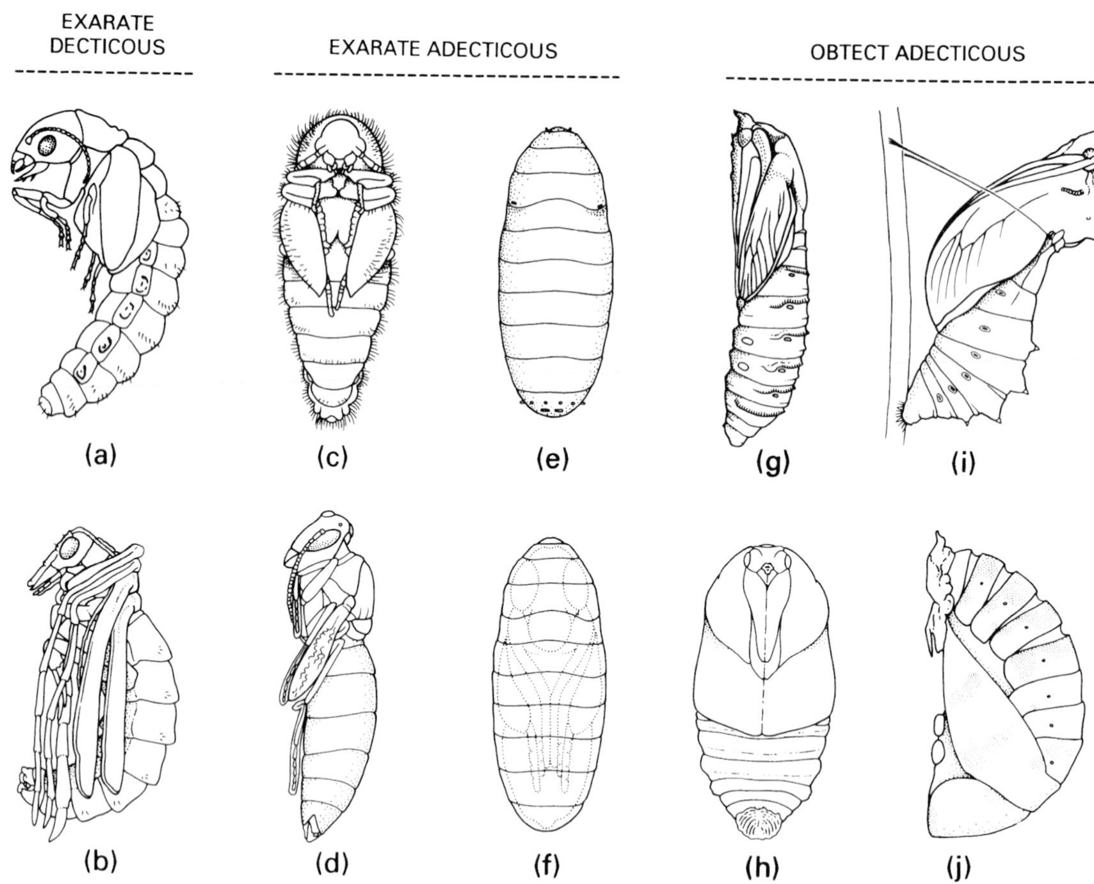

Fig. 6.7 Examples of pupal types. Exarate decticous pupae: (a) Megaloptera: Sialidae; (b) Mecoptera: Bittacidae. Exarate adecticous pupae: (c) Coleoptera: Dermestidae; (d) Hymenoptera: Vespidae; (e,f) Diptera: Calliphoridae, puparium and pupa within. Obtect adecticous pupae: (g) Lepidoptera: Cossidae; (h) Lepidoptera: Saturniidae; (i) Lepidoptera: Papilionidae, chrysalis; (j) Coleoptera: Coccinellidae. ((a) After Evans 1978; (b,c,e,g) after CSIRO 1970; (d) after Chu 1949; (h) after Common 1990; (i) after Common & Waterhouse 1972; (j) after Palmer 1914.)

adult may have rested in the pupal cuticle until the appropriate environmental trigger for emergence. Changes in temperature or light and perhaps chemical signals may synchronize adult emergence in most species.

Hormonal control of emergence has been studied most comprehensively in Lepidoptera, especially in the tobacco hornworm, *Manduca sexta* (Lepidoptera: Sphingidae), notably by James Truman, Lynn Riddiford, and colleagues. The description of the following events at eclosion are based largely on *M. sexta* but are believed to be similar in other insects and at other molts. At least five hormones are involved in eclosion (see also section 6.3). A few days prior to eclosion the ecdysteroid level declines, and a series of physiological and behavioral events are initiated in preparation for ecdysis, including the release of two neuropeptides. **Ecdysis triggering hormone** (ETH), from epitracheal glands called Inka cells, and **eclosion hormone** (EH), from neurosecretory cells in the brain, act in concert to trigger pre-eclosion behavior, such as seeking a site suitable for ecdysis and movements to aid later extrication from the old cuticle. ETH is released first and ETH and EH stimulate each other's release, forming a positive feedback loop. The build-up of EH also releases **crustacean cardioactive peptide** (CCAP) from cells

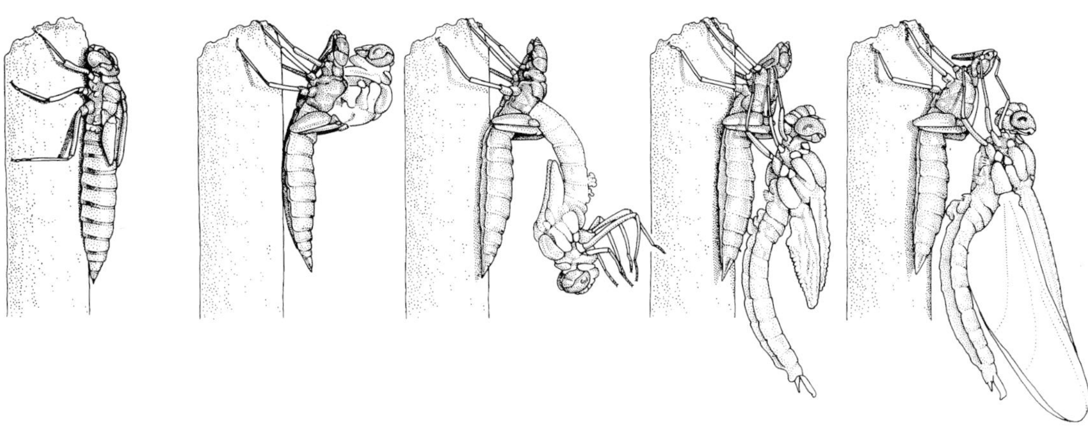

Fig. 6.8 The nymphal–imaginal molt of a male dragonfly of *Aeshna cyanea* (Odonata: Aeshnidae). The final-instar nymph climbs out of the water prior to the shedding of its cuticle. The old cuticle splits mid-dorsally, the teneral adult frees itself, swallows air and must wait many hours for its wings to expand and dry. (After Blaney 1976.)

in the ventral nerve cord. CCAP switches off pre-eclosion behavior and switches on eclosion behavior, such as abdominal contraction and wing-base movements, and accelerates heartbeat. EH appears also to permit the release of further neurohormones – **bursicon** and **cardiopeptides** – that are involved in wing expansion after ecdysis. The cardiopeptides stimulate the heart, facilitating movement of hemolymph into the thorax and thus into the wings. Bursicon induces a brief increase in cuticle plasticity to permit wing expansion, followed by sclerotization of the cuticle in its expanded form.

The newly emerged, or **teneral**, adult has soft cuticle, which permits expansion of the body surface by swallowing air, by taking air into the tracheal sacs, and by locally increasing hemolymph pressure by muscular activity. The wings normally hang down (Fig. 6.8; see also Plate 3.4), which aids their inflation. Pigment deposition in the cuticle and epidermal cells occurs just before or after emergence and is either linked to, or followed by, sclerotization of the body cuticle under the influence of the neurohormone bursicon.

Following emergence from the pupal cuticle, many holometabolous insects void a fecal fluid called the **meconium**. This represents the metabolic wastes that have accumulated during the pupal stadium. Sometimes the teneral adult retains the meconium in the rectum until sclerotization is complete, thus aiding increase in body size.

Reproduction is the main function of adult life and the length of the imaginal stadium, at least in the female, is related to the duration of egg production. Reproduction is discussed in detail in Chapter 5. Senescence correlates with termination of reproduction and death may be predetermined in the ontogeny of an insect. Females may die after egg deposition and males may die after mating. An extended post-reproductive life is important in distasteful, aposematic insects to allow predators to learn the distastefulness of the prey at a developmental period when prey individuals are expendable (section 14.4).

6.3 PROCESS AND CONTROL OF MOLTING

For practical reasons an instar is defined from ecdysis to ecdysis (Fig. 6.1), but morphologically and physiologically a new instar comes into existence at the time of **apolysis** when the epidermis separates from the cuticle of the previous stage. Apolysis is difficult to detect in most insects but knowledge of its occurrence may be important because many insects spend a substantial period in the pharate state (cloaked within the cuticle of the previous instar) awaiting conditions favorable for emergence as the next stage. Insects often survive adverse conditions as pharate pupae or pharate adults (e.g. some diapausing adult moths) because in this state the double cuticular layer restricts water loss during a developmental period during which metabolism is reduced and requirements for gaseous exchange are minimal.

Molting is a complex process involving hormonal, behavioral, epidermal, and cuticular changes that lead up to the shedding of the old cuticle. The epidermal cells are actively involved in molting – they are responsible for partial breakdown of the old cuticle and formation of the new cuticle. The molt commences with the retraction of the epidermal cells from the inner surface of the old cuticle, usually in an antero-posterior direction. This separation is not total because muscles and sensory nerves retain their connection with the old cuticle. Apolysis is either correlated with or followed by mitotic division of the epidermal cells leading to increases in the volume and surface area of the epidermis. The subcuticular or **apolysial space** formed after apolysis becomes filled with the secreted but inactive molting fluid. The chitinolytic and proteolytic enzymes of the molting fluid are not activated until the epidermal cells have laid down the protective outer layer of a new cuticle. Then the inner part of the old cuticle (the endocuticle) is lysed and presumably resorbed, while the new pharate cuticle continues to be deposited as an undifferentiated procuticle. Ecdysis commences with the remnants of the old cuticle splitting along the dorsal midline as a result of increase in hemolymph pressure. The cast cuticle consists of the indigestible protein, lipid, and chitin of the old epicuticle and exocuticle. Once free of the constraints of this previous "skin", the newly ecdysed insect expands the new cuticle by swallowing air or water and/or by increasing hemolymph pressure in different body parts to smooth out the wrinkled and folded epicuticle and stretch the procuticle. After cuticular expansion, some or much of the body surface may become sclerotized by the chemical stiffening and darkening of the procuticle to form exocuticle (section 2.1). However, in larval insects most of the body cuticle remains membranous and exocuticle is confined to the head capsule. Following ecdysis, more proteins and chitin are secreted from the epidermal cells thus adding to the inner part of the procuticle, the endocuticle, which may continue to be deposited well into the intermolt period. Sometimes the endocuticle is partially sclerotized during the stadium and frequently the outer surface of the cuticle is covered in wax secretions. Finally, the stadium draws to an end and apolysis is initiated once again.

The above events are effected by hormones acting on the epidermal cells to control the cuticular changes and also on the nervous system to co-ordinate the behaviors associated with ecdysis. Hormonal regulation of molting has been studied most thoroughly at metamorphosis, when endocrine influences on molting *per se* are difficult to separate from those involved in the control of morphological change. The classical view of the hormonal regulation of molting and metamorphosis is presented schematically in Fig. 6.9; the endocrine centers and their hormones are described in more detail in Chapter 3. Three major types of hormones control molting and metamorphosis:

1 neuropeptides, including prothoracicotropic hormone (PTTH), ETH, and EH;
2 ecdysteroids;
3 juvenile hormone (JH), which may occur in several different forms even in the same insect.

Neurosecretory cells in the brain secrete PTTH, which passes down nerve axons to the corpora cardiaca, a pair of neuroglandular bodies that store and later release PTTH into the hemolymph. The PTTH stimulates ecdysteroid synthesis and secretion by the prothoracic or molting glands. Ecdysteroid release then initiates the changes in the epidermal cells that lead to the production of new cuticle. The characteristics of the molt are regulated by JH from the corpora allata; JH inhibits the expression of adult features so that a high hemolymph level (titer) of JH is associated with a larval–larval molt, and a lower titer with a larval–pupal molt; JH is absent at the pupal–adult molt.

Ecdysis is mediated by ETH and EH, and EH at least appears to be important at every molt in the life history of perhaps all insects. This neuropeptide acts on a steroid-primed central nervous system to evoke the co-ordinated motor activities associated with escape from the old cuticle. Eclosion hormone derives its name from the pupal–adult ecdysis, or eclosion, for which its importance was first discovered and before its wider role was realized. Indeed, the association of EH with molting appears to be ancient, as other arthropods (e.g. crustaceans) have EH homologues. In the well-studied tobacco hornworm (section 6.2.4), the more recently discovered ETH is as important to ecdysis as EH, with ETH and EH stimulating each other's release, but the taxonomic distribution of ETH is not yet known. In many insects, another neuropeptide, bursicon, controls sclerotization of the exocuticle and postmolt deposition of endocuticle.

The relationship between the hormonal environment and the epidermal activities that control molting and cuticular deposition in a lepidopteran, the tobacco hornworm *Manduca sexta*, are presented in Fig. 6.10.

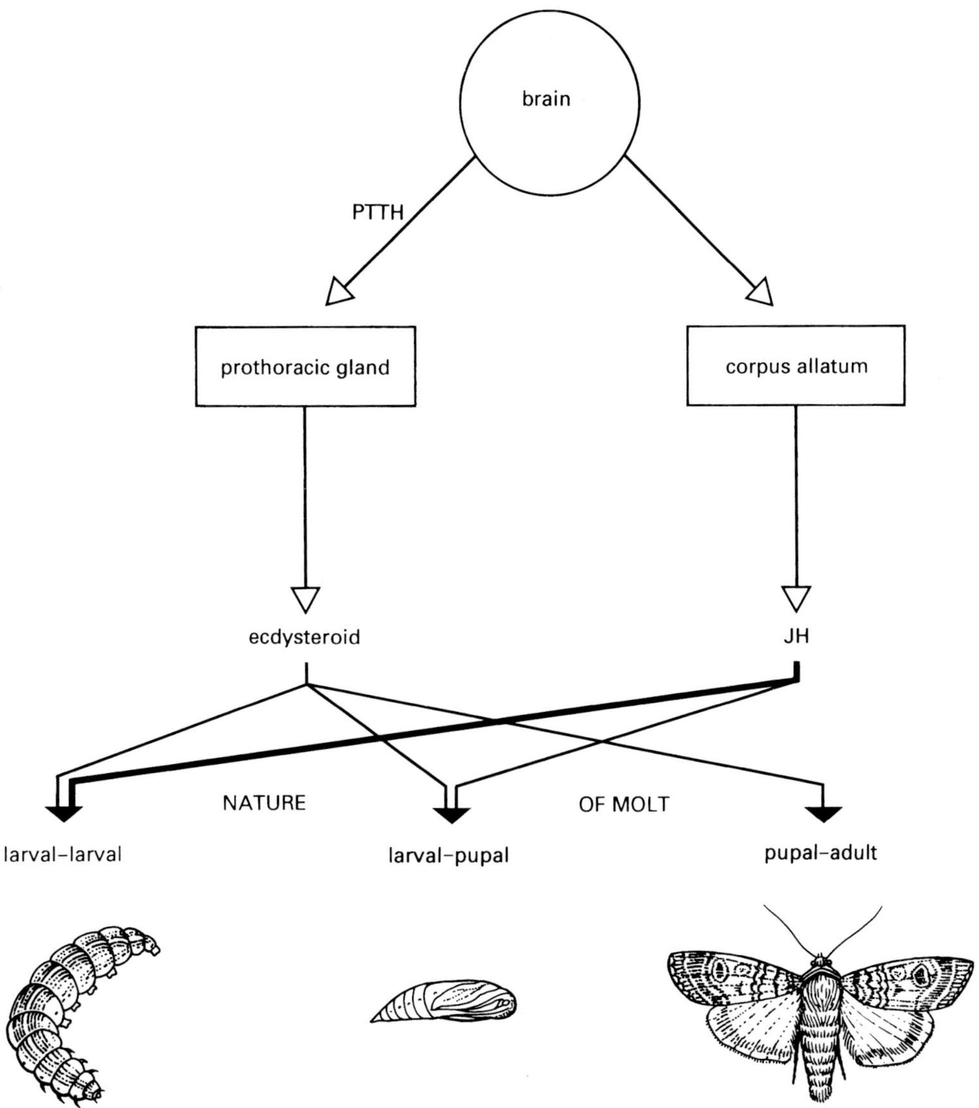

Fig. 6.9 Schematic diagram of the classical view of endocrine control of the epidermal processes that occur in molting and metamorphosis in an endopterygote insect. This scheme simplifies the complexity of ecdysteroid and JH secretion and does not indicate the influence of neuropeptides such as eclosion hormone. JH, juvenile hormone; PTTH, prothoracicotropic hormone. (After Richards 1981.)

Only now are we beginning to understand how hormones regulate molting and metamorphosis at the cellular and molecular levels. However, detailed studies on the tobacco hornworm clearly show the correlation between the ecdysteroid and JH titers and the cuticular changes that occur in the last two larval instars and in prepupal development. During the molt at the end of the fourth larval instar, the epidermis responds to the surge of ecdysteroid by halting synthesis of endocuticle and the blue pigment insecticyanin. A new epicuticle is synthesized, much of the old cuticle is digested, and resumption of endocuticle and insecticyanin production

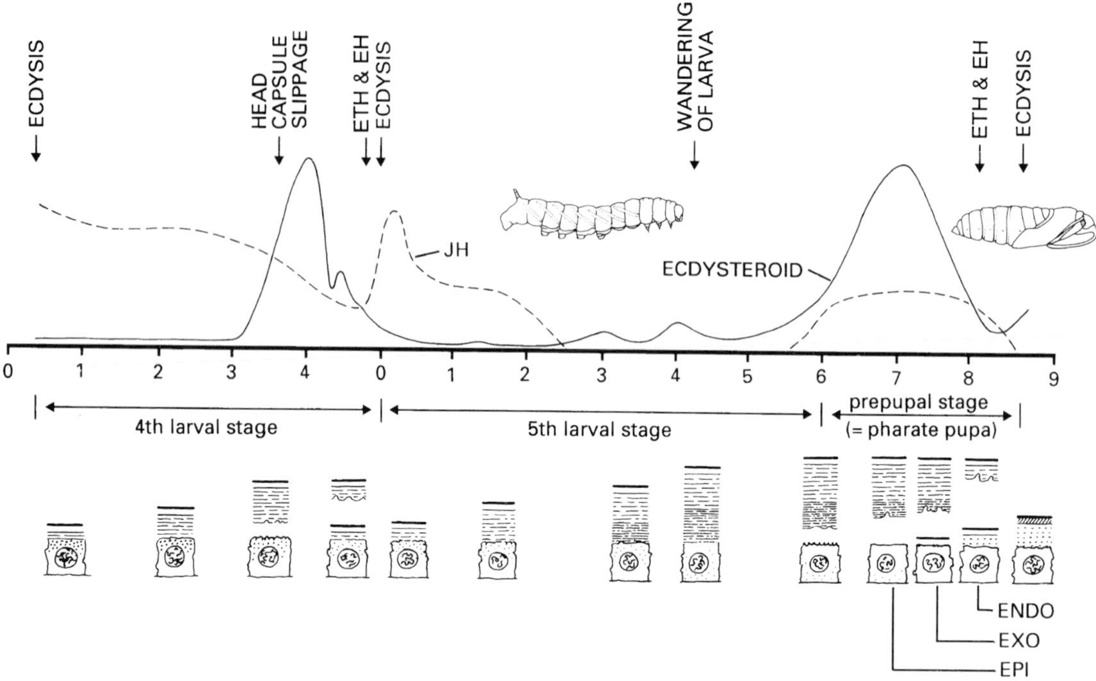

Fig. 6.10 Diagrammatic view of the changing activities of the epidermis during the fourth and fifth larval instars and prepupal (= pharate pupal) development in the tobacco hornworm, *Manduca sexta* (Lepidoptera: Sphingidae) in relation to the hormonal environment. The dots in the epidermal cells represent granules of the blue pigment insecticyanin. ETH, ecdysis triggering hormone; EH, eclosion hormone; JH, juvenile hormone; EPI, EXO, ENDO, deposition of pupal epicuticle, exocuticle, and endocuticle, respectively. The numbers on the *x*-axis represent days. (After Riddiford 1991.)

occurs by the time of ecdysis. In the final larval instar the JH declines to undetectable levels, allowing small rises in ecdysteroid that first stimulate the epidermis to produce a stiffer cuticle with thinner lamellae and then elicit wandering in the larva. When ecdysteroid initiates the next molt, the epidermal cells produce pupal cuticle as a result of the activation of many new genes. The decline in ecdysteroid level towards the end of each molt seems to be essential for, and may be the physiological trigger causing, ecdysis to occur. It renders the tissues sensitive to EH and permits the release of EH into the hemolymph (see section 6.2.4 for further discussion of the actions of eclosion hormone). Apolysis at the end of the fifth larval instar marks the beginning of a prepupal period when the developing pupa is pharate within the larval cuticle. Differentiated exocuticle and endocuticle appear at this larval–pupal molt. During larval life, the epidermal cells covering most of the body do not produce exocuticle, so the caterpillar's cuticle is soft and flexible allowing considerable growth within an instar as a result of feeding.

6.4 VOLTINISM

Insects are short-lived creatures, whose lives can be measured by their **voltinism** – the numbers of generations per year. Most insects take a year or less to develop, with either one generation per year (**univoltine** insects), or two (**bivoltine** insects), or more than two (**multivoltine**, or **polyvoltine**, insects). Generation times in excess of one year (**semivoltine** insects) are found, for example, amongst some inhabitants of the polar extremes, where suitable conditions for development may exist for only a few weeks in each year. Large insects that rely upon nutritionally poor diets also develop slowly over many years. For example, periodic cicadas feeding on sap from tree roots may take either

13 or 17 years to mature, and beetles that develop within dead wood have been known to emerge after more than 20 years' development.

Most insects do not develop continuously throughout the year, but arrest their development during unfavorable times by quiescence or diapause (section 6.5). Many univoltine and some bivoltine species enter diapause at some stage, awaiting suitable conditions before completing their life cycle. For some univoltine insects, many social insects, and others that take longer than a year to develop, adult longevity may extend to several years. In contrast, the adult life of multivoltine insects may be as little as a few hours at low tide for marine midges such as *Clunio* (Diptera: Chironomidae), or a single evening for many Ephemeroptera.

Multivoltine insects tend to be small and fast-developing, using resources that are more evenly available throughout the year. Univoltinism is common amongst temperate insects, particularly those that use resources that are seasonally restricted. These might include insects whose aquatic immature stages rely on spring algal bloom, or phytophagous insects using short-lived annual plants. Bivoltine insects include those that develop slowly on evenly spread resources, and those that track a bimodally distributed factor, such as spring and fall temperature. Some species have fixed voltinism patterns, whereas others may vary with geography, particularly in insects with broad latitudinal or elevational ranges.

6.5 DIAPAUSE

The developmental progression from egg to adult often is interrupted by a period of dormancy. This occurs particularly in temperate areas when environmental conditions become unsuitable, such as in seasonal extremes of high or low temperatures, or drought. Dormancy may occur in summer (**aestivation** (estivation)) or in winter (**hibernation**), and may involve either quiescence or diapause. **Quiescence** is a halted or slowed development as a direct response to unfavorable conditions, with development resuming immediately favorable conditions return. In contrast, **diapause** involves arrested development combined with adaptive physiological changes, with development recommencing not necessarily on return of suitable conditions, but only following particular physiological stimuli. Distinguishing between quiescence and diapause requires detailed study.

Diapause at a fixed time regardless of varied environmental conditions is termed **obligatory**. Univoltine insects (those with one generation per year) often have obligatory diapause to extend an essentially short life cycle to one full year. Diapause that is optional is termed **facultative**, and this occurs widely in insects, including many bi- or multivoltine insects in which diapause occurs only in the generation that must survive the unfavorable conditions. Facultative diapause can be food induced: thus when summer aphid prey populations are low the ladybird beetles *Hippodamia convergens* and *Semidalia unidecimnotata* aestivate, but if aphids remain in high densities, as in irrigated crops, the predators will continue to develop without diapause.

Diapause can last from days to months or in rare cases years, and can occur in any life-history stage from egg to adult. The diapausing stage predominantly is fixed within any species and can vary between close relatives. Egg and/or pupal diapause is common, probably because these stages are relatively closed systems, with only gases being exchanged during embryogenesis and metamorphosis, respectively, allowing better survival during environmental stress. In the adult stage, **reproductive diapause** describes the cessation or suspension of reproduction in mature insects. In this state metabolism may be redirected to migratory flight (section 6.7), production of cryoprotectants (section 6.6.1), or simply reduced during conditions inclement for the survival of adult (and/or immature) stages. Reproduction commences post-migration or when conditions for successful oviposition and immature stage development return.

Much research on diapause has been carried out in Japan in relation to silk production from cultured silkworms (*Bombyx mori*). Optimal silk production comes from the generation with egg diapause, but this conflicts with a commercial need for continuous production, which comes from individuals reared from non-diapausing eggs. The complex mechanisms that promote and break diapause in this species are now well understood. However, these mechanisms may not apply generally, and as the example of *Aedes* below indicates, several different mechanisms may be at play in different, even closely related, insects, and much is still to be discovered.

Major environmental cues that induce and/or terminate diapause are photoperiod, temperature, food quality, moisture, pH, and chemicals including oxygen, urea, and plant secondary compounds. Identification of the contribution of each may be difficult, as for example in species of the mosquito genus *Aedes* that lay diapausing eggs into seasonally dry pools or containers. Flooding of the oviposition site at any time may terminate embryonic diapause in some *Aedes* species. In other species, many successive inundations may be required to break diapause, with the cues apparently including chemical changes such as lowering of pH by microbial decomposition of pond detritus. Furthermore, one environmental cue may enhance or override a previous one. For example, if an appropriate diapause-terminating cue of inundation occurs while the photoperiod and/or temperature is "wrong", then diapause may not break, or only a small proportion of eggs may hatch.

Photoperiod is significant in diapause because alteration in day length predicts much about future seasonal environmental conditions, with photoperiod increasing as summer heat approaches and diminishing towards winter cold (section 6.10.2). Insects can detect day-length or night-length changes (photoperiodic stimuli), sometimes with extreme accuracy, through brain photoreceptors rather than compound eyes or ocelli. The insect brain also stores the "programming" for diapause, such that transplant of a diapausing moth pupal brain into a non-diapausing pupa will induce diapause in the recipient. The reciprocal operation causes resumption of development in a diapausing recipient. This programming may long precede the diapause and even span a generation, such that maternal conditions can govern the diapause in the developing stages of her offspring.

Many studies have shown endocrine control of diapause, but substantial variation in mechanisms for the regulation of diapause reflects the multiple independent evolution of this phenomenon. Generally in diapausing larvae, the production of ecdysteroid molting hormone from the prothoracic gland ceases, and JH plays a role in termination of diapause. Resumption of ecdysteroid secretion from the prothoracic glands appears essential for the termination of pupal diapause. JH is important in diapause regulation in adult insects but, as with the immature stages, may not be the only regulator. In larvae, pupae, and adults of *Bombyx mori*, complex antagonistic interactions occur between a **diapause hormone**, originating from paired neurosecretory cells in the suboesophageal ganglion, and JH from the corpora allata. The adult female produces diapause eggs when the ovariole is under the influence of diapause hormone, whereas in the absence of this hormone and in the presence of juvenile hormone, non-diapause eggs are produced.

6.6 DEALING WITH ENVIRONMENTAL EXTREMES

The most obvious environmental variables that confront an insect are seasonal fluctuations in temperature and humidity. The extremes of temperatures and humidities experienced by insects in their natural environments span the range of conditions encountered by terrestrial organisms, with only the suite of deep oceanic hydrothermic vent taxa encountering higher temperatures. For reasons of human interest in cryobiology (revivable preservation) the responses to extremes of cold and desiccation have been better studied than those to high temperatures alone.

The options available for avoidance of the extremes are behavioral avoidance, such as by burrowing into soil of a more equable temperature, migration (section 6.7), diapause (section 6.5), and *in situ* tolerance/survival in a very altered physiological condition, the topic of the following sections.

6.6.1 Cold

Biologists have long been interested in the occurrence of insects at the extremes of the Earth, in surprising diversity and sometimes in large numbers. Holometabolous insects are abundant in refugial sites within 3° of the North Pole, although fewer, notably a chironomid midge and some penguin and seal lice, are found on the Antarctic proper. Freezing, high elevations, including glaciers, sustain resident insects, such as the Himalayan *Diamesa* glacier midge (Diptera: Chironomidae), which sets a record for cold activity, being active at an air temperature of −16°C. Snowfields also support seasonally cold-active insects such as grylloblattids, and *Chionea* (Diptera: Tipulidae) and *Boreus* (Mecoptera), the snow "fleas". Low-temperature environments pose physiological problems that resemble dehydration in the reduction of available water, but clearly also include the need to avoid freezing of body fluids. Expansion and ice crystal formation typically kill mammalian cells and tissues, but perhaps some insect cells can tolerate freezing. Insects may possess one or several of a suite of mechanisms – collectively termed **cryoprotection** – that allows survival of cold extremes. These mechanisms may apply in any life-history stage, from resistant eggs to adults. Although they form a continuum, the following categories can aid understanding.

Freeze tolerance

Freeze-tolerant insects include some of the most cold-hardy species, mainly occurring in Arctic, sub-Arctic, and Antarctic locations that experience the most extreme winter temperatures (e.g. −40 to −80°C). Protection is provided by seasonal production of ice-nucleating agents (INA) under the induction of falling temperatures and prior to onset of severe cold. These proteins, lipoproteins, and/or endogenous crystalline substances such as urates, act as sites where (safe) freezing is encouraged outside cells, such as in the hemolymph, gut, or Malpighian tubules. Controlled and gentle extracellular ice formation acts also to gradually dehydrate cell contents, in which state freezing is avoided. In addition, substances such as glycerol and/or related polyols, and sugars including sorbitol and trehalose, allow **supercooling** (remaining liquid at subzero temperature without ice formation) and also protect tissues and cells prior to full INA activation and after freezing. Antifreeze proteins may also be produced; these fulfill some of the same protective roles, especially during freezing conditions in fall and during the spring thaw, outside the core deep-winter freeze. Onset of internal freezing often requires body contact with external ice to trigger ice nucleation, and may take place with little or no internal supercooling. Freeze tolerance does not guarantee survival, which depends not only on the actual minimum temperature experienced but also upon acclimation before cold onset, the rapidity of onset of extreme cold, and perhaps also the range and fluctuation in temperatures experienced during thawing. In the well-studied galling tephritid fly *Eurosta solidaginis*, all these mechanisms have been demonstrated, plus tolerance of cell freezing, at least in fat body cells.

Freeze avoidance

Freeze avoidance describes both a survival strategy and a species' physiological ability to survive low temperatures without internal freezing. In this definition, insects that avoid freezing by supercooling can survive extended periods in the supercooled state and show high mortality below the supercooling point, but little above it, and are freeze avoiders. Mechanisms for encouraging supercooling include evacuation of the digestive system to remove the promoters of ice nucleation, plus pre-winter synthesis of polyols and antifreeze agents. In these insects cold hardiness (potential to survive cold) can be calculated readily by comparison of the supercooling point (below which death occurs) and the lowest temperature the insect experiences. Freeze avoidance has been studied in the autumnal moth, *Epirrita autumnata*, and goldenrod gall moth, *Epiblema scudderiana*.

Chill tolerance

Chill-tolerant species occur mainly from temperate areas polewards, where insects survive frequent encounters with subzero temperatures. This category contains species with extensive supercooling ability (see above) and cold tolerance, but is distinguished from these by mortality that is dependent on duration of cold exposure and low temperature (above the supercooling point), i.e. the longer and the colder the freezing spell, the more deaths are attributable to freezing-induced cellular and tissue damage. A notable ecological grouping that demonstrates high chill tolerance are species that survive extreme cold (lower than supercooling point) by relying on snow cover, which provides "milder" conditions where chill tolerance permits survival. Examples of studied chill-tolerant species include the beech weevil, *Rhynchaenus fagi*, in Britain, and the bertha armyworm, *Mamestra configurata*, in Canada.

Chill susceptibility

Chill-susceptible species lack cold hardiness, and although they may supercool, death is rapid on exposure to subzero temperatures. Such temperate insects tend to vary in summer abundances according to the severity of the preceding winter. Thus, several studied European pest aphids (*Myzus persicae*, *Sitobion avenae*, and *Rhopalosiphum padi*) can supercool to −24°C (adults) or −27°C (nymphs) yet show high mortality when held at subzero temperatures for just a minute or two. Eggs show much greater cold hardiness than nymphs or adults. As overwintering eggs are produced only by sexual (**holocyclic**) species or clones, aphids with this life cycle predominate at increasingly high latitudes in comparison with those in which overwintering is in a nymphal or adult stage (**anholocyclic** species or clones).

Opportunistic survival

Opportunistic survival is observed in insects living in stable, warm climates in which cold hardiness is little developed. Even though supercooling is possible, in species that lack avoidance of cold through diapause or quiescence (section 6.5), mortality occurs when an irreversible lower threshold for metabolism is reached. Survival of predictable or sporadic cold episodes for these species depends upon exploitation of favorable sites, for example by migration (section 6.7) or by local opportunistic selection of appropriate microhabitats.

Clearly, low-temperature tolerance is acquired convergently, with a range of different mechanisms and chemistries involved. A unifying feature may be that the mechanisms for cryoprotection are rather similar to those shown for avoidance of dehydration which may be preadaptive for cold tolerance. Although each of the above categories contains a few unrelated species, amongst the terrestrial bembidiine Carabidae (Coleoptera) the Arctic and sub-Arctic regions contain a radiation of cold-tolerant species. A preadaptation to aptery (wing loss) has been suggested for these beetles, as it is too cold to warm flight muscles. Nonetheless, the summer Arctic is plagued by actively flying, biting dipterans that warm themselves by their resting orientation towards the sun.

6.6.2 Heat

The hottest inhabited places on Earth occur in the ocean, where suboceanic thermal vents support a unique assemblage of organisms based on thermophilous bacteria, and insects are absent. In contrast, in a terrestrial equivalent, vents in thermally active areas support a few specialist insects. The hottest waters in thermal springs of Yellowstone National Park are too hot to touch, but by selection of slightly cooler microhabitats amongst the cyanobacteria/blue-green algal mats, a brine fly, *Ephydra bruesi* (Ephydridae), can survive at 43°C. At least some other species of ephydrids, stratiomyiids, and chironomid larvae (all Diptera) tolerate nearly 50°C in Iceland, New Zealand, South America, and perhaps other sites where volcanism provides hot-water springs. The other aquatic temperature-tolerant taxa are found principally amongst the Odonata and Coleoptera.

High temperatures tend to kill cells by denaturing proteins, altering membrane and enzyme structures and properties, and by loss of water (dehydration). Inherently, the stability of non-covalent bonds that determine the complex structure of proteins determines the upper limits, but below this threshold there are many different but interrelated temperature-dependent biochemical reactions. Exactly how insects tolerant of high temperature cope biochemically is little known. **Acclimation**, in which a gradual exposure to increasing (or decreasing) temperatures takes place, certainly provides a greater disposition to survival at extreme temperatures compared with instantaneous exposure. When comparisons of effects of temperature are made, acclimation conditioning should be considered.

Options of dealing with high air temperatures include behaviors such as use of a burrow during the hottest times. This activity takes advantage of the buffering of soils, including desert sands, against temperature extremes so that near-stable temperatures occur within a few centimeters of the fluctuations of the exposed surface. Overwintering pupation of temperate insects frequently takes place in a burrow made by a late-instar larva, and in hot, arid areas night-active insects such as predatory carabid beetles may pass the extremes of the day in burrows. Arid-zone ants, including Saharan *Cataglyphis*, Australian *Melophorus*, and Namibian *Ocymyrmex*, show several behavioral features to maximize their ability to use some of the hottest places on Earth. Long legs hold the body in cooler air above the substrate, they can run as fast as 1 m s^{-1}, and are good navigators to allow rapid return to the burrow. Tolerance of high temperature is an advantage to *Cataglyphis* because they scavenge upon insects that have died from heat stress. However, *Cataglyphis bombycina* suffers predation from a lizard that also has a high temperature tolerance, and predator avoidance restricts the aboveground activity of *Cataglyphis* to a very narrow temperature band, between that at which the lizard ceases activity and its own upper lethal thermal threshold. *Cataglyphis* minimizes exposure to high temperatures using the strategies outlined above, and adds thermal respite activity – climbing and pausing on grass stems above the desert substrate, which may exceed 46°C. Physiologically, *Cataglyphis* may be amongst the most thermally tolerant land animals because they can accumulate high levels of "heat-shock proteins" in advance of their departure to forage from their (cool) burrow to the ambient external heat. The few minutes duration of the foraging frenzy is too short for synthesis of these protective proteins after exposure to the heat.

The proteins once termed "heat-shock proteins" (abbreviated as "hsp") may be best termed stress-induced proteins when involved in temperature-related activities, as at least some of the suite can be induced also by desiccation and cold. Their function at higher temperatures appears to be to act as molecular chaperones assisting in protein folding. In cold conditions, protein folding is not the problem, but rather it is loss of membrane fluidity, which can be restored by fatty acid changes and by denaturing of membrane phospholipids, perhaps also under some control of stress proteins.

The most remarkable specialization involves a larval chironomid midge, *Polypedilum vanderplanki*, which lives in West Africa on granite outcrops in temporary pools, such as those that form in depressions made by native people when grinding grain. The larvae do not form cocoons when the pools dry, but their bodies lose water until they are almost completely dehydrated. In this condition of **cryptobiosis** (alive but with all metabolism ceased), the larvae can tolerate temperature extremes, including artificially imposed temperatures in dry air from more than 100°C down to −27°C. On wetting, the larvae revive rapidly, feed and continue development until the onset of another cycle of desiccation or until pupation and emergence.

6.6.3 Aridity

In terrestrial environments, temperature and humidity are intimately linked, and responses to high temperatures are inseparable from concomitant water stress. Although free water may be unavailable in the arid tropics for long periods, many insects are active year-round in places such as the Namib Desert, an essentially rain-free desert in southwestern Africa. This desert has provided a research environment for the study of water relations in arid-zone insects ever since the discovery of "fog basking" amongst some tenebrionid beetles. The cold oceanic current that abuts the hot Namib Desert produces daily fog that sweeps inland. This provides a source of aerial moisture that can be precipitated onto the bodies of beetles that present a head-down stance on the slip face of sand dunes, facing the fog-laden wind. The precipitated moisture then runs to the mouth of the beetle. Such atmospheric water gathering is just one from a range of insect behaviors and morphologies that allow survival under these stressful conditions. Two different strategies exemplified by different beetles can be compared and contrasted: detritivorous tenebrionids and predaceous carabids, both of which have many aridity-tolerant species.

The greatest water loss by most insects occurs via evaporation from the cuticle, with lesser amounts lost through respiratory gas exchange at the spiracles and through excretion. Some arid-zone beetles have reduced their water loss 100-fold by one or more strategies including extreme reduction in evaporative water loss through the cuticle (section 2.1), reduction in spiracular water loss, reduction in metabolism, and extreme reduction of excretory loss. In the studied arid-zone species of tenebrionids and carabids, cuticular water permeability is reduced to almost zero such that water loss is virtually a function of metabolic rate alone – i.e. loss is by the respiratory pathway, predominantly related to variation in the local humidity around the spiracles. Enclosure of the spiracles in a humid subelytral space is an important mechanism for reduction of such losses. Observation of unusually low levels of sodium in the hemolymph of studied tenebrionids compared with levels in arid-zone carabids (and most other insects) implies reduced sodium pump activity, reduced sodium gradient across cell membranes, a concomitantly inferred reduction in metabolic rate, and reduced respiratory water loss. Uric acid precipitation when water is reabsorbed from the rectum allows the excretion of virtually dry urine (section 3.7.2), which, with retention of free amino acids, minimizes loss of everything except the nitrogenous wastes. All these mechanisms allow the survival of a tenebrionid beetle in an arid environment with seasonal food and water shortage. In contrast, desert carabids include species that maintain a high sodium pump activity and sodium gradient across cell membranes, implying a high metabolic rate. They also excrete more dilute urine, and appear less able to conserve free amino acids. Behaviorally, carabids are active predators, needing a high metabolic rate for pursuit, which would incur greater rates of water loss. This may be compensated for by the higher water content of their prey, compared with the desiccated detritus that forms the tenebrionid diet.

To test if these distinctions are different "adaptive" strategies, or if tenebrionids differ more generally from carabids in their physiology, irrespective of any arid tolerance, will require wider sampling of taxa, and some appropriate tests to determine whether the observed physiological differences are correlated with taxonomic relationships (i.e. are preadaptive for life in low-humidity environments) or ecology of the species. Such tests have not been undertaken.

6.7 MIGRATION

Diapause, as described above, allows an insect to track

its resources in time – when conditions become inclement, development ceases until diapause breaks. An alternative to shutdown is to track resources in space by directed movement. The term **migration** was formerly restricted to the to-and-fro major movements of vertebrates, such as wildebeest, salmonid fish, and migratory birds including swallows, shorebirds, and maritime terns. However, there are good reasons to expand this to include organisms that fulfill some or all of the following criteria, in and around specific phases of movement:

- persistent movement away from an original home range;
- relatively straight movement in comparison with station-tending or zig-zagging within a home range;
- undistracted by (unresponsive to) stimuli from home range;
- distinctive pre- and post-movement behaviors;
- reallocation of energy within the body.

All migrations in this wider sense are attempts to provide a homogeneous suitable environment despite temporal fluctuations in a single home range. Criteria such as length of distance traveled, geographical area in which migration occurs, and whether or not the outward-bound individual undertakes a return are unimportant to this definition. Furthermore, thinning out of a population (dispersal) or advance across a similar habitat (range extension) are not migrations. According to this definition, seasonal movements from the upper mountain slopes of the Sierra Nevada to the Central Valley by the convergent ladybird beetle (*Hippodamia convergens*) is as much a migratory activity as is a transcontinental movement of a monarch butterfly (*Danaus plexippus*). Pre-migration behaviors in insects include redirecting metabolism to energy storage, cessation of reproduction, and production of wings in polymorphic species in which winged and wingless forms coexist (polyphenism; section 6.8.2). Feeding and reproduction are resumed post-migration. Some responses are under hormonal control, whereas others are environmentally induced. Evidently, pre-migration changes must anticipate the altered environmental conditions that migration has evolved to avoid. As with induction of diapause (above), principal amongst these cues is change in day length (photoperiod). A strong linkage exists between the several cues for onset and termination of reproductive diapause and induction and cessation of migratory response in studied species, including monarch butterflies and milkweed bugs (*Oncopeltus fasciatus*). From their extensive range associated with North American host milkweed plants (Asclepiadaceae), individuals of both species migrate south. At least in this migrant generation of monarchs, a magnetic compass complements solar navigation in deriving the bearings towards the overwintering site. Shortening day length induces a reproductive diapause in which flight inhibition is removed and energy is transferred to flight instead of reproduction. The overwintering generation of both species (monarch butterflies at their winter roost are shown in Plate 3.5) is in diapause, which ends with a two- (or more) stage migration from south to north that essentially tracks the sequential development of subtropical to temperate annual milkweeds as far as southern Canada. The first flight in early spring from the overwintering area is short, with both reproduction and flight effort occurring during days of short length, but the next generation extends far northwards in longer days, either as individuals or by consecutive generations. Few if any of the returning individuals are the original outward migrants. In the milkweed bugs there is a circadian rhythm (Box 4.4) with oviposition and migration temporally segregated in the middle of the day, and mating and feeding concentrated at the end of the daylight period. As both milkweed bugs and monarch butterflies have non-migratory multivoltine relatives that remain in the tropics, it seems that the ability to diapause and thus escape in the fall has allowed just these two species to invade summer milkweed stands of the temperate region. In contrast, amongst noctuid moths of the genus *Spodoptera* (armyworms) a number of species show a diapause-related migration and others a variable pre-reproductive period.

It is a common observation that insects living in "temporary" habitats of limited duration have a higher proportion of flighted species, and within polymorphic taxa, a higher proportion of flighted individuals. In longer-lasting habitats loss of flightedness, either permanently or temporarily, is more common. Thus, amongst European water-striders (Hemiptera: Gerridae) species associated with small ephemeral water bodies are winged and regularly migrate to seek new water bodies; those associated with large lakes tend to winglessness and sedentary life histories. Evidently, flightedness relates to the tendency (and ability) to migrate in locusts, as exemplified in *Chortoicetes terminifera* (the Australian migratory locust) and *Locusta migratoria* which demonstrate adaptive migration to exploit ephemerally available favorable conditions in arid regions (see section 6.10.5 for *L. migratoria* behavior).

Although the massed movements described above are very conspicuous, even the "passive dispersal" of small and lightweight insects can fulfill many of the criteria of migration. Thus, even reliance upon wind (or water) currents for movement may involve the insect being capable of any or all of the following:

- changing behavior to embark, such as young scale insects crawling to a leaf apex and adopting a posture there to enhance the chances of extended aerial movement;
- being in appropriate physiological and developmental condition for the journey, as in the flighted stage of otherwise apterous aphids;
- sensing appropriate environmental cues to depart, such as seasonal failure of the host plant of many aphids;
- recognizing environmental cues on arrival, such as odors or colors of a new host plant, and making controlled departure from the current.

Naturally, embarkation on such journeys does not always bring success and there are many strandings of migratory insects in unsuitable habitat, such as icefields and in open oceans. Nonetheless, it is clear that some fecund insects that can make use of predictable meteorological conditions can make long journeys in a consistent direction, depart from the air current and establish in a suitable, novel habitat. Aphids are a prime example, but certain thrips and scale insects and other agriculturally damaging pests are capable of locating new host plants by this means.

6.8 POLYMORPHISM AND POLYPHENISM

The existence of several generations per year often is associated with morphological change between generations. Similar variation may occur contemporaneously within a population, such as the existence simultaneously of both winged and flightless forms ("**morphs**"). Sexual differences between males and females and the existence of strong differentiation in social insects such as ants and bees are further obvious examples of the phenomenon. The term **polymorphism** encompasses all such discontinuities, which occur in the same life-history phase at a frequency greater than might be expected from repeated mutation alone. It is defined as the simultaneous or recurrent occurrence of distinct morphological differences, reflecting and often including physiological, behavioral, and/or ecological differences among conspecific individuals.

6.8.1 Genetic polymorphism

The distinction between the sexes is an example of a particular polymorphism, namely sexual dimorphism, which in insects is almost totally under genetic determination. Environmental factors may affect sexual expression, as in castes of some social insects or in feminization of genetically male insects by mermithid nematode infections. Aside from the dimorphism of the sexes, different genotypes may co-occur within a single species, maintained by natural selection at specific frequencies that vary from place to place and time to time throughout the range. For example, adults of some gerrid bugs are fully winged and capable of flight, whereas other coexisting individuals of the same species are brachypterous and cannot fly. Intermediates are at a selective disadvantage and the two genetically determined morphs coexist in a balanced polymorphism. Some of the most complex, genetically based, polymorphisms have been discovered in butterflies that mimic chemically protected butterflies of another species (the model) for purposes of defense from predators (section 14.5). Some butterfly species may mimic more than one model and, in these species, the accuracy of the several distinct mimicry patterns is maintained because inappropriate intermediates are not recognized by predators as being distasteful and are eaten. Mimetic polymorphism predominantly is restricted to the females, with the males generally monomorphic and non-mimetic. The basis for the switching between the different mimetic morphs is relatively simple Mendelian genetics, which may involve relatively few genes or supergenes.

It is a common observation that some individual species with a wide range of latitudinal distributions show different life-history strategies according to location. For example, populations living at high latitudes (nearer the pole) or high elevation may be univoltine, with a long dormant period, whereas populations nearer the equator or lower in elevation may be multivoltine, and develop continuously without dormancy. Dormancy is environmentally induced (sections 6.5 & 6.10.2), but the ability of the insect to recognize and respond to these cues is programmed genetically. In addition, at least some geographical variation in life histories results from genetic polymorphism.

6.8.2 Environmental polymorphism, or polyphenism

A phenotypic difference between generations that lacks a genetic basis and is determined entirely by the environment often is termed **polyphenism**. An example is the temperate to tropical Old World pierid butterfly *Eurema hecabe*, which shows a seasonal change in wing color between summer and fall morphs. Photoperiod induces morph change, with a dark-winged summer morph induced by a long day of greater than 13 h. A short day of less than 12 h induces the paler-winged fall morph, particularly at temperatures of under 20°C, with temperature affecting males more than females.

Amongst the most complex polyphenisms are those seen in the aphids. Within parthenogenetic lineages (i.e. in which there is absolute genetic identity) the females may show up to eight distinct phenotypes, in addition to polymorphisms in sexual forms. These female aphids may vary in morphology, physiology, fecundity, offspring timing and size, development time, longevity, and host-plant choice and utilization. Environmental cues responsible for alternative morphs are similar to those that govern diapause and migration in many insects (sections 6.5 & 6.7), including photoperiod, temperature, and maternal effects, such as elapsed time (rather than number of generations) since the winged founding mother. Overcrowding triggers many aphid species to produce a winged dispersive phase. Crowding also is responsible for one of the most dramatic examples of polyphenism, the phase transformation from the solitary young locusts (hoppers) to the gregarious phase (section 6.10.5). Studies on the physiological mechanisms that link environmental cues to these phenotype changes have implicated JH in many aphid morph shifts.

If aphids show the greatest number of polyphenisms, the social insects come a close second, and undoubtedly have a greater degree of morphological differentiation between morphs, termed **castes**. This is discussed in more detail in Chapter 12; suffice it to say that maintenance of the phenotypic differences between castes as different as queens, workers, and soldiers includes physiological mechanisms such as pheromones transferred with food, olfactory and tactile stimuli, and endocrine control including JH and ecdysone. Superimposed on these polyphenisms are the dimorphic differences between the sexes, which impose some limits on variation.

6.9 AGE-GRADING

Identification of the growth stages or ages of insects in a population is important in ecological or applied entomology. Information on the proportion of a population in different developmental stages and the proportion of the adult population at reproductive maturity can be used to construct time-specific life-tables or budgets to determine factors that cause and regulate fluctuations in population size and dispersal rate, and to monitor fecundity and mortality factors in the population. Such data are integral to predictions of pest outbreaks as a result of climate and to the construction of models of population response to the introduction of a control program.

Many different techniques have been proposed for estimating either the growth stage or the age of insects. Some provide an estimate of chronological (calendar) age within a stadium, whereas most estimate either instar number or relative age within a stadium, in which case the term **age-grading** is used in place of age determination.

6.9.1 Age-grading of immature insects

For many population studies it is important to know the number of larval or nymphal instars in a species and to be able to recognize the instar to which any immature individual belongs. Generally, such information is available or its acquisition is feasible for species with a constant and relatively small number of immature instars, especially those with a lifespan of a few months or less. However, it is logistically difficult to obtain such data for species with either many or a variable number of instars, or with overlapping generations. The latter situation may occur in species with many asynchronous generations per year or in species with a life cycle of longer than one year. In some species there are readily discernible qualitative (e.g. color) or meristic (e.g. antennal segment number) differences between consecutive immature instars. More frequently, the only obvious difference between successive larval or nymphal instars is the increase in size that occurs after each molt (the molt increment). Thus, it should be possible to determine the actual number of instars in the life history of a species from a frequency histogram of measurements of a sclerotized body part (Fig. 6.11).

Entomologists have sought to quantify this size progression for a range of insects. One of the earliest

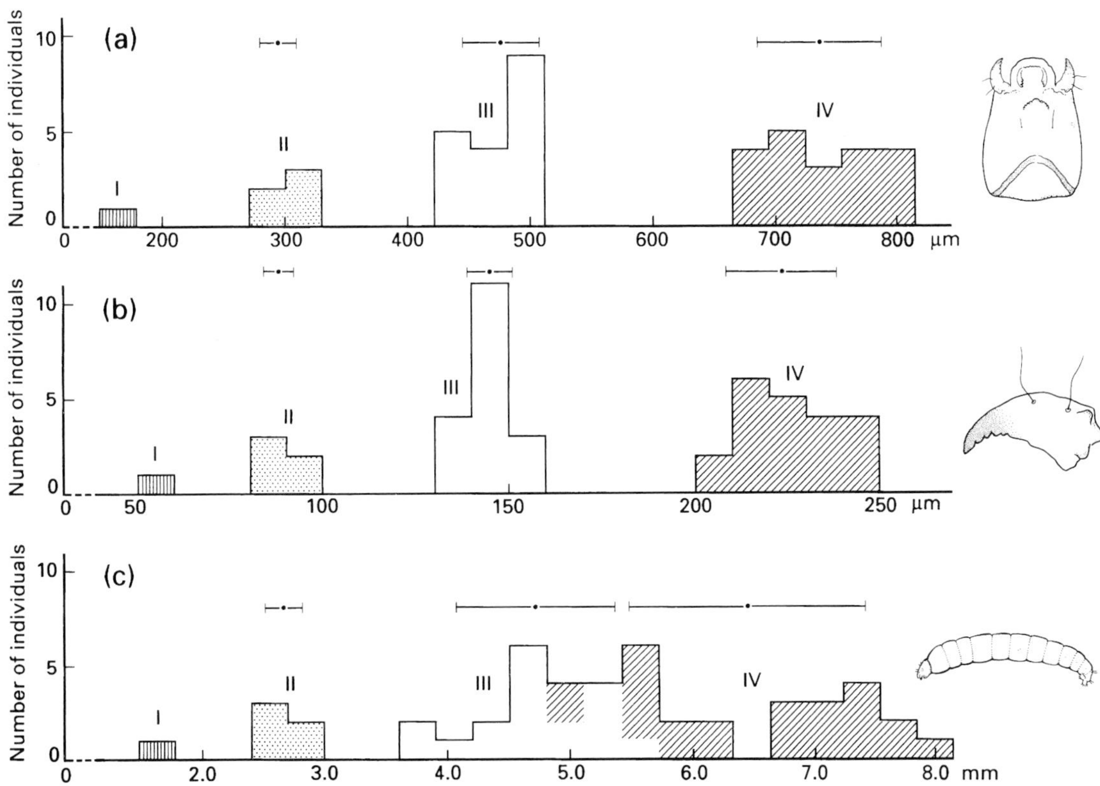

Fig. 6.11 Growth and development in a marine midge, *Telmatogeton* (Diptera: Chironomidae), showing increases in: (a) head capsule length; (b) mandible length; and (c) body length between the four larval instars (I–IV). The dots and horizontal lines above each histogram represent the means and standard deviations of measurements for each instar. Note that the lengths of the sclerotized head and mandible fall into discrete size classes representing each instar, whereas body length is an unreliable indicator of instar number, especially for separating the third- and fourth-instar larvae.

attempts was that of H.G. Dyar, who in 1890 established a "rule" from observations on the caterpillars of 28 species of Lepidoptera. Dyar's measurements showed that the width of the head capsule increased in a regular linear progression in successive instars by a ratio (range 1.3–1.7) that was constant for a given species. **Dyar's rule** states that:

postmolt size/premolt size (or molt increment) = constant

Thus, if logarithms of measurements of some sclerotized body part in different instars are plotted against the instar number, a straight line should result; any deviation from a straight line indicates a missing instar. In practice, however, there are many departures from Dyar's rule, as the progression factor is not always constant, especially in field populations subject to variable conditions of food and temperature during growth.

A related empirical "law" of growth is Przibram's rule, which states that an insect's weight is doubled during each instar and at each molt all linear dimensions are increased by a ratio of 1.26. The growth of most insects shows no general agreement with this rule, which assumes that the dimensions of a part of the insect body should increase at each molt by the same ratio as the body as a whole. In reality, growth in most insects is allometric, i.e. the parts grow at rates peculiar to themselves, and often very different from the growth rate of the body as a whole. The horned adornments on the head and thorax of *Onthophagus* dung beetles discussed in section 5.3 exemplify the trade-offs associated with allometric growth.

6.9.2 Age-grading of adult insects

The age of an adult insect is not determined easily. However, adult age is of great significance, particularly in the insect vectors of disease. For instance, it is crucial to epidemiology that the age (longevity) of an adult female mosquito be known, as this relates to the number of blood meals taken and therefore the number of opportunities for pathogen transmission. Most techniques for assessing the age of adult insects estimate relative (not chronological) age and hence age-grading is the appropriate term.

Three general categories of age assessment have been proposed, relating to:

1 age-related changes in physiology and morphology of the reproductive system;

2 changes in somatic structures;

3 external wear and tear.

The latter approach has proved unreliable but the other methods have wide applicability.

In the first method, age is graded according to reproductive physiology in a technique applicable only to females. Examination of an ovary of a **parous** insect (one that has laid at least one egg) shows that evidence remains after each egg is laid (or even resorbed) in the form of a **follicular relic** that denotes an irreversible change in the epithelium. The deposition of each egg, together with contraction of the previously distended membrane, leaves one follicular relic per egg. The actual shape and form of the follicular relic varies between species, but one or more residual dilations of the lumen, with or without pigment or granules, is common in the Diptera. Females that have no follicular relic have not developed an egg and are termed **nulliparous**.

Counting follicular relics can give a comparative measure of the physiological age of a female insect, for example allowing discrimination of parous from nulliparous individuals, and often allowing further segregation within parous individuals according to the number of ovipositions. The chronological age can be calculated if the time between successive ovipositions (the **ovarian cycle**) is known. However, if there is one ovarian cycle per blood meal, as in many medically significant flies, it is the physiological age (number of cycles) that is of greater significance than the precise chronological age.

The second generally applicable method of age determination has a more direct relationship with chronology, and most of the somatic features that allow age estimation are present in both sexes. Estimates of age can be made from measures of cuticle growth, fluorescent pigments, fat body size, cuticular hardness and, in females only, color and/or patterning of the abdomen. Cuticular growth estimates of age rely upon there being a daily rhythm of deposition of the endocuticle. In exopterygotes, cuticular layers are more reliable, whereas in endopterygotes, the apodemes (internal skeletal projections upon which muscles attach) are more dependable. The daily layers are most distinctive when the temperature for cuticle formation is not attained for part of each day. This use of growth rings is confounded by development temperatures too cold for deposition, or too high for the daily cycle of deposition and cessation. A further drawback to the technique is that deposition ceases after a certain age is attained, perhaps only 10–15 days after eclosion. Physiological age can be determined by measuring the pigments that accumulate in the aging cells of many animals, including insects. These pigments fluoresce and can be studied by fluorescence microscopy. Lipofuscin from postmitotic cells in most body tissues, and pteridine eye pigments have been measured in this way, especially in flies.

6.10 ENVIRONMENTAL EFFECTS ON DEVELOPMENT

The rate or manner of insect development or growth may depend upon a number of factors. These include the type and amount of food, the amount of moisture (for terrestrial species) and heat (measured as temperature), or the presence of environmental signals (e.g. photoperiod), mutagens and toxins, or other organisms, either predators or competitors. Two or more of these factors may interact to complicate interpretation of growth characteristics and patterns.

6.10.1 Temperature

Most insects are poikilothermic, that is with body temperature more or less directly varying with environmental temperature, thus heat is the force driving the rate of growth and development when food is unlimited. A rise in temperature, within a favorable range, will speed up the metabolism of an insect and consequently increase its rate of development. Each species and each stage in the life history may develop at its own rate in relation to temperature. Thus, **physiological time**, a measure of the amount of heat required

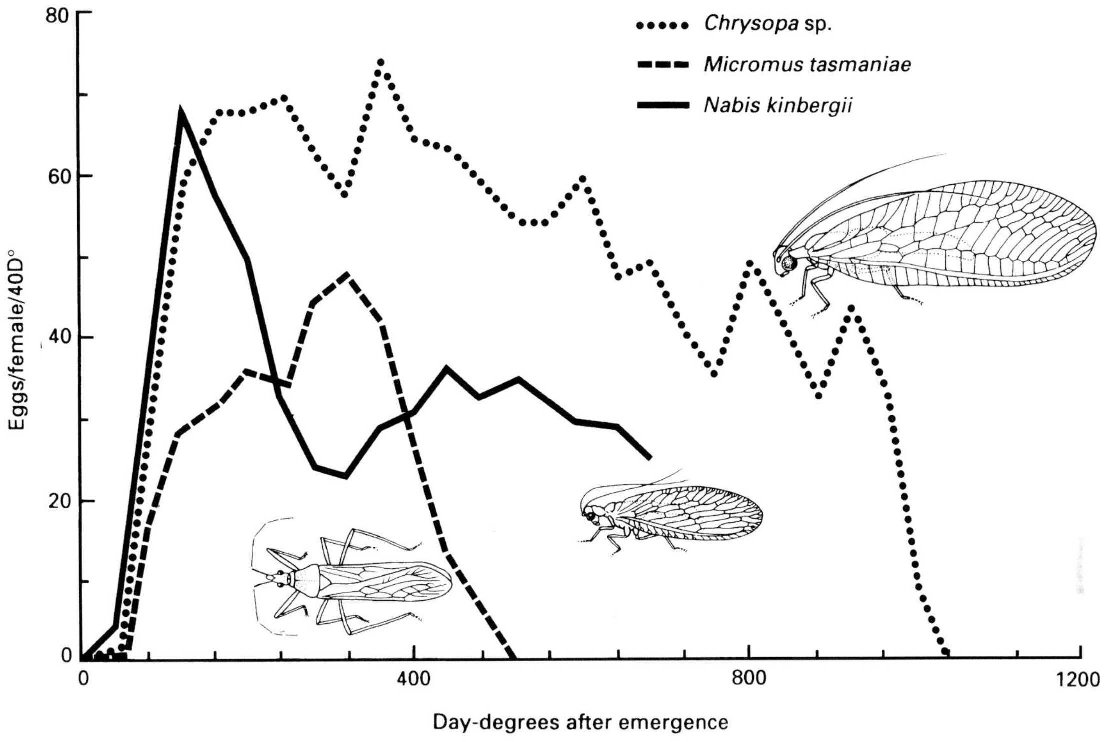

Fig. 6.12 Age-specific oviposition rates of three predators of cotton pests, *Chrysopa* sp. (Neuroptera: Chrysopidae), *Micromus tasmaniae* (Neuroptera: Hemerobiidae), and *Nabis kinbergii* (Hemiptera: Nabidae), based on physiological time above respective development thresholds of 10.5°C, −2.9°C, and 11.3°C. (After Samson & Blood 1979.)

over time for an insect to complete development or a stage of development, is more meaningful as a measure of development time than age in calendar time. Knowledge of temperature–development relationships and the use of physiological time allow comparison of the life cycles and/or fecundity of pest species in the same system (Fig. 6.12), and prediction of the larval feeding periods, generation length, and time of adult emergence under variable temperature conditions that exist in the field. Such predictions are especially important for pest insects, as control measures must be timed carefully to be effective.

Physiological time is the cumulative product of total development time (in hours or days) multiplied by the temperature (in degrees) above the **developmental (or growth) threshold**, or the temperature below which no development occurs. Thus, physiological time is commonly expressed as **day-degrees** (D°) or hour-degrees (h°). Normally, physiological time is estimated for a species by rearing a number of individuals of the life-history stage(s) of interest under different constant temperatures in several identical growth cabinets. The developmental threshold is estimated by the linear regression *x*-axis method, as outlined in Box 6.2, although more accurate threshold estimates can be obtained by more time-consuming methods.

In practice, the application of laboratory-estimated physiological time to natural populations may be complicated by several factors. Under fluctuating temperatures, especially if the insects experience extremes, growth may be retarded or accelerated compared with the same number of day-degrees under constant temperatures. Furthermore, the temperatures actually experienced by the insects, in their often sheltered microhabitats on plants or in soil or litter, may be several degrees different from the temperatures recorded at a meteorological station even just a few meters away. Insects may select microhabitats that ameliorate cold

Box 6.2 Calculation of day-degrees

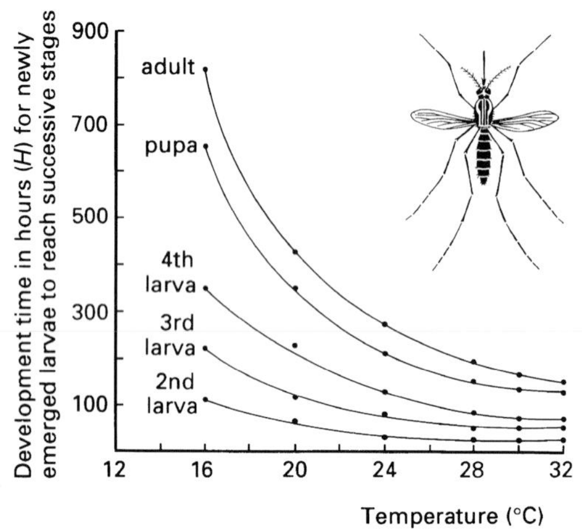

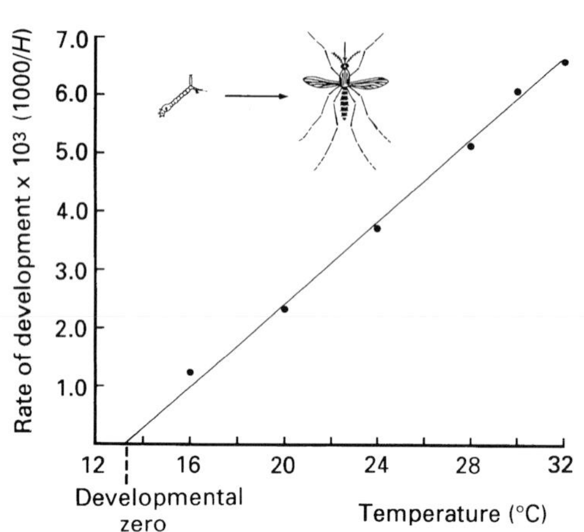

An outline of a simple method to estimate day-degrees (after Daly et al. 1978) is exemplified by data on the relationship between temperature and development in the yellow-fever mosquito, *Aedes aegypti* (Diptera: Culicidae) (after Bar-Zeev 1958).

1 In the laboratory, establish the average time required for each stage to develop at different constant temperatures. The graph on the left shows the time in hours (H) for newly hatched larvae of *Ae. aegypti* to reach successive stages of development when incubated at various temperatures.

2 Plot the reciprocal of development time ($1/H$), the development rate, against temperature to obtain a sigmoid curve with the middle part of the curve approximately linear. The graph on the right shows the linear part of this relationship for the total development of *Ae. aegypti* from the newly hatched larva to the adult stage. A straight line would not be obtained if extreme development temperatures (e.g. higher than 32°C or lower than 16°C) had been included.

3 Fit a linear regression line to the points and calculate the slope of this line. The slope represents the amount in hours by which development rates are increased for each 1 degree of increased temperature. Hence, the reciprocal of the slope gives the number of hour-degrees, above threshold, required to complete development.

4 To estimate the developmental threshold, the regression line is projected to the x-axis (abscissa) to give the developmental zero, which in the case of *Ae. aegypti* is 13.3°C. This zero value may differ slightly from the actual developmental threshold determined experimentally, probably because at low (or high) temperatures the temperature–development relationship is rarely linear. For *Ae. aegypti*, the developmental threshold actually lies between 9 and 10°C.

5 The equation of the regression line is $1/H = k(T^{\circ} - T^{t})$, where H = development period, T° = temperature, T^{t} = development threshold temperature, and k = slope of line.

Thus, the physiological time for development is $H(T^{\circ} - T^{t}) = 1/k$ hour-degrees, or $H(T^{\circ} - T^{t})/24 = 1/k = K$ day-degrees, with K = thermal constant, or K-value.

By inserting the values of H, T°, and T^{t} for the data from *Ae. aegypti* in the equation given above, the value of K can be calculated for each of the experimental temperatures from 14 to 36°C:

Temperature (°C)	14	16	20	24	28	30	32	34	36
K	1008	2211	2834	2921	2866	2755	2861	3415	3882

Thus, the K-value for *Ae. aegypti* is approximately independent of temperature, except at extremes (14 and 34–36°C), and averages about 2740 hour-degrees or 114 day-degrees between 16 and 32°C.

night conditions or reduce or increase daytime heat. Thus, predictions of insect life-cycle events based on extrapolation from laboratory to field temperature records may be inaccurate. For these reasons, the laboratory estimates of physiological time should be corroborated by calculating the hour-degrees or day-degrees required for development under more natural conditions, but using the laboratory-estimated developmental threshold, as follows.

1 Place newly laid eggs or newly hatched larvae in their appropriate field habitat and record temperature each hour (or calculate a daily average – a less accurate method).

2 Estimate the time for completion of each instar by discarding all temperature readings below the developmental threshold of the instar and subtracting the developmental threshold from all other readings to determine the effective temperature for each hour (or simply subtract the development threshold temperature from the daily average temperature). Sum the degrees of effective temperature for each hour from the beginning to the end of the stadium. This procedure is called thermal summation.

3 Compare the field-estimated number of hour-degrees (or day-degrees) for each instar with that predicted from the laboratory data. If there are discrepancies, then microhabitat and/or fluctuating temperatures may be influencing insect development or the developmental zero read from the graph may be a poor estimate of the developmental threshold.

Another problem with laboratory estimation of physiological time is that insect populations maintained for lengthy periods under laboratory conditions frequently undergo acclimation to constant conditions or even genetic change in response to the altered environment or as a result of population reductions that produce genetic "bottle-necks". Therefore, insects maintained in rearing cages may exhibit different temperature–development relationships from individuals of the same species in wild populations.

For all of the above reasons any formula or model that purports to predict insect response to environmental conditions must be tested carefully for its fit with natural population responses.

6.10.2 Photoperiod

Many insects, perhaps most, do not develop continuously all year round, but avoid some seasonally adverse conditions by a resting period (section 6.5) or migration (section 6.7). Summer dormancy (aestivation) and winter dormancy (hibernation) provide two examples of avoidance of seasonal extremes. The most predictable environmental indicator of changing seasons is **photoperiod** – the length of the daily light phase or, more simply, day length. Near the equator, although sunrise to sunset of the longest day may be only a few minutes longer than on the shortest day, if the period of twilight is included then total day length shows more marked seasonal change. The photoperiod response is to duration rather than intensity and there is a critical threshold intensity of light below which the insect does not respond; this threshold is often as dim as twilight, but rarely as low as bright moonlight. Many insects appear to measure the duration of the light phase in the 24 h period, and some have been shown experimentally to measure the duration of dark. Others recognize long days by light falling within the "dark" half of the day.

Most insects can be described as "long-day" species, with growth and reproduction in summer and with dormancy commencing with decreasing day length. Others show the reverse pattern, with "short-day" (often fall and spring) activity and summer aestivation. In some species the life-history stage in which photoperiod is assessed is in advance of the stage that reacts, as is the case when the photoperiodic response of the maternal generation of silkworms affects the eggs of the next generation.

The ability of insects to recognize seasonal photoperiod and other environmental cues requires some means of measuring time between the cue and the subsequent onset or cessation of diapause. This is achieved through a "biological clock" (Box 4.4), which may be driven by internal (endogenous) or external (exogenous) daily cycles, called **circadian rhythms**. Interactions between the short time periodicity of circadian rhythms and longer-term seasonal rhythms, such as photoperiod recognition, are complex and diverse, and have probably evolved many times within the insects.

6.10.3 Humidity

The high surface area : volume ratio of insects means that loss of body water is a serious hazard in a terrestrial environment, especially a dry one. Low moisture content of the air can affect the physiology and thus the development, longevity, and oviposition of many insects. Air holds more water vapor at high than at low temperatures. The relative humidity (RH) at a particular temperature is the ratio of actual water vapor present to that necessary for saturation of the air at that temperature. At low relative humidities, development may be retarded, for example in many pests of stored products; but at high relative humidities or in saturated air (100% RH), insects or their eggs may drown or be infected more readily by pathogens. The fact that stadia may be greatly lengthened by unfavorable humidity has serious implications for estimates of development times, whether calendar or physiological time is used. The complicating effects of low, and sometimes even high, air moisture levels should be taken into account when gathering such data.

6.10.4 Mutagens and toxins

Stressful conditions induced by toxic or mutagenic chemicals may affect insect growth and form to varying degrees, ranging from death at one extreme to slight phenotypic modifications at the other end of the spectrum. Some life-history stages may be more sensitive to mutagens or toxins than others, and sometimes the phenotypic effects may not be easily measured by crude estimates of stress, such as percentage survival. One sensitive and efficient measure of the amount of genetic or environmental stress experienced by insects during development is the incidence of **fluctuating asymmetry**, or the quantitative differences between the left and right sides of each individual in a sample of the population. Insects are usually bilaterally symmetrical if grown under ideal conditions, so the left and right halves of their bodies are mirror images (except for obvious differences in structures such as the genitalia of some male insects). If grown under stressful conditions, however, the degree of asymmetry tends to increase.

The measurement of fluctuating asymmetry has many potential uses in theoretical and economic entomology and in assessment of environmental quality. For example, it can be used as an indicator of developmental stability to determine the effect on non-target organisms of exposure to insecticides or vermicides, such as avermectins. Bush flies (*Musca vetustissima*) breeding in the dung of cattle treated for nematode control with Avermectin B_1 are significantly more asymmetric for two morphometric wing characters than flies breeding in the dung of untreated cattle. Fluctuating asymmetry has been used as a measure of environmental quality. For example, water quality has been assessed by comparing the amount of asymmetry in aquatic insects reared in polluted and clean water. In industrially polluted waters, particular bloodworms (larvae of chironomid midges) may survive but often exhibit gross developmental abnormalities. However, at lower levels of pollutants, more subtle effects may be detected as deviations from symmetry compared with clean-water controls. In addition, measures of developmental effects on non-target insects have been used to assess the specificity of biocides prior to marketing. The technique is not completely reliable, with doubts having been raised about interpretation (variation in response between different organ systems measured) and concerning the underlying mechanism causing any responses measured.

6.10.5 Biotic effects

In most insect orders, adult size has a strong genetic component and growth is strongly determinate. In many Lepidoptera, for example, final adult size is relatively constant within a species; reduction in food quality or availability delays caterpillar growth rather than causing reduced final adult size, although there are exceptions. In contrast, in flies that have limited or ephemeral larval resources, such as a dung pat or temporary pool, cessation of larval growth would result in death as the habitat shrinks. Thus larval crowding and/or limitation of food supply tend to shorten development time and reduce final adult size. In some mosquitoes and midges, success in short-lived pool habitats is attained by a small proportion of the larval population developing with extreme rapidity relative to their slower siblings. In pedogenetic gall midges (section 5.10.1), crowding with reduced food supply terminates larva-only reproductive cycles and induces the production of adults, allowing dispersal to more favorable habitats.

Food quality appears important in all these cases, but there may be related effects, for example as a result of crowding. Clearly, it can be difficult to segregate out food effects from other potentially limiting factors. In the California red scale, *Aonidiella aurantii* (Hemiptera: Diaspididae), development and reproduction on orange trees is fastest on fruit, intermediate on twigs, and slowest on leaves. Although these differences may reflect

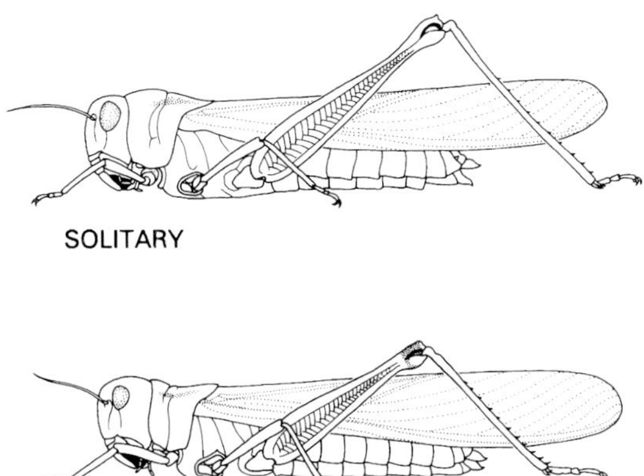

Fig. 6.13 Solitary and gregarious females of the migratory locust, *Locusta migratoria* (Orthoptera: Acrididae). The solitaria adults have a pronounced pronotal crest and the femora are larger relative to the body and wing than in the gregaria adults. Intermediate morphologies occur in the transiens (transient stage) during the transformation from solitaria to gregaria or the reverse.

differing nutritional status, a microclimatic explanation cannot be excluded, as fruit may retain heat longer than the relatively smaller-volumed stems and leaves, and such slight temperature differences might affect the development of the insects.

The effects of crowding on development are well understood in some insects, as in locusts in which two extreme phases, termed **solitary** and **gregarious** (Fig. 6.13), differ in morphometrics, color, and behavior. At low densities locusts develop into the solitary phase, with a characteristic uniform-colored "hopper" (nymph), and large-sized adult with large hind femora. As densities increase, induced in nature by high survivorship of eggs and young nymphs under favorable climatic conditions, graded changes occur and a darker-striped nymph develops to a smaller locust with shorter hind femora. The most conspicuous difference is behavioral, with more solitary individuals shunning each other's company but making concerted nocturnal migratory movements that result eventually in aggregations in one or a few places of gregarious individuals, which tend to form enormous and mobile swarms. The behavioral shift is induced by crowding, as can be shown by splitting a single locust egg pod into two: rearing the offspring at low densities induces solitary locusts, whereas their siblings reared under crowded conditions develop into gregarious locusts. The response to high population density results from the integration of several cues, including the sight, touch, and sometimes the odor (pheromone) of conspecifics, which lead to endocrine and neuroendocrine (ecdysteroid) changes associated with developmental transformation.

Under certain circumstances biotic effects can override growth factors. Across much of the eastern USA, 13- and 17-year periodic cicadas (*Magicicada* spp.) emerge highly synchronously. At any given time, nymphal cicadas are of various sizes and in different instars according to the nutrition they have obtained from feeding on the phloem from roots of a variety of trees. Whatever their growth condition, after the elapse of 13 or 17 years since the previous emergence and egg-laying, the final molt of all nymphs prepares them for synchronous emergence as adults. In a very clever experiment host plants were induced to flush twice in one year, inducing adult cicada emergence one year early compared to controls on the roots of single-flushing trees. This implies that synchronized timing for cicadas depends on an ability to "count off" annual events – the predictable flush of sap with the passing of each spring once a year (except when experimenters manipulate it!).

6.11 CLIMATE AND INSECT DISTRIBUTIONS

Earlier in this chapter we saw how environmental factors may affect insect development. Here we examine some predictive models of how insect abundance and distribution change with abiotic factors. These models have application to past climate reconstruction, and increasingly are tested for veracity against range changes modulated by present, on-going climate change.

6.11.1 Modeling climatic effects on insect distributions

The abundance of any poikilothermic species is determined largely by proximate ecological factors including the population densities of predators and competitors (section 13.4) and interactions with habitat, food availability, and climate. Although the distributions of insect species result from these ecological factors, there is also a historical component. Ecology

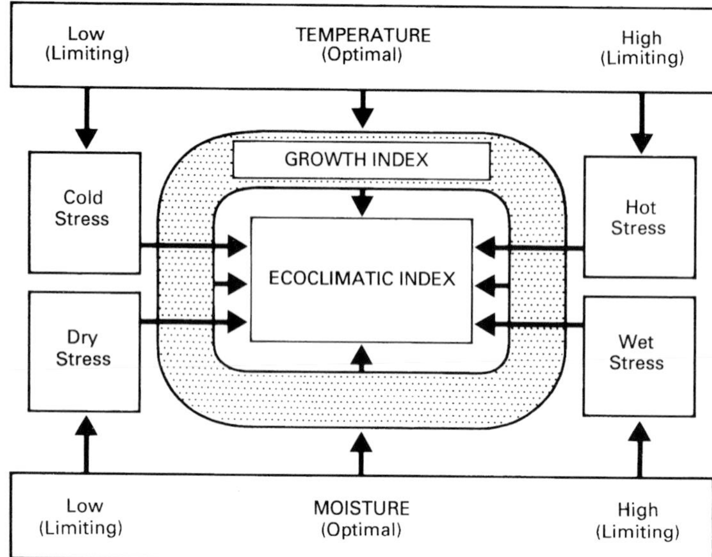

Fig. 6.14 Flow diagram depicting the derivation of the "ecoclimatic index" (EI) as the product of population growth index and four stress indices. The EI value describes the climatic favorability of a given location for a given species. Comparison of EI values allows different locations to be assessed for their relative suitability to a particular species. (After Sutherst & Maywald 1985.)

determines whether a species can continue to live in an area; history determines whether it does, or ever had the chance to live there. This difference relates to timing; given enough time, an ecological factor becomes a historical factor. In the context of present-day studies of where invasive insects occur and what the limits of their spread might be, history may account for the original or native distribution of a pest. However, knowledge of ecology may allow prediction of potential or future distributions under changed environmental conditions (e.g. as a result of the "greenhouse effect"; see section 6.11.2) or as a result of accidental (or intentional) dispersal by humans. Thus, ecological knowledge of insect pests and their natural enemies, especially information on how climate influences their development, is vital for the prediction of pest outbreaks and for successful pest management.

There are many models pertaining to the population biology of economic insects, especially those affecting major crop systems in western countries. One example of a climatic model of arthropod distribution and abundance is the computer-based system called CLIMEX (developed by R.W. Sutherst and G.F. Maywald), which allows the prediction of an insect's potential relative abundance and distribution around the world, using ecophysiological data and the known geographical distribution. An annual "ecoclimatic index" (EI), describing the climatic favorability of a given location for permanent colonization of an insect species, is derived from a climatic database combined with estimates of the response of the organisms to temperature, moisture, and day length. The EI is calculated as follows (Fig. 6.14). First, a population growth index (GI) is determined from weekly values averaged over a year to obtain a measure of the potential for population increase of the species. The GI is estimated from data on the seasonal incidence and relative abundance in different parts of the species' range. Second, the GI is reduced by incorporation of four stress indices, which are measures of the deleterious effects of cold, heat, dry, and wet.

Commonly, the existing geographical distribution and seasonal incidence of a pest species are known but biological data pertaining to climatic effects on development are scanty. Fortunately, the limiting effects of climate on a species usually can be estimated reliably from observations on the geographical distribution. The climatic tolerances of the species are inferred from the climate of the sites where the species is known to occur and are described by the stress indices of the CLIMEX model. The values of the stress indices are progressively adjusted until the CLIMEX predictions agree with the observed distribution of the species. Naturally, other information on the climatic tolerances of the species should be incorporated where possible because the above procedure assumes that the present distribution is climate limited, which might be an oversimplification.

Such climatic modeling based on world data has been carried out for tick species and for insects such as the Russian wheat aphid, the Colorado potato beetle, screw-worm flies, biting flies of *Haematobia* species, dung beetles, and fruit flies (Box 6.3). The output has great utility in applied entomology, namely in epidemiology, quarantine, management of insect pests, and entomological management of weeds and animal pests (including other insects).

In reality, detailed information on ecological performances may never be attained for many taxa, although such data are essential for the autecological-based distribution models described above. Nonetheless, there are demands for models of distribution in the absence of ecological performance data. Given these practical constraints, a class of modeling has been developed that accepts distribution point data as surrogates for "performance (process) characteristics" of organisms. These points are defined bioclimatically, and potential distributions can be modeled using some flexible procedures. Analyses assume that current species distributions are restricted (constrained) by bioclimatic factors. A suite of models developed in Australia (e.g. BIOCLIM, developed by Henry Nix and colleagues) allow estimation of potential constraints on species distribution in a stepwise process. First, the sites at which a species occurs are recorded and the climate estimated for each data point, using a set of bioclimatic measures based on the existing irregular network of weather stations across the region under consideration. Factors such as annual precipitation, seasonality of precipitation, precipitation of the driest quarter, minimum temperature of the coldest period, maximum temperature of the warmest period, and elevation appear to be particularly influential and are likely to have wide significance in determining the distribution of poikilothermic organisms. From this information a bioclimatic profile is developed from the pooled climate per site estimates, providing a profile of the range of climatic conditions at all sites for the species. Next, the bioclimatic profiles so produced are matched with climate estimates at other mapped sites across a regional grid to identify all other locations with similar climates. Specialized software then can be used to measure similarity of sites, with comparison being made via a digital elevation model with fine resolution. All locations within the grid with similar climates to the species-profile form a predicted bioclimatic domain. This is represented spatially (mapped) as a "predicted potential distribution" for the taxon under consideration, in which isobars (or colors) represent different degrees of confidence in the prediction of presence.

The estimated potential distribution of the chironomid midge genus *Austrochlus* (Diptera) based on data points in south-western Australia is shown in Fig. 6.15. Based on climatic (predominantly seasonal rainfall) parameters, dark locations show high probability of occurrence and light grey show less likelihood. The model, based on two well-surveyed, partially sympatric species from south-western Australia, predicts the occurrence of an ecologically related taxon in central Australia, which has been since discovered within the predicted range. The effectiveness of bioclimatic modeling in predicting distributions of sister taxa, as shown here and in other studies, implies that much speciation has been by vicariance, with little or no ecological divergence (section 8.6).

6.11.2 Climatic change and insect distributions

The modeling techniques above lend themselves to back-tracking, allowing reconstruction of past species distributions based on models of previous climate and/or reconstruction of past climates based on postglacial fossil remains representing past distributional information. Such studies were based originally on pollen remains (palynology) from lake benthic cores, in which rather broad groups of pollens, with occasional indicative species, were used to track vegetational changes through time, across landscapes, and even associated with previous climates. More refined data came from preserved ostracods, beetles (especially their elytra), and the head capsules of larval chironomids. These remnants of previous inhabitants derive from short-lived organisms that appear to respond rapidly to climatic events. Extrapolation from inferred bioclimatic controls governing the present-day distributional range of insect species and their assemblages to those same taxa preserved at time of deposition allows reconstructions of previous climates. For example, major features from the late Quaternary period include a rapid recovery from extreme conditions at the peak of last glaciation (14,500 years ago), with intermittent reversal to colder periods in a general warming trend. Verification for such insect-based reconstructions has come from independent chemical signals and congruence with a Younger Dryas cold period (11,400–10,500 years ago), and documented records in human

Box 6.3 Climatic modeling for fruit flies

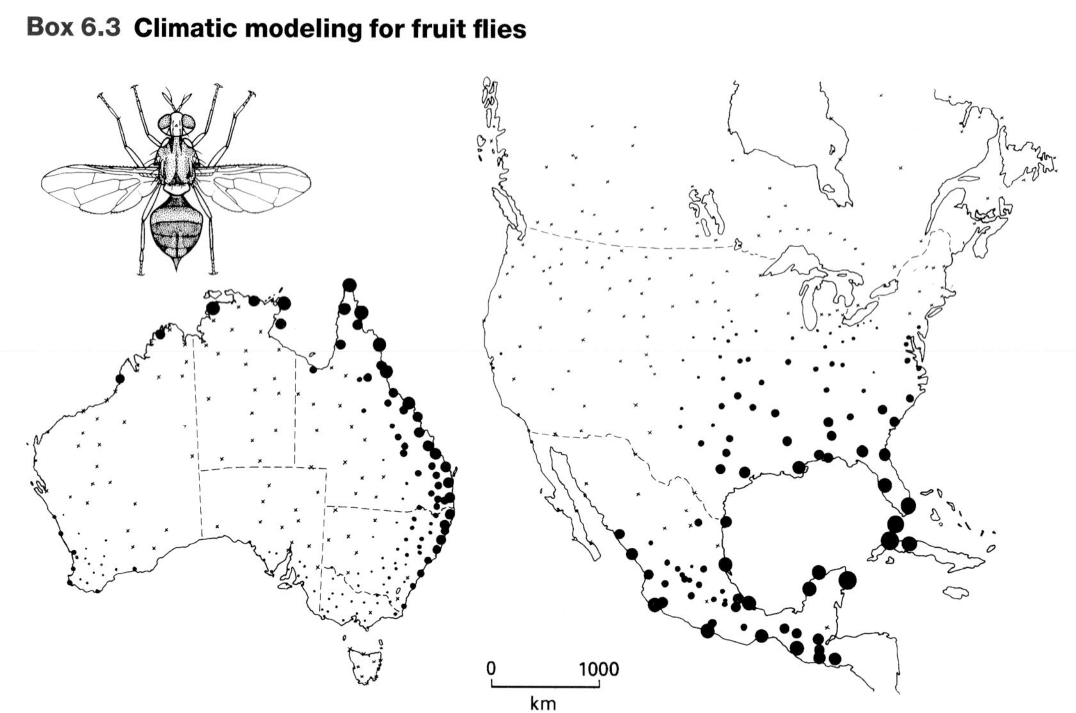

The Queensland fruit fly, *Bactrocera tryoni*, is a pest of most commercial fruits. The females oviposit into the fruit and larval feeding followed by rotting quickly destroys it. Even if damage in an orchard is insignificant, any infestation is serious because of restrictions on interstate and overseas marketing of fruit-fly-infested fruit.

CLIMEX has been used by R.W. Sutherst and G.F. Maywald to describe the response of *B. tryoni* to Australia's climate. The growth and stress indices of CLIMEX were estimated by inference from maps of the geographical distribution and from estimates of the relative abundance of this fly in different parts of its range in Australia. The map of Australia depicts the ecoclimatic indices (EI) describing the favorableness of each site for permanent colonization by *B. tryoni*. The area of each circle is proportional to its EI. Crosses indicate that the fly could not permanently colonize the site.

The potential survival of *B. tryoni* as an immigrant pest in North America can be predicted using CLIMEX by climate-matching with the fly's native range. Accidental transport of this fly could lead to its establishment at the point of entry or it might be taken to other areas with climates more favorable to its persistence. Should *B. tryoni* become established in North America, the eastern seaboard from New York to Florida and west to Kansas, Oklahoma, and Texas in the USA, and much of Mexico are most at risk. Canada and most of the central and western USA are unlikely to support permanent colonization. Thus, only certain regions of the continent are at high risk of infestation by *B. tryoni* and quarantine authorities in those places should maintain appropriate vigilance. (After Sutherst & Maywald 1991.)

history such as a medieval 12th century warm event and the 17th century Little Ice Age when "Ice Fairs" were held on the frozen River Thames. Inferred changes in temperatures range from 1 to 6°C, sometimes over just a few decades.

Confirmation of past temperature-associated biotic changes leads to the advocacy of such models to predict future range changes. For example, estimates for disease-transmitting mosquitoes and biting midges under different climate-change scenarios have ranged

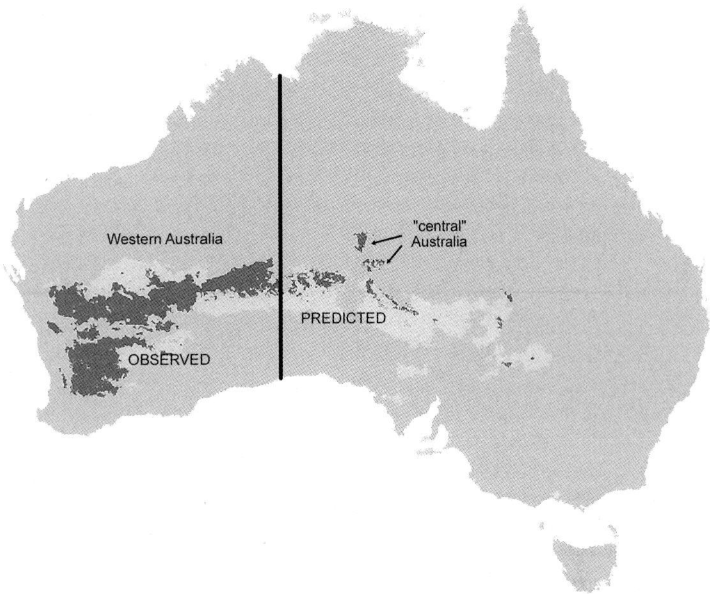

Fig. 6.15 Modeled distribution for *Austrochlus* species (Diptera: Chironomidae) based on presence data. Black, predicted presence within 98% confidence limits; pale grey, within 95% confidence. (After Cranston et al. 2002.)

from naïve estimates of increased range of disease vectors into populated areas currently disease-free (where vectors actually already exist in the absence of the virus) to sophisticated models accounting for altered development rates for vector and arbovirus, and altered environments for larval development. Future levels of predicted climate change remain unclear, allowing certain policy makers to deny its existence or its biotic significance. However, by the turn of the millennium Europe had warmed 0.8°C in the 20th century and realistic expectations are for a further increase of between 2.1 and 4.6°C mean global change in this century, along with commensurate variation in other climatic factors such as seasonality and rainfall. That predicted changes in distributions of insects are occurring is evident from studies of individual species, but the generality of these examples has been unclear. However, a study of species of western European butterflies (limited to non-migrants and excluding monophagous and/or geographically restricted taxa) is quite conclusive. Significant northward extension of ranges is demonstrated for many taxa (65% of 52 species), with some stasis (34%), and retraction south from an earlier northern limit for only one species. Data for the southern boundary, limited to 40 species, revealed retraction northward for 22%, stasis for 72%, and southward extension for only one species. The subset of the data for which sufficient historical detail was known for both northern and southern boundaries comprised 35 species: of these 63% shifted northward, 29% were stable at both boundaries, 6% shifted southwards, and one species extended both boundaries. For the many species whose boundaries moved, an observed range shift of from 35 to 240 km in the past 30–100 years coincides quite closely with the (north) polewards movement of the isotherms over the period. That such range changes have been induced by a modest temperature increase of <1°C surely is a warning of the dramatic effects of the ongoing "global warming" over the next century.

FURTHER READING

Binnington, K. & Retnakaran, A. (eds.) (1991) *Physiology of the Insect Epidermis.* CSIRO Publications, Melbourne.

Carroll, S.B. (1995) Homeotic genes and the evolution of arthropods and chordates. *Nature* **376**, 479–85.

Chapman, R.F. (1998) *The Insects. Structure and Function*, 4th edn. Cambridge University Press, Cambridge.

Daly, H.V. (1985) Insect morphometrics. *Annual Review of Entomology* **30**, 415–38.

Danks, H.V. (ed.) (1994) *Insect Life Cycle Polymorphism: Theory, Evolution and Ecological Consequences for Seasonality and Diapause Control.* Kluwer Academic, Dordrecht.

Dingle, H. (1996) *Migration. The Biology of Life on the Move.* Oxford University Press, New York.

Hayes, E.J. & Wall, R. (1999) Age-grading adult insects: a review of techniques. *Physiological Entomology* **24**, 1–10.

Heming, B.-S. (2003) *Insect Development and Evolution.* Cornell University Press, Ithaca, NY.

Kerkut, G.A. & Gilbert, L.I. (eds.) (1985) *Comprehensive Insect Physiology, Biochemistry, and Pharmacology*, Vols. 7 & 8, *Endocrinology I and II*. Pergamon Press, Oxford.

Moran, N.A. (1992) The evolution of aphid life cycles. *Annual Review of Entomology* **37**, 321–48.

Nagy, L. (1998) Changing patterns of gene regulation in the evolution of arthropod morphology. *American Zoologist* **38**, 818–28.

Nijhout, H.F. (1999) Control mechanisms of polyphenic development in insects. *BioScience* **49**, 181–92.

Parmesan, C., Ryrholm, N., Stefanescu, C. et al. (1999) Poleward shifts in geographical ranges of butterfly species associated with regional warming. *Nature* **399**, 579–83.

Pender, M.P. (1991) Locust phase polymorphism and its endocrine relations. *Advances in Insect Physiology* **23**, 1–79.

Resh, V.H. & Cardé, R.T. (eds.) (2003) *Encyclopedia of Insects.* Academic Press, Amsterdam. [Particularly see articles on cold/heat protection; development, hormonal control of; diapause; embryogenesis; imaginal discs; molting.]

Sander, K., Gutzeit, H.O. & Jäckle, H. (1985) Insect embryogenesis: morphology, physiology, genetical and molecular aspects. In: *Comprehensive Insect Physiology, Biochemistry, and Pharmacology*, Vol. 1: *Embryogenesis and Reproduction* (eds. G.A. Kerkut & L.I. Gilbert), pp. 319–85. Pergamon Press, Oxford.

Sehnal, F. (1985) Growth and life cycles. In: *Comprehensive Insect Physiology, Biochemistry, and Pharmacology*, Vol. 2: *Postembryonic Development* (eds. G.A. Kerkut & L.I. Gilbert), pp. 1–87. Pergamon Press, Oxford.

Sutherst, R.W. & Maywald, G.F. (1985) A computerized system for matching climates in ecology. *Agriculture, Ecosystems and Environment* **13**, 281–99.

Worner, S.P. (1988) Ecoclimatic assessment of potential establishment of exotic pests. *Journal of Economic Entomology* **81**, 9731–83.

Papers and references in *European Journal of Entomology* (1996) **93**(3), especially: Bale, J.S., Insect cold hardiness: A matter of life and death (pp. 369–82); Lee, R.E., Costanzo, J.P., & Mugano, J.A., Regulation of supercooling and ice nucleation in insects (pp. 405–18).

Chapter 7

INSECT SYSTEMATICS: PHYLOGENY AND CLASSIFICATION

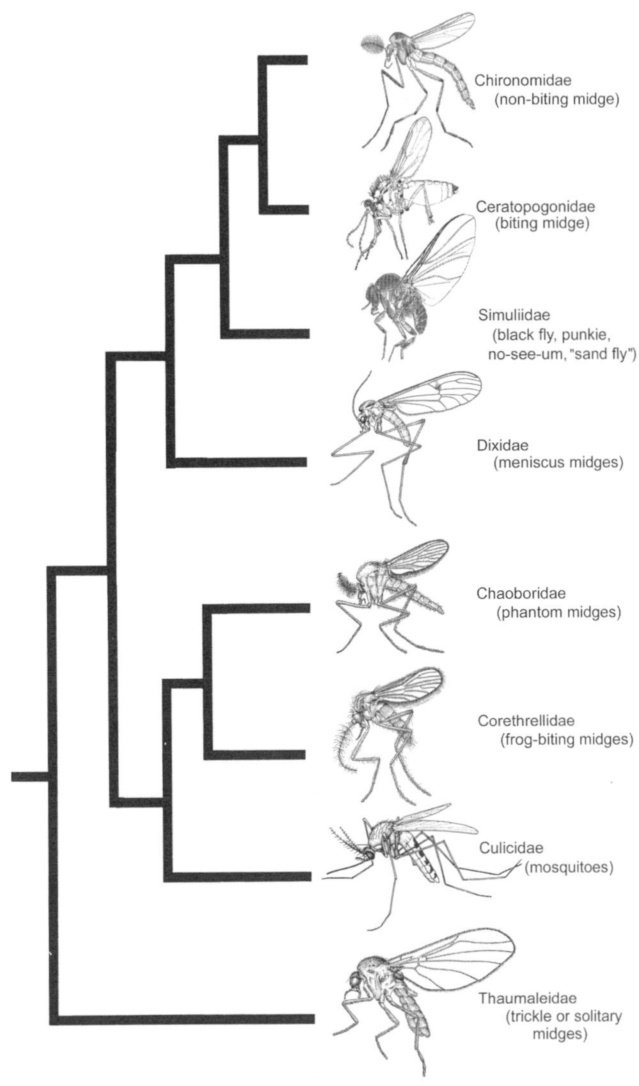

Tree showing proposed relationships between mosquitoes, midges, and their relatives. (After various sources.)

Because there are so many guides to the identity and classification of birds, mammals, and flowers, it is tempting to think that every organism in the living world is known. However, if we compared different books, treatments will vary, perhaps concerning the taxonomic status of a geographical race of bird, or of the family to which a species of flowering plant belongs. Scientists do not change and confuse such matters perversely. Differences can reflect uncertainty concerning relationships and the most appropriate classification may be elusive. Changes may arise from continuing acquisition of knowledge concerning relationships, perhaps through the addition of molecular data to previous anatomical studies. For insects, **taxonomy** – the basic work of recognizing, describing, naming, and classification – is incomplete because there are so many species, with much variation.

The study of the kinds and diversity of organisms and their inter-relationships – **systematics** – has been portrayed sometimes as dull and routine. Certainly, taxonomy involves time-consuming activities, including exhaustive library searches and specimen study, curation of collections, measurements of features from specimens, and sorting of perhaps thousands of individuals into morphologically distinctive and coherent groups (which are first approximations to species), and perhaps hundreds of species into higher groupings. These essential tasks require considerable skill and are fundamental to the wider science of systematics, which involves the investigation of the origin, diversification, and distribution, both historical and current, of organisms. Modern systematics has become an exciting and controversial field of research, due largely to the accumulation of increasing amounts of nucleotide sequence data and the application of explicit analytical methods to both morphological and DNA data, and partly to increasing interest in the documentation and preservation of biological diversity.

Taxonomy provides the database for systematics. The collection of these data and their interpretation once was seen as a matter of personal taste, but recently has been the subject of challenging debate. Entomological systematists have featured as prominent participants in this vital biological enterprise. In this chapter the methods of interpreting relationships are reviewed briefly, followed by details of the current ideas on a classification based on the postulated evolutionary relationships within the Hexapoda, of which the Insecta forms the largest of four classes.

7.1 PHYLOGENETICS

The unraveling of evolutionary history, **phylogenetics**, is a stimulating and contentious area of biology, particularly for the insects. Although the various groups (**taxa**), especially the orders, are fairly well defined, the phylogenetic relationships among insect taxa are a matter of much conjecture, even at the level of orders. For example, the order Strepsiptera is a discrete group that is recognized easily by having the fore wings modified as balancing organs, yet the identity of its close relatives is not obvious. Stoneflies (Plecoptera) and mayflies (Ephemeroptera) somewhat resemble each other, but this resemblance is superficial and misleading as an indication of relationship. The stoneflies are more closely related to the orthopteroids (cockroaches, termites, mantids, earwigs, grasshoppers, crickets, and their allies) than to mayflies. Resemblance may not indicate evolutionary relationships. Similarity may derive from being related, but equally it can arise through **homoplasy**, meaning convergent or parallel evolution of structures either by chance or by selection for similar functions. Only similarity as a result of common ancestry (**homology**) provides information regarding phylogeny. Two criteria for homology are:

1 **similarity** in outward appearance, development, composition, and position of features (characters);

2 **conjunction** – two homologous features (characters) cannot occur simultaneously in the same organism.

A test for homology is **congruence** (correspondence) with other homologies.

In segmented organisms such as insects (section 2.2), features may be repeated on successive segments, for example each thoracic segment has a pair of legs, and the abdominal segments each have a pair of spiracles. **Serial homology** refers to the correspondence of an identically derived feature of one segment with the feature on another segment (Chapter 2).

Traditionally, morphology (external anatomy) provided most data upon which insect relationships were reconstructed. Some of the ambiguity and lack of clarity regarding insect phylogeny was blamed on inherent deficiencies in the phylogenetic information provided by these morphological characters. After investigations of the utility of chromosomes and then differences in electrophoretic mobility of proteins, molecular sequence data from the mitochondrial and the nuclear genomes have become the most prevalent tools used to solve many unanswered questions, including those con-

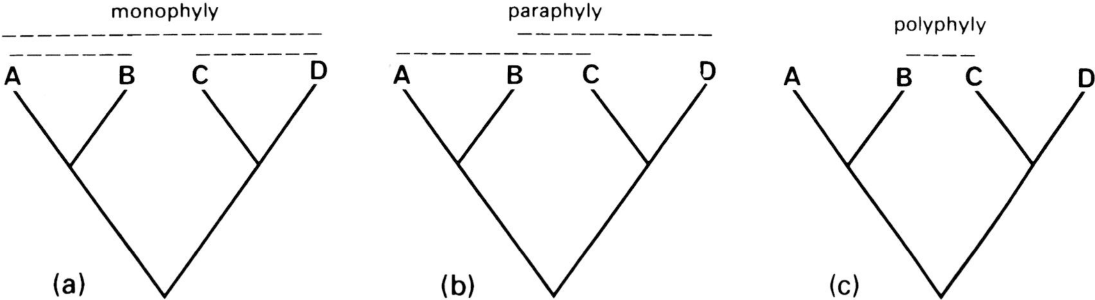

Fig. 7.1 A cladogram showing the relationships of four species, A, B, C, and D, and examples of (a) the three monophyletic groups, (b) two of the four possible (ABC, ABD, ACD, BCD) paraphyletic groups, and (c) one of the four possible (AC, AD, BC, and BD) polyphyletic groups that could be recognized based on this cladogram.

cerning higher relationships among insects. However, molecular data are not foolproof; as with all data sources the signal can be obscured by homoplasy. Nevertheless, with appropriate choice of taxa and genes, molecules do help resolve certain phylogenetic questions that morphology has been unable to answer. Another source of useful data for inferring the phylogenies of some insect groups derives from the DNA of their bacterial symbionts. For example, the primary endosymbionts (but not the secondary endosymbionts) of aphids, mealybugs, and psyllids co-speciate with their hosts, and bacterial relationships can be used (with caution) to estimate host relationships. Evidently, the preferred approach to estimating phylogenies is a holistic one, using data from as many sources as possible and retaining an awareness that not all similarities are equally informative in revealing phylogenetic pattern.

7.1.1 Systematic methods

The various methods that attempt to recover the pattern produced by evolutionary history rely on observations on living and fossil organisms. As a simplification, three differing methods can be identified: phenetics, cladistics, and evolutionary systematics.

The **phenetic** method (phenetics) relies on estimates of overall similarity, usually derived from morphology, but sometimes from behavior and other traits, and increasingly from molecular evidence. Many of those who have applied phenetics have claimed that evolution is unknowable and the best that we can hope for are patterns of resemblance; however, other scientists believe that the phenetic pattern revealed is as good an estimate of evolutionary history as can be obtained. Alternative methods to phenetics are based on the premise that the pattern produced by evolutionary processes can be estimated, and, furthermore, ought to be reflected in the classification. Overall similarity, the criterion of phenetics, may not recover this pattern of evolution and phenetic classifications are therefore artificial.

The **cladistic** method (cladistics) seeks patterns of special similarity based only on shared, evolutionarily novel features (**synapomorphies**). Synapomorphies are contrasted with shared ancestral features (**plesiomorphies** or **symplesiomorphies**), which do not indicate closeness of relationship. Furthermore, features that are unique to a particular group (**autapomorphies**) but unknown outside the group do not indicate inter-group relationships, although they are very useful for diagnosing the group. Construction of a **cladogram** (Fig. 7.1), a treelike diagram portraying the phylogenetic branching pattern, is fundamental to cladistics. From this tree, **monophyletic** groups, or **clades**, their relationships to each other, and a classification, can be inferred directly. **Sister groups** are taxa that are each other's closest relatives. A monophyletic group contains a hypothetical ancestor and *all* of its descendants.

Further groupings can be identified from Fig. 7.1: **paraphyletic** groups lack one clade from amongst the descendants of a common ancestor, and often are created by the recognition (and removal) of a derived subgroup; **polyphyletic** groups fail to include two or more clades from amongst the descendants of a common ancestor (e.g. A and D in Fig. 7.1c). Thus, when we recognize the monophyletic Pterygota (winged or secondarily apterous insects), a grouping of the remainder of the Insecta, the non-monophyletic "apterygotes", is rendered paraphyletic. If we were to recognize a group of flying insects with wings restricted to the mesothorax (dipterans, male coccoids, and a few ephemeropterans), this would be a polyphyletic grouping. Paraphyletic groups should be avoided if possible because their only defining features are ancestral ones shared with other indirect relatives. Thus, the absence of wings in the paraphyletic apterygotes is an ancestral feature shared by many other invertebrates. The mixed ancestry of polyphyletic groups means that they are biologically uninformative and such artificial taxa should never be included in any classification.

Evolutionary systematics also uses estimates of derived similarity but, in contrast to cladistics, estimates of the amount of evolutionary change are included with the branching pattern in order to produce a classification. Thus, an evolutionary approach emphasizes distinctness, granting higher taxonomic status to taxa separated by "gaps". These gaps may be created by accelerated morphological innovation in a lineage, and/or by extinction of intermediate, linking forms. Thus, ants once were given superfamily rank (the Formicoidea) within the Hymenoptera because ants are highly specialized with many unique features that make them look very different from their nearest relatives. However, phylogenetic studies show ants belong in the superfamily Vespoidea, and are given the rank of family, the Formicidae (Fig. 12.2).

Current classifications of insects mix all three practices, with most orders being based on groups (taxa) with distinctive morphology. It does not follow that these groups are monophyletic, for instance Blattodea, Psocoptera, and Mecoptera almost certainly are each paraphyletic (see below). However, it is unlikely that any higher-level groups are polyphyletic. In many cases, the present groupings coincide with the earliest colloquial observations on insects, for example the term "beetles" for Coleoptera. However, in other cases, such old colloquial names cover disparate modern groupings, as with the old term "flies", now seen to encompass unrelated orders from mayflies (Ephemeroptera) to true flies (Diptera). Refinements continue as classification is found to be out of step with our developing understanding of the phylogeny. Thus, current classifications increasingly combine traditional views with recent ideas on phylogeny.

7.1.2 Taxonomy and classification

Difficulties with attaining a comprehensive, coherent classification of the insects arise when phylogeny is obscured by complex evolutionary diversifications. These include radiations associated with adoption of specialized plant or animal feeding (phytophagy and parasitism; section 8.6) and radiations from a single founder on isolated islands (section 8.7). Difficulties arise also because of conflicting evidence from immature and adult insects, but, above all, they derive from the immense number of species (section 1.3.2).

Scientists who study the taxonomy of insects – i.e. describe, name, and classify them – face a daunting task. Virtually all the world's vertebrates are described, their past and present distributions verified and their behaviors and ecologies studied at some level. In contrast, perhaps only 5–20% of the estimated number of insect species have been described formally, let alone studied biologically. The disproportionate allocation of taxonomic resources is exemplified by Q.D. Wheeler's report for the USA of seven described mammal species per mammal taxonomist in contrast to 425 described insects per insect taxonomist. These ratios, which probably have worldwide application, become even more alarming if we include estimates of undescribed species. There are very few unnamed mammals, but estimates of global insect diversity may involve millions of undescribed species.

Despite these problems, we are moving towards a consensus view on many of the internal relationships of Insecta and their wider grouping, the Hexapoda. These are discussed below.

7.2 THE EXTANT HEXAPODA

The Hexapoda (usually given the rank of superclass) contains all six-legged arthropods. Traditionally, the closest relatives of hexapods have been considered to be the myriapods (centipedes, millipedes, and their allies). However, as shown in Box 7.1, molecular sequence and developmental data plus some morphology (especially of the compound eye and nervous system) suggest a more recent shared ancestry for hexapods and crustaceans than for hexapods and myriapods.

Diagnostic features of the Hexapoda include the possession of a unique **tagmosis** (section 2.2), which is the specialization of successive body segments that more or less unite to form sections or tagmata, namely

Box 7.1 Relationships of the Hexapoda to other Arthropoda

The immense phylum Arthropoda, the joint-legged animals, includes several major lineages: the myriapods (centipedes, millipedes, and their relatives), the chelicerates (horseshoe crabs and arachnids), the crustaceans (crabs, shrimps, and relatives), and the hexapods (the six-legged arthropods – the Insecta and their relatives). The onychophorans (velvet worms, lobopods) have been included in the Arthropoda, but are considered now to lie outside, amongst probable sister groups. Traditionally, each major arthropod lineage has been considered monophyletic, but at least some investigations have revealed non-monophyly of one or more groups. Analyses of molecular data (some of which were naïve in sampling and analytical methods) suggested paraphyly, possibly of myriapods and/or crustaceans. Even accepting monophyly of arthropods, estimation of inter-relationships has been contentious with almost every possible relationship proposed by someone. A once-influential view of the late Sidnie Manton proposed three groups of arthropods, namely the Uniramia (lobopods, myriapods, and insects, united by having single-branched legs), Crustacea, and Chelicerata, each derived independently from a different (but unspecified) non-arthropod group. More recent morphological and molecular studies reject this hypothesis, asserting monophyly of arthropodization, although proposed internal relationships cover a range of possibilities. Part of Manton's Uniramia group – the Atelocerata (also known as Tracheata) comprising myriapods plus hexapods – is supported by some morphology. These features include the presence (in at least some groups) of a tracheal system, Malpighian tubules, unbranched limbs, eversible coxal vesicles, postantennal organs, and anterior tentorial arms. Furthermore, there is no second antenna (or homolog) as seen in crustaceans. Proponents of this myriapod plus hexapod relationship saw Crustacea either grouping with the chelicerates and the extinct trilobites, distinct from the Atelocerata, or forming its sister group in a clade termed the Mandibulata. In all these schemes, the closest relatives of the Hexapoda always were the Myriapoda or a subordinate group within Myriapoda.

In contrast, certain shared morphological features, including ultrastructure of the nervous system (e.g. brain structure, neuroblast formation, and axon development), the visual system (e.g. fine structure of the ommatidia, optic nerves), and developmental processes, especially segmentation, argued for a closer relationship of Hexapoda to Crustacea. Such a grouping, termed the Pancrustacea, excludes myriapods. Molecular sequence data alone, or combined with morphology, tend to support Pancrustacea over Atelocerata. However, not all analyses actually recover Pancrustacea and certain genes evidently fail to retain phylogenetic signal from what was clearly a very ancient divergence.

If the Pancrustacea hypothesis of relationship is correct, then features understood previously to support the monophyly of Atelocerata need re-consideration. Postantennal organs occur only in Collembola and Protura in Hexapoda, and may be convergent with similar organs in Myriapoda or homologous with the second antenna of Crustacea. The shared absence of features such as the second antenna provides poor evidence of relationship. Malpighian tubules of hexapods must exist convergently in arachnids and evidence for homology between their structure and development in hexapods and myriapods remains inadequately studied. Coxal vesicles are not always developed and may not be homologous in the Myriapoda and those Hexapoda (apterygotes) possessing these structures. Thus, morphological characters supporting Atelocerata may be non-homologous and may have been convergently acquired in association with the adoption of a terrestrial mode of life.

A major finding from molecular embryology is that the developmental expression of the homeotic (developmental regulatory) gene *Dll* (*Distal-less*) in the mandible of studied insects resembled that observed in sampled crustaceans. This finding refutes Manton's argument for arthropod polyphyly and the claim that hexapod mandibles were derived independently from those of crustaceans. Data derived from the neural, visual, and developmental systems, although sampled across few taxa, may reflect more accurately the phylogeny than did many earlier-studied morphological features. Whether the Crustacea in totality or a component thereof constitute the sister group to the Hexapoda is still debatable. Morphology generally supports a monophyletic Crustacea, but inferences from some molecular data imply paraphyly, including a suggestion that Malacostraca alone form the sister taxon to Hexapoda. Given that analysis of combined morphological and molecular data supports monophyly of Crustacea and Pancrustacea, a single origin of Crustacea seems most favored. Nonetheless, some data imply a quite radically different relationship of Collembola to Crustacea, implying a polyphyletic Hexapoda. In this view, aberrant collembolan morphology (entognathy, unusual abdominal segmentation, lack of Malpighian tubules, single claw, unique furcula, unique embryology) derives from an early-branching pancrustacean ancestry, with terrestriality acquired independently of Hexapoda. Such a view deserves further study – evidently there remain many questions in the unraveling of the evolution of the Hexapoda and Insecta.

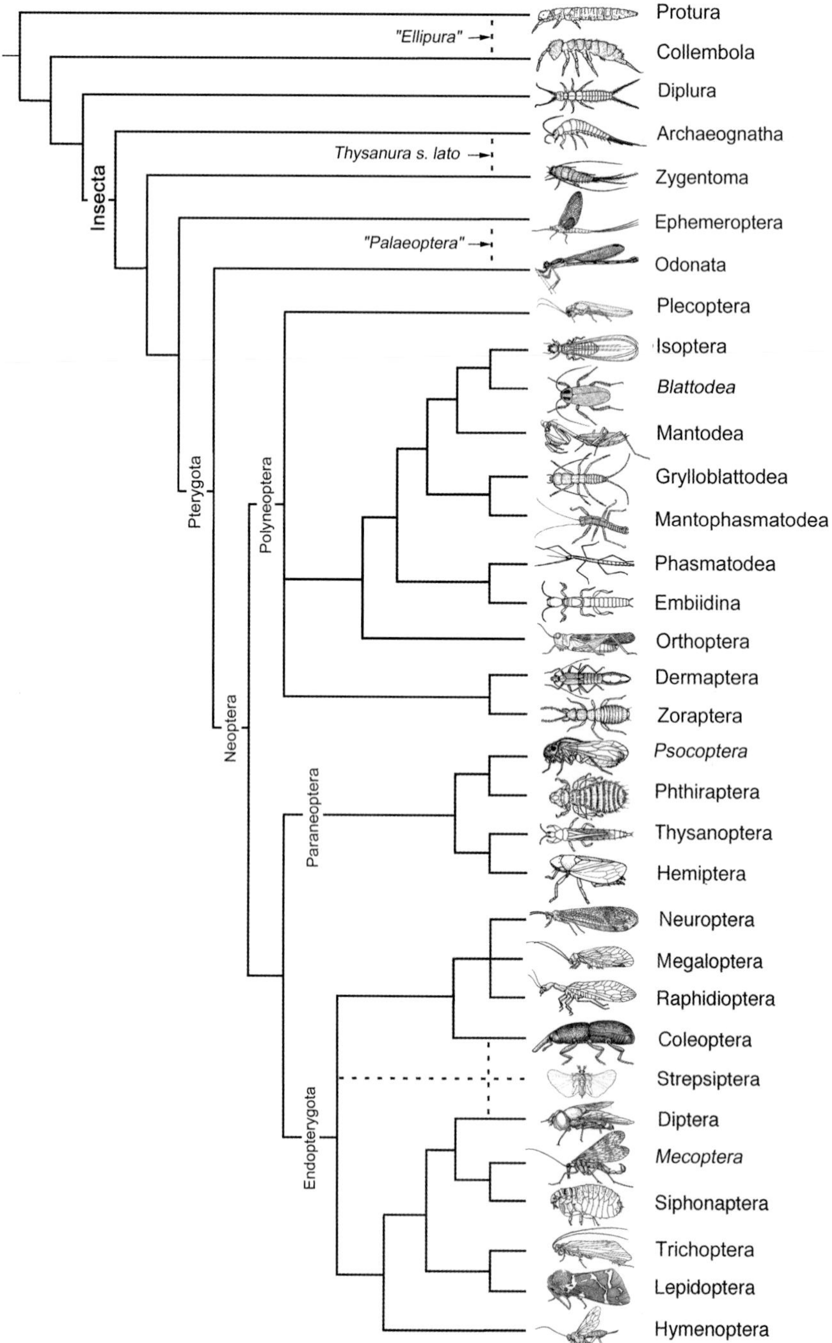

Fig. 7.2 Cladogram of postulated relationships of extant hexapods, based on combined morphological and nucleotide sequence data. Italicized names indicate paraphyletic taxa. Broken lines indicate uncertain relationships. Thysanura *sensu lato* refers to Thysanura in the broad sense. (Data from several sources.)

the head, thorax, and abdomen. The head is composed of a pregnathal region (usually considered to be three segments) and three gnathal segments bearing mandibles, maxillae, and labium, respectively; the eyes are variously developed, and may be lacking. The thorax comprises three segments, each of which bears one pair of legs, and each thoracic leg has a maximum of six segments in extant forms, but was primitively 11-segmented with up to five **exites** (outer appendages of the leg), a coxal **endite** (an inner appendage of the leg) and two terminal claws. The abdomen originally had 11 segments plus a telson or some homologous structure; if abdominal limbs are present, they are smaller and weaker than those on the thorax, and primitively were present on all except the tenth segment.

The earliest branches in the hexapod phylogeny undoubtedly involve organisms whose ancestors were terrestrial (non-aquatic) and wingless. However, any combined grouping of these taxa is not monophyletic, being based on evident symplesiomorphies or otherwise doubtfully derived characters. Included orders are Protura, Collembola, Diplura, Archaeognatha, and Zygentoma (= Thysanura). The Insecta proper comprise Archaeognatha, Zygentoma, and the huge radiation of Pterygota (the primarily winged hexapods). As a consequence of the Insecta being ranked as a class, the successively more distant sister groups Diplura, Collembola, and Protura, which are considered to be of equal rank, are treated as classes.

Some relationships among the component taxa of Hexapoda are uncertain, although the cladograms shown in Figs. 7.2 and 7.3, and the classification presented in the following sections reflect our current synthetic view. Previously, Collembola, Protura, and Diplura were grouped as "Entognatha", based on resemblance in mouthpart morphology. Entognathan mouthparts are enclosed in folds of the head, in contrast to mouthparts of the Insecta (Archaeognatha + Zygentoma + Pterygota) which are exposed (ectognathous). However, two different types of entognathy have been recognized, one type apparently shared by Collembola and Protura, and the second seemingly unique to Diplura. Other morphological evidence and some molecular data analyses indicate that Diplura may be closer to Insecta than to the other entognathans, rendering Entognatha paraphyletic (as indicated by broken lines in Fig. 7.3). Some highly controversial studies indicate derivation of Collembola (and perhaps Protura) from within the Crustacea, independently from other hexapods.

7.3 PROTURA (PROTURANS), COLLEMBOLA (SPRINGTAILS), AND DIPLURA (DIPLURANS)

7.3.1 Class and order Protura (proturans)

(see also Box 9.2)

Proturans are small, delicate, elongate, mostly unpigmented hexapods, lacking eyes and antennae, with entognathous mouthparts consisting of slender mandibles and maxillae that slightly protrude from the mouth cavity. Maxillary and labial palps are present. The thorax is poorly differentiated from the 12-segmented abdomen. Legs are five-segmented. A gonopore lies between segments 11 and 12, and the anus is terminal. Cerci are absent. Larval development is anamorphic, that is with segments added posteriorly during development. Protura either is sister to Collembola, forming Ellipura in a weakly supported relationship based on entognathy and lack of cerci, or is sister to all remaining Hexapoda.

7.3.2 Class and order Collembola (springtails) (see also Box 9.2)

Collembolans are minute to small and soft bodied, often with rudimentary eyes or ocelli. The antennae are four- to six-segmented. The mouthparts are entognathous, consisting predominantly of elongate maxillae and mandibles enclosed by lateral folds of head, and lacking maxillary and labial palps. The legs are four-segmented. The abdomen is six-segmented with a sucker-like ventral tube or collophore, a retaining hook and a furcula (forked jumping organ) on segments 1, 3, and 4, respectively. A gonopore is present on segment 5, the anus on segment 6. Cerci are absent. Larval development is epimorphic, that is with segment number constant through development. Certain controversial studies suggest that Collembola may have a different evolutionary origin to the rest of the Hexapoda (see Box 7.1). If Collembola do belong to the Hexapoda, then they form either the sister group to Protura comprising the clade Ellipura or alone form the sister to Diplura + Insecta.

7.3.3 Class and order Diplura (diplurans)

(see also Box 9.2)

Diplurans are small to medium sized, mostly

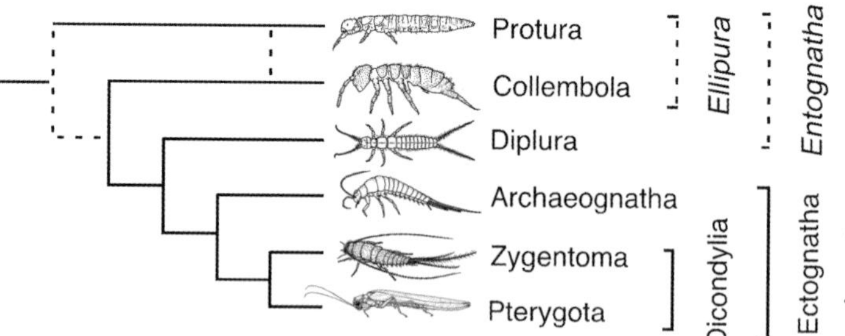

Fig. 7.3 Cladogram of postulated relationships of early-branching hexapod orders, based on morphological data. Italicized names indicate likely paraphyletic taxa. Broken lines indicate uncertain relationships. (Data from several sources.)

unpigmented, possess long, moniliform antennae (like a string of beads), but lack eyes. The mouthparts are entognathous, with tips of well-developed mandibles and maxillae protruding from the mouth cavity, and maxillary and labial palps reduced. The thorax is poorly differentiated from the 10-segmented abdomen. The legs are five-segmented and some abdominal segments have small styles and protrusible vesicles. A gonopore lies between segments 8 and 9, the anus is terminal. Cerci are slender to forceps-shaped. The tracheal system is relatively well developed, whereas it is absent or poorly developed in other entognath groups. Larval development is epimorphic, with segment number constant through development. Diplura undoubtedly forms the sister group to Insecta.

7.4 CLASS INSECTA (TRUE INSECTS)

Insects range from minute to large (0.2 mm to 30 cm long) with very variable appearance. Adult insects typically have ocelli and compound eyes, and the mouthparts are exposed (ectognathous) with the maxillary and labial palps usually well developed. The thorax may be weakly developed in immature stages but is distinct in flighted adult stages, associated with development of wings and the required musculature; it is weakly developed in wingless taxa. Thoracic legs have more than five segments. The abdomen is primitively 11-segmented with the gonopore nearly always on segment 8 in the female and segment 9 in the male. Cerci are primitively present. Gas exchange is predominantly tracheal with spiracles present on both the thorax and abdomen, but may be variably reduced or absent as in some immature stages. Larval/nymphal development is epimorphic, that is, with the number of body segments constant during development.

The 30 orders of insects traditionally have been divided into two groups. Monocondylia is represented by just one small order, Archaeognatha, in which each mandible has a single posterior articulation with the head. Dicondylia (Fig. 7.3), which contains all of the other orders and the overwhelming majority of species, has mandibles characterized by a secondary anterior articulation in addition to the primary posterior one. The traditional group Apterygota for the primitively wingless taxa Archaeognatha + Zygentoma appears paraphyletic on most (but not all) modern analyses (Figs. 7.2 & 7.3).

7.4.1 Archaeognatha and Zygentoma (Thysanura *sensu lato*)

Order Archaeognatha (archaeognathans, bristletails) (see also Box 9.3)

Archaeognathans are medium sized, elongate-cylindrical, and primitively wingless ("apterygotes"). The head bears three ocelli and large compound eyes that are in contact medially. The antennae are multisegmented. The mouthparts project ventrally, can be partially retracted into the head, and include elongate mandibles with two neighboring condyles each and elongate seven-segmented maxillary palps. Often a coxal style occurs on coxae of legs 2 and 3, or 3 alone. Tarsi are two- or three-segmented. The abdomen continues in an even contour from the humped thorax, and bears ventral muscle-containing styles (representing reduced limbs) on segments 2–9, and generally one or two pairs of eversible vesicles medial to the styles on segments 1–7. Cerci are multisegmented and shorter than the median caudal appendage. Development occurs without change in body form.

The fossil taxon Monura belongs in Thysanura

sensu lato. The two families of recent Archaeognatha, Machilidae and Meinertellidae, form an undoubted monophyletic group. The order probably is placed as the earliest branch of the Insecta, and as sister group to Zygentoma + Pterygota (Fig. 7.3). Alternatively, a potentially influential recent molecular analysis revived the concept of Archaeognatha as sister to Zygentoma, in a grouping that should be called Thysanura (*sensu lato* – meaning in the broad sense in which the name was first used for apterous insects with "bristle tails").

Order Zygentoma (Thysanura, silverfish)
(see also Box 9.3)

Zygentomans (thysanurans) are medium sized, dorsoventrally flattened, and primitively wingless ("apterygotes"). Eyes and ocelli are present, reduced or absent, the antennae are multisegmented. The mouthparts are ventrally to slightly forward projecting and include a special form of double-articulated (dicondylous) mandibles, and five-segmented maxillary palps. The abdomen continues the even contour of the thorax, and includes ventral muscle-containing styles (representing reduced limbs) on at least segments 7–9, sometimes on 2–9, and with eversible vesicles medial to the styles on some segments. Cerci are multisegmented and subequal to the length of the median caudal appendage. Development occurs without change in body form.

There are four extant families. Zygentoma is the sister group of the Pterygota (Fig. 7.3) alone, or perhaps with Archaeognatha in Thysanura *sensu lato* (see above under Archaeognatha).

7.4.2 Pterygota

Pterygota, treated as an infraclass, are the winged or secondarily wingless (apterous) insects, with thoracic segments of adults usually large and with the meso- and metathorax variably united to form a pterothorax. The lateral regions of the thorax are well developed. Abdominal segments number 11 or fewer, and lack styles and vesicular appendages like those of apterygotes. Most Ephemeroptera have a median terminal filament. The spiracles primarily have a muscular closing apparatus. Mating is by copulation. Metamorphosis is hemi- to holometabolous, with no adult ecdysis, except for the subimago (subadult) stage in Ephemeroptera.

Informal grouping "Palaeoptera"

Insect wings that cannot be folded against the body at rest, because articulation is via axillary plates that are fused with veins, have been termed "palaeopteran" (old wings). Living orders with such wings typically have triadic veins (paired main veins with intercalated longitudinal veins of opposite convexity/concavity to the adjacent main veins) and a network of cross-veins (figured in Boxes 10.1 and 10.2). This wing venation and articulation, together with paleontological studies of similar features, was taken to imply that Odonata and Ephemeroptera form a monophyletic group, termed Palaeoptera. The group was argued to be sister to Neoptera which comprises all remaining extant and primarily winged orders. However, reassessment of morphology of extant early-branching lineages and recent nucleotide sequence evidence fails to provide strong support for monophyly of Palaeoptera. Here we treat Ephemeroptera as sister group to Odonata + Neoptera, giving a higher classification of Pterygota into three divisions.

Division (and order) Ephemeroptera (mayflies)
(see also Box 10.1)
Ephemeroptera has a fossil record dating back to the Carboniferous and is represented today by a few thousand species. In addition to their "palaeopteran" wing features mayflies display a number of unique characteristics including the non-functional, strongly reduced adult mouthparts, the presence of just one axillary plate in the wing articulation, a hypertrophied costal brace, and male fore legs modified for grasping the female during copulatory flight. Retention of a subimago (subadult stage) is unique. Nymphs (larvae) are aquatic and the mandible articulation, which is intermediate between monocondyly and the dicondylous ball-and-socket joint of all higher Insecta, may be diagnostic. Historic contraction of ephemeropteran diversity and remnant high levels of homoplasy render phylogenetic reconstruction difficult. Ephemeroptera traditionally has been divided into two suborders: Schistonota (with nymphal fore-wing pads separate from each other for over half their length) containing superfamilies Baetoidea, Heptagenioidea, Leptophlebioidea, and Ephemeroidea, and Pannota ("fused back" – with more extensively fused fore-wing pads) containing Ephemerelloidea and Caenoidea. Recent studies suggest this concept of Schistonota is paraphyletic, but no robust alternative scheme has been proposed.

Division (and order) Odonata (dragonflies and damselflies) (see also Box 10.2)
Odonates have "palaeopteran" wings as well as many additional unique features, including the presence of two axillary plates (humeral and posterior axillary) in the wing articulation and many features associated with specialized copulatory behavior, including possession of secondary copulatory apparatus on ventral segments 2–3 of the male and the formation of a tandem wheel during copulation (Box 5.3). The immature stages are aquatic and possess a highly modified prehensile labium for catching prey (Fig. 13.4).

Odonatologists (those that study odonates) traditionally recognized three groups generally ranked as suborders: Zygoptera (damselflies), Anisozygoptera and Anisoptera (dragonflies). Anisozygoptera is minor, containing fossil taxa but only one extant genus with two species. Assessment of the monophyly or paraphyly of each suborder has relied very much on interpretation of the very complex wing venation. Interpretation of wing venation within the odonates and between them and other insects has been prejudiced by prior ideas about relationships. Thus the Comstock and Needham naming system for wing veins implies that the common ancestor of modern Odonata was anisopteran, and the venation of zygopterans is reduced. In contrast, the Tillyard-named venational system implies that Zygoptera is a grade (is paraphyletic) to Anisozygoptera, which itself is a grade on the way to a monophyletic Anisoptera. A well-supported view, incorporating information from the substantial fossil record, has Zygoptera probably paraphyletic, Anisozygoptera undoubtedly paraphyletic, and Anisoptera as monophyletic sister to some extinct anisozygopterans.

Zygoptera contains three broad superfamilial groupings, the Coenagrionoidea, Lestoidea, and Calopterygoidea. Amongst Anisoptera four major lineages can be recognized, but their relationships to each other are obscure.

Division Neoptera
Neopteran ("new wing") insects diagnostically have wings capable of being folded back against their abdomen when at rest, with wing articulation that derives from separate movable sclerites in the wing base, and wing venation with none to few triadic veins and mostly lacking anastomosing (joining) cross-veins (Fig. 2.21).

The phylogeny (and hence classification) of the neopteran orders remains subject to debate, mainly concerning (a) the placement of many extinct orders described only from fossils of variably adequate preservation, (b) the relationships among the Polyneoptera (orthopteroid plus plecopteroid orders), and (c) the relationships of the highly derived Strepsiptera.

Here we summarize the most recent research findings, based on both morphology and molecules. No single or combined data set provides unambiguous resolution of insect order-level phylogeny and there are several areas of controversy. Some questions arise from inadequate data (insufficient or inappropriate taxon sampling) and character conflict within existing data (support for more than one relationship). In the absence of a robust phylogeny, ranking is somewhat subjective and "informal" ranks abound.

A group of 11 orders is termed the Polyneoptera (if monophyletic and considered to be sister to the remaining Neoptera) or Orthopteroid–Plecopteroid assemblage (if monophyly is uncertain). The remaining neopterans can be divided readily into two monophyletic groups, namely Paraneoptera (hemipteroid assemblage) and Endopterygota (= Holometabola). These three clades may be given the rank of subdivision. Polyneoptera and Paraneoptera both have plesiomorphic hemimetabolous development in contrast to the complete metamorphosis of Endopterygota.

Subdivision Polyneoptera (or Orthopteroid–Plecopteroid assemblage)

This grouping comprises the orders Plecoptera, Mantodea, Blattodea, Isoptera, Grylloblattodea, Mantophasmatodea, Orthoptera, Phasmatodea, Embiidina, Dermaptera, and Zoraptera.

Some early-branching events amongst the neopteran orders are becoming better understood, but some relationships remain poorly resolved, and often contradictory between those suggested by morphology and those from molecular data. The 11 included orders may form a monophyletic Polyneoptera based on the shared presence of tarsal plantulae (lacking only in Zoraptera) and certain analyses of nucleotide sequences. Within Polyneoptera, the grouping comprising Blattodea (cockroaches), Isoptera (termites), and Mantodea (mantids) – the Dictyoptera (Fig. 7.4) – is robust. All three orders within Dictyoptera share distinctive features of the head skeleton (perforated tentorium), mouthparts (paraglossal musculature), digestive system (toothed proventriculus), and female genitalia (shortened ovipositor above a large subgen-

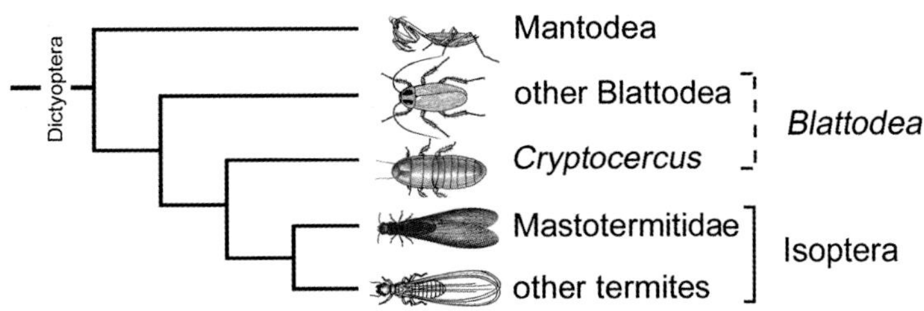

Fig. 7.4 Cladogram of postulated relationships within Dictyoptera, based on combined morphological and nucleotide sequence data. The broken line indicates a paraphyletic taxon. (Data from several sources.)

ital plate) which demonstrate monophyly substantiated by nearly all analyses based on nucleotide sequences. Dermaptera (the earwigs) and Zoraptera (zorapterans) form an unexpected higher clade based on recent nucleotide sequence data: some analyses place this group outside the Polyneoptera as sister to the remaining Neoptera, but the position is best represented as unresolved at the base of the assemblage (Fig. 7.2). The Grylloblattodea (the ice crawlers or rock crawlers; now apterous, but with winged fossils) forms a well-supported clade with the newly established order Mantophasmatodea.

Some data suggested that Orthoptera (crickets, katydids, grasshoppers, locusts, etc.), Phasmatodea (stick-insects or phasmids), and Embiidina (webspinners) may be closely related in a grouping called Orthopteroidea, although recent investigations suggest an earlier-branching position for Orthoptera. The relationships of Plecoptera (stoneflies) to other groupings are poorly understood.

Order Plecoptera (stoneflies) (see also Box 10.3)

Plecoptera are mandibulate in the adult, with filiform antennae, bulging compound eyes, two to three ocelli and subequal thoracic segments. The fore and hind wings are membranous and similar except that the hind wings are broader; aptery and brachyptery are frequent. The abdomen is 10-segmented, with remnants of segments 11 and 12 present, including cerci. Nymphs are aquatic.

Monophyly of the order is supported by few morphological characters, including in the adult the looping and partial fusion of gonads and male seminal vesicles, and the absence of an ovipositor. In nymphs the presence of strong, oblique, ventro-longitudinal muscles running intersegmentally allowing lateral undulating swimming, and the probably widespread "cercus heart", an accessory circulatory organ associated with posterior abdominal gills, support the monophyly of the order. Nymphal plecopteran gills may occur on almost any part of the body, or may be absent. This varied distribution causes problems of homology of gills between families, and between those of Plecoptera and other orders. Whether Plecoptera are ancestrally aquatic or terrestrial is debatable. The phylogenetic position of Plecoptera is certainly amongst "lower Neoptera", early in the diversification of the assemblage, possibly as sister group to the remainder of Polyneoptera, but portrayed here as unresolved (Fig. 7.2).

Internal relationships have been proposed as two predominantly vicariant suborders, the austral (southern hemisphere) Antarctoperlaria and northern Arctoperlaria. The monophyly of Antarctoperlaria is argued based on the unique sternal depressor muscle of the fore trochanter, lack of the usual tergal depressor, and presence of floriform chloride cells which may have a sensory function. Some included taxa are the large-sized Eustheniidae and Diamphipnoidae, the Gripopterygidae, and Austroperlidae – all southern hemisphere families. Some nucleotide sequence studies support this clade.

The sister group Arctoperlaria lacks defining morphology, but is united by a variety of mechanisms associated with drumming (sound production) associated with mate-finding. Component families Scopuridae, Taeniopterygidae, Capniidae, Leuctridae, and Nemouridae (including Notonemouridae) are essentially northern hemisphere with a lesser radiation of Notonemouridae into the southern hemisphere. Some nucleotide sequence analyses suggest paraphyly of Arctoperlaria, with most elements of Notonemouridae forming the sister group to the remainder of the families. Relationships amongst extant Plecoptera have been used in hypothesizing origins of wings from "thoracic gills", and in tracing the possible development of aerial flight from surface flapping with legs trailing on the water surface, and forms of gliding. Current views of the phylogeny suggest these traits are secondary and reductional.

Order Isoptera (termites, white ants) (see also Box 12.3)
Isoptera forms a small order of eusocial insects with a polymorphic caste system of reproductives, workers, and soldiers. Mouthparts are blattoid and mandibulate. Antennae are long and multisegmented. The fore and hind wings generally are similar, membranous, and with restricted venation; but *Mastotermes* (Mastotermitidae) with complex wing venation and a broad hind-wing anal lobe is exceptional. The male external genitalia are weakly developed and symmetrical, in contrast to the complex, asymmetrical genitalia of Blattodea and Mantodea. Female *Mastotermes* have a reduced blattoid-type ovipositor.

The Isoptera has always been considered to belong in Dictyoptera close to Blattodea, but precise relationships have been uncertain. A long-held view that Mastotermitidae is the earliest extant branch in the Isoptera is upheld by all studies – the distinctive features mentioned above evidently are plesiomorphies. Recent studies that included structure of the proventriculus and nucleotide sequence data suggest that termites arose from within the cockroaches, thereby rendering Blattodea paraphyletic (Fig. 7.4). Under this scenario, the (wingless) woodroaches of North America and eastern Asia (genus *Cryptocercus*) are sister group to Isoptera. Alternative suggestions of the independent origin (hence convergence) of the semisociality (parental care and transfer of symbiotic gut flagellates between generations) of *Cryptocercus* and the sociality of termites (section 12.4.2) no longer seem likely.

Order Blattodea (cockroaches) (see also Box 9.8)
Cockroaches are dorsoventrally flattened insects with filiform, multisegmented antennae and mandibulate, ventrally projecting mouthparts. The prothorax has an enlarged, shield-like pronotum, that often covers the head; the meso- and metathorax are rectangular and subequal. The fore wings are sclerotized tegmina protecting membranous hind wings folded fan-like beneath. Hind wings often may be reduced or absent, and if present characteristically have many vein branches and a large anal lobe. The legs may be spiny and the tarsi are five-segmented. The abdomen has 10 visible segments, with a subgenital plate (sternum 9), bearing in the male well-developed asymmetrical genitalia, with one or two styles, and concealing the reduced 11th segment. Cerci have one or usually many segments; the female ovipositor valves are small, concealed beneath tergum 10.

Although long considered an order (and hence monophyletic) convincing evidence shows the termites arose from within the cockroaches, and the "order" thus is rendered paraphyletic. The sister group of the Isoptera appears to be *Cryptocercus*, undoubtedly a cockroach (Fig. 7.4). Other internal relationships of the Blattodea are not well understood, with apparent conflict between morphology and limited molecular data. Usually from five to eight families are recognized. Blatellidae and Blaberidae (the largest families) are thought to be sister groups. The many early fossils allocated to Blattodea that possess a well-developed ovipositor are considered best as belonging to a blattoid stemgroup, that is, from prior to the ordinal diversification of the Dictyoptera.

Order Mantodea (mantids) (see also Box 13.2)
Mantodea are predatory, with males generally smaller than females. The small, triangular head is mobile, with slender antennae, large, widely separated eyes and mandibulate mouthparts. The prothorax is narrow and elongate, with the meso- and metathorax shorter. The fore wings form leathery tegmina with a reduced anal area; the hind wings are broad and membranous, with long unbranched veins and many cross-veins, but often are reduced or absent. The fore legs are raptorial, whereas the mid and hind legs are elongate for walking. The abdomen has a visible 10th segment, bearing variably segmented cerci. The ovipositor predominantly is internal and the external male genitalia are asymmetrical.

Mantodea forms the sister group to Blattodea + Isoptera (Fig. 7.4), and shares many features with Blattodea such as strong direct flight muscles and weak indirect (longitudinal) flight muscles, asymmetrical male genitalia and multisegmented cerci. Derived features of Mantodea relative to Blattodea involve modifications associated with predation, including leg morphology, an elongate prothorax, and features associated with visual predation, namely the mobile head with large, separated eyes. Internal relationships of the eight families of Mantodea are uncertain and little studied.

Order Grylloblattodea (= Grylloblattaria, Notoptera) (grylloblattids, ice crawlers or rock crawlers)
(see also Box 9.4)
Grylloblattids are moderate-sized, soft-bodied insects with anteriorly projecting mandibulate mouthparts and the compound eyes are either reduced or absent. The antennae are multisegmented and the mouthparts mandibulate. The quadrate prothorax is larger than

the meso- or metathorax, and wings are absent. The legs have large coxae and five-segmented tarsi. Ten abdominal segments are visible with rudiments of segment 11, including five- to nine-segmented cerci. The female has a short ovipositor, and the male genitalia are asymmetrical.

Several ordinal names have been used for these insects but Grylloblattodea is preferred because this name has the widest usage in published work and its ending matches the names of some related orders. Most of the rules of nomenclature do not apply to names above the family group and thus there is no name priority at ordinal level. The phylogenetic placement of Grylloblattodea also has been controversial, generally being argued to be relictual, either "bridging the cockroaches and orthopterans", or "primitive amongst orthopteroids". The antennal musculature resembles that of mantids and embiids, mandibular musculature resembles Dictyoptera, and the maxillary muscles those of Dermaptera. Embryologically grylloblattids are confirmed as orthopteroids. Molecular phylogenetic study emphasizing grylloblattids strongly supports a sister-group relationship to the newly discovered Mantophasmatodea, and these combined are sister to Dictyoptera.

Order Mantophasmatodea (see also Box 13.3)

Mantophasmatodea is the most recently recognized order, comprising three families from Africa, and Baltic amber specimens. Mantophasmatodeans all are apterous, without even wing rudiments. The head is hypognathous with generalized mouthparts and long, slender, multisegmented antennae. Coxae are not enlarged, the fore and mid femora are broadened and have bristles or spines ventrally; hind legs are elongate; tarsi are five-segmented, with euplanulae on the basal four; the ariolum is very large and the distal tarsomere is held off the substrate. Male cerci are prominent, clasping and not differentially articulated with tergite 10; female cerci are short and one-segmented. A distinct short ovipositor projects beyond a short subgenital lobe, lacking any protective operculum (plate below ovipositor) as seen in phasmids. Based on morphology, placement of the new order was difficult, but relationships with phasmids (Phasmatodea) and/or ice crawlers (Grylloblattodea) were suggested. Nucleotide sequencing data have justified the rank of order, and strongly confirmed a sister-group relationship to Grylloblattodea. This grouping may be the extant remnants of radiation in the distant geological past represented by fossil taxa such as Titanoptera, Caloneuridea, and Cnemidolestodea (perhaps an earlier name for Mantophasmatodea).

Order Orthoptera (grasshoppers, locusts, katydids, crickets) (see also Box 11.5)

Orthopterans are medium-sized to large insects with hind legs enlarged for jumping (saltation). The compound eyes are well developed, the antennae are elongate and multisegmented, and the prothorax is large with a shield-like pronotum curving downwards laterally. The fore wings form narrow, leathery tegmina, and the hind wings are broad, with numerous longitudinal and cross-veins, folded beneath the tegmina by pleating; aptery and brachyptery are frequent. The abdomen has eight or nine annular visible segments, with the two or three terminal segments reduced, and one-segmented cerci. The ovipositor is well developed, formed from highly modified abdominal appendages.

Virtually all morphological evidence and some molecular data suggested that the Orthoptera were closely related to Phasmatodea, to the extent that some entomologists united the orders. However, different wing bud development, egg morphology, and lack of auditory organs in phasmatids suggest distinction. Recent intensive molecular data place the Orthoptera as an early branch in the assemblage as shown in Fig. 7.2, but this requires further study.

The division of Orthoptera into two monophyletic suborders, Caelifera (grasshoppers and locusts – predominantly day-active, fast-moving, visually acute, terrestrial herbivores) and Ensifera (katydids and crickets – often night-active, camouflaged or mimetic, predators, omnivores, or phytophages), is supported on morphological and molecular evidence. Grylloidea probably form the sister group to all other ensiferan taxa but they are highly divergent. On grounds of some molecular and morphological data, Tettigoniidae and Haglidae form a monophyletic group, sister to Stenopelmatidae and relatives (Mormon crickets, wetas, Cooloola monsters, and the like), but alternative analyses suggest different or unresolved relationships. For Caelifera a well-supported recent proposal for four superfamilies, namely (Tridactyloidea (Tetragoidea (Eumastacoidea + "higher Caelifera"))) reconciles molecular evidence with certain earlier suggestions from morphology. The major grouping of acridoid grasshoppers (Acridoidea) lies in the unnamed clade "higher Caelifera", which contains also several less-speciose superfamilies.

Order Phasmatodea (phasmatids, phasmids, stick-insects or walking sticks) (see also Box 11.6)

Phasmatodea exhibit body shapes that are variations on elongate cylindrical and stick-like or flattened, or often leaf-like. The mouthparts are mandibulate. The compound eyes are relatively small and placed anterolaterally, with ocelli only in winged species, and often only in males. The wings, if present, are functional in males, but often reduced in females, and many species are apterous in both sexes. Fore wings form short leathery tegmina, whereas the hind wings are broad with a network of numerous cross-veins and with the anterior margin toughened to protect the folded wing. The legs are elongate, slender, and adapted for walking, with five-segmented tarsi. The abdomen is 11-segmented, with segment 11 often forming a concealed supra-anal plate in males or a more obvious segment in females.

Phasmatodea have long been considered as sister to Orthoptera within the orthopteroid assemblage. Recent evidence from morphology in support of this grouping comes from neurophysiological studies, namely the dorsal position of the cell body of salivary neuron 1 in the suboesophageal ganglion and presence of serotonin in salivary neuron 2. Phasmatodea are distinguished from the Orthoptera by their body shape, asymmetrical male genitalia, proventricular structure, and lack of rotation of nymphal wing pads during development. Recent evidence for a sister-group relationship to Embiidina (as in Fig. 7.2) comes from combined morphological and nucleotide sequence data from several genes. Phasmatodea conventionally have been classified in three families (although some workers raise many subfamilies to family rank). The only certainty in internal relationships is that plesiomorphic western North American *Timema* is sister to the remaining extant members of the order (termed Euphasmida). An interpretation of recent nucleotide sequence data suggests that Phasmatodea ancestrally were wingless and flightedness may have re-evolved several to many times in the radiation of the order.

Order Embiidina (= Embioptera) (embiids, webspinners) (see also Box 9.5)

Embiidina have an elongate, cylindrical body, somewhat flattened in the male. The head has kidney-shaped compound eyes that are larger in males than females, and lacks ocelli. The antennae are multi-segmented and the mandibulate mouthparts project forwards (prognathy). All females and some males are apterous; but if present, the wings are characteristically soft and flexible, with blood sinus veins stiffened for flight by blood pressure. The legs are short, with three-segmented tarsi, and the basal segment of each fore tarsus is swollen because it contains silk glands. The hind femora are swollen by strong tibial muscles. The abdomen is 10-segmented with rudiments of segment 11 and with two-segmented cerci. The female external genitalia are simple (no ovipositor), and those of males are complex and asymmetrical.

Embiids are undoubtedly monophyletic based above all on the ability to produce silk from unicellular glands in the anterior basal tarsus. A general morphological resemblance to Plecoptera based on reduced phallomeres, a trochantin-episternal sulcus, and separate coxopleuron and premental lobes is not supported by nucleotide sequences that instead imply a sister-group relationship with Phasmatodea. Internal relationships amongst the described higher taxa of Embiidina suggest that the prevailing classification into eight families includes many non-monophyletic groups. Evidently, much further study is needed to understand relationships within Embiidina, and among it and other neopterans.

Order Dermaptera (earwigs) (see also Box 9.7)

Adult earwigs are elongate and dorsoventrally flattened with mandibulate, forward-projecting mouthparts, compound eyes ranging from large to absent, no ocelli, and short annulate antennae. The tarsi are three-segmented with a short second tarsomere. Many species are apterous or, if winged, the fore wings are small, leathery, and smooth, forming unveined tegmina, and the hind wings are large, membranous, semi-circular, and dominated by an anal fan of radiating vein branches connected by cross-veins.

The five species commensal or ectoparasitic on bats in south-east Asia were placed in suborder Arixeniina. A few species semi-parasitic on African rodents were placed in suborder Hemimerina. Earwigs in both of these groups are blind, apterous, and exhibit pseudo-placental viviparity. Recent morphological study of Hemimerina suggests derivation from within Forficulina, rendering that suborder paraphyletic. The relationships of Arixeniina to more "typical" earwigs (Forficulina) are uninvestigated. Within Forficulina, only four (Karshiellidae, Apachyidae, Chelisochidae, and Forficulidae) of eight or nine families proposed appear to be supported by synapomorphies. Other families may not be monophyletic, as much weight has been placed on plesiomorphies, especially of the penis specifically and genitalia more generally, or homoplasies (convergences) in furcula form and wing reduction.

A sister-group relationship to Dictyoptera that is well supported on morphology, including many features of the wing venation, is not supported by nucleotide sequences that demonstrate an earlier-branching sister-group relationship to Zoraptera (Fig. 7.2). Whether the pair of orders is considered part of Polyneoptera or sister to the remainder of Neoptera is as yet unclear, and the relationship is best shown as unresolved.

Order Zoraptera (zorapterans) (see also Box 9.6)
Zoraptera is one of the smallest and probably the least known pterygote order. Zorapterans are small, rather termite-like insects, with simple morphology. They have biting, generalized mouthparts, including five-segmented maxillary palps and three-segmented labial palps. Sometimes both sexes are apterous, and in alate forms the hind wings are smaller than the fore wings; the wings are shed as in ants and termites. Wing venation is highly specialized and reduced.

Traditionally the order contained only one family (Zorotypidae) and one genus (*Zorotypus*), but has been divided into several genera of uncertain monophyly, delimited predominantly on wing venation. The phylogenetic position of Zoraptera based on morphology has been controversial, ranging through membership of the hemipteroid orders, sister to Isoptera, an orthopteroid, or a blattoid. Wing shape and venation resembles that of narrow-winged Isoptera, and analysis of major wing structures and musculature imply Zoraptera belong in a wide "blattoid" lineage. Hind-leg musculature revealed a derived condition shared only by Embiidina. Cephalic, abdominal, and nucleotide characters indicate an early divergence, perhaps as sister to Dermaptera, originating before the origin of the Dictyoptera clade.

Subdivision Paraneoptera (Acercaria, or Hemipteroid assemblage)

This subdivision comprises the orders Psocoptera, Phthiraptera, Thysanoptera, and Hemiptera. This group is defined by derived features of the mouthparts, including the slender, elongate maxillary lacinia separated from the stipes and a swollen postclypeus containing an enlarged cibarium (sucking pump), and the reduction in tarsomere number to three or less.

Within Paraneoptera, the monophyletic superorder Psocodea contains Phthiraptera (parasitic lice) and Psocoptera (booklice). Phthiraptera is monophyletic, but the clade arose from within Psocoptera, rendering that group paraphyletic. Although sperm morphology and some molecular sequence data imply that Hemiptera is sister to Psocodea + Thysanoptera, a grouping of Thysanoptera + Hemiptera (called superorder Condylognatha) is supported by derived head and mouthparts including the stylet mouthparts, features of the wing base, and the presence of a sclerotized ring between antennal flagellomeres. Condylognatha thus forms the sister group to Psocodea.

Order Psocoptera (psocids, barklice, booklice) (see also Box 11.9)
Psocoptera is a worldwide order of cryptic small insects, with a large, mobile head, bulbous postclypeus, and membranous wings held roof-like over the abdomen. Evidently, Psocoptera belong with Phthiraptera in a monophyletic clade Psocodea. However, Psocoptera is rendered paraphyletic by a postulated relationship of Phthiraptera to the psocopteran family Liposcelidae. Internal relationships of the more than 30 families of psocids are poorly known and of the three suborders, Troctomorpha, Trogiomorpha, and Psocomorpha, there is support only for the monophyly of Psocomorpha.

Order Phthiraptera (parasitic lice) (see also Box 15.3)
Phthirapterans are wingless obligate ectoparasites of birds and mammals. Monophyly of, and relationships among, traditional suborders Anoplura, Amblycera, Ischnocera, and Rhyncophthirina are poorly understood and nearly all possible arrangements have been proposed. The latter three suborders have been treated as a monophyletic Mallophaga (biting and chewing lice) based on their feeding mode and morphology, in contrast to the piercing and blood-feeding Anoplura. Cladistic analysis of morphology has disputed mallophagan monophyly, suggesting the relationship Amblycera (Ischnocera (Anoplura + Rhyncophthirina)). Ignorance of robust estimates of relationship restricts estimation of evolutionary interactions, such as co-speciation, between lice and their bird and mammal hosts.

Order Thysanoptera (thrips) (see also Box 11.7)
The development of Thysanoptera is intermediate between hemi- and holometabolous. Their head is elongate and the mouthparts are unique in that the maxillary laciniae form grooved stylets, the right

mandible is atrophied, but the left mandible forms a stylet; all three stylets together form the feeding apparatus. The tarsi are one- or two-segmented, and the pretarsus has an apical protrusible adhesive ariolum (bladder or vesicle). Reproduction in thrips is haplodiploid.

Limited molecular evidence supports a traditional morphological division of the Thysanoptera into two suborders, Tubulifera containing a sole, speciose, family Phlaeothripidae, and Terebrantia. Terebrantia includes one speciose family, Thripidae, and seven smaller families. Relationships among families in Terebrantia are poorly resolved, although phylogenies are being generated at lower levels concerning aspects of the evolution of sociality, especially the origins of gall-inducing thrips, and of "soldier" castes in Australian gall-inducing Thripidae.

Order Hemiptera (bugs, cicadas, leafhoppers, planthoppers, spittle bugs, aphids, jumping plant lice, scale insects, whiteflies, moss bugs) (see also Boxes 10.6 & 11.8)

Hemiptera, the largest non-endopterygote order, has diagnostic mouthparts, with mandibles and maxillae modified as needle-like stylets, lying in a beak-like, grooved labium, collectively forming a rostrum or proboscis. Within this, the stylet bundle contains two canals, one delivering saliva and the other uptaking fluid. Hemiptera lack maxillary and labial palps. The prothorax and mesothorax usually are large and the metathorax small. Venation of both pairs of wings can be reduced; some species are apterous, and male scale insects have only one pair of wings. Legs often have complex pretarsal adhesive structures. Cerci are lacking.

Hemiptera and Thysanoptera are sister groups within Paraneoptera. Hemiptera once was divided into two groups, Heteroptera (true bugs) and "Homoptera" (cicadas, leafhoppers, planthoppers, spittle bugs, aphids, psylloids, scale insects, and whiteflies), treated as either suborders or as orders. All "homopterans" are terrestrial plant feeders and many share a common biology of producing honeydew and being ant-attended. Although sharing defining features, such as wings held roof-like over the abdomen, fore wings either membranous or in the form of tegmina of uniform texture, and with the rostrum arising ventrally close to the anterior of the thorax, "Homoptera" represents a paraphyletic grade rather than a clade (Fig. 7.5). This view finds support in re-interpreted morphological data and from analyses of nucleotide sequences, which also suggest more complicated relationships among the higher groups of hemipterans (Fig. 7.5).

The rank of hemipteran clades has been much disputed. We follow a system of five suborders recognized on phylogenetic grounds. Fulgoromorpha, Cicadomorpha, Coleorrhyncha, and Heteroptera (collectively termed the Euhemiptera) form the sister group to

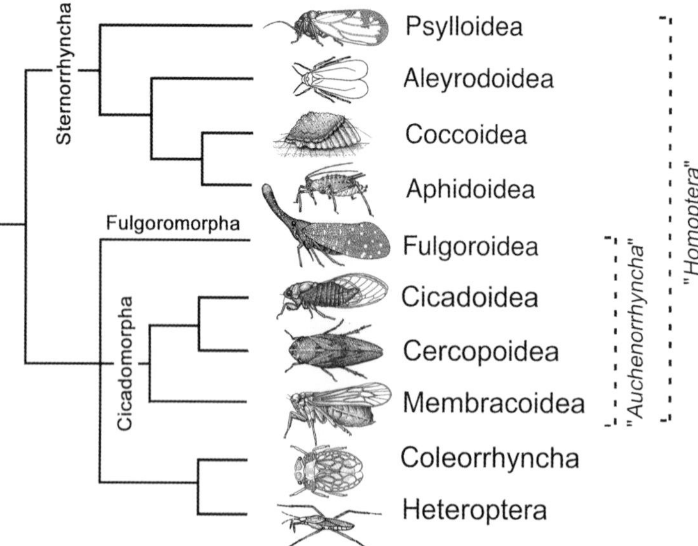

Fig. 7.5 Cladogram of postulated relationships within Hemiptera, based on combined morphological and nucleotide sequence data. Broken lines indicate paraphyletic taxa, with names italicized. (After Bourgoin & Campbell 2002.)

suborder Sternorrhyncha. The latter contains the aphids (Aphidoidea), jumping plant lice (Psylloidea), scale insects (Coccoidea), and whiteflies (Aleyrodoidea), which are characterized principally by their possession of a particular kind of gut filter chamber, a rostrum that appears to arise between the bases of their front legs and, if winged, by absence of the vannus and vannal fold in the hind wings. Some relationships among Euhemiptera are unsettled. A traditional grouping called the Auchenorrhyncha, morphologically defined by their possession of a tymbal acoustic system, an aristate antennal flagellum, and reduction of the proximal median plate in the wing base, contains the Fulgoromorpha (planthoppers) and Cicadomorpha (cicadas, leafhoppers, and spittle bugs). Paleontological data combined with nucleotide sequences suggest that Cicadomorpha is sister to Coleorrhyncha + Heteroptera (sometimes called Prosorrhyncha), which would render Auchenorrhyncha paraphyletic. However, relationships among Cicadomorpha, Fulgoromorpha, and Coleorrhyncha + Heteroptera are still disputed and thus are portrayed here as an unresolved trichotomy (Fig. 7.5).

Heteroptera (true bugs, including assassin bugs, backswimmers, lace bugs, stink bugs, waterstriders, and others) has as its sister group the Coleorrhyncha, containing only one family, Peloridiidae or moss bugs. Although small, cryptic and rarely collected, moss bugs have generated considerable phylogenetic interest due to their combination of ancestral and derived hemipteran features, and their exclusively "relictual" Gondwanan distribution. Heteropteran diversity is distributed amongst about 80 families, forming the largest hemipteran clade. Heteroptera is diagnosed most easily by the presence of metapleural scent glands, and monophyly is undisputed.

Subdivision Endopterygota (= Holometabola)

Endopterygota comprise insects with holometabolous development in which immature (larval) instars are very different from their respective adults. The adult wings and genitalia are internalized in their pre-adult expression, developing in imaginal discs that are evaginated at the penultimate molt. Larvae lack true ocelli. The "resting stage" or pupa is non-feeding, and precedes an often active pharate ("cloaked" in pupal cuticle) adult. Unique derived features are less evident in adults than in earlier developmental stages, but the clade is recovered consistently from all phylogenetic analyses.

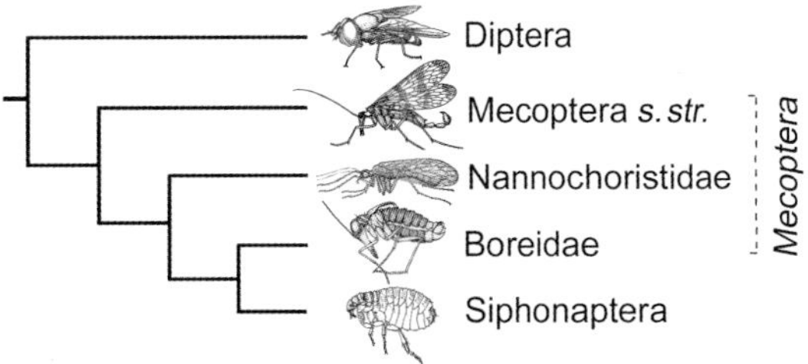

Fig. 7.6 Cladogram of postulated relationships of Antliophora, based on a combination of morphological and nucleotide sequence data. The broken lines indicate a paraphyletic taxon, with its name italicized; *s. str.* refers to the restricted sense. (After Whiting 2002.)

Two or three groups currently are proposed amongst the endopterygotes, of which one of the strongest is a sister-group relationship termed Amphiesmenoptera between the Trichoptera (caddisflies) and Lepidoptera (butterflies and moths). A plausible scenario of an ancestral amphiesmenopteran taxon envisages a larva living in damp soil amongst liverworts and mosses followed by radiation into water (Trichoptera) or into terrestrial plant-feeding (Lepidoptera).

A second strongly supported relationship is between three orders: Neuroptera, Megaloptera, and Raphidioptera, called Neuropterida and sometimes treated as a group of ordinal rank, which shows a sister-group relationship to Coleoptera.

A third, postulated relationship – Antliophora – unites Diptera (true flies), Siphonaptera (fleas), and Mecoptera (scorpionflies and hangingflies). Their relationships, particularly concerning Siphonaptera, have been debated. Fleas were considered as sister group to Diptera, but anatomical and nucleotide sequence evidence increasingly points to a relationship with the curious-looking mecopterans, the snow fleas of the family Boreidae (Fig. 7.6).

Strepsiptera is phylogenetically enigmatic, but resemblance of their first-instar larvae (called triungulins) to those of certain Coleoptera, notably parasitic Rhipiphoridae, and some wing-base features have been cited as indicative of a close relationship. This suggested placement is becoming less likely, as molecular evidence (and haltere development) suggests alternatives, either with Diptera or distant from either Diptera or Coleoptera. Strepsiptera has undergone much morphological and molecular evolution, and is highly derived with few features shared with any other taxon. Such long-isolated evolution of the genome can create a problem known as "long-branch attraction", in which nucleotide sequences may converge by chance mutations alone with those of an unrelated taxon with a similarly long independent evolution, for the strepsipteran notably with Diptera. The issue of relationship remains unresolved, although morphological study of wing-base morphology suggests that proximity to neither Diptera nor Coleoptera is likely.

The relationships of two major orders of endopterygotes, Coleoptera and Hymenoptera, remain to be considered. Several positions have been proposed for Coleoptera but current evidence derived from female genitalia and ambivalent evidence from eye structure supports a sister-group relationship to Neuropterida. This group is sister to the remaining Endopterygota in many analyses. Hymenoptera may be the sister to Antliophora + Amphiesmenoptera, but the many highly derived features of adults and reductions in larvae limit morphological justification for this position.

Within the limits of uncertainty, the relationships within Endopterygota are summarized in Fig. 7.2, in which uncertain or ambiguous associations are shown by interrupted lines and suspect paraphyletic taxon names are italicized.

Order Coleoptera (beetles) (see also Boxes 10.6 & 11.10)
Coleoptera undoubtedly lie amongst early branches of the Endopterygota. The major shared derived feature of Coleoptera is the development of the fore wings as sclerotized rigid elytra, which extend to cover some or many of the abdominal segments, and beneath which the propulsive hind wings are elaborately folded when at rest. Some molecular studies show Coleoptera polyphyletic or paraphyletic with respect to some or all of Neuropterida. However, this is impossible to reconcile with the morphological support for coleopteran monophyly, and we accept that a sister-group relationship to Neuropterida is most probable.

Within Coleoptera, four modern lineages (treated as suborders) are recognized: Archostemata, Adephaga, Polyphaga, and Myxophaga. Archostemata includes only the small families Ommatidae, Crowsoniellidae, Cupedidae, and Micromalthidae, and probably forms the sister group to the remaining extant Coleoptera. The few known larvae are wood-miners with a sclerotized ligula and a large mola on each mandible. Adults have movable hind coxae with usually visible trochantins, and five (not six) ventral abdominal plates (ventrites), but share with Myxophaga and Adephaga wing folding features, lack of any cervical sclerites, and an external prothoracic pleuron. In contrast to Myxophaga, the pretarsus and tarsus are unfused.

Adephaga is diverse, second in size only to Polyphaga, and includes ground beetles, tiger beetles, whirligigs, predaceous diving beetles, and wrinkled bark beetles, amongst others. Larval mouthparts are adapted for liquid-feeding, with a fused labrum and no mandibular mola. Adults have the notopleural sutures visible on the prothorax and have six visible abdominal sterna with the first three fused into a single ventrite which is divided by the hind coxae. Pygidial defense glands are widespread in adults. The most speciose included family is Carabidae, or ground beetles, with a predominantly predaceous feeding habit, but Adephaga also includes the aquatic families, Dytiscidae, Gyrinidae, Haliplidae and Noteridae, and the mycophagous Rhysodidae, or wrinkled bark beetles. Morphology suggests that Adephaga is sister group to the combined Myxophaga and Polyphaga, although some nucleotide sequences suggest Adephaga as sister to Polyphaga, with Myxophaga sister to the two combined.

Myxophaga is a clade of small, primarily riparian aquatic beetles, comprising families Lepiceridae, Torridincolidae, Hydroscaphidae, and Microsporidae, united by the synapomorphic fusion of the pretarsus and tarsus. The three-segmented larval antenna, five-segmented larval legs with a single pretarsal claw, fusion of trochantin with the pleuron, and ventrite structure support a sister-group relationship of Myxophaga with the Polyphaga. This has been challenged by some workers, notably because some interpretations of wing venation and folding support Polyphaga (Archostemata (Myxophaga + Adephaga)).

Polyphaga contains the majority (>90% of species) of beetle diversity, with about 300,000 described species. The suborder includes rove beetles (Staphylinoidea), scarabs and stag beetles (Scarabaeoidea), metallic wood-boring beetles (Buprestoidea), click beetles and fireflies (Elateroidea), as well as the diverse Cucujiformia, including fungus beetles, grain beetles, ladybird beetles, darkling beetles, blister beetles, longhorn beetles, leaf beetles, and weevils. The prothoracic pleuron is not visible externally, but is fused with the trochantin and remnant internally as a "cryptopleuron". Thus, one suture between the notum and the sternum is visible in the prothorax in polyphagans, whereas two sutures (the sternopleural and notopleural) often are visible externally in other suborders (unless secondary fusion between the sclerites obscures the sutures, as in *Micromalthus*). The transverse fold of the hind wing never crosses the media posterior (MP) vein, cervical sclerites are present, and hind coxae are mobile and do not divide the first ventrite. Female polyphagan beetles have telotrophic ovarioles, which is a derived condition within beetles.

The internal classification of Polyphaga involves several superfamilies or series, whose constituents are relatively stable, although some smaller families (whose rank even is disputed) are allocated to different clades by different authors. Large superfamilies include Hydrophiloidea, Staphylinoidea, Scarabaeoidea, Buprestoidea, Byrrhoidea, Elateroidea, Bostrichoidea, and the grouping Cucujiformia. This latter includes the vast majority of phytophagous (plant-eating) beetles, united by cryptonephric Malpighian tubules of the normal type, the eye with a cone ommatidium with open rhabdom, and lack of functional spiracles on the eighth abdominal segment. Constituent superfamilies of Cucujiformia are Cleroidea, Cucujoidea, Tenebrionoidea, Chrysomeloidea, and Curculionoidea. Evidently, adoption of a phytophagous lifestyle correlates with speciosity in beetles, with Cucujiformia, especially weevils (Curculionoidea), forming a major radiation (see section 8.6).

Neuropterida, or neuropteroid orders

Orders Megaloptera (alderflies, dobsonflies, fishflies), Raphidioptera (snakeflies), and Neuroptera (lacewings, antlions, owlflies) (see also Boxes 10.6 & 13.4)
Neuropterida comprise three minor (species-poor) orders, whose adults have multisegmented antennae, large, separated eyes, and mandibulate mouthparts. The prothorax may be larger than either the meso- or metathorax, which are about equal in size. Legs sometimes are modified for predation. The fore and hind wings are quite similar in shape and venation, with folded wings often extending beyond the abdomen. The abdomen lacks cerci.

Megalopterans are predatory only in the aquatic larval stage; although adults have strong mandibles, they are not used in feeding. Adults closely resemble neuropterans, except for the presence of an anal fold in the hind wing. Raphidiopterans are terrestrial predators as adults and larvae. The adult is mantid-like, with an elongate prothorax, and the head is mobile and used to strike, snake-like, at prey. The larval head is large and forwardly directed. Many adult neuropterans are predators, and have wings typically characterized by numerous cross-veins and "twigging" at the ends of veins. Neuropteran larvae usually are active predators with slender, elongate mandibles and maxillae combined to form piercing and sucking mouthparts.

Megaloptera, Raphidioptera, and Neuroptera may be treated as separate orders, united in Neuropterida, or Raphidioptera may be included in Megaloptera. Neuropterida undoubtedly is monophyletic with new support from morphology of the wing-base sclerites. This latter feature also supports the long-held view that Neuropterida forms a sister group to Coleoptera. Each component appears monophyletic, although a doubt remains concerning megalopteran monophyly. There remains uncertainty about internal relationships, which traditionally have Megaloptera and Raphidioptera as sister groups. Recent reanalyses with some new character suites propose Megaloptera as sister to Neuroptera with a novel scenario of ancestral aquatic larvae (as seen in Sisyridae within Neuroptera, and in all Megaloptera) in Neuropterida.

Order Strepsiptera (see also Box 13.6)
Strepsiptera form an enigmatic order showing extreme sexual dimorphism. The male's head has bulging eyes comprising few large facets and lacks ocelli; the antennae are flabellate or branched, with four to seven segments. The fore wings are stubby and lack veins, whereas the hind wings are broadly fan-shaped, with few radiating veins; the legs lack trochanters and often also claws. Females are either coccoid-like or larviform, wingless, and usually retained in a pharate (cloaked) state, protruding from the host. The first-instar larva is a triungulin, without antennae and mandibles, but with three pairs of thoracic legs; subsequent instars are maggot-like, lacking mouthparts or appendages. The pupa, which has immovable mandibles but appendages free from its body, develops within a puparium formed from the last larval instar.

The phylogenetic position of Strepsiptera has been subject to much speculation because modifications associated with their endoparasitic lifestyle mean that few characteristics are shared with possible relatives. In having posteromotor flight (using only metathoracic wings) they resemble Coleoptera, but other putative synapomorphies with Coleoptera appear suspect or mistaken. The fore-wing-derived halteres of strepsipterans are gyroscopic organs of equilibrium with the same functional role as the halteres of Diptera (although the latter are derived from the hind wing). Nucleotide sequence studies indicate that Strepsiptera might be a sister group to Diptera, which is one relationship indicated on Fig. 7.2 by the broken line.

Order Mecoptera (scorpionflies, hangingflies) (see also Box 13.5)
Mecopteran adults have an elongate, ventrally projecting rostrum, containing elongate, slender mandibles and maxillae, and an elongate labium. The eyes are large and separated, the antennae filiform and multisegmented. The fore and hind wings are narrow, similar in size, shape, and venation, but often are reduced or absent. The legs may be modified for predation. Larvae have a heavily sclerotized head capsule, are mandibulate, and may have compound eyes comprising three to 30 ocelli (absent in Panorpidae, indistinct in Nannochoristidae). The thoracic segments are about equal, and have short thoracic legs with fused tibia and tarsus and a single claw. Prolegs usually are present on abdominal segments 1–8, and the terminal segment (10) has either paired hooks or a suction disk. The pupa is immobile, mandibulate, and with appendages free.

Although some adult Mecoptera resemble neuropterans, strong evidence supports a relationship to Diptera. Intriguing recent morphological studies, plus robust evidence from molecular sequences, suggest that Siphonaptera arose from within Mecoptera, as a sister group to the "snow fleas" (Boreidae) (Fig. 7.6). The phylogenetic position of Nannochoristidae, a southern hemisphere mecopteran taxon currently treated as being of subfamily rank, has a significant bearing on internal relationships within Antliophora. Nucleotide sequence data suggest that it is sister to Boreidae + Siphonaptera, and therefore is of equivalent rank to the boreids, fleas, and the residue of Mecoptera (*sensu stricto*) – and logically each should be treated as orders, or Siphonaptera reduced in rank within Mecoptera.

Order Siphonaptera (fleas) (see also Box 15.4)
Siphonaptera are bilaterally compressed, apterous ectoparasites, with mouthparts specialized for piercing and sucking, lacking mandibles but with an unpaired labral stylet and two elongate serrate, lacinial stylets that together lie within a maxillary sheath. A salivary pump injects saliva into the wound, and cibarial and pharyngeal pumps suck up the blood meal. Fleas lack compound eyes and the antennae lie in deep lateral grooves. The body is armed with many posteriorly directed setae and spines, some of which form combs, especially on the head and anterior thorax. The metathorax houses very large muscles associated with the long and strong hind legs, which power the prodigious leaps made by these insects.

After early suggestions that the fleas arose from a mecopteran, the weight of evidence suggested they formed the sister group to Diptera. However, increasing molecular and novel morphological evidence now points to a sister-group relationship to only part of Mecoptera, specifically the Boreidae (snow fleas) (Fig. 7.6). Internal relationships of the fleas are under study and preliminary results imply that monophyly of many families is uncertain.

Order Diptera (true flies) (see also Boxes 5.4, 10.5, & 15.5)
Diptera are readily recognized by the development of hind (metathoracic) wings as balancers, or halteres (halters), and in the larval stages by a lack of true legs and the often maggot-like appearance. Venation of the fore (mesothoracic), flying wings ranges from complex to extremely simple. Mouthparts range from biting-and-sucking (e.g. biting midges and mosquitoes) to "lapping"-type with a pair of pseudotracheate labella functioning as a sponge (e.g. house flies). Dipteran larvae lack true legs, although various kinds of locomotory apparatus range from unsegmented pseudolegs to creeping welts on maggots. The larval head capsule may be complete, partially undeveloped, or completely absent in a maggot head that consists only of the internal sclerotized mandibles ("mouth hooks") and supporting structures.

Traditionally, Diptera had two suborders, Nematocera (crane flies, midges, mosquitoes, and gnats) with a slender, multisegmented antennal flagellum, and heavier-built Brachycera ("higher flies" including hover flies, blow flies, and dung flies) with a shorter, stouter, and fewer-segmented antenna. However, Brachycera is sister to only part of "Nematocera", and thus Nematocera is paraphyletic.

Internal relationships amongst Diptera are becoming better understood, although with some notable exceptions. Ideas concerning early branches in dipteran phylogeny are inconsistent. Traditionally, Tipulidae (or Tipulomorpha) is a first-branching clade on evidence from the wing and other morphology. Such an arrangement is difficult to reconcile with the much more derived larva, in which the head capsule is variably reduced. Furthermore, some molecular evidence casts doubt on this position for the crane flies, but as yet does not produce a robust estimate for any alternative early-branching pattern. Alternative views based on morphology have suggested that the relictual family Tanyderidae, with complex ("primitive") wing venation, arose early in the diversification of the order. Support comes also from the tanyderid larval morphology, and putative placement in Psychodomorpha, considered a probable early-branching clade.

There is strong support for a grouping called Culicomorpha, comprising mosquitoes (Culicidae) and their relatives (Corethrellidae, Chaoboridae, Dixidae) and their sister group the black flies, midges, and relatives (Simuliidae, Thaumaleidae, Ceratopogonidae, Chironomidae), and for Bibionomorpha, comprising the fungus gnats (Mycetophilidae, Bibionidae, Anisopodidae, and possibly Cecidomyiidae (gall midges)).

Monophyly of Brachycera, comprising "higher flies", is established by features including the larva having a posterior elongate head contained within the prothorax, a divided mandible and loss of premandible, and in the adult by eight or fewer antennal flagellomeres, two or fewer palp segments, and separation of the male genitalia into two parts (epandrium and hypandrium). All relationships of Brachycera are to a subgroup within "Nematocera", perhaps as sister to Psychodomorpha or even to Culicomorpha (molecular data only), but strong support is provided for a sister relationship to the Bibionomorpha, or to a group within the Anisopodidae. Brachycera contains four equivalent groups with internally unresolved relationships: Tabanomorpha (with a brush on the larval mandible and the larval head retractile); Stratiomyomorpha (with larval cuticle calcified and pupation in last-larval instar exuviae); Xylophagomorpha (with a distinctive elongate, conical, strongly sclerotized larval head capsule, and abdomen posteriorly ending in a sclerotized plate with terminal hooks); and Muscomorpha (adults with tibial spurs absent, flagellum with no more than four flagellomeres, and female cercus single-segmented). This latter speciose group contains Asiloidea (robber flies, bee flies, and relatives) and Eremoneura (Empidoidea and Cyclorrhapha). Eremoneura is a strongly supported clade based on wing venation (loss or fusion of vein M_4 and closure of anal cell before margin), presence of ocellar setae, unitary palp and genitalic features, plus larval stage with only three instars and maxillary reduction. Cyclorrhaphans, united by metamorphosis in a puparium formed by the last instar larval skin, include a heterogeneous group including Syrphidae (hover flies) and the Schizophora defined by the presence of a balloon-like ptilinum that everts from the frons to assist the adult escape the puparium. Within Schizophora, the "higher" cyclorrhaphans include the ecologically very diverse acalypterates, and the blow flies and relatives (Calypteratae).

Order Hymenoptera (ants, bees, wasps, sawflies, and wood wasps) (see also Box 12.2)

The mouthparts of adults are directed ventrally to forward projecting, ranging from generalized mandibulate to sucking and chewing, with mandibles often used for killing and handling prey, defense, and nest building. The compound eyes often are large; the antennae are long, multisegmented, and often prominently held forwardly or recurved dorsally. "Symphyta" (wood wasps and sawflies) has a conventional three-segmented thorax, but in Apocrita (ants, bees, and wasps) the propodeum (abdominal segment 1) is included with the thorax to form a mesosoma. The wing venation is relatively complete in large sawflies, and reduced in Apocrita in correlation with body size, such that very small species of 1–2 mm have only one divided vein, or none. In Apocrita, the second abdominal segment (and sometimes also the third) forms a constriction, or petiole (Box 12.2). Female genitalia include an ovipositor, comprising three valves and two major basal sclerites, which in aculeate Hymenoptera is modified as a sting associated with a venom apparatus.

Symphytan larvae are eruciform (caterpillar-like), with three pairs of thoracic legs bearing apical claws and with some abdominal legs. Apocritan larvae are apodous, with the head capsule frequently reduced but with prominent strong mandibles.

Hymenoptera forms the sister group to Amphiesmenoptera (= Trichoptera + Lepidoptera) + Antliophora (= Diptera + Mecoptera/Siphonaptera) (Fig. 7.2), although an earlier-branching position in the Holometabola has been advocated. Hymenoptera often are treated as containing two suborders, Symphyta (wood wasps and sawflies) and Apocrita (wasps, bees, and ants). However, Apocrita appears to be sister to one family of symphytan only, the Orussidae, and thus "symphytans" form a paraphyletic group.

Within Apocrita, aculeate (Aculeata) and parasitic (Parasitica or terebrant) wasp groups were considered each to be monophyletic, but aculeates evidently originated from within a paraphyletic Parasitica. Internal relationships of aculeates, including vespids (paper wasps, yellow jackets, etc.), formicids (ants), and apids (bees), and the monophyly of subordinate groups are under scrutiny. Apidae evidently arose as sister to, or from within, Sphecidae (digger wasps), but the precise relationships of another significant group of aculeates, Formicidae (ants), within Vespoidea are less certain (Fig. 12.2).

Order Trichoptera (caddisflies) (see also Box 10.4)
The moth-like adult trichopteran has reduced mouthparts lacking any proboscis, but with three- to five-segmented maxillary palps and three-segmented labial palps. The antennae are multisegmented and filiform and often as long as the wings. The compound eyes are large, and there are two to three ocelli. The wings are haired or less often scaled, and differentiated from all but the most basal Lepidoptera by the looped anal veins in the fore wing, and absence of a discal cell. The larva is aquatic, has fully developed mouthparts, three pairs of thoracic legs (each with at least five segments), and lacks the ventral prolegs characteristic of lepidopteran larvae. The abdomen terminates in hook-bearing prolegs. The tracheal system is closed, and associated with tracheal gills on most abdominal segments. The pupa also is aquatic, enclosed in a retreat often made of silk, with functional mandibles that aid in emergence from the sealed case.

Amphiesmenoptera (Trichoptera + Lepidoptera) is now unchallenged, despite earlier suggestions that Trichoptera may have originated within Lepidoptera. Proposed internal relationships within the Trichoptera range from stable and well supported, to unstable and anecdotal. Monophyly of suborder Annulipalpia (comprising families Hydropsychidae, Polycentropodidae, Philopotamidae, and some close relatives) is well supported by larval and adult morphology – including presence of an annulate apical segment of both adult maxillary and larval palp, absence of male phallic parameres, presence of papillae lateral to the female cerci, and in the larva by the presence of elongate anal hooks and reduced abdominal tergite 10.

The monophyly of the case-making suborder Integripalpia (comprising families Phryganeidae, Limnephilidae, Leptoceridae, Sericostomatidae, and relatives) is supported by the absence of the *m* cross-vein, hind wings broader than fore wings especially in the anal area, female lacking both segment 11 and cerci, and larval character states including usually complete sclerotization of the mesonotum, hind legs with lateral projection, lateral and mid-dorsal humps on abdominal segment 1, and short and stout anal hooks.

Monophyly of a third putative suborder, Spicipalpia, is more contentious. Defined for a grouping of families Glossosomatidae, Hydroptilidae, and Rhyacophilidae (and perhaps the Hydrobiosidae), uniting features are the spiculate apex of the adult maxillary and labial palps, the ovoid second segment of the maxillary palp, and an eversible oviscapt (egg-laying appendage). Morphological and molecular evidence fail to confirm Spicipalpia monophyly, unless at least Hydroptilidae is removed.

All possible relationships between Annulipalpia, Integripalpia, and Spicipalpia have been proposed, sometimes associated with scenarios concerning the evolution of case-making. An early idea that Annulipalpia are sister to a paraphyletic Spicipalpia + monophyletic Integripalpia finds support from some morphological and molecular data.

Order Lepidoptera (moths and butterflies)
(see also Box 11.11)
Adult heads bear a long, coiled proboscis formed from greatly elongated maxillary galeae; large labial palps usually are present, but other mouthparts are absent, except that mandibles are present primitively in some groups. The compound eyes are large, and ocelli usually are present. The multisegmented antennae often are pectinate in moths and knobbed or clubbed in butterflies. The wings are covered completely with a double layer of scales (flattened modified macrotrichia), and the hind and fore wings are linked by either a frenulum, a jugum, or simple overlap. Lepidopteran larvae have a sclerotized head capsule with mandibulate mouthparts, usually six lateral ocelli, and short three-segmented antennae. The thoracic legs are five-segmented with single claws, and the abdomen has 10 segments with short prolegs on some segments. Silk gland products are extruded from a characteristic spinneret at the median apex of the labial prementum. The pupa usually is contained within a silken cocoon.

The early-branching events in the radiation of this large order is considered well-enough resolved to serve as a test for the ability of particular nucleotide sequences to recover the expected phylogeny. Although more than 98% of the species of Lepidoptera belong in Ditrysia, the morphological diversity is concentrated in a small non-ditrysian grade. Three of the four suborders are species-poor early branches, each with just a single family (Micropterigidae, Agathiphagidae, Heterobathmiidae); these lack the synapomorphy of the mega-diverse fourth suborder Glossata, namely the characteristically developed coiled proboscis formed from the fused galea (Fig. 2.12). The

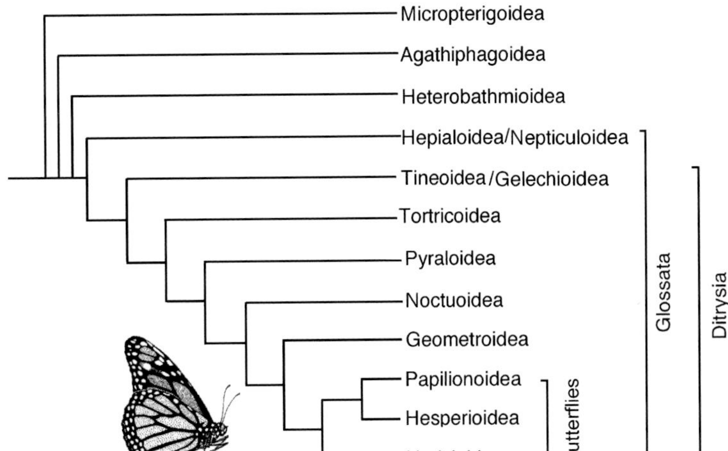

Fig. 7.7 Cladogram of postulated relationships of selected lepidopteran higher taxa, based on morphological data. (After Kristensen & Skalski 1999.)

highly speciose Glossata contains a comb-like branching pattern of many species-poor taxa, plus a species-rich grouping united by the larva (caterpillar) having abdominal prolegs with muscles and apical crochets (hooklets). This latter group contains the diverse Ditrysia, defined by the unique two genital openings in the female, one the ostium bursae on sternite 8, the other the genitalia proper on sternites 9 and 10. Additionally, the wing coupling is always frenulate or amplexiform and not jugate, and the wing venation tends to be heteroneuran (with venation dissimilar between fore and hind wings). Trends in the evolution of Ditrysia include elaboration of the proboscis and the reduction to loss of maxillary palpi. One of the best-supported relationships in Ditrysia is the grouping of Hesperioidea (skippers) and Papilionoidea (butterflies), united by their clubbed, dilate antennae, lack of frenulum in the wing and large humeral lobe on the hind wing. To this the neotropical Hedyloidea has been added to form the clade known as the butterflies (Fig. 7.7).

FURTHER READING

Beutel, R.G. & Haas, F. (2000) Phylogenetic relationships of the suborders of Coleoptera (Insecta). *Cladistics* **16**, 103–41.

Bitsch, C. & Bitsch, J. (2000) The phylogenetic interrelationships of the higher taxa of apterygote hexapods. *Zoologica Scripta* **29**, 131–56.

Caterino, M.S., Cho, S. & Sperling, F.A.H. (2000) The current state of insect molecular systematics: a thriving Tower of Babel. *Annual Review of Entomology* **45**, 1–54.

Cranston, P.S. & Gullan, P.J. (2003) Phylogeny of insects. In: *Encyclopedia of Insects* (eds. V.H. Resh & R.T. Cardé), pp. 882–98. Academic Press, Amsterdam.

Cranston, P.S., Gullan, P.J. & Taylor, R.W. (1991) Principles and practice of systematics. In: *The Insects of Australia*, 2nd edn. (CSIRO), pp. 109–24. Melbourne University Press, Carlton.

Felsenstein, J. (2004) *Inferring Phylogenies*. Sinauer Associates, Sunderland, MA.

Hall, B.G. (2004) *Phylogenetic Trees Made Easy; A How-To Manual*, 2nd edn. Sinauer Associates, Sunderland, MA.

Klass, K.-D., Zompro, O., Kristensen, N.P. & Adis, J. (2002) Mantophasmatodea: a new insect order with extant members in the Afrotropics. *Science* **296**, 1456–9.

Kristensen, N.P. (1991) Phylogeny of extant hexapods. In: *The Insects of Australia*, 2nd edn. (CSIRO), pp. 125–40. Melbourne University Press, Carlton.

Kristensen, N.P. (1997) Early evolution of the Lepidoptera + Trichoptera lineage: phylogeny and the ecological scenario. In: *The Origin of Biodiversity in Insects: Phylogenetic Tests of Evolutionary Scenarios* (ed. P. Grandcolas). *Mémoires du Muséum National d'Histoire Naturelle* **173**, 253–71.

Kristensen, N.P. (1999) Phylogeny of endopterygote insects, the most successful lineage of living organisms. *European Journal of Entomology* **96**, 237–53.

Kristensen, N.P. & Skalski, A.W. (1999) Phylogeny and paleontology. In: *Lepidoptera: Moths and Butterflies 1. Handbuch der Zoologie/Handbook of Zoology*, Vol. IV, Part 35 (ed. N.P. Kristensen), pp. 7–25. Walter de Gruyter, Berlin.

Lo, N., Tokuda, J., Watanabe, H. et al. (2000) Evidence from multiple gene sequences indicates that termites evolved from wood-feeding termites. *Current Biology* **10**, 801–4.

Ronquist, F. (1999) Phylogeny of the Hymenoptera (Insecta): the state of the art. *Zoologica Scripta* **28**, 3–11.
Schuh, R.T. (2000) *Biological Systematics: Principles and Applications*. Cornell University Press, Ithaca.
Skelton, P. & Smith, A. (2002) *Cladistics: A Practical Primer on CD-ROM*. Cambridge University Press, Cambridge.
Whiting, M.F. (1998) Phylogenetic position of the Strepsiptera: review of molecular and morphological evidence. *International Journal of Morphology and Embryology* **27**, 53–60.
Whiting, M.F. (2002) Phylogeny of the holometabolous insect orders: molecular evidence. *Zoologica Scripta* **31**, 3–15.
Whiting, M.F. (2002) Mecoptera is paraphyletic: multiple genes and phylogeny of Mecoptera and Siphonaptera. *Zoologica Scripta* **312**, 93–104.
Yeates, D.K. & Wiegmann, B.M. (1999) Congruence and controversy: toward a higher-level phylogeny of Diptera. *Annual Review of Entomology* **44**, 397–428.

Chapter 8

INSECT BIOGEOGRAPHY AND EVOLUTION

Reconstructions of giant Carboniferous insects. (Inspired by a drawing by Mary Parrish in Labandeira 1998.)

The insects have had a long history since the divergence of the Hexapoda from the Crustacea many millions of years ago. In this time the Earth has undergone much evolution itself, from droughts to floods, from ice ages to arid heat. Extra-terrestrial objects have collided with the Earth, and major extinction events have occurred periodically. Through this long time insects have changed their ranges, and evolved to display the enormous modern diversity outlined in our opening chapter.

In this chapter we review patterns and causes for the distribution of insects on the planet – their biogeography – then introduce fossil and contemporary evidence for their age. We ask what evidence there is for aquatic or terrestrial origins of the group, then address in detail some aspects of insect evolution that have been proposed to explain their success – the origin of wings (and hence flight) and of metamorphosis. We summarize explanations for their diversification and conclude with a review of insects on Pacific islands, highlighting the role of patterns seen there as a more general explanation of insect radiations.

8.1 INSECT BIOGEOGRAPHY

Viewers of television nature documentaries, biologically alert visitors to zoos or botanic gardens, and global travelers will be aware that different plants and animals live in different parts of the world. This is more than a matter of differing climate and ecology. Thus, Australia has suitable trees but no woodpeckers, tropical rainforests but no monkeys, and prairie grasslands without native ungulates. American deserts have cacti, but arid regions elsewhere have a range of ecological analogs including succulent euphorbs, but no native cacti. The study of the distributions and the past historical and current ecological explanations for these distributions is the discipline of biogeography. Insects, no less than plants and vertebrates, show patterns of restriction to one geographic area (endemism) and entomologists have been, and remain, amongst the most prominent biogeographers. Our ideas on the biological relationships between the size of an area, the number of species that the area can support, and changes in species (turnover) in ecological time – called island biogeography – have come from the study of island insects (see section 8.7). Researchers note that islands can be not only oceanic but also habitats isolated in metaphorical "oceans" of unsuitable habitat – such as mountain tops in lowlands, or isolated forest remnants in agro-landscapes.

Entomologists have been prominent amongst those who have studied dispersal between areas, across land bridges, and along corridors, with ground beetle specialists being especially prominent. Since the 1950s the paradigm of a static-continent Earth has shifted to one of dynamic movement powered by plate tectonics. Much of the evidence for faunas drifting along with their continents came from entomologists studying the distribution and evolutionary relationships of taxa shared exclusively between the modern disparate remnants of the once-united southern continental land mass (Gondwana). Amongst this cohort, those studying aquatic insects were especially prominent, perhaps because the adult stages are ephemeral and the immature stages so tied to freshwater habitats, that long-distance trans-oceanic dispersal seemed an unlikely explanation for the many observed disjunct distributions. Stoneflies, mayflies, dragonflies, and aquatic flies including midges (Diptera: Chironomidae) show southern hemisphere disjunct associations, even at low taxonomic levels (species groups, genera). Current distributions imply that their direct ancestors must have been around and subjected to Earth history events in the Upper Jurassic and Cretaceous. Such findings imply that a great many groups must have been around for at least 130 million years. Such time-scales appear to be confirmed by increasing amounts of fossil material, and by some estimations of the purported clock-like acquisition of mutations in molecules.

On the finer scale, insect studies have played a major role in understanding the role of geography in processes of species formation and maintenance of local differentiation. Naturally, the genus *Drosophila* figures prominently with its Hawai'ian radiation having provided valuable data. Studies of parapatric speciation – divergence of spatially separated populations that share a boundary – have involved detailed understanding of orthopteran, especially grasshopper, genetics and micro-distributions. Experimental evidence for sympatric speciation has been derived from research on tephritid fruit flies. The range modeling analyses outlined in section 6.11.1 exemplify some potential applications of ecological biogeographic rationales to relatively recent historical, environmental, and climatic events that influence distributions. Entomologists using these tools to interpret recent fossil material from lake sediments have played a vital role in recognizing how insect distributions have tracked past environmental change, and allowed estimation of past climate fluctuations.

Strong biogeographic patterns in the modern fauna

are becoming more difficult to recognize and interpret since humans have been responsible for the expansion of ranges of certain species and the loss of much endemism, such that many of our most familiar insects are cosmopolitan (that is, virtually worldwide) in distribution. There are at least five explanations for this expansion of so many insects of previously restricted distribution.

1 Human-loving (anthropophilic) insects such as many cockroaches, silverfish, and house flies accompany humans virtually everywhere.

2 Humans create disturbed habitats wherever they live and some synanthropic (human-associated) insects act rather like weedy plants and are able to take advantage of disturbed conditions better than native species can. Synanthropy is a weaker association with humans than anthropophily.

3 Insect (and other arthropod) external parasites (ectoparasites) and internal parasites (endoparasites) of humans and domesticated animals are often cosmopolitan.

4 Humans rely on agriculture and horticulture, with a few food crops cultivated very widely. Plant-feeding (phytophagous) insects associated with plant species that were once localized but now disseminated by humans can follow the introduced plants and may cause damage wherever the host plants grow. Many insects have been distributed in this way.

5 Insects have expanded their ranges by deliberate anthropogenic (aided by humans) introduction of selected species as biological control agents to control pest plants and animals, including other insects.

Attempts are made to restrict the shipment of agricultural, horticultural, forestry, and veterinary pests through quarantine regulations, but much of the mixing of insect faunas took place before effective measures were implemented. Thus, pest insects tend to be identical throughout climatically similar parts of the world meaning that applied entomologists must take a worldwide perspective in their studies.

8.2 THE ANTIQUITY OF INSECTS

8.2.1 The insect fossil record

Until recently, the oldest fossil hexapods were Collembola, including *Rhyniella praecursor*, known from about 400 mya (million years ago) in the Lower Devonian of Rhynie, Scotland, and slightly younger archaeognathans from North America (Fig. 8.1). Reinterpretation of another Rhynie fossil, *Rhyniognatha hirsti*, known only from its mouthparts, suggests that it is the oldest "ectognathous" insect. Tantalizing evidence from Lower Devonian fossil plants shows damage resembling that caused by the piercing-and-sucking mouthparts of insects or mites. Any earlier fossil evidence for Insecta or their relatives will be difficult to find because appropriate freshwater fossiliferous deposits are scarce prior to the Devonian.

In the Carboniferous, an extensive radiation is evidenced by substantial Upper Carboniferous fossils. Lower Carboniferous fossils are unknown, again because of lack of freshwater deposits. By some 300 mya a probably monophyletic grouping of Palaeodictyopteroidea comprising three now-extinct ordinal groups, the Megasecoptera, Palaeodictyoptera (Fig. 8.2), and Diaphanopterodea, was diverse. Palaeodictyopteroideans varied in size (with wingspans up to 56 cm), diversity (over 70 genera in 21 families are known), and in morphology, notably in mouthparts and wing articulation and venation. An apparently paraphyletic "Protodonata", perhaps a stemgroup of Odonata, had prothorax winglets as did Palaeodictyopteroidea, and included Permian insects with the largest wingspans ever recorded. Extant orders represented unambiguously by Carboniferous fossils are limited to the Orthoptera; putative Ephemeroptera, fossil hemipteroids, and blattoids are treated best as paraphyletic groups lacking the defining features of any extant clade. The beak-like, piercing mouthparts and expanded clypeus of some Carboniferous insects indicate an early origin of plant-feeding, although it was not until the Permian that gymnosperms (conifers and allies) became abundant in the previously fern-dominated flora. Concurrently, a dramatic increase took place in ordinal diversity, with some 30 orders known from the Permian. The evolution of the plant-sucking Hemiptera may have been associated with the newly available plants with thin cortex and sub-cortical phloem. Other Permian insects included those that fed on pollen, another resource of previously restricted supply.

Certain Carboniferous and Permian insects were very large, exemplified by the giant Bolsover dragonfly and palaeodictyopteran on a *Psaronius* tree fern, depicted in the vignette for this chapter. Fossils of some Ephemeroptera and dragonfly-like Protodonata had wingspans of up to 45 cm and 71 cm respectively. A plausible explanation for this gigantism is that the respiratory restriction on insect size (section 3.5.1) might have been alleviated by elevated atmospheric oxygen levels during the late Palaeozoic (between

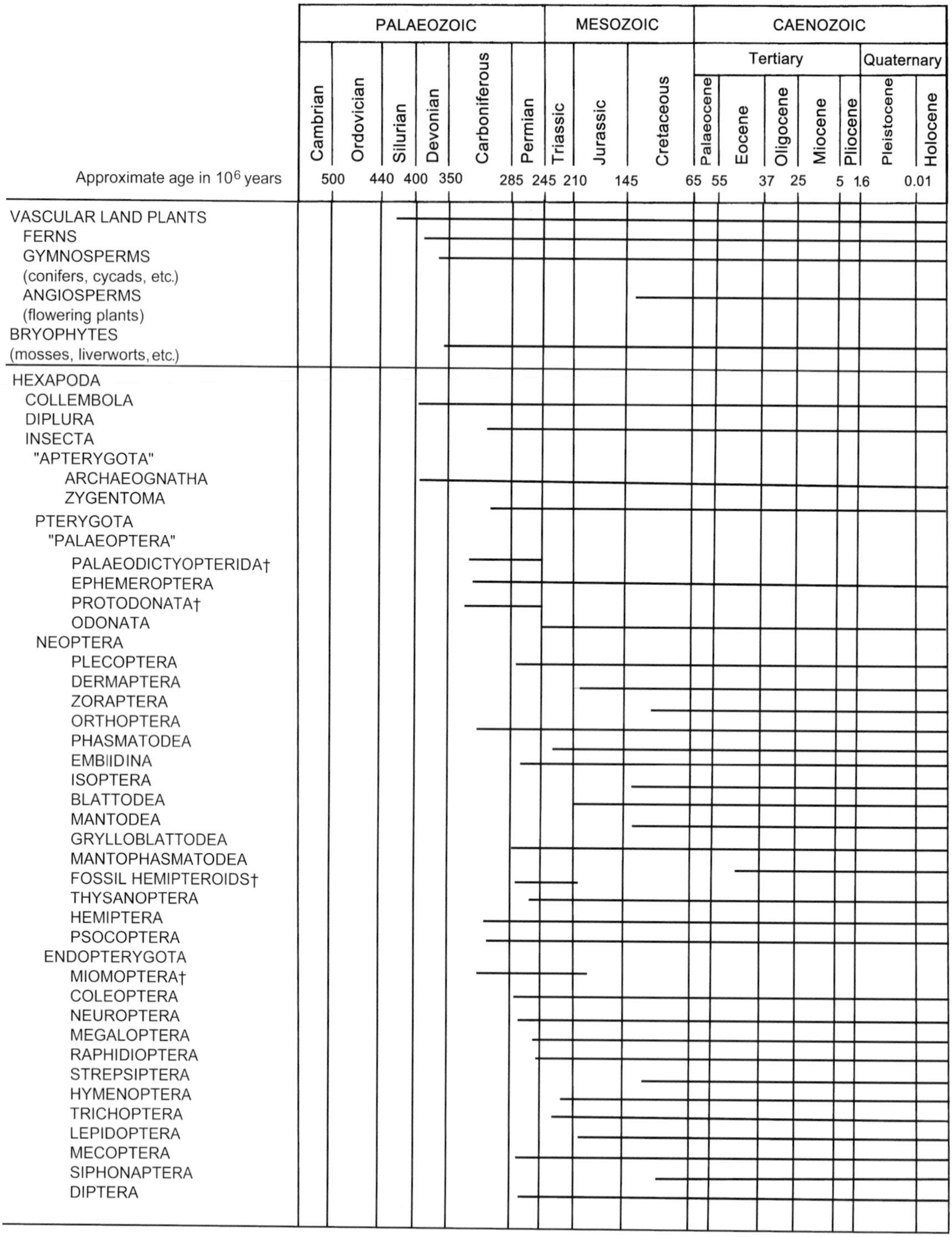

	PALAEOZOIC						MESOZOIC			CAENOZOIC							
										Tertiary					Quaternary		
	Cambrian	Ordovician	Silurian	Devonian	Carboniferous	Permian	Triassic	Jurassic	Cretaceous	Palaeocene	Eocene	Oligocene	Miocene	Pliocene	Pleistocene	Holocene	
Approximate age in 10^6 years	500	440	400	350	285	245	210	145	65	55	37	25	5	1.6	0.01		

VASCULAR LAND PLANTS
FERNS
GYMNOSPERMS
(conifers, cycads, etc.)
ANGIOSPERMS
(flowering plants)
BRYOPHYTES
(mosses, liverworts, etc.)

HEXAPODA
COLLEMBOLA
DIPLURA
INSECTA
"APTERYGOTA"
ARCHAEOGNATHA
ZYGENTOMA
PTERYGOTA
"PALAEOPTERA"
PALAEODICTYOPTERIDA†
EPHEMEROPTERA
PROTODONATA†
ODONATA
NEOPTERA
PLECOPTERA
DERMAPTERA
ZORAPTERA
ORTHOPTERA
PHASMATODEA
EMBIIDINA
ISOPTERA
BLATTODEA
MANTODEA
GRYLLOBLATTODEA
MANTOPHASMATODEA
FOSSIL HEMIPTEROIDS†
THYSANOPTERA
HEMIPTERA
PSOCOPTERA
ENDOPTERYGOTA
MIOMOPTERA†
COLEOPTERA
NEUROPTERA
MEGALOPTERA
RAPHIDIOPTERA
STREPSIPTERA
HYMENOPTERA
TRICHOPTERA
LEPIDOPTERA
MECOPTERA
SIPHONAPTERA
DIPTERA

Fig. 8.1 The geological history of insects in relation to plant evolution. Taxa that contain only fossils are indicated by the symbol †. The fossil record of Protura, Phthiraptera, and Zoraptera dates from only the Holocene, presumably because of the small size and/or rarity of these organisms. Miomoptera represent the earliest fossil endopterygotes. The placement of *Rhyniognatha* is unknown. (Insect records after Kukalová-Peck 1991; and Conrad Labandeira (Smithsonian Institution) pers. comm.)

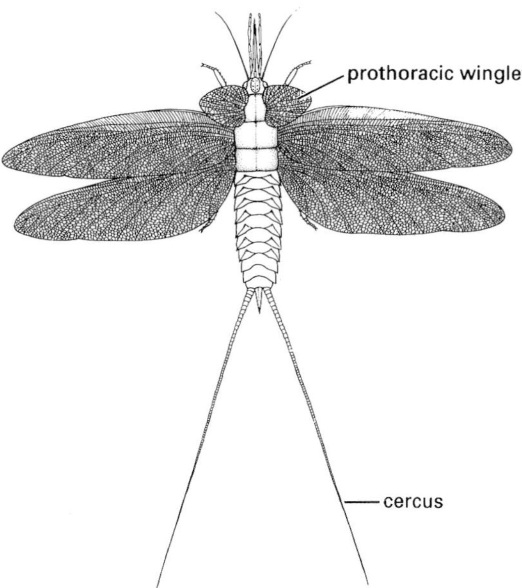

Fig. 8.2 Reconstruction of *Stenodictya lobata* (Palaeodictyoptera: Dictyoneuridae). (After Kukalová 1970.)

370 and 250 mya), which would have promoted greater oxygen diffusion in the tracheae. Furthermore, if other gases were unchanged, then any extra atmospheric oxygen may have facilitated flight in denser air. Despite the obvious appeal of this hypothesis to account for Palaeozoic gigantism, such a climatic reconstruction remains controversial and the physiological and morphological consequences of alterations of gaseous composition have been little studied for insects.

Many groups present in the Permian, including Ephemeroptera, Odonata, Plecoptera, Grylloblattodea, Phasmatodea, Orthoptera, Dictyoptera and other Polyneoptera, plus Coleoptera and Neuropterida amongst other early originating Holometabola, survived the period. However, early lineages such as Palaeodictyopteroidea (including Diaphanopteroidea, Megasecoptera, and Palaeodictyoptera) disappeared at the end of the Permian. This Permian–Triassic boundary was a time of major extinction that particularly affected marine biota, and coincided with a dramatic reduction in diversity in taxa and feeding types within surviving insect orders.

The Triassic period (commencing about 245 mya) is famed for the "dominance" by dinosaurs and pterosaurs, and the origin of the mammals; but the insects were radiating apace. The major orders of modern insects, excepting Hymenoptera and Lepidoptera, were well established by the beginning of the Triassic. Hymenoptera are seen first in this period, but represented only by symphytans. The oldest living families appeared, together with diversified taxa with aquatic immature (and some adult) stages, including modern Odonata, Heteroptera, and many families of nematocerous Diptera. The Jurassic saw the first appearance of aculeate Hymenoptera, many early-branching Diptera, and the first Brachycera. Triassic and Jurassic fossils include some excellent preserved material in fine-grained deposits such as those of Solenhofen, the site of beautifully preserved insects and *Archaeopteryx*. The origin of birds (Aves) and their subsequent diversification marked the first aerial competition for insects since the evolution of flight.

In the Cretaceous (145–65 mya) and throughout the subsequent Tertiary period (65–1.6 mya), excellent arthropod specimens were preserved in amber – a resinous plant secretion that trapped insects and hardened into a clear preservative (see Plate 3.6, facing p. 14). The excellence of preservation of whole insects in amber contrasts favorably with compression fossils that may comprise little more than crumpled wings. Many early fossil records of extant higher taxa (groups above species level) derive from these well-preserved amber specimens, but inherent sampling biases must be recognized. Smaller (more easily trapped) and forest-dwelling taxa are over-represented. Amber of Cretaceous origin occurs in France, Lebanon, Burma, Siberia, Canada, and New Jersey. The biota of this period shows a numerical dominance of insects coincident with angiosperm (flowering plant) diversification. However, major mouthpart types of extant insects evolved prior to the angiosperm radiation, associated with insect feeding on earlier terrestrial plant radiations. The fossil record indicates the great antiquity of certain insect–plant associations. The lower Cretaceous of China (130 mya) has revealed both early angiosperms and a distinctive fly belonging to Nemestrinidae with a characteristic long proboscis associated with angiosperm pollination. Elsewhere, a fossil leaf of an ancestral sycamore has the highly characteristic mine of the extant genus *Ectoedemia* (Lepidoptera: Nepticulidae), suggesting at least a 97 million year association between the nepticulid moth and particular plants. Both Coleoptera and Lepidoptera, which are primarily phytophagous orders, commenced their massive radiations in the Cretaceous. By 65 mya, the insect fauna looks rather modern, with some fossils able to be allocated to extant genera. For many animals, notably the dinosaurs, the Cretaceous–Tertiary ("K–T") boundary marked a major extinction event. Although it is generally believed that the insects entered the Tertiary with little extinction, recent studies show that although generalized insect–plant interactions survived, the prior high diversity of specialist insect–plant associations was greatly attenuated. At least in the paleobiota of south-western North Dakota, at 65 mya a major ecological perturbation setback specialized insect–plant associations.

Our understanding of Tertiary insects increasingly comes from amber from the Dominican Republic (see Plate 3.6) dated to the Oligocene/Miocene (37–22 mya) to complement the abundant and well-studied Baltic amber that derives from Eocene/Oligocene deposits (50–30 mya). Baltic ambers have been preserved and are now partially exposed beneath the northern European Baltic Sea and, to a lesser extent, the southern North Sea, brought to shore by periodic storms. Many attempts have been made to extract, amplify, and sequence ancient DNA from fossil insects preserved in amber – an idea popularized by the movie *Jurassic Park*. Amber resin is argued to dehydrate specimens and thus protect their DNA from bacterial degradation. Success in sequencing ancient DNA has been claimed for a variety of amber-preserved insects, including a termite (30 myo (million years old)), a stingless bee (25–40 myo), and a weevil (120–135 myo); however, attempts to authenticate these ancient sequences by repetition have failed. Degradation of DNA from amber fossils and contamination by fresh DNA are unresolved issues, but the extraction of material from amber insects leads to certain destruction of the specimen. Even if authentic DNA sequences have been obtained from amber insects, as yet they do not provide more useful information than that which derives from the often exquisitely preserved morphology.

An unchallengeable outcome of recent studies of fossil insects is that many species and higher taxa, especially genera and families, are revealed as much older than thought previously. At species level, all northern temperate, sub-Arctic, and Arctic zone fossil insects dating from the last million years or so appear to be morphologically identical to existing species. Many of these fossils belong to beetles (particularly their elytra), but the situation seems no different amongst other insects. Pleistocene climatic fluctuations (glacial and interglacial cycles) evidently caused taxon range changes, via movements and extinctions of individuals, but resulted in the genesis of few, if any, new species, at least as defined on their morphology. The implication is that if species of insect are typically greater than a million years old, then insect higher taxa such as genera and families may be of immense age.

Modern microscopic paleontological techniques can allow inference of age for insect taxa based on recognition of specific types of feeding damage caused to plants which are fossilized. As seen above, leaf mining in ancient sycamore leaves is attributable to a genus of extant nepticulid moth, despite the absence of any preserved remains of the actual insect. In like manner, a hispine beetle (Chrysomelidae) causing a unique type of grazing damage on young leaves of ginger plants (Zingiberaceae) has never been seen preserved as contemporary with the leaf fossils. The characteristic damage caused by their leaf chewing is recognizable, however, in the Late Cretaceous (c. 65 mya) deposits from Wyoming, some 20 million years before any body fossil of the culprit hispine. To this day, these beetles specialize in feeding on young leaves of gingers and heliconias of the modern tropics.

Despite these valuable contributions made by fossils, several tempting inferences concerning their use in phylogenetic reconstruction ought to be resisted, namely that:

1 all character states in the fossil are in the ancestral condition;

2 any fossil represents an actual ancestral form of later taxa;

3 the oldest fossil of a group necessarily represents the phylogenetically earliest taxon.

Notwithstanding this last point, fossil insects tend to demonstrate that the stratigraphic (time) sequence of earliest-dated fossils reflects the branching sequence of the phylogeny. Early-branching (so-called "basal") taxa within each order are represented first – for example, the Mastotermitidae before the higher termites (Fig. 7.4), midges, gnats, and sand flies before house and blow flies, and primitive moths before the butterflies (Fig. 7.7). Furthermore, insect fossils can show that taxa currently restricted (or perhaps disjunct) in distribution once were distributed more widely. Some such taxa include the following.

• Mastotermitidae (Isoptera), now represented by one northern Australian species, is diverse and abundant in ambers from the Dominican Republic (Caribbean), Brazil, Mexico, USA, France, Germany, and Poland dating from Cenomanian Cretaceous (c. 100 mya) to lower Miocene (some 20 mya);

• the biting-midge subfamily Austroconopinae (Diptera: Ceratopogonidae), now restricted to one extant species of *Austroconops* in Western Australia, is represented by several species in Lower Cretaceous Lebanese amber (Neocomian, 120 mya) and Upper Cretaceous Siberian amber (90 mya);
• *Leptomyrmex*, an ant genus now distributed in the western Pacific (eastern Australia, New Guinea, New Caledonia), is known from 30–40 myo Dominican amber.

An emerging pattern stemming from ongoing study of amber insects, especially those dating from the Cretaceous, is the former presence in the north of groups now restricted to the south. We might infer that the modern distributions, often involving Australia, are relictual due to differential extinction in the north. The question of whether such patterns relate in any way to northern extinction at the K–T boundary due to bolide ("meteorite") impact deserves further study. Did the differential extinction of specialist insect–plant associations in South Dakota (above) extend to the southern hemisphere? Evidently, some insect taxa presently restricted to the southern hemisphere but reported as having been present 30–40 mya in Dominican and subsequent Baltic ambers did survive the event and regional extinction has occurred more recently.

The relationship of fossil insect data to phylogeny derivation is complex. Although early fossil taxa seem often to precede phylogenetically later-branching ("more derived") taxa, it is methodologically unsound to assume so. Although phylogenies can be reconstructed from the examination of characters observed in extant material alone, fossils provide important information, not least allowing dating of the minimum age of origin of diagnostic derived characters and clades. Optimally, all data, fossil and extant, can and should be reconciled into a single estimate of evolutionary history.

8.2.2 Living insect distributions as evidence for antiquity

Evidence from the current distribution (biogeography) confirms the antiquity of many insect lineages. The disjunct distribution, specific ecological requirements, and restricted vagility of insects in a number of genera suggest that their constituent species were derived from ancestors that existed prior to the continental movements of the Jurassic and Cretaceous periods (commencing some 155 mya). For example, the occurrence of several closely related species from several lineages of chironomid midges (Diptera) only in southern Africa and Australia suggests that the ancestral taxon ranges were fragmented by separation of the continental masses during the breakup of the supercontinent Gondwana, giving a minimum age of 120 myo. Such estimates are substantiated by related Cretaceous amber fossil specimens, dating from only slightly later than commencement of the southern continental breakup.

The aphid subtribe Melaphidina, all species of which have complex life cycles involving gall induction on sumac (*Rhus*) species, has a distribution disjunct between Asia and North America. This biogeographic evidence provides an estimated minimum age for this aphid–sumac association of 48 million years, based on the date of climatically driven vicariance of the Asian and American plant lineages. The possibility of recent dispersal is refuted by appropriately aged fossils of melaphane aphids from both continents.

A similarly intimate relationship shown by the fig–fig wasp association (Box 11.4) has been subjected to molecular phylogenetic analysis for host figs and wasp pollinators. The radiations of both show episodes of colonization and radiation that largely track each other (co-speciation). The origin of the mutualism is dated to c. 90 mya (after Africa separated from Gondwana) with subsequent evolutionary radiations associated with continental fragmentation including the northward movement of India. Disjunctions in mutualistic relationships, such as the above two cases, strongly suggest concerted vicariant distributions, since both partners in the relationship must relocate simultaneously – which is highly unlikely under a dispersal interpretation.

The woodroaches (*Cryptocercus*) have a disjunct distribution in Eurasia, western USA, and the Appalachians of south-eastern USA where there is cryptic diversity. The species of *Cryptocercus* are near indistinguishable morphologically, but are distinctive in their chromosome number, mitochondrial and nuclear sequences, and in their endosymbionts. *Cryptocercus* harbor endosymbiont bacteria in bacteriocytes of their fat bodies (see section 3.6.5). Phylogenetic analysis of the bacterial RNA sequences shows that they follow faithfully the branching pattern of their host cockroaches. Using an existing estimate of a clock-like model of molecular evolution, dates have been reconstructed for the major disjunctions in woodroach evolution. The earliest branch, the North American/east Asian separation, was dated at 70–115 mya, and the separation of western from eastern North American clades at 53–88 mya. Though doubts always will exist concerning reconstructions based on assumed clock-like change in nucleotides, these patterns evidently are rather old, and did not result from Pleistocene glaciations with recent dispersal. Morphological stasis is evident in the lack of obvious differentiation over this long time period.

Such morphological conservatism and yet great antiquity of many insect species needs to be reconciled with the obvious species and genetic diversity discussed in Chapter 1. The occurrence of species assemblages in Pleistocene deposits that resemble those seen today (although not necessarily at the same geographical location) suggests considerable physiological, ecological, and morphological constancy of species. In comparative terms, insects display slower rates of morphological evolution than is apparent in many larger animals such as mammals. For example, *Homo sapiens* is a mere 100,000 years old; and if we classify (correctly) humans as morphologically highly derived chimpanzees, then any grouping of humans and the two chimpanzee species is some 5 myo. Perhaps, therefore, the difference from insects lies in mammals having undergone a recent radiation and yet already suffered major extinctions including significant losses in the Pleistocene. In contrast, insects underwent early and many subsequent radiations, each followed by relative stasis and persistence of lineages (see section 8.6).

8.3 WERE THE FIRST INSECTS AQUATIC OR TERRESTRIAL?

Arthropods evolved in the sea. This belief is based on evidence from the variety of forms preserved in Cambrian-age marine-derived deposits, such as the Burgess Shale in Canada and the Qiongzhusi Formation at Chengjiang in southern China. It is commonly believed that insects evolved after their hexapod ancestors become terrestrial, rather than insects making the ancestral transition from the ocean to land via estuaries and freshwater. The main evidence in support of a terrestrial origin for the Insecta derives from the fact that all extant non-pterygote insects (the apterygotes) and the other hexapods (Diplura, Collembola, and Protura) are terrestrial. That is, all the early-branching taxa in the hexapod phylogenetic tree (Fig. 7.2) live on land and there is no evidence from fossils (either by their possessing aquatic features or from details of preservation site) to suggest that the ancestors of these groups were not terrestrial (although they may have been associated with the margins of aquatic habitats). In contrast, the juveniles of five pterygote orders (Ephemeroptera, Odonata, Plecoptera, Megaloptera, and Trichoptera) live almost exclusively in freshwater. Given the positions of the Ephemeroptera and Odonata in Fig. 7.2, the ancestral condition for the protopterygotes probably involved immature development in freshwater.

Another line of evidence against an aquatic origin for the earliest insects is the difficulty in envisaging how a tracheal system could have evolved in water. In an aerial environment, simple invagination of external respiratory surfaces and subsequent internal elaboration could have given rise to a tracheal system (as shown in Fig. 8.3a) that later served as a preadaptation for tracheal gas exchange in the gills of aquatic insects (as shown in Fig. 8.3b). Thus, gill-like structures could assume an efficient oxygen uptake function (more than just diffusion across the cuticle) but only *after* the evolution of tracheae in a terrestrial ancestor.

There is no single explanation as to why virtually all insects with aquatic immature development have retained an aerial adult stage. Certainly, retention has occurred independently in several lineages (such as a number of times within both the Coleoptera and Diptera). The suggestion that a flighted adult is a predator-avoidance mechanism seems unlikely as predation could be avoided by a motile aquatic adult, as with so many crustaceans. It is conceivable that an aerial stage is retained to facilitate mating – perhaps there are mechanical disadvantages to underwater copulation in insects, or perhaps mate recognition systems may not function in water, especially if they are pheromonal or auditory.

8.4 EVOLUTION OF WINGS

As we have seen, much of the success of insects can be attributed to the wings, found in the numerically dominant pterygotes. Pterygotes are unusual among winged animals in that no limbs lost their pre-existing function as a result of the acquisition of flight. As we cannot observe the origins of flight, and fossils (although relatively abundant) have not greatly assisted in interpretation, any hypotheses of the origins of flight must be speculative. Several ideas have been promoted, and the area remains controversial.

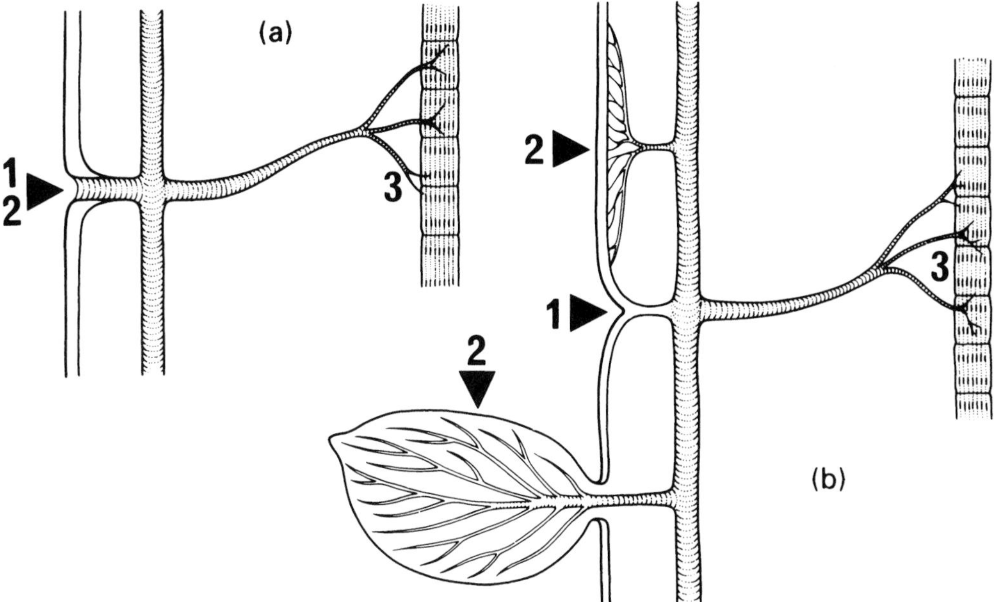

Fig. 8.3 Stylized tracheal system. (a) Oxygen uptake through invagination. (b) Invagination closed, with tracheal gas exchange through gill. 1, indicates point of invagination of the tracheal system; 2, indicates point for oxygen uptake; 3, indicates point for oxygen delivery, such as muscles. (After Pritchard et al. 1993.)

One of the longest-standing hypotheses attributes the origin of the wings to postulated lobes, derivations from the thoracic terga, called **paranota**. These lobes were not articulated and thus tracheation, innervation, venation, and musculature would have been of secondary derivation. The paranotal lobe hypothesis has been displaced in favor of one inferring wing origination from serially repeated, pre-existing, mobile structures of the pleuron. These were most probably the outer appendage (**exite**) and inner appendage (**endite**) of a basal leg segment, the **epicoxa** (Fig. 8.4a). Each "protowing" or winglet was formed by the fusion of an exite and endite lobe of the respective ancestral leg, with exites and endites having tracheation and articulation. Fossil evidence indicates the presence of articulated winglets on all body segments, best developed on the thorax (Fig. 8.4b). Molecular studies of development (Box 6.1) seem to substantiate the exite–endite model for wing origins.

The exite–endite hypothesis of wing origin can be reconciled with another recurring view: that wings derive from tracheal gills of an ancestral aquatic "protopterygote". Although the earliest insects undoubtedly were terrestrial, the earliest pterygote insects probably had aquatic immature stages (section 8.3). The abdominal gills of aquatic mayfly nymphs may be homologous with the abdominal winglets of the protopterygote, and be serially homologous with thoracic wings. Winglets are postulated in aquatic juveniles to have functioned in gas exchange and/or ventilation or even to assist in swimming, with the terrestrial adult co-opting them for an aerodynamic function.

All hypotheses concerning early wings make a common assumption that winglets originally had a non-flight function, as small winglets could have little or no use in flapping flight. Suggestions for preadaptive functions have included any (or all!) of the following:

- protection of the legs;
- covers for the spiracles;
- thermoregulation;
- sexual display;
- aid in concealment by breaking up the outline;
- predator avoidance by extension of escape jump by gliding.

Aerodynamic function came only after enlargement. However, aquatic nymphal gills may even have been

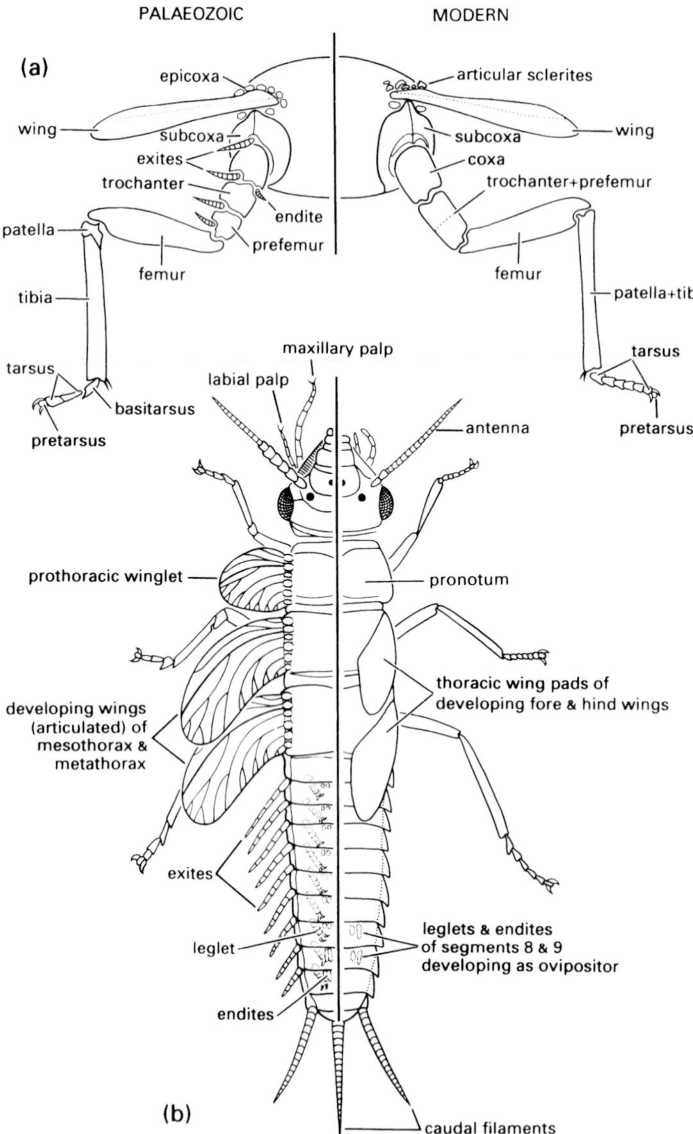

Fig. 8.4 Appendages of hypothetical primitive Palaeozoic (left of each diagram) and modern (right of each diagram) pterygotes (winged insects): (a) thoracic segment of adult showing generalized condition of appendages; (b) dorsal view of nymphal morphology. (Modified from Kukalová-Peck 1991; to incorporate ideas of J.W.H. Trueman (unpublished).)

large enough to give some immediate significant aerial advantage if retained in a terrestrial adult.

The manner in which flight evolved is also highly speculative and contentious but, whatever the origin of winglets, they came to assume some aerodynamic function. Four routes to flight have been argued, via:

1 floating, in which small insects were assisted in passive dispersal by convection;

2 paragliding, in which winglets assisted in stable gliding or parachuting from trees and tall vegetation;

3 running–jumping to flying;

4 surface sailing, in which the raised winglets allowed the adults of aquatic insects to skim across the water surface.

The first two hypotheses apply equally to fixed, non-articulated winglets and to articulated but rigidly extended winglets. Articulated winglets and flapping flight can most easily be incorporated into the running–jumping scenario of developing flight. The "floating" route to flight suffers from the flaw that wings actually hinder passive dispersal, and selection would tend to favor diminution in body size and reduction in the wings with commensurate increase in features such as long hairs. The third, running-jump route is unlikely, as no insect could attain the necessary velocity for flight originating from the ground, and only the scenario of a powered leap to allow limited gliding or flight is at all plausible.

The surface-sailing hypothesis requires articulated winglets and can account for the loss of the abdominal winglets, which would be downwind of the thoracic ones and thus barely contribute to sailing performance. Some extant stoneflies (Plecoptera) can sail across water in this manner. Perhaps surface sailing drove the evolution of wing length more than aerial gliding did, and when winglets had reached certain dimensions then gliding or flapping flight may have been facilitated greatly.

Aerodynamic theory has been applied to the problem of how large winglets had to be to give some aerodynamic advantage, and model insects have been constructed for wind-tunnel testing. Although a size-constrained and fixed-wing model lacks realism, the following evidence has been produced. An unwinged 1 cm-long insect model lacks control in a glide, and even small winglets give an immediate advantage by allowing some retarding of velocity. The possession of caudal filaments and/or paired cerci would give greater glide stability, particularly when associated with the reduction and eventual loss of posterior abdominal winglets. Additional control over gliding or flight would come with increase in body and winglet size.

There is a basic structural division of the ptergygotes into "Palaeoptera", with movable, non-folding wings, and the Neoptera, with complex wing articulation that allows folding of the wings backwards along the body (section 7.4.2). Some authors have suggested that the two wing-base types are so different that wings must have originated at least twice. However, it can be demonstrated that there is a basic venational pattern common to all pterygotes irrespective of the articulation, implying monophyly (a single origin) of wings, but not necessarily of flight. The primitive pterygote wing base apparently involved many articulated sclerites: such a system is seen in fossil palaeopterans and, in a variably modified form, in extant neopterans. In the Ephemeroptera and Odonata the basal sclerites have undergone extensive fusion that prevents the wing from flexing backwards. However, the nature of these fusions, and others that have occurred within the neopteran lineages, indicates that many different pathways have been used, and fusion *per se* does not indicate monophyly of the Palaeoptera. The likelihood that the primitive winglet had complex articulation provides a major criticism of the aerodynamic hypothesis above and the supposition of fixed wings renders suspect the conclusions from these unrealistic model insects.

The traditional proposal for the origin of venation involves tracheated, supporting or strengthening ridges on the protowing. Alternatively, the veins arose along the courses of hemolymph canals that supplied the winglets, in a manner seen in the gills of some aquatic insects. The basic venational pattern (section 2.4.2, Fig. 2.21) consists of eight veins each arising from a basal blood sinus, named from anterior to posterior: precosta, costa, subcosta, radius, media, cubitus, anal, and jugal. Each vein (perhaps excepting the media) branched basally into anterior concave and posterior convex components, with additional dichotomous branching away from the base, and a polygonal pattern of cells. Evolution of the insect wing has involved frequent reduction in the number of cells, development of bracing struts (cross-veins), selected increase in division of some veins, and reduction in complexity or complete loss of others. Furthermore, there have been changes in the muscles used for powered flight and in the phases of wing beat (section 3.1.4). Alteration in function has taken place, including the protection of the hind pair of wings by the modified fore wings (tegmina or elytra) in some groups. Increased flight control has been gained in some other groups by coupling the fore and hind wings as a single unit, and in Diptera by reduction of the metathoracic wings to halteres that function like gyroscopes.

8.5 EVOLUTION OF METAMORPHOSIS

As we have seen earlier, the evolution of metamorphosis – which allows larval immature stages to be separated ecologically from the adult stage to avoid competition – seems to have been an important factor encouraging diversification.

Although how holometaboly (with larval juvenile instars highly differentiated from adults by metamorphosis) evolved from incremental hemimetaboly has been debated, strong support for one candidate hypothesis is now available. This involves a proposal that a pronymph (hatchling or pre-hatching stage, distinct from subsequent nymphal stages) is the evolutionary precursor to the holometabolous larva, and the holometabolous pupa is the sole nymphal stage.

In ametabolous taxa, which form the earliest branches in the hexapodan phylogeny, at each molt the subsequent instar is a larger version of the previous, and development is linear, progressive, and continuous. Even early flying insects, such as the Palaeodictyoptera (Fig. 8.2), in all stages (sizes) of fossil nymphs had proportionally scaled winglets, and thus were ametabolous. A distinctive earliest developmental stage, the **pronymph**, forms an exception to the proportionality of nymphal development. The pronymph stage, which feeds only on yolk reserves, can survive independently and move for some days after hatching. Hemimetabolous insects, which differ from ametabolous taxa in that the adult instar with fully formed genitalia and wings undergoes no further molting, also have a recognizably distinct pronymph.

The body proportions of the pronymph differ from those of subsequent nymphal stages, perhaps constrained by confinement inside the egg and by the need to assist in hatching (if indeed this stage is that which hatches). Clearly, the pterygote pronymph is not just a highly miniaturized first-instar nymph. In certain orders (Blattodea, Phthiraptera, Hemiptera) the hatchling may be a pharate first-instar nymph, inside the pronymphal cuticle. At hatching the nymph emerges from the egg, since the first molt occurs concurrently with eclosion. In Odonata and Orthoptera the hatchling is the actual pronymph which can undertake limited, often specialized, post-hatching movement to locate a potential nymphal development site before molting to the first true nymph.

The larval stages of Holometabola are theorized to be homologous to this pronymphal stage, and the hemimetabolous nymphal stages are contracted into the holometabolous pupa, which is the only nymphal stage. Supporting evidence for this view comes from recognition of differences between pronymphal, nymphal, and larval cuticle, the timing of different cuticle formations relative to embryogenetic stages (**katatrepsis** – adoption of the final position in egg – and dorsal closure; see section 6.2.1), and interruption of neuroblast-induced neuron production during larval stages, which resumes in the nymph.

The mechanism that could cause such dramatic changes in development is **heterochrony** – alteration in the timing of activation of different controls involved in developmental cascades (Box 6.1). Metamorphosis is controlled by the interplay between neuropeptides, ecdysteroids, and especially juvenile hormone (JH) titers, as seen in section 6.3. The balance between controlling factors commences in the egg, and continues throughout development: subtle differences in timing of events lead to very different outcomes. Earlier appearance of elevated JH in the embryo prevents maturity of some aspects of the nymph, leading to development of a prolarva, which then is maintained in larval form by continued high JH which suppresses maturation. Pupation (entry to the nymphal stage) takes place when JH is reduced, and maturation then requires increased JH. Holometabolous development occurs because JH remains high, with the JH-free period delayed until the end of immature growth (metamorphosis). This contrasts with hemimetaboly, in which postembryonic, continuous low JH exposure allows nymphal development to progress evenly towards the adult form.

In early-branching ("primitive") Holometabola, JH prevents any precocious production of adult features in the larva until the pupa. However, in later-branching orders and more derived families some adult features can escape suppression by JH and may commence development in early larval instars. Such features include wings, legs, antennae, eyes, and genitalia: their early expression is seen in groups of primordial cells that become imaginal discs – differentiated already for their final adult function (section 6.2, Fig. 6.4). With scope to vary the onset of differentiation of each adult organ in the larvae, great variation and flexibility in life-cycle evolution is permitted, including capacity to greatly shorten them.

Timing of developmental control is evident also in the basic larval shape, especially the variety of larval leg forms shown in Fig. 6.6. Onset of JH expression can retard development of the pronymph/prolarva at any stage in leg expression, from apodous (no expression) to essentially fully developed. Such legs, although termed prolegs in larvae (section 2.4.1), have innervation similar to, but less developed than, that found in adult legs. Evidently, immature and imaginal legs are homologous – since adult leg imaginal discs develop within the prolegs, no matter how well developed the prolegs are.

The major unanswered question in this view of the evolution of holometaboly is that if larval evolution results from a protracted equivalent of the pronymphal stage, how did the pronymph become able to feed? Although crustacean pronymphs (e.g. the nauplius stage of decapods) feed, the equivalent stage in extant hexapods apparently cannot do so. The pronymph has a short post-hatching existence, but if it finds itself in a suitable microhabitat by female oviposition-site selection or its own limited ability to search, and could feed, then there would be a tendency to select for this ability in ensuing instars. Selection is seen as continuing, because decoupling of larval and adult food resources reduces competition between juvenile and adult for food, thus separating resources used in growth from those for reproduction. The evident success of Holometabola derives not least from this segregation.

8.6 INSECT DIVERSIFICATION

An estimated half of all insect species chew, suck, gall, or mine the living tissues of higher plants (**phytophagy**), yet only nine (of 30) extant insect orders are primarily phytophagous. This imbalance suggests that when a barrier to phytophagy (e.g. plant defenses) is breached, an asymmetry in species number occurs, with the phytophagous lineage being much more speciose than the lineage of its closest relative (the sister group) of different feeding mode. For example, the tremendous diversification of the almost universally phytophagous Lepidoptera can be compared with that of its sister group, the relatively species-poor, non-phytophagous Trichoptera. Likewise, the enormous phytophagous beetle group Phytophaga (Chrysomeloidea plus Curculionoidea) is overwhelmingly more diverse than the entire Cucujoidea, the whole or part of which forms the sister group to the Phytophaga. Clearly, the diversifications of insects and flowering plants are related in some way, which we explore further in Chapter 11. By analogy, the diversification of phytophagous insects should be accompanied by the diversification of their insect parasites or parasitoids, as discussed in Chapter 13. Such parallel species diversifications clearly require that the phytophage or parasite be able to seek out and recognize its host(s). Indeed, the high level of host-specificity observed for insects is possible only because of their highly developed sensory and neuromotor systems.

An asymmetry, similar to that of phytophagy compared with non-phytophagy, is seen if flightedness is contrasted to aptery. The monophyletic Pterygota (winged or secondarily apterous insects) are vastly more speciose than their immediate sister group, the Zygentoma (silverfish), or the totality of primitively wingless apterygotes. The conclusion is unavoidable: the gain of flight correlates with a radiation under any definition of the term. Secondary aptery occurs in some pterygotes – amongst many isolated species, some genera, and in all members of the Phthiraptera (parasitic lice) and Siphonaptera (fleas), two small orders showing an ectoparasitic radiation. The Phthiraptera are less diverse than their flighted sister group (part or all of the Psocoptera), although the Siphonaptera are more speciose than their likely sister group (part or all of the Mecoptera). Many Phasmatodea are flightless, and there are indications that the order originated from unflighted ancestors. Radiation within the phasmids thus seems to have included sporadic regain of wings. This seemingly anomalous hypothesis deserves further study, not least regarding how developmental pathways regulating wing development function.

Flight allows insects the increased mobility necessary to use patchy food resources and habitats and to evade non-winged predators. These abilities may enhance species survival by reducing the threats of extinction, but wings also allow insects to reach novel habitats by dispersal across a barrier and/or by expansion of their range. Thus, vagile pterygotes may be more prone to species formation by the two modes of geographical (**allopatric**) speciation: small isolated populations formed by the vagaries of chance dispersal by winged adults may be the progenitors of new species, or the continuous range of widely distributed species may become fragmented into isolates by **vicariance** (range division) events such as vegetation fragmentation or geological changes.

New species arise as the genotypes of isolated populations diverge from those of parental populations. Such isolation may be phenological (temporal or behavioral) as **sympatric** speciation, as well as spatial or geographical, and host transfers or changes in breeding times are documented better for insects than for any other organisms.

The Endopterygota (see section 7.4.2) contains the numerically large orders Diptera, Lepidoptera, Hymenoptera, and Coleoptera (section 1.3). An explanation for their success lies in their metamorphosis, discussed in detail above, which allows the adult and larval stages to differ or overlap in phenology, depending upon timing of suitable conditions. Alternative food resources and/or habitats may be used by a sedentary larva and vagile adult, enhancing species survival by avoidance of intraspecific competition. Furthermore, deleterious conditions for some life-history stages, such as extreme temperatures, low water levels, or shortage of food, may be tolerated by a less susceptible life-history stage, for example a diapausing larva, non-feeding pupa, or migratory adult.

No single factor explains the astonishing diversification of the insects. An early origin and an elevated rate of species formation in association with the angiosperm radiation, combined with high species persistence through time, leave us with the great number of living species. We can obtain some ideas on the processes involved by study of selected cases of insect radiations in which the geological framework for their evolution is well known, as on some Pacific islands.

8.7 INSECT EVOLUTION IN THE PACIFIC

Study of the evolution of insects (and other arthropods such as spiders) of oceanic islands such as Hawai'i and the Galapagos is comparable in importance to those of the perhaps more famous plants (e.g. Hawai'an silverswords), birds (Hawai'ian honeycreepers and "Darwin's finches" of the Galapagos), land snails (Hawai'i), and lizards (Galapagos iguanas). The earliest and most famous island evolutionary studies of insects involved the Hawai'ian fruit flies (Diptera: Drosophilidae). This radiation has been revisited many times, but recent research has included evolutionary studies of certain crickets, microlepidopterans, carabid beetles, pipunculid flies, mirid bugs, and damselflies.

Why this interest in the fauna of isolated island chains in the mid-Pacific? The Hawai'ian fauna is highly endemic, with an estimated 99% of its native arthropod species found nowhere else. The Pacific is an immense ocean, in which lies Hawai'i, an archipelago (island chain) some 3800 km distant from the nearest continental land mass (North America) or high islands (the Marquesas). The geological history, which is quite well known, involves continued production of new volcanic material at an oceanic hotspot located in the south-east of the youngest island, Hawai'i, whose maximum age is 0.43 myo. Islands lying to the north-west are increasingly older, having been transported to their current locations (Fig. 8.5) by the north-western movement of the Pacific plate. The production of islands in this way is likened to a "conveyor belt" carrying islands away from the hotspot (which stays in the same relative position). Thus, the oldest existing above-water "high islands" (that is of greater elevation than a sand bar/atoll) are Niihau (aged 4.9 myo) and Kauia (c. 5.1 myo) positioned to the north-west. Between these two and Hawai'i lie Oahu (aged 3.7 myo), Molokai (1.9 myo), Maui and Lanai (1.3 myo), and Kahoolawe (1.0 myo). Undoubtedly there have been older islands – some estimates are that the chain originated some 80 mya – but only since about 23 mya have there been continuous high islands for colonization.

Since the islands are mid-oceanic and volcanic, they originated without any terrestrial biota, and so the present inhabitants must have descended from colonists. The great distance from source areas (other islands, the continents) implies that colonization is a rare event – and this is borne out in nearly all studies. The biota of islands is quite discordant (unbalanced) compared to that of continents. Major groups are missing, presumably by chance failure to arrive and flourish. Those that did arrive successfully and found viable populations often speciated, and may exhibit quite strange biologies with respect to their ancestors. Thus, some Hawai'ian damselflies have terrestrial larvae, in contrast to aquatic larvae elsewhere; Hawai'ian geometrid moth caterpillars are predaceous, not phytophagous; otherwise marine midge larvae are found in freshwater torrents.

As a consequence of the rarity of founding events, most insect radiations have been identified as monophyletic, that is the complete radiation belongs to a clade derived from one founder individual or population. For some clades each species of the radiation is restricted to one island, whereas other ("widespread") species can be found on more than one island. Fundamental to understanding the history of the colonization and subsequent diversification is a phylogeny of relationships between the species in the clade. The Hawai'ian Drosophilidae lack any widespread species (i.e. all are single island endemics) and their relationships have been studied, first with morphology and more recently with molecular techniques. Interpretation of the history of this clade is rather straightforward; species distributions generally are congruent with the geology such that the colonists of older islands and older volcanoes (those of Oahu and Molokai) gave rise to descendants that have radiated more recently on the younger islands and younger volcanoes of Maui and Hawai'i. Similar scenarios of an older colonization with more recent radiation associated with island age are seen in the Hawai'ian Pipunculidae, damselflies, and mirids, and this probably is typical for all the diversified biota. Where estimates have been made to date the colonization and radiation, it seems few if any originated prior to the currently oldest high island (c. 5 mya), and sequential colonization seems to have been approximately contemporaneous with each newly formed island.

Aquatic insects, such as black flies (Diptera:

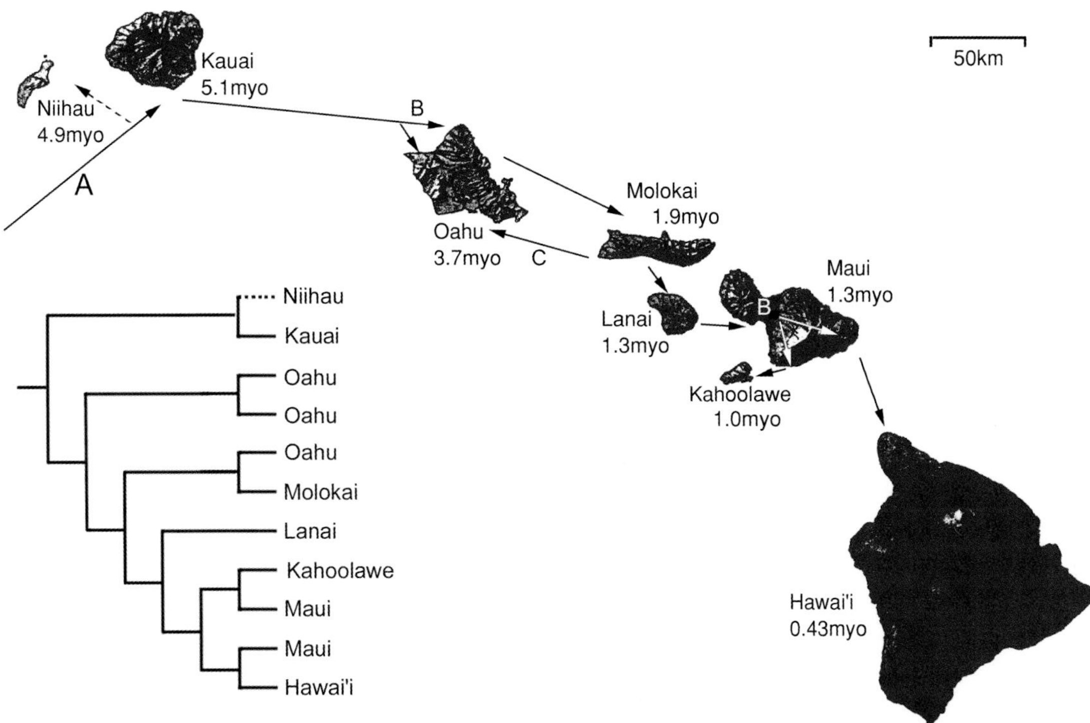

Fig. 8.5 Area cladogram showing phylogenetic relationships of hypothetical insect taxa with taxon names replaced by their areas of endemism in the Hawai'ian archipelago. The pattern of colonization and speciation of the insects on the islands is depicted by arrows showing the sequence and direction of events: A, founding; B, diversification within an island; C, back-colonization event; myo, million years old. Broken line, extinct lineage.

Simuliidae) whose larvae live in running water, cannot colonize islands until persistent streams and seepages form. As islands age in geological time, greatest environmental heterogeneity with maximum aquatic habitat diversity may occur in middle-age, until senescence-induced erosion and loss of elevated areas cause extinction. In this "middle-age", speciation may occur on a single island as specializations in different habitats, as in Hawai'ian *Megalagrion* damselflies. In this clade, most speciation has been associated with existing ecological larval-habitat specialists (fast-running water, seepages, plant axils, or even terrestrial habitats) colonizing and subsequently differentiating on newly formed islands as they arose from the ocean and suitable habitats became available. However, on top of this pattern there can be radiations associated with different habitats on the same island, perhaps very rapidly after initial colonization. Furthermore, examples exist showing recolonization from younger islands to older (back-founder events) that indicate substantial complexity in the evolution of some insect radiations on islands.

Sources for the original colonizers sometimes have been difficult to find because the offspring of Hawai'ian radiations often are very distinct from any prospective non-Hawai'ian relatives; however, the western or south-western Pacific is a likely source for platynine carabids, *Megalagrion* damselflies, and several other groups, and North America for some mirid bugs. In contrast, the evolution of the insect fauna of the Galapagos on the eastern side of the Pacific Ocean presents a rather different story to that of Hawai'i. Widespread insect species on the Galapagos predominantly are shared with Central or South America, and endemic species often have sister-group relationships with the nearest South American mainland, as is proposed for much of the fauna. The biting-midge (Ceratopogonidae) fauna derives apparently from many independent founding events, and similar findings come from other families of flies. Evidently, long-distance dispersal from the nearest continent outweighs *in-situ* speciation in generating the diversity of the Galapagos compared to Hawai'i. Nonetheless, some estimates of arrival of founders are earlier than the currently oldest islands.

Orthopteroids of the Galapagos and Hawai'i show another evolutionary feature associated with island-living – wing loss or reduction (aptery or brachyptery) in one or both sexes. Similar losses are seen in carabid beetles, with multiple losses proposed from phylogenetic analyses. Furthermore, there are extensive radiations of certain insects in the Galapagos and Hawai'i associated with underground habitats such as larva tubes and caves. Studies of the role of sexual selection – primarily female choice of mating partner (section 5.3) – suggest that this may have played an important role in species differentiation on islands, at least of crickets and fruit flies. Whether this is enhanced relative to continental situations is unclear.

All islands of the Pacific are highly impacted by the arrival and establishment of non-native species, through introductions perhaps by continued over-water colonizations, but certainly associated with human commerce, including well-meaning biological control activities. Some accidental introductions, such as of tramp ant species (Box 1.2) and a mosquito vector of avian malaria, have affected Hawai'ian native ecosystems detrimentally across many taxa. Even parasitoids introduced to control agricultural pests have spread to native moths in remote natural habitats (section 16.5). Our unique natural laboratories for the study of evolutionary processes are being destroyed apace.

FURTHER READING

Austin, J.J., Ross, A.J., Smith, A.B., Fortey, R.A. & Thomas, R.H. (1997) Problems of reproducibility – does geologically ancient DNA survive in amber-preserved insects? *Proceedings of the Royal Society of London B* **264**, 467–74.

Clark, J.W., Hossain, S., Burnside, C.A. & Kambhampati, S. (2001) Coevolution between a cockroach and its bacterial endosymbiont: a biogeographic perspective. *Proceedings of the Royal Society of London B* **298**, 393–8.

Coope, R. (1991) The study of the "nearly fossil". *Antenna* **15**, 158–63.

Cranston, P.S. & Naumann, I. (1991) Biogeography. In: *The Insects of Australia*, 2nd edn. (CSIRO), pp. 181–97. Melbourne University Press, Carlton.

Dudley, R. (1998) Atmospheric oxygen, giant Palaeozoic insects and the evolution of aerial locomotor performance. *Journal of Experimental Biology* **201**, 1043–50.

Elias, S.A. (1994) *Quaternary Insects and their Environments*. Smithsonian Institution Press, Washington, DC.

Gillespie, R.G. & Roderick, G.K. (2002) Arthropods on islands: colonization, speciation, and conservation. *Annual Review of Entomology* **47**, 595–632.

Grimaldi, D. (2003) Fossil record. In: *Encyclopedia of Insects* (eds. V.H. Resh & R.T. Cardé), pp. 882–8. Academic Press, Amsterdam.

Jordan, S., Simon, C. & Polhemus, D. (2003) Molecular systematics and adaptive radiation of Hawaii's endemic damselfly genus *Megalagrion* (Odonata: Coenagrionidae). *Systematic Biology* **52**, 89–109.

Kukalová-Peck, J. (1983) Origin of the insect wing and wing articulation from the arthropodan leg. *Canadian Journal of Zoology* **61**, 1618–69.

Kukalová-Peck, J. (1987) New Carboniferous Diplura, Monura, and Thysanura, the hexapod ground plan, and the role of thoracic side lobes in the origin of wings (Insecta). *Canadian Journal of Zoology* **65**, 2327–45.

Kukalová-Peck, J. (1991) Fossil history and the evolution of hexapod structures. In: *The Insects of Australia*, 2nd edn. (CSIRO), pp. 141–79. Melbourne University Press, Carlton.

Labandeira, C.C., Dilcher, D.L., Davis, D.R. & Wagner, D.L. (1994) Ninety-seven million years of angiosperm–insect association: palaeobiological insights into the meaning of coevolution. *Proceedings of the National Academy of Sciences of the USA* **91**, 12278–82.

Machado, C.A., Jousselin, E., Kjellberg, F., Compton, S.G. & Herre, E.A. (2001) Phylogenetic relationships, historical biogeography and character evolution of fig-pollinating wasps. *Proceedings of the Royal Society of London B* **268**, 685–94.

Mitter, C., Farrell, B. & Wiegmann, B. (1988) The phylogenetic study of adaptive zones: has phytophagy promoted insect diversification? *American Naturalist* **132**, 107–28.

Moran, N.A. (1989) A 48 million year old aphid–host plant association and complex life cycle: biogeographic evidence. *Science* **245**, 173–5.

Pritchard, G., McKee, M.H., Pike, E.M., Scrimgeour, G.J. & Zloty, J. (1993) Did the first insects live in water or in air? *Biological Journal of the Linnean Society* **49**, 31–44.

Resh, V.H. & Cardé, R.T. (eds.) (2003) *Encyclopedia of Insects*. Academic Press, Amsterdam. [Particularly see articles on biogeographical patterns; fossil record; island biogeography and evolution.]

Trueman, J.W.H. (1990) Evolution of insect wings: a limb-exite-plus-endite model. *Canadian Journal of Zoology* **68**, 1333–5.

Truman, J.W. & Riddiford, L.M. (1999) The origins of insect metamorphosis. *Nature* **401**, 447–52.

Truman, J.W. & Riddiford, L.M. (2002) Endocrine insights into the evolution of metamorphosis in insects. *Annual Review of Entomology* **33**, 467–500.

Chapter 9

GROUND-DWELLING INSECTS

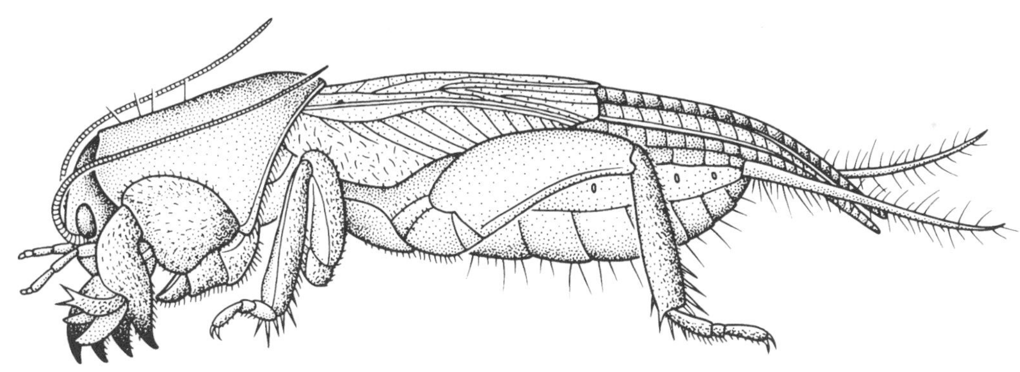

A mole cricket. (After Eisenbeis & Wichard 1987.)

A profile of a typical soil shows an upper layer of recently derived vegetational material, termed **litter**, overlying more decayed material that intergrades with **humus**-enriched organic soils. These organic materials lie above mineralized soil layers, which vary with local geology and climate, such as rainfall and temperature. Particle size and soil moisture are important influences on the microdistributions of subterranean organisms.

The decompositional habitat, comprising decaying wood, leaf litter, carrion, and dung, is an integral part of the soil system. The processes of decay of vegetation and animal matter and return of nutrients to the soil involve many organisms, notably fungi. Fungal hyphae and fruiting bodies provide a medium exploited by many insects, and all faunas associated with decompositional substrates include insects and other hexapods.

In this chapter we consider the ecology and taxonomic range of soil and decompositional faunas in relation to the differing macrohabitats of soil and decaying vegetation and humus, dead and decaying wood, dung, and carrion. We survey the importance of insect–fungal interactions and examine two intimate associations. A description of a specialized subterranean habitat (caves) is followed by a discussion of some uses of terrestrial hexapods in environmental monitoring. The chapter concludes with seven taxonomic boxes that deal with: non-insect hexapods (Collembola, Protura, and Diplura); primitively wingless bristletails and silverfish (Archaeognatha and Zygentoma); three small hemimetabolous orders, the Grylloblattodea, Embiidina, and Zoraptera; earwigs (Dermaptera); and cockroaches (Blattodea).

9.1 INSECTS OF LITTER AND SOIL

Litter is fallen vegetative debris, comprising materials such as leaves, twigs, wood, fruit, and flowers in various states of decay. The processes that lead to the incorporation of recently fallen vegetation into the humus layer of the soil involve degradation by microorganisms, such as bacteria, protists, and fungi. The actions of nematodes, earthworms, and terrestrial arthropods, including crustaceans, mites, and a range of hexapods (Fig. 9.1), mechanically break down large particles and deposit finer particles as feces. Acari (mites), termites (Isoptera), ants (Formicidae), and many beetles (Coleoptera) are important arthropods of litter and humus-rich soils. The immature stages of many insects, including beetles, flies (Diptera), and moths (Lepidoptera), may be abundant in litter and soils. For example, in Australian forests and woodlands, the eucalypt leaf litter is consumed by larvae of many oecophorid moths and certain chrysomelid leaf beetles. The soil fauna also includes many non-insect hexapods (Collembola, Protura, and Diplura) and primitively wingless insects, the Archaeognatha and Zygentoma. Many Blattodea, Orthoptera, and Dermaptera occur only in terrestrial litter – a habitat to which several of the minor orders of insects, the Zoraptera, Embiidina, and Grylloblattodea, are restricted. Soils that are permanently or regularly waterlogged, such as marshes and riparian (stream marginal) habitats, intergrade into the fully aquatic habitats described in Chapter 10 and show faunal similarities.

In a soil profile, the transition from the upper, recently fallen litter to the lower well-decomposed litter to the humus-rich soil below may be gradual. Certain arthropods may be confined to a particular layer or depth and show a distinct behavior and morphology appropriate to the depth. For example, amongst the Collembola, *Onychurus* lives in deep soil layers and has reduced appendages, is blind and white, and lacks a furcula, the characteristic collembolan springing organ. At intermediate soil depths, *Hypogastrura* has simple eyes, and short appendages with the furcula shorter than half the body length. In contrast, Collembola such as *Orchesella* that live amongst the superficial leaf litter have larger eyes, longer appendages, and an elongate furcula, more than half as long as the body.

A suite of morphological variations can be seen in soil insects. Larvae often have well-developed legs to permit active movement through the soil, and pupae frequently have spinose transverse bands that assist movement to the soil surface for eclosion. Many adult soil-dwelling insects have reduced eyes and their wings are protected by hardened fore wings, or are reduced (brachypterous), or lost altogether (apterous) or, as in the reproductives of ants and termites, shed after the dispersal flight (deciduous, or caducous). Flightlessness (that is either through primary absence or secondary loss of wings) in ground-dwelling organisms may be countered by jumping as a means of evading predation: the collembolan furcula is a spring mechanism and the alticine Coleoptera ("flea-beetles") and terrestrial Orthoptera can leap to safety. However, jumping is of little value in subterranean organisms. In these insects, the fore legs may be modified for digging (Fig. 9.2) as **fossorial** limbs, seen in groups that construct tunnels,

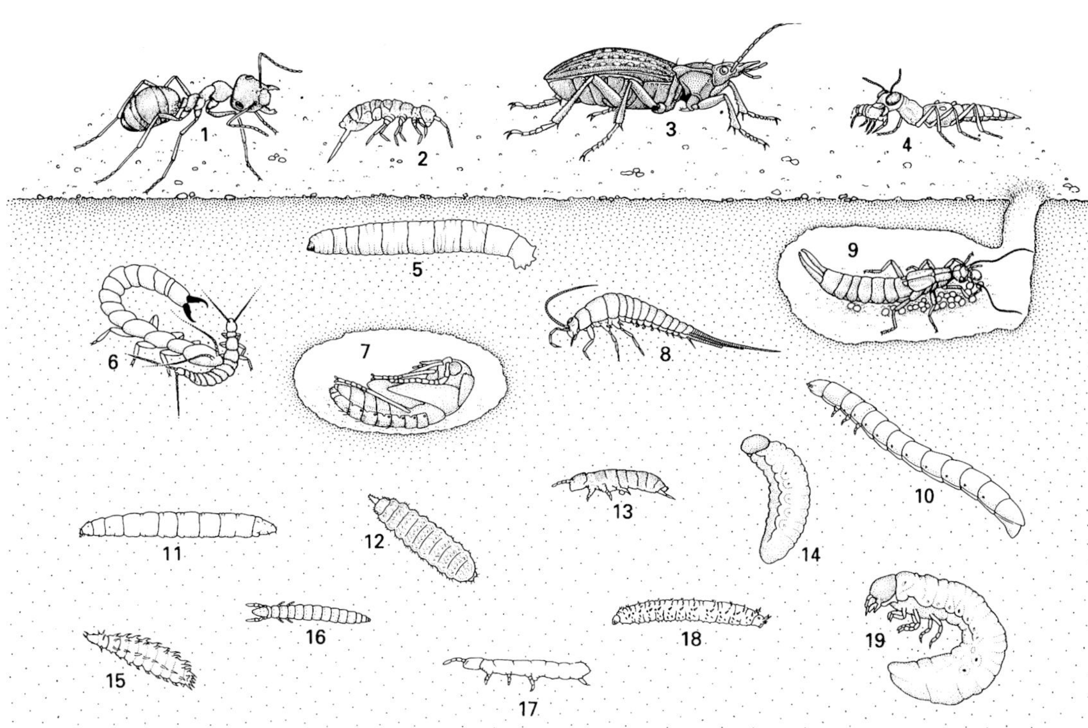

Fig. 9.1 Diagrammatic view of a soil profile showing some typical litter and soil insects and other hexapods. Note that organisms living on the soil surface and in litter have longer legs than those found deeper in the ground. Organisms occurring deep in the soil usually are legless or have reduced legs; they are unpigmented and often blind. The organisms depicted are: (1) worker of a wood ant (Hymenoptera: Formicidae); (2) springtail (Collembola: Isotomidae); (3) ground beetle (Coleoptera: Carabidae); (4) rove beetle (Coleoptera: Staphylinidae) eating a springtail; (5) larva of a crane fly (Diptera: Tipulidae); (6) japygid dipluran (Diplura: Japygidae) attacking a smaller campodeid dipluran; (7) pupa of a ground beetle (Coleoptera: Carabidae); (8) bristletail (Archaeognatha: Machilidae); (9) female earwig (Dermaptera: Labiduridae) tending her eggs; (10) wireworm, larva of a tenebrionid beetle (Coleoptera: Tenebrionidae); (11) larva of a robber fly (Diptera: Asilidae); (12) larva of a soldier fly (Diptera: Stratiomyidae); (13) springtail (Collembola: Isotomidae); (14) larva of a weevil (Coleoptera: Curculionidae); (15) larva of a muscid fly (Diptera: Muscidae); (16) proturan (Protura: Sinentomidae); (17) springtail (Collembola: Isotomidae); (18) larva of a March fly (Diptera: Bibionidae); (19) larva of a scarab beetle (Coleoptera: Scarabaeidae). (Individual organisms after various sources, especially Eisenbeis & Wichard 1987.)

such as mole crickets (as depicted in the vignette of this chapter), immature cicadas, and many beetles.

The distribution of subterranean insects changes seasonally. The constant temperatures at greater soil depths are attractive in winter as a means of avoiding low temperatures above ground. The level of water in the soil is important in governing both vertical and horizontal distributions. Frequently, larvae of subterranean insects that live in moist soils will seek drier sites for pupation, perhaps to reduce the risks of fungal disease during the immobile pupal stage. The subterranean nests of ants usually are located in drier areas, or the nest entrance is elevated above the soil surface to prevent flooding during rain, or the whole nest may be elevated to avoid excess ground moisture. Location and design of the nests of ants and termites is very important to the regulation of humidity and temperature because, unlike social wasps and bees, they cannot ventilate their nests by fanning, although they can migrate within nests or, in some species, between them. The passive regulation of the internal nest environment is exemplified by termites of *Amitermes* (see Fig. 12.9) and *Macrotermes* (see Fig. 12.10), which maintain an internal environment suitable for the

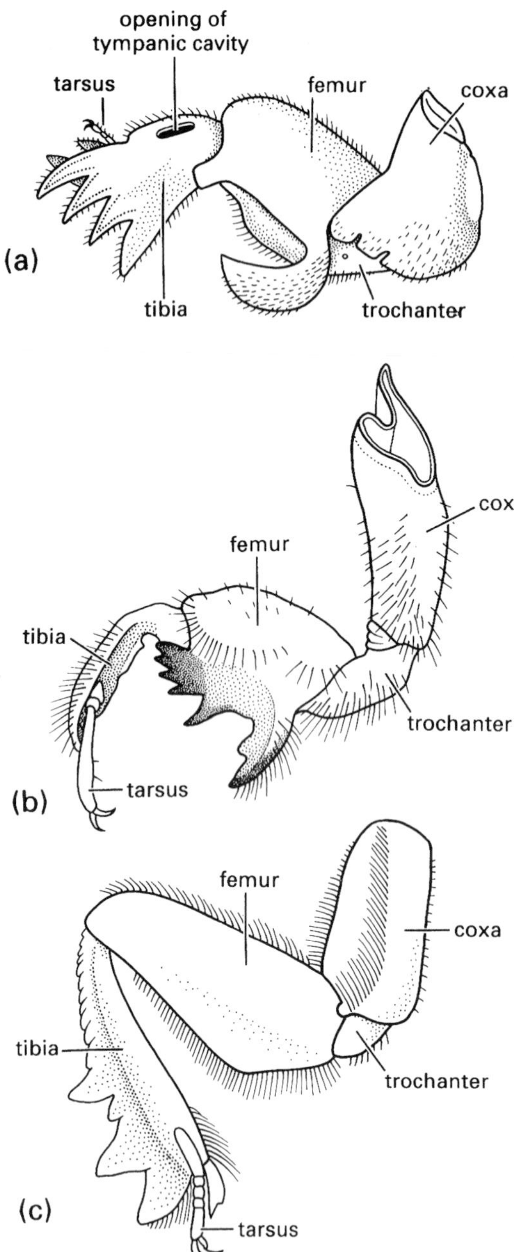

Fig. 9.2 Fossorial fore legs of: (a) a mole cricket of *Gryllotalpa* (Orthoptera: Gryllotalpidae); (b) a nymphal periodical cicada of *Magicicada* (Hemiptera: Cicadidae); and (c) a scarab beetle of *Canthon* (Coleoptera: Scarabaeidae). ((a) After Frost 1959; (b) after Snodgrass 1967; (c) after Richards & Davies 1977.)

growth of particular fungi that serve as food (section 12.2.4).

Many soil-dwelling hexapods derive their nutrition from ingesting large volumes of soil containing dead and decaying vegetable and animal debris and associated microorganisms. These bulk-feeders, known as **saprophages** or **detritivores**, include hexapods such as some Collembola, beetle larvae, and certain termites (Isoptera: Termitinae, including *Termes* and relatives). Although these have not been demonstrated to possess symbiotic gut protists they appear able to digest cellulose from the humus layers of the soil. Copious excreta (feces) is produced and these organisms clearly play a significant role in structuring soils of the tropics and subtropics.

For arthropods that consume humic soils, the subsoil parts of plants (the roots) will be encountered frequently. The fine parts of roots often have particular associations with fungal mycorrhizae and rhizobacteria, forming a zone called the **rhizosphere**. Bacterial and fungal densities are an order of magnitude higher in soil close to the rhizosphere compared with soil distant from roots, and microarthropod densities are correspondingly higher close to the rhizosphere. The selective grazing of Collembola, for example, can curtail growth of fungi that are pathogenic to plants, and their movements aid in transport of beneficial fungi and bacteria to the rhizosphere. Furthermore, interactions between microarthropods and fungi in the rhizosphere and elsewhere may aid in mineralization of nitrogen and phosphates, making these elements available to plants; but further experimental evidence is required to quantify these beneficial roles.

9.1.1 Root-feeding insects

Out-of-sight herbivores feeding on the roots of plants have been neglected in studies of insect–plant interactions, although it is recognized that 50–90% of plant biomass may be below ground. Root-feeding activities have been difficult to quantify in space and time, even for charismatic taxa like the periodic cicadas (*Magicicada* spp.). The damaging effects caused by root chewers and miners such as larvae of hepialid and ghost moths, and beetles including wireworms (Elateridae), false wireworms (Tenebrionidae), weevils (Curculionidae), scarabaeids, flea-beetles, and galerucine chrysomelids may become evident only if the above-ground plants collapse. However, lethality is one end of a spectrum of responses, with some plants responding with increased above-ground growth to root grazing, others neutral (perhaps through resistance), and others sustaining subcritical damage. Sap-sucking insects on the plant roots such as some aphids (Box 11.2) and scale insects (Box 9.1) cause loss of plant vigor, or death, especially if insect-damaged necrotized tissue is invaded secondarily by fungi and bacteria. Although when the nymphs of periodic cicadas occur in orchards they can cause serious damage, the nature of the relationship with the roots upon which they feed remains poorly known (see also section 6.10.5).

Soil-feeding insects probably do not selectively avoid the roots of plants. Thus, where there are high densities of fly larvae that eat soil in pastures, such as Tipulidae (leatherjackets), Sciaridae (black fungus gnats), and Bibionidae (March flies), roots are damaged by their activities. There are frequent reports of such activities causing economic damage in managed pastures, golf courses, and turf-production farms.

The use of insects as biological control agents for control of alien/invasive plants has emphasized phytophages of above-ground parts such as seeds and leaves (see section 11.2.6) but has neglected root-damaging taxa. Even with increased recognition of their importance, 10 times as many above-ground control agents are released compared to root feeders. By the year 2000, over 50% of released root-feeding biological control agents contributed to the suppression of target invasive plants; in comparison about 33% of the above-ground biological control agents contributed some suppression of their host plant. Coleoptera, particularly Curculionidae and Chrysomelidae, appear to be most successful in control, whereas Lepidoptera and Diptera are less so.

9.2 INSECTS AND DEAD TREES OR DECAYING WOOD

The death of trees may involve insects that play a role in the transmission of pathogenic fungi amongst trees. Thus, wood wasps of the genera *Sirex* and *Urocercus* (Hymenoptera: Siricidae) carry *Amylostereum* fungal spores in invaginated intersegmental sacs connected to the ovipositor. During oviposition, spores and mucus are injected into the sapwood of trees, notably *Pinus* species, causing mycelial infection. The infestation causes locally drier conditions around the xylem, which is optimal for development of larval *Sirex*. The fungal disease in Australia and New Zealand can cause death of fire-damaged trees or those stressed by drought conditions. The role of bark beetles (*Scolytus* spp., Coleoptera: Curculionidae: Scolytinae) in the spread of Dutch elm disease is discussed in section 4.3.3. Other insect-borne fungal diseases transmitted to live trees may result in tree mortality, and continued decay of these and those that die of natural causes often involves further interactions between insects and fungi.

The ambrosia beetles (Curculionidae: Platypodinae and some Scolytinae) are involved in a notable association with ambrosia fungus and dead wood, which has been popularly termed "the evolution of agriculture" in beetles. Adult beetles excavate tunnels (often called galleries), predominantly in dead wood (Fig. 9.3), although some attack live wood. Beetles mine in the phloem, wood, twigs, or woody fruits, which they infect with wood-inhabiting ectosymbiotic "ambrosia" fungi that they transfer in special cuticular pockets called **mycangia**, which store the fungi during the insects' aestivation or dispersal. The fungi, which come from a wide taxonomic range, curtail plant defenses and break down wood making it more nutritious for the beetles. Both larvae and adults feed on the conditioned wood and directly on the extremely nutritious fungi. The association between ambrosia fungus and beetles appears to be very ancient, perhaps originating as long ago as 60 million years with gymnosperm host trees, but with subsequent increased diversity associated with multiple transfers to angiosperms.

Some mycophagous insects, including beetles of the families Lathridiidae and Cryptophagidae, are strongly attracted to recently burned forest to which they carry fungi in mycangia. The cryptophagid beetle *Henoticus serratus*, which is an early colonizer of burned forest in some areas of Europe, has deep depressions on the underside of its pterothorax (Fig. 9.4), from which glandular secretions and material of the ascomycete fungus *Trichoderma* have been isolated. The beetle probably uses its legs to fill its mycangia with fungal material, which it transports to newly burnt habitats as an inoculum. Ascomycete fungi are important food sources for many **pyrophilous** insects, i.e. species strongly attracted to burning or newly burned areas or which occur mainly in burned forest for a few years after the fire. Some predatory and wood-feeding insects are also pyrophilous. A number of pyrophilous heteropterans (Aradidae), flies (Empididae and Platypezidae), and beetles (Carabidae and Buprestidae) have been shown to be attracted to the heat or smoke of fires, and

Box 9.1 Ground pearls

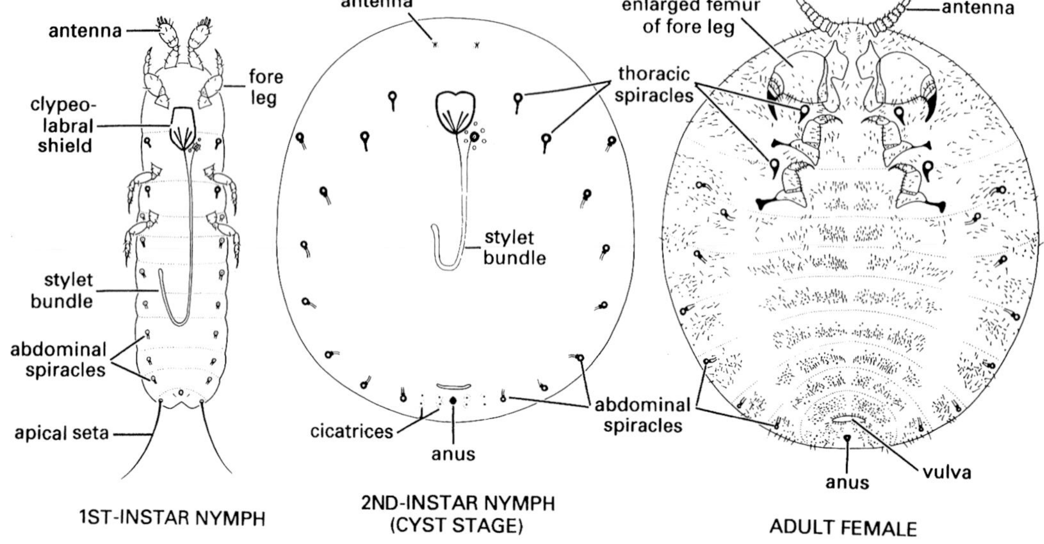

In parts of Africa, the encysted nymphs ("ground pearls") of certain subterranean scale insects are sometimes made into necklaces by the local people. These nymphal insects have few cuticular features, except for their spiracles and sucking mouthparts. They secrete a transparent or opaque, glassy or pearly covering that encloses them, forming spherical to ovoid "cysts" of greatest dimension 1–8 mm, depending on species. Ground pearls belong to several genera of Margarodinae (Hemiptera: Margarodidae), including *Eumargarodes*, *Margarodes*, *Neomargarodes*, *Porphyrophora*, and *Promargarodes*. They occur worldwide in soils among the roots of grasses, especially sugarcane, and grape vines (*Vitis vinifera*). They may be abundant and their nymphal feeding can cause loss of plant vigor and death; in lawns, feeding results in brown patches of dead grass. In South Africa they are serious vineyard pests; in Australia different species reduce sugarcane yield; and in the south-eastern USA one species is a grass pest.

Plant damage mostly is caused by the female insects because many species are parthenogenetic, or at least males have never been found, and when males are present they are smaller than the females. There are three female instars (as illustrated here for *Margarodes* (= *Sphaeraspis*) *capensis*, after De Klerk et al. 1982): the first-instar nymph disperses in the soil seeking a feeding site on roots, where it molts to the second-instar or cyst stage; the adult female emerges from the cyst between spring and fall (depending on species) and, in species with males, comes to the soil surface where mating occurs. The female then buries back into the soil, digging with its large fossorial fore legs. The foreleg coxa is broad, the femur is massive, and the tarsus is fused with the strongly sclerotized claw. In parthenogenetic species, females may never leave the soil. Adult females have no mouthparts and do not feed; in the soil, they secrete a waxy mass of white filaments – an ovisac, which surrounds their several hundred eggs.

Although ground pearls can feed via their thread-like stylets, which protrude from the cyst, second-instar nymphs of most species are capable of prolonged dormancy (up to 17 years has been reported for one species). Often the encysted nymphs can be kept dry in the laboratory for one to several years and still be capable of "hatching" as adults. This long life and ability to rest dormant in the soil, together with resistance to desiccation, mean that they are difficult to eradicate from infested fields and even crop rotations do not eliminate them effectively. Furthermore, the protection afforded by the cyst wall and subterranean existence makes insecticidal control largely inappropriate. Many of these curious pestiferous insects are probably African and South American in origin and, prior to quarantine restrictions, may have been transported within and between countries as cysts in soil or on rootstocks.

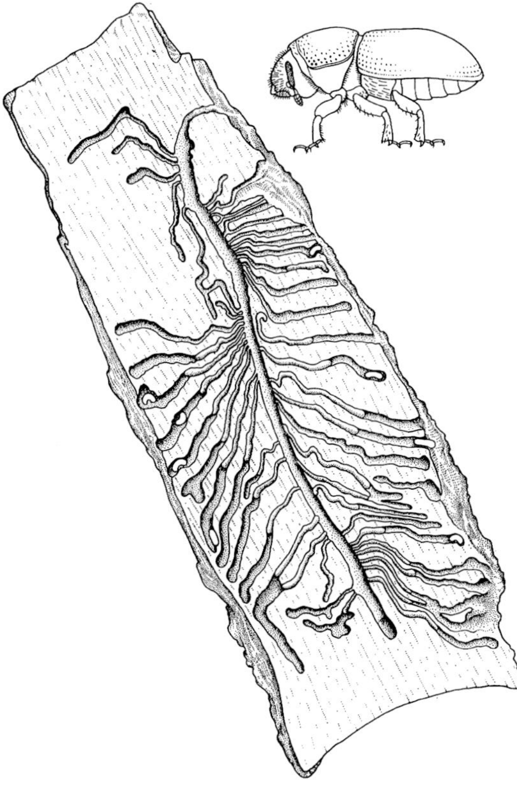

Fig. 9.3 A plume-shaped tunnel excavated by the bark beetle *Scolytus unispinosus* (Coleoptera: Scolytidae) showing eggs at the ends of a number of galleries; enlargement shows an adult beetle. (After Deyrup 1981.)

often from a great distance. Species of jewel beetle (Buprestidae: *Melanophila* and *Merimna*) locate burnt wood by sensing the infrared radiation typically produced by forest fires (section 4.2.1).

Fallen, rotten timber provides a valuable resource for a wide variety of detritivorous insects if they can overcome the problems of living on a substrate rich in cellulose and deficient in vitamins and sterols. Termites are able to live entirely on this diet, either through the possession of cellulase enzymes in their digestive systems and the use of gut symbionts (section 3.6.5) or with the assistance of fungi (section 9.5.3). Cockroaches and termites have been shown to produce endogenous cellulase that allows digestion of cellulose from the diet of rotting wood. Other **xylophagous** (wood-eating) strategies of insects include very long life cycles with slow development and probably the use of xylophagous microorganisms and fungi as food.

9.3 INSECTS AND DUNG

The excreta or dung produced by vertebrates may be a rich source of nutrients. In the grasslands and rangelands of North America and Africa, large ungulates produce substantial volumes of fibrous and nitrogen-rich dung that contains many bacteria and protists. Insect **coprophages** (dung-feeding organisms) utilize this resource in a number of ways. Certain higher flies – such as the Scathophagidae, Muscidae (notably the worldwide house fly, *Musca domestica*, the Australian *M. vetustissima*, and the widespread tropical buffalo fly,

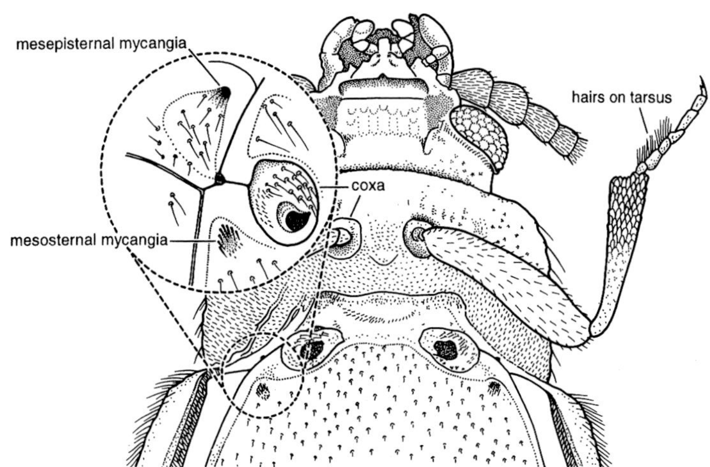

Fig. 9.4 Underside of the thorax of the beetle *Henoticus serratus* (Coleoptera: Cryptophagidae) showing the depressions, called mycangia, which the beetle uses to transport fungal material that inoculates new substrate on recently burnt wood. (After drawing by Göran Sahlén in Wikars 1997.)

Haematobia irritans), Faniidae, and Calliphoridae – oviposit or larviposit into freshly laid dung. Development can be completed before the medium becomes too desiccated. Within the dung medium, predatory fly larvae (notably other species of Muscidae) can seriously reduce survival of coprophages. However, in the absence of predators or disturbance of the dung, nuisance-level populations of flies can be generated from larvae developing in dung in pastures.

The insects primarily responsible for disturbing dung, and thereby limiting fly breeding in the medium, are dung beetles, belonging to the family Scarabaeidae. Not all larvae of scarabs use dung: some ingest general soil organic matter, whereas some others are herbivorous on plant roots. However, many are coprophages. In Africa, where many large herbivores produce large volumes of dung, several thousand species of scarabs show a wide variety of coprophagous behaviors. Many can detect dung as it is deposited by a herbivore, and from the time that it falls to the ground invasion is very rapid. Many individuals arrive, perhaps up to many thousands for a single fresh elephant dropping. Most dung beetles excavate networks of tunnels immediately beneath or beside the pad (also called a pat), and pull down pellets of dung (Fig. 9.5). Other beetles excise a chunk of dung and move it some distance to a dug-out chamber, also often within a network of tunnels. This movement from pad to nest chamber may occur either by head-butting an unformed lump, or by rolling molded spherical balls over the ground to the burial site. The female lays eggs into the buried pellets, and the larvae develop within the fecal food ball, eating fine and coarse particles. The adult scarabs also may feed on dung, but only on the fluids and finest particulate matter. Some scarabs are generalists and utilize virtually any dung encountered, whereas others specialize according to texture, wetness, pad size, fiber content, geographical area, and climate; a range of scarab activities ensures that all dung is buried within a few days at most.

In tropical rainforests, an unusual guild of dung beetles has been recorded foraging in the tree canopy on every subcontinent. These specialist coprophages have been studied best in Sabah, Borneo, where a few species of *Onthophagus* collect the feces of primates (such as gibbons, macaques, and langur monkeys) from the foliage, form it into balls and push the balls over the edge of leaves. If the balls catch on the foliage below, then the dung-rolling activity continues until the ground is reached.

In Australia, a continent in which native ungulates are absent, native dung beetles cannot exploit the volume and texture of dung produced by introduced domestic cattle, horses, and sheep. As a result, dung once lay around in pastures for prolonged periods, reducing the quality of pasture and allowing the development of prodigious numbers of nuisance flies. A program to introduce alien dung beetles from Africa and Mediterranean Europe has been successful in accelerating dung burial in many regions.

9.4 INSECT–CARRION INTERACTIONS

In places where ants are important components of the fauna, the corpses of invertebrates are discovered and removed rapidly, by widely scavenging and efficient ants. In contrast, vertebrate corpses (carrion) support a wide diversity of organisms, many of which are insects. These form a **succession** – a non-seasonal, directional, and continuous sequential pattern of populations of species colonizing and being eliminated as carrion decay progresses. The nature and timing of the succession depends upon the size of the corpse, seasonal and ambient climatic conditions, and the surrounding non-biological (edaphic) environment, such as soil type. The organisms involved in the succession vary according to whether they are upon or within the carrion, in the substrate immediately below the corpse, or in the soil at an intermediate distance below or away from the corpse. Furthermore, each succession will comprise different species in different geographical areas, even in places with similar climates. This is because few species are very widespread in distribution, and each biogeographic area has its own specialist carrion faunas. However, the broad taxonomic categories of cadaver specialists are similar worldwide.

The first stage in carrion decomposition, **initial decay**, involves only microorganisms already present in the body, but within a few days the second stage, called **putrefaction**, begins. About two weeks later, amidst strong odors of decay, the third, **black putrefaction** stage begins, followed by a fourth, **butyric fermentation** stage, in which the cheesy odor of butyric acid is present. This terminates in an almost dry carcass and the fifth stage, slow **dry decay**, completes the process, leaving only bones.

The typical sequence of corpse **necrophages**, saprophages, and their parasites is often referred to as following "waves" of colonization. The first wave

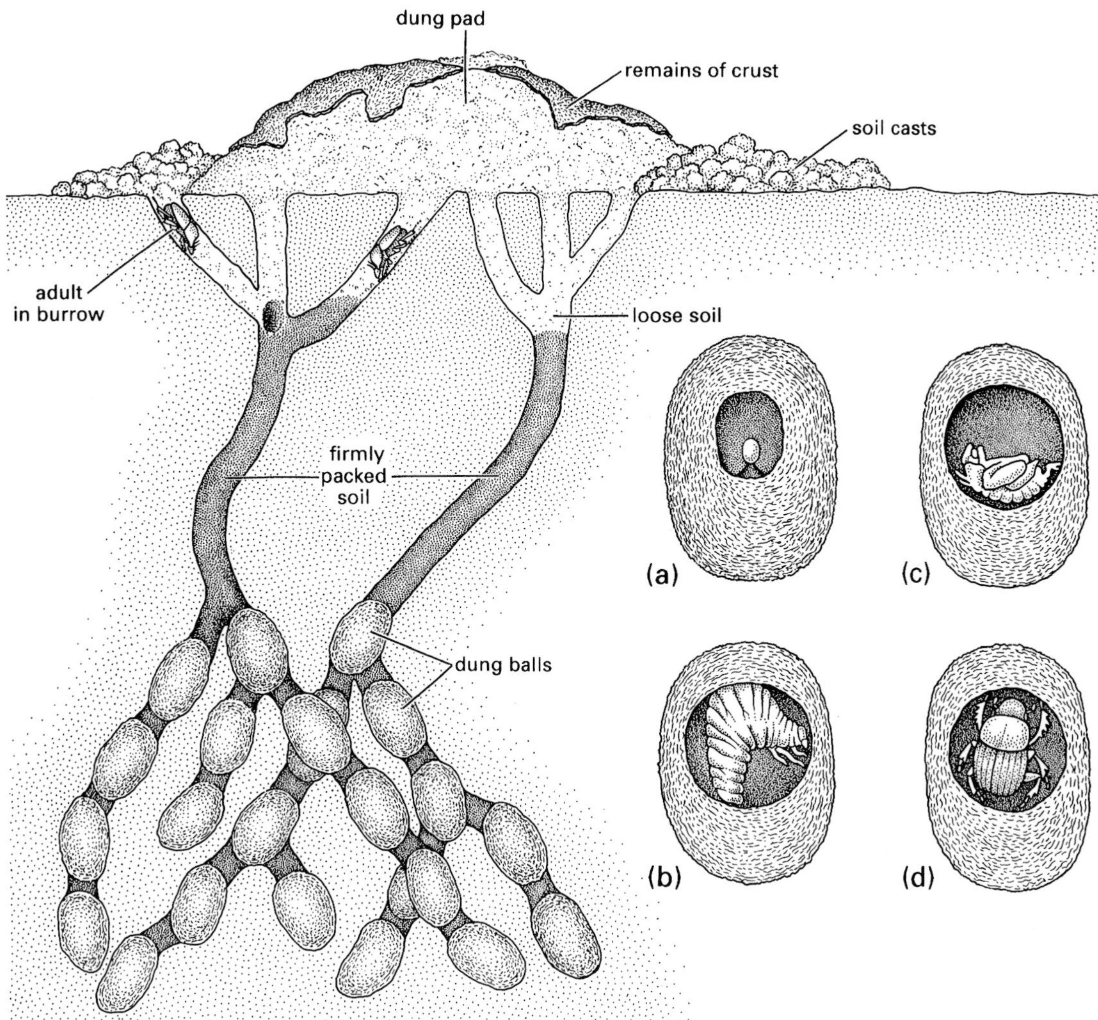

Fig. 9.5 A pair of dung beetles of *Onthophagus gazella* (Coleoptera: Scarabaeidae) filling in the tunnels that they have excavated below a dung pad. The inset shows an individual dung ball within which beetle development takes place: (a) egg; (b) larva, which feeds on the dung; (c) pupa; and (d) adult just prior to emergence. (After Waterhouse 1974.)

involves certain blow flies (Diptera: Calliphoridae) and house flies (Muscidae) that arrive within hours or a few days at most. The second wave is of sarcophagids (Diptera) and additional muscids and calliphorids that follow shortly thereafter, as the corpse develops an odor. All these flies either lay eggs or larviposit on the corpse. The principal predators on the insects of the corpse fauna are staphylinid, silphid, and histerid beetles, and hymenopteran parasitoids may be entomophagous on all the above hosts. At this stage, blow fly activity ceases as their larvae leave the corpse and pupate in the ground. When the fat of the corpse turns rancid, a third wave of species enters this modified substrate, notably more dipterans, such as certain Phoridae, Drosophilidae, and *Eristalis* rat-tailed maggots (Syrphidae) in the liquid parts. As the corpse becomes butyric, a fourth wave of cheese-skippers (Diptera: Piophilidae) and related flies use the body. A fifth wave occurs as the ammonia-smelling carrion dries out, and adult and larval Dermestidae and

Cleridae (Coleoptera) become abundant, feeding on keratin. In the final stages of dry decay, some tineid larvae ("clothes moths") feed on any remnant hair.

Immediately beneath the corpse, larvae and adults of the beetle families Staphylinidae, Histeridae, and Dermestidae are abundant during the putrefaction stage. However, the normal, soil-inhabiting groups are absent during the carrion phase, and only slowly return as the corpse enters late decay. The rather predictable sequence of colonization and extinction of carrion insects allows forensic entomologists to estimate the age of a corpse, which can have medico-legal implications in homicide investigations (section 15.6).

9.5 INSECT–FUNGAL INTERACTIONS

9.5.1 Fungivorous insects

Fungi and, to a lesser extent, slime molds are eaten by many insects, termed **fungivores** or **mycophages**, which belong to a range of orders. Amongst insects that use fungal resources, Collembola and larval and adult Coleoptera and Diptera are numerous. Two feeding strategies can be identified: **microphages** gather small particles such as spores and hyphal fragments (see Plate 3.7, facing p. 14) or use more liquid media; whereas **macrophages** use the fungal material of fruiting bodies, which must be torn apart with strong mandibles. The relationship between fungivores and the specificity of their fungus feeding varies. Insects that develop as larvae in the fruiting bodies of large fungi are often obligate fungivores, and may even be restricted to a narrow range of fungi; whereas insects that enter such fungi late in development or during actual decomposition of the fungus are more likely to be saprophagous or generalists than specialist mycophages. Longer-lasting macrofungi such as the pored mushrooms, Polyporaceae, have a higher proportion of mono- or oligophagous associates than ephemeral and patchily distributed mushrooms such as the gilled mushrooms (Agaricales).

Smaller and more cryptic fungal food resources also are used by insects, but the associations tend to be less well studied. Yeasts are naturally abundant on live and fallen fruits and leaves, and **fructivores** (fruit-eaters) such as larvae of certain nitidulid beetles and drosophilid fruit flies are known to seek and eat yeasts. Apparently, fungivorous drosophilids that live in decomposing fruiting bodies of fungi also use yeasts, and specialization on particular fungi may reflect variations in preferences for particular yeasts. The fungal component of lichens is probably used by grazing larval lepidopterans and adult plecopterans.

Amongst the Diptera that utilize fungal fruiting bodies, the Mycetophilidae (fungus gnats) are diverse and speciose, and many appear to have oligophagous relationships with fungi from amongst a wide range used by the family. The use by insects of subterranean fungal bodies in the form of mycorrhizae and hyphae within the soil is poorly known. The phylogenetic relationships of the Sciaridae (Diptera) to the mycetophilid "fungus gnats" and evidence from commercial mushroom farms all suggest that sciarid larvae normally eat fungal mycelia. Other dipteran larvae, such as certain phorids and cecidomyiids, feed on commercial mushroom mycelia and associated microorganisms, and may also use this resource in nature.

9.5.2 Fungus farming by leaf-cutter ants

The subterranean ant nests of the genus *Atta* (15 species) and the rather smaller colonies of *Acromyrmex* (24 species) are amongst the major earthen constructions in neotropical rainforest. Calculations suggest that the largest nests of *Atta* species involve excavation of some 40 tonnes of soil. Both these genera are members of a tribe of myrmecine ants, the Attini, in which the larvae have an obligate dependence on symbiotic fungi for food. Other genera of Attini have monomorphic workers (of a single morphology) and cultivate fungi on dead vegetable matter, insect feces (including their own and, for example, caterpillar "frass"), flowers, and fruit. In contrast, *Atta* and *Acromyrmex*, the more derived genera of Attini, have polymorphic workers of several different kinds or castes (section 12.2.3) that exhibit an elaborate range of behaviors including cutting living plant tissues, hence the name "leaf-cutter ants". In *Atta*, the largest worker ants excise sections of live vegetation with their mandibles (Fig. 9.6a) and transport the pieces to the nest (Fig. 9.6b). During these processes, the working ant has its mandibles full, and may be the target of attack by a particular parasitic phorid fly (illustrated in the top right of Fig. 9.6a). The smallest worker is recruited as a defender, and is carried on the leaf fragment.

When the material reaches the nest, other individuals lick any waxy cuticle from the leaves and macerate the plant tissue with their mandibles. The mash is then

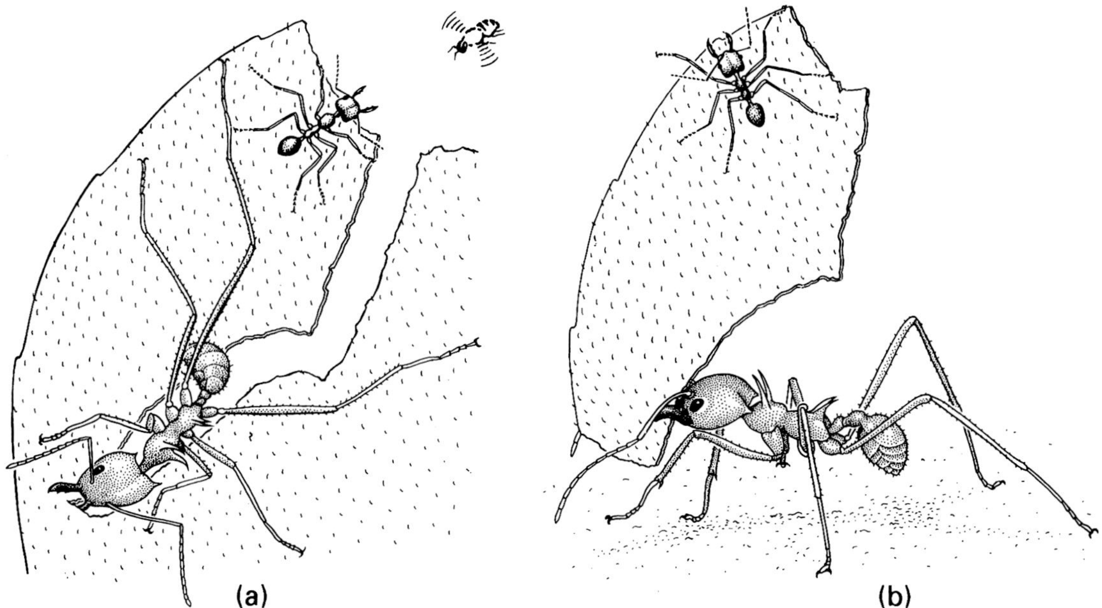

Fig. 9.6 The fungus gardens of the leaf-cutter ant, *Atta cephalotes* (Formicidae), require a constant supply of leaves. (a) A medium-sized worker, called a media, cuts a leaf with its serrated mandibles while a minor worker guards the media from a parasitic phorid fly (*Apocephalus*) that lays its eggs on living ants. (b) A guarding minor hitchhikes on a leaf fragment carried by a media. (After Eibl-Eibesfeldt & Eibl-Eibesfeldt 1967.)

inoculated with a fecal cocktail of enzymes from the hindgut. This initiates digestion of the fresh plant material, which acts as an incubation medium for a fungus, known only from these "fungus gardens" of leaf-cutter ants. Another specialized group of workers tends the gardens by inoculating new substrate with fungal hyphae and removing other species of undesirable fungi in order to maintain a monoculture. Control of alien fungi and bacteria is facilitated by pH regulation (4.5–5.0) and by antibiotics, including those produced by mutualistic *Pseudonocardia* bacteria associated with ant cuticle. In darkness, and at optimal humidity and a temperature close to 25°C, the cultivated fungal mycelia produce nutritive hyphal bodies called **gongylidia**. These are not sporophores, and appear to have no function other than to provide food for ants in a mutualistic relationship in which the fungus gains access to the controlled environment. Gongylidia are manipulated easily by the ants, providing food for adults, and are the exclusive food eaten by larval attine ants. Digestion of fungi requires specialized enzymes, which include chitinases produced by ants from their labial glands.

A single origin of fungus domestication might be expected given the vertical transfer of fungi by transport in the mouth of the founding gyne (new queen) and regurgitation at the new site. However, molecular phylogenetic studies of the fungi show domestication from free-living stocks has taken place several times, although the ancestral symbiosis is at least 50 million years old. All but one domesticate belongs to the Basidiomycetes of the tribe Leucocoprini in the family Lepiotaceae, propagated as a mycelium or occasionally as a unicellular yeast. Although each attine nest has a single species of fungus, amongst different nests of a single species a range of fungus species are tended. Obviously, some ant species can change their fungus when a new nest is constructed, perhaps when colony foundation is by more than one queen (pleiometrosis). Lateral (horizontal) transfer was observed when a Central American *Atta* species introduced to Florida rapidly adopted the local attine-tended fungus for its gardens.

Leaf-cutter ants dominate the ecosystems in which they occur; some grassland *Atta* species consume as much vegetation per hectare as domestic cattle, and certain rainforest species are estimated to cause up to 80% of all leaf damage and to consume up to 17% of all leaf production. The system is an effective converter of plant cellulose to usable carbohydrate, with at least 45% of the original cellulose of fresh leaves converted by the time the spent substrate is ejected into a dung store as refuse from the fungus garden. However, fungal gongylidia contribute only a modest fraction of the metabolic energy of the ants, because about 95% of the respiratory requirements of the colony is provided by adults feeding on plant sap from chewed leaf fragments.

Leaf-cutter ants may be termed highly polyphagous, as studies have shown them to utilize between 50 and 70% of all neotropical rainforest plant species. However, as the adults feed on the sap of fewer species, and the larvae are monophagous on fungus, the term polyphagy strictly may be incorrect. The key to the relationship is the ability of the worker ants to harvest from a wide variety of sources, and the cultivated fungus to grow on a wide range of hosts. Coarse texture and latex production by leaves can discourage attines, and chemical defenses may play a role in deterrence. However, leaf-cutter ants have adopted a strategy to evade plant defensive chemicals that act on the digestive system: they use the fungus to digest the plant tissue. The ants and fungus co-operate to break down plant defenses, with the ants removing protective leaf waxes that deter fungi, and the fungi in turn producing carbohydrates from cellulose indigestible to the ants.

9.5.3 Fungus cultivation by termites

The terrestrial microfauna of tropical savannas (grasslands and open woodlands) and some forests of the Afrotropical and Oriental (Indo-Malayan) zoogeographic regions can be dominated by a single subfamily of Termitidae, the Macrotermitinae. These termites may form conspicuous above-ground mounds up to 9 m high, but more often their nests consist of huge underground structures. Abundance, density, and production of macrotermitines may be very high and, with estimates of a live biomass of 10 g m^{-2}, termites consume over 25% of all terrestrial litter (wood, grass, and leaf) produced annually in some west African savannas.

The litter-derived food resources are ingested, but not digested by the termites: the food is passed rapidly through the gut and, upon defecation, the undigested feces are added to comb-like structures within the nest. The combs may be located within many small subterranean chambers or one large central hive or brood chamber. Upon these combs of feces, a *Termitomyces* fungus develops. The fungi are restricted to Macrotermitinae nests, or occur within the bodies of termites. The combs are constantly replenished and older parts eaten, on a cycle of 5–8 weeks. Fungus action on the termite fecal substrate raises the nitrogen content of the substrate from about 0.3% until in the asexual stages of *Termitomyces* it may reach 8%. These asexual spores (mycotêtes) are eaten by the termites, as well as the nutrient-enriched older comb. Although some species of *Termitomyces* have no sexual stage, others develop above-ground basidiocarps (fruiting bodies, or "mushrooms") at a time that coincides with colony-founding forays of termites from the nest. A new termite colony is inoculated with the fungus by means of asexual or sexual spores transferred in the gut of the founder termite(s).

Termitomyces lives as a monoculture on termite-attended combs, but if the termites are removed experimentally or a termite colony dies out, or if the comb is extracted from the nest, many other fungi invade the comb and *Termitomyces* dies. Termite saliva has some antibiotic properties but there is little evidence for these termites being able to reduce local competition from other fungi. It seems that *Termitomyces* is favored in the fungal comb by the remarkably constant microclimate at the comb, with a temperature of 30°C and scarcely varying humidity together with an acid pH of 4.1–4.6. The heat generated by fungal metabolism is regulated appropriately via a complex circulation of air through the passageways of the nest, as illustrated for the above-ground nest of the African *Macrotermes natalensis* in Fig. 12.10.

The origin of the mutualistic relationship between termite and fungus seems not to derive from joint attack on plant defenses, in contrast to the ant–fungus interaction seen in section 9.5.2. Termites are associated closely with fungi, and fungus-infested rotting wood is likely to have been a primitive food preference. Termites can digest complex substances such as pectins and chitins, and there is good evidence that they have endogenous cellulases, which break down dietary cellulose. However, the Macrotermitinae have shifted some of their digestion to *Termitomyces* outside of the gut. The fungus facilitates conversion of plant compounds to more nutritious products and probably allows a wider range of cellulose-containing foods to be consumed by the termites. Thus, the macrotermitines successfully utilize the abundant resource of dead vegetation.

9.6 CAVERNICOLOUS INSECTS

Caves often are perceived as extensions of the subterranean environment, resembling deep soil habitats in the lack of light and the uniform temperature, but differing in the scarcity of food. Food sources in shallow caves include roots of terrestrial plants, but in deeper caves there is no plant material other than that originating from any stream-derived debris. In many caves nutrient supplies come from fungi and the feces (guano) of bats and certain cave-dwelling birds, such as swiftlets in the Orient.

Cavernicolous (cave-dwelling) insects include those that seek refuge from adverse external environmental conditions – such as moths and adult flies, including mosquitoes, that hibernate to avoid winter cold, or aestivate to avoid summer heat and desiccation. **Troglobiont** or **troglobite** insects are restricted to caves, and often are phylogenetically related to soil-dwelling ones. The troglobite assemblage may be dominated by Collembola (especially the family Entomobryidae), and other important groups include the Diplura (especially the family Campodeidae), orthopteroids (including cave crickets, Rhaphidophoridae), and beetles (chiefly carabids, but including fungivorous silphids).

In Hawai'i, past and present volcanic activity produces a spectacular range of "lava tubes" of different isolation in space and time from other volcanic caves. Here, studies of the wide range of troglobitic insects and spiders living in lava tubes have helped us to gain an understanding of the possible rapidity of morphological divergence rates under these unusual conditions. Even caves formed by very recent lava flows such as on Kilauea have endemic or incipient species of *Caconemobius* cave crickets.

Dermaptera and Blattodea may be abundant in tropical caves, where they are active in guano deposits. In south-east Asian caves a troglobite earwig is ectoparasitic on roosting bats. Associated with cavernicolous vertebrates there are many more conventional ectoparasites, such as hippoboscid, nycteribid, and streblid flies, fleas, and lice.

9.7 ENVIRONMENTAL MONITORING USING GROUND-DWELLING HEXAPODS

Human activities such as agriculture, forestry, and pastoralism have resulted in the simplification of many terrestrial ecosystems. Attempts to quantify the effects of such practices – for the purposes of conservation assessment, classification of land-types, and monitoring of impacts – have tended to be phytosociological, emphasizing the use of vegetational mapping data. More recently, data on vertebrate distributions and communities have been incorporated into surveys for conservation purposes.

Although arthropod diversity is estimated to be very great (section 1.3), it is rare for data derived from this group to be available routinely in conservation and monitoring. There are several reasons for this neglect. Firstly, when "flagship" species elicit public reaction to a conservation issue, such as loss of a particular habitat, these organisms are predominantly furry mammals, such as pandas and koalas, or birds; rarely are they insects. Excepting perhaps some butterflies, insects often lack the necessary charisma in the public perception.

Secondly, insects generally are difficult to sample in a comparable manner within and between sites. Abundance and diversity fluctuate on a relatively short time-scale, in response to factors that may be little understood. In contrast, vegetation often shows less temporal variation; and with knowledge of mammal seasonality and of the migration habits of birds, the seasonal variations of vertebrate populations can be taken into account.

Thirdly, arthropods often are more difficult to identify accurately, because of the numbers of taxa and some deficiencies in taxonomic knowledge (alluded to for insects in Chapter 8 and discussed more fully in Chapter 17). Whereas competent mammalogists, ornithologists, or field botanists might expect to identify to species level, respectively, all mammals, birds, and plants of a geographically restricted area (outside the tropical rainforests), no entomologist could aspire to do so.

Nonetheless, aquatic biologists routinely sample and identify all macroinvertebrates (mostly insects) in regularly surveyed aquatic ecosystems, for purposes including monitoring of deleterious change in environmental quality (section 10.5). Comparable studies of terrestrial systems, with objectives such as establishment of rationales for conservation and the detection of pollution-induced changes, are undertaken in some countries. The problems outlined above have been addressed in the following ways.

Some charismatic insect species have been highlighted, usually under "endangered-species" legislation designed with vertebrate conservation in mind. These species predominantly have been lepidopterans and much has been learnt of the biology of selected species. However, from the perspective of site classification for conservation purposes, the structure of selected soil and litter communities has greater realized and potential value than any single-species study. Sampling problems are alleviated by using a single collection method, often that of pit-fall trapping, but including the extraction of arthropods from litter samples by a variety of means (see section 17.1.2). Pitfall traps collect mobile terrestrial arthropods by capturing them in containers filled with preserving fluid and sunken level with the substrate. Traps can be aligned along a transect, or dispersed according to a standard quadrat-based sampling regime. According to the sample size required, they can be left *in situ* for several days or for up to a few weeks. Depending on the sites surveyed, arthropod collections may be dominated by Collembola, Formicidae, and Coleoptera, particularly ground

Box 9.2 Non-insect hexapods (Collembola, Protura, and Diplura)

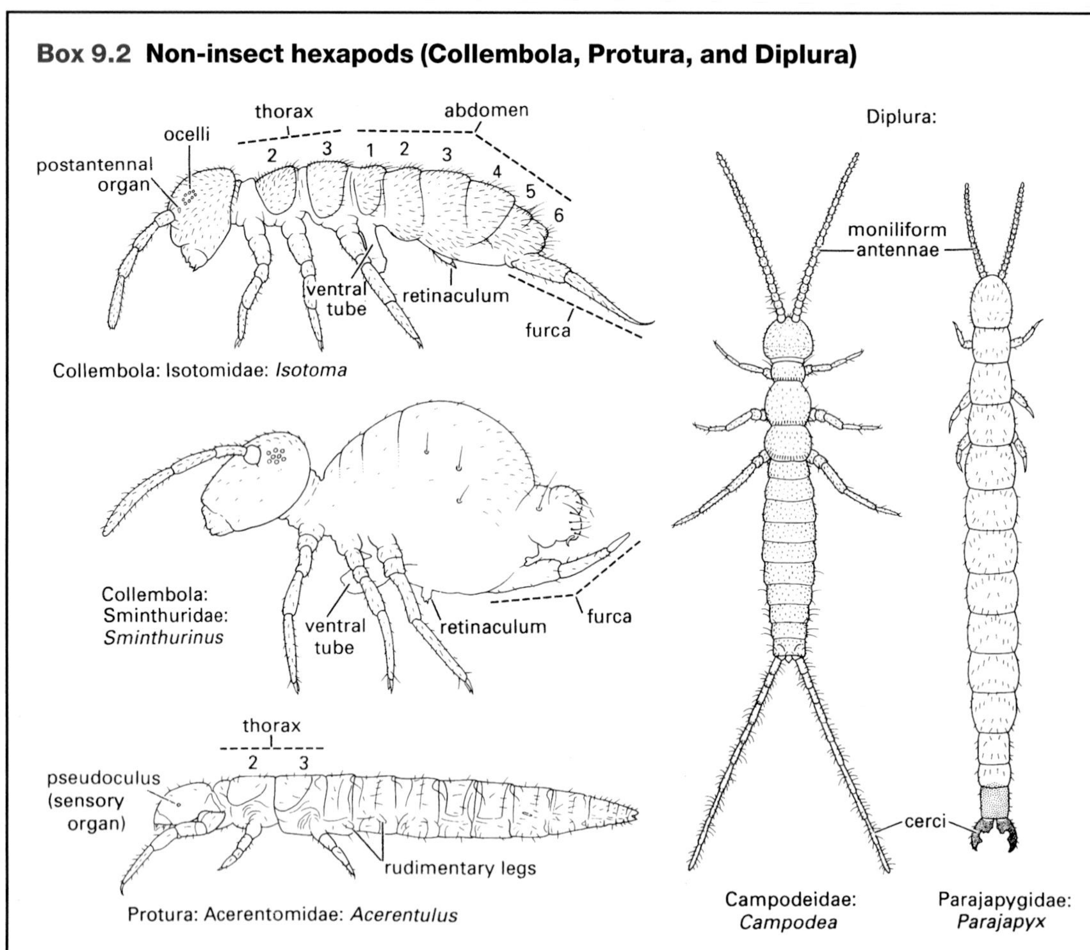

The Collembola, Protura, and Diplura have been united as the "Entognatha", based on a similar mouthpart morphology in which mandibles and maxillae are enclosed in folds of the head, except when everted for feeding. Although the entognathy of Diplura is thought not to be homologous with that of Collembola and Protura (section 7.2), it is convenient to treat these classes together here. All have indirect fertilization – males deposit sperm bundles or stalked spermatophores, which are picked up from the substrate by unattended females. For phylogenetic considerations concerning these three classes, see sections 7.2 and 7.3.

Protura (proturans)

The proturans are non-insect hexapods, with over 600 species in eight families. They are small (<2 mm long) to very small, delicate, elongate, pale to white, with a fusiform body and conically shaped head. The thorax is poorly differentiated from the abdomen. Eyes and antennae are lacking, and the mouthparts are entognathous, consisting of slender mandibles and maxillae, slightly protruding from a pleural fold cavity; maxillary and labial palps are present. The thorax is weakly developed, and bears legs each comprising five segments; the anterior legs are held forward (as shown here for *Acerentulus*, after Nosek 1973), fulfilling an antennal sensory function. The adult abdomen is 12-segmented with the gonopore between segments 11 and 12, and a terminal anus; cerci are absent. Immature development is anamorphic (with segments added posteriorly during development). Proturans are cryptic, found exclusively in soil, moss, and leaf litter. Their biology is little known, but some species are known to feed on mycorrhizal fungi.

Collembola (springtails)

The springtails are treated as non-insect hexapods, but intriguing evidence suggests an alternative, independent origin from Crustacea (see section 7.3). There are about 9000 described species in some 27 families, but the true species diversity may be much higher. Small (usually 2–3 mm, but up to 12 mm) and soft-bodied, their body varies in shape from globular to elongate (as illustrated here for *Isotoma* and *Sminthurinus*, after Fjellberg 1980), and is pale or often characteristically pigmented grey, blue, or black. The eyes and/or ocelli are often poorly developed or absent; the antennae have four to six segments. Behind the antennae usually there is a pair of postantennal organs, which are specialized sensory structures (believed by some to be the remnant apex of the second antenna of crustaceans). The entognathous mouthparts comprise elongate maxillae and mandibles enclosed by pleural folds of the head; maxillary and labial palps are absent. The legs each comprise four segments. The six-segmented abdomen has a sucker-like ventral tube (the collophore), a retaining hook (the retinaculum), and a furca (sometimes called furcula; forked jumping organ, usually three-segmented) on segments 1, 3, and 4, respectively, with the gonopore on segment 5 and the anus on segment 6; cerci are absent. The ventral tube is the main site of water and salt exchange and thus is important to fluid balance, but also can be used as an adhesive organ. The springing organ or furca, formed by fusion of a pair of appendages, is longer in surface-dwelling species than those living within the soil. In general, jump length is correlated with furca length, and some species can spring up to 10 cm. Amongst hexapods, collembolan eggs uniquely are microlecithal (lacking large yolk reserves) and holoblastic (with complete cleavage). The immature instars are similar to the adults, developing epimorphically (with a constant segment number); maturity is attained after five molts, but molting continues for life. Springtails are most abundant in moist soil and litter, where they are major consumers of decaying vegetation, but also they occur in caves, in fungi, as commensals with ants and termites, on still water surfaces, and in the intertidal zone. Most species feed on fungal hyphae or dead plant material, some species eat other small invertebrates, and only a very few species are injurious to living plants. For example, the "lucerne flea" *Sminthurus viridis* (Sminthuridae) damages the tissues of crops such as lucerne and clover and can cause economic injury. Springtails can reach extremely high densities (e.g. 10,000–100,000 individuals m^{-2}) and are ecologically important in adding nutrients to the soil via their feces and in facilitating decomposition processes, for example by stimulating and inhibiting the activities of different microorganisms.

Diplura (diplurans)

The diplurans are non-insect hexapods, with some 1000 species in eight or nine families. They are small to medium sized (2–5 mm, exceptionally up to 50 mm), mostly unpigmented, and weakly sclerotized. They lack eyes, and their antennae are long, moniliform, and multi-segmented. The mouthparts are entognathous, and the mandibles and maxillae are well developed, with their tips visible protruding from the pleural fold cavity; the maxillary and labial palps are reduced. The thorax is little differentiated from the abdomen, and bears legs each comprising five segments. The abdomen is 10-segmented, with some segments having small styles and protrusible vesicles; the gonopore is between segments 8 and 9, and the anus is terminal; the cerci are filiform (as illustrated here for *Campodea*, after Lubbock 1873) to forceps-like (as in *Parajapyx* shown here, after Womersley 1939). Development of the immature forms is epimorphic, with molting continuing through life. Some species are gregarious, and females of certain species tend the eggs and young. Diplurans are generally omnivorous, some feed on live and decayed vegetation, and japygid diplurans are predators.

Box 9.3 Archaeognatha (bristletails) and Zygentoma (Thysanura; silverfish)

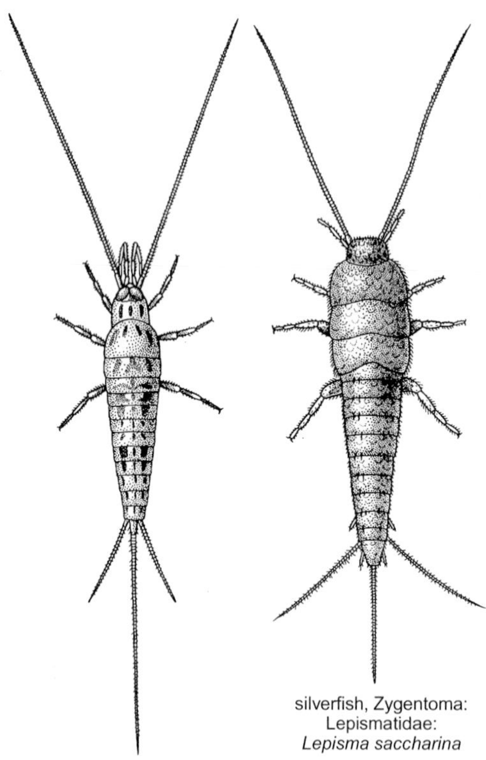

silverfish, Zygentoma: Lepismatidae: *Lepisma saccharina*

bristletail, Archaeognatha: Machilidae: *Petrobius maritima*

The Archaeognatha and Zygentoma represent the surviving remnants of a wider radiation of primitively flightless insects. These two apterygote orders superficially are similar, but differ in pleural structures and quite fundamentally in their mouthpart morphology. The thoracic segments are subequal and unfused, with poorly developed pleura. The abdomen is 11-segmented, with styles and often protrusible vesicles on some segments; it bears a long, multisegmented caudal appendix dorsalis, located mediodorsally on the tergum of segment 11, forming an epiproct extension lying between the paired cerci and dorsal to the genitalia. In females the gonapophyses of segments 8 and 9 form an ovipositor. Fertilization is indirect, by transfer of a spermatophore or sperm droplets. Development is ametabolous and molting continues for life. For phylogenetic considerations see section 7.4.1.

Archaeognatha (bristletails)

The bristletails are primitively wingless insects, with some 500 species in two extant families. They are moderate sized, 6–25 mm long, elongate, and cylindrical. The head is hypognathous, and bears large compound eyes that are in contact dorsally; three ocelli are present; the antennae are multisegmented. The mouthparts are partially retracted into the head, and include elongate, monocondylar (single-articulated) mandibles, and elongate, seven-segmented maxillary palps. The thorax is humped, and the legs have large coxae each bearing a style and the tarsi are two- or three-segmented. The abdomen continues the thoracic contour; segments 2–9 bear ventral muscle-containing styles (representing limbs), whereas segments 1–7 have one or two pairs of protrusible vesicles medial to the styles (fully developed only in mature individuals). The paired multisegmented cerci are shorter than the median caudal appendage (as shown here for *Petrobius maritima*, after Lubbock 1873).

Fertilization is indirect, with sperm droplets attached to silken lines produced from the male gonapophyses, or stalked spermatophores are deposited on the ground, or more rarely sperm are deposited on the female's ovipositor. Bristletails often are active nocturnally, feeding on litter, detritus, algae, lichens and mosses, and sheltering beneath bark or in litter during the day. They can run fast and jump, using the arched thorax and flexed abdomen to spring considerable distances.

Zygentoma (Thysanura; silverfish)

Silverfish are primitively wingless insects, with some 400 species in five extant families. Their bodies are moderately sized (5–30 mm long) and dorsoventrally flattened, often with silvery scales. The head is hypognathous to slightly prognathous; compound eyes are absent or reduced to isolated ommatidia, and there may be one to three ocelli present; the antennae are multisegmented. The mouthparts are mandibulate, and include dicondylar (double-articulated) mandibles, and five-segmented maxillary palps. The legs have large coxae and two- to five-segmented tarsi. The abdomen continues the taper of the thorax, with segments 7–9 at least, but sometimes 2–9, bearing ventral muscle-containing styles; mature individuals may have a pair of protrusible vesicles medial to the styles on segments 2–7, although these are often reduced or absent. The paired elongate multisegmented cerci are nearly as long as the median caudal appendage (as shown here for *Lepisma saccharina*, after Lubbock 1873).

Fertilization is indirect, via flask-shaped spermatophores that females pick up from the substrate. Many silverfish live in litter or under bark; some are subterranean or are cavernicolous, but some species can tolerate low humidity and high temperatures of arid areas; for example, there are desert-living lepismatid silverfish in the sand dunes of the Namib Desert in south-western Africa, where they are important detritivores. Some other zygentoman species live in mammal burrows, a few are commensals in nests of ants and termites, and several species are familiar synanthropic insects, living in human dwellings. These include *L. saccharina*, *Ctenolepisma longicauda* (silverfishes), and *Lepismodes inquilinus* (= *Thermobia domestica*) (the firebrat), which eat materials such as paper, cotton, and plant debris, using their own cellulase to digest the cellulose.

Box 9.4 Grylloblattodea (Grylloblattaria, Notoptera; grylloblattids, ice or rock crawlers)

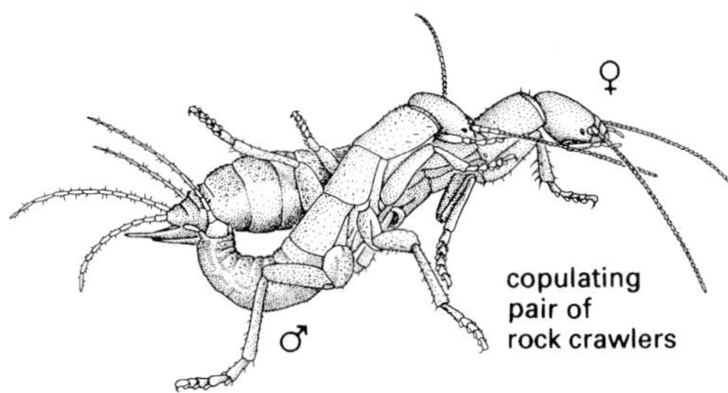

Grylloblattodea comprise a single family, Grylloblattidae, containing some 25 described species, restricted to western North America and central to eastern Asia including Japan. North American species are particularly tolerant of cold and may live at high elevations on glaciers and snow banks; East Asian species may live at sea level in temperate forest. Grylloblattodea are moderately sized insects (20–35 mm long) with an elongate, pale, cylindrical body that is soft and pubescent. The head is prognathous, and the compound eyes are reduced or absent; ocelli are absent. The antennae are multisegmented, and the mouthparts mandibulate. The quadrate prothorax is larger than the meso- or metathorax; wings are absent. The legs are cursorial, with large coxae and five-segmented tarsi. The abdomen has 10 visible segments and the rudiments of segment 11, with five- to nine-segmented cerci. The female has a short ovipositor, and the male genitalia are asymmetrical.

Copulation takes place side-by-side with the male on the right, as illustrated here for a common Japanese species, *Galloisiana nipponensis* (after Ando 1982). Eggs may diapause up to a year in damp wood or soil under stones. Nymphs, which resemble adults, develop slowly through eight instars. North American rock crawlers are active by day and night at low temperatures, feeding on dead arthropods and organic material, notably from the surface of ice and snow in spring snow-melt, within caves (including ice caves), in alpine soil, and damp places such as beneath rocks.

Phylogenetic relations are discussed in section 7.4.2 and depicted in Fig. 7.2.

beetles (Carabidae), Tenebrionidae, Scarabaeidae, and Staphylinidae, with some terrestrial representatives of many other orders.

Taxonomic difficulties often are alleviated by selecting (from amongst the organisms collected) one or more higher taxonomic groups for species-level identification. The carabids are often selected for study because of the diversity of species sampled, the pre-existing ecological knowledge, and availability of taxonomic keys to species level, although these are largely restricted to temperate northern hemisphere taxa.

Studies to date are ambivalent concerning correlates between species diversity (including taxon richness) established from vegetational survey and those from terrestrial insect trapping. Evidence from the well-documented British biota suggests that vegetational diversity does not predict insect diversity. However, a study in more natural, less human-affected environments in southern Norway showed congruence between carabid faunal indices and those obtained by vegetational and bird surveys. Further studies are required into the nature of any relationships between terrestrial insect richness and diversity data obtained by conventional biological survey of selected plants and vertebrates.

Box 9.5 Embiidina or Embioptera (embiids, webspinners)

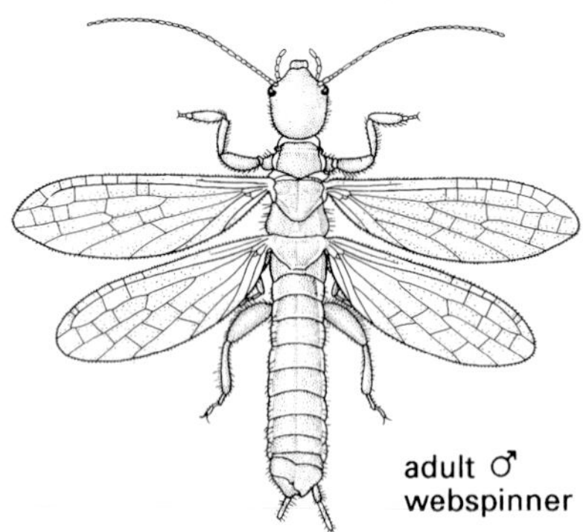

There are some 300 described species of embiids (perhaps up to an order of magnitude more remain undescribed) in at least eight families, worldwide. Small to moderately sized, they have an elongate, cylindrical body, somewhat flattened in males. The head is prognathous, and the compound eyes are reniform (kidney-shaped), larger in males than females; ocelli are absent. The antennae are multisegmented, and the mouthparts are mandibulate. The quadrate prothorax is larger than the meso- or metathorax. All females and some males are apterous, and, if present, the wings (illustrated here for *Embia major*, after Imms 1913) are characteristically soft and flexible, with blood sinus veins stiffened for flight by hemolymph pressure. The legs are short, with three-segmented tarsi; the basal segment of each fore tarsus is swollen and contains silk glands, whereas the hind femora are swollen with strong tibial muscles. The abdomen is 10-segmented, with only the rudiments of segment 11; the cerci comprise two segments and are responsive to tactile stimuli. The female external genitalia are simple, whereas the male genitalia are complex and asymmetrical.

During copulation, the male holds the female with his prognathous mandibles and/or his asymmetrical cerci. The eggs and early nymphal stages are tended by the female parent, and the immature stages resemble the adults except for their wings and genitalia. Embiids live gregariously in silken galleries, spun with the tarsal silk glands (present in all instars); their galleries occur in leaf litter, beneath stones, on rocks, on tree trunks (see Plate 4.1, facing p. 14), or in cracks in bark and soil, often around a central retreat. Their food comprises litter, moss, bark, and dead leaves. The galleries are extended to new food sources, and the safety of the gallery is left only when mature males disperse to new sites, where they mate, do not feed, and sometimes are cannibalized by females (see Plate 4.2). Webspinners readily reverse within their galleries, for example when threatened by a predator.

Phylogenetic relations are discussed in section 7.4.2 and depicted in Fig. 7.2.

Box 9.6 Zoraptera

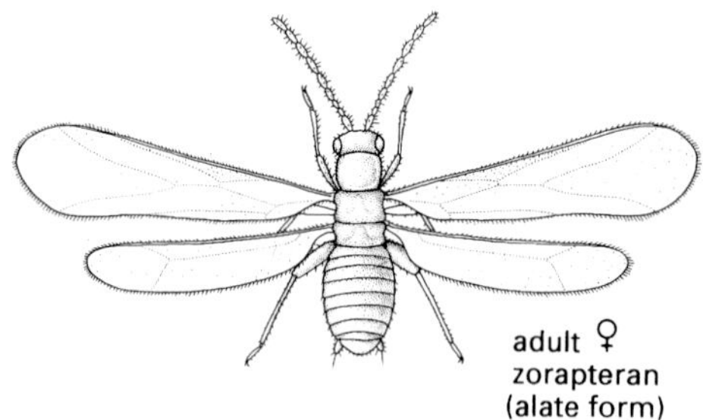

These insects comprise the single genus *Zorotypus*, sometimes subdivided into several genera, containing just over 30 described species found worldwide in tropical and warm temperate regions except Australia. They are small (<4 mm long) and rather termite-like. The head is hypognathous, and compound eyes and ocelli are present in winged species but absent in apterous species. The antennae are moniliform and nine-segmented, and the mouthparts are mandibulate, with five-segmented maxillary palps and three-segmented labial palps. The subquadrate prothorax is larger than the similar-shaped meso- and metathorax. The wings are polymorphic; some forms are apterous in both sexes, whereas other forms are alate, with two pairs of paddle-shaped wings with reduced venation and smaller hind wings (as illustrated here for *Zorotypus hubbardi*, after Caudell 1920). The wings are shed as in ants and termites. The legs have well-developed coxae, expanded hind femora bearing stout ventral spines, and two-segmented tarsi, each with two claws. The 11-segmented abdomen is short and rather swollen, with cerci comprising just a single segment. The male genitalia are asymmetric.

The immature stages are polymorphic according to wing development. Zorapterans are gregarious, occurring in leaf litter, rotting wood, or near termite colonies, eating fungi and perhaps small arthropods. Phylogenetically they are enigmatic, with a probable relationship within the Polyneoptera (see section 7.4.2 and Fig. 7.2).

Box 9.7 Dermaptera (earwigs)

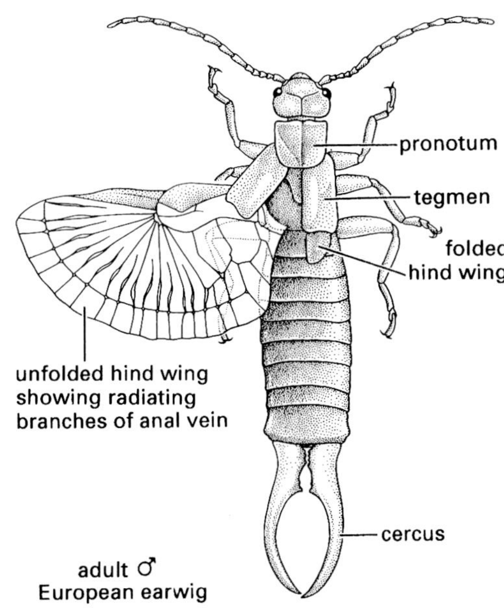

The earwigs comprise an order containing some 2000 species in about 10 families found worldwide. They are hemimetabolous, with small to moderately sized (4–25 mm long) elongate bodies that are dorsoventrally flattened. The head is prognathous; the compound eyes may be large, small, or absent, and ocelli are absent. The antennae are short to moderate length and filiform with segments elongate; there are fewer antennal segments in immature individuals than in the adult. The mouthparts are mandibulate (section 2.3.1; Fig. 2.10). The legs are relatively short, and the tarsi are three-segmented with the second tarsomeres short. The prothorax has a shield-like pronotum, and the meso- and metathoracic sclerites are of variable size. Earwigs are apterous or, if winged, their fore wings are small and leathery, with smooth, unveined tegmina; the hind wings are large, membranous, and semi-circular (as illustrated here for an adult male of the common European earwig, *Forficula auricularia*) and when at rest are folded fan-like and then longitudinally, protruding slightly from beneath the tegmina; hind-wing venation is dominated by the anal fan of branches of A_1 and cross-veins. The abdominal segments are telescoped (terga overlapping), with 10 visible segments in the male and eight in the female, terminating in prominent cerci modified into forceps; the latter are often heavier, larger, and more curved in males than in females.

Copulation is end-to-end, and male spermatophores may be retained in the female for some months prior to fertilization. Oviparous species lay eggs often in a burrow in debris (Fig. 9.1), guard the eggs and lick them to remove fungus. The female may assist the nymphs to hatch from the eggs, and may care for them until the second or third instar, after which she may cannibalize her offspring. Maturity is attained after four or five molts. The two parasitic groups (of uncertain rank), Arixeniina and Hemimerina, exhibit pseudoplacental viviparity (section 5.9).

Earwigs are mostly cursorial and nocturnal, with most species rarely flying. Feeding is predominantly on dead and decaying vegetable and animal matter, with some predation and some damage to living vegetation, especially in gardens. Some are commensals or ectoparasites of bats in south-east Asia (Arixeniina) or semi-parasites of South African rodents (Hemimerina): earwigs in both tribes are blind, apterous, and with rod-like forceps. The forceps of free-living earwigs are used for manipulating prey, for defense and offense, and in some species for grasping the partner during copulation. The common name "earwig" may derive from a supposed predilection for entering ears, or from a corruption of "ear wing" referring to the shape of the wing, but these are unsupported.

Phylogenetic relations are discussed in section 7.4.2 and depicted in Fig. 7.2.

Box 9.8 Blattodea (Blattaria; cockroaches, roaches)

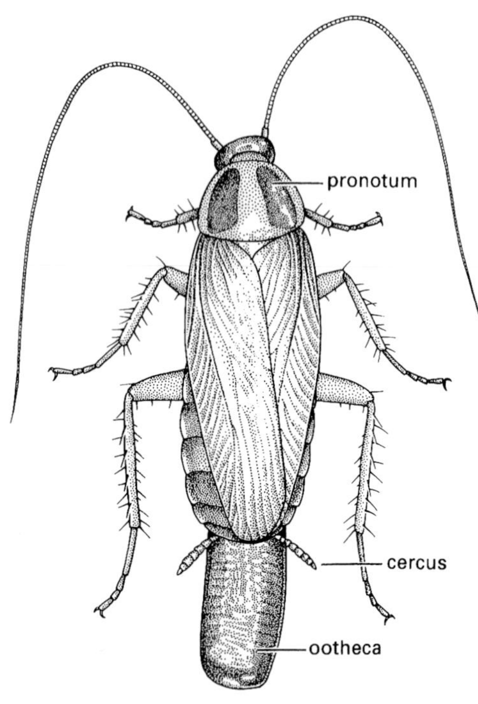

adult ♀ cockroach

The cockroaches make up an order of over 3500 species in at least eight families worldwide. They are hemimetabolous, with small to large (<3 mm to >100 mm), dorsoventrally flattened bodies. The head is hypognathous, and the compound eyes may be moderately large to small, or absent in cavernicolous species; ocelli are represented by two pale spots. The antennae are filiform and multisegmented, and the mouthparts are mandibulate. The prothorax has an enlarged, shield-like pronotum, often covering the head; the meso- and metathorax are rectangular and subequal. The fore wings (Fig. 2.22c) are sclerotized as tegmina, protecting the membranous hind wings; each tegmen lacks an anal lobe, and is dominated by branches of veins R and CuA. In contrast, the hind wings have a large anal lobe, with many branches in the R, CuA, and anal sectors; at rest they lie folded fan-like beneath the tegmina. Wing reduction is frequent. The legs are often spinose (Fig. 2.19) and have five-segmented tarsi. The large coxae abut each other and dominate the ventral thorax. The abdomen has 10 visible segments, with the subgenital plate (sternum 9) often bearing one or a pair of styles in the male, and concealing segment 11 that is represented only by paired paraprocts. The cerci comprise from one to usually many segments. The male genitalia are asymmetrical, and the female's ovipositor valves are concealed inside a genital atrium.

Mating may involve stridulatory courtship, both sexes may produce sex pheromones, and the female may mount the male prior to end-to-end copulation. Eggs generally are laid in a purse-shaped ootheca comprising two parallel rows of eggs with a leathery enclosure (section 5.8), which may be carried externally by the female (as illustrated here for a female of *Blatella germanica*, after Cornwell 1968). Certain species demonstrate a range of forms of ovoviviparity in which a variably reduced ootheca is retained within the reproductive tract in a "uterus" (or brood sac) during embryogenesis, often until nymphal hatching; true viviparity is rare. Parthenogenesis occurs in a few species. Nymphs develop slowly, resembling small apterous adults.

Cockroaches are amongst the most familiar insects, owing to the widespread human-associated habits of some 30 species, including *Periplaneta americana* (the American cockroach), *B. germanica* (the German cockroach), and *B. orientalis* (the Oriental cockroach). These nocturnal, malodorous, disease-carrying, refuge-seeking, peridomestic roaches are unrepresentative of the wider diversity. Typically, cockroaches are tropical, either nocturnal or diurnal, and sometimes arboreal, with some cavernicolous species. Cockroaches include solitary and gregarious species; *Cryptocercus* (the woodroach) lives in family groups. Cockroaches mostly are saprophagous scavengers, but some eat wood and use enteric amoebae to break it down. *Cryptocercus*, uniquely in Blattodea, utilizes flagellate internal protists to digest cellulose.

Phylogenetic relations are discussed in section 7.4.2 and depicted in Figs. 7.2 and 7.4. Evidence presented in earlier editions of this textbook suggested that *Cryptocercus* convergently acquired its termite-like features, such as sociality and digestion of cellulose, via protists. Now this similarity appears to reflect actual relationships, with Isoptera having arisen from within Blattodea (Fig. 7.4). Although this renders the latter paraphyletic, we continue to use "Blattodea" until these relationships are confirmed.

FURTHER READING

Blossey, B. & Hunt-Joshi, T.R. (2003) Belowground herbivory by insects: influence on plants and aboveground herbivores. *Annual Review of Entomology* **48**, 521–47.

Dindal, D.L. (ed.) (1990) *Soil Biology Guide*. John Wiley & Sons, Chichester.

Edgerly, J.S. (1997) Life beneath silk walls: a review of the primitively social Embiidina. In: *The Evolution of Social Behaviour in Insects and Arachnids* (eds. J.C. Choe & B.J. Crespi), pp. 14–25. Cambridge University Press, Cambridge.

Eisenbeis, G. & Wichard, W. (1987) *Atlas on the Biology of Soil Arthropods*, 2nd edn. Springer-Verlag, Berlin.

Hopkin, S.P. (1997) *Biology of Springtails*. Oxford University Press, Oxford.

Lövei, G.L. & Sunderland, K.D. (1996) Ecology and behaviour of ground beetles (Coleoptera: Carabidae). *Annual Review of Entomology* **41**, 231–56.

Lussenhop, J. (1992) Mechanisms of microarthropod–microbial interactions in soil. *Advances in Ecological Research* **23**, 1–33.

McGeoch, M.A. (1998) The selection, testing and application of terrestrial insects as bioindicators. *Biological Reviews* **73**, 181–201.

Mueller, U.G., Rehner, S.A. & Schultz, T.R. (1998) The evolution of agriculture in ants. *Science* **281**, 203–9.

New, T.R. (1998) *Invertebrate Surveys for Conservation*. Oxford University Press, Oxford.

North, R.D., Jackson, C.W. & Howse, P.E. (1997) Evolutionary aspects of ant–fungus interactions in leaf-cutting ants. *Trends in Ecology and Evolution* **12**, 386–9.

Paine, T.D., Raffia, K.F. & Harrington, T.C. (1997) Interactions among scolytid bark beetles, their associated fungi, and live host conifers. *Annual Review of Entomology* **42**, 179–206.

Resh, V.H. & Cardé, R.T. (eds.) (2003) *Encyclopedia of Insects*. Academic Press, Amsterdam. [Particularly see articles on cave insects; soil habitats.]

Stork, N.E. (ed.) (1990) *The Role of Ground Beetles in Ecological and Environmental Studies*. Intercept, Andover.

Villani, M.G. & Wright, R.J. (1990) Environmental influences on soil macroarthropod behavior in agricultural systems. *Annual Review of Entomology* **35**, 249–69.

Chapter 10

AQUATIC INSECTS

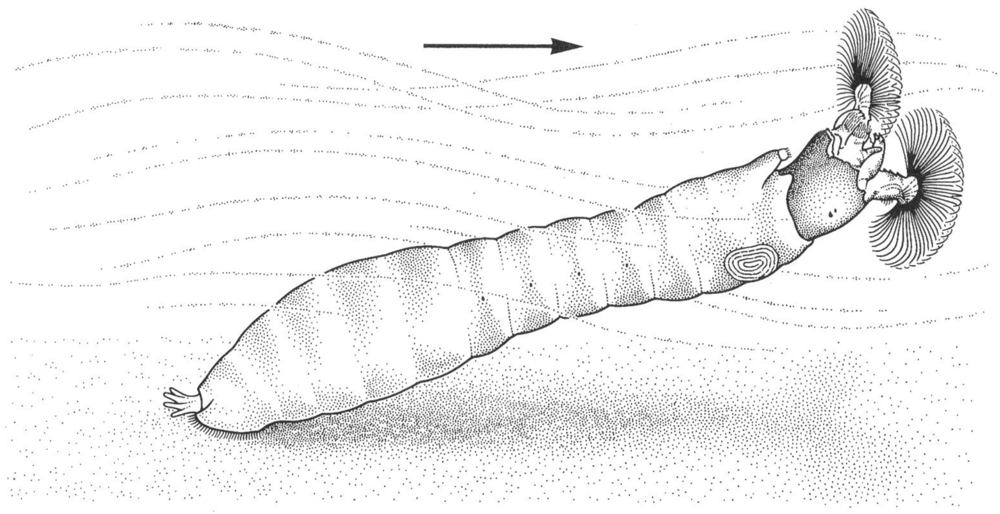

A black-fly larva in the typical filter-feeding posture. (After Currie 1986.)

Every inland waterbody, whether a river, stream, seepage, or lake, supports a biological community. The most familiar components often are the vertebrates, such as fish and amphibians. However, at least at the macroscopic level, invertebrates provide the highest number of individuals and species, and the highest levels of biomass and production. In general, the insects dominate freshwater aquatic systems, where only nematodes can approach the insects in terms of species numbers, biomass, and productivity. Crustaceans may be abundant, but are rarely diverse in species, in saline (especially temporary) inland waters. Some representatives of nearly all orders of insects live in water, and there have been many invasions of freshwater from the land. Insects have been almost completely unsuccessful in marine environments, with a few sporadic exceptions such as some water-striders (Hemiptera: Gerridae) and larval dipterans.

This chapter surveys the successful insects of aquatic environments and considers the variety of mechanisms they use to obtain scarce oxygen from the water. Some of their morphological and behavioral modifications to life in water are described, including how they resist water movement, and a classification based on feeding groups is presented. The use of aquatic insects in biological monitoring of water quality is reviewed and the few insects of the marine and intertidal zones are discussed. Taxonomic boxes summarize information on mayflies (Ephemeroptera), dragonflies and damselflies (Odonata), stoneflies (Plecoptera), caddisflies (Trichoptera), and other orders of importance in aquatic ecosystems.

10.1 TAXONOMIC DISTRIBUTION AND TERMINOLOGY

The orders of insects that are almost exclusively aquatic in their immature stages are the Ephemeroptera (mayflies; Box 10.1), Odonata (damselflies and dragonflies; Box 10.2), Plecoptera (stoneflies; Box 10.3), and Trichoptera (caddisflies; Box 10.4). Amongst the major insect orders, Diptera (Box 10.5) have many aquatic representatives in the immature stages, and a substantial number of Hemiptera and Coleoptera have at least some aquatic stages (Box 10.6), and in the less speciose minor orders two families of Megaloptera and some Neuroptera develop in freshwater (Box 10.6). Some Hymenoptera parasitize aquatic prey but these, together with certain collembolans, orthopteroids, and other predominantly terrestrial frequenters of damp places, are considered no further in this chapter.

Aquatic entomologists often (correctly) restrict use of the term **larva** to the immature (i.e. postembryonic and prepupal) stages of holometabolous insects; **nymph** (or **naiad**) is used for the pre-adult hemimetabolous insects, in which the wings develop externally. However, for the odonates, the terms larva, nymph, and naiad have been used interchangeably, perhaps because the sluggish, non-feeding, internally reorganizing, final-instar odonate has been likened to the pupal stage of a holometabolous insect. Although the term "larva" is being used increasingly for the immature stages of all aquatic insects, we accept new ideas on the evolution of metamorphosis (section 8.5) and therefore use the terms larva and nymphs in their strict sense, including for immature odonates.

Some aquatic adult insects, including notonectid bugs and dytiscid beetles, can use atmospheric oxygen when submerged. Other adult insects are fully aquatic, such as several naucorid bugs and hydrophilid and elmid beetles, and can remain submerged for extended periods and obtain respiratory oxygen from the water. However, by far the greatest proportion of the adults of aquatic insects are aerial, and it is only their nymphal or larval (and often pupal) stages that live permanently below the water surface, where oxygen must be obtained whilst out of direct contact with the atmosphere. The ecological division of life history allows the exploitation of two different habitats, although there are a few insects that remain aquatic throughout their lives. Exceptionally, *Helichus*, a genus of dryopid beetles, has terrestrial larvae and aquatic adults.

10.2 THE EVOLUTION OF AQUATIC LIFESTYLES

Hypotheses concerning the origin of wings in insects (section 8.4) have different implications regarding the evolution of aquatic lifestyles. The paranotal theory suggests that the "wings" originated in adults of a terrestrial insect for which immature stages may have been aquatic or terrestrial. Some proponents of the preferred exite–endite theory speculate that the progenitor of the pterygotes had aquatic immature stages. Support for the latter hypothesis appears to come from the fact that the two extant basal groups of Pterygota (mayflies and odonates) are aquatic, in contrast to the terrestrial apterygotes; but the aquatic habits of Ephemeroptera and Odonata cannot have been primary, as the tracheal system indicates a preceding terrestrial stage (section 8.3).

Whatever the origins of the aquatic mode of life, all proposed phylogenies of the insects demonstrate that it must have been adopted, adopted and lost, and readopted in several lineages, through geological time. The multiple independent adoptions of aquatic lifestyles are particularly evident in the Coleoptera and Diptera, with aquatic taxa distributed amongst many families across each of these orders. In contrast, all species of Ephemeroptera and Plecoptera are aquatic, and in the Odonata, the only exceptions to an almost universal aquatic lifestyle are the terrestrial nymphs of a few species.

Movement from land to water causes physiological problems, the most important of which is the requirement for oxygen. The following section considers the physical properties of oxygen in air and water, and the mechanisms by which aquatic insects obtain an adequate supply.

10.3 AQUATIC INSECTS AND THEIR OXYGEN SUPPLIES

10.3.1 The physical properties of oxygen

Oxygen comprises 200,000 ppm (parts per million) of air, but in aqueous solution its concentration is only about 15 ppm in saturated cool water. Energy at the cellular level can be provided by anaerobic respiration but it is inefficient, providing 19 times less energy per unit of substrate respired than aerobic respiration. Although insects such as bloodworms (certain chironomid midge larvae) survive extended periods of almost anoxic conditions, most aquatic insects must obtain oxygen from their surroundings in order to function effectively.

The proportions of gases dissolved in water vary according to their solubilities: the amount is inversely proportional to temperature and salinity, and proportional to pressure, decreasing with elevation. In **lentic** (standing) waters, diffusion through water is very slow; it would take years for oxygen to diffuse several meters from the surface in still water. This slow rate, combined with the oxygen demand from microbial breakdown of submerged organic matter, can totally deplete the oxygen on the bottom (**benthic anoxia**). However, the oxygenation of surface waters by diffusion is enhanced by turbulence, which increases the surface area, forces aeration, and mixes the water. If this turbulent mixing is prevented, such as in a deep lake with a small surface area or one with extensive sheltering vegetation or under extended ice cover, anoxia can be prolonged or permanent. Living under these circumstances, benthic insects must tolerate wide annual and seasonal fluctuations in oxygen availability.

Oxygen levels in **lotic** (flowing) conditions can reach 15 ppm, especially in cold water. Equilibrium concentrations may be exceeded if photosynthesis generates locally abundant oxygen, such as in macrophyte- and algal-rich pools in sunlight. However, when this vegetation respires at night oxygen is consumed, leading to a decline in dissolved oxygen. Aquatic insects must cope with a diurnal range of oxygen tensions.

10.3.2 Gaseous exchange in aquatic insects

The gaseous exchange systems of insects depend upon oxygen diffusion, which is rapid through the air, slow through water, and even slower across the cuticle. Eggs of aquatic insects absorb oxygen from water with the assistance of a chorion (section 5.8). Large eggs may have the respiratory surface expanded by elaborated horns or crowns, as in water-scorpions (Hemiptera: Nepidae). Oxygen uptake by the large eggs of giant water bugs (Hemiptera: Belostomatidae) is assisted by unusual male parental tending of the eggs (Box 5.5).

Although insect cuticle is very impermeable, gas diffusion across the body surface may suffice for the smallest aquatic insects, such as some early-instar larvae or all instars of some dipteran larvae. Larger aquatic insects, with respiratory demands equivalent to spiraculate air-breathers, require either augmentation of gas-exchange areas or some other means of obtaining increased oxygen, because the reduced surface area to volume ratio precludes dependence upon cutaneous gas exchange.

Aquatic insects show several mechanisms to cope with the much lower oxygen levels in aqueous solutions. Aquatic insects may have open tracheal systems with spiracles, as do their air-breathing relatives. These may be either polypneustic (8–10 spiracles opening on the body surface) or oligopneustic (one or two pairs of open, often terminal spiracles), or closed and lacking direct external connection (section 3.5, Fig. 3.11).

10.3.3 Oxygen uptake with a closed tracheal system

Simple cutaneous gaseous exchange in a closed tracheal system suffices for only the smallest aquatic insects, such as early-instar caddisflies (Trichoptera). For larger insects, although cutaneous exchange can account for a substantial part of oxygen uptake, other mechanisms are needed.

A prevalent means of increasing surface area for gaseous exchange is by **gills** – tracheated cuticular lamellar extensions from the body. These are usually abdominal (ventral, lateral, or dorsal) or caudal, but may be located on the mentum, maxillae, neck, at the base of the legs, around the anus in some Plecoptera (Fig. 10.1), or even within the rectum, as in dragonfly nymphs. Tracheal gills are found in the immature stages of Odonata, Plecoptera, Trichoptera, aquatic Megaloptera and Neuroptera, some aquatic Coleoptera, a few Diptera and pyralid lepidopterans, and probably reach their greatest morphological diversity in the Ephemeroptera.

In interpreting these structures as gills, it is important to demonstrate that they do function in oxygen uptake. In experiments with nymphs of *Lestes* (Odonata: Lestidae), the huge caudal gill-like lamellae of some individuals were removed by being broken at the site of natural autotomy. Both gilled and ungilled individuals were subjected to low-oxygen environments in closed-bottle respirometry, and survivorship was assessed. The three caudal lamellae of this odonate met all criteria for gills, namely:

- large surface area;
- moist and vascular;
- able to be ventilated;
- responsible normally for 20–30% of oxygen uptake.

However, as temperature rose and dissolved oxygen fell, the gills accounted for increased oxygen uptake, until the maximum uptake reached 70%. At this high level, the proportion equaled the proportion of gill surface to total body surface area. At low temperatures (<12°C) and with dissolved oxygen at the environmental maximum of 9 ppm, the gills of the lestid accounted for very little oxygen uptake; cuticular uptake was presumed to be dominant. When *Siphlonurus* mayfly nymphs were tested similarly, at 12–13°C the gills accounted for 67% of oxygen uptake, which was proportional to their fraction of the total surface area of the body.

Dissolved oxygen can be extracted using respiratory pigments. These pigments are almost universal in vertebrates but also are found in some invertebrates and even in plants and protists. Amongst the aquatic insects, some larval chironomids (bloodworms) and a few notonectid bugs possess hemoglobins. These

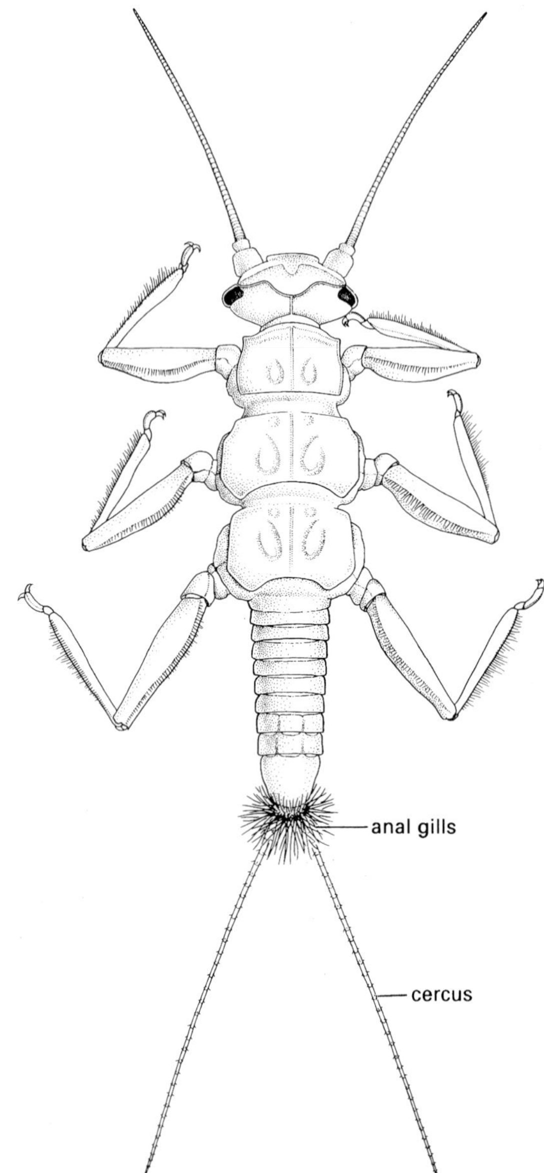

Fig. 10.1 A stonefly nymph (Plecoptera: Gripopterygidae) showing filamentous anal gills.

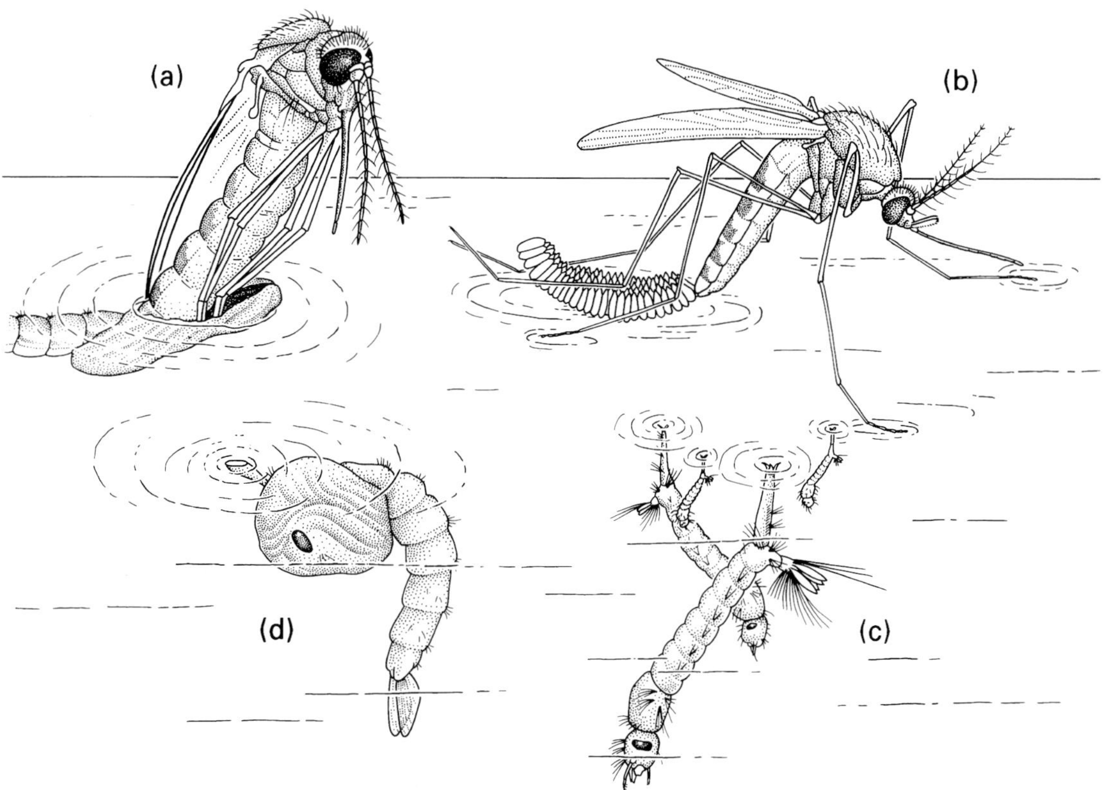

Fig. 10.2 The life cycle of the mosquito *Culex pipiens* (Diptera: Culicidae): (a) adult emerging from its pupal exuviae at the water surface; (b) adult female ovipositing, with her eggs adhering together as a floating raft; (c) larvae obtaining oxygen at the water surface via their siphons; (d) pupa suspended from the water meniscus, with its respiratory horn in contact with the atmosphere. (After Clements 1992.)

molecules are homologous (same derivation) to the hemoglobin of vertebrates such as ourselves. The hemoglobins of vertebrates have a low affinity for oxygen; i.e. oxygen is obtained from a high-oxygen aerial environment and unloaded in muscles in an acid (carbonic acid from dissolved carbon dioxide) environment – the Bohr effect. Where environmental oxygen concentrations are consistently low, as in the virtually anoxic and often acidic sediments of lakes, the Bohr effect would be counterproductive. In contrast to vertebrates, chironomid hemoglobins have a high affinity for oxygen. Chironomid midge larvae can saturate their hemoglobins through undulating their bodies within their silken tubes or substrate burrows to permit the minimally oxygenated water to flow over the cuticle. Oxygen is unloaded when the undulations stop, or when recovery from anaerobic respiration is needed. The respiratory pigments allow a much more rapid oxygen release than is available by diffusion alone.

10.3.4 Oxygen uptake with an open spiracular system

For aquatic insects with open spiracular systems, there is a range of possibilities for obtaining oxygen. Many immature stages of Diptera can obtain atmospheric oxygen by suspending themselves from the water meniscus, in the manner of a mosquito larva and pupa (Fig. 10.2). There are direct connections between the atmosphere and the spiracles in the terminal respiratory siphon of the larva, and in the thoracic respiratory organ of the pupa. Any insect that uses atmospheric oxygen is independent of low dissolved oxygen levels, such as occur in rank or stagnant waters. This independence from dissolved oxygen is particularly prevalent amongst larvae of flies, such as ephydrids, one species of which can live in oil-tar ponds, and certain pollution-tolerant hover flies (Syrphidae), the "rat-tailed maggots".

Several other larval Diptera and psephenid beetles have cuticular modifications surrounding the spiracular openings, which function as gills, to allow an increase in the extraction rate of dissolved oxygen without spiracular contact with the atmosphere. An unusual source of oxygen is the air stored in roots and stems of aquatic macrophytes. Aquatic insects including the immature stages of some mosquitoes, hover flies, and *Donacia*, a genus of chrysomelid beetles, can use this source. In *Mansonia* mosquitoes, the spiracle-bearing larval respiratory siphon and pupal thoracic respiratory organ both are modified for piercing plants.

Temporary air stores (**compressible gills**) are common means of storing and extracting oxygen. Many adult dytiscid, gyrinid, helodid, hydraenid, and hydrophilid beetles, and both nymphs and adults of many belostomatid, corixid, naucorid, and pleid hemipterans use this method of enhancing gaseous exchange. The gill is a bubble of stored air, in contact with the spiracles by various means, including subelytral retention in adephagan water beetles (Fig. 10.3), and fringes of specialized hydrofuge hairs on the body and legs, as in some polyphagan water beetles. When the insect dives from the surface, air is trapped in a bubble in which all gases start at atmospheric equilibrium. As the submerged insect respires, oxygen is used up and the carbon dioxide produced is lost due to its high solubility in water. Within the bubble, as the partial pressure of oxygen drops, more diffuses in from solution in water but not rapidly enough to prevent continued depletion in the bubble. Meanwhile, as the proportion of nitrogen in the bubble increases, it diffuses outwards, causing diminution in the size of the bubble. This contraction in size gives rise to the term "compressible gill". When the bubble has become too small, it is replenished by the insect returning to the surface.

The longevity of the bubble depends upon the relative rates of consumption of oxygen and of gaseous diffusion between the bubble and the surrounding water. A maximum of eight times more oxygen can be supplied from the compressible gill than was in the original bubble. However, the available oxygen varies according to the amount of exposed surface area of the bubble and the prevailing water temperature. At low temperatures the metabolic rate is lower, more gases remain dissolved in water, and the gill is long lasting. Conversely, at higher temperatures metabolism is higher, less gas is dissolved, and the gill is less effective.

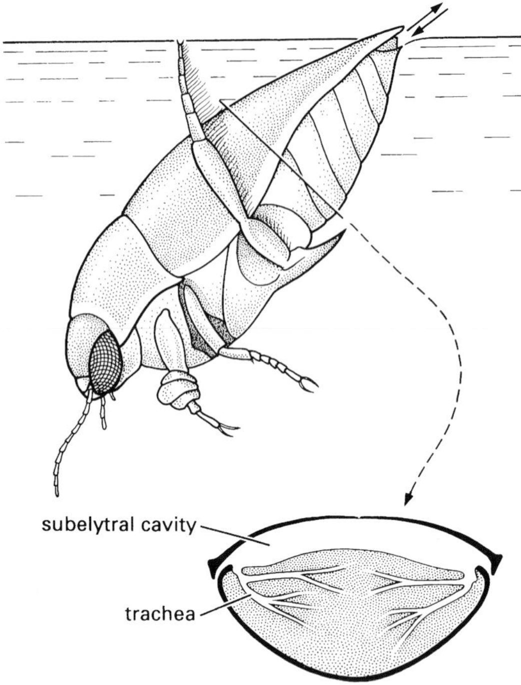

Fig. 10.3 A male water beetle of *Dytiscus* (Coleoptera: Dytiscidae) replenishing its store of air at the water surface. Below is a transverse section of the beetle's abdomen showing the large air store below the elytra and the tracheae opening into this air space. Note: the tarsi of the fore legs are dilated to form adhesive pads that are used to hold the female during copulation. (After Wigglesworth 1964.)

A further modification of the air-bubble gill, the **plastron**, allows some insects to use permanent air stores, termed an "incompressible gill". Water is held away from the body surface by hydrofuge hairs or a cuticular mesh, leaving a permanent gas layer in contact with the spiracles. Most of the gas is relatively insoluble nitrogen but, in response to metabolic use of oxygen, a gradient is set up and oxygen diffuses from water into the plastron. Most insects with such a gill are relatively sedentary, as the gill is not very effective in responding to high oxygen demand. Adults of some curculionid, dryopid, elmid, hydraenid, and hydrophilid beetles, nymphs and adults of naucorid bugs, and pyralid moth larvae use this mode of oxygen extraction.

10.3.5 Behavioral ventilation

A consequence of the slow diffusion rate of oxygen through water is the development of an oxygen-depleted layer of water that surrounds the gaseous uptake surface, whether it be the cuticle, gill, or spiracle. Aquatic insects exhibit a variety of ventilation behaviors that disrupt this oxygen-depleted layer. Cuticular gaseous diffusers undulate their bodies in tubes (Chironomidae), cases (young caddisfly nymphs), or under shelters (young lepidopteran larvae) to produce fresh currents across the body. This behavior continues even in later-instar caddisflies and lepidopterans in which gills are developed. Many ungilled aquatic insects select their positions in the water to allow maximum aeration by current flow. Some dipterans, such as blephariceriid (Fig. 10.4) and deuterophlebiid larvae, are found only in torrents; ungilled simuliids, plecopterans, and case-less caddisfly larvae are found commonly in high-flow areas. The very few sedentary aquatic insects with gills, notably black-fly (simuliid) pupae, some adult dryopid beetles, and the immature stages of a few lepidopterans, maintain local high oxygenation by positioning themselves in areas of well-oxygenated flow. For mobile insects, swimming actions, such as leg movements, prevent the formation of a low-oxygen boundary layer.

Although most gilled insects use natural water flow to bring oxygenated water to them, they may also undulate their bodies, beat their gills, or pump water in and out of the rectum, as in anisopteran nymphs. In lestid zygopteran nymphs (for which gill function is discussed in section 10.3.3), ventilation is assisted by "pull-downs" (or "push-ups") that effectively move oxygen-reduced water away from the gills. When dissolved oxygen is reduced through a rise in temperature, *Siphlonurus* nymphs elevate the frequency and increase the percentage of time spent beating gills.

10.4 THE AQUATIC ENVIRONMENT

The two different aquatic physical environments, the lotic (flowing) and lentic (standing), place very different constraints on the organisms living therein. In the following sections, we highlight these conditions and discuss some of the morphological and behavioral modifications of aquatic insects.

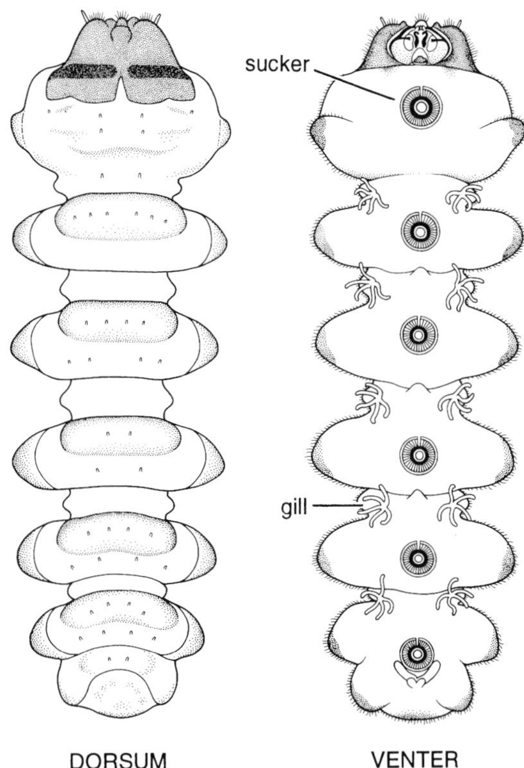

Fig. 10.4 Dorsal (left) and ventral (right) views of the larva of *Edwardsina polymorpha* (Diptera: Blephariceridae); the venter has suckers which the larva uses to adhere to rock surfaces in fast-flowing water.

10.4.1 Lotic adaptations

In lotic systems, the velocity of flowing water influences:

- substrate type, with boulders deposited in fast-flow and fine sediments in slow-flow areas;
- transport of particles, either as a food source for filter-feeders or, during peak flows, as scouring agents;
- maintenance of high levels of dissolved oxygen.

A stream or river contains heterogeneous microhabitats, with riffles (shallower, stony, fast-flowing sections) interspersed with deeper natural pools. Areas of erosion of the banks alternate with areas where

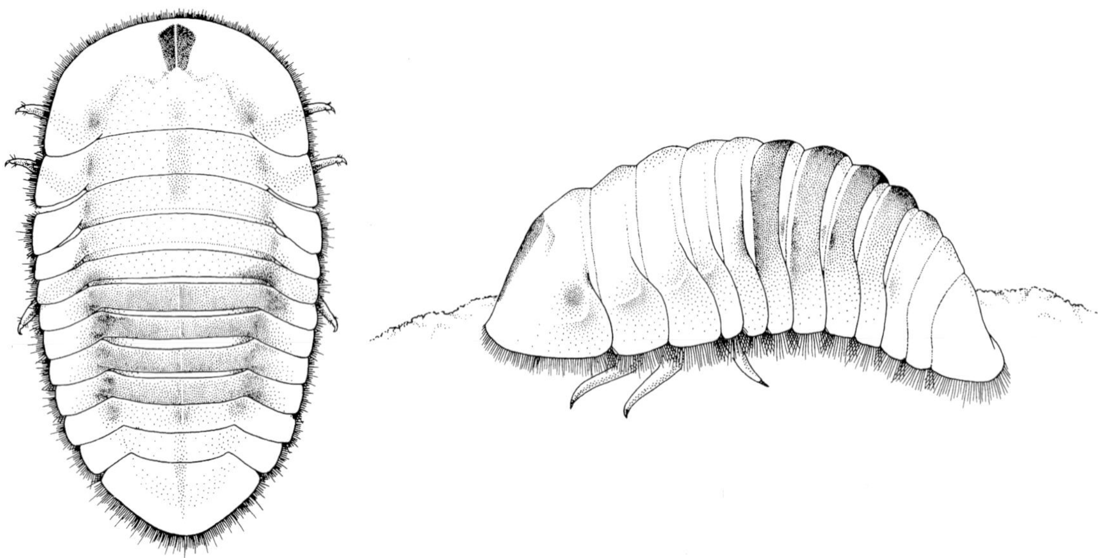

Fig. 10.5 Dorsal and lateral views of the larva of a species of water penny (Coleoptera: Psephenidae).

sediments are deposited, and there may be areas of unstable, shifting sandy substrates. The banks may have trees (a vegetated **riparian** zone) or be unstable, with mobile deposits that change with every flood. Typically, where there is riparian vegetation, there will be local accumulations of drifted **allochthonous** (external to the stream) material such as leaf packs and wood. In parts of the world where extensive pristine, forested catchments remain, the courses of streams often are periodically blocked by naturally fallen trees. Where the stream is open to light, and nutrient levels allow, **autochthonous** (produced within the stream) growth of plants and macroalgae (**macrophytes**) will occur. Aquatic flowering plants may be abundant, especially in chalk streams.

Characteristic insect faunas inhabit these various substrates, many with particular morphological modifications. Thus, those that live in strong currents (**rheophilic** species) tend to be dorsoventrally flattened (Fig. 10.5), sometimes with laterally projecting legs. This is not strictly an adaptation to strong currents, as such modification is found in many aquatic insects. Nevertheless, the shape and behavior minimizes or avoids exposure by allowing the insect to remain within a boundary layer of still water close to the surface of the substrate. However, the fine-scale hydraulic flow of natural waters is much more complex than once believed, and the relationship between body shape, streamlining, and current velocity is not simple.

The cases constructed by many rheophilic caddisflies assist in streamlining or otherwise modifying the effects of flow. The variety of shapes of the cases (Fig. 10.6) must act as ballast against displacement. Several aquatic larvae have suckers (Fig. 10.4) that allow the insect to stick to quite smooth exposed surfaces, such as rock-faces on waterfalls and cascades. Silk is widely produced, allowing maintenance of position in fast flow. Black-fly larvae (Simuliidae) (see the vignette to this chapter) attach their posterior claws to a silken pad that they spin on a rock surface. Others, including hydropsychid caddisflies (Fig. 10.7) and many chironomid midges, use silk in constructing retreats. Some spin silken mesh nets to trap food brought into proximity by the stream flow.

Many lotic insects are smaller than their counterparts in standing waters. Their size, together with flexible body design, allows them to live amongst the cracks and crevices of boulders, stones, and pebbles in the bed (**benthos**) of the stream, or even in unstable, sandy substrates. Another means of avoiding the current is to live in accumulations of leaves (leaf packs) or to mine in immersed wood – substrates that are used by many

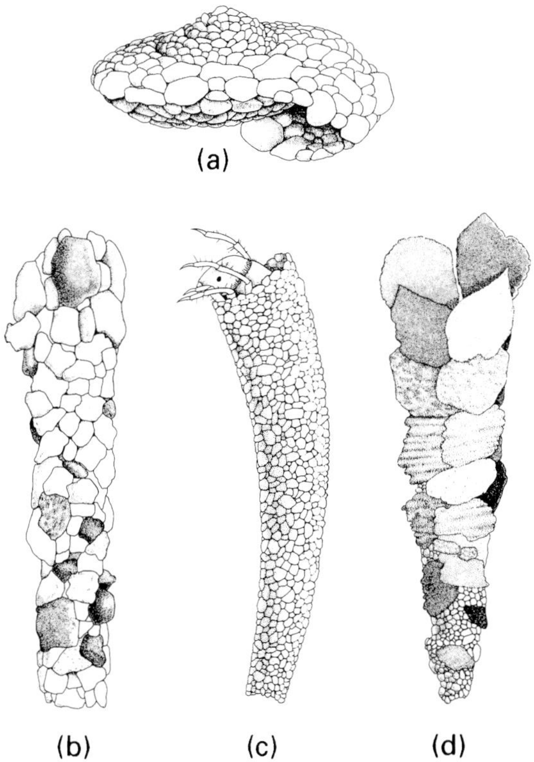

Fig. 10.6 Portable larval cases of representative families of caddisflies (Trichoptera): (a) Helicopsychidae; (b) Philorheithridae; (c) and (d) Leptoceridae.

beetles and specialist dipterans, such as crane-fly larvae (Diptera: Tipulidae).

Two behavioral strategies are more evident in running waters than elsewhere. The first is the strategic use of the current to allow **drift** from an unsuitable location, with the possibility of finding a more suitable patch. Predatory aquatic insects frequently drift to locate aggregations of prey. Many other insects, such as stoneflies and mayflies, notably *Baetis* (Ephemeroptera: Baetidae), may show a diurnal periodic pattern of drift. "Catastrophic" drift is a behavioral response to physical disturbance, such as pollution or severe flow episodes. An alternative response, of burrowing deep into the substrate (the **hyporheic** zone), is a second particularly lotic behavior. In the hyporheic zone, the vagaries of flow regime, temperature, and perhaps predation can be avoided, although food and oxygen availability may be diminished.

10.4.2 Lentic adaptations

With the exception of wave action at the shore of larger bodies of water, the effects of water movement cause little or no difficulty for aquatic insects that live in lentic environments. However, oxygen availability is more of a problem and lentic taxa show a greater variety of mechanisms for enhanced oxygen uptake compared with lotic insects.

The lentic water surface is used by many more species (the **neustic** community of **semi-aquatic** insects) than the lotic surface, because the physical properties of surface tension in standing water that can support an insect are disrupted in turbulent flowing water. Water-striders (Hemiptera: Gerromorpha: Gerridae, Veliidae) are amongst the most familiar neustic insects that exploit the surface film (Box 10.6). They use hydrofuge (water-repellent) hair piles on the legs and venter to avoid breaking the film. Water-striders move with a rowing motion and they locate prey items (and in some species, mates) by detecting vibratory ripples on the water surface. Certain staphylinid beetles use chemical means to move around the meniscus, by discharging from the anus a detergent-like substance that releases local surface tension and propels the beetle forwards. Some elements of this neustic community can be found in still-water areas of streams and rivers, and related species of Gerromorpha can live in estuarine and even oceanic water surfaces (section 10.8).

Underneath the meniscus of standing water, the larvae of many mosquitoes feed (Fig. 2.16), and hang suspended by their respiratory siphons (Fig. 10.2), as do certain crane flies and stratiomyiids (Diptera). Whirligig beetles (Gyrinidae) (Fig. 10.8) also are able to straddle the interface between water and air, with an upper unwettable surface and a lower wettable one. Uniquely, each eye is divided such that the upper part can observe the aerial environment, and the lower half can see underwater.

Between the water surface and the benthos, planktonic organisms live in a zone divisible into an upper **limnetic** zone (i.e. penetrated by light) and a deeper **profundal** zone. The most abundant planktonic insects belong to *Chaoborus* (Diptera: Chaoboridae); these "phantom midges" undergo diurnal vertical migration, and their predation on *Daphnia* is discussed in section 13.4. Other insects such as diving beetles (Dytiscidae) and many hemipterans, such as Corixidae, dive and swim actively through this zone in search of

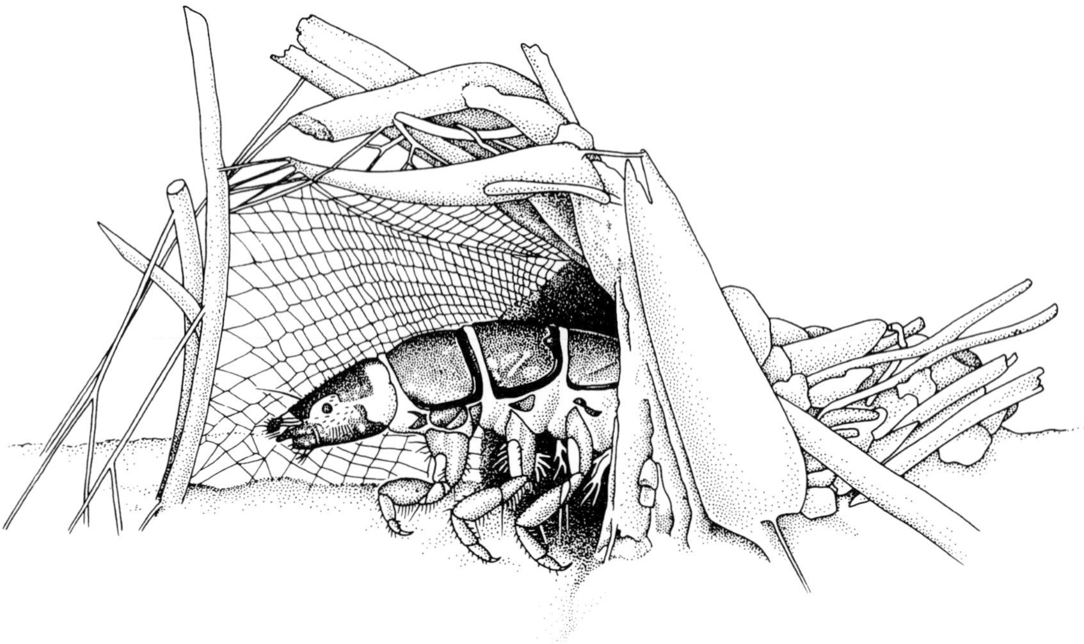

Fig. 10.7 A caddisfly larva (Trichoptera: Hydropsychidae) in its retreat; the silk net is used to catch food. (After Wiggins 1978.)

prey. The profundal zone generally lacks planktonic insects, but may support an abundant benthic community, predominantly of chironomid midge larvae, most of which possess hemoglobin. Even the profundal benthic zone of some deep lakes, such as Lake Baikal in Siberia, supports some midges, although at eclosion the pupa may have to rise more than 1 km to the water surface.

In the **littoral** zone, in which light reaches the benthos and macrophytes can grow, insect diversity is at its maximum. Many differentiated microhabitats are available and physico-chemical factors are less restricting than in the dark, cold, and perhaps anoxic conditions of the deeper waters.

10.5 ENVIRONMENTAL MONITORING USING AQUATIC INSECTS

Aquatic insects form assemblages that vary with their geographical location, according to historical biogeographic and ecological processes. Within a more restricted area, such as a single lake or river drainage, the community structure derived from within this pool of locally available organisms is constrained largely by physico-chemical factors of the environment. Amongst the important factors that govern which species live in a particular waterbody, variations in oxygen availability obviously lead to different insect communities. For example, in low-oxygen conditions, perhaps caused by oxygen-demanding sewage pollution, the community is typically species-poor and differs in composition from a comparable well-oxygenated system, as might be found upstream of a pollution site. Similar changes in community structure can be seen in relation to other physico-chemical factors such as temperature, sediment, and substrate type and, of increasing concern, pollutants such as pesticides, acidic materials, and heavy metals.

All of these factors, which generally are subsumed under the term "water quality", can be measured physico-chemically. However, physico-chemical monitoring requires:

• knowledge of which of the hundreds of substances to monitor;

• understanding of the synergistic effects when two or more pollutants interact (which often exacerbates or multiplies the effects of any compound alone);

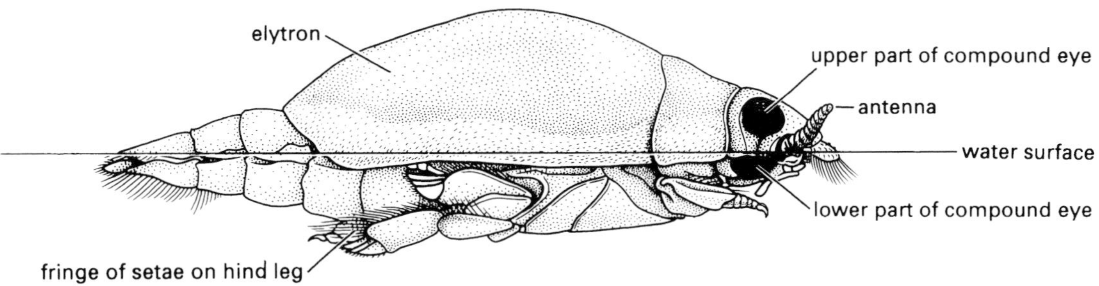

Fig. 10.8 The whirligig beetle, *Gyretes* (Coleoptera: Gyrinidae), swimming on the water surface. Note: the divided compound eye allows the beetle to see both above and below water simultaneously; hydrofuge hairs on the margin of the elytra repel water. (After White et al. 1984.)

- continuous monitoring to detect pollutants that may be intermittent, such as nocturnal release of industrial waste products.

The problem is that we often do not know in advance which of the many substances released into waterways are significant biologically; even with such knowledge, continuous monitoring of more than a few is difficult and expensive. If these impediments could be overcome, the important question remains: what are the biological effects of pollutants? Organisms and communities that are exposed to aquatic pollutants integrate multiple present and immediate-past environmental effects. Increasingly, insects are used in the description and classification of aquatic ecosystems and in the detection of deleterious effects of human activities. For the latter purpose, aquatic insect communities (or a subset of the animals that comprise an aquatic community) are used as surrogates for humans: their observed responses give early warning of damaging changes.

In this **biological monitoring** of aquatic environments, the advantages of using insects include:

- ability to select amongst the many insect taxa in any aquatic system, according to the resolution required;
- availability of many ubiquitous or widely distributed taxa, allowing elimination of non-ecological reasons why a taxon might be missing from an area;
- functional importance of insects in aquatic ecosystems, ranging from secondary producers to top predators;
- ease and lack of ethical constraints in sampling aquatic insects, giving sufficient numbers of individuals and taxa to be informative, and yet still be able to be handled;
- ability to identify most aquatic insects to a meaningful level;
- predictability and ease of detection of responses of many aquatic insects to disturbances, such as particular types of pollution.

Typical responses observed when aquatic insect communities are disturbed include:

- increased abundance of certain mayflies, such as Caenidae with protected abdominal gills, and caddisflies including filter-feeders such as Hydropsychidae, as particulate material (including sediment) increases;
- increase in numbers of hemoglobin-possessing bloodworms (Chironomidae) as dissolved oxygen is reduced;
- loss of stonefly nymphs (Plecoptera) as water temperature increases;
- substantial reduction in diversity with pesticide run-off;
- increased abundance of a few species but general loss of diversity with elevated nutrient levels (organic enrichment, or **eutrophication**).

More subtle community changes can be observed in response to less overt pollution sources, but it can be difficult to separate environmentally induced changes from natural variations in community structure.

10.6 FUNCTIONAL FEEDING GROUPS

Although aquatic insects are used widely in the context of applied ecology (section 10.5) it may not be possible, necessary, or even instructive, to make detailed species-level identifications. Sometimes the taxonomic framework is inadequate to allow identification to this level, or time and effort do not permit resolution. In most aquatic entomological studies there is a necessary trade-off between maximizing ecological information and reducing identification time. Two solutions to this dilemma involve summary by subsuming taxa into (i) more readily identified higher taxa (e.g. families, genera), or (ii) functional groupings based on feeding mechanisms ("functional feeding groups").

The first strategy assumes that a higher taxonomic category summarizes a consistent ecology or behavior amongst all member species, and indeed this is evident from some of the broad summary responses noted above. However, many closely related taxa diverge in their ecologies, and higher-level aggregates thus contain a diversity of responses. In contrast, functional groupings need make no taxonomic assumptions but use mouthpart morphology as a guide to categorizing feeding modes. The following categories are generally recognized, with some further subdivisions used by some workers:

- **shredders** feed on living or decomposing plant tissues, including wood, which they chew, mine, or gouge;
- **collectors** feed on fine particulate organic matter by filtering particles from suspension (see the chapter vignette of a filter-feeding black-fly larva of the *Simulium vittatum* complex with body twisted and cephalic feeding fans open) or fine detritus from sediment;
- **scrapers** feed on attached algae and diatoms by grazing solid surfaces;
- **piercers** feed on cell and tissue fluids from vascular plants or larger algae, by piercing and sucking the contents;
- **predators** feed on living animal tissues by engulfing and eating the whole or parts of animals, or piercing prey and sucking body fluids;
- **parasites** feed on living animal tissue as external or internal parasites of any stage of another organism.

Functional feeding groups traverse taxonomic ones; for example, the grouping "scrapers" includes some convergent larval mayflies, caddisflies, lepidopterans, and dipterans, and within Diptera there are examples of each functional feeding group.

One important ecological observation associated with such functional summary data is the often observed sequential downstream changes in proportions of functional feeding groups. This aspect of the **river continuum concept** relates the sources of energy inputs into the flowing aquatic system to its inhabitants. In riparian tree-shaded headwaters where light is low, photosynthesis is restricted and energy derives from high inputs of allochthonous materials (leaves, wood, etc.). Here, shredders such as some stoneflies and caddisflies tend to predominate, because they can break up large matter into finer particles. Further downstream, collectors such as larval black flies (Simuliidae) and hydropsychid caddisflies filter the fine particles generated upstream and themselves add particles (feces) to the current. Where the waterway becomes broader with increased available light allowing photosynthesis in the mid-reaches, algae and diatoms (periphyton) develop and serve as food on hard substrates for scrapers, whereas macrophytes provide a resource for piercers. Predators tend only to track the localized abundance of food resources. There are morphological attributes broadly associated with each of these groups, as grazers in fast-flowing areas tend to be active, flattened, and current-resisting, compared with the sessile, clinging filterers; scrapers have characteristic robust, wedge-shaped mandibles.

Changes in functional groups associated with human activities include:

- reduction in shredders with loss of riparian habitat, and consequent reduction in autochthonous inputs;
- increase in scrapers with increased periphyton development resulting from enhanced light and nutrient entry;
- increase in filtering collectors below impoundments, such as dams and reservoirs, associated with increased fine particles in upstream standing waters.

10.7 INSECTS OF TEMPORARY WATERBODIES

In a geological time-scale, all waterbodies are temporary. Lakes fill with sediment, become marshes, and eventually dry out completely. Erosion reduces the catchments of rivers and their courses change. These historical changes are slow compared with the lifespan of insects and have little impact on the aquatic fauna, apart from a gradual alteration in environmental conditions. However, in certain parts of the world, waterbodies may fill and dry on a much shorter time-scale. This is particularly evident where rainfall is very seasonal or intermittent, or where high temperatures cause elevated evaporation rates. Rivers may run during periods of predictable seasonal rainfall, such as the "winterbournes" on chalk downland in southern England that flow only during, and immediately following, winter rainfall. Others may flow only intermittently after unpredictable heavy rains, such as streams of the arid zone of central Australia and deserts of the western USA. Temporary bodies of standing waters may last for as little as a few days, as in water-filled footprints of animals, rocky depressions, pools beside a falling river, or in impermeable clay-lined pools filled by flood or snow-melt.

Even though temporary, these habitats are very productive and teem with life. Aquatic organisms appear almost immediately after the formation of such habitats. Amongst the macroinvertebrates, crustaceans are numerous and many insects thrive in ephemeral waterbodies. Some insects lay eggs into a newly formed aquatic habitat within hours of its filling, and it seems that gravid females of these species are transported to such sites over long distances, associated with the frontal meteorological conditions that bring the rainfall. An alternative to colonization by the adult is the deposition by the female of desiccation-resistant eggs into the dry site of a future pool. This behavior is seen in some odonates and many mosquitoes, especially of the genus *Aedes*. Development of the diapausing eggs is induced by environmental factors that include wetting, perhaps requiring several consecutive immersions (section 6.5).

A range of adaptations is shown amongst insects living in ephemeral habitats compared with their relatives in permanent waters. First, development to the adult often is more rapid, perhaps because of increased food quality and lowered interspecific competition. Second, development may be staggered or asynchronous, with some individuals reaching maturity very rapidly, thereby increasing the possibility of at least some adult emergence from a short-lived habitat. Associated with this is a greater variation in size of adult insects from ephemeral habitats – with metamorphosis hastened as a habitat diminishes. Certain larval midges (Diptera: Chironomidae and Ceratopogonidae) can survive drying of an ephemeral habitat by resting in silk- or mucus-lined cocoons amongst the debris at the bottom of a pool, or by complete dehydration (section 6.6.2). In a cocoon, desiccation of the body can be tolerated and development continues when the next rains fill the pool. In the dehydrated condition temperature extremes can be withstood.

Persistent temporary pools develop a fauna of predators, including immature beetles, bugs, and odonates, which are the offspring of aerial colonists. These colonization events are important in the genesis of faunas of newly flowing intermittent rivers and streams. In addition, immature stages present in remnant water beneath the streambed may move into the main channel, or colonists may be derived from permanent waters with which the temporary water connects. It is a frequent observation that novel flowing waters are colonized initially by a single species, often otherwise rare, that rapidly attains high population densities and then declines rapidly with the development of a more complex community, including predators.

Temporary waters are often saline, because evaporation concentrates salts, and this type of pool develops communities of specialist saline-tolerant organisms. However, few if any species of insect living in saline inland waters also occur in the marine zone – nearly all of the former have freshwater relatives.

10.8 INSECTS OF THE MARINE, INTERTIDAL, AND LITTORAL ZONES

The estuarine and subtropical and tropical mangrove zones are transitions between fresh and marine waters. Here, the extremes of the truly marine environment, such as wave and tidal actions, and some osmotic effects, are ameliorated. Mangroves and "saltmarsh" communities (such as *Spartina*, *Sarcocornia*, *Halosarcia*, and *Sporobolus*) support a complex phytophagous insect fauna on the emergent vegetation. In intertidal substrates and tidal pools, biting flies (mosquitoes and biting midges) are abundant and may be diverse. At the littoral margin, species of any of four families of hemipterans stride on the surface, some venturing onto the open water. A few other insects, including some *Bledius* staphylinid beetles, cixid fulgoroid bugs, and root-feeding *Pemphigus* aphids, occupy the zone of prolonged inundation by salt water. This fauna is restricted compared with freshwater and terrestrial ecosystems.

Splash-zone pools on rocky shores have salinities that vary because of rainwater dilution and solar concentration. They can be occupied by many species of corixid bugs and several larval mosquitoes and crane flies. Flies and beetles are diverse on sandy and muddy marine shores, with some larvae and adults feeding along the strandline, often aggregated on and under stranded seaweeds.

Within the intertidal zone, which lies between high and low neap-tide marks, the period of tidal inundation varies with the location within the zone. The insect fauna of the upper level is indistinguishable from the strandline fauna. At the lower end of the zone, in

Text continues on p. 260.

Box 10.1 Ephemeroptera (mayflies)

The mayflies constitute a small order of some 3000 described species, with highest diversity in temperate areas. Adults have reduced mouthparts and large compound eyes, especially in males, and three ocelli. Their antennae are filiform, sometimes multisegmented. The thorax, particularly the mesothorax, is enlarged for flight, with large triangular fore wings and smaller hind wings (as illustrated here for an adult male of the ephemerid *Ephemera danica*, after Stanek 1969; Elliott & Humpesch 1983), which are sometimes much reduced or absent. Males have elongate fore legs used to seize the female during the mating flight. The abdomen is 10-segmented, typically with three long, multisegmented, caudal filaments consisting of a pair of lateral cerci and usually a median terminal filament. Nymphs have 12–45 aquatic instars, with fully developed mandibulate mouthparts. Developing wings are visible in older nymphs (as shown here for a leptophlebiid nymph). Respiration is aided by a closed tracheal system lacking spiracles, with abdominal lamellar gills on some segments, sometimes elsewhere, including on the maxillae and labium. Nymphs have three usually filiform caudal filaments consisting of paired cerci and a variably reduced (rarely absent) median terminal filament. The penultimate instar or subimago (subadult) is fully winged, and flies or crawls.

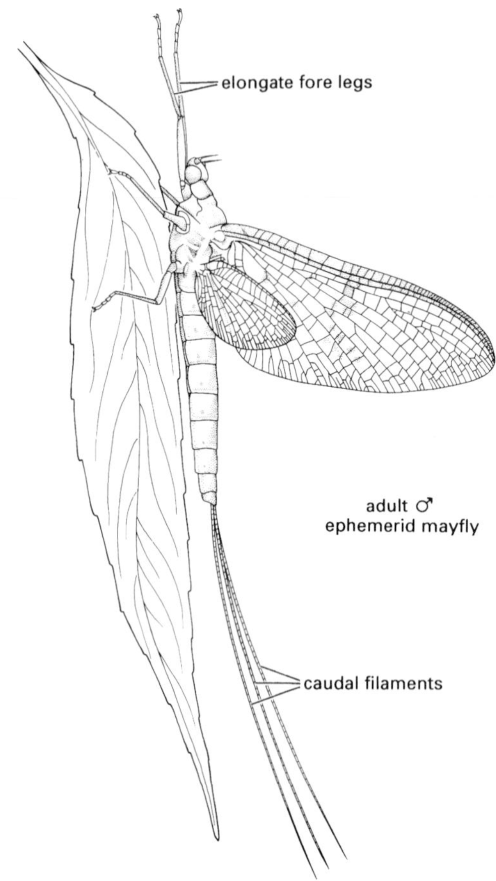

The subimago and adult are non-feeding and short-lived. Exceptionally, the subimagos mate and the adult stage is omitted. Imagos typically form mating swarms, sometimes of thousands of males, over water or nearby landmarks. Copulation usually takes place in flight, and eggs are laid in water by the female either dipping her abdominal apex below the surface or crawling under the water.

Nymphs graze on periphyton (algae, diatoms, aquatic fungi) or collect fine detritus; some are predatory on other aquatic organisms. Development takes from 16 days, to over one year in cold and high-latitude waters; some species are multivoltine. Nymphs occur predominantly in well-oxygenated cool fast-flowing streams, with fewer species in slower rivers and cool lakes; some tolerate elevated temperatures, organic enrichment, or increased sediment loads.

Phylogenetic relations are discussed in section 7.4.2 and depicted in Fig. 7.2.

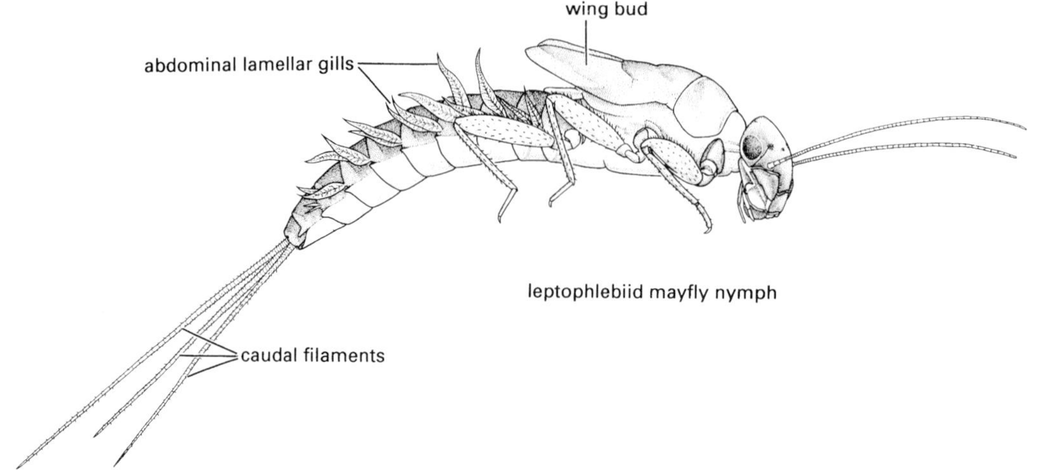

Box 10.2 Odonata (damselflies and dragonflies)

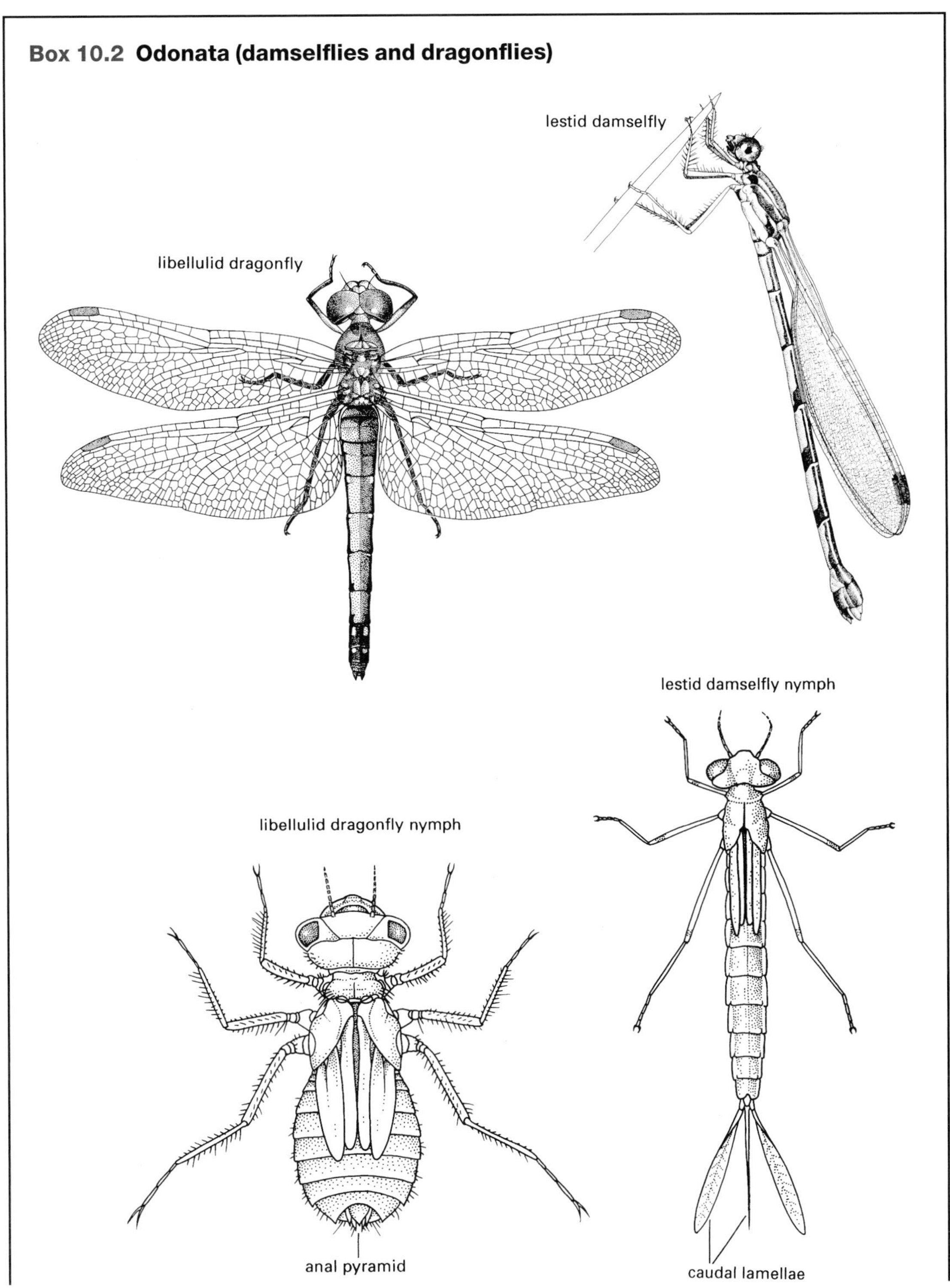

These conspicuous insects comprise a small, largely tropical order containing about 5500 described species, with about one-half belonging to the suborder Zygoptera (damselflies), and the remaining half to the suborder Anisoptera (dragonflies). Two Oriental species have been placed in a third suborder, the Anisozygoptera, which is likely invalid (section 7.4.2). The adults are medium to large (from <2 cm to >15 cm long, with a maximum wingspan of 17 cm in the South American giant damselfly (Pseudostigmatidae: *Mecistogaster*)). They have a mobile head with large, multifaceted compound eyes, three ocelli, short bristle-like antennae, and mandibulate mouthparts. The thorax is enlarged to accommodate the flight muscles of two pairs of elongate membranous wings that are richly veined. The slender 10-segmented abdomen terminates in clasping organs in both sexes; males possess secondary genitalia on the venter of the second to third abdominal segments; females often have an ovipositor at the ventral apex of the abdomen. In adult zygopterans the eyes are widely separated and the fore and hind wings are equal in shape with narrow bases (as illustrated on p. 253 in the top right figure for a lestid, *Austrolestes*, after Bandsma & Brandt 1963). Anisopteran adults have eyes either contiguous or slightly separated, and their wings have characteristic closed cells called the triangle (T) and hypertriangle (ht) (Fig. 2.22b); the hind wings are considerably wider at the base than the fore wings (as illustrated in the top left figure for a libellulid dragonfly, *Sympetrum*, after Gibbons 1986). Odonate nymphs have a variable number of up to 20 aquatic instars, with fully developed mandibulate mouthparts, including an extensible grasping labium or "mask" (Fig. 13.4). The developing wings are visible in older nymphs. The tracheal system is closed and lacks spiracles, but specialized gas-exchange surfaces are present on the abdomen as external gills (Zygoptera) or internal folds in the rectum (Anisoptera; Fig. 3.11f). Zygopteran nymphs (such as the lestid illustrated on the lower right, after CSIRO 1970) are slender, with the head wider than the thorax, and the apex of the abdomen with three (rarely two) elongate tracheal gills (caudal lamellae). Anisopteran nymphs (such as the libellulid illustrated on the lower left, after CSIRO 1970) are more stoutly built, with the head rarely much broader than the thorax, and the abdominal apex characterized by an anal pyramid consisting of three short projections and a pair of cerci in older nymphs. Many anisopteran nymphs rapidly eject water from their anus – "jet propulsion" – as an escape mechanism.

Prior to mating, the male fills his secondary genitalia with sperm from the primary genital opening on the ninth abdominal segment. At mating, the male grasps the female by her neck or prothorax and the pair fly in tandem, usually to a perch. The female then bends her abdomen forwards to connect to the male's secondary genitalia, thus forming the "wheel" position (as illustrated in Box 5.3). The male may displace sperm of a previous male before transferring his own (Box 5.3), and mating may last from seconds to several hours, depending on species. Egg-laying may take place with the pair still in tandem. The eggs (Fig. 5.10) are laid onto a water surface, into water, mud, or sand, or into plant tissue, depending on species. After eclosion, the hatchling ("pronymph") immediately molts to the first true nymph, which is the first feeding stage.

The nymphs are predatory on other aquatic organisms, whereas the adults catch terrestrial aerial prey. At metamorphosis (Fig. 6.8), the pharate adult moves to the water/land surface where atmospheric gaseous exchange commences; then it crawls from the water, anchors terrestrially, and the imago emerges from the cuticle of the final-instar nymph. The imago is long-lived, active, and aerial. Nymphs occur in all waterbodies, particularly in well-oxygenated, standing waters, but elevated temperatures, organic enrichment, or increased sediment loads are tolerated by many species.

Phylogenetic relations are discussed in section 7.4.2 and depicted in Fig. 7.2.

Box 10.3 Plecoptera (stoneflies)

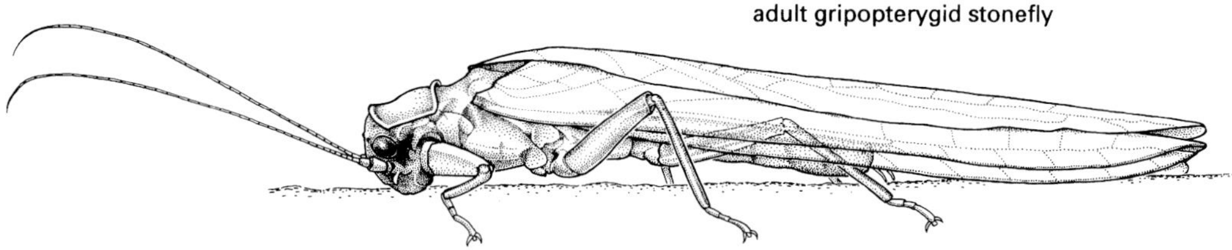

The stoneflies constitute a minor and often cryptic order of 16 families, with more than 2000 species worldwide, predominantly in temperate and cool areas. They are hemimetabolous, with adults resembling winged nymphs. The adult (see Plate 4.3, facing p. 14) is mandibulate with filiform antennae, bulging compound eyes, and two or three ocelli. The thoracic segments are subequal, and the fore and hind wings are membranous and similar (except the hind wings are broader), with the folded wings partly wrapping the abdomen and extending beyond the abdominal apex (as illustrated for an adult of the Australian gripopterygid, *Illiesoperla*); however, aptery and brachyptery are frequent. The legs are unspecialized, and the tarsi comprise three segments. The abdomen is soft and 10-segmented, with vestiges of segments 11 and 12 serving as paraprocts, cerci, and epiproct, a combination of which serve as male accessory copulatory structures, sometimes in conjunction with the abdominal sclerites of segments 9 and 10. The nymphs have 10–24, rarely as many as 33, aquatic instars, with fully developed mandibulate mouthparts; the wings pads are first visible in half-grown nymphs. The tracheal system is closed, with simple or plumose gills on the basal abdominal segments or near the anus (Fig. 10.1) – sometimes extrusible from the anus – or on the mouthparts, neck, or thorax, or lacking altogether. The cerci are usually multisegmented, and there is no median terminal filament.

Stoneflies usually mate during daylight; some species drum the substrate with their abdomen prior to mating. Eggs are dropped into water, laid in a jelly on water, or laid underneath stones in water or into damp crevices near water. Eggs may diapause. Nymphal development may take several years in some species.

Nymphs may be omnivores, detritivores, herbivores, or predators. Adults feed on algae, lichen, higher plants, and/or rotten wood; some may not eat. Mature nymphs crawl to the water's edge where adult emergence takes place. Nymphs occur predominantly on stony or gravelly substrates in cool water, mostly in well-aerated streams, with fewer species in lakes. Generally they are very intolerant of organic and thermal pollution.

Phylogenetic relations are discussed in section 7.4.2 and depicted in Fig. 7.2.

Box 10.4 Trichoptera (caddisflies)

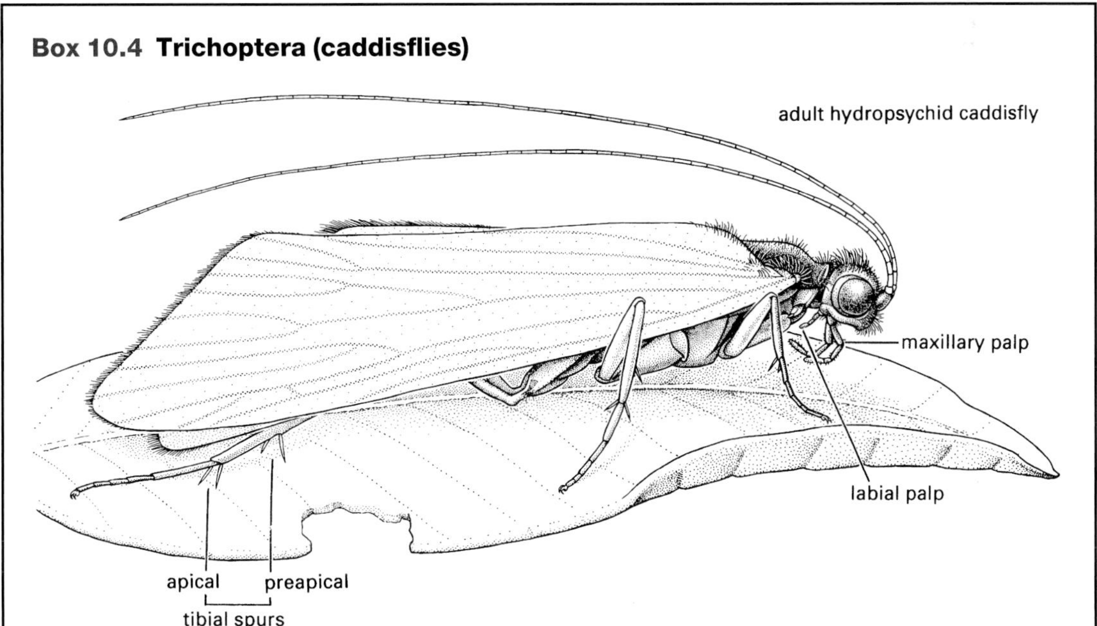

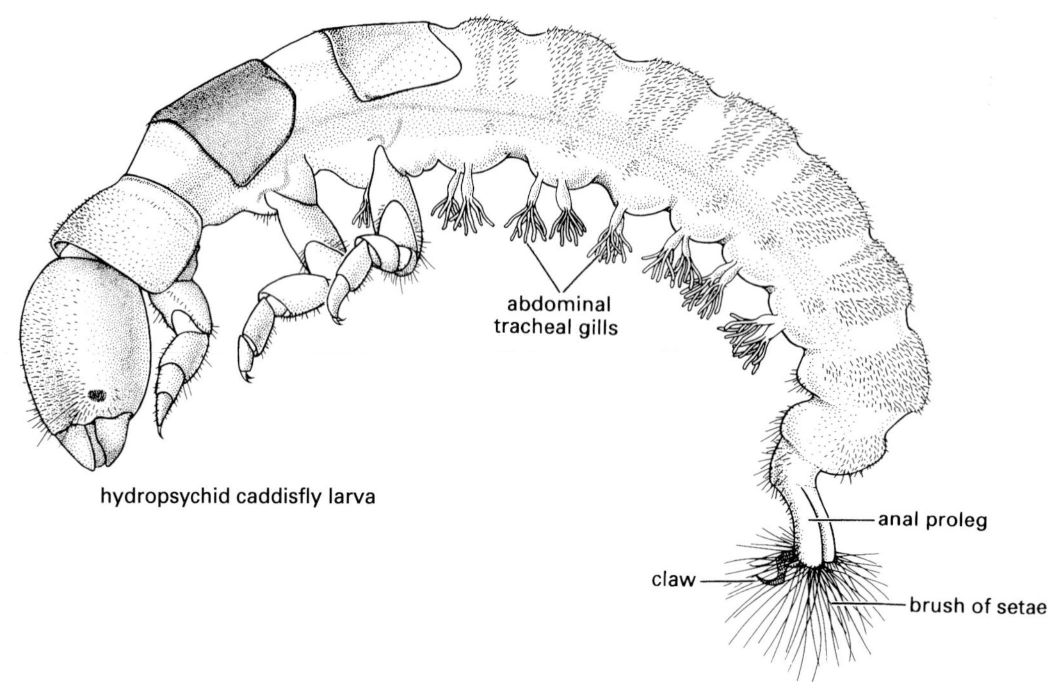

Caddisflies comprise an order of over 11,000 described species and more than 40 families found worldwide. They are holometabolous, with a moth-like adult (as illustrated here for a hydropsychid) usually covered in hairs; setal warts (setose protuberances) often occur on the dorsum of the head and thorax. The head has reduced mouthparts, but with three- to five-segmented maxillary palps and three-segmented labial palps (cf. the proboscis of most Lepidoptera). The antennae are multisegmented and filiform, often as long as or longer than the wings. There are large compound eyes and two or three ocelli. The prothorax is smaller than the meso- or metathorax, and the wings are haired or sometimes scaled, although they can be distinguished from lepidopteran wings by their different wing venation, including anal veins looped in the fore wings and no discal cell. The abdomen typically is 10-segmented, with the male terminalia more complex (often with claspers) than in the female.

The larvae have five to seven aquatic instars, with fully developed mouthparts and three pairs of thoracic legs, each with at least five segments, and without the ventral prolegs characteristic of lepidopteran larvae. The abdomen terminates in hook-bearing prolegs. The tracheal system is closed, with tracheal gills often on most or all nine abdominal segments (as illustrated here for a hydropsychid, *Cheumatopsyche* sp.), and sometimes associated with the thorax or anus. Gas exchange is also cuticular, enhanced by ventilatory undulation of the larva in its tubular case. The pupa is aquatic, enclosed in a silken retreat or case, with large functional mandibles to chew free from the pupal case or cocoon; it also has free legs with setose mid-tarsi to swim to the water surface; its gills coincide with the larval gills. Eclosion involves the pharate adult swimming to the water surface, where the pupal cuticle splits; the exuviae are used as a floating platform.

Caddisflies are predominantly univoltine, with development exceeding one year at high latitudes and elevations. The larvae are saddle-, purse-, or tube-case-making (Fig. 10.6), or free-living, net-spinning (Fig. 10.7); they exhibit diverse feeding habits and include predators, filterers, and/or shredders of organic matter, and some grazers on macrophytes. Net-spinners are restricted to flowing waters, with case-makers frequent also in standing waters. Adults may ingest nectar or water, but often do not feed.

Phylogenetic relations are discussed in section 7.4.2 and depicted in Fig. 7.2.

Box 10.5 Diptera (true flies)

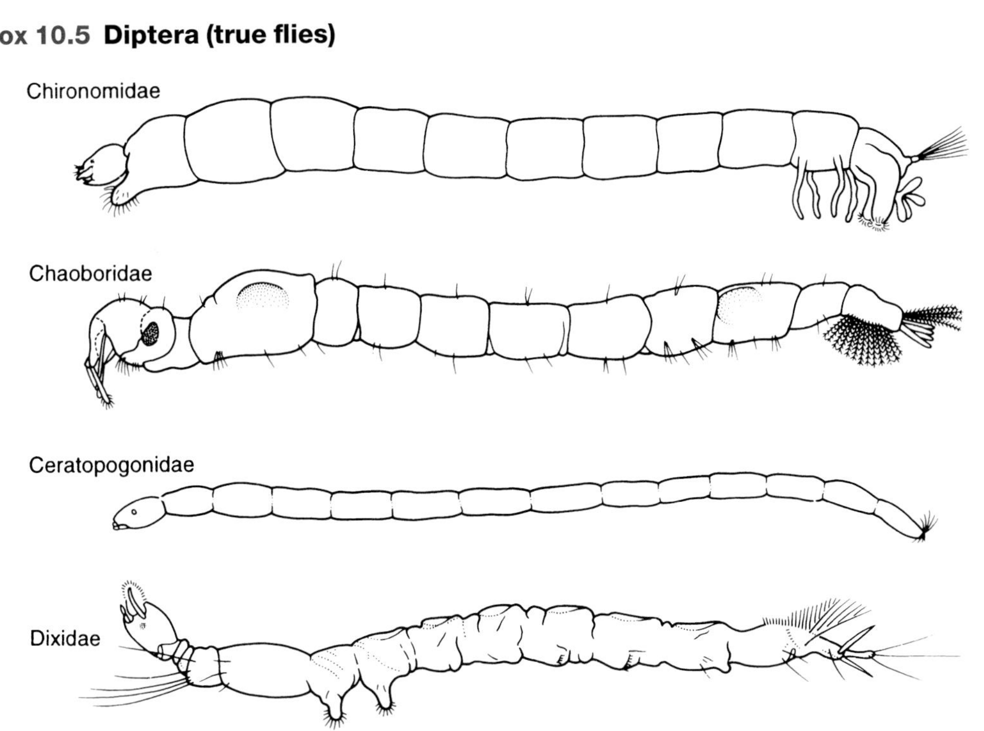

Amongst the Diptera, aquatic larvae are typical of many Nematocera, with over 10,000 aquatic species in several families, including the speciose Chironomidae (non-biting midges), Ceratopogonidae (biting midges), Culicidae (mosquitoes; Fig. 10.2), and Simuliidae (black flies) (see vignette for this chapter). Dipterans are holometabolous, and the adults are terrestrial and aerial (diagnosed in Box 15.5). The larvae are commonly vermiform (as illustrated here for the third-instar larvae of (from top to bottom) *Chironomus*, *Chaoborus*, a ceratopogonid, and *Dixa*, after Lane & Crosskey 1993), diagnostically with unsegmented prolegs, variably distributed on the body. Primitively the larvae have a sclerotized head and horizontally operating mandibles, whereas in more derived groups the head is progressively reduced, ultimately (in the maggot) with the head and mouthparts atrophied to a cephalopharyngeal skeleton. The larval tracheal system is open (amphi- or meta-, rarely propneustic) or closed, with cuticular gaseous exchange through spiracular gills or a terminal, elongate respiratory siphon with a spiracular connection to the atmosphere. There are usually three or four (in black flies up to 10) larval instars (Fig. 6.1). Pupation predominantly occurs underwater: the pupa is non-mandibulate, with appendages fused to the body; a puparium is formed in derived groups (few of which are aquatic) from the tanned retained third-instar larval cuticle. Emergence at the water surface may involve use of the cast exuviae as a platform (Chironomidae and Culicidae), or through the adult rising to the surface in a bubble of air secreted within the pupa (Simuliidae).

Development time varies from 10 days to over one year, with many multivoltine species; adults may be ephemeral to long-lived. At least some dipteran species occur in virtually every aquatic habitat, from the marine coast, salt lagoons, and sulfurous springs to fresh and stagnant waterbodies, and from temporary containers to rivers and lakes. Temperatures tolerated range from 0°C for some species up to 55°C for a few species that inhabit thermal pools (section 6.6.2). The environmental tolerance to pollution shown by certain taxa is of value in biological indication of water quality.

The larvae show diverse feeding habits, ranging from filter feeding (as shown in Fig. 2.16 and the vignette of this chapter), through algal grazing and saprophagy to micropredation.

Phylogenetic relations are discussed in section 7.4.2 and depicted in Fig. 7.2.

Box 10.6 Other aquatic orders

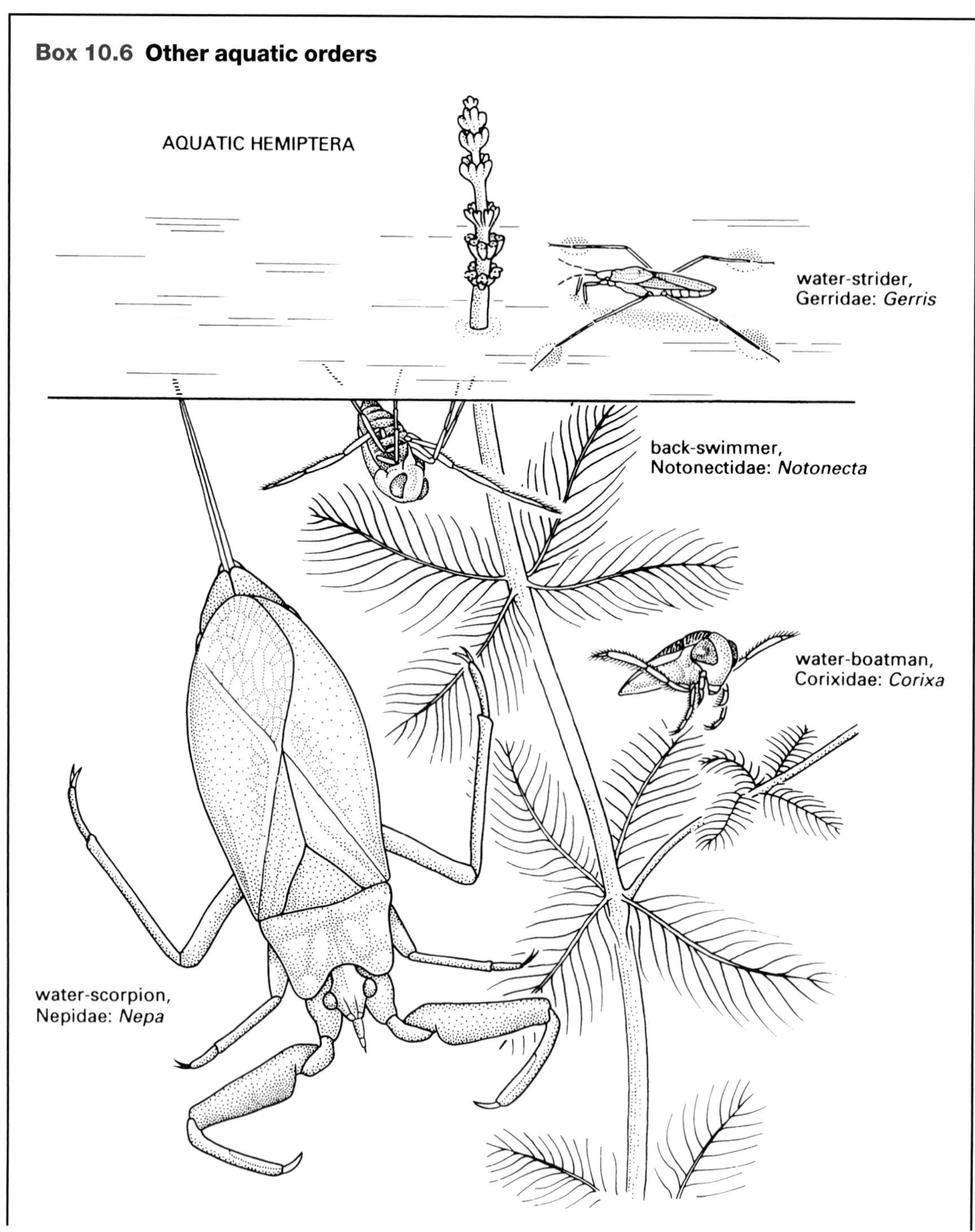

Hemiptera (bugs)

Amongst these hemimetabolous insects, there are about 4000 aquatic and semi-aquatic (including marine) species in about 20 families worldwide, belonging to three heteropteran infraorders (Gerromorpha, Leptopodomorpha, and Nepomorpha). These possess the subordinal characteristics (Box 11.8) of mouthparts modified as a rostrum (beak) and fore wings as hemelytra. They are spiraculate with various gaseous-exchange mechanisms. Nymphs have one and adults have two or more tarsal segments. The three- to five-segmented antennae are inconspicuous in aquatic groups but obvious in semi-aquatic ones. There is often reduction, loss, and/or polymorphism of wings. There are five (rarely four) nymphal instars and species are often univoltine. Gerromorphs (water-striders, represented here by *Gerris*) scavenge or are predatory on the water surface. Diving taxa are either predatory – for example back-swimmers (Notonectidae) such as *Notonecta*, water-scorpions (Nepidae) such as *Nepa*, and giant water bugs (Belostomatidae) (Box 5.5) – or phytophagous detritivores – for example some water-boatmen (Corixidae) such as *Corixa*.

Coleoptera (beetles)

The Coleoptera is a diverse order, and contains over 5000 aquatic species (although these form less than 2% of the world's described beetle species). About 10 families are exclusively aquatic as larvae and adults, an additional few are predominantly aquatic as larvae and terrestrial as adults or vice versa, and several more have sporadic aquatic representation. They are holometabolous, and adults diagnostically have the mesothoracic wings modified as rigid elytra (Fig. 2.22d and Box 11.10). Gaseous exchange in adults is usually by temporary or permanent air stores. The larvae are very variable, but all have a distinct sclerotized head with strongly developed mandibles and two- or three-segmented antennae. They have three pairs of jointed thoracic legs, and lack abdominal prolegs. The tracheal system is open and peripneustic with nine pairs of spiracles, but there is a variably reduced spiracle number in most aquatic larvae; some have lateral and/or ventral abdominal gills, sometimes hidden beneath the terminal sternite. Pupation is terrestrial (except in some Psephenidae), and the pupa lacks functional mandibles. Aquatic Coleoptera exhibit diverse feeding habits, but both larvae and adults of most species are predatory.

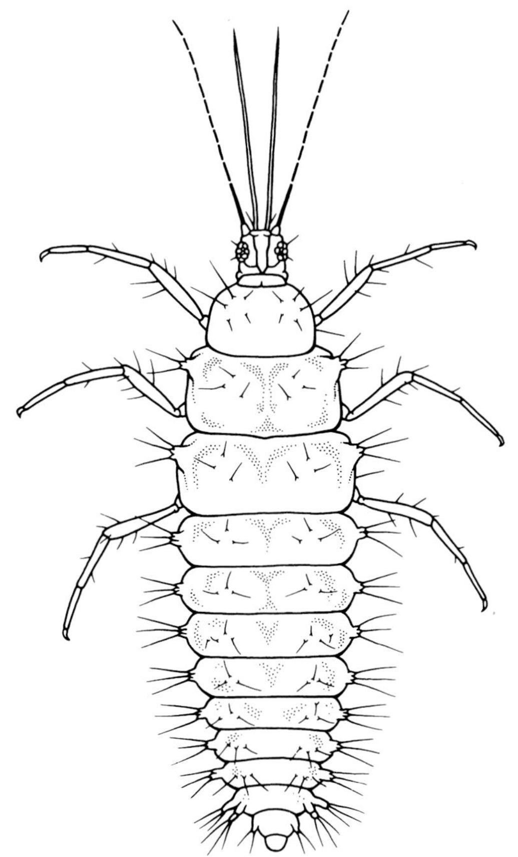

sisyrid larva

Neuroptera (lacewings)

The lacewings are holometabolous, predominantly terrestrial predators (Box 13.4), but the approximately 50 species of spongillaflies (Sisyridae) have aquatic larvae. Sisyrid larvae (as illustrated here, after CSIRO 1970) have elongate stylet-like mandibles, filamentous antennae, paired ventral abdominal gills, and lack terminal prolegs. The pupa has functional mandibles. Adults are small and soft-bodied with subequal wings lacking an anal lobe on the hind wing. The eggs are laid in trees overhanging running water, and hatching larvae drop into the water where they seek out and feed upon sponges by sucking out the living cells. There are three larval instars, with rapid development, and they may be multivoltine. Pupation takes place in a silken cocoon out of water.

Megaloptera: Corydalidae: *Archichauliodes*

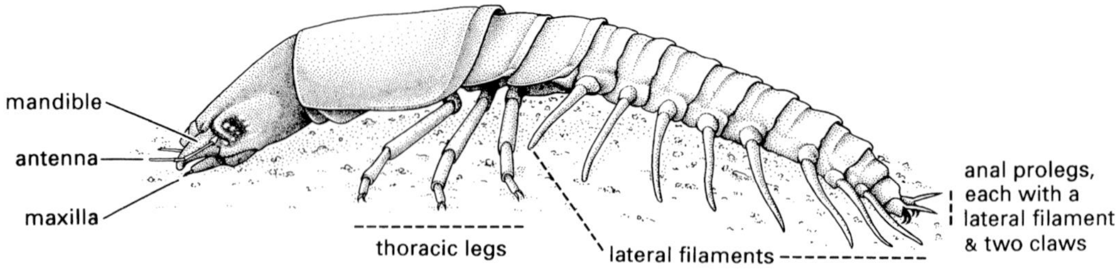

mandible
antenna
maxilla
thoracic legs
lateral filaments
anal prolegs, each with a lateral filament & two claws

Megaloptera (alderflies, dobsonflies, fishflies)

Megalopterans are holometabolous, with about 300 species in two families worldwide – Sialidae (alderflies, with adults 10–15 mm long) and the larger Corydalidae (dobsonflies and fishflies, with adults up to 75 mm long). Adults (illustrated in Box 13.4) have unspecialized mouthparts with strong mandibles. The wings are unequal, with a large pleated anal field on the hind wing that infolds when the wings are at rest over the back. The abdomen is soft. The larvae are prognathous, with well-developed mouthparts, including three-segmented labial palps (similar-looking gyrinid beetle larvae have one- or two-segmented palps). They are spiraculate, with gills consisting of four- to five-segmented (Sialidae) or two-segmented lateral filaments on the abdominal segments. The larval abdomen terminates in an unsegmented median caudal filament (Sialidae) or a pair of anal prolegs (as shown here for a species of *Archichauliodes* (Corydalidae)). The pupa is beetle-like (Fig. 6.7a), except that it has mobility due to its free legs and has a head similar to that of the larva, including functional mandibles. The larvae (sometimes called hellgrammites) have 10–12 instars and take at least one year, usually two or more, to develop. Pupation is away from water, often in chambers in damp soil under stones, or in damp timber. The larvae are predatory and scavenging, in lotic and lentic waters, and are intolerant of pollution.

conditions that are essentially marine, crane flies, chironomid midges, and species of several families of beetles occur. The female of a remarkable Australasian marine trichopteran (Chathamiidae: *Philanisus plebeius*) lays its eggs in a starfish coelom. The early-instar caddisflies feed on starfish tissues, but later free-living instars construct cases of algal fragments.

Three lineages of chironomid midges are amongst the few insects that have diversified in the marine zone. *Telmatogeton* (Fig. 6.11) is common in mats of green algae, such as *Ulva*, and occurs worldwide, including many isolated oceanic islands. In Hawai'i the genus has re-invaded freshwater. The ecologically convergent *Clunio* also is found worldwide. In some species, adult emergence from marine rock pools is synchronized by the lunar cycle to coincide with the lowest tides. A third lineage, *Pontomyia*, ranges from intertidal to oceanic, with larvae found at depths of up to 30 m on coral reefs.

The only insects on open oceans are pelagic water-striders (*Halobates*), which have been sighted hundreds of kilometers from shore in the Pacific Ocean. The distribution of these insects coincides with mid-oceanic accumulations of flotsam, where food of terrestrial origin supplements a diet of marine chironomid midges.

Physiology is unlikely to be a factor restraining diversification in the marine environment because so many different taxa are able to live in inland saline waters and in various marine zones. When living in highly saline waters, submerged insects can alter their osmoregulation to reduce chloride uptake and increase the concentration of their excretion through Malpighian tubules and rectal glands. In the pelagic water-striders, which live on the surface film, contact with saline waters must be limited.

As physiological adaptation appears to be a surmountable problem, explanations for the failure of insects to diversify in the sea must be sought elsewhere. The most likely explanation is that the insects originated well after other invertebrates, such as the Crustacea and Mollusca, had already dominated the sea. The advantages to terrestrial (including freshwater) insects of internal fertilization and flight are superfluous in the marine environment, where gametes can be shed directly into the sea and the tide and oceanic currents aid dispersal. Notably, of the few successful marine insects, many have modified wings or have lost them altogether.

FURTHER READING

Andersen, N.M. (1995) Cladistic inference and evolutionary scenarios: locomotory structure, function, and performance in water striders. *Cladistics* **11**, 279–95.

Dudgeon, D. (1999) *Tropical Asian Streams. Zoobenthos, Ecology and Conservation.* Hong Kong University Press, Hong Kong.

Eriksen, C.H. (1986) Respiratory roles of caudal lamellae (gills) in a lestid damselfly (Odonata: Zygoptera). *Journal of the North American Benthological Society* **5**, 16–27.

Eriksen, C.H. & Moeur, J.E. (1990) Respiratory functions of motile tracheal gills in Ephemeroptera nymphs, as exemplified by *Siphlonurus occidentalis* Eaton. In: *Mayflies and Stoneflies: Life Histories and Biology* (ed. I.C. Campbell), pp. 109–18. Kluwer Academic Publishers, Dordrecht.

Merritt, R.W. & Cummins, K.W. (eds.) (1996) *An Introduction to the Aquatic Insects of North America*, 3rd edn. Kendall/Hunt Publishing, Dubuque, IA.

Resh, V. & Rosenberg, D. (eds.) (1984) *The Ecology of Aquatic Insects.* Praeger, New York.

Rosenberg, D.M. & Resh, V.H. (eds.) (1993) *Freshwater Biomonitoring and Benthic Macroinvertebrates.* Chapman & Hall, London.

Ward, J.V. (1992) *Aquatic Insect Ecology.* John Wiley & Sons, Chichester.

Chapter 11

INSECTS AND PLANTS

Specialized, plant-associated neotropical insects. (After various sources.)

Insects and plants share ancient associations that date from the Carboniferous, some 300 million years ago (Fig. 8.1). Evidence preserved in fossilized plant parts of insect damage indicates a diversity of types of **phytophagy** (plant-feeding) by insects, which are presumed to have had different mouthparts, and associated with tree and seed ferns from Late Carboniferous coal deposits. Prior to the origin of the now dominant angiosperms (flowering plants), the diversification of other seed-plants, namely conifers, seed ferns, cycads, and (extinct) bennettiales, provided the template for radiation of insects with specific plant-feeding associations. Some of these, such as weevils and thrips with cycads, persist to this day. However, the major diversification of insects became manifest later, in the Cretaceous period. At this time, angiosperms dramatically increased in diversity in a radiation that displaced the previously dominant plant groups of the Jurassic period. Interpreting the early evolution of the angiosperms is contentious, partly because of the paucity of fossilized flowers prior to the period of radiation, and also because of the apparent rapidity of the origin and diversification within the major angiosperm families. However, according to estimates of their phylogeny, the earliest angiosperms may have been insect-pollinated, perhaps by beetles. Many living representatives of primitive families of beetles feed on fungi, fern spores, or pollen of other non-angiosperm taxa such as cycads. As this feeding type preceded the angiosperm radiation, it can be seen as a preadaptation for angiosperm pollination. The ability of flying insects to transport pollen from flower to flower on different plants is fundamental to cross-pollination. Other than the beetles, the most significant and diverse present-day pollinator taxa belong to three orders – the Diptera (flies), Hymenoptera (wasps and bees), and Lepidoptera (moths and butterflies). Pollinator taxa within these orders are unrepresented in the fossil record until late in the Cretaceous. Although insects almost certainly pollinated cycads and other primitive plants, insect pollinators may have promoted speciation in angiosperms, through pollinator-mediated isolating mechanisms.

As seen in Chapter 9, many modern-day non-insect hexapods and apterygote insects scavenge in soil and litter, predominantly feeding on decaying plant material. The earliest true insects probably fed similarly. This manner of feeding certainly brings soil-dwelling insects into contact with plant roots and subterranean storage organs, but specialized use of plant aerial parts by sap sucking, leaf chewing, and other forms of phytophagy arose later in the phylogeny of the insects. Feeding on living tissues of higher plants presents problems that are experienced neither by the scavengers living in the soil or litter, nor by predators. First, to feed on leaves, stems, or flowers a phytophagous insect must be able to gain and retain a hold on the vegetation. Second, the exposed phytophage may be subject to greater desiccation than an aquatic or litter-dwelling insect. Third, a diet of plant tissues (excluding seeds) is nutritionally inferior in protein, sterol, and vitamin content compared with food of animal or microbial origin. Last, but not least, plants are not passive victims of phytophages, but have evolved a variety of means to deter herbivores. These include physical defenses, such as spines, spicules or sclerophyllous tissue, and/or chemical defenses that may repel, poison, reduce food digestibility, or otherwise adversely affect insect behavior and/or physiology. Despite these barriers, about half of all living insect species are phytophagous, and the exclusively plant-feeding Lepidoptera, Curculionidae (weevils), Chrysomelidae (leaf beetles), Agromyzidae (leaf-mining flies), and Cynipidae (gall wasps) are very speciose. Plants represent an abundant resource and insect taxa that can exploit this have flourished in association with plant diversification (section 1.3.4).

This chapter begins with a consideration of the evolutionary interactions among insects and their plant hosts, amongst which a euglossine bee pollinator at work on the flower of a *Stanhopea* orchid, a chrysomelid beetle feeding on the orchid leaf, and a pollinating bee fly hovering nearby are illustrated in the chapter vignette. The vast array of interactions of insects and living plants can be grouped into three categories, defined by the effects of the insects on the plants. Phytophagy (herbivory) includes leaf chewing, sap sucking, seed predation, gall induction, and mining the living tissues of plants (section 11.2). The second category of interactions is important to plant reproduction and involves mobile insects that transport pollen between conspecific plants (pollination) or seeds to suitable germination sites (myrmecochory). These interactions are mutualistic because the insects obtain food or some other resource from the plants that they service (section 11.3). The third category of insect–plant interaction involves insects that live in specialized plant structures and provide their host with either nutrition or defense against herbivores, or both (section 11.4). Such mutualisms, like the nutrient-producing fly larvae that live unharmed within the pitchers of carnivorous plants, are unusual but provide fascinating opportunities for evolutionary and ecological studies. There is a vast literature dealing with insect–plant interactions and the interested reader should consult the reading list at the end of this chapter.

The chapter concludes with seven taxonomic boxes that summarize the morphology and biology of the primarily phytophagous orders Orthoptera, Phasmatodea, Thysanoptera, Hemiptera, Psocoptera, Coleoptera, and Lepidoptera.

11.1 COEVOLUTIONARY INTERACTIONS BETWEEN INSECTS AND PLANTS

Reciprocal interactions over evolutionary time between phytophagous insects and their food plants, or between pollinating insects and the plants they pollinate, have been described as **coevolution**. This term, coined by P.R. Ehrlich and P.H. Raven in 1964 from a study of butterflies and their host plants, was defined broadly, and now several modes of coevolution are recognized. These differ in the emphasis placed on the specificity and reciprocity of the interactions.

Specific or **pair-wise coevolution** refers to the evolution of a trait of one species (such as an insect's ability to detoxify a poison) in response to a trait of another species (such as the elaboration of the poison by the plant), which in turn evolved originally in response to the trait of the first species (i.e. the insect's food preference for that plant). This is a strict mode of coevolution, as reciprocal interactions between specific pairs of species are postulated. The outcomes of such coevolution may be evolutionary "arms races" between eater and eaten, or convergence of traits in mutualisms so that both members of an interacting pair appear perfectly adapted to each other. Reciprocal evolution between the interacting species may contribute to at least one of the species becoming subdivided into two or more reproductively isolated populations (as exemplified by figs and fig wasps; Box 11.4), thereby generating species diversity.

Another mode, **diffuse** or **guild coevolution**, describes reciprocal evolutionary change among groups, rather than pairs, of species. Here the criterion of specificity is relaxed so that a particular trait in one or more species (e.g. of flowering plants) may evolve in response to a trait or suite of traits in several other species (e.g. as in several different, perhaps distantly related, pollinating insects).

These are the main modes of coevolution that relate to insect–plant interactions, but clearly they are not mutually exclusive. The study of such interactions is beset by the difficulty in demonstrating unequivocally that any kind of coevolution has occurred. Evolution takes place over geological time and hence the selection pressures responsible for changes in "coevolving" taxa can be inferred only retrospectively, principally from correlated traits of interacting organisms. Specificity of interactions among living taxa can be demonstrated or refuted far more convincingly than can historical reciprocity in the evolution of the traits of these same taxa. For example, by careful observation, a flower bearing its nectar at the bottom of a very deep tube may be shown to be pollinated exclusively by a particular fly or moth species with a proboscis of appropriate length (e.g. Fig. 11.8), or a hummingbird with a particular length and curvature of its beak. Specificity of such an association between any individual pollinator species and plant is an observable fact, but flower tube depth and mouthpart morphology are mere correlation and only suggest coevolution (section 11.3.1).

11.2 PHYTOPHAGY (OR HERBIVORY)

The majority of plant species support complex faunas of herbivores, each of which may be defined in relation to the range of plant taxa used. Thus, **monophages** are specialists that feed on one plant taxon, **oligophages** feed on few, and **polyphages** are generalists that feed on many plant groups. The adjectives for these feeding categories are **monophagous**, **oligophagous**, and **polyphagous**. Gall-inducing cynipid wasps (Hymenoptera) exemplify monophagous insects as nearly all species are host-plant specific; furthermore, all cynipid wasps of the tribe Rhoditini induce their galls only on roses (*Rosa*) (Fig. 11.5d) and almost all species of Cynipini form their galls only on oaks (*Quercus*) (Fig. 11.5c). The monarch or wanderer butterfly, *Danaus plexippus* (Nymphalidae), is an example of an oligophagous insect, with larvae that feed on various milkweeds, predominantly species of *Asclepias*. The polyphagous gypsy moth, *Lymantria dispar* (Lymantriidae), feeds on a wide range of tree genera and species, and the Chinese wax scale, *Ceroplastes sinensis* (Hemiptera: Coccidae), is truly polyphagous with its recorded host plants belonging to about 200 species in at least 50 families. In general, most phytophagous insect groups, except Orthoptera, tend to be specialized in their feeding.

Many plants appear to have broad-spectrum defenses against a very large suite of enemies, including insect and vertebrate herbivores and pathogens. These primarily physical or chemical defenses are discussed in section 16.6 in relation to host-plant resistance to insect pests. Spines or pubescence on stems and leaves, silica or sclerenchyma in leaf tissue, or leaf shapes that aid camouflage are amongst the physical attributes of plants that may deter some herbivores. Furthermore, in addition to the chemicals considered essential to plant function, most plants contain compounds whose role generally is assumed to be defensive, although these chemicals may have, or once may have had, other metabolic functions or simply be metabolic waste products. Such chemicals are often called **secondary plant compounds**, **noxious phytochemicals**, or **allelochemicals**. A huge array exists, including phenolics (such as tannins), terpenoid compounds (essential oils), alkaloids, cyanogenic glycosides, and sulfur-containing glucosinolates. The anti-herbivore action of many of these compounds has been demonstrated or inferred. For example, in *Acacia*, the loss of the otherwise widely distributed cyanogenic glycosides in those species that harbor mutualistic stinging ants implies that the secondary plant chemicals do have an anti-herbivore function in those many species that lack ant defenses.

In terms of plant defense, secondary plant compounds may act in one of two ways. At a behavioral level, these chemicals may repel an insect or inhibit feeding and/or oviposition. At a physiological level, they may poison an insect or reduce the nutritional content of its food. However, the same chemicals that repel some insect species may attract others, either for oviposition or feeding (thus acting as kairomones; section 4.3.3). Such insects, thus attracted, are said to be adapted to the chemicals of their host plants, either by tolerating, detoxifying, or even sequestering them. An example is the monarch butterfly, *D. plexippus*, which usually oviposits on milkweed plants, many of which contain toxic cardiac glycosides (cardenolides), which the feeding larva can sequester for use as an anti-predator device (sections 14.4.3 & 14.5.1).

Secondary plant compounds have been classified into two broad groups based on their inferred biochemical actions: (i) qualitative or toxic, and (ii) quantitative. The former are effective poisons in small quantities (e.g. alkaloids, cyanogenic glycosides), whereas the latter are believed to act in proportion to their concentration, being more effective in greater amounts (e.g. tannins, resins, silica). In practice, there probably is a continuum of biochemical actions, and tannins are not simply digestion-reducing chemicals but have more complex anti-digestive and other physiological effects. However, for insects that are specialized to feed on particular plants containing any secondary plant compound(s), these chemicals actually can act as phagostimulants. Furthermore, the narrower the host-plant range of an insect, the more likely that it will be repelled or deterred by non-host-plant chemicals, even if these substances are not noxious if ingested.

The observation that some kinds of plants are more susceptible to insect attack than others also has been explained by the relative **apparency** of the plants. Thus, large, long-lived, clumped trees are very much more apparent to an insect than small, annual, scattered herbs. Apparent plants tend to have quantitative secondary compounds, with high metabolic costs in their production. Unapparent plants often have qualitative or toxic secondary compounds, produced at little metabolic cost. Human agriculture often turns unapparent plants into apparent ones, when monocultures of annual plants are cultivated, with corresponding increases in insect damage.

Another consideration is the predictability of resources sought by insects, such as the suggested predictability of the presence of new leaves on a eucalypt tree or creosote bush in contrast to the erratic spring flush of new leaves on a deciduous tree. However, the question of what is predictability (or apparency) of plants to insects is essentially untestable. Furthermore, insects can optimize the use of intermittently abundant resources by synchronizing their life cycles to environmental cues identical to those used by the plant.

A third correlate of variation in herbivory rates concerns the nature and quantities of resources (i.e. light, water, nutrients) available to plants. One hypothesis is that insect herbivores feed preferentially on stressed plants (e.g. affected by water-logging, drought, or nutrient deficiency), because stress can alter plant physiology in ways beneficial to insects. Alternatively, insect herbivores may prefer to feed on vigorously growing plants (or plant parts) in resource-rich habitats. Evidence for and against both is available. Thus, gall-forming phylloxera (Box 11.2) prefers fast-growing meristematic tissue found in rapidly extending shoots of its healthy native vine host. In apparent contrast, the larva of *Dioryctria albovitella* (the pinyon pine cone and shoot boring moth; Pyralidae) attacks the growing shoots of nutrient-deprived and/or water-stressed pinyon pine (*Pinus edulis*) in preference to adjacent, less-stressed trees. Experimental alleviation of water stress has been shown to reduce rates of infestation, and enhance pine growth. Examination of a wide range of resource studies leads to the following partial explanation: boring and sucking insects seem to perform better on stressed plants, whereas gall inducers and chewing insects are adversely affected by plant stress. Additionally, performance of chewers may be reduced more on stressed, slow-growing plants than on stressed, fast growers.

The presence in Australia of a huge radiation of oecophorid moths whose larvae specialize in feeding on fallen eucalypt leaves suggests that even well-defended food resources can become available to the specialist herbivore. Evidently, no single hypothesis (model) of herbivory is consistent with all observed patterns of temporal and spatial variation within plant individuals, populations, and communities. However, all models of current herbivory theory make two assumptions, both of which are difficult to substantiate. These are:

1 damage by herbivores is a dominant selective force on plant evolution;

2 food quality has a dominant influence on the abundance of insects and the damage they cause.

Even the substantial evidence that hybrid plants may incur much greater damage from herbivores than either adjacent parental population is not unequivocal evidence of either assumption. Selection against hybrids clearly could affect plant evolution; but any such herbivore preference for hybrids would be expected to constrain rather than promote plant genetic diversification. The food quality of hybrids arguably is higher than that of the parental plants, as a result of less efficient chemical defenses and/or higher nutritive value of the genetically "impure" hybrids. It remains unclear whether the overall population abundance of herbivores is altered by the presence of hybrids (or by food quality *per se*) or merely is redistributed among the plants available. Furthermore, the role of natural enemies in regulating herbivore populations often is overlooked in studies of insect–plant interactions.

Many studies have demonstrated that phytophagous insects can impair plant growth, both in the short term and the long term. These observations have led to the suggestion that host-specific herbivores may affect the relative abundances of plant species by reducing the competitive abilities of host plants. The occurrence of induced defenses (Box 11.1) supports the idea that it is advantageous for plants to deter herbivores. In contrast with this view is the controversial hypothesis that "normal" levels of herbivory may be advantageous or selectively neutral to plants. Some degree of pruning, pollarding, or mowing may increase (or at least not reduce) overall plant reproductive success by altering growth form or longevity and thus lifetime seed set. The important evolutionary factor is lifetime reproductive success, although most assessments of herbivore effects on plants involve only measurements of plant production (biomass, leaf number, etc.).

A major problem with all herbivory theories is that they have been founded largely on studies of leaf-chewing insects, as the damage caused by these insects is easier to measure and factors involved in defoliation are more amenable to experimentation than for other types of herbivory. The effects of sap-sucking, leaf-mining, and gall-inducing insects may be as important although, except for some agricultural and horticultural pests such as aphids, they are generally poorly understood.

11.2.1 Leaf chewing

The damage caused by leaf-chewing insects is readily visible compared, for example, with that of many sap-sucking insects. Furthermore, the insects responsible for leaf tissue loss are usually easier to identify than the small larvae of species that mine or gall plant parts. By far the most diverse groups of leaf-chewing insects are the Lepidoptera and Coleoptera. Most moth and butterfly caterpillars and many beetle larvae and adults feed on leaves, although plant roots, shoots, stems, flowers, or fruits often are eaten as well. Certain Australian adult scarabs, especially species of *Anoplognathus* (Coleoptera: Scarabaeidae; commonly called Christmas beetles) (Fig. 11.1), can cause severe defoliation of eucalypt trees. The most important foliage-eating pests in north temperate forests are lepidopteran larvae, such as those of the gypsy moth, *Lymantria dispar* (Lymantriidae). Other important groups of leaf-chewing insects worldwide are the Orthoptera (most species) and Hymenoptera (most Symphyta). The stick-insects (Phasmatodea) generally have only minor impact as leaf chewers, although outbreaks of the spur-legged stick-insect, *Didymuria violescens* (Box 11.6), can defoliate eucalypts in Australia.

High levels of herbivory result in economic losses to forest trees and other plants, so reliable and repeatable methods of estimating damage are desirable. Most

Box 11.1 Induced defenses

Plants contain various chemicals that may deter, or at least reduce their suitability to, some herbivores. These are the secondary plant compounds (noxious phytochemicals, or allelochemicals). Depending on plant species, such chemicals may be present in the foliage at all times, only in some plant parts, or only in some parts during particular stages of ontogeny, such as during the growth period of new leaves. Such **constitutive** defenses provide the plant with continuous protection, at least against non-adapted phytophagous insects. If defense is costly (in energetic terms) and if insect damage is intermittent, plants would benefit from being able to turn on their defenses only when insect feeding occurs. There is good experimental evidence that, in some plants, damage to the foliage induces chemical changes in the existing or future leaves, which adversely affect insects. This phenomenon is called **induced defense** if the induced chemical response benefits the plant. However, sometimes the induced chemical changes may lead to greater foliage consumption by lowering food quality for herbivores, which thus eat more to obtain the necessary nutrients.

Both short-term (or rapidly induced) and long-term (or delayed) chemical changes have been observed in plants as responses to herbivory. For example, proteinase-inhibitor proteins are produced rapidly by some plants in response to wounds caused by chewing insects. These proteins can significantly reduce the palatability of the plant to some insects. In other plants, the production of phenolic compounds may be increased, either for short or prolonged periods, within the wounded plant part or sometimes the whole plant. Alternatively, the longer-term carbon–nutrient balance may be altered to the detriment of herbivores.

Such induced chemical changes have been demonstrated for some but not all studied plants. Even when they occur, their function(s) may not be easy to demonstrate, especially as herbivore feeding is not always deterred. Sometimes induced chemicals may benefit the plant indirectly, not by reducing herbivory but by attracting natural enemies of the insect herbivores (section 4.3.3). Moreover, the results of studies on induced responses may be difficult to interpret because of large variation in foliage quality between and within individual plants, as well as the complication that minor variations in the nature of the damage can lead to different outcomes. In addition, insect herbivore populations in the field are regulated by an array of factors and the effects of plant chemistry may be ameliorated or exacerbated depending on other conditions.

An even more difficult area of study involves what the popular literature refers to as "talking trees", to describe the controversial phenomenon of damaged plants releasing signals (volatile chemicals) that elicit increased resistance to herbivory in undamaged neighbors. Whether such interplant communication is important in nature is unclear but within-plant responses to herbivory certainly can occur at some distance from the site of insect damage, as a result of intraplant chemical signals. The nature and control of these systemic signals have been little studied in relation to herbivory and yet manipulation of such chemicals may provide new opportunities for increasing plant resistance to herbivorous insect pests.

methods rely on estimating leaf area lost due to leaf-chewing insects. This can be measured directly from foliar damage, either by once-off sampling, or monitoring marked branches, or by destructively collecting separate samples over time ("spot sampling"), or indirectly by measuring the production of insect **frass** (feces). These sorts of measurements have been undertaken in several forest types, from rainforests to xeric (dry) forests, in many countries worldwide. Herbivory levels tend to be surprisingly uniform. For temperate forests, most values of proportional leaf area missing range from 3 to 17%, with a mean value of 8.8 ± 5.0% ($n = 38$) (values from Landsberg & Ohmart 1989). Data collected from rainforests and mangrove forests reveal similar levels of leaf area loss (range 3–15%, with mean 8.8 ± 3.5%). However, during outbreaks, especially of introduced pest species, defoliation levels may be very high and even lead to plant death. For some plant taxa, herbivory levels may be high (20–45%) even under natural, non-outbreak conditions.

Levels of herbivory, measured as leaf area loss, differ among plant populations or communities for a number of reasons. The leaves of different plant species vary in their suitability as insect food because of variations in nutrient content, water content, type and concentrations of secondary plant compounds, and degree of sclerophylly (toughness). Such differences may occur because of inherent differences among plant taxa and/or may relate to the maturity and growing conditions of the individual leaves and/or the plants sampled.

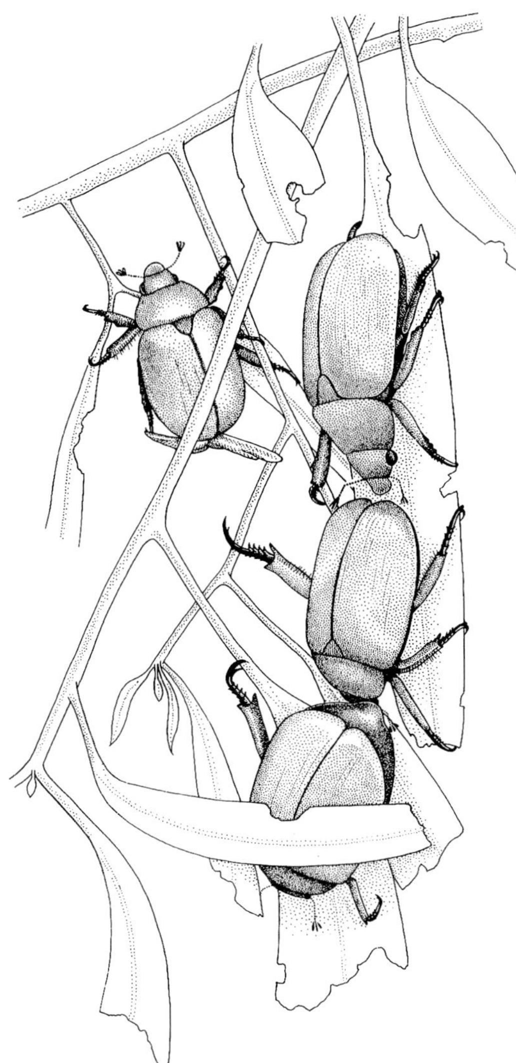

Fig. 11.1 Christmas beetles of *Anoplognathus* (Coleoptera: Scarabaeidae) on the chewed foliage of a eucalypt tree (Myrtaceae).

Assemblages in which the majority of the constituent tree species belong to different families (such as in many north temperate forests) may suffer less damage from phytophages than those that are dominated by one or a few genera (such as Australian eucalypt/acacia forests). In the latter systems, specialist insect species may be able to transfer relatively easily to new, closely related plant hosts. Favorable conditions thus may result in considerable insect damage to all or most tree species in a given area. In diverse (multigenera) forests, oligophagous insects are unlikely to switch to species unrelated to their normal hosts. Furthermore, there may be differences in herbivory levels within any given plant population over time as a result of seasonal and stochastic factors, including variability in weather conditions (which affects both insect and plant growth) or plant defenses induced by previous insect damage (Box 11.1). Such temporal variation in plant growth and response to insects can bias herbivory estimates made over a restricted time period.

11.2.2 Plant mining and boring

A range of insect larvae reside within and feed on the internal tissues of living plants. **Leaf-mining** species live between the two epidermal layers of a leaf and their presence can be detected externally after the area that they have fed upon dies, often leaving a thin layer of dry epidermis. This leaf damage appears as tunnels, blotches, or blisters (Fig. 11.2). Tunnels may be straight (linear) to convoluted and often widen throughout their course (Fig. 11.2a), as a result of larval growth during development. Generally, larvae that live in the confined space between the upper and lower leaf epidermis are flattened. Their excretory material, frass, is left in the mine as black or brown pellets (Fig. 11.2a,b,c,e) or lines (Fig. 11.2f).

The leaf-mining habit has evolved independently in only four holometabolous orders of insects: the Diptera, Lepidoptera, Coleoptera, and Hymenoptera. The commonest types of leaf miners are larval flies and moths. Some of the most prominent leaf mines result from the larval feeding of agromyzid flies (Fig. 11.2a–d). Agromyzids are virtually ubiquitous; there are about 2500 species, all of which are exclusively phytophagous. Most are leaf miners, although some mine stems and a few occur in roots or flower heads. Some anthomyiids and a few other fly species also mine leaves. Lepidopteran leaf miners (Fig. 11.2e–g) mostly belong to the families Gracillariidae, Gelechiidae, Incurvariidae, Lyonetiidae, Nepticulidae, and Tisheriidae. The habits of leaf-mining moth larvae are diverse, with many variations in types of mines, methods of feeding, frass disposal, and larval morphology. Generally, the larvae are more specialized than those of other leaf-mining orders and are very dissimilar to their non-mining relatives. A number of moth species have habits that intergrade

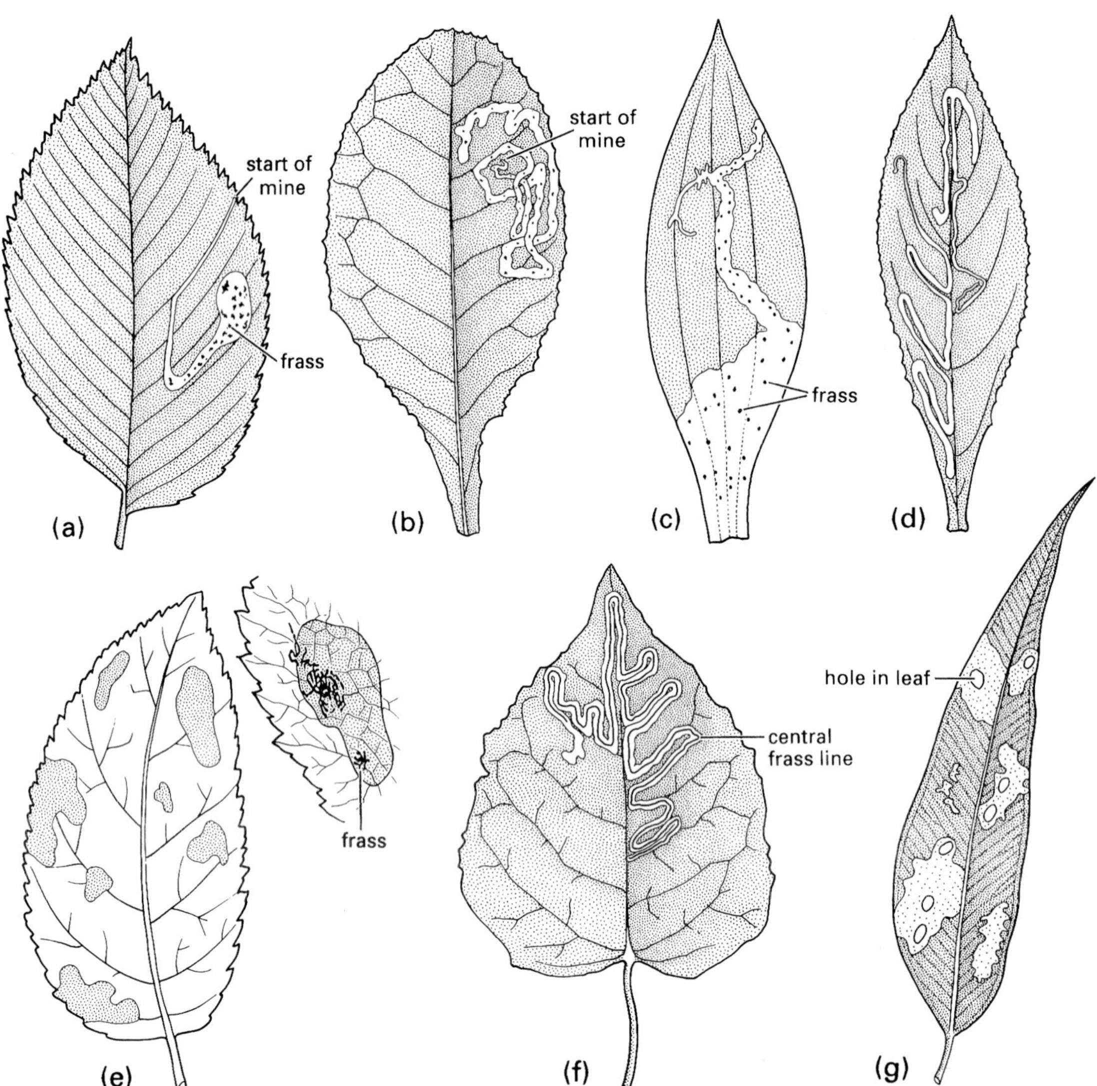

Fig. 11.2 Leaf mines: (a) linear-blotch mine of *Agromyza aristata* (Diptera: Agromyzidae) in leaf of an elm, *Ulmus americana* (Ulmaceae); (b) linear mine of *Chromatomyia primulae* (Agromyzidae) in leaf of a primula, *Primula vulgaris* (Primulaceae); (c) linear-blotch mine of *Chromatomyia gentianella* (Agromyzidae) in leaf of a gentian, *Gentiana acaulis* (Gentianaceae); (d) linear mine of *Phytomyza senecionis* (Agromyzidae) in leaf of a ragwort, *Senecio nemorensis* (Asteraceae); (e) blotch mines of the apple leaf miner, *Lyonetia speculella* (Lepidoptera: Lyonetiidae), in leaf of apple, *Malus* sp. (Rosaceae); (f) linear mine of *Phyllocnistis populiella* (Lepidoptera: Gracillariidae) in leaf of poplar, *Populus* (Salicaceae); (g) blotch mines of jarrah leaf miner, *Perthida glyphopa* (Lepidoptera: Incurvariidae), in leaf of jarrah, *Eucalyptus marginata* (Myrtaceae). ((a,e–f) After Frost 1959; (b–d) after Spencer 1990.)

with gall inducing and leaf rolling. Leaf-mining Hymenoptera principally belong to the sawfly superfamily Tenthredinoidea, with most leaf-mining species forming blotch mines. Leaf-mining Coleoptera are represented by certain species of jewel beetles (Buprestidae), leaf beetles (Chrysomelidae), and weevils (Curculionoidea).

Leaf miners can cause economic damage by attacking the foliage of fruit trees, vegetables, ornamental plants, and forest trees. The spinach leaf miner (or mangold fly), *Pegomya hyoscyami* (Diptera: Anthomyiidae), causes commercial damage to the leaves of spinach and beet. The larvae of the birch leaf miner, *Fenusa pusilla* (Hymenoptera: Tenthredinidae), produce blotch mines in birch foliage in north-eastern North America, where this sawfly is considered a serious pest. In Australia, certain eucalypts are prone to the attacks of leaf miners, which can cause unsightly damage. The leaf blister sawflies (Hymenoptera: Pergidae: *Phylacteophaga*) tunnel in and blister the foliage of some species of *Eucalyptus* and related genera of Myrtaceae. The larvae of the jarrah leaf miner, *Perthida glyphopa* (Lepidoptera: Incurvariidae), feed in the leaves of jarrah, *Eucalyptus marginata*, causing blotch mines and then holes after the larvae have cut leaf discs for their pupal cases (Fig. 11.2g). Jarrah is an important timber tree in Western Australia and the feeding of these leaf miners can cause serious leaf damage in vast areas of eucalypt forest.

Mining sites are not restricted to leaves, and some insect taxa display a diversity of habits. For example, different species of *Marmara* (Lepidoptera: Gracillariidae) not only mine leaves but some burrow below the surface of stems, or in the joints of cacti, and a few even mine beneath the skin of fruit. One species that typically mines the cambium of twigs even extends its tunnels into leaves if conditions are crowded. **Stem mining**, or feeding in the superficial layer of twigs, branches, or tree trunks, can be distinguished from **stem boring**, in which the insect feeds deep in the plant tissues. Stem boring is just one form of **plant boring**, which includes a broad range of habits that can be subdivided according to the part of the plant eaten and whether the insects are feeding on living or dead and/or decaying plant tissues. The latter group of saprophytic insects is discussed in section 9.2 and is not dealt with further here. The former group includes larvae that feed in buds, fruits, nuts, seeds, roots, stalks, and wood. **Stalk borers**, such as the wheat stem sawflies (Hymenoptera: Cephidae: *Cephus* species) and the European corn borer (Lepidoptera: Pyralidae: *Ostrinia nubilalis*) (Fig. 11.3a), attack grasses and more succulent plants, whereas **wood borers** feed in the twigs, stems, and/or trunks of woody plants where they may eat the bark, phloem, sapwood, or heartwood. The wood-boring habit is typical of many Coleoptera, especially the larvae of jewel beetles (Buprestidae), longicorn (or longhorn) beetles (Cerambycidae), and weevils (Curculionoidea), and some Lepidoptera (e.g. Hepialidae and Cossidae; Fig. 1.3) and Hymenoptera. The root-boring habit is well developed in the Lepidoptera, but many moth larvae do not differentiate between the wood of trunks, branches, or roots. Many species damage plant storage organs by boring into tubers, corms, and bulbs.

The reproductive output of many plants is reduced or destroyed by the feeding activities of larvae that bore into and eat the tissues of fruits, nuts, or seeds. **Fruit borers** include:

- Diptera (especially Tephritidae, such as the apple maggot, *Rhagoletis pomonella*, and the Mediterranean fruit fly, *Ceratitis capitata*);
- Lepidoptera (e.g. some tortricids, such as the oriental fruit moth, *Grapholita molesta*, and the codling moth, *Cydia pomonella*; Fig. 11.3b);
- Coleoptera (particularly certain weevils, such as the plum curculio, *Conotrachelus nenuphar*).

Weevil larvae also are common occupants of seeds and nuts and many species are pests of stored grain (section 11.2.5).

11.2.3 Sap sucking

The feeding activities of insects that chew or mine leaves and shoots cause obvious damage. In contrast, structural damage caused by sap-sucking insects often is inconspicuous, as the withdrawal of cell contents from plant tissues usually leaves the cell walls intact. Damage to the plant may be difficult to quantify even though the sap sucker drains plant resources (by removing phloem or xylem contents), causing loss of condition such as retarded root growth, fewer leaves, or less overall biomass accumulation compared with unaffected plants. These effects may be detectable with confidence only by controlled experiments in which the growth of infested and uninfected plants is compared. Certain sap-sucking insects do cause conspicuous tissue necrosis either by transmitting diseases, especially viral ones, or by injecting toxic saliva, whereas others induce obvious tissue distortion or growth abnormalities called galls (section 11.2.4).

Most sap-sucking insects belong to the Hemiptera. All hemipterans have long, thread-like mouthparts consisting of appressed mandibular and maxillary stylets forming a bundle lying in a groove in the labium (Box 11.8). The maxillary stylet contains a salivary canal that directs saliva into the plant, and a food canal through which plant juice or sap is sucked up into the insect's gut. Only the stylets enter the tissues of the host plant (Fig. 11.4a). They may penetrate

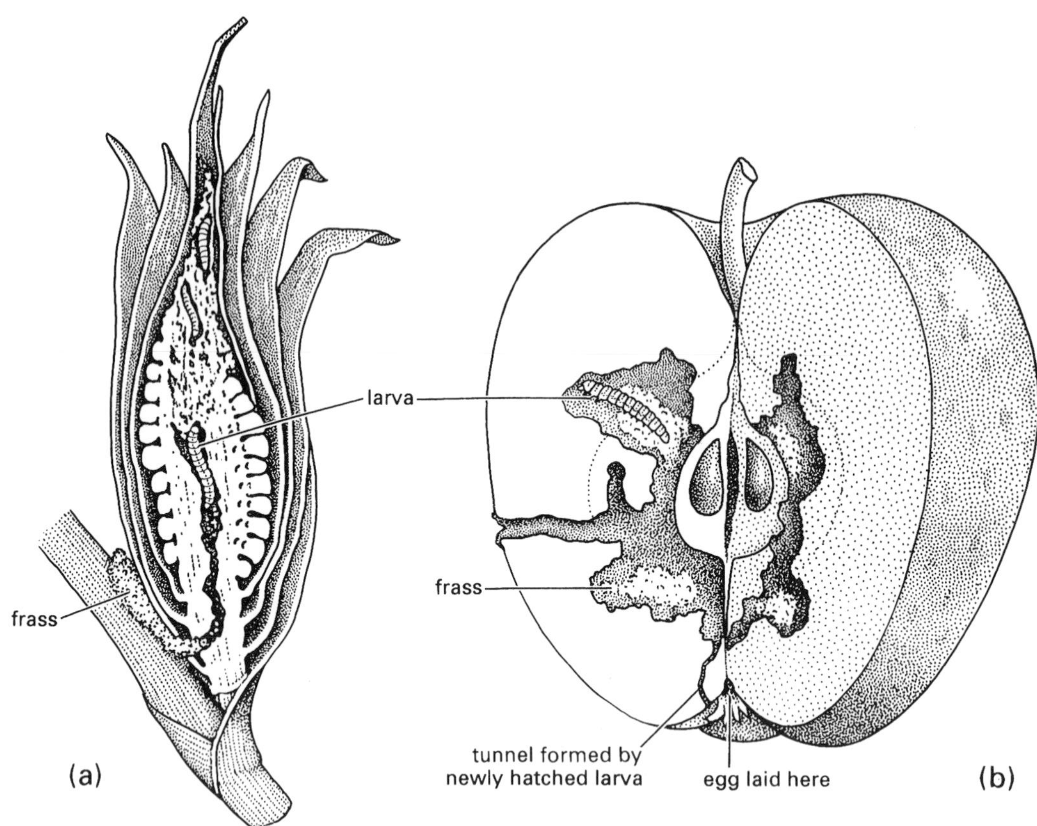

Fig. 11.3 Plant borers: (a) larvae of the European corn borer, *Ostrinia nubilalis* (Lepidoptera: Pyralidae), tunneling in a corn stalk; (b) a larva of the codling moth, *Cydia pomonella* (Lepidoptera: Tortricidae), inside an apple. (After Frost 1959.)

superficially into a leaf or deeply into a plant stem or leaf midrib, following either an intracellular or intercellular path, depending on species. The feeding site reached by the stylet tips may be in the parenchyma (e.g. some immature scale insects, many Heteroptera), the phloem (e.g. most aphids, mealybugs, soft scales, psyllids, and leafhoppers), or the xylem (e.g. spittle bugs and cicadas). In addition to a hydrolyzing type of saliva, many species produce a solidifying saliva that forms a sheath around the stylets as they enter and penetrate the plant tissue. This sheath can be stained in tissue sections and allows the feeding tracks to be traced to the feeding site (Fig. 11.4b,c). The two feeding strategies of hemipterans, stylet-sheath and macerate-and-flush feeding, are described in section 3.6.2, and the gut specializations of hemipterans for dealing with a watery diet are discussed in Box 3.3. Many species of plant-feeding Hemiptera are considered serious agricultural and horticultural pests. Loss of sap leads to wilting, distortion, or stunting of shoots. Movement of the insect between host plants can lead to the efficient transmission of plant viruses and other diseases, especially by aphids and whiteflies. The sugary excreta (**honeydew**) of phloem-feeding Hemiptera, particularly coccoids, is used by black sooty molds, which soil leaves and fruits and can impair photosynthesis.

Thrips (Thysanoptera) that feed by sucking plant juices penetrate the tissues using their stylets (Fig. 2.13) to pierce the epidermis and then rupture individual cells below. Damaged areas discolor and the leaf, bud, flower, or shoot may wither and die. Plant damage typically is concentrated on rapidly growing tissues, so that flowering and leaf flushing may be seriously disrupted. Some thrips inject toxic saliva during feeding or

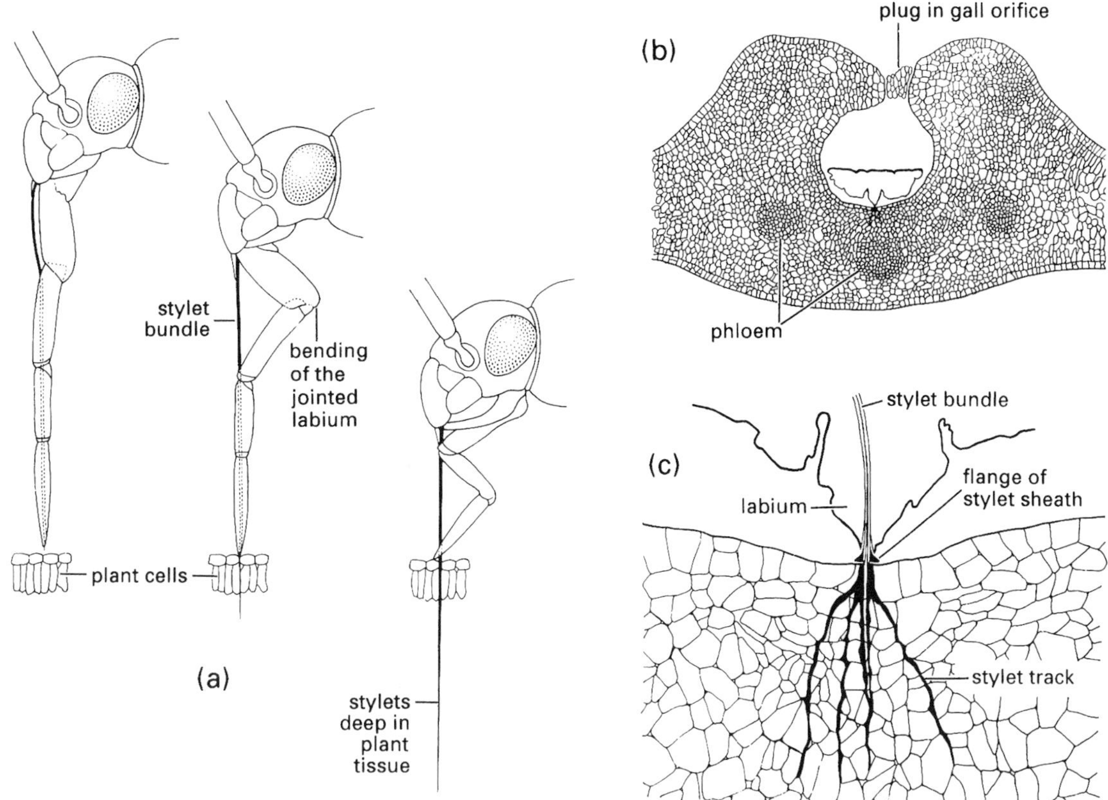

Fig. 11.4 Feeding in phytophagous Hemiptera: (a) penetration of plant tissue by a mirid bug showing bending of the labium as the stylets enter the plant; (b) transverse section through a eucalypt leaf gall containing a feeding nymph of a scale insect, *Apiomorpha* (Eriococcidae); (c) enlargement of the feeding site of (b) showing multiple stylet tracks (formed of solidifying saliva) resulting from probing of the parenchyma. ((a) After Poisson 1951.)

transmit viruses, such as the *Tospovirus* (Bunyaviridae) carried by the pestiferous western flower thrips, *Frankliniella occidentalis*. A few hundred thrips species have been recorded attacking cultivated plants, but only 10 species transmit tospoviruses.

Outside the Hemiptera and Thysanoptera, the sap-sucking habit is rare in extant insects. Many fossil species, however, had a rostrum with piercing-and-sucking mouthparts. Palaeodictyopteroids (Fig. 8.2), for example, probably fed by imbibing juices from plant organs.

11.2.4 Gall induction

Insect-induced plant **galls** result from a very specialized type of insect–plant interaction in which the morphology of plant parts is altered, often substantially and characteristically, by the influence of the insect. Generally, galls are defined as pathologically developed cells, tissues, or organs of plants that have arisen by hypertrophy (increase in cell size) and/or hyperplasia (increase in cell number) as a result of stimulation from foreign organisms. Some galls are induced by viruses, bacteria, fungi, nematodes, and mites, but insects cause many more. The study of plant galls is called **cecidology**, gall-causing animals (insects, mites, and nematodes) are **cecidozoa**, and galls induced by cecidozoa are referred to as **zoocecidia**. Cecidogenic insects account for about 2% of all described insect species, with perhaps 13,000 species known. Although galling is a worldwide phenomenon across most plant groups, global survey shows an eco-geographical pattern with gall incidence more frequent in vegetation with a sclerophyllous habit, or at least living on plants in wet–dry seasonal environments.

On a world basis, the principal cecidozoa in terms of number of species are representatives of just three orders of insects – the Hemiptera, Diptera, and Hymenoptera. In addition, about 300 species of mostly tropical Thysanoptera (thrips) are associated with galls, although not necessarily as inducers, and some species of Coleoptera (mostly weevils) and microlepidoptera (small moths) induce galls. Most hemipteran galls are elicited by Sternorrhyncha, in particular aphids, coccoids, and psyllids; their galls are structurally diverse and those of gall-inducing eriococcids (Coccoidea: Eriococcidae) often exhibit spectacular sexual dimorphism, with galls of female insects much larger and more complex than those of their conspecific males (Fig. 11.5a,b). Worldwide, there are several hundred gall-inducing coccoid species in about 10 families, about 350 gall-forming Psylloidea, mostly in two families, and perhaps 700 gall-inducing aphid species distributed among the three families, Phylloxeridae (Box 11.2), Adelgidae, and Aphididae.

The Diptera contains the highest number of gall-inducing species, perhaps thousands, but the probable number is uncertain because many dipteran gall inducers are poorly known taxonomically. Most cecidogenic flies belong to one family of at least 4500 species, the Cecidomyiidae (gall midges), and induce simple or complex galls on leaves, stems, flowers, buds, and even roots. The other fly family that includes some important cecidogenic species is the Tephritidae, in which gall inducers mostly affect plant buds, often of the Asteraceae. Galling species of both cecidomyiids and tephritids are of actual or potential use for biological control of some weeds. Three superfamilies of wasps contain large numbers of gall-inducing species: Cynipoidea contains the gall wasps (Cynipidae, at least 1300 species), which are among the best-known gall insects in Europe and North America, where hundreds of species form often extremely complex galls, especially on oaks and roses (Fig. 11.5c,d); Tenthredinoidea has a number of gall-forming sawflies, such as *Pontania* species (Tethredinidae) (Fig. 11.5g); and Chalcidoidea includes several families of gall inducers, especially species in the Agaonidae (fig wasps; Box 11.4), Eurytomidae, and Pteromalidae.

There is enormous diversity in the patterns of development, shape, and cellular complexity of insect galls (Fig. 11.5). They range from relatively undifferentiated masses of cells ("indeterminate" galls) to highly organized structures with distinct tissue layers ("determinate" galls). Determinate galls usually have a shape that is specific to each insect species. Cynipids, cecidomyiids, and eriococcids form some of the most histologically complex and specialized galls; these galls have distinct tissue layers or types that may bear little resemblance to the plant part from which they are derived. Among the determinate galls, different shapes correlate with mode of gall formation, which is related to the initial position and feeding method of the insect (as discussed below). Some common types of galls are:

• **covering galls**, in which the insect becomes enclosed within the gall, either with an opening (ostiole) to the exterior, as in coccoid galls (Fig. 11.5a,b), or without any ostiole, as in cynipid galls (Fig. 11.5c);
• **filz galls**, which are characterized by their hairy epidermal outgrowths (Fig. 11.5d);
• **roll** and **fold galls**, in which differential growth provoked by insect feeding results in rolled or twisted leaves, shoots, or stems, which are often swollen, as in many aphid galls (Fig. 11.5e);
• **pouch galls**, which develop as a bulge of the leaf blade, forming an invaginated pouch on one side and a prominent bulge on the other, as in many psyllid galls (Fig. 11.5f);
• **mark galls**, in which the insect egg is deposited inside stems or leaves so that the larva is completely enclosed throughout its development, as in sawfly galls (Fig. 11.5g);
• **pit galls**, in which a slight depression, sometimes surrounded by a swelling, is formed where the insect feeds;
• **bud** and **rosette galls**, which vary in complexity and cause enlargement of the bud or sometimes multiplication and miniaturization of new leaves, forming a pine-cone-like gall.

Gall formation may involve two separate processes: (i) initiation and (ii) subsequent growth and maintenance of structure. Usually, galls can be stimulated to develop only from actively growing plant tissue. Therefore, galls are initiated on young leaves, flower buds, stems, and roots, and rarely on mature plant parts. Some complex galls develop only from undifferentiated meristematic tissue, which becomes molded into a distinctive gall by the activities of the insect. Development and growth of insect-induced galls (including, if present, the nutritive cells upon which some insects feed) depend upon continued stimulation of the plant cells by

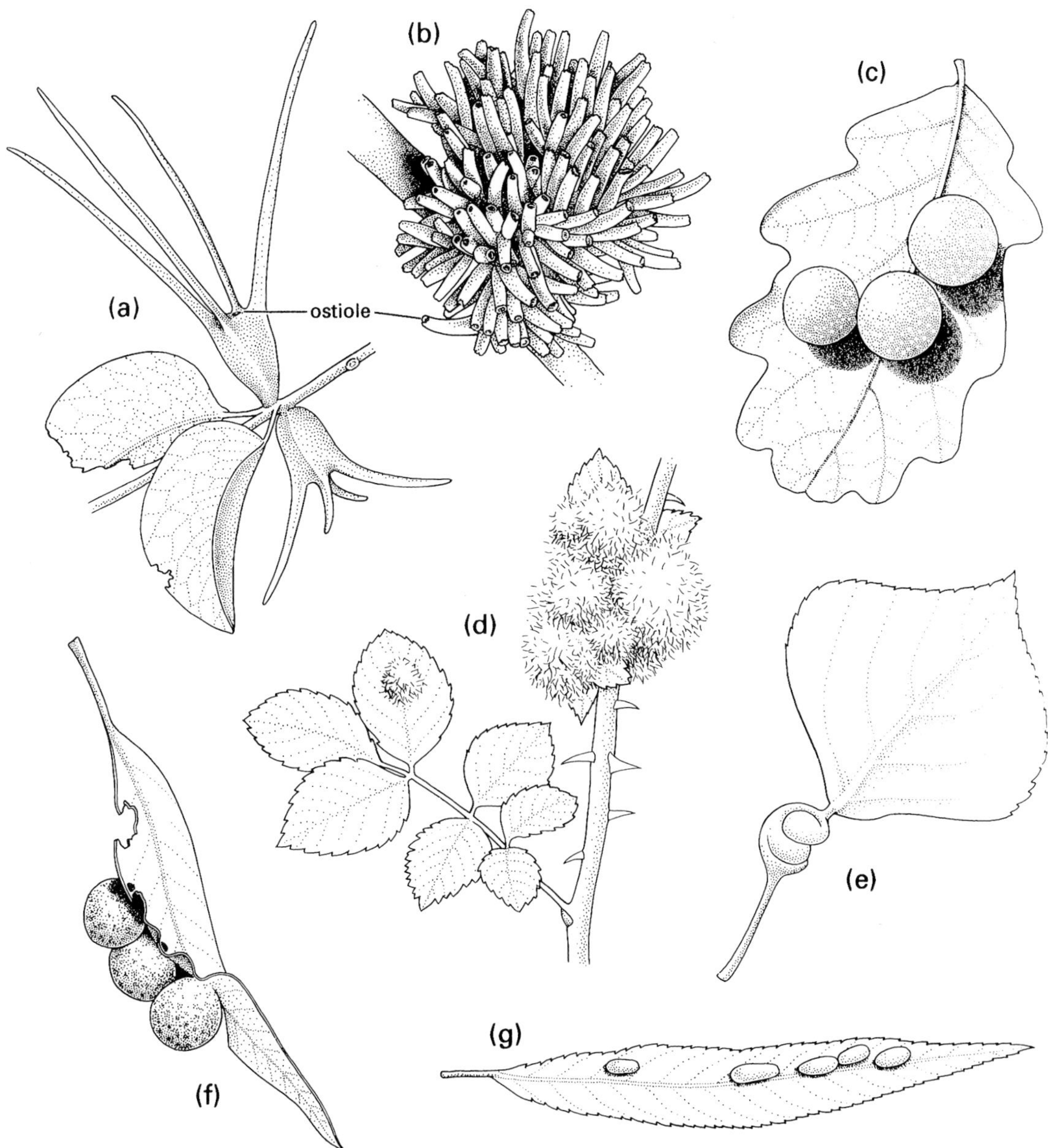

Fig. 11.5 A variety of insect-induced galls: (a) two coccoid galls, each formed by a female of *Apiomorpha munita* (Hemiptera: Eriococcidae) on the stem of *Eucalyptus melliodora*; (b) a cluster of galls each containing a male of *A. munita* on *E. melliodora*; (c) three oak cynipid galls formed by *Cynips quercusfolii* (Hymenoptera: Cynipidae) on a leaf of *Quercus* sp.; (d) rose bedeguar galls formed by *Diplolepis rosae* (Hymenoptera: Cynipidae) on *Rosa* sp.; (e) a leaf petiole of lombardy poplar, *Populus nigra*, galled by the aphid *Pemphigus spirothecae* (Hemiptera: Aphididae); (f) three psyllid galls, each formed by a nymph of *Glycaspis* sp. (Hemiptera: Psyllidae) on a eucalypt leaf; (g) willow bean galls of the sawfly *Pontania proxima* (Hymenoptera: Tenthredinidae) on a leaf of *Salix* sp. ((d–g) After Darlington 1975.)

Box 11.2 The grape phylloxera

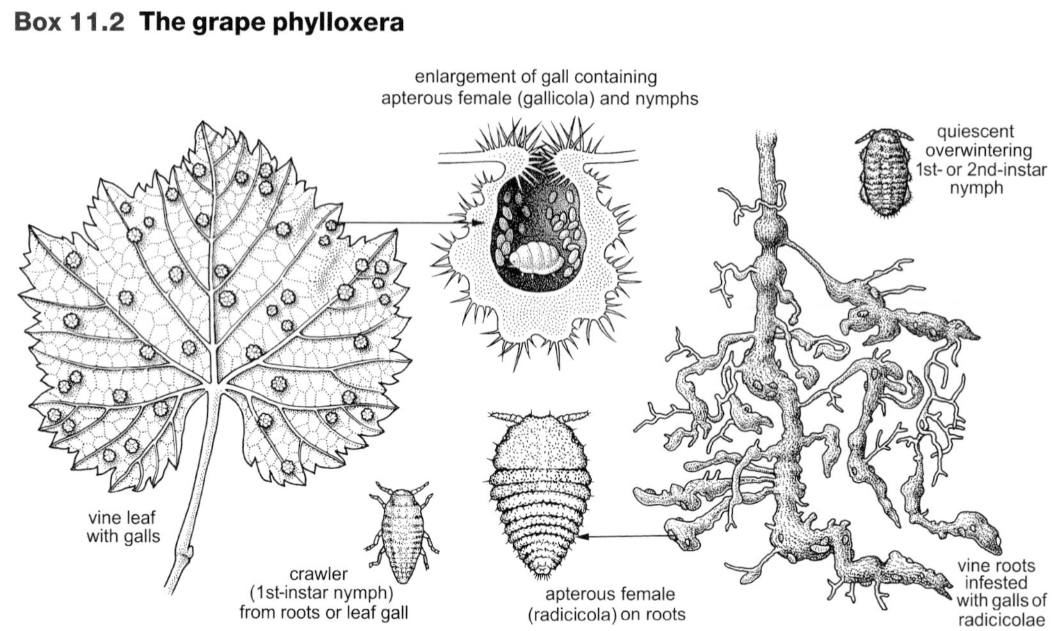

An example of the complexity of a galling life cycle, host-plant resistance, and even naming of an insect is provided by the grape phylloxera, sometimes called the grape louse. This aphid's native range and host is temperate–subtropical from eastern North America and the south-west including Mexico, on a range of species of wild grapes (Vitaceae: *Vitis* spp.). Its complete life cycle is holocyclic (restricted to a single host). In its native range, its life cycle commences with the hatching of an overwintering egg, which develops into a **fundatrix** that crawls from the vine bark to a developing leaf where a pouch gall is formed in the rapidly growing meristematic tissue (as shown here, after several sources). Numerous generations of further apterous offspring are produced, most of which are **gallicolae** – gall inhabitants that either continue to use the maternal gall or induce their own. Some of the apterae, termed **radicicolae**, migrate downwards to the roots. In warm climate regions such as California, South Africa, and Australia where the phylloxera is introduced, it is radicicolae that survive the winter when vine leaves are shed along with their gallicolae. In the soil, radicicolae form nodose and tuberose galls (swellings) on the subapices of young roots (as illustrated here for the asexual life cycle). In fall, in those biotypes with sexual stages, alates (**sexuparae**) are produced that fly from the soil to the stems of the vine, where they give rise to apterous, non-feeding **sexuales**. These mate, and each female lays a single overwintering egg. Within the natural range of aphid and host, the plants appear to show little damage from phylloxera, except perhaps in the late season in which limited growth provides only a little new meristematic tissue for the explosive increase in gallicolae.

This straightforward (for an aphid) life cycle shows modifications outside the natural range, involving loss of the sexual and aerial stages, with persistence owing to entirely parthenogenetic radicicolae. Also involved are dramatic deleterious effects on the host vine by phylloxera feeding. This is of major economic importance when the host is *Vitis vinifera*, the native grape vine of the Mediterranean and Middle East. In the mid-19th century American vines carrying phylloxera were imported into Europe; these devastated European grapes, which had no resistance to the aphid. Damage is principally through roots rotting under heavy loads of radicicolae rather than sucking *per se*, and generally there is no aerial gall-inducing stage. The shipment from eastern USA to France by Charles Valentine Riley of a natural enemy, the mite *Tyroglyphus phylloxerae*, in 1873 was the first intercontinental attempt to control a pest insect. However, eventual control was achieved by grafting the already very diverse range of European grape cultivars (cépages such as Cabernet, Pinot Noir, or Merlot) onto phylloxera-resistant rootstocks of North American *Vitis* species. Some *Vitis* species are not attacked by phylloxera, and in others the infestation starts and is either tolerated at a low level or rejected. Resistance (section 16.6) is mainly a matter of the speed at which the plant can produce inhibitory complex compounds from naturally produced phenolics that can isolate each developing tuberose gall. Recently it seems that some genotypes of phylloxera have circumvented certain resistant rootstocks, and resurgence may be expected.

The history of the scientific name of grape phylloxera is nearly as complicated as the life cycle – phylloxera may now refer only to the family Phylloxeridae, in which species of *Phylloxera* are mainly on *Juglans* (walnuts), *Carya* (pecans), and relatives. The grape phylloxera has been known as *Phylloxera vitifoliae* and also as *Viteus vitifoliae* (under which name it is still known in Europe), but it is increasingly accepted that the genus name should be *Daktulosphaira* if a separate genus is warranted. Whether there is a single species (*D. vitifoliae*) with a very wide range of behaviors associated with different host species and cultivars is an open question. There certainly is wide geographical variation in responses and host tolerances but as yet no morphometric, molecular, or behavioral traits correlate well with any of the reported "biotypes" of *D. vitifoliae*.

the insect. Gall growth ceases if the insect dies or reaches maturity. It appears that gall insects, rather than the plants, control most aspects of gall formation, largely via their feeding activities.

The mode of feeding differs in different taxa as a consequence of fundamental differences in mouthpart structure. The larvae of gall-inducing beetles, moths, and wasps have biting and chewing mouthparts, whereas larval gall midges and nymphal aphids, coccoids, psyllids, and thrips have piercing and sucking mouthparts. Larval gall midges have vestigial mouthparts and largely absorb nourishment by suction. Thus, these different insects mechanically damage and deliver chemicals (or perhaps genetic material, see below) to the plant cells in a variety of ways.

Little is known about what stimulates gall induction and growth. Wounding and plant hormones (such as cytokinins) appear important in indeterminate galls, but the stimuli are probably more complex for determinate galls. Oral secretions, anal excreta, and accessory gland secretions have been implicated in different insect–plant interactions that result in determinate galls. The best-studied compounds are the salivary secretions of Hemiptera. Salivary substances, including amino acids, auxins (and other plant growth regulators), phenolic compounds, and phenol oxidases, in various concentrations, may have a role either in gall initiation and growth or in overcoming the defensive necrotic reactions of the plant. Plant hormones, such as auxins and cytokinins, must be involved in cecidogenesis but it is equivocal whether these hormones are produced by the insect, by the plant as a directed response to the insect, or are incidental to gall induction. In certain complex galls, such as those of eriococcoids and cynipids, it is conceivable that the development of the plant cells is redirected by semiautonomous genetic entities (viruses, plasmids, or transposons) transferred from the insect to the plant. Thus, the initiation of such galls may involve the insect acting as a DNA or RNA donor, as in some wasps that parasitize insect hosts (Box 13.1). Unfortunately, in comparison with anatomical and physiological studies of galls, genetic investigations are in their infancy.

The gall-inducing habit may have evolved either from plant mining and boring (especially likely for Lepidoptera, Hymenoptera, and certain Diptera) or from sedentary surface feeding (as is likely for Hemiptera, Thysanoptera, and cecidomyiid Diptera). It is believed to be beneficial to the insects, rather than a defensive response of the plant to insect attack. All gall insects derive their food from the tissues of the gall and also some shelter or protection from natural enemies and adverse conditions of temperature or moisture. The relative importance of these environmental factors to the origin of the galling habit is difficult to ascertain because current advantages of gall living may differ from those gained in the early stages of gall evolution. Clearly, most galls are "sinks" for plant assimilates – the nutritive cells that line the cavity of wasp and fly galls contain higher concentrations of sugars, protein, and lipids than ungalled plant cells. Thus, one advantage of feeding on gall rather than normal plant tissue is the availability of high-quality food. Moreover, for sedentary surface feeders, such as aphids, psyllids, and coccoids, galls furnish a more protected microenvironment than the normal plant surface. Some cecidozoa may "escape" from certain parasitoids and predators that are unable to penetrate galls, particularly galls with thick woody walls.

Other natural enemies, however, specialize in feeding on gall-living insects or their galls and sometimes it is difficult to determine which insects were the original inhabitants. Some galls are remarkable for the association of an extremely complex community of species, other than the gall causer, belonging to diverse insect groups. These other species may be either parasitoids of the gall former (i.e. parasites that cause the eventual death of their host; Chapter 13) or inquilines ("guests" of the gall former) that obtain their nourishment from tissues of the gall. In some cases, gall inquilines cause the original inhabitant to die through abnormal growth of the gall; this may obliterate the cavity in which the gall former lives or prevent emergence from the gall. If two species are obtained from a single gall or a single type of gall, one of these insects must be a parasitoid, an inquiline, or both. There are even cases of hyperparasitism, in which the parasitoids themselves are subject to parasitization (section 13.3.1).

11.2.5 Seed predation

Plant seeds usually contain higher levels of nutrients than other tissues, providing for the growth of the seedling. Specialist seed-eating insects use this resource. Notable seed-eating insects are many beetles (below), harvester ants (especially species of *Messor*, *Monomorium*, and *Pheidole*), which store seeds in underground granaries, bugs (many Coreidae, Lygaeidae, Pentatomidae, Pyrrhocoridae, and Scutelleridae) that suck out the contents of developing or mature seeds, and a few moths (such as some Gelechiidae and Oecophoridae).

Harvester ants are ecologically significant seed predators. These are the dominant ants in terms of biomass and/or colony numbers in deserts and dry grasslands in many parts of the world. Usually, the species are highly polymorphic, with the larger individuals possessing powerful mandibles capable of cracking open seeds. Seed fragments are fed to larvae, but probably many harvested seeds escape destruction either by being abandoned in stores or by germinating quickly within the ant nests. Thus, seed harvesting by ants, which could be viewed as exclusively detrimental, actually may carry some benefits to the plant through dispersal and provision of local nutrients to the seedling.

An array of beetles (especially Curculionidae and bruchine Chrysomelidae) develop entirely within individual seeds or consume several seeds within one fruit. Some bruchine seed beetles, particularly those attacking leguminous food plants such as peas and beans, are serious pests. Species that eat dried seeds are preadapted to be pests of stored products such as pulses and grains. Adult beetles typically oviposit onto the developing ovary or the seeds or fruits, and some larvae then mine through the fruit and/or seed wall or coat. The larvae develop and pupate inside seeds, thus destroying them. Successful development usually occurs only in the final stages of maturity of seeds. Thus, there appears to be a "window of opportunity" for the larvae; a mature seed may have an impenetrable seed coat but if young seeds are attacked, the plant can abort the infected seed or even the whole fruit or pod if little investment has been made in it. Aborted seeds and those shed to the ground (whether mature or not) generally are less attractive to seed beetles than those retained on the plant, but evidently stored-product pests have no difficulty in developing within cast (i.e. harvested and stored) seeds. The larvae of the granary weevil, *Sitophilus granarius* (Box 11.10), and rice weevil, *S. oryzae*, develop inside dry grains of corn, wheat, rice, and other plants.

Plant defense against seed predation includes the provision of protective seed coatings or toxic chemicals (allelochemicals), or both. Another strategy is the synchronous production by a single plant species of an abundance of seeds, often separated by long intervals of time. Seed predators either cannot synchronize their life cycle to the cycle of glut and scarcity, or are overwhelmed and unable to find and consume the total seed production.

11.2.6 Insects as biological control agents for weeds

Weeds are simply plants that are growing where they are not wanted. Some weed species are of little economic or ecological consequence, whereas the presence of others results in significant losses to agriculture or causes detrimental effects in natural ecosystems. Most plants are weeds only in areas outside their native distribution, where suitable climatic and edaphic conditions, usually in the absence of natural enemies, favor their growth and survival. Sometimes exotic plants that have become weeds can be controlled by introducing host-specific phytophagous insects from the area of origin of the weed. This is called classical biological control of weeds and it is analogous to the classical biological control of insect pests (as explained in detail in section 16.5). Another form of biological control, called augmentation (section 16.5), involves increasing the natural level of insect enemies of a weed and thus requires mass rearing of insects for inundative release. This method of controlling weeds is unlikely to be cost-effective for most insect–plant systems. The tissue damage caused by introduced or augmented insect enemies of weeds may limit or reduce vegetative growth (as shown for the weed discussed in Box 11.3), prevent or reduce reproduction, or make the weed less competitive than other plants in the environment.

A classical biological control program involves a sequence of steps that include biological as well as sociopolitical considerations. Each program is initiated with a review of available data (including taxonomic and distributional information) on the weed, its plant relatives, and any known natural enemies. This forms the basis for assessment of the nuisance status of the target weed and a strategy for collecting, rearing, and testing the utility of potential insect enemies. Regulatory authorities must then approve the proposal to attempt control of the weed. Next, foreign exploration and local surveys must determine the potential control agents attacking the weed both in its native and introduced ranges. The weed's ecology, especially in relation to its natural enemies, must be studied in its native range. The host-specificity of potential control agents must be tested, either inside or outside the country of introduction and, in the former case, always in quarantine. The results of these tests will determine whether the regulatory authorities approve the importation of the agents for subsequent release or only for further testing, or refuse approval. After importation, there is a period of rearing in quarantine to eliminate any imported diseases or parasitoids, prior to mass rearing in preparation for field release. Release is dependent on the quarantine procedures being approved by the regulatory authorities. After release, the establishment, spread, and effect of the insects on the weed must be monitored. If weed control is attained at the initial release site(s), the spread of the insects is assisted by manual distribution to other sites.

There have been some outstandingly successful cases of deliberately introduced insects controlling invasive weeds. The control of the water weed salvinia by a *Cyrtobagous* weevil (as outlined in Box 11.3), and of prickly pear cacti, *Opuntia* species, by the larvae of the *Cactoblastis* moth are just two examples. On the whole, however, the chances of successful biological control of weeds by released phytophagous organisms are not high (Fig. 11.6) and vary in different circumstances, often unpredictably. Furthermore, biological control systems that are highly successful and appropriate for weed control in one geographical region may be potentially disastrous in another region. For example, in Australia, which has no native cacti, *Cactoblastis* was used safely and effectively to almost completely destroy vast infestations of *Opuntia* cactus. However, this moth also was introduced into the West Indies and from there spread to Cuba and Florida, where it has increased the likelihood of extinction of native cactus species, and it now threatens North America's (and Mexico's) unique cacti-dominated ecosystems.

In general, perennial weeds of uncultivated areas are well suited to classical biological control, as long-lived plants, which are predictable resources, are generally associated with host-specific insect enemies. Cultivation, however, can disrupt these insect populations. In contrast, augmentation of insect enemies of a weed may be best suited to annual weeds of cultivated land, where mass-reared insects could be released to control the plant early in its growing season. Sometimes it is claimed that highly variable, genetically outcrossed weeds are hard to control and that insects "newly associated" (in an evolutionary sense) with a weed have greater control potential because of their infliction of greater damage. However, the number of studies for which control assessment is possible is limited and the reasons for variation or failure in control of weeds are diverse. Currently, prediction of the success or failure of control in terms of weed or phytophage ecology and/or behavior is unsatisfactory. The interactions of plants, insects, and environmental factors are complicated and likely to be case-specific.

In addition to the uncertainty of success of classical biological control programs, the control of certain weeds can cause potential conflicts of interest. Sometimes not everyone may consider the target a weed. For example, in Australia, the introduced *Echium plantagineum* (Boraginaceae) is called "Paterson's curse" by those who consider it an agricultural weed and "Salvation Jane" by some pastoralists and beekeepers who regard it as a source of fodder for livestock and nectar for bees. A second type of conflict may arise if the natural phytophages of the weed are oligophagous rather than monophagous, and thus may feed on a few species other than the target weed. In this case, the introduction of insects that are not strictly host-specific may pose a risk

Box 11.3 Salvinia and phytophagous weevils

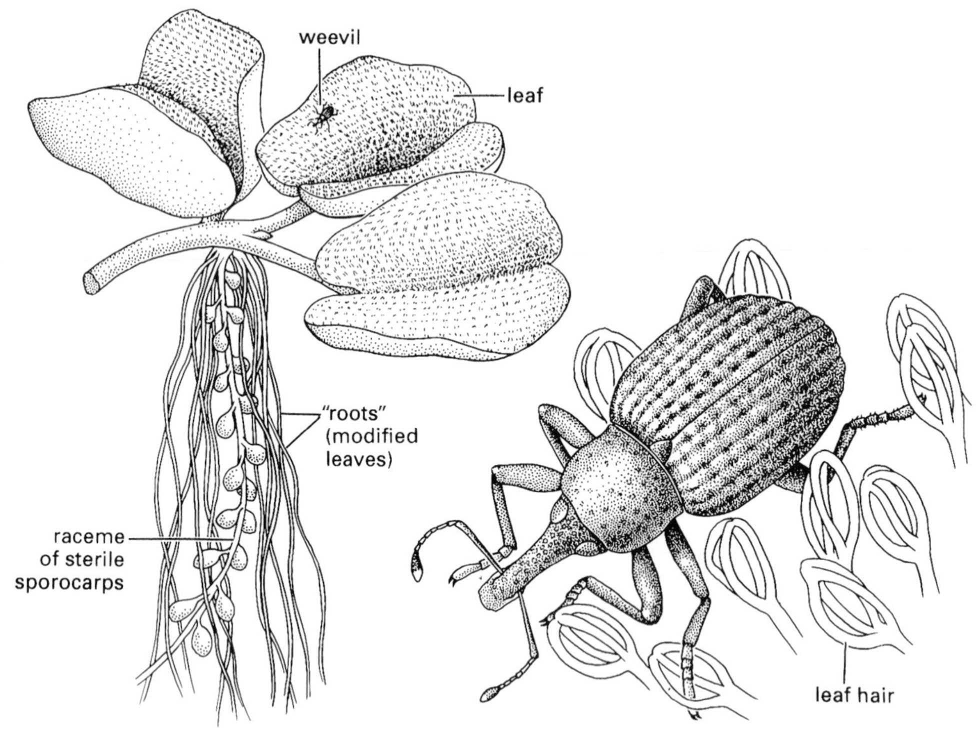

The floating aquatic fern salvinia (Salviniaceae: *Salvinia molesta*) (illustrated here, after Sainty & Jacobs 1981) has spread by human agency since 1939 to many tropical and subtropical lakes, rivers, and canals throughout the world. Salvinia colonies consist of ramets (units of a clone) connected by horizontal branching rhizomes. Growth is favored by warm, nitrogen-rich water. Conditions suitable for vegetative propagation and the absence of natural enemies in its non-native range have allowed very rapid colonization of large expanses of freshwater. Salvinia becomes a serious weed because its thick mats completely block waterways, choking the flow and disrupting the livelihood of people who depend on them for transport, irrigation, and food (especially fish, rice, sago palms, etc.). This problem was especially acute in parts of Africa, India, south-east Asia, and Australasia, including the Sepik River in Papua New Guinea. Expensive manual and mechanical removal and herbicides could achieve limited control, but some 2000 km^2 of water surface were covered by this invasive plant by the early 1980s. The potential of biological control was recognized in the 1960s, although it was slow to be used (for reasons outlined below) until the 1980s, when outstanding successes were achieved in most areas where biological control was attempted. Choked lakes and rivers became open water again.

The phytophagous insect responsible for this spectacular control of *S. molesta* is a tiny (2 mm long) weevil (Curculionidae) called *Cyrtobagous salviniae* (shown enlarged in the drawing on the right, after Calder & Sands 1985). Adult weevils feed on salvinia buds, whereas larvae tunnel through buds and rhizomes as well as feeding externally on roots. The weevils are host-specific, have a high searching efficiency for salvinia, and can live at high population densities without intraspecific interference stimulating emigration. These characteristics allow the weevils to control salvinia effectively.

Initially, biological control of salvinia failed because of unforeseen taxonomic problems with the weed and the weevil. Prior to 1972, the weed was thought to be *Salvinia auriculata*, which is a South American species fed upon by the weevil *Cyrtobagous singularis*. Even when the weed's correct identity was established as *Salvinia molesta*, it was not until 1978 that its native range was discovered to be south-eastern Brazil. Weevils feeding there on *S. molesta* were believed to be conspecific with *C. singularis* feeding on *S. auriculata*. However, after preliminary testing and subsequent success in controlling *S. molesta*, the weevil was recognized as specific to *S. molesta*, new to science, and named as *C. salviniae*.

The benefits of control to people living in Africa, Asia, the Pacific, and other warm regions are substantial, whether measured in economic terms or as savings in human health and social systems. For example, villages in Papua New Guinea that were abandoned because of salvinia have been reoccupied. Similarly, the environmental benefits of eliminating salvinia infestations are great, as this weed is capable of reducing a complex aquatic ecosystem to a virtual monoculture. Now, control by this weevil is benefiting aquatic systems in the USA, especially the south-eastern states where *S. molesta* was introduced in the 1990s through the aquarium and landscape trades.

The economics of salvinia control have been studied only in Sri Lanka, where a cost–benefit analysis showed returns on investment of 53 : 1 in terms of cash and 1678 : 1 in terms of hours of labor. Appropriately, the team responsible for the ecological research that led to biological control of salvinia was recognized by the award of the UNESCO Science Prize in 1985. Taxonomists made essential contributions by establishing the true identities of the salvinias and the weevils.

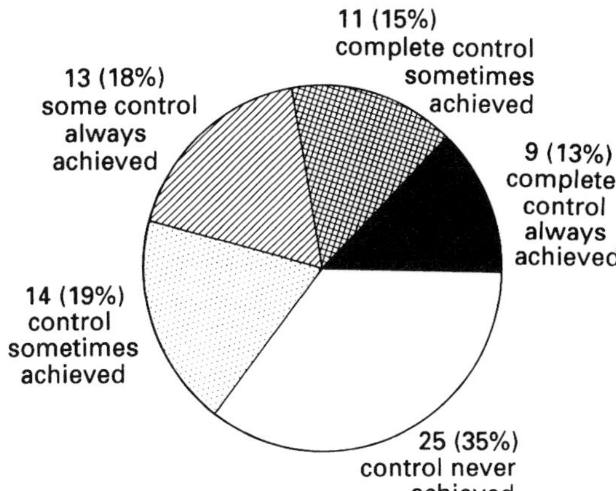

Fig. 11.6 Pie chart showing the possible outcomes of releases of alien phytophagous organisms against invasive plants for the biological control of these weeds. The data include 72 weed species that have agents introduced and established long enough to permit control assessment. (After Sheppard 1992; based on data from Julien 1992.)

for beneficial and/or native plants in the proposed area of introduction of the control agent(s). For example, some of the insects that can be or have been introduced into Australia as control agents for *E. plantagineum* also feed on other boraginaceous plants. The risks of damage to such non-target species must be assessed carefully prior to releasing foreign insects for the biological control of a weed. Some introduced phytophagous insects may become pests in their new habitat.

11.3 INSECTS AND PLANT REPRODUCTIVE BIOLOGY

Insects are intimately associated with plants. Agriculturalists, horticulturalists, and gardeners are aware of their role in damage and disease dispersal. However, certain insects are vitally important to many plants, assisting in their reproduction, through pollination, or their dispersal, through spreading their seeds.

11.3.1 Pollination

Sexual reproduction in plants involves **pollination** – the transfer of pollen (male germ cells in a protective covering) from the anthers of a flower to the stigma (Fig. 11.7a). A pollen tube grows from the stigma down the style to an ovule in the ovary where it fertilizes the egg. Pollen generally is transferred either by an animal pollinator or by the wind. Transfer may be from anthers to stigma of the same plant (either of the same flower or a different flower) (self-pollination), or between flowers on different plants (with different genotypes) of the same species (cross-pollination). Animals, especially insects, pollinate most flowering plants. It is argued that the success of the angiosperms relates to the development of these interactions. The benefits of insect pollination (**entomophily**) over wind pollination (**anemophily**) include:

- increase in pollination efficiency, including reduction of pollen wastage;
- successful pollination under conditions unsuitable for wind pollination;

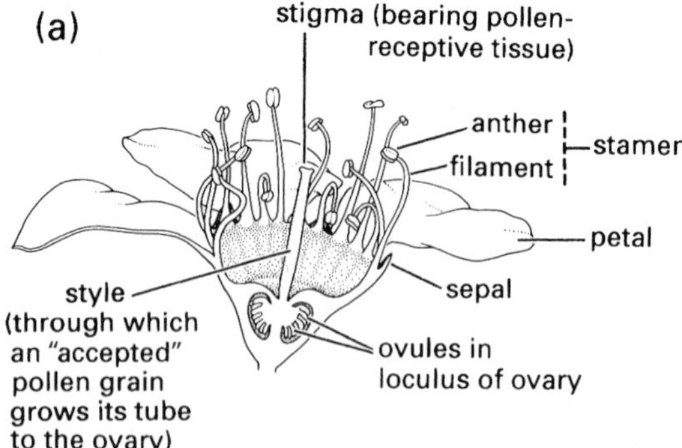

Fig. 11.7 Anatomy and pollination of a tea-tree flower, *Leptospermum* (Myrtaceae): (a) diagram of a flower showing the parts; (b) a jewel beetle, *Stigmodera* sp. (Coleoptera: Buprestidae), feeding from a flower.

• maximization of the number of plant species in a given area (as even rare plants can receive conspecific pollen carried into the area by insects).

Within-flower self-pollination also brings some of these advantages, but continued selfing induces deleterious homozygosity, and rarely is a dominant fertilization mechanism.

Generally, it is advantageous to a plant for its pollinators to be specialist visitors that faithfully pollinate only flowers of one or a few plant species. Pollinator constancy, which may initiate the isolation of small plant populations, is especially prevalent in the Orchidaceae – the most speciose family of vascular plants.

The major **anthophilous** (flower-frequenting) taxa among insects are the beetles (Coleoptera), flies (Diptera), wasps, bees, and ants (Hymenoptera), thrips (Thysanoptera), and butterflies and moths (Lepidoptera). These insects visit flowers primarily to obtain nectar and/or pollen, but even some predatory insects may pollinate the flowers that they visit. Nectar primarily consists of a solution of sugars, especially glucose, fructose, and sucrose. Pollen often has a high protein content plus sugar, starch, fat, and traces of vitamins and inorganic salts. In the case of a few bizarre interactions, male hymenopterans are attracted neither by pollen nor by nectar but by the resemblance of certain orchid flowers in shape, color, and odor to their conspecific females (see Plate 4.4, facing p. 14). In attempting to mate (**pseudocopulate**) with the insect-mimicking flower (see Plate 4.5), the male inadvertently pollinates the orchid with pollen that adhered to his body during previous pseudocopulations. Pseudocopulatory pollination is common among Australian thynnine wasps (Tiphiidae), but occurs in a few other wasp groups, some bees, and rarely in ants.

Cantharophily (beetle pollination) may be the oldest form of insect pollination. Beetle-pollinated flowers often are white or dull colored, strong smelling, and regularly bowl- or dish-shaped (Fig. 11.7). Beetles mostly visit flowers for pollen, although nutritive tissue or easily accessible nectar may be utilized, and the plant's ovaries usually are well protected from the biting mouthparts of their pollinators. The major beetle families that commonly or exclusively contain anthophilous species are the Buprestidae (jewel beetles; Fig. 11.7b), Cantharidae (soldier beetles), Cerambycidae (longicorn or longhorn beetles), Cleridae (checkered beetles), Dermestidae, Lycidae (net-winged beetles), Melyridae (soft-winged flower beetles), Mordellidae (tumbling flower beetles), Nitidulidae (sap beetles), and Scarabaeidae (scarabs).

Myophily (fly pollination) occurs when flies visit flowers to obtain nectar (see Plate 4.6), although hover flies (Syrphidae) feed chiefly on pollen rather than nectar. Fly-pollinated flowers tend to be less showy than other insect-pollinated flowers but may have a strong smell, often malodorous. Flies generally utilize many different sources of food and thus their pollinating activity is irregular and unreliable. However, their sheer abundance and the presence of some flies throughout the year mean that they are important pollinators for many plants. Both dipteran groups contain anthophilous species. Among the Nematocera, mosquitoes and bibionids are frequent blossom visitors, and predatory midges, principally of *Forcipomyia* species (Ceratopogonidae), are essential pollinators of cocoa flowers. Pollinators are more numerous in the Brachycera, in which at least 30 families are known to contain anthophilous species. Major pollinator taxa are the Bombyliidae (bee flies), Syrphidae, and muscoid families.

Many members of the large order Hymenoptera visit flowers for nectar and/or pollen. The Apocrita, which contains most of the wasps (as well as bees and ants), is more important than the Symphyta (sawflies) in terms of **sphecophily** (wasp pollination). Many pollinators are found in the superfamilies Ichneumonoidea and Vespoidea. Fig wasps (Chalcidoidea: Agaonidae) are highly specialized pollinators of the hundreds of species of figs (discussed in Box 11.4). Ants (Vespoidea: Formicidae) are rather poor pollinators, although **myrmecophily** (ant pollination) is known for a few plant species. Ants are commonly anthophilous (flower loving), but rarely pollinate the plants that they visit. Two hypotheses, perhaps acting together, have been postulated to explain the paucity of ant pollination. First, ants are flightless, often small, and their bodies frequently are smooth, thus they are unlikely to facilitate cross-pollination because the foraging of each worker is confined to one plant, they often avoid contact with the anthers and stigmas, and pollen does not adhere easily to them. Second, the metapleural glands of ants produce secretions that spread over the integument and inhibit fungi and bacteria, but also can affect pollen viability and germination. Some plants actually have evolved mechanisms to deter ants; however, a few, especially in hot dry habitats, appear to have evolved adaptations to ant pollination.

Generally, bees are regarded as the most important group of insect pollinators. They collect nectar and pollen for their brood as well as for their own consumption. There are over 20,000 species of bees worldwide and all are anthophilous. Plants that depend on **melittophily** (bee pollination) often have bright (yellow or blue), sweet-smelling flowers with nectar guides – lines (often visible only as ultraviolet light) on the petals that direct pollinators to the nectar. The main bee pollinator worldwide is the honey bee, *Apis mellifera* (Apidae). The pollination services provided by this bee are extremely important for many crop plants (section 1.2), but in natural ecosystems serious problems can be caused. Honey bees compete with native insect pollinators by depleting nectar and pollen supplies and may disrupt pollination by displacing the specialist pollinators of native plant species.

Most members of the Lepidoptera feed from flowers using a long, thin proboscis. In the speciose Ditrysia (the "higher" Lepidoptera) the proboscis is retractile (Fig. 2.12), allowing feeding and drinking from sources distant from the head. Such a structural innovation may have contributed to the radiation of this successful group, which contains 98% of all lepidopteran species. Flowers pollinated by butterflies and moths often are regular, tubular, and sweet smelling. **Phalaenophily** (moth pollination) typically is associated with light-colored, pendant flowers that have nocturnal or crepuscular anthesis (opening of flowers); whereas **psychophily** (butterfly pollination) is typified by red, yellow, or blue upright flowers that have diurnal anthesis.

Insect–plant interactions associated with pollination are clearly mutualistic. The plant is fertilized by appropriate pollen, and the insect obtains food (or sometimes fragrances) supplied by the plant, often specifically to attract the pollinator. It is clear that plants may experience strong selection as a result of insects. In contrast, in most pollination systems, evolution of the pollinators may have been little affected by the plants that they visited. For most insects, any particular plant is just another source of nectar or pollen and even insects that appear to be faithful pollinators over a short observation period may utilize a range of plants in their lifetime. Nevertheless, symmetrical influences do occur in some insect–plant pollination systems, as evidenced by the specializations of each fig wasp species to the fig species that it pollinates (Box 11.4), and by correlations between moth proboscis (tongue) lengths and flower depths for a range of orchids and some other plants. For example, the Malagasy star orchid, *Angraecum sesquipedale*, has floral spurs that may exceed 30 cm in length, and has a pollinator with a tongue length of some 22 cm, a giant hawkmoth, *Xanthopan morgani praedicta* (Sphingidae) (Fig. 11.8). Only this moth can reach the nectar at the apex of the floral spurs and, during the process of pushing its head into the flower, it pollinates the orchid. This is cited often as a spectacular example of a coevolved "long-tongued" pollinator, whose existence had been predicted by Charles Darwin and Alfred Russel Wallace, who knew of the long-spurred flower but not the hawkmoth. However, the

Box 11.4 Figs and fig wasps

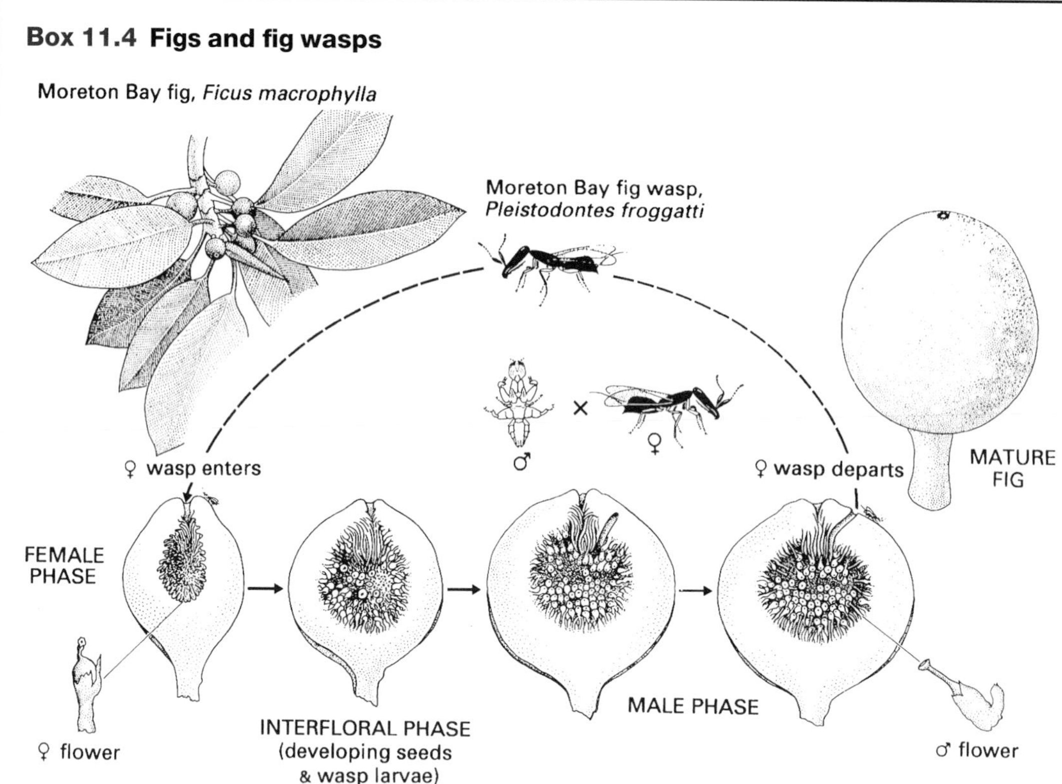

Figs belong to the large, mostly tropical genus *Ficus* (Moraceae) of about 900 species. Each species of fig (except for the self-fertilizing cultivated edible fig) has a complex obligatory mutualism with usually only one species of pollinator. These pollinators all are fig wasps belonging to the hymenopteran family Agaonidae, which comprises numerous species in 20 genera. Each fig tree produces a large crop of 500–1,000,000 fruit (syconia) as often as twice a year, but each fruit requires the action of at least one wasp in order to set seeds. Fig species are either dioecious (with male syconia on separate plants to those bearing the female syconia) or monoecious (with both male and female flowers in the same syconium), with monoecy being the ancestral condition. The following description of the life cycle of a fig wasp in relation to fig flowering and fruiting applies to monoecious figs, such as *F. macrophylla* (illustrated here, after Froggatt 1907; Galil & Eisikowitch 1968).

The female wasp enters the fig syconium via the ostiole (small hole), pollinates the female flowers, which line the spheroidal cavity inside, oviposits in some of them (always short-styled ones), and dies. Each wasp larva develops within the ovary of a flower, which becomes a gall flower. Female flowers (usually long-styled ones) that escape wasp oviposition form seeds. About a month after oviposition, wingless male wasps emerge from their seeds and mate with female wasps still within the fig ovaries. Shortly after, the female wasps emerge from their seeds, gather pollen from another lot of flowers within the syconium (which is now in the male phase), and depart the mature fig to locate another conspecific fig tree in the phase of fig development suitable for oviposition. Different fig trees in a population are in different sexual stages, but all figs on one tree are synchronized. Species-specific volatile attractants produced by the trees allow very accurate, error-free location of another fig tree by the wasps.

Phylogenetic studies suggest that the fig and fig wasp mutualism arose only once because both interacting groups are monophyletic. Significant co-speciation has been inferred but rarely tested. For any given fig and wasp pair, reciprocal selection pressures presumably result in matching of fig and fig wasp traits. For example, the sensory receptors of the wasp respond only to the volatile chemicals of its host fig, and the size and morphology of the guarding scales of the fig ostiole allow entry only to a fig wasp of the "correct" size and shape. It is likely that divergence in a local population of either fig or fig wasp, whether by genetic drift or selection, will induce coevolutionary change in the other. Host-specificity provides reproductive isolation among both wasps and figs, so coevolutionary divergence among populations is likely to lead to speciation. The amazing diversity of *Ficus* and Agaonidae may be a consequence of this coevolution.

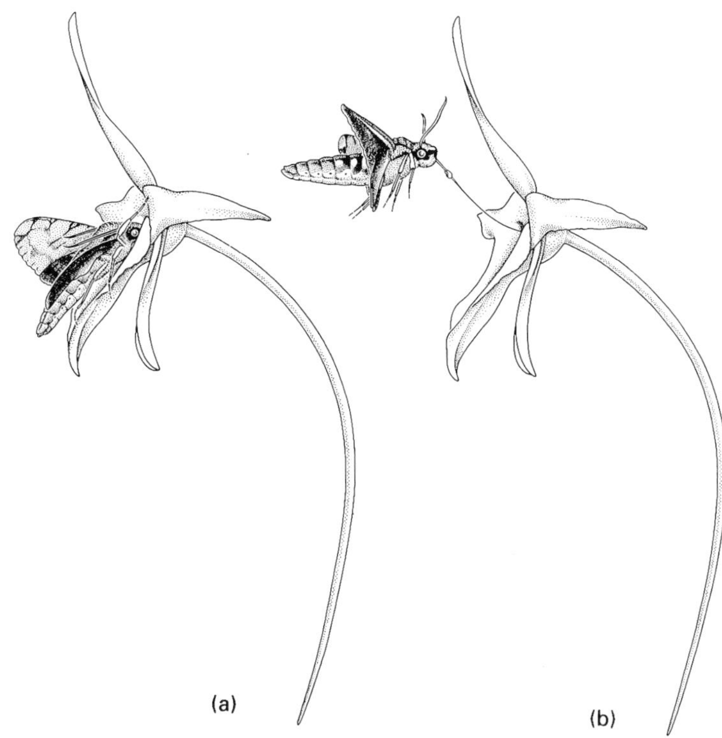

Fig. 11.8 A male hawkmoth of *Xanthopan morgani praedicta* (Lepidoptera: Sphingidae) feeding from the long floral spur of a Malagasy star orchid, *Angraecum sesquipedale*: (a) full insertion of the moth's proboscis; (b) upward flight during withdrawal of the proboscis with the orchid pollinium attached. (After Wasserthal 1997.)

interpretation of this relationship as coevolution has been challenged with the suggestion that the long tongue evolved in the nectar-feeding moth to evade (by distance-keeping and feeding in hovering flight) ambushing predators (e.g. spiders) lurking in other less-specialized flowers frequented by *X. morgani*. In this interpretation, pollination of *A. sesquipedale* follows a host-shift of the preadapted pollinator, with only the orchid showing adaptive evolution. The specificity of location of pollinia (pollen masses) on the tongue of *X. morgani* seems to argue against the pollinator-shift hypothesis, but detailed field study is required to resolve the controversy. Unfortunately, this rare Malagasy insect–plant system is threatened because its natural rainforest habitat is being destroyed.

11.3.2 Myrmecochory: seed dispersal by ants

Many ants are seed predators that harvest and eat seeds (section 11.2.5). Seed dispersal may occur when seeds are accidentally lost in transport or seed stores are abandoned. Some plants, however, have very hard seeds that are inedible to ants and yet many ant species actively collect and disperse them, a phenomenon called **myrmecochory**. These seeds have food bodies, called **elaiosomes**, with special chemical attractants that stimulate ants to collect them. Elaiosomes are seed appendages that vary in size, shape, and color and contain nutritive lipids, proteins, and carbohydrates in varying proportions. These structures have diverse derivations from various ovarian structures in different plant groups. The ants, gripping the elaiosome with their mandibles (Fig. 11.9), carry the entire seed back to their nest, where the elaiosomes are removed and typically fed to the ant larvae. The hard seeds are then discarded, intact and viable, either in an abandoned gallery of the nest, or close to the nest entrance in a refuse pile.

Myrmecochory is a worldwide phenomenon, but is disproportionately prevalent in three plant assemblages: early flowering herbs in the understorey of north temperate mesic forests; perennials in Australian and southern African sclerophyll vegetation; and an eclectic assemblage of tropical plants. Myrmecochorous

Fig. 11.9 An ant of *Rhytidoponera tasmaniensis* (Hymenoptera: Formicidae) carrying a seed of *Dillwynia juniperina* (Fabaceae) by its elaiosome (seed appendage).

plants number more than 1500 species in Australia and about 1300 in South Africa, whereas only about 300 species occur in the rest of the world. They are distributed amongst more than 20 plant families and thus represent an ecological, rather than a phylogenetic, group, although they are predominantly legumes.

This association is of obvious benefit to the ants, for which the elaiosomes represent food; and the mere existence of the elaiosomes is evidence that the plants have become adapted for interactions with ants. Myrmecochory may reduce intraspecific and/or interspecific competition amongst plants by removing seeds to new sites. Seed removal to underground ant nests provides protection from fire or seed predators, such as some birds, small mammals, and other insects. Post-fire South African fynbos (plant) community structure varies according to the presence of different seed dispersing ants (Box 1.2). Furthermore, ant nests are rich in plant nutrients, making them better microsites for seed germination and seedling establishment. However, no universal explanation for myrmecochory should be expected, as the relative importance of factors responsible for myrmecochory must vary according to plant species and geographical location.

Myrmecochory can be called a mutualism, but specificity and reciprocity do not characterize the association. There is no evidence that any myrmecochorous plant relies on a single ant species to collect its seeds. Similarly, there is no evidence that any ant species has adapted to collect the seeds of one particular myrmecochorous species. Of course, ants that harvest elaiosome-bearing seeds could be called a guild, and the myrmecochorous plants of similar form and habitat also could represent a guild. However, it is highly unlikely that myrmecochory represents an outcome of diffuse or guild coevolution, as no reciprocity can be inferred. Elaiosomes are just food items to ants, which display no obvious adaptations to myrmecochory. Thus, this fascinating form of seed dispersal appears to be the result of plant evolution, as a result of selection from ants in general, and not of coevolution of plants and specific ants.

11.4 INSECTS THAT LIVE MUTUALISTICALLY IN SPECIALIZED PLANT STRUCTURES

A great many insects live within plant structures, in bored-out stems, leaf mines, or galls, but these insects create their own living spaces by destruction or physiological manipulation. In contrast, some plants have specialized structures or chambers, which house mutualistic insects and form in the absence of these guests. Two types of these special insect–plant interactions are discussed below.

11.4.1 Ant–plant interactions involving domatia

Domatia (little houses) may be hollow stems, tubers, swollen petioles, or thorns, which are used by ants either for feeding or as nest sites, or both. True domatia are cavities that form independently of ants, such as in plants grown in glasshouses from which ants are excluded. It may be difficult to recognize true domatia in the field because ants often take advantage of natural hollows and crevices such as tunnels bored by beetle or moth larvae. Plants with true domatia, called ant plants or **myrmecophytes**, often are trees, shrubs, or vines of the secondary regrowth or understorey of tropical lowland rainforest.

Ants benefit from association with myrmecophytes through provision of shelter for their nests and readily available food resources. Food comes either directly from the plant through food bodies or extrafloral nectaries (Fig. 11.10a), or indirectly via honeydew-excreting hemipterans living within the domatia (Fig. 11.10b). Food bodies are small nutritive nodules on the foliage or stems of ant plants. Extrafloral nectaries (EFNs) are glands that produce sugary secretions (possibly also containing amino acids) attractive to ants and other insects. Plants with EFNs often occur in temperate areas and lack domatia, for example many Australian *Acacia* species, whereas plants with food

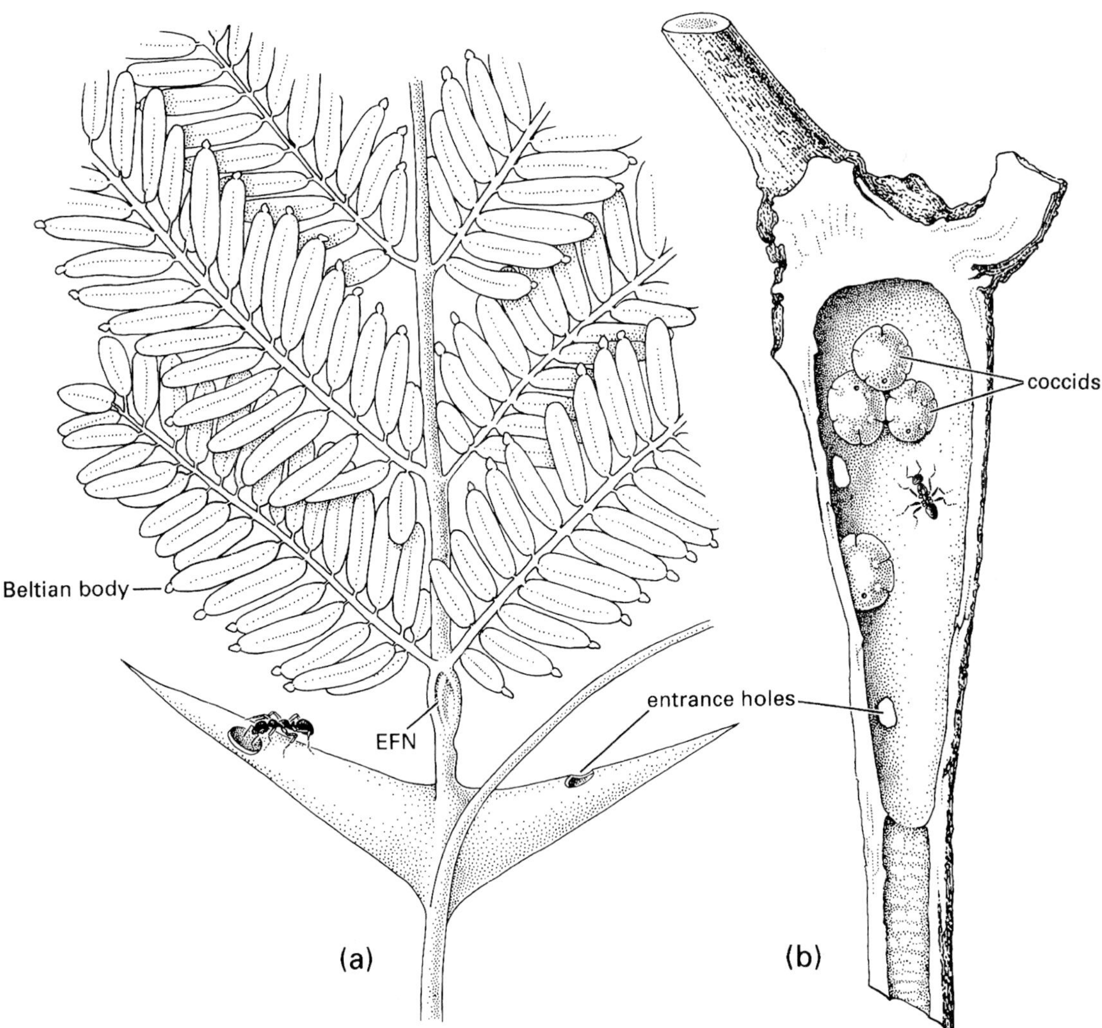

Fig. 11.10 Two myrmecophytes showing the domatia (hollow chambers) that house ants and the food resources available to the ants: (a) a neotropical bull's-horn acacia, *Acacia sphaerocephala* (Fabaceae), with hollow thorns, food bodies, and extrafloral nectaries (EFNs) that are used by the resident *Pseudomyrmex* ants; (b) a hollow swollen internode of *Kibara* (Monimiaceae) with scale insects of *Myzolecanium kibarae* (Hemiptera: Coccidae) that excrete honeydew that is eaten by the resident ants of *Anonychomyrma scrutator*. ((a) After Wheeler 1910; (b) after Beccari 1877.)

bodies nearly always have domatia, and some plants have both EFNs and food bodies. Many myrmecophytes, however, lack both of the latter structures and instead the ants "farm" soft scales or mealybugs (Coccoidea: Coccidae or Pseudococcidae) for their honeydew (sugary excreta derived from phloem on which they feed) and possibly cull them to obtain protein. Like EFNs and food bodies, coccoids can draw the ants into a closer relationship with the plant by providing a resource on that plant.

Obviously, myrmecophytes receive some benefits from ant occupancy of their domatia. The ants may provide protection from herbivores and plant competitors or supply nutrients to their host plant. Some

Fig. 11.11 A tuber of the epiphytic myrmecophyte *Myrmecodia beccarii* (Rubiaceae), cut open to show the chambers inhabited by ants. Ants live in smooth-walled chambers and deposit their refuse in warted tunnels, from which nutrients are absorbed by the plant. (After Monteith 1990.)

ants aggressively defend their plant against grazing mammals, remove herbivorous insects, and prune or detach other plants, such as epiphytes and vines that grow on their host. This extremely aggressive tending is demonstrated by ants of *Pseudomyrmex* that protect *Acacia* in tropical America. Rather than protection, some myrmecophytes derive mineral nutrients and nitrogen from ant-colony waste via absorption through the inner surfaces of the domatia. Such plant "feeding" by ants, called **myrmecotrophy**, can be documented by following the fate of a radioactive label placed in ant prey. Prey is taken into the domatia, eaten, and the remains are discarded in refuse tunnels; the label ends up in the leaves of the plant. Myrmecotrophy occurs in the epiphytic *Myrmecodia* (Rubiaceae) (Fig. 11.11), species of which occur in the Malaysian and Australian regions, especially in New Guinea.

The majority of ant–plant associations may be opportunistic and unspecialized, although some tropical and subtropical ants (e.g. some *Pseudomyrmex* and *Azteca* species) are totally dependent on their particular host plants (e.g. *Acacia* or *Triplaris* and *Cecropia* species, respectively) for food and shelter. Likewise, if deprived of their attendant ants, these myrmecophytes decline. These relationships clearly are obligatorily mutualistic and, no doubt, others remain to be documented.

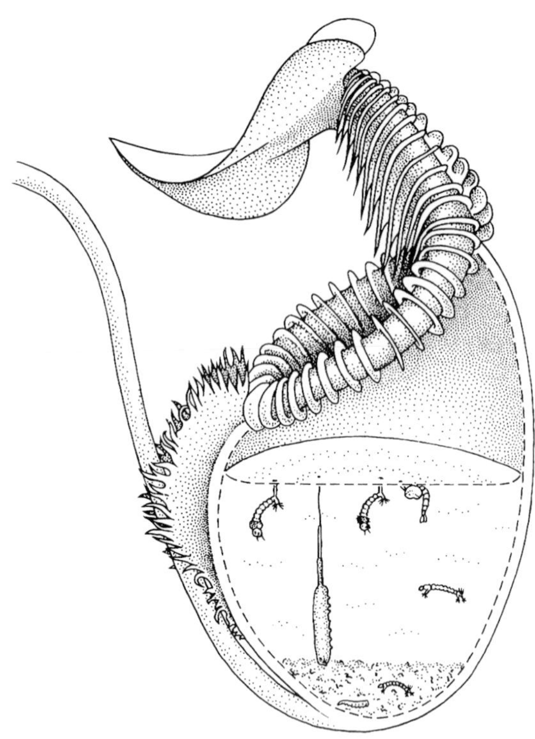

Fig. 11.12 A pitcher of *Nepenthes* (Nepenthaceae) cut open to show fly inquilines in the fluid: (clockwise from the top left) two mosquito larvae, a mosquito pupa, two chironomid midge larvae, a small maggot, and a large rat-tailed maggot.

11.4.2 Phytotelmata: plant-held water containers

Many plants support insect communities in structures that retain water. The containers formed by water retained in leaf axils of many bromeliads ("tank-plants"), gingers, and teasels, for example, or in rot-holes of trees, appear incidental to the plants. Others, namely the pitcher plants, have a complex architecture, designed to lure and trap insects, which are digested in the container liquid (Fig. 11.12).

The pitcher plants are a convergent grouping of the American Sarraceniaceae, Old World Nepenthaceae, and Australian endemic Cephalotaceae. They generally live in nutrient-poor soils. Odor, color, and nectar entice insects, predominantly ants, into modified leaves – the "pitchers". Guard hairs and slippery walls prevent exit and thus the prey cannot escape and drowns in the pitcher liquid, which contains digestive enzymes secreted by the plant.

Text continues on p. 294.

Box 11.5 Orthoptera (grasshoppers, locusts, katydids, and crickets)

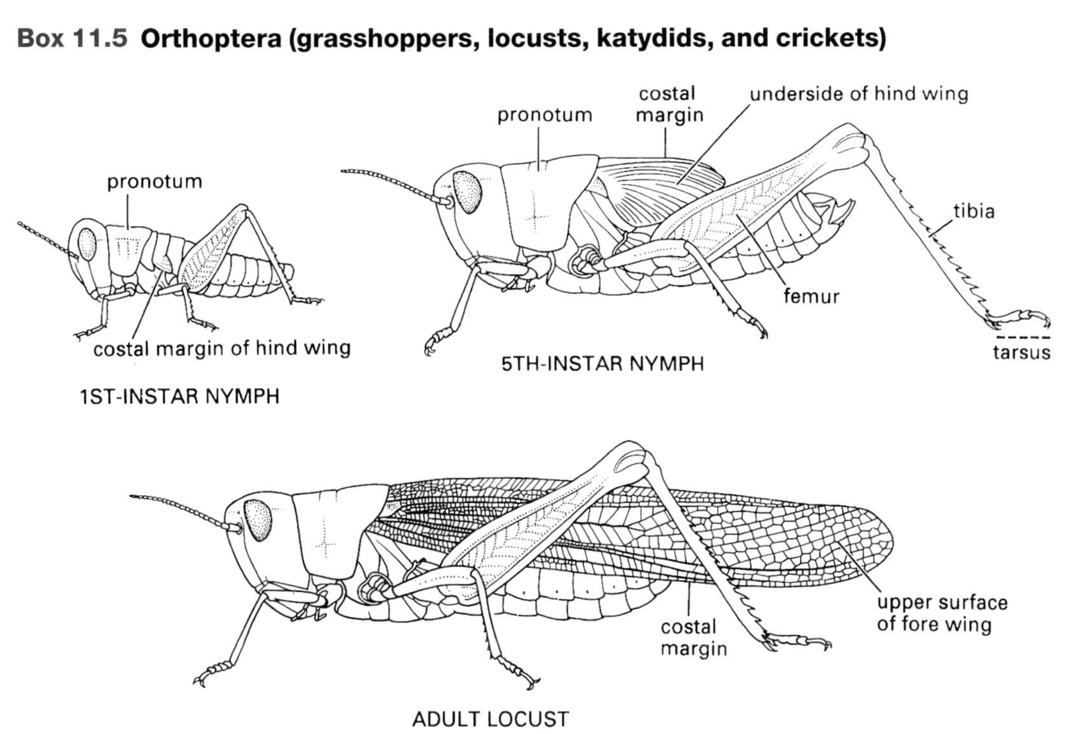

The Orthoptera is a worldwide order of more than 20,000 species in some 13 to 30 families (the classification is unstable), comprising two suborders: Caelifera (grasshoppers and locusts) and Ensifera (katydids and crickets). Orthopterans have hemimetabolous development, and are typically elongate cylindrical, medium-sized to large (up to 12 cm long), with enlarged hind legs for jumping. They are hypognathous and mandibulate, and have well-developed compound eyes; ocelli may be present or absent. The antennae are multisegmented. The prothorax is large, with a shield-like pronotum curving over the pleura; the mesothorax is small, and the metathorax large. The fore wings form narrow, leathery tegmina; the hind wings are broad, with numerous longitudinal and cross-veins, folded beneath the tegmina by pleating. Aptery and brachyptery are frequent. The legs are often elongate and slender, and the hind legs large, usually saltatorial; the tarsi have 1–4 segments. The abdomen has 8–9 annular visible segments, with two or three terminal segments reduced. Females have a well-developed appendicular ovipositor (Fig. 2.23b,c; Box 5.2). The cerci each consist of a single segment.

Courtship may be elaborate and often involves communication by sound production and reception (sections 4.1.3 & 4.1.4). In copulation the male is astride the female, with mating sometimes prolonged for many hours. Ensiferan eggs are laid singly into plants or soil, whereas Caelifera use their ovipositor to bury batches of eggs in soil chambers. Egg diapause is frequent. Nymphs resemble small adults except in the lack of development of wings and genitalia, but apterous adults may be difficult to distinguish from nymphs. In all winged species, nymphal wing pad orientation changes between molts (as illustrated here for a locust); in early instars the wing pad rudiments are laterally positioned with the costal margin ventral, until prior to the penultimate nymphal instar (actually the third last molt) they rotate about their base so that the costal margin is dorsal and the morphological ventral surface is external; the hind wing then overlaps the fore wing (as in the fifth-instar nymph illustrated here). During the molt to the adult, the wings resume their normal position with the costal margin ventral. This wing pad "rotation", otherwise known only in Odonata, is unique to the Orthoptera amongst the orthopteroid orders.

Caelifera are predominantly day-active, fast-moving, visually acute, terrestrial herbivores, and include some destructive insects such as migratory locusts (section 6.10.5; Fig. 6.13). Ensifera are more often night-active, camouflaged or mimetic, and are predators, omnivores, or phytophages.

Phylogenetic relationships are considered in section 7.4.2 and relationships of the order are depicted in Fig. 7.2.

Box 11.6 Phasmatodea (phasmatids, phasmids, stick-insects or walking sticks)

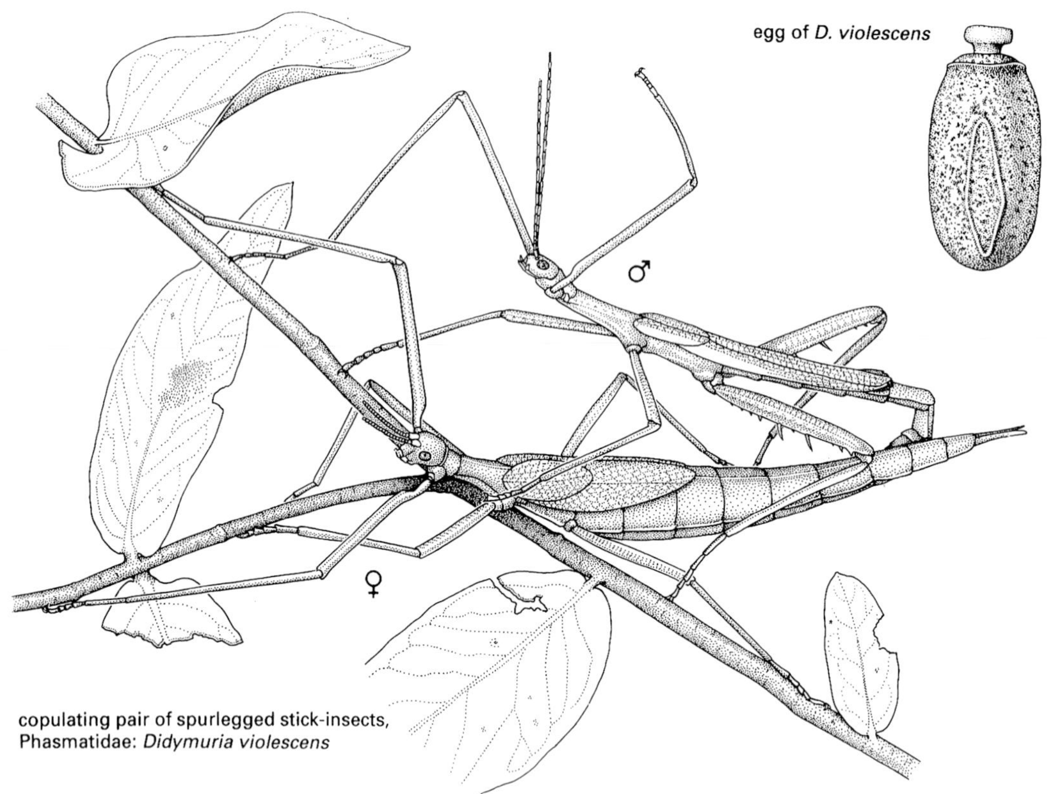

copulating pair of spurlegged stick-insects, Phasmatidae: *Didymuria violescens*

The Phasmatodea is a worldwide, predominantly tropical order of more than 3000 species that is lacking a phylogenetically based classification. They have hemimetabolous development, and are elongate cylindrical and stick-like or flattened and often leaf-like in form, up to >30 cm in body length (the longest species has a total length, including legs, of about 50 cm and is from Borneo). They have mandibulate mouthparts. The compound eyes are anterolaterally placed and relatively small, and ocelli occur only in winged species, often only in males. The antennae range from short to long, with 8–100 segments. The prothorax is small, and the mesothorax and metathorax are elongate if winged, shorter if apterous. The wings, when present, are functional in males but are often reduced in females; many species are apterous in both sexes. The fore wings form leathery tegmina, whereas the hind wings are broad, with a network of numerous cross-veins and the anterior margin toughened as a remigium that protects the folded wing. The legs are elongate, slender, gressorial, with five-segmented tarsi; they can be shed in defense (section 14.3) and may be regenerated at a nymphal molt. The abdomen is 11-segmented, with segment 11 often forming a concealed supra-anal plate in males or a more obvious segment in females; the male genitalia are concealed and asymmetrical. The cerci are variably lengthened and consist of a single segment.

In the often prolonged copulation the smaller male is astride the female, as illustrated here for the spurlegged stick-insect, *Didymuria violescens* (Phasmatidae). The eggs often resemble seeds (as shown here in the enlargement of the egg of *D. violescens*, after CSIRO 1970) and are deposited singly, glued on vegetation or dropped to the ground; there may be lengthy egg diapause. Nymphal phasmatids mostly resemble adults except in their lack of wing and genitalia development, the absence of ocelli, and the fewer antennal segments.

Phasmatodea are phytophagous and predominantly resemble (mimic) various vegetational features such as stems, sticks, and leaves. In conjunction with crypsis, phasmatids demonstrate an array of anti-predator defenses ranging from general slow movement, grotesque and often asymmetrical postures, to death feigning (sections 14.1 & 14.2).

Phylogenetic relationships are considered in section 7.4.2 and depicted in Fig. 7.2.

Box 11.7 Thysanoptera (thrips)

A slide-mounted thrips

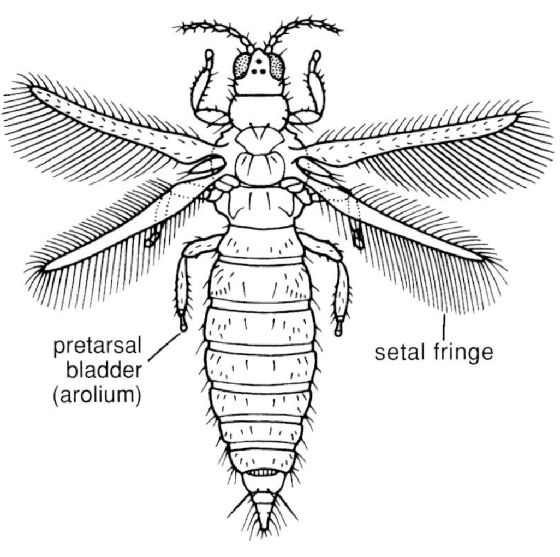

Terebrantia: Thripidae

Living thrips, at rest

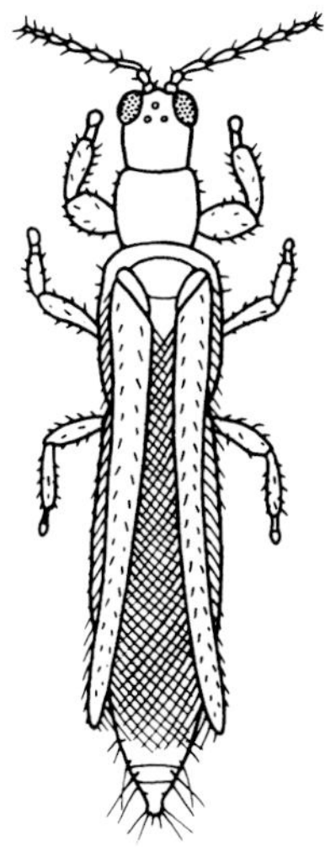

Terebrantia: Thripidae

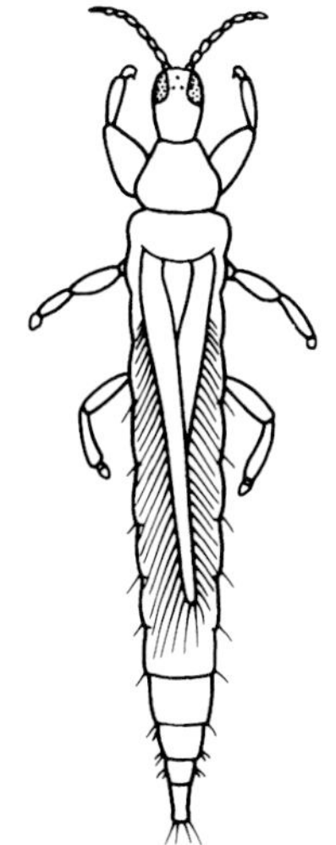

Tubulifera: Phlaeothripidae

The Thysanoptera is a worldwide order of minute to small insects (from 0.5 mm to a maximum length of 15 mm), comprising about 5000 species in two suborders: Terebrantia with seven families (including the speciose Thripidae); and Tubulifera with one family (the speciose Phlaeothripidae). Their development is intermediate between hemi- and holometabolous. The body is slender and elongate, and the head is elongate and usually hypognathous. The mouthparts (Fig. 2.13a) comprise the maxillary laciniae formed as grooved stylets, with the right mandible atrophied and the left mandible formed as a further stylet; the maxillary stylets form a feeding tube. The compound eyes range from small to large, and there are three ocelli in fully winged forms. The antennae are four- to nine-segmented and anteriorly directed. Thoracic development varies according to the presence of wings; fore and hind wings are similar and narrow with a long setal fringe (as illustrated on the left for a terebrantian thrips, after Lewis 1973). At rest the wings are parallel in Terebrantia (middle figure) but overlap in Tubulifera (right figure); microptery and aptery occur. The legs are short and gressorial, sometimes with the fore legs raptorial and the hind legs saltatory; the tarsi are one- or two-segmented, and the pretarsus has an apical protrusible adhesive arolium (bladder or vesicle). The abdomen is 11-segmented (though with only 10 segments visible). In males the genitalia are concealed and symmetrical. In females the cerci are absent; the ovipositor is serrate in Terebrantia, very reduced in Tubulifera.

Eggs are laid into plant tissue (Terebrantia) or into crevices or exposed vegetation (Tubulifera). The first- and second-instar nymphs resemble small adults except with regard to their wings and genitalia; however, instars 3–4 (Terebrantia) or 3–5 (Tubulifera) are resting or pupal stages, during which significant tissue reconstruction takes place. Female thrips are diploid, whereas males (if present) are haploid, produced from unfertilized eggs. Arrhenotokous parthenogenesis is common; thelytoky is rare (section 5.10.1).

The primitive feeding mode of thrips probably was fungal feeding, and about half of the species feed only on fungi, mostly hyphae. Most other thrips primarily are phytophages, feeding on flowers or leaves and including some gall inducers, and there are a few predators. Plant-feeding thrips use their single mandibular stylet to pierce a hole through which the maxillary stylets are inserted. The contents of single cells are sucked out one at a time; pollen- or spore-feeding thrips similarly remove the contents of individual pollen grains or spores. Several cosmopolitan thrips species (e.g. western flower thrips, *Frankliniella occidentalis*) act as vectors of viruses that damage plants. Thrips may aggregate in flowers, where they may act as pollinators. Subsocial behavior, including parental care, is exhibited by a few thrips (section 12.1.1).

Phylogenetic relationships are considered in section 7.4.2 and depicted in Fig. 7.2.

Box 11.8 Hemiptera (bugs, cicadas, leafhoppers, spittle bugs, planthoppers, aphids, jumping plant lice, scale insects, whiteflies)

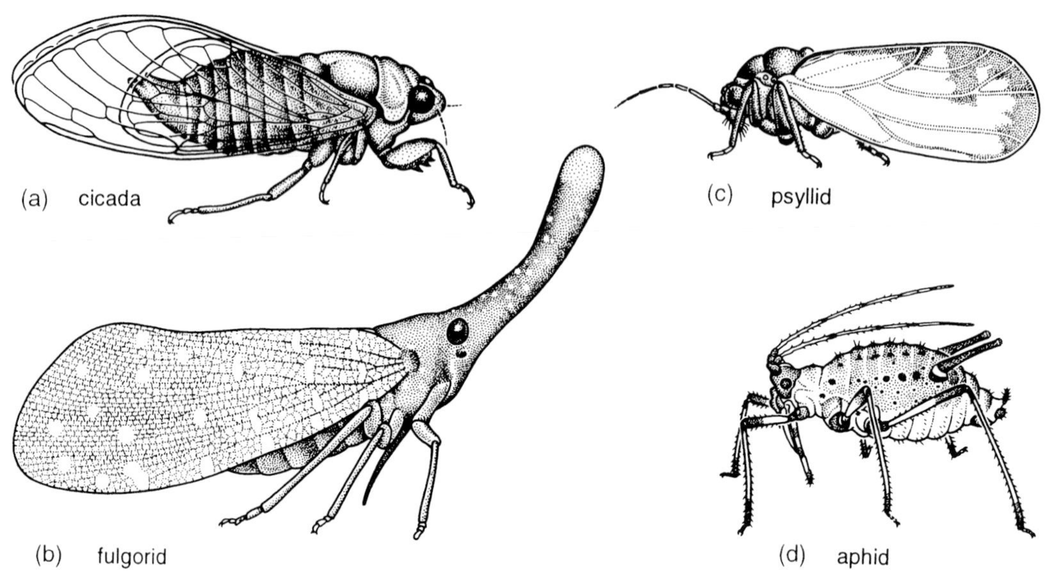

The Hemiptera is distributed worldwide, and is the most diverse of the non-endopterygote orders, with more than 90,000 species in about 140 families. Historically, it was divided into two suborders (sometimes treated as orders): Heteroptera (bugs) and "Homoptera" (cicadas, leafhoppers, spittle bugs, planthoppers, aphids, jumping plant lice (= psylloids), scale insects (= coccoids), and whiteflies). However, homopterans represent a grade of organization (a paraphyletic rather than a monophyletic group). Currently, five suborders can be recognized (Fig. 7.5): (i) Heteroptera, the "true" bugs; (ii) Coleorrhyncha, the moss bugs (family Peloridiidae); (iii) Cicadomorpha (cicadas, leafhoppers, and spittle bugs); (iv) Fulgoromorpha (planthoppers); and (v) Sternorrhyncha (aphids, jumping plant lice, scale insects, and whiteflies). Sometimes the Cicadomorpha and Fulgoromorpha are collectively called the Auchenorrhyncha, but this grouping may also be paraphyletic. Four hemipterans are illustrated here in lateral view: (a) *Cicadetta montana* (Cicadidae), the only British cicada (after drawing by Jon Martin in Dolling 1991); (b) a green lantern bug, *Pyrops sultan* (Fulgoridae), from Borneo (after Edwards 1994); (c) the psyllid *Psyllopsis fraxini* (Psyllidae), which deforms leaflets of ash trees in Britain (after drawing by Jon Martin in Dolling 1991); and (d) an apterous viviparous female of the aphid *Macromyzus woodwardiae* (Aphididae) (after Miyazaki 1987a).

Hemipteran compound eyes are often large, and ocelli may be present or absent. Antennae vary from short with few segments to filiform and multisegmented. The mouthparts comprise mandibles and maxillae modified as needle-like stylets, lying in a beak-like grooved labium (as shown for a pentatomid heteropteran in (e) and (f)), collectively forming a **rostrum** or proboscis. The stylet bundle contains two canals, one delivering saliva, the other uptaking fluid (as shown in (f)); there are no palps. The thorax often consists of large pro- and mesothorax, but a small metathorax. Both pairs of wings often have reduced venation; some hemipterans are apterous, and rarely there may be just one pair of wings (in male scale insects). The legs are frequently gressorial, sometimes raptorial, often with complex pretarsal adhesive structures. The abdomen is variable, and cerci are absent.

Most Heteroptera hold their head horizontally, with the rostrum anteriorly distinct from the prosternum (although the rostrum may be in body contact at coxal bases and on anterior abdomen). When at rest the wings are usually folded flat over the abdomen (Fig. 5.8). The fore wings usually are thickened basally and

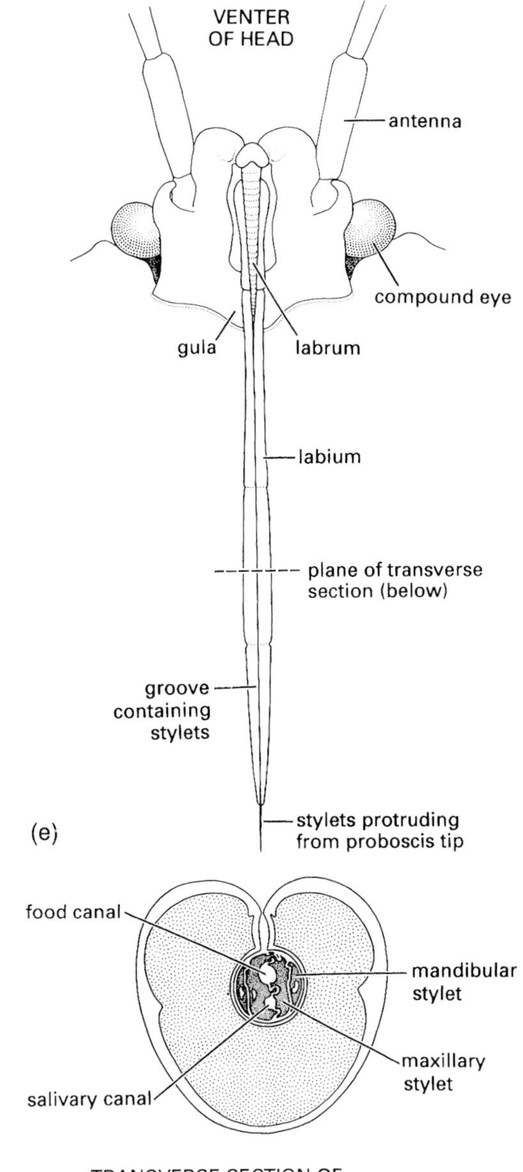

membranous apically to form hemelytra (Fig. 2.22e). Heteroptera mostly have abdominal scent glands. Apterous heteropterans can be identified by the rostrum arising from the anteroventral region of the head and the presence of a large gula. Non-heteropterans hold the head deflexed with the complete length of the rostrum appressed to the prosternum, directed posteriorly often between the coxal bases. They have membranous wings that rest roof-like over the abdomen; apterous species are identified by the absence of a gula and the rostrum arising from the posteroventral head or near the prosternum. Mouthparts are absent in some aphids, and in some female and all male scale insects.

Nymphal Heteroptera (Fig. 6.2) resemble adults except in the lack of development of the wings and genitalia. However, immature Sternorrhyncha show much variation in a range of complex life cycles. Many aphids exhibit parthenogenesis (section 5.10.1), usually alternating with seasonal sexual reproduction. The immature stages of Aleyrodoidea (whiteflies) and Coccoidea (scale insects) may differ greatly from adults, with larviform stages followed by a quiescent, non-feeding "pupal" stage, in convergently acquired holometaboly.

The primitive feeding mode is piercing and sucking plant tissue (Fig. 11.4), and many species induce galls on their host plants (section 11.2.4; Box 11.2). All hemipterans have large salivary glands and an alimentary canal modified for absorption of liquids, with a filter chamber to remove water (Box 3.3). Many hemipterans rely exclusively on living plant sap (from phloem or xylem and sometimes parenchyma). Elimination of large quantities of honeydew by phloem-feeding Sternorrhyncha provides the basis for mutualistic relationships with ants. Many hemipterans exude waxes (Fig. 2.5), which form powdery (see Plate 5.1, facing p. 14) or plate-like protective covers. Non-phytophagous Heteroptera comprise many predators, some scavengers, a few hematophages (blood-feeders), and some necrophages (consumers of dead prey), with the last trophic group including successful colonizers of aquatic environments (Box 10.6) and some of the few insects to live on the oceans (section 10.8).

Phylogenetic relationships are considered in section 7.4.2 and depicted in Fig. 7.5.

Box 11.9 Psocoptera (booklice, barklice, or psocids)

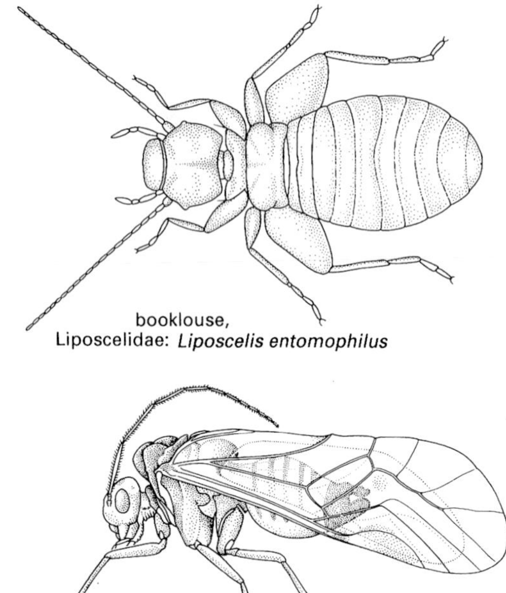

booklouse,
Liposcelidae: *Liposcelis entomophilus*

psocid,
Psocidae: *Amphigerontia contaminata*

The Psocoptera is a worldwide order of common but cryptic minute to small insects (1–10 mm long), with over 3000 species in 36 families. Development is hemimetabolous with five or six nymphal instars. They have a large and mobile head, and large compound eyes; three ocelli are present in winged species, but absent in apterous ones. The antennae are usually 13-segmented and filiform. The mouthparts have asymmetrical chewing mandibles, rod-shaped maxillary laciniae, and reduced labial palps. The thorax varies according to the presence of wings. The pronotum is small, whereas the meso- and metanotum are larger. The legs are gressorial and slender. The wings are often reduced or absent (as shown here for the booklouse *Liposcelis entomophilus* (Liposcelidae), after Smithers 1982). When present the wings are membranous, with reduced venation, with the hind wing coupled to the larger fore wing in flight and at rest, when the wings are held roof-like over the abdomen (as shown here for the psocid *Amphigerontia contaminata* (Psocidae), after Badonnel 1951). The abdomen has 10 visible segments, with the 11th represented by a dorsal epiproct and paired lateral paraprocts. Cerci are always absent.

Courtship often involves a nuptial dance, followed by spermatozoa transfer via a spermatophore. Eggs are laid in groups or singly onto vegetation or under bark, in sites where nymphs subsequently develop. Parthenogenesis is common, and may be obligatory or facultative. Viviparity is known in at least one genus.

Adults and nymphs feed on fungi (hyphae and spores), lichens, algae, insect eggs, or are scavengers on dead organic matter. Some species are solitary; others may be communal, forming small groups of adults and nymphs beneath webs.

Phylogenetic relationships are considered in section 7.4.2 and depicted in Fig. 7.5.

This apparently inhospitable environment provides the home for a few specialist insects that live above the fluid, and many more living as larvae within. The adults of these insects can move in and out of the pitchers with impunity. Mosquito and midge larvae are the most common inhabitants, but other fly larvae of more than 12 families have been reported worldwide, and odonates, spiders, and even a stem-mining ant occur in south-east Asian pitchers. Many of these insect inquilines live in a mutualistic relationship with the plant, digesting trapped prey and microorganisms and excreting nutrients in a readily available form to the plant. Another unusual pitcher plant associate is a *Camponotus* ant that nests in the hollow tendrils of the pitcher plant *Nepenthes bicalcarata* in Borneo. The ants feed on large, trapped prey or mosquito larvae, which they haul from the pitchers, and thereby benefit the plant by preventing the accumulation of excess prey, which can lead to putrefaction of pitcher contents.

Box 11.10 Coleoptera (beetles)

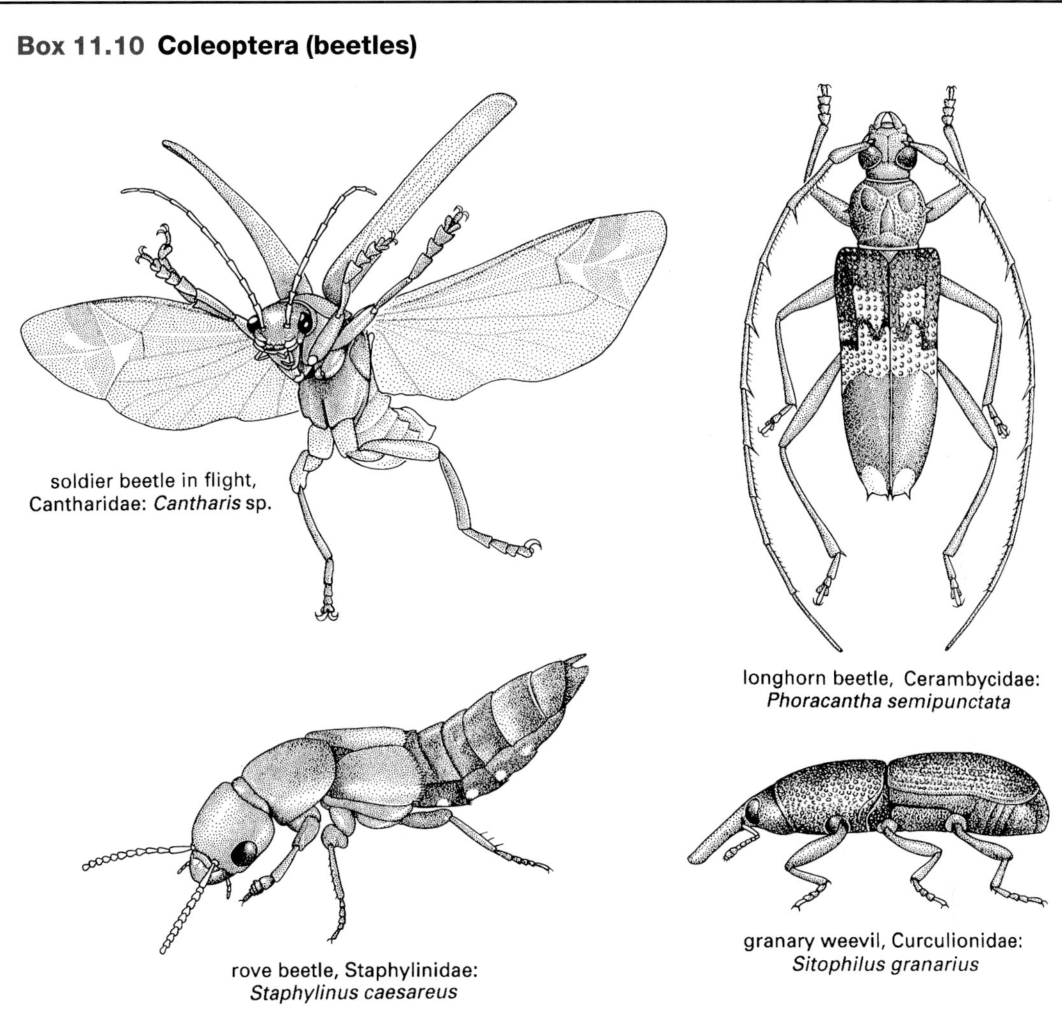

soldier beetle in flight, Cantharidae: *Cantharis* sp.

longhorn beetle, Cerambycidae: *Phoracantha semipunctata*

rove beetle, Staphylinidae: *Staphylinus caesareus*

granary weevil, Curculionidae: *Sitophilus granarius*

The Coleoptera is probably the largest order of insects, with some 350,000 described species in four suborders (Archostemata, Myxophaga, Adephaga, and the speciose Polyphaga). Although the family-level classification is unstable, some 500 families and subfamilies are recognized. Adult beetles range from small to very large, but are usually heavily sclerotized, sometimes even armored, and often compact. Development is holometabolous. The mouthparts are mandibulate, and compound eyes range from well developed (sometimes even meeting medially) to absent; ocelli are usually absent. The antennae comprise 11 or frequently fewer segments (exceptionally with 20 segments in male Rhipiceridae). The prothorax is distinct, large, and extends laterally beyond the coxae; the mesothorax is small (at least dorsally), and fused to the metathorax to form the wing-bearing pterothorax. The fore wings are modified as sclerotized, rigid **elytra** (Fig. 2.22d & Plate 1.2, facing p. 14), whose movement may assist in lift or may be restricted to opening and closing before and after flight; the elytra cover the hind wings and abdominal spiracles allowing control of water loss. The hind wings are longer than the elytra when extended for flight (as illustrated on the upper left for a soldier beetle, *Cantharis* sp. (Cantharidae), after Brackenbury 1990), and have variably reduced venation, much of which is associated with complex pleating to allow the wings to be folded longitudinally and transversely beneath the elytra even if the latter are reduced in size, as in rove beetles (Staphylinidae) such as *Staphylinus caesareus* (illustrated on the lower left, after Stanek 1969). The legs are very variably developed, with coxae that are sometimes large and mobile; the tarsi are primitively five-segmented, although often with a reduced number of segments, and bear variously shaped claws and adhesive structures (Fig. 10.3). Sometimes the legs are fossorial (Fig. 9.2c) for digging in soil or wood, or modified for swimming (Figs. 10.3 & 10.8) or jumping. The abdomen is primitively nine-segmented in females, and 10-segmented in males, with at least one terminal segment retracted; the sterna are usually strongly sclerotized, often more so than the terga. Females have a substitutional ovipositor, whereas the male external genitalia are primitively trilobed (Fig. 2.24b). Cerci are absent.

Larvae exhibit a wide range of morphologies, but most can be recognized by the sclerotized head capsule with opposable mandibles and their usually five-segmented thoracic legs, and can be distinguished from similar lepidopteran larvae by the lack of ventral abdominal crochet-bearing prolegs and lack of a median labial silk gland. Similar symphytan wasp larvae have prolegs on abdominal segments 2–7. Beetle larvae vary in body shape and leg structure; some are **apodous** (lacking any thoracic legs; Fig. 6.6g), whereas legged larvae may be **campodeiform** (prognathous with long thoracic legs; Fig. 6.6e), **eruciform** (grub-like with short legs), or **scarabaeiform** (grub-like but long-legged; Fig. 6.6f). Pupation is often in a specially constructed cell or chamber (Fig. 9.1), rarely in a cocoon spun from silk from Malpighian tubules, or exposed as in coccinellids (Fig. 6.7j).

Beetles occupy virtually every conceivable habitat, including freshwater (Box 10.6), a few marine and intertidal habitats, and, above all, every vegetational microhabitat from external foliage (Fig. 11.1), flowers, buds, stems, bark, and roots, to internal sites such as in galls in any living plant tissue or in any kind of dead material in all its various states of decomposition. Saprophagy and fungivory are fairly common, and dung and carrion are exploited (sections 9.3 & 9.4, respectively). Few beetles are parasitic but carnivory is frequent, occurring in nearly all Adephaga and many Polyphaga, including Lampyridae (fireflies) and many Coccinellidae (ladybird beetles; vignette to Chapter 16, Fig. 5.9). Herbivorous Chrysomelidae and Curculionidae are widely introduced as biological control agents of weedy plants, and Coccinellidae have been used as biological control agents for aphid and coccoid pests of plants (Box 16.2). Some beetles are significant pests of roots in pastures and crops (especially larval Scarabaeidae), of timber (especially Cerambycidae such as *Phoracantha semipunctata*, illustrated on the upper right, after Duffy 1963), and of stored products (such as the granary weevil, *Sitophilus granarius* (Curculionidae), illustrated on the lower right). These last beetles tend to be adapted to dry conditions and thrive on stored grains, cereals, pulses, and dried animal material such as skins and leather.

Phylogenetic relationships are considered in section 7.4.2 and depicted in Fig. 7.2.

Box 11.11 Lepidoptera (butterflies and moths)

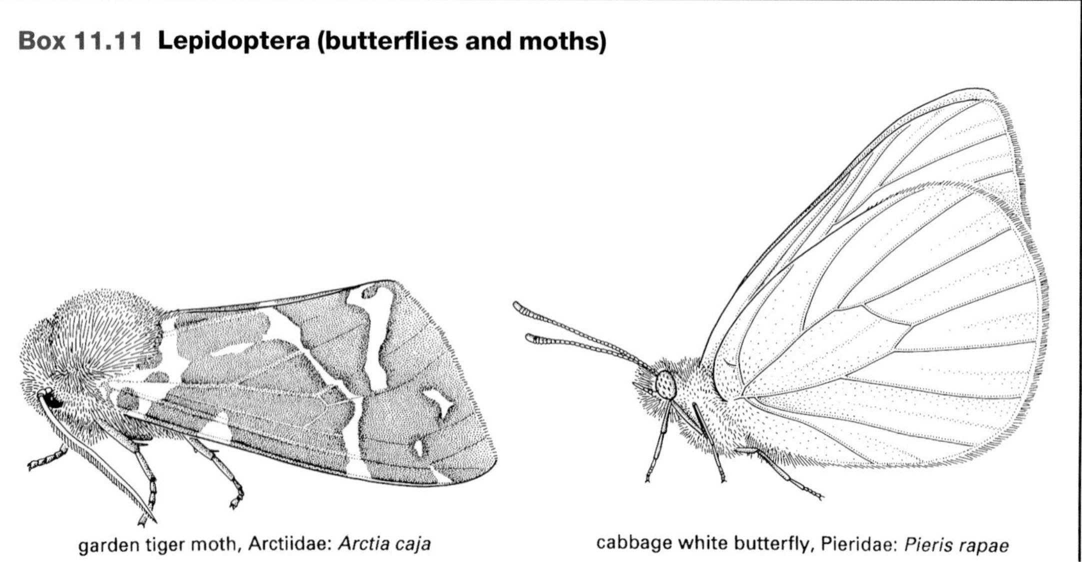

garden tiger moth, Arctiidae: *Arctia caja*

cabbage white butterfly, Pieridae: *Pieris rapae*

The Lepidoptera is one of the major insect orders, both in terms of size, with some 160,000 described species in more than 120 families, and in terms of popularity, with many amateur and professional entomologists studying the order, particularly the butterflies. Three of the four suborders contain few species and lack the characteristic proboscis of the largest suborder, Glossata, which contains the speciose series Ditrysia, defined by unique abdominal features especially in the genitalia. Adult lepidopterans range in size from very small (some microlepidopterans) to large (see Plates 1.1 & 1.3, facing p. 14), with wingspans up to 30 cm. Development is holometabolous (vignette to Chapter 6). The head is hypognathous, bearing a long coiled proboscis (Fig. 2.12) formed from greatly elongated maxillary galeae; large labial palps are usually present, whereas other mouthparts are absent, although mandibles are primitively present. The compound eyes are large, and ocelli and/or **chaetosemata** (paired sensory organs lying dorsolateral on the head) are frequent. The antennae are multisegmented, often pectinate in moths (Fig. 4.6) and knobbed or clubbed in butterflies. The prothorax is small, with paired dorsolaterally placed plates (**patagia**), whereas the mesothorax is large and bears a scutum and scutellum, and a lateral tegula protects the base of each fore wing. The metathorax is small. The wings are completely covered with a double layer of scales (flattened modified macrotrichia), and hind and fore wings are linked by a frenulum, jugum, or simple overlap. Wing venation consists predominantly of longitudinal veins with few cross-veins and some large cells, notably the discal (Fig. 2.22a). The legs are long and usually gressorial, with five tarsomeres. The abdomen is 10-segmented, with segment 1 variably reduced, and segments 9 and 10 modifed as external genitalia (Fig. 2.23a). Internal female genitalia are very complex.

Premating behavior including courtship often involves pheromones (Figs. 4.7 & 4.8). Encounter between the sexes is often aerial, but copulation is on the ground or a perch. Eggs are laid on, close to or, more rarely, within a larval host plant. Egg numbers and degree of aggregation are very variable. Diapause is common.

Lepidopteran larvae can be recognized by their sclerotized, hypognathous or prognathous head capsule, mandibulate mouthparts, usually six lateral stemmata (Fig. 4.9a), short three-segmented antennae, five-segmented thoracic legs with single claws, and 10-segmented abdomen with short prolegs on some segments (usually on 3–6 and 10, but may be reduced) (Figs. 6.6a,b & 14.6). Silk-gland products are extruded from a characteristic **spinneret** at the median apex of the labial prementum. The pupa is usually contained within a silken cocoon, typically adecticous and obtect (a **chrysalis**) (Fig. 6.7g–i), with only some abdominal segments unfused; the pupa is exarate in primitive groups.

Adult lepidopterans that feed utilize nutritious liquids, such as nectar, honeydew, and other seepages from live and decaying plants, and a few species pierce fruit. However, none suck sap from the vessels of live plants. Many species supplement their diet by feeding on nitrogenous animal wastes. Most larvae feed exposed on higher plants and form the major insect phytophages; a few "primitive" species feed on non-angiosperm plants, and some feed on fungi. Several are predators and others are scavengers, notably amongst the Tineidae (wool moths).

The larvae are often cryptic (see Plate 5.2), particularly when feeding in exposed positions, or warningly colored (aposematic) to alert predators to their toxicity (Chapter 14). Toxins derived from larval food plants often are retained by adults, which show anti-predator devices including advertisement of non-palatability and defensive mimicry (section 14.5).

Although the butterflies popularly are considered to be distinct from the moths, they form a clade that lies deep within the phylogeny of the Lepidoptera: butterflies are not the sister group to all moths. Butterflies are day-flying whereas most moths are active at night or dusk. In life, butterflies hold their wings together vertically above the body (as shown here on the right for a cabbage white butterfly) in contrast to moths, which hold their wings flat or wrapped around the body (as shown on the left for the garden tiger moth); a few lepidopteran species have brachypterous adults and sometimes completely wingless adult females (see Plate 4.7).

Phylogenetic relationships are considered in section 7.4.2 and depicted in Figs. 7.2 and 7.7.

FURTHER READING

Bernays, E.A. (1998) Evolution of feeding behaviour in insect herbivores. *BioScience* **48**, 35–44.

Cook, J.M. & Rasplus, J.-Y. (2003) Mutualists with attitude: co-evolving fig wasps and figs. *Trends in Ecology and Evolution* **18**, 241–8.

Farrell, B.D., Mitter, C. & Futuyma, D.J. (1992) Diversification at the insect–plant interface. *BioScience* **42**, 34–42.

Herrera, C.M. & Pellmyr, O. (eds.) (2002) *Plant–Animal Interactions*. Blackwell Publishing, Oxford.

Huxley, C.R. & Cutler, D.F. (eds.) (1991) *Ant–Plant Interactions*. Oxford University Press, Oxford.

Johnson, W.T. & Lyon, H.H. (1991) *Insects that Feed on Trees and Shrubs*, 2nd edn. Comstock Publishing Associates of Cornell University Press, Ithaca, NY.

Karban, R. & Agrawal, A.A. (2002) Herbivore offense. *Annual Review of Ecology and Systematics* **33**, 641–4.

Karban, R. & Baldwin, I.T. (1997) *Induced Responses to Herbivory*. University of Chicago Press, Chicago, IL.

Landsberg, J. & Ohmart, C. (1989) Levels of insect defoliation in forests: patterns and concepts. *Trends in Ecology and Evolution* **4**, 96–100.

McFadyen, R.E.C. (1998) Biological control of weeds. *Annual Review of Entomology* **43**, 369–93.

Miles, P.W. (1999) Aphid saliva. *Biological Reviews* **74**, 41–85.

Nilsson, L.A. (1998) Deep flowers for long tongues. *Trends in Ecology and Evolution* **13**, 259–60.

Pellmyr, O. (1992) Evolution of insect pollination and angiosperm diversification. *Trends in Ecology and Evolution* **7**, 46–9.

Price, P.W. (1997) *Insect Ecology*, 3rd edn. John Wiley & Sons, New York.

Price, P.W. (2003) *Macroevolutionary Theory on Macroecological Patterns*. Cambridge University Press, Cambridge.

Raman, A., Schaefer, C.W. & Withers, T.M. (eds.) (2004) *Biology, Ecology, and Evolution of Gall-Inducing Arthropods*. Oxford & IBH Publishing, New Delhi.

Resh, V.H. & Cardé, R.T. (eds.) (2003) *Encyclopedia of Insects*. Academic Press, Amsterdam. [Particularly see articles on phytophagous insects; plant–insect interactions; pollination and pollinators.]

Room, P.M. (1990) Ecology of a simple plant–herbivore system: biological control of *Salvinia*. *Trends in Ecology and Evolution* **5**, 74–9.

Rosenthal, G.A. & Berenbaum, M.R. (eds.) (1992) *Herbivores, their Interactions with Secondary Plant Metabolites*, 2nd edn. Academic Press, San Diego, CA.

Speight, M.R., Hunter, M.D. & Watt, A.D. (1999) *Ecology of Insects. Concepts and Applications*. Blackwell Science, Oxford.

Thompson, J.N. (1989) Concepts of coevolution. *Trends in Ecology and Evolution* **4**, 179–83.

Wasserthal, L.T. (1997) The pollinators of the Malagasy star orchids *Angraecum sesquipedale*, *A. sororium* and *A. compactum* and the evolution of extremely long spurs by pollinator shift. *Botanica Acta* **110**, 343–59.

Watt, A.D., Stork, N.E. & Hunter, M.D. (eds.) (1997) *Forests and Insects*. Chapman & Hall, London.

White, T.C.R. (1993) *The Inadequate Environment. Nitrogen and the Abundance of Animals*. Springer-Verlag, Berlin.

Whitham, T.G., Martinsen, G.D., Floate, K.D., Dungey, H.S., Potts, B.M. & Keim, P. (1999) Plant hybrid zones affect biodiversity: tools for a genetic-based understanding of community structure. *Ecology* **80**, 416–28.

Willmer, P.G. & Stone, G.N. (1997) How aggressive ant-guards assist seed-set in *Acacia* flowers. *Nature* **388**, 165–7.

Chapter 12

INSECT SOCIETIES

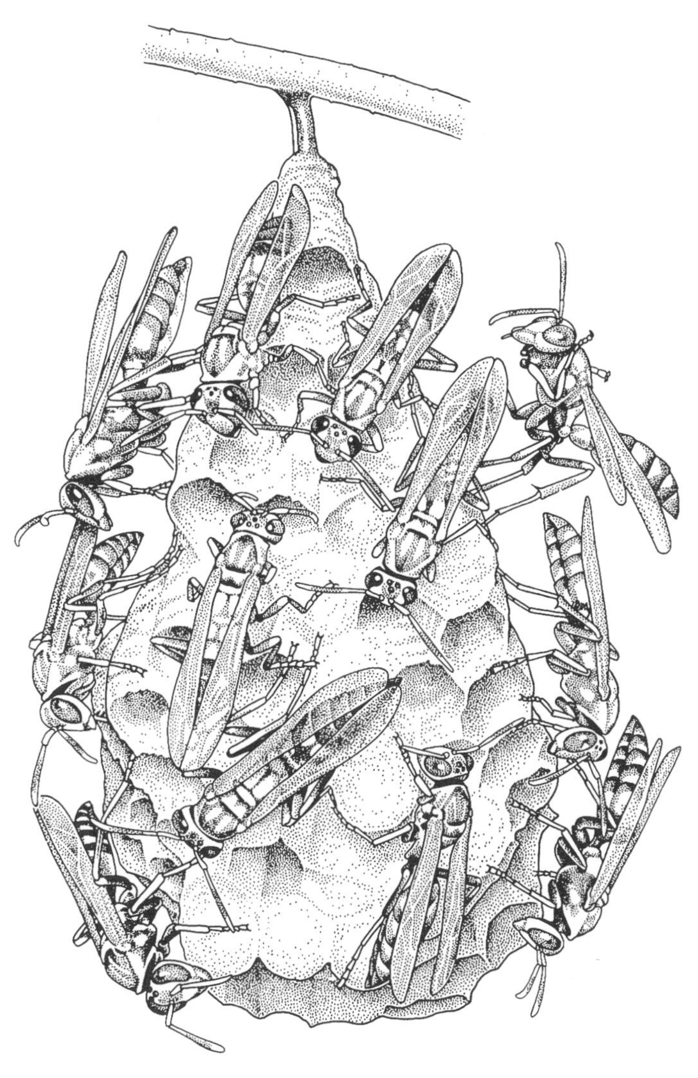

Vespid wasp nest. (After Blaney 1976.)

The study of insect social behaviors is a popular entomological topic and there is a voluminous literature, ranging from the popular to the highly theoretical. The proliferation of some insects, notably the ants and termites, is attributed to the major change from a solitary lifestyle to a social one.

Social insects are ecologically successful and have important effects on human life. Leaf-cutter ants (*Atta* spp.) are the major herbivores in the Neotropics, and in south-western US deserts, harvester ants take as many seeds as do mammals. Ecologically dominant "tramp" ants can threaten our agriculture, outdoor behavior, and biodiversity (Box 1.2). Termites turn over at least as much soil as do earthworms in many tropical regions. The numerical dominance of social insects can be astonishing, with a Japanese supercolony of *Formica yessensis* estimated at 306 million workers and over 1 million queens dispersed over 2.7 km^2 amongst 45,000 interconnected nests. In West African savanna, densities of up to 20 million resident ants per hectare have been estimated, and single nomadic colonies of driver ants (*Dorylus* sp.) may attain 20 million workers. Estimates of the value of honey bees in commercial honey production, as well as in pollination of agricultural and horticultural crops, run into many billions of dollars per annum in the USA alone. Social insects clearly affect our lives.

A broad definition of social behavior could include all insects that interact in any way with other members of their species. However, entomologists limit **sociality** to a more restricted range of **co-operative** behaviors. Amongst the social insects, we can recognize **eusocial** ("true social") **insects**, which co-operate in reproduction and have division of reproductive effort, and **subsocial** ("below social") **insects**, which have less strongly developed social habits, falling short of extensive co-operation and reproductive partitioning. **Solitary** insects exhibit no social behaviors.

Eusociality is defined by three traits:

1 Division of labor, with a **caste system** involving sterile or non-reproductive individuals assisting those that reproduce.

2 Co-operation among colony members in tending the young.

3 Overlap of generations capable of contributing to colony functioning.

Eusociality is restricted to all ants and termites and some bees and wasps, such as the vespine paper wasps depicted in the vignette of this chapter. Subsociality is a more widespread phenomenon, known to have arisen independently in 13 orders of insects, including some cockroaches, embiids, thysanopterans, hemipterans, beetles, and hymenopterans. As insect lifestyles become better known, forms of subsociality may be found in yet more orders. The term "presociality" often is used for social behaviors that do not fulfill the strict definition of eusociality. However, the implication that presociality is an evolutionary precursor to eusociality is not always correct and the term is best avoided.

In this chapter we discuss subsociality prior to detailed treatment of eusociality in bees, wasps, ants, and termites. We conclude with some ideas concerning the origins and success of eusociality.

12.1 SUBSOCIALITY IN INSECTS

12.1.1 Aggregation

Non-reproductive aggregations of insects, such as the gregarious overwintering of monarch butterflies at specific sites in Mexico and California (see Plate 3.5, facing p. 14), are social interactions. Many tropical butterflies form roosting aggregations, particularly in **aposematic** species (distasteful and with warning signals including color and/or odor). Aposematic phytophagous insects often form conspicuous feeding aggregations, sometimes using pheromones to lure conspecific individuals to a favorable site (section 4.3.2). A solitary aposematic insect runs a greater risk of being encountered by a naïve predator (and being eaten by it) than if it is a member of a conspicuous group. Belonging to a conspicuous social grouping, either of the same or several species, provides benefits by the sharing of protective warning coloration and the education of local predators.

12.1.2 Parental care as a social behavior

Parental care may be considered to be a social behavior; although few insects, if any, show a complete lack of parental care: eggs are not deposited randomly. Females select an appropriate oviposition site, affording protection to the eggs and ensuring an appropriate food resource for the hatching offspring. The ovipositing female may protect the eggs in an ootheca, or deposit them directly into suitable substrate with her ovipositor, or modify the environment, as in nest construction. Parental care conventionally is seen as postoviposition and/or posthatching attention, including the provision and protection of food resources for the young. A convenient basis for discussing parental care is to distinguish between care with and without nest construction.

Parental care without nesting

For most insects, the highest mortality occurs in the egg and first instar, and many insects tend these stages until the more mature larvae or nymphs can better fend for themselves. The orders of insects in which tending of eggs and young is most frequent are the Blattodea, Orthoptera, and Dermaptera (orthopteroid orders), Embiidina, Psocoptera, Thysanoptera, Hemiptera, Coleoptera, and Hymenoptera. There has been a tendency to assume that subsociality is a precursor of isopteran eusociality, as the eusocial termites are related to cockroaches. The phylogenetic position (Fig. 7.4) and social behavior, including parental care, of the subsocial cockroach family Cryptocercidae has provoked speculation on the origin of sociality, discussed in more detail in section 12.4.2.

Egg and early-instar attendance is predominantly a female role; yet paternal guarding is known in some Hemiptera, notably amongst some tropical assassin bugs (Reduviidae) and giant water bugs (Belostomatidae). The female belostomatid oviposits onto the dorsum of the male, which receives eggs in small batches after each copulation. The eggs, which die if neglected, are tended in various ways by the male (Box 5.5). There is no tending of belostomatid nymphs, unlike some other hemipterans in which the female (or in some reduviids, the male) may guard at least the early-instar nymphs. In these species, experimental removal of the tending adult increases losses of eggs and nymphs as a result of parasitization and/or predation. Other functions of parental care include keeping the eggs free from fungi, maintaining appropriate conditions for egg development, herding the young, and sometimes actually feeding them.

In an unusual case, certain treehoppers (Hemiptera: Membracidae) have "delegated" parental care of their young to ants. Ants obtain honeydew from treehoppers, which are protected from their natural enemies by the presence of the ants. In the presence of protective ants, brooding females prematurely may cease to tend a first brood and raise a second one. Another species of membracid will abandon its eggs in the absence of ants and seek a larger treehopper aggregation, where ants are in attendance, before laying another batch of eggs.

Many wood-mining beetles show advanced subsocial care that verges on the nesting described in the following section and on eusociality. For instance, all Passalidae (Coleoptera) live in communities of larvae and adults, with the adults chewing dead wood to form a substrate for the larvae to feed upon. Some ambrosia beetles (Curculionidae: Platypodinae) prepare galleries for their offspring (section 9.2), where the larvae feed on cultivated fungus and are defended by a male that guards the tunnel entrance. Whether or not these feeding galleries are called nests is a matter of semantics.

Parental care with solitary nesting

Nesting is a social behavior in which eggs are laid in a pre-existing or newly constructed structure to which the parent(s) bring food supplies for the young. Nesting, as thus defined, is seen in only five insect orders. Nest builders amongst the subsocial Orthoptera, Dermaptera, Coleoptera, and subsocial Hymenoptera are discussed below; the nests of eusocial Hymenoptera and the prodigious mounds of the eusocial termites are discussed later in this chapter.

Earwigs of both sexes overwinter in a nest. In spring, the male is ejected when the mother starts to tend the eggs (Fig. 9.1). In some species mother earwigs forage and provide food for the young nymphs. Mole crickets and other ground-nesting crickets exhibit somewhat similar behavior. A greater range of nesting behaviors is seen in the beetles, particularly in the dung beetles (Scarabaeidae) and carrion beetles (Silphidae). For these insects the attractiveness of the short-lived, scattered, but nutrient-rich dung (and carrion) food resource induces competition. Upon location of a fresh source, dung beetles bury it to prevent drying out or being ousted by a competitor (section 9.3; Fig. 9.5). Some scarabs roll the dung away from its source; others coat the dung with clay. Both sexes co-operate, but the female is mostly responsible for burrowing and preparation of the larval food source. Eggs are laid on the buried dung and in some species no further interest is taken. However, parental care is well developed in others, commonly with maternal attention to fungus reduction, and removal or exclusion of conspecifics and ants by paternal defense.

Amongst the Hymenoptera, subsocial nesting is restricted to some **aculeate** Apocrita within the superfamilies Chrysidoidea, Vespoidea, and Apoidea (Fig. 12.2); these wasps and bees are the most prolific and diverse nest builders amongst the insects. Excepting bees, nearly all these insects are parasitoids, in which adults attack and immobilize arthropod prey upon which the young feed. Wasps demonstrate a series of increasingly complex prey handling and nesting strategies, from using the prey's own burrow (e.g. many Pompilidae), to building a simple burrow following prey capture (a few Sphecidae), to construction of a nest burrow before prey-capture (most Sphecidae). In bees and masarine wasps, pollen replaces arthropod prey as the food source that is collected and stored for the larvae. Nest complexity in the aculeates ranges from a single burrow provisioned with one food item for one developing egg, to linearly or radially arranged multicellular nests. The primitive nest site was probably a pre-existing burrow, with the construction medium later being soil or sand. Further specializations involved the use of plant material – stems, rotten wood, and even solid wood by carpenter bees (Xylocopini) – and free-standing constructions of chewed vegetation (Megachilinae), mud (Eumeninae), and saliva (Colletinae). A range of natural materials are used in making and sealing cells, including mud, plants, and saps, resins, and oils secreted by plants as rewards for pollination, and even the wax adorning soft scale insects. In some subsocial nesters such as mason wasps (Eumeninae), many individuals of one species may aggregate, building their nests close together.

Parental care with communal nesting

When favorable conditions for nest construction are scarce and scattered throughout the environment, communal nesting may occur. Even under apparently favorable conditions, many subsocial and all eusocial hymenopterans share nests. Communal nesting may arise if daughters nest in their natal nest, enhancing utilization of nesting resources and encouraging mutual defense against parasites. However, communal nesting in subsocial species allows "antisocial" or selfish behavior, with frequent theft or takeover of nest and prey, so that extended time defending the nest against others of the same species may be required. Furthermore, the same cues that lead the wasps and bees to communal nesting sites easily can direct specialized nest parasites to the location. Examples of communal nesting in subsocial species are known or presumed in the Sphecidae, and in bees among Halictinae, Megachilinae, and Andreninae.

After oviposition, female bees and wasps remain in their nests, often until the next generation emerges as adults. They generally guard, but they also may remove feces and generally maintain nest hygiene. The supply of provisions to the nest may be through mass provisioning, as in many communal sphecids and subsocial bees, or replenishment, as seen in the many vespid wasps that return with new prey as their larvae develop.

Subsocial aphids and thrips

Certain aphids belonging to the subfamilies Pemphiginae and Hormaphidinae (Hemiptera: Aphididae) have a sacrificial sterile soldier caste, consisting of some first- or second-instar nymphs that exhibit aggressive behavior and never develop into adults. Soldiers are pseudoscorpion-like, as a result of body sclerotization and enlarged anterior legs, and will attack intruders using either their frontal horns (anterior cuticular projections) (Fig. 12.1) or feeding stylets (mouthparts) as piercing weapons. These modified individuals may defend good feeding sites against competitors or defend their colony against predators. As the offspring are produced by parthenogenesis, soldiers and normal nymphs from the same mother aphid should be genetically identical, favoring the evolution of these non-reproductive and apparently altruistic soldiers (as a result of increased inclusive fitness via kin selection; section 12.4). A similar phenomenon occurs in other related aphid species, but in this case all nymphs become temporary soldiers, which later molt into normal, non-aggressive individuals that reproduce. These unusual aphid polymorphisms have led some researchers to claim that the Hemiptera is a third insect order displaying eusociality. Although these few aphid species clearly have a reproductive division of labor, they do not appear to fulfill the other attributes of eusocial insects, as overlap of generations capable of contributing to colony labor is equivocal and tending of offspring does not occur. Here we consider these aphids to exhibit subsocial behavior.

A range of subsocial behaviors is seen in a few species of several genera of thrips (Thysanoptera: Phlaeothripidae). At least in the gall thrips, the level of sociality appears to be similar to that of the aphids discussed above. Thrips sociality is well developed in a bark-dwelling species of *Anactinothrips* from Panama, in which thrips live communally, co-operate in brood care, and forage with their young in a highly co-ordinated fashion. However, this species has no obvious non-reproductive females and all adults may disappear before the young are fully grown. Evolution of subsocial

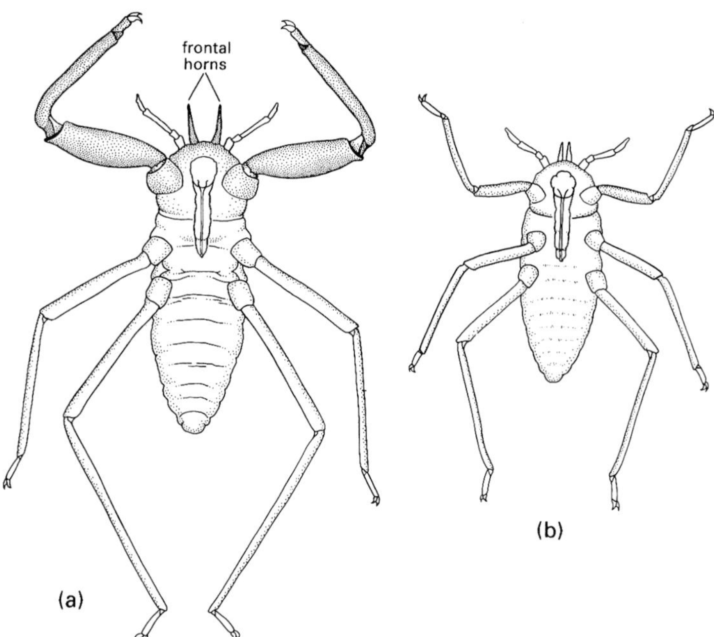

Fig. 12.1 First-instar nymphs of the subsocial aphid *Pseudoregma alexanderi* (Hemiptera: Hormaphidinae): (a) pseudoscorpion-like soldier; (b) normal nymph. (After Miyazaki 1987b.)

behaviors in *Anactinothrips* may bring advantages to the young in group foraging, as feeding sites, although stable over time, are patchy and difficult to locate. In several species of Australian gall thrips, females show polymorphic wing reduction associated in some species with very enlarged fore legs. This "soldier" morph is more frequent amongst the first young to develop, which are differentially involved in defending the gall against intrusion by other species of thrips, and appear to be incapable of dispersing or of inducing galls. As in most thrips, sex determination is via haplodiploidy, with gall foundation by a single female producing polymorphic offspring, and with establishment of multiple generations. Self-sacrificing defense by some individuals is favored by demonstrated high relatedness of the offspring (altruism; section 12.4). Generational overlap is modest, and soldiers defend their siblings and their offspring rather than their mother (who has died). Soldiers reproduce, but at a much lower rate than the foundress. Such examples are valuable in showing the circumstances under which co-operation might have evolved.

Quasisociality and semisociality

Division of reproductive labor is restricted to the subsocial aphids amongst the insect groups discussed above: all females of all the other subsocial insects can reproduce. Within the social Hymenoptera, females show variation in fecundity, or reproductive division of labor. This variation ranges from fully reproductive (the subsocial species described above), through reduced fecundity (many halictine bees), the laying of only male eggs (workers of *Bombus*), sterility (workers of *Aphaenogaster*), to super-reproductives (queens of *Apis*). This range of female behaviors is reflected in the classification of social behaviors in the Hymenoptera. Thus, in **quasisocial** behavior, a communal nest consists of members of the same generation all of which assist in brood rearing, and all females are able to lay eggs, even if not necessarily at the same time. In **semisocial** behavior, the communal nest similarly contains members of the same generation co-operating in brood care, but there is division of reproductive labor, with some females (queens) laying eggs, whereas their sisters act as workers and rarely lay eggs. This differs from eusociality only in that the workers are sisters to the egg-laying queens, rather than daughters, as is the case in eusociality. As in primitive eusocial hymenopterans there is no morphological (size or shape) difference between queens and workers.

Any or all the subsocial behaviors discussed above may be evolutionary precursors of eusociality. It is clear that solitary nesting is the primitive behavior, with communal nesting (and additional subsocial behaviors) having arisen independently in many lineages of aculeate hymenopterans.

12.2 EUSOCIALITY IN INSECTS

Eusocial insects have a division of labor in their colonies, involving a caste system comprising a restricted reproductive group of one or several **queens**, aided by **workers** – non-reproductive individuals that assist the reproducers – and in termites and many ants, an additional defensive **soldier** group. There may be further division into subcastes that perform specific tasks. At their most specialized, members of some castes, such as queens and soldiers, may lack the ability to feed themselves. The tasks of workers therefore include bringing food to these individuals as well as to the **brood** – the developing offspring.

The primary differentiation is female from male. In eusocial Hymenoptera, which have a **haplodiploid** genetic system, queens control the sex of their offspring. Releasing stored sperm fertilizes haploid eggs, which develop into diploid female offspring, whereas unfertilized eggs produce male offspring. At most times of the year, reproductive females (queens, or **gynes**) are rare compared with sterile female workers. Males do not form castes and may be infrequent and short-lived, dying soon after mating. In termites (Isoptera), males and females may be equally represented, with both sexes contributing to the worker caste. A single male termite, the **king**, may permanently attend the gyne.

Members of different castes, if derived from a single pair of parents, are close genetically and may be morphologically similar, or, as a result of environmental influence, may be morphologically very different, in an environmental polymorphism termed **polyphenism**. Individuals within a caste (or subcaste) often differ behaviorally, in what is termed **polyethism**, either by an individual performing different tasks at different times in its life (age polyethism), or by individuals within a caste specializing on certain tasks during their lives. The intricacies of social insect caste systems can be considered in terms of the increasing complexity demonstrated in the Hymenoptera, but concluding with the remarkable systems of the termites (Isoptera). The characteristics of these two orders, which contain the majority of the eusocial species, are given in Boxes 12.2 and 12.3.

12.2.1 The primitively eusocial hymenopterans

Hymenopterans exhibiting primitive eusociality include polistine vespids (paper wasps of the genus *Polistes*), stenogastrine wasps, and even one sphecid (Fig. 12.2). In these wasps, all individuals are morphologically similar and live in colonies that seldom last more than one year. The colony is often founded by more than one gyne, but rapidly becomes **monogynous**, i.e. dominated by one queen with other foundresses either departing the nest or remaining but reverting to a worker-like state. The queen establishes a dominance hierarchy physically by biting, chasing, and begging for food, with the winning queen gaining monopoly rights to egg-laying and initiation of cell construction. Dominance may be incomplete, with non-queens laying some eggs: the dominant queen may eat these eggs or allow them to develop as workers to assist the colony. The first brood of females produced by the colony is of small workers, but subsequent workers increase in size as nutrition improves and as worker assistance in rearing increases. Sexual retardation in subordinates is reversible: if the queen dies (or is removed experimentally) either a subordinate foundress takes over, or if none is present, a high-ranking worker can mate (if males are present) and lay fertile eggs. Some other species of primitively eusocial wasps are **polygynous**, retaining several functional queens throughout the duration of the colony; whereas others are **serially polygynous**, with a succession of functional queens.

Primitively eusocial bees, such as certain species of Halictinae (Fig. 12.2), have a similar breadth of behaviors. In female castes, differences in size between queens and workers range from little or none to no overlap in their sizes. Bumble bees (Apidae: *Bombus* spp.) found colonies through a single gyne, often after a fight to the death between gynes vying for a nest site. The first brood consists only of workers that are dominated by the queen physically, by aggression and by eating of any worker eggs, and by means of pheromones that modify the behavior of the workers. In the absence of the queen, or late in the season as the queen's physical and chemical influence wanes, workers can undergo ovarian development. The queen eventually fails to maintain dominance over those workers that have commenced ovarian development, and the queen either is killed or driven from the nest. When this happens workers are unmated, but they can produce

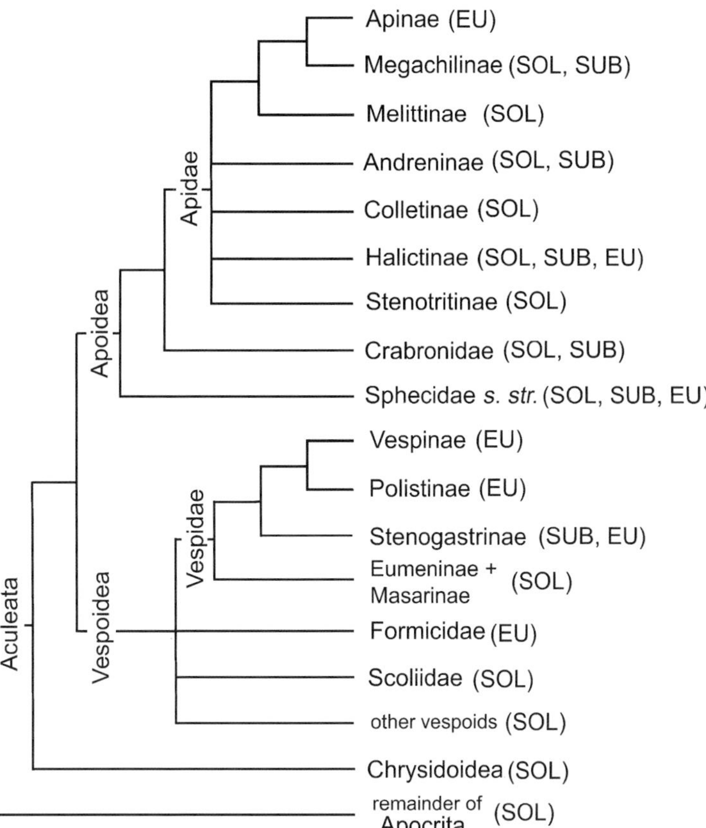

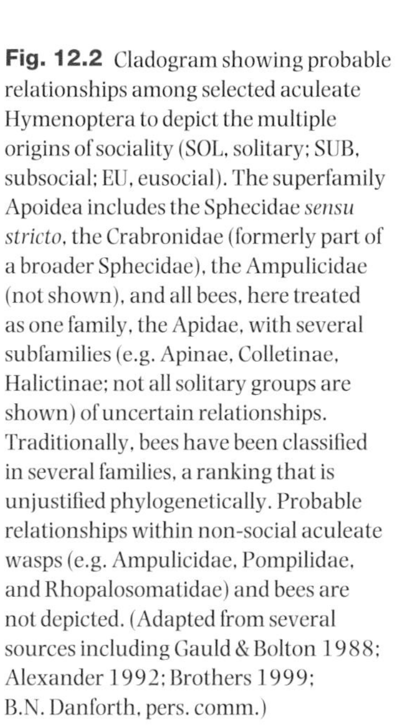
Fig. 12.2 Cladogram showing probable relationships among selected aculeate Hymenoptera to depict the multiple origins of sociality (SOL, solitary; SUB, subsocial; EU, eusocial). The superfamily Apoidea includes the Sphecidae *sensu stricto*, the Crabronidae (formerly part of a broader Sphecidae), the Ampulicidae (not shown), and all bees, here treated as one family, the Apidae, with several subfamilies (e.g. Apinae, Colletinae, Halictinae; not all solitary groups are shown) of uncertain relationships. Traditionally, bees have been classified in several families, a ranking that is unjustified phylogenetically. Probable relationships within non-social aculeate wasps (e.g. Ampulicidae, Pompilidae, and Rhopalosomatidae) and bees are not depicted. (Adapted from several sources including Gauld & Bolton 1988; Alexander 1992; Brothers 1999; B.N. Danforth, pers. comm.)

male offspring from their haploid eggs. Gynes are thus derived solely from the fertilized eggs of the queen.

12.2.2 Specialized eusocial hymenopterans: wasps and bees

The highly eusocial hymenopterans comprise the ants (family Formicidae) and some wasps, notably Vespinae, and many bees, including most Apinae (Figs. 12.2 & 12.3). Bees are derived from sphecid wasps and differ from wasps in anatomy, physiology, and behavior in association with their dietary specialization. Most bees provision their larvae with nectar and pollen rather than animal material. Morphological adaptations of bees associated with pollen collection include plumose (branched) hairs, and usually a widened hind basitarsus adorned with hairs in the form of a brush (**scopa**) or a fringe surrounding a concavity (the **corbicula**, or pollen basket) (Fig. 12.4). Pollen collected on the body hairs is groomed by the legs and transferred to the mouthparts, scopae, or corbiculae. The diagnostic features and the biology of all hymenopterans are dealt with in Box 12.2, which includes an illustration of the morphology of a worker vespine wasp and a worker ant.

Colony and castes in eusocial wasps and bees

The female castes are dimorphic, differing markedly in their appearance. Generally, the queen is larger than any worker, as in vespines such as the European wasps (*Vespula vulgaris* and *V. germanica*), and honey bees (*Apis* spp.). The typical eusocial wasp queen has a differentially (allometrically) enlarged gaster (abdomen). In worker wasps the bursa copulatrix is small, preventing mating, even though in the absence of a queen their ovaries will develop.

In the vespine wasps, the colony-founding queen, or gyne, produces only workers in the first brood.

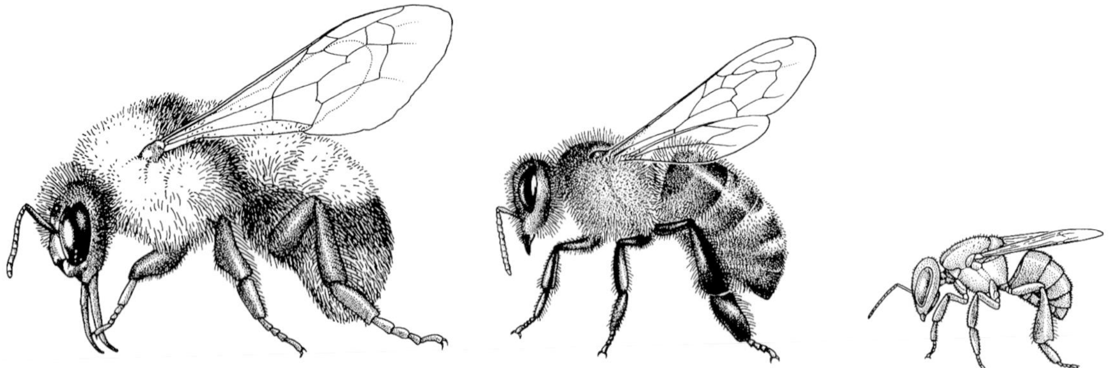

Fig. 12.3 Worker bees from three eusocial genera, from left, *Bombus*, *Apis*, and *Trigona* (Apidae: Apinae), superficially resemble each other in morphology, but they differ in size and ecology, including their pollination preferences and level of eusociality. (After various sources, especially Michener 1974.)

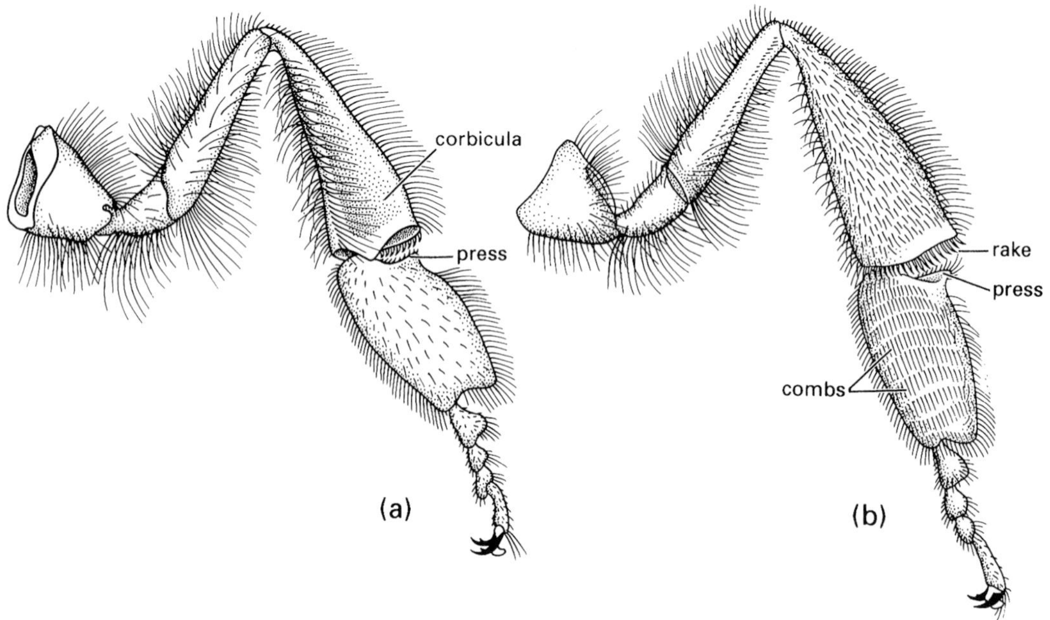

Fig. 12.4 The hind leg of a worker honey bee, *Apis mellifera* (Hymenoptera: Apidae): (a) outer surface showing corbicula, or pollen basket (consisting of a depression fringed by stiff setae), and the press that pushes the pollen into the basket; (b) the inner surface with the combs and rakes that manipulate pollen into the press prior to packing. (After Snodgrass 1956; Winston 1987.)

Immediately after these are hatched, the queen wasp ceases to forage and devotes herself exclusively to reproduction. As the colony matures, subsequent broods include increasing proportions of males, and finally gynes are produced late in the season from larger cells than those from which workers are produced.

The tasks of vespine workers include:

- distribution of protein-rich food to larvae and carbohydrate-rich food to adult wasps;
- cleaning cells and disposal of dead larvae;
- ventilation and air-conditioning of the nest by wing-fanning;

- nest defense by guarding entrances;
- foraging outside for water, sugary liquids, and insect prey;
- construction, extension, and repair of the cells and inner and outer nest walls with wood pulp, which is masticated to produce paper.

Each worker is capable of carrying out any of these tasks, but often there is an age polyethism: newly emerged workers tend to remain in the nest engaged in construction and food distribution. A middle-aged foraging period follows, which may be partitioned into wood-pulp collection, predation, and fluid-gathering phases. In old age, guarding duties dominate. As newly recruited workers are produced continuously, the age structure allows flexibility in performing the range of tasks required by an active colony. There are seasonal variations, with foraging occupying much of the time of the colony in the founding period, with fewer resources – or a lesser proportion of workers' time – devoted to these activities in the mature colony. Male eggs are laid in increasing numbers as the season progresses, perhaps by queens, or by workers on whom the influence of the queen has waned.

The biology of the honey bee, *Apis mellifera*, is extremely well studied because of the economic significance of honey and the relative ease of observing honey-bee behavior (Box 12.1). Workers differ from queens in being smaller, possessing wax glands, having a pollen-collecting apparatus comprising pollen combs and a corbicula on each hind leg, in having a barbed sting that cannot be retracted after use, and in some other features associated with the tasks that workers perform. The queen's sting is scarcely barbed and is retractible and reusable, allowing repeated assaults on pretenders to the queen's position. Queens have a shorter proboscis than workers and lack several glands.

Honey-bee workers are more or less monomorphic, but exhibit polyethism. Thus, young workers tend to be "hive bees", engaged in within-hive activities, such as nursing larvae and cleaning cells, and older workers are foraging "field bees". Seasonal changes are evident, such as the 8–9-month longevity of winter bees, compared with the 4–6-week longevity of summer workers. Juvenile hormone (JH) is involved in these behavioral changes, with levels of JH rising from winter to spring, and also in the change from hive bees to field bees. Honey-bee worker activities correlate with seasons, notably in the energy expenditure involved in thermoregulation of the hive.

Caste differentiation in honey bees, as in eusocial hymenopterans generally, is largely **trophogenic**, i.e. determined by the quantity and quality of the larval diet. In species that provision each cell with enough food to allow the egg to develop to the pupa and adult without further replenishment, differences in the food quantity and quality provided to each cell determine how the larva will develop. In honey bees, although cells are constructed according to the type of caste that is to develop within them, the caste is determined neither by the egg laid by the queen, nor by the cell itself, but by food supplied by workers to the developing larva (Fig. 12.5). The type of cell guides the queen as to whether to lay fertilized or unfertilized eggs, and identifies to the worker which type of rearing (principally food) to be supplied to the occupant. Food given to future queens is known as "royal jelly" and differs from worker food in having a high sugar content and being composed predominantly of mandibular gland products, namely pantothenic acid and biopterin. Eggs and larvae up to three days old can differentiate into queens or workers according to upbringing. However, by the third day a potential queen has been fed royal jelly at up to 10 times the rate of less-rich food supplied to a future worker. At this stage, if a future queen is transferred to a worker cell for further development, she will become an intercaste, a worker-like queen. The opposite transfer, of a three-day-old larva reared as a worker into a queen cell, gives rise to a queen-like worker, still retaining the pollen baskets, barbed stings, and mandibles of a worker. After four days of appropriate feeding, the castes are fully differentiated and transfers between cell types result in either retention of the early determined outcome or failure to develop.

Trophogenic effects cannot always be separated from endocrine effects, as nutritional status is linked to corpora allata activity. It is clear that JH levels correlate with polymorphic caste differentiation in eusocial insects. However, there seems to be much specific and temporal variation in JH titers and no common pattern of control is yet evident.

The queen maintains control over the workers' reproduction principally through pheromones. The mandibular glands of queens produce a compound identified as (*E*)-9-oxodec-2-enoic acid (9-ODA), but the intact queen inhibits worker ovarian development more effectively than this active compound. A second pheromone has been found in the gaster of the queen, and this, together with a second component of the mandibular gland, effectively inhibits ovarian development. Queen recognition by the rest of the colony

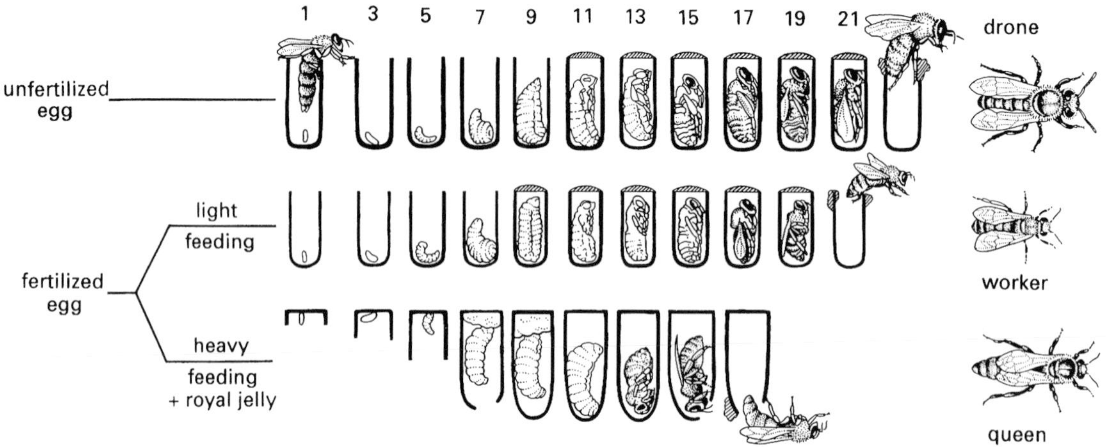

Fig. 12.5 Development of the honey bee, *Apis mellifera* (Hymenoptera: Apidae), showing the factors that determine differentiation of the queen-laid eggs into drones, workers, and queens (on the left) and the approximate developmental times (in days) and stages for drones, workers, and queens (on the right). (After Winston 1987.)

involves a pheromone disseminated by attendant workers that contact the queen and then move about the colony as messenger bees. Also, as the queen moves around on the comb whilst ovipositing into the cells, she leaves a trail of **footprint pheromone**. Production of queens takes place in cells that are distant from the effects of the queen's pheromone control, as occurs when nests become very large. Should the queen die, the volatile pheromone signal dissipates rapidly, and the workers become aware of the absence. Honey bees have very strongly developed chemical communication, with specific pheromones associated with mating, alarm, and orientation as well as colony recognition and regulation. Physical threats are rare, and are used only by young gynes towards workers.

Males, termed **drones**, are produced throughout the life of the honey-bee colony, either by the queen or perhaps by workers with developed ovaries. Males contribute little to the colony, living only to mate: their genitalia are ripped out after copulation and they die.

Nest construction in eusocial wasps

The founding of a new colony of eusocial vespid wasps takes place in spring, following the emergence of an overwintering queen. After her departure from the natal colony the previous fall, the new queen mates, but her ovarioles remain undeveloped during the temperature-induced winter quiescence or facultative diapause. As spring temperatures rise, queens leave hibernation and feed on nectar or sap, and the ovarioles grow. The resting site, which may be shared by several overwintering queens, is not a prospective site for foundation of the new colony. Each queen scouts individually for a suitable cavity and fighting may occur if sites are scarce.

Nest construction begins with the use of the mandibles to scrape wood fibers from sound or, more rarely, rotten wood. The wasp returns to the nest site using visual cues, carrying the wood pulp masticated with water and saliva in the mandibles. This pulpy paper is applied to the underside of a selected support at the top of the cavity. From this initial buttress, the pulp is formed into a descending pillar, upon which is suspended ultimately the embryonic colony of 20–40 cells (Fig. 12.6). The first two cells, rounded in cross-section, are attached and then an umbrella-like envelope is formed over the cells. The envelope is elevated by about the width of the queen's body above the cells, allowing the queen to rest there, curled around the pillar. The developing colony grows by the addition of further cells, now hexagonal in cross-section and wider at the open end, and by either extension of the envelope or construction of a new one. The queen forages only for building material at the start of nest construction. As the larvae develop from the first cells, both liquid and insect prey are sought to nourish the developing larvae, although wood pulp continues to be collected for

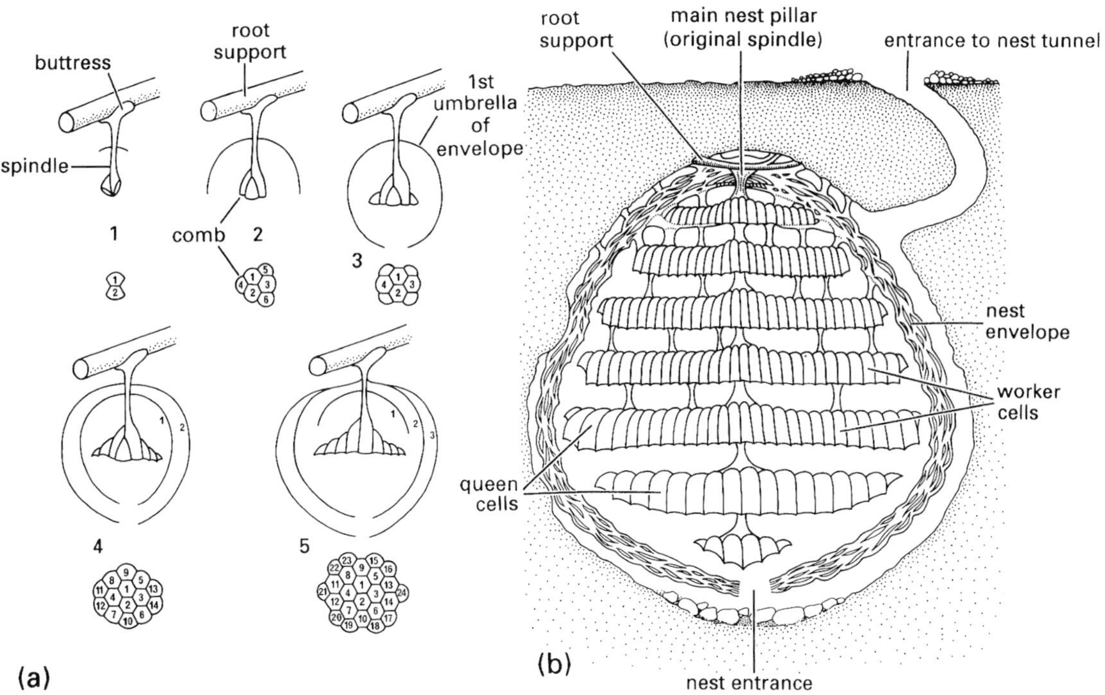

Fig. 12.6 The nest of the common European wasp, *Vespula vulgaris* (Hymenoptera: Vespidae): (a) initial stages (1–5) of nest construction by the queen (the embryonic phase of the colony's life); (b) a mature nest. (After Spradbery 1973.)

further cell construction. This first embryonic phase of the life of the colony ceases as the first workers emerge.

As the colony grows, further pillars are added, providing support to more lateral areas where brood-filled cells are aligned in **combs** (series of adjoining cells aligned in parallel rows). The early cells and envelopes become overgrown, and their materials may be reused in later construction. In a subterranean nest, the occupants may have to excavate soil and even small stones to allow colony expansion, resulting in a mature nest (as in Fig. 12.6), which may contain as many as 12,000 cells. The colony has some independence from external temperature, as thoracic heating through wing beating and larval feeding can raise temperature, and high temperature can be lowered by directional fanning or by evaporation of liquid applied to the pupal cells.

At the end of the season, males and gynes (potential queens) are produced and are fed with larval saliva and prey brought into the nest by workers. As the old queen fails and dies, and gynes emerge from the nest, the colony declines rapidly and the nest is destroyed as workers fight and larvae are neglected. Potential queens and males mate away from the nest, and the mated female seeks a suitable overwintering site.

Nesting in honey bees

In honey bees, new colonies are initiated if the old one becomes too crowded. When a bee colony becomes too large and the population density too high, a founder queen, accompanied by a swarm of workers, seeks a new nest site. Because workers cannot survive long on the honey reserves carried in their stomachs, a suitable site must be found quickly. Scouts may have started the search several days before formation of the swarm. If a suitable cavity is found, the scout returns to the cluster and communicates the direction and quality of the site by a dance (Box 12.1). Optimally, a new site should be beyond the foraging territory of the old nest, but not so distant that energy is expended in long-distance flight. Bees from temperate areas select enclosed nest sites in cavities of about 40 liters in volume, whereas more tropical bees choose smaller cavities or nest outside.

Box 12.1 The dance language of bees

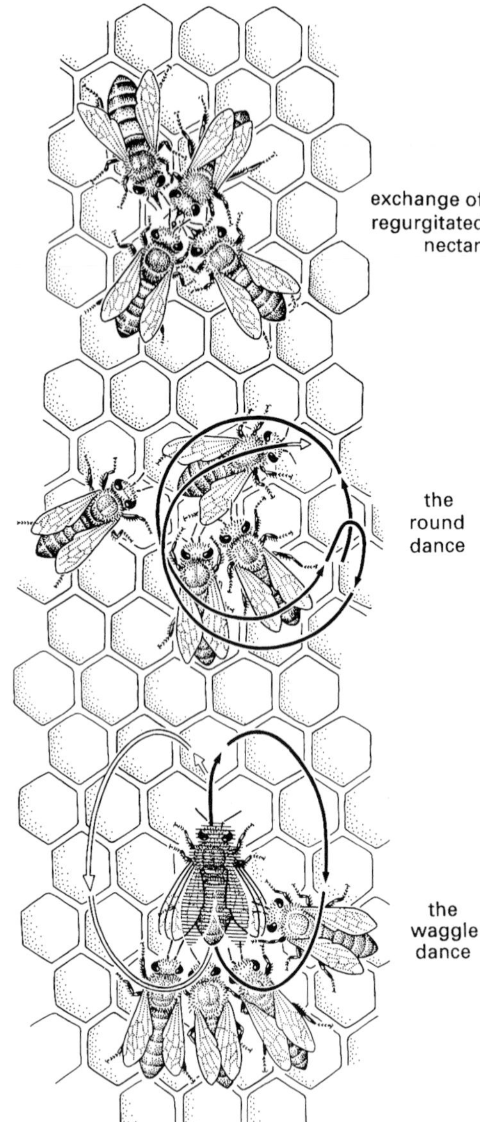

Honey bees have impressive communication abilities. Their ability to communicate forage sites to their nestmates first was recognized when a marked worker provided with an artificial food source was allowed to return to its hive, and then prevented from returning to the food. The rapid appearance at the food source of other workers indicated that information concerning the resource had been transferred within the hive. Subsequent observations using a glass-fronted hive showed that foragers often performed a dance on return to the nest. Other workers followed the dancer, made antennal contact and tasted regurgitated food, as depicted here in the upper illustration (after Frisch 1967). Olfactory communication alone could be discounted by experimental manipulation of food sources, and the importance of dancing became recognized. Variations within different dances allow communication and recruitment of workers to close or distant food sources, and to food versus prospective nest sites. The purpose and messages associated with three dances – the round, waggle, and dorsoventral abdominal vibrating (DVAV) – have become well understood.

Nearby food is communicated by a simple **round dance** involving the incoming worker exchanging nectar and making tight circles, with frequent reversals, for a few seconds to a few minutes, as shown in the central illustration (after Frisch 1967). The quality of nectar or pollen from the source is communicated by the vigor of the dance. Although no directionality is conveyed, 89% of 174 workers contacted by the dancer during a round dance were able to find the novel food source within five minutes, probably by flying in ever increasing circles until the local source is found.

More distant food sources are identified by a **waggle dance**, which involves abdomen-shaking during a figure-of-eight circuit, shown in the lower illustration (after Frisch 1967), as well as food sharing. Informative characteristics of the dance include the length of the straight part (measured by the number of comb cells traversed), the dance tempo (number of dances per unit time), the duration of waggling and noise production (buzzing) during the straight-line section, and the orientation of the straight run relative to gravity. Messages conveyed are the energy required to get to the source (rather than absolute distance), quality of the forage, and direction relative to the sun's position (Box 4.4). This interpretation of the significance and information content of the waggle dance was challenged by some experimentalists, who ascribed food-site location entirely to odors particular to the site and borne by the dancer. The claim mainly centered on the protracted time taken for the worker observers of the dance to locate a specific site. The duration matches more the time expected for a bee to take to locate an odor plume and subsequently zig-zag up the plume (Fig. 4.7), compared with direct flight from bearings provided in the dance. Following some well-designed studies, it is now evident that food finding is as effective and efficient when the experimental source is placed upwind as when it is placed downwind. Furthermore, although experienced workers can locate food by odor, the waggle dance serves to communicate information to naïve workers that allows them to head in the correct general direction. Close to the food source, specific odor does appear to be significant, and the final stages of orientation may be the slow part of location (particularly in experimental set-ups, with non-authentic food sources stationed beside human observers).

The function of the **vibration dance** (DVAV) differs from the round and waggle dances in regulating the daily and seasonal foraging patterns in relation to fluctuating food supply. Workers vibrate their bodies, particularly their abdomens, in a dorsoventral plane, usually whilst in contact with another bee. Vibration dances peak at times of the day and season when the colony needs to be primed for increased foraging, and these dances act to recruit workers into the waggle-dance area. Vibration dancing with queen contact appears to lessen the inhibitory capacity of the queen, and is used during the period when queen rearing is taking place. Cessation of this kind of vibration dancing may result in the queen departing with a swarm, or in the mating flight of new queens.

Communication of a suitable site for a new nest differs somewhat from communication of a food source. The returning scout dances without any nectar or pollen and the dance lasts for 15–30 minutes rather than the 1–2 minutes' duration of the forager's dance. At first, several scouts returning from various prospective new sites will all dance, with differences in tempo, angle, and duration that indicate the different directions and quality of the sites, as in a waggle dance. More scouts then fly out to prospect and some sites are rejected. Gradually, a consensus is attained, as shown by one dance that indicates the agreed site.

Following consensus over the nest site, workers start building a nest using wax. **Wax** is unique to social bees and is produced by workers that metabolize honey in fat cells located close to the wax glands. These modified epidermal cells lie beneath **wax mirrors** (overlapping plates) ventrally on the fourth to seventh abdominal segments. Flakes of wax are extruded beneath each wax mirror and protrude slightly from each segment of a worker that is actively producing wax. Wax is quite malleable at the ambient nest temperature of 35°C, and when mixed with saliva can be manipulated for cell construction. At nest foundation, workers already may have wax protruding from the abdominal wax glands. They start to construct combs of back-to-back hexagonal cells in a parallel series, or comb. Combs are separated from one another by pillars and bridges of wax. A thick cell base of wax is extended into a thin-walled cell of remarkably constant dimensions, despite a series of workers being involved in construction. In contrast to other social insects such as the vespids described above, cells do not hang downwards but are angled at about 13° above the horizontal, thereby preventing loss of honey. The precise orientation of the cells and comb derives from the bees' ability to detect gravity through the proprioceptor hair plates at the base of their necks. Although removal of the hair plates prevents cell construction, worker bees could construct serviceable cells under conditions of weightlessness in space.

Unlike most other bees, honey bees do not chew up and reuse wax: once a cell is constructed it is permanently part of the nest, and cells are reused after the brood has emerged or the food contents have been used. Cell sizes vary, with small cells used to rear workers, and larger ones for drones (Fig. 12.5). Later in the life of the nest, elongate conical cells in which queens are reared are constructed at the bottom and sides of the nest. The brood develops and pollen is stored in lower and more central cells, whereas honey is stored in upper and peripheral cells. Workers form honey primarily from nectar taken from flowers, but also from secretions from extrafloral nectaries, or insect-produced honeydew. Workers carry nectar to the hive in honey stomachs, from which it may be fed directly to the brood and to other adults. However, most often it is converted to honey by enzymatic digestion of the sugars to simpler forms and reduction of the water content by evaporation before storage in wax-sealed cells until required to feed adults or larvae. It has been calculated that in 66,000 bee-hours of labor, 1 kg of beeswax can be formed into 77,000 cells, which can support the weight of 22 kg of honey. An average colony requires about 60–80 kg of honey per annum.

Unlike wasps, honey bees do not hibernate with the arrival of the lower temperatures of temperate winter. Colonies remain active through the winter, but foraging is curtailed and no brood is reared. Stored honey provides an energy source for activity and heat generation within the nest. As outside temperatures drop, the workers cluster together, heads inwards, forming an inactive layer of bees on the outside, and warmer, more active, feeding bees on the inside. Despite the prodigious stores of honey and pollen, a long or extremely cold winter may kill many bees.

Beehives are artificial constructions that resemble feral honey-bee nests in some dimensions, notably the distance between the combs. When given wooden frames separated by an invariable natural spacing interval of 9.6 mm honey bees construct their combs within the frame without formation of the internal waxen bridges needed to separate the combs of a feral nest. This width between combs is approximately the space required for bees to move unimpeded on both combs. The ability to remove frames allows the apiculturalist (beekeeper) to examine and remove the honey, and replace the frames in the hive. The ease of construction allows the building of several ranks of boxes. The hives can be transported to suitable locations without damaging the combs. Although the apiculture industry has developed through commercial production of honey, lack of native pollinators in monocultural agricultural systems has led to increasing reliance on the mobility of hive bees to ensure the pollination of crops as diverse as canola, nuts, soybeans, fruits, clover, alfalfa, and other fodder crops. In the USA alone, in 1998 some 2,500,000 bee colonies were rented for pollination purposes and the value to US agriculture attributable to honey-bee pollination was about US$15 billion in 2000. Yield losses of over 90% of fruit, seed, and nut crops would occur without honey-bee pollination. The role of the many species of eusocial native bees is little recognized, but may be important in areas of natural vegetation.

12.2.3 Specialized hymenopterans: ants

Ants (Formicidae) form a well-defined, highly specialized group within the superfamily Vespoidea (Fig. 12.2). The morphology of a worker ant of *Formica* is illustrated in Box 12.2.

Colony and castes in ants

All ants are social and their species are polymorphic. There are two major female castes, the reproductive queen and the workers, usually with complete dimorphism between them. Many ants have monomorphic workers, but others have distinct subcastes called, according to their size, **minor**, **media**, or **major** workers. Although workers may form clearly different morphs, more often there is a gradient in size. Workers are never winged, but queens have wings that are shed after mating, as do males, which die after mating. Winged individuals are called **alates**. Polymorphism in ants is accompanied by polyethism, with the queen's role restricted to oviposition, and the workers performing all other tasks. If workers are monomorphic, there may be temporal or age polyethism, with young workers undertaking internal nurse duties and older ones foraging outside the nest. If workers are polymorphic, the subcaste with the largest individuals, the major workers, usually has a defensive or soldier role.

The workers of certain ants, such as the fire ants (*Solenopsis*), have reduced ovaries and are irreversibly sterile. In others, workers have functional ovaries and may produce some or all of the male offspring by laying haploid (unfertilized) eggs. In some species, when the queen is removed, the colony continues to produce gynes from fertilized eggs previously laid by the queen, and males from eggs laid by workers. The inhibition by the queen of her daughter workers is quite striking in the African weaver ant, *Oecophylla longinoda*. A mature colony of up to half a million workers, distributed amongst as many as 17 nests, is prevented from reproduction completely by a single queen. Workers, however, do produce male offspring in nests that lie outside the influence (or territory) of the queen. Queens prevent the production of reproductive eggs by workers, but may allow the laying of specialized **trophic** eggs that are fed to the queen and/or larvae. By this means the queen not only prevents any reproductive competition, but directs much of the protein in the colony towards her own offspring.

Caste differentiation is largely trophogenic (diet-determined), involving biased allocation of volume and quality of food given to the larvae. A high-protein diet promotes differentiation of gyne/queen and a less rich, more dilute diet leads to differentiation of workers. The queen generally inhibits the development of gynes indirectly by modifying the feeding behavior of workers towards female larvae, which have the potential to differentiate as either gynes or workers. In *Myrmica*, large, slowly developing larvae will become gynes, so stimulation of rapid development and early metamorphosis of small larvae, or food deprivation and irritating of large larvae by biting to accelerate development, both induce differentiation as workers. When queen influence wanes, either through the increased size of the colony, or because the inhibitory pheromone is impeded in its circulation throughout the colony,

Fig. 12.7 Weaver ants of *Oecophylla* making a nest by pulling together leaves and binding them with silk produced by larvae that are held in the mandibles of worker ants. (After CSIRO 1970; Hölldobler 1984.)

gynes are produced at some distance from the queen. There is also a role for JH in caste differentiation. JH tends to induce queen development during egg and larval stages, and induces production of major workers from already differentiated workers.

According to a seasonal cycle, ant gynes mature to winged reproductives, or alates, and remain in the nest in a sexually inactive state until external conditions are suitable for departing the nest. At the appropriate time they make their nuptial flight, mate, and attempt to found a new colony.

Nesting in ants

The subterranean soil nests of *Myrmica* and the mounds of plant debris of *Formica* are typical temperate ant nests. Colonies are founded when a mated queen sheds her wings and overwinters, sealed into a newly dug nest that she will never leave. In spring, the queen lays some eggs and feeds the hatched larvae by **stomodeal** or **oral trophallaxis**, i.e. regurgitation of liquid food from her internal food reserves. Colonies develop slowly whilst worker numbers build up, and a nest may be many years old before alates are produced.

Colony foundation by more than one queen, known as **pleometrosis**, appears to be fairly widespread, and the digging of the initial nest may be shared, as in the honeypot ant *Myrmecocystus mimicus*. In this species and others, multi-queen nests may persist as polygynous colonies, but monogyny commonly arises through dominance of a single queen, usually following rearing of the first brood of workers. Polygynous nests often are associated with opportunistic use of ephemeral resources, or persistent but patchy resources.

The woven nests of *Oecophylla* species are well-known, complex structures (Fig. 12.7). These African and Asian/Australian weaver ants have extended territories that workers continually explore for any leaf that can be bent. A remarkable collaborative construction effort follows, in which leaves are manipulated into a tent-shape by linear ranks of workers, often involving "living chains" of ants that bridge wide gaps between the leaf edges. Another group of workers take larvae from existing nests and carry them held delicately between their mandibles to the construction site. There, larvae are induced to produce silk threads from their well-developed silk glands and a nest is woven linking the framework of leaves.

Living plant tissues provide a location for nests of ants such as *Pseudomyrmex ferrugineus*, which nests in the expanded thorns of the Central American bull's-horn acacia trees (Fig. 11.10a). In such mutualisms involving plant defense, plants benefit by deterrence of phytophagous animals by the ants, as discussed in section 11.4.1.

Foraging efficiency of ants can be very high. A typical mature colony of European red ants (*Formica polyctena*) is estimated to harvest about 1 kg of arthropod food per day. The legionary, or army, and driver ants are popularly known for their voracious predatory activities. These ants, which predominantly belong to the subfamilies Ecitoninae and Dorylinae, alternate cyclically between sedentary (**statary**) and migratory or nomadic phases. In the latter phase, a nightly **bivouac** is formed, which often is no more than an exposed cluster of the entire colony. Each morning, the millions-strong colony moves *in toto*, bearing the larvae. The advancing edge of this massive group raids and forages on a wide range of terrestrial arthropods, and group predation allows even large prey items to be overcome. After some two weeks of nomadism, a statary period commences, during which the queen lays 100,000–300,000 eggs in a statary bivouac. This is more sheltered than a typical overnight bivouac, perhaps within an old ants' nest, or beneath a log. In the three weeks before the eggs hatch, larvae of the previous oviposition complete their development to emerge as new workers, thus stimulating the next migratory period.

Not all ants are predatory. Some ants harvest grain and seeds (myrmecochory; section 11.3.2) and others, including the extraordinary honeypot ants, feed almost exclusively on insect-produced honeydew, including that of scale insects tended inside nests (section 11.4.1). Workers of honeypot ants return to the nest with crops filled with honeydew, which is fed by oral trophallaxis to selected workers called **repletes**. The abdomen of repletes are so distensible that they become virtually immobile "honey pots" (Fig. 2.4), which act as food reserves for all in the nest.

12.2.4 Isoptera (termites)

All termites (Isoptera) are eusocial. Their diagnostic features and biology are summarized in Box 12.3.

Colony and castes in termites

In contrast to the adult and female-only castes of holometabolous eusocial Hymenoptera, the castes of the hemimetabolous Isoptera involve immature stages and equal representation of the sexes. However, before castes are discussed further, terms for termite immature stages must be clarified. Termitologists refer to the developmental instars of reproductives as nymphs, more properly called brachypterous nymphs; and the instars of sterile lineages as larvae, although strictly the latter are apterous nymphs.

The termites may be divided into two groups – the "lower" and "higher" termites. The species-rich higher termites (Termitidae) differ from lower termites in the following manners:

- Members of the Termitidae lack the symbiotic flagellates found in the hind gut of lower termites; these protists (protozoa) secrete enzymes (including cellulases) that may contribute to the breakdown of gut contents. One subfamily of Termitidae uses a cultivated fungus to predigest food.
- Termitidae have a more elaborate and rigid caste system. For example, in most lower termites there is little or no distension of the queen's abdomen, whereas termitid queens undergo extraordinary **physogastry**, in which the abdomen is distended to 500–1000% of its original size (Fig. 12.8; see Plate 5.4).

All termite colonies contain a pair of **primary reproductives** – the queen and king (Plate 5.4), which are former alate (winged) adults from an established colony. Upon loss of the primary reproductives, potential replacement reproductives occur (in some species a small number may be ever-present). These individuals, called **supplementary reproductives**, or **neotenics**, are arrested in their development, either with wings present as buds (brachypterous neotenics) or without wings (apterous neotenics, or **ergatoids**), and can take on the reproductive role if the primary reproductives die.

In contrast to these reproductives, or potentially reproductive castes, the colony is dominated numerically by sterile termites that function as workers and soldiers of both sexes. Soldiers have distinctive heavily sclerotized heads, with large mandibles or with a

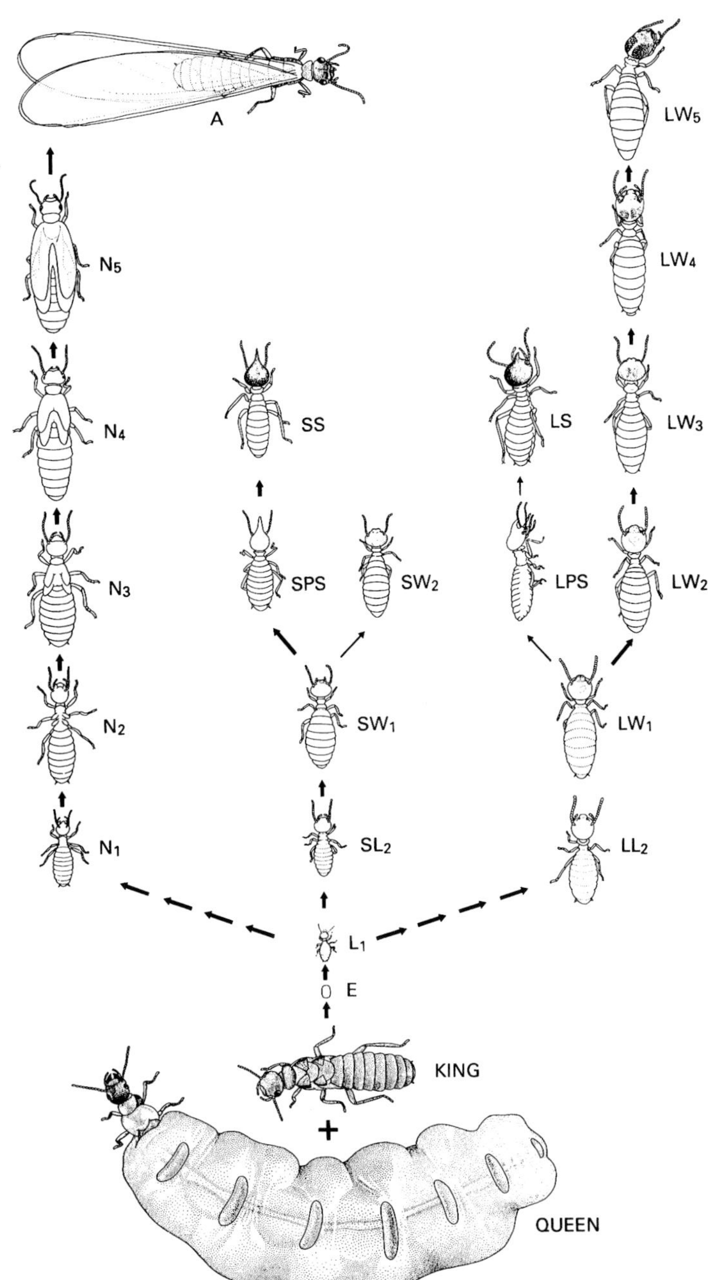

Fig. 12.8 Developmental pathways of the termite *Nasutitermes exitiosus* (Isoptera: Termitidae). Heavy arrows indicate the main lines of development, light arrows the minor lines. A, alate; E, egg; L, larva; LL, large larva; LPS, large presoldier; LS, large soldier; LW, large worker; N, nymph; SL, small larva; SPS, small presoldier; SS, small soldier; SW, small worker. The numbers indicate the stages. (Pathways based on Watson & Abbey 1985.)

strongly produced snout (or **nasus**) through which sticky defensive secretions are ejected. Two classes, major and minor soldiers, may occur in some species. Workers are unspecialized, weakly pigmented and poorly sclerotized, giving rise to the popular name of "white ants".

Caste differentiation pathways are portrayed best in the more rigid system of the higher termites

(Termitidae), which can then be contrasted with the greater plasticity of the lower termites. In *Nasutitermes exitiosus* (Termitidae: subfamily Nasutitermitinae) (Fig. 12.8), two different developmental pathways exist; one leads to reproductives and the other (which is further subdivided) gives rise to sterile castes. This differentiation may occur as early as the first larval stage, although some castes may not be recognizable morphologically until later molts. The reproductive pathway (on the left in Fig. 12.8) is relatively constant between termite taxa and typically gives rise to alates – the winged reproductives that leave the colony, mate, disperse, and found new colonies. In *N. exitiosus* no neotenics are formed; replacement for lost primary reproductives comes from amongst alates retained in the colony. Other *Nasutitermes* show great developmental plasticity.

The sterile (neuter) lineages are complex and variable between different termite species. In *N. exitiosus*, two categories of second-instar larvae can be recognized according to size differences probably relating to sexual dimorphism, although which sex belongs to which size category is unclear. In both lineages a subsequent molt produces a third-instar nymph of the worker caste, either small or large according to the pathway. These third-instar workers have the potential (**competency**) to develop into a soldier (via an intervening **presoldier** instar) or remain as workers through several more molts. The sterile pathway of *N. exitiosus* involves larger workers continuing to grow at successive molts, whereas the small worker ceases to molt beyond the fourth instar. Those that molt to become presoldiers and then soldiers develop no further.

The lower termites are more flexible, exhibiting more routes to differentiation. Lower termites have no true worker caste, but employ a functionally equivalent "child-labor" **pseudergate** caste composed of either nymphs whose wing buds have been eliminated (regressed) by molting or, less frequently, brachypterous nymphs or even undifferentiated larvae. Unlike the "true" workers of the higher termites, pseudergates are developmentally plastic and retain the capacity to differentiate into other castes by molting. In lower termites, differentiation of nymphs from larvae, and reproductives from pseudergates, may not be possible until a relatively late instar is reached. If there is sexual dimorphism in the sterile line, the larger workers are often male, but workers may be monomorphic. This may be through the absence of sexual dimorphism, or more rarely, because only one sex is represented. Molts in species of lower termites may give:

- morphological change within a caste;
- no morphological advance (stationary molt);
- change to a new caste (such as a pseudergate to a reproductive);
- saltation to a new morphology, missing a normal intermediate instar;
- supplementation, adding an instar to the normal route;
- reversion to an earlier morphology (such as a pseudergate from a reproductive), or a presoldier from any nymph, late-instar larva, or pseudergate.

Instar determination is impossibly difficult in the light of these molting potentialities. The only inevitability is that a presoldier must molt to a soldier.

Certain unusual termites lack soldiers. Even the universal presence of only one pair of reproductives has exceptions; multiple primary queens cohabit in some colonies of some Termitidae.

Individuals in a termite colony are derived from one pair of parents. Therefore, genetic differences existing between castes either must be sex-related or due to differential expression of the genes. Gene expression is under complex multiple and synergistic influences entailing hormones (including neurohormones), external environmental factors, and interactions between colony members. Termite colonies are very structured and have high homeostasy – caste proportions are restored rapidly after experimental or natural disturbance, by recruitment of individuals of appropriate castes and elimination of individuals excess to colony needs. Homeostasis is controlled by several pheromones that act specifically upon the corpora allata and more generally on the rest of the endocrine system. In the well-studied *Kalotermes*, primary reproductives inhibit differentiation of supplementary reproductives and alate nymphs. Presoldier formation is inhibited by soldiers, but stimulated through pheromones produced by reproductives.

Pheromones that inhibit reproduction are produced inside the body by reproductives and disseminated to pseudergates by **proctodeal trophallaxis**, i.e. by feeding on anal excretions. Transfer of pheromones to the rest of the colony is by oral trophallaxis. This was demonstrated experimentally in a *Kalotermes* colony by removing reproductives and dividing the colony into two halves with a membrane. Reproductives were reintroduced, orientated within the membrane such that their abdomens were directed into one half of the colony and their heads into the other. Only in the "head-end" part of the colony did pseudergates differentiate as reproductives: inhibition continued at the "abdomen-end". Painting the protruding abdomen with varnish eliminated any cuticular chemical messengers but failed to remove the inhibition on pseudergate development. In constrast, when the anus was blocked, pseudergates became reproductive, thereby verifying anal transfer. The inhibitory pheromones produced by both queen and king have complementary or synergistic effects: a female pheromone stimulates the male to release inhibitory pheromone, whereas the male pheromone has a lesser stimulatory effect on the female. Production of primary and supplementary reproductives involves removal of these pheromonal inhibitors produced by functioning reproductives.

Increasing recognition of the role of JH in caste differentiation comes from observations such as the differentiation of pseudergates into soldiers after injection or topical application of JH or implantation of the corpora allata of reproductives. Some of the effects of pheromones on colony composition may be due to JH production by the primary reproductives. Caste determination in Termitidae originates as early as the egg, during maturation in the ovary of the queen. As the queen grows, the corpora allata undergoes hypertrophy and may attain a size 150 times greater than the gland of the alate. The JH content of eggs also varies, and it is possible that a high JH level in the egg causes differentiation to follow the sterile lineage. This route is enforced if the larvae are fed proctodeal foods (or trophic eggs) that are high in JH, whereas a low level of JH in the egg allows differentiation along the reproductive pathway. In higher and lower termites, worker and soldier differentiation from the third-instar larva is under further hormonal control, as demonstrated by the induction of individuals of these castes by JH application.

Nesting in termites

In the warmer parts of the temperate northern hemisphere, drywood termites (Kalotermitidae, especially *Cryptotermes*) are most familiar because of the structural damage that they cause to timber in buildings. Termites are pests of drywood and dampwood in the subtropics and tropics, but in these regions termites may be more familiar through their spectacular mound nests. In the timber pests, colony size may be no greater than a few hundred termites, whereas in the mound formers (principally species of Termitidae and some Rhinotermitidae), several million individuals may be involved. The Formosan subterranean termite (*Coptotermes formosanus*, Rhinotermitidae) which mostly lives in underground nests, and is a serious pest in the south-eastern USA, can form huge colonies of up to 8 million individuals.

In all cases, a new nest is founded by a male and female following the nuptial flight of alates. A small cavity is excavated into which the pair seal themselves. Copulation takes place in this royal cell, and egg-laying commences. The first offspring are workers, which are fed on regurgitated wood or other plant matter, primed with gut symbionts, until they are old enough to feed themselves and enlarge the nest. Early in the life of the colony, production is directed towards workers, with later production of soldiers to defend the colony. As the colony matures, but perhaps not until it is 5–10 years old, production of reproductives commences. This involves differentiation of alate sexual forms at the appropriate season for swarming and foundation of new colonies.

Tropical termites can use virtually all cellulose-rich food sources, above and below the ground, from grass tussocks and fungi to living and dead trees. Workers radiate from the mound, often in subterranean tunnels, less often in above-ground, pheromone-marked trails, in search of materials. In the subfamily Macrotermitinae (Termitidae), fungi are raised in combs of termite feces within the mound, and the complete culture of fungus and excreta is eaten by the colony (section 9.5.3). These fungus-tending termites form the largest termite colonies known, with estimated millions of inhabitants in some East African species.

The giant mounds of tropical termites mostly belong to species in the Termitidae. As the colony grows through production of workers, the mound is enlarged by layers of soil and termite feces until mounds as much as a century old attain massive dimensions. Diverse mound architectures characterize different termite species; for example, the "magnetic mounds" of *Amitermes meridionalis* in northern Australia have a narrow north–south and broad east–west orientation, and can be used like a compass (Fig. 12.9). Orientation relates to thermoregulation, as the broad face of the mound receives maximum exposure to the warming of the early and late sun, and the narrowest face presented to the high and hot midday sun. Aspect is not the only means of temperature regulation: intricate internal design, especially in fungus-farming *Macrotermes*

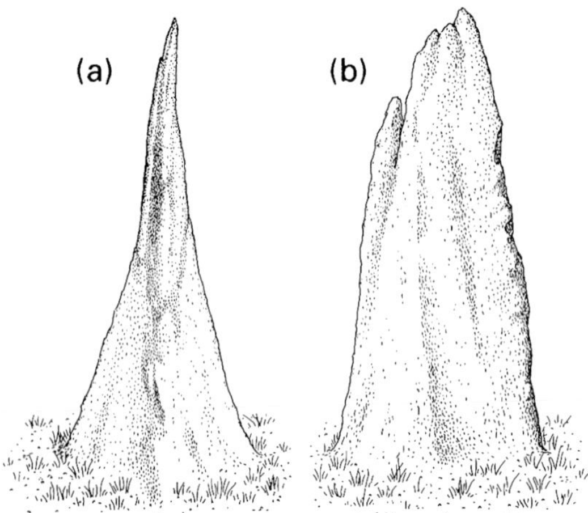

Fig. 12.9 A "magnetic" mound of the debris-feeding termite *Amitermes meridionalis* (Isoptera: Termitidae) showing: (a) the north–south view, and (b) the east–west view. (After Hadlington 1987).

species, allows circulation of air to give microclimatic control of temperature and carbon dioxide (Fig. 12.10).

12.3 INQUILINES AND PARASITES OF SOCIAL INSECTS

The abodes of social insects provide many other insects with a hospitable place for their development. The term **inquiline** refers to an organism that shares a home of another. This covers a vast range of organisms that have some kind of obligate relationship with another organism, in this case a social insect. Complex classification schemes involve categorization of the insect host and the known or presumed ecological relationship between inquiline and host (e.g. myrmecophile, termitoxene). However, two alternative divisions appropriate to this discussion involve the degree of integration of the inquiline lifestyle with that of the host. Thus, **integrated** inquilines are incorporated into their hosts' social lives by behavioral modification of both parties, whereas **non-integrated** inquilines are adapted ecologically to the nest, but do not interact socially with the host. Predatory inquilines may negatively affect the host, whereas other inquilines may merely shelter within the nest, or give benefit, such as by feeding on nest debris.

Integration may be achieved by mimicking the chemical cues used by the host in social communication (such as pheromones), or by tactile signaling that releases social behavioral responses, or both. The term **Wasmannian mimicry** covers some or all chemical or tactile mimetic features that allow the mimic to be accepted by a social insect, but the distinction from other forms of mimicry (notably Batesian; section 14.5.1) is unclear. Wasmannian mimicry may, but need not, include imitation of the body form. Conversely, mimicry of a social insect may not imply inquilinism – the ant mimics shown in Fig. 14.12 may gain some protection from their natural enemies as a result of their ant-like appearance, but are not symbionts or nest associates.

The breaking of the social insect chemical code occurs through the ability of an inquiline to produce appeasement and/or adoption chemicals – the messengers that social insects use to recognize one another and to distinguish themselves from intruders. Caterpillars of *Maculinea arion* (the large blue butterfly) and congeners that develop in the nests of red ants (*Myrmica* spp.) as inquilines or parasites evidently surmount the nest defenses (Box 1.1). Certain staphylinid beetles also can do this, for example *Atemeles pubicollis*, which lives as a larva in the nest of the European ant, *Formica rufa*. The staphylinid larva produces a glandular secretion that induces brood-tending ants to groom the alien. Food is obtained by adoption of the begging posture of an ant larva, in which the larva rears up and contacts the adult ant mouthparts, provoking a release of regurgitated food. The diet of the staphylinid is supplemented by predation on larvae of ants and of their own species. Pupation and adult eclosion take place in the *Formica rufa* nest. However, this ant ceases activity in winter and during this period the staphylinid seeks alternative shelter. Adult beetles leave the wooded *Formica* habitat and migrate to the more open grassland habitat of *Myrmica* ants. When a *Myrmica* ant is encountered, secretions from the "appeasement glands" are offered that suppress the aggression of the ant, and then the products of glands on the lateral abdomen attract the ant. Feeding on these secretions appears to facilitate "adoption", as the ant subsequently carries the beetle back to its nest where the immature adult overwinters as a tolerated food-thief. In spring, the reproductively mature adult beetle departs for the woods to seek out another *Formica* nest for oviposition.

Amongst the inquilines of termites, many show convergence in shape in terms of physogastry (dilation of

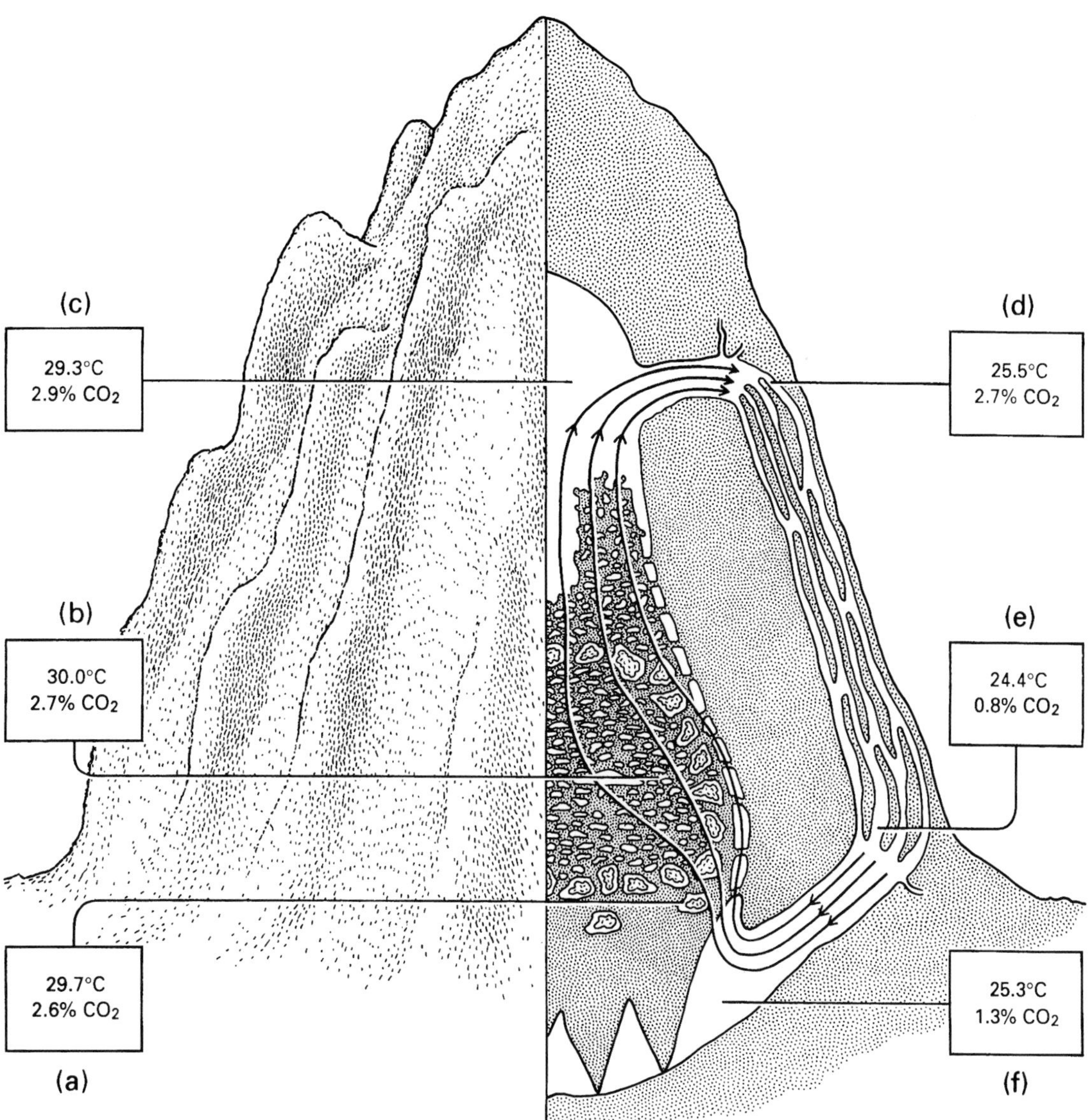

Fig. 12.10 Section through the mound nest of the African fungus-farming termite *Macrotermes natalensis* (Isoptera: Termitidae) showing how air circulating in a series of passageways maintains favorable culture conditions for the fungus at the bottom of the nest (a) and for the termite brood (b). Measurements of temperature and carbon dioxide are shown in the boxes for the following locations: (a) the fungus combs; (b) the brood chambers; (c) the attic; (d) the upper part of a ridge channel; (e) the lower part of a ridge channel; and (f) the cellar. (After Lüscher 1961.)

the abdomen), seen also in queen termites. In the curious case of flies of *Termitoxenia* and relatives (Diptera: Phoridae), the physogastric females from termite nests were the only stage known for so long that published speculation was rife that neither larvae nor males existed. It was suggested that the females hatched directly from huge eggs, were brachypterous throughout their lives (hitching a ride on termites for dispersal), and, uniquely amongst the endopterygotes, the flies were believed to be protandrous hermaphrodites, functioning first as males, then as females. The truth is more prosaic: sexual dimorphism in the group is so great that wild-caught, flying males had been unrecognized and placed in a different taxonomic group. The females are winged, but shed all but the stumps of the anterior veins after mating, before entering the termitarium. Although the eggs are large, short-lived larval stages exist. As the postmated female is **stenogastrous** (with a small abdomen), physogastry must develop whilst in the termitarium. Thus, *Termitoxenia* is only a rather unconventional fly, well adapted to the rigors of life in a termite nest, in which its eggs are treated by the termites as their own, and with attenuation of the vulnerable larval stage, rather than the possessor of a unique suite of life-history features.

Inquilinism is not restricted to non-social insects that breach the defenses (section 12.4.3) and abuse the hospitality of social insects. Even amongst the social Hymenoptera some ants may live as temporary or even permanent social parasites in the nests of other species. A reproductive female inquiline gains access to a host nest and usually kills the resident queen. In some cases, the intruder queen produces workers, which eventually take over the nest. In others, the inquiline usurper produces only males and reproductives – the worker caste is eliminated and the nest survives only until the workers of the host species die off.

In a further twist of the complex social lives of ants, some species are slave-makers; they capture pupae from the nests of other species and take them to their own nest where they are reared as slave workers. This phenomenon, known as **dulosis**, occurs in several inquiline species, all of which found their colonies by parasitism.

The phylogenetic relationships between ant hosts and ant inquilines reveal an unexpectedly high proportion of instances in which host and inquiline belong to sister species (i.e. each other's closest relatives), and many more are congeneric close relatives. One possible explanation envisages the situation in which daughter species formed in isolation come into secondary contact after mating barriers have developed. If no differentiation of colony-identifying chemicals has taken place, it is possible for one species to invade the colony of the other undetected, and parasitization is facilitated.

Non-integrated inquilines are exemplified by hover flies of the genus *Volucella* (Diptera: Syrphidae), the adults of which are Batesian mimics of either *Polistes* wasps or of *Bombus* bees. Female flies appear free to fly in and out of hymenopteran nests, and lay eggs whilst walking over the comb. Hatching larvae drop to the bottom of the nest where they scavenge on fallen detritus and fallen prey. Another syrphid, *Microdon*, has a myrmecophilous larva so curious that it was described first as a mollusk, then as a coccoid. It lives unscathed amongst nest debris (and perhaps sometimes as a predator on young ant larvae), but the emerged adult is recognized as an intruder. Non-integrated inquilines include many predators and parasitoids whose means of circumventing the defenses of social insects are largely unknown.

Social insects also support a few parasitic arthropods. For example, varroa and tracheal mites (Acari) and the bee louse, *Braula coeca* (Diptera: Braulidae; section 13.3.3), all live on honey bees (Apidae: *Apis* spp.). The extent of colony damage caused by the tracheal mite *Acarapis woodi* is controversial, but infestations of *Varroa* are resulting in serious declines in honey-bee populations in most parts of the world. Varroa mites feed externally on the bee brood (see Plate 5.5) leading to deformation and death of the bees. Low levels of mite infestation are difficult to detect and it can take several years for a mite population to build to a level that causes extensive damage to the hive. Some *Apis* species, such as *A. cerana*, appear more resistant to varroa but interpretation is complicated by the existence of a sibling species complex of varroa mites with distinct biogeographic and virulence patterning. This suggests that great care should be taken to avoid promiscuous mixing of different bee and mite genotypes.

12.4 EVOLUTION AND MAINTENANCE OF EUSOCIALITY

At first impression the complex social systems of hymenopterans and termites bear a close resemblance and it is tempting to suggest a common origin. However, examination of the phylogeny presented in Chapter 7 (Fig. 7.2) shows that these two orders, and the social aphids and thrips, are distantly related and a single evolutionary origin is inconceivable. Thus, the possible routes for the origin of eusociality in Hymenoptera and Isoptera are examined separately, followed by a discussion on the maintenance of social colonies.

12.4.1 The origins of eusociality in Hymenoptera

According to estimates derived from the proposed phylogeny of the Hymenoptera, eusociality has arisen independently in wasps, bees, and ants (Fig. 12.2) with multiple origins within wasps and bees, and some losses by reversion to solitary behavior. Comparisons of life histories between living species with different degrees of social behavior allow extrapolation to possible historical pathways from solitariness to sociality. Three possible routes have been suggested and in each case, communal living is seen to provide benefits through sharing the costs of nest construction and defense of offspring.

The first suggestion envisages a monogynous (single queen) subsocial system with eusociality developing through the queen remaining associated with her offspring through increased maternal longevity.

In the second scenario, involving semisociality and perhaps applicable only to certain bees, several unrelated females of the same generation associate and establish a colonial nest in which there is some reproductive division of labor, with an association that lasts only for one generation.

The third scenario involves elements of the previous two, with a communal group comprising related females (rather than unrelated) and multiple queens (in a polygynous system), within which there is increasing reproductive division. The association of queens and daughters arises through increased longevity.

These life-history-based scenarios must be considered in relation to genetic theories concerning eusociality, notably concerning the origins and maintenance by selection of **altruism** (or self-sacrifice in reproduction). Ever since Darwin, there has been debate about altruism – why should some individuals (non-reproductive workers) sacrifice their reproductive potential for the benefit of others?

Four proposals for the origins of the extreme reproductive sacrifice seen in eusociality are discussed below. Three proposals are partially or completely compatible with one another, but **group selection**, the first considered, seems incompatible. In this case, selection is argued to operate at the level of the group: an efficient colony with an altruistic division of reproductive labor will survive and produce more offspring than one in which rampant individual self-interest leads to anarchy. Although this scenario aids in understanding the maintenance of eusociality once it is established, it contributes little if anything to explaining the origin(s) of reproductive sacrifice in non-eusocial or subsocial insects. The concept of group selection operating on pre-eusocial colonies runs counter to the view that selection operates on the genome, and hence the origin of altruistic individual sterility is difficult to accept under group selection. It is amongst the remaining three proposals, namely kin selection, maternal manipulation, and mutualism, that the origins of eusociality are more usually sought.

The first, **kin selection**, stems from recognition that **classical** or **Darwinian fitness** – the direct genetic contribution to the gene pool by an individual through its offspring – is only part of the contribution to an individual's total, or **inclusive**, or **extended**, **fitness**. An additional indirect contribution, termed the **kinship component**, must be included. This is the contribution to the gene pool made by an individual that assists and enhances the reproductive success of its kin. Kin are individuals with similar or identical genotypes derived from the relatedness due to having the same parents. In the Hymenoptera, kin relatedness is enhanced by the haplodiploid genetic system. In this system, males are haploid so that each sperm (produced by mitosis) contains 100% of the paternal genes. In contrast, the egg (produced by meiosis) is diploid, containing only half the maternal genes. Thus, daughter offspring, produced from fertilized eggs, share all their father's genes, but only half of their mother's genes. Because of this, full sisters (i.e. those with the same father) share on average three-quarters of their genes. Therefore, sisters share more genes with each other than they would with their own female offspring (50%). Under these conditions, the inclusive fitness of a sterile female (worker) is greater than its classical fitness. As selection operating on an individual should maximize its inclusive fitness, a worker should invest in the survival of her sisters, the queen's offspring, rather than in the production of her own female young.

However, haplodiploidy alone is an inadequate explanation for the origin of eusociality, because altruism does not arise solely from relatedness. Haplodiploidy is universal in hymenopterans and kinship has encouraged repeated eusociality, but eusociality is not universal in the Hymenoptera. Furthermore, other haplodiploid insects such as thrips are not eusocial, although there may be social behavior. Other factors promoting eusociality are recognized in Hamilton's rule, which emphasizes the ratio of costs and benefits of altruistic behavior as well as relatedness. The

	sister	half-sister	own son	son of full sister	queen's son (brother)	son of half-sister
worker	0.75	0.375	0.5	0.375	0.25	0.125

Fig. 12.11 Relatedness of a given worker to other possible occupants of a hive. (After Whitfield 2002.)

conditions under which selection will favor altruism can be expressed as follows:

$$rB - C > 0$$

where r is the coefficient of relatedness, B is the benefit gained by the recipient of altruism, and C is the cost suffered by the donor of altruism. Thus, variations in benefits and costs modify the consequences of the particular degrees of relatedness expressed in Fig. 12.11, although these factors are difficult to quantify.

Kinship calculations assume that all offspring of a single mother in the colony share an identical father, and this assumption is implicit in the kinship scenario for the origin of eusociality. At least in higher eusocial insects, queens may mate multiply with different males, and thus r values are less than predicted by the monogamous model. This effect impinges on maintenance of an already existing eusocial system, discussed below in section 12.4.3. Whatever, the opportunity to help relatives, in combination with high relatedness through haplodiploidy, predisposes insects to eusociality.

At least two further ideas concern the origins of eusociality. The first involves maternal manipulation of offspring (both behaviorally and genetically), such that by reducing the reproductive potential of certain offspring, parental fitness may be maximized by assuring reproductive success of a few select offspring. Most female Aculeata can control the sex of offspring through fertilizing the egg or not, and are able to vary offspring size through the amount of food supplied, making maternal manipulation a plausible option for the origin of eusociality.

A further well-supported scenario emphasizes the roles of competition and mutualism. This envisages individuals acting to enhance their own classical fitness with contributions to the fitness of neighbors arising only incidentally. Each individual benefits from colonial life through communal defense by shared vigilance against predators and parasites. Thus, mutualism (including the benefits of shared defense and nest construction) and kinship encourage the establishment of group living. Differential reproduction within a familial-related colony confers significant fitness advantages on all members through their kinship. In conclusion, the three scenarios are not mutually exclusive, but are compatible in combination, with kin selection, female manipulation, and mutualism acting in concert to encourage evolution of eusociality.

The Vespinae illustrate a trend to eusociality commencing from a solitary existence, with nest-sharing and facultative labor division being a derived condition. Further evolution of eusocial behavior is envisaged as developing through a dominance hierarchy that arose from female manipulation and reproductive competition among the nest-sharers: the "winners" are queens and the "losers" are workers. From this point onwards, individuals act to maximize their fitness and the caste system becomes more rigid. As the queen and colony acquire greater longevity and the number of generations retained increases, short-term monogynous societies (those with a succession of queens) become long-term, monogynous, **matrifilial** (mother–daughter) colonies. Exceptionally, a derived polygynous condition may arise in large colonies, and/or in colonies where queen dominance is relaxed.

The evolution of sociality from solitary behavior should not be seen as unidirectional, with the eusocial bees and wasps at a "pinnacle". Recent phylogenetic studies show many reversions from eusocial to semisocial and even to solitary lifestyles. Such reversions have occurred in halictine and allodapine bees. These losses demonstrate that even with haplodiploidy predisposing towards group living, unsuitable environmental conditions can counter this trend, with selection able to act against eusociality.

12.4.2 The origins of eusociality in Isoptera

In contrast to the haplodiploidy of Hymenoptera, termite sex is determined universally by an XX–XY chromosome system and thus there is no genetic predisposition toward kinship-based eusociality. Furthermore, and in contrast to the widespread subsociality of hymenopterans, the lack of any intermediate stages on the route to termite eusociality has obscured its origin. Subsocial behaviors in some mantids and cockroaches (the nearest relatives of the termites) have been proposed to be an evolutionary precursor to the eusociality in Isoptera. Notably, behavior in the family Cryptocercidae, which is sister branch to the termite lineage (Fig. 7.4), demonstrates how reliance on a nutrient-poor food source and adult longevity might predispose to social living. The internal symbiotic organisms needed to aid the digestion of a cellulose-rich, but nutrient-poor, diet of wood is central to this argument. The need to transfer symbionts to replenish supplies lost at each molt encourages unusual levels of intracolony interaction through trophallaxis. Furthermore, transfer of symbionts between members of successive generations requires overlapping generations. Trophallaxis, slow growth induced by the poor diet, and parental longevity, act together to encourage group cohesion. These factors, together with patchiness of adequate food resources such as rotting logs, can lead to colonial life, but do not readily explain altruistic caste origins. When an individual gains substantial benefits from successful foundation of a colony, and where there is a high degree of intracolony relatedness (as is found in some termites), eusociality may arise. However, the origin of eusociality in termites remains much less clear-cut than in eusocial hymenopterans.

12.4.3 Maintenance of eusociality – the police state

As we have seen, workers in social hymenopteran colonies forgo their reproduction and raise the brood of their queen, in a system that depends upon kinship – proximity of relatedness – to "justify" their sacrifice. Once non-reproductive castes have evolved (theoretically under conditions of single paternity), the requirement for high relatedness may be relaxed if workers lack any opportunity to reproduce, through mechanisms such as chemical control by the queen. Nonetheless, sporadically, and especially when the influence of the queen wanes, some workers may lay their own eggs. These "non-queen" eggs are not allowed to survive: the eggs are detected and eaten by a "police force" of other workers. This is known from honey bees, certain wasps, and some ants, and may be quite widespread although uncommon. For example, in a typical honey-bee hive of 30,000 workers, on average only three have functioning ovaries. Although these individuals are threatened by other workers, they can be responsible for up to 7% of the male eggs in any colony. Because these eggs lack chemical odors produced by the queen, they can be detected and are eaten by the policing workers with such efficiency that only 0.1% of a honey-bee colony's males derive from a worker as a mother.

Hamilton's rule (section 12.4.1) provides an explanation for the policing behavior. The relatedness of a sister to her sister (worker to worker) is $r = 0.75$, which is reduced to $r = 0.375$ if the queen has multiply mated (as happens). An unfertilized egg of a worker, if allowed to develop, becomes a son to which his mother's relatedness is $r = 0.5$. This kinship value is greater than to her half-sisters ($0.5 > 0.375$), thus providing an incentive to escape queen control. However, from the perspective of the other workers, their kinship to the son of another worker is only $r = 0.125$, "justifying" the killing of a half-nephew (another worker's son), and tending the development of her sisters ($r = 0.75$) or half-sisters ($r = 0.375$) (relationships portrayed in Fig. 12.11). The evolutionary benefits to any worker derive from raising the queen's eggs and destroying her sisters'. However, when the queen's strength wanes or she dies, the pheromonal repression of the colony ceases, anarchy breaks out and the workers all start to lay eggs.

Outside the extreme rigidity of the honey-bee colony, a range of policing activities can be seen. In colonies of ants that lack clear division into queens and workers, a hierarchy exists with only certain individuals' reproduction tolerated by nestmates. Although enforcement involves violence towards an offender, such regimes have some flexibility, since there is regular ousting of the reproductives. Even for honey bees, as the queen's performance diminishes and her pheromonal control wanes, workers' ovaries develop and rampant egg-laying takes place. Workers of some vespids discriminate between offspring of a singly-mated or a promiscuous queen, and behave according to kinship. Presumably, polygynous colonies at some stage have allowed additional queens to develop, or to return and be tolerated, providing possibilities for invasiveness by relaxed internest interactions (Box 1.2). The inquilines discussed in section 12.3 and Box 1.1 evidently evade policing efforts, but the mechanisms are poorly known as yet.

In an unusual development in southern Africa, anarchistic behavior has taken hold in hives of African honey bees (*Apis mellifera scutellatus*) that are being invaded by a different parasitic subspecies, Cape honey bee (*A. m. capensis*). The invader workers, which do little work, produce diploid female eggs that are clones of themselves. These evade the regular policing of the colony, presumably by chemical mimicry of the queen pheromone. The colony is destroyed rapidly by these social parasites, which then move on to invade another hive.

12.5 SUCCESS OF EUSOCIAL INSECTS

As we saw in the introduction to this chapter, social insects can attain numerical and ecological dominance in some regions. In Box 1.2 we describe some examples in which ants can become a nuisance by their dominance. Social insects tend to abundance at low latitudes and low elevations, and their activities are conspicuous in summer in temperate areas, or year-round in subtropical to tropical climates. As a generalization, the most abundant and dominant social insects are the most derived phylogenetically and have the most complex social organization.

Three qualities of social insects contribute to their competitive advantage, all of which derive from the caste system that allows multiple tasks to be performed. Firstly, the tasks of foraging, feeding the queen, caring for offspring, and maintenance of the nest can be performed simultaneously by different groups rather than sequentially as in solitary insects. Performing tasks in parallel means that one activity does not jeopardize another, thus the nest is not vulnerable to predators or parasites whilst foraging is taking place. Furthermore, individual errors have little or no consequence in parallel operations compared with those performed serially. Secondly, the ability of the colony to marshal all workers can overcome serious difficulties that a solitary insect cannot deal with, such as defense against a much larger or more numerous predator, or construction of a nest under unfavorable conditions. Thirdly, the specialization of function associated with castes allows some homeostatic regulation, including holding of food reserves in some castes (such as honeypot ants) or in developing larvae, and behavioral control of temperature and other microclimatic conditions within the nest. The ability to vary the proportion of individuals allocated to a particular caste allows appropriate distribution of community resources according to the differing demands of season and colony age. The widespread use of a variety of pheromones allows a high level of control to be exerted, even over millions of individuals. However, within this apparently rigid eusocial system, there is scope for a wide variety of different life histories to have evolved, from the nomadic army ants to the parasitic inquilines.

FURTHER READING

Billen, J. (ed.) (1992) *Biology and Evolution of Social Insects.* Leuven University Press, Leuven.

Carpenter, J. (1989) Testing scenarios: wasp social behavior. *Cladistics* **5**, 131–44.

Choe, J.C. & Crespi, B.J. (eds.) (1997) *Social Behavior in Insects and Arachnids.* Cambridge University Press, Cambridge.

Crozier, R.H. & Pamilo, P. (1996) *Evolution of Social Insect Colonies: Sex Allocation and Kin Selection.* Oxford University Press, Oxford.

Danforth, B.N. (2002) Evolution of sociality in a primitively eusocial lineage of bees. *Proceedings of the National Academy of Science (USA)* **99**, 286–90.

de Wilde, J. & Beetsma, J. (1982) The physiology of caste development in social insects. *Advances in Insect Physiology* **16**, 167–246.

Dyer, F.C. (2002) The biology of the dance language. *Annual Review of Entomology* **47**, 917–49.

Fletcher, D.J.C. & Ross, K.G. (1985) Regulation of reproduction in eusocial Hymenoptera. *Annual Review of Entomology* **30**, 319–43.

Hardie, J. & Lees, A.D. (1985) Endocrine control of polymorphism and polyphenism. In: *Comprehensive Insect Physiology, Biochemistry, and Pharmacology*, Vol. 8: *Endocrinology II* (eds. G.A. Kerkut & L.I. Gilbert), pp. 441–90. Pergamon Press, Oxford.

Hölldobler, B. & Wilson, E.O. (1990) *The Ants.* Springer-Verlag, Berlin.

Itô, Y. (1989) The evolutionary biology of sterile soldiers in aphids. *Trends in Ecology and Evolution* **4**, 69–73.

Kiester, A.R. & Strates, E. (1984) Social behavior in a thrips from Panama. *Journal of Natural History* **18**, 303–14.

Kranz, B.D., Schwarz, M.P., Morris, D.C. & Crespi, B.J. (2002) Life history of *Kladothrips ellobus* and *Oncothrips rodwayi*: Insight into the origin and loss of soldiers in gall-inducing thrips. *Ecological Entomology* **27**, 49–57.

Lenior, A., D'Ettorre, P., Errard, C. & Hefetz, A. (2001) Chemical ecology and social parasitism in ants. *Annual Review of Entomology* **46**, 573–99.

Quicke, D.L.J. (1997) *Parasitic Wasps.* Chapman & Hall, London.

Resh, V.H. & Cardé, R.T. (eds.) (2003) *Encyclopedia of Insects.* Academic Press, Amsterdam. [Particularly see articles on *Apis* species; beekeeping; caste; dance language; division of labor in insect societies; Hymenoptera; Isoptera; sociality.]

Retnakaran, A. & Percy, J. (1985) Fertilization and special modes of reproduction. In: *Comprehensive Insect Physiology, Biochemistry, and Pharmacology*, Vol. 1: *Embryogenesis and*

Box 12.2 Hymenoptera (bees, ants, wasps, sawflies, and wood wasps)

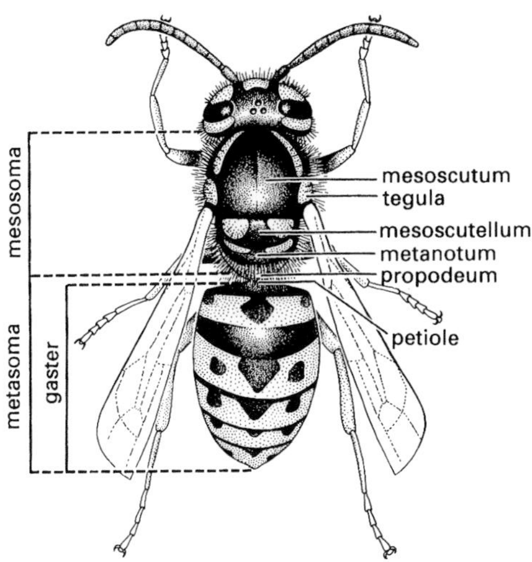

worker of the European wasp, *Vespula germanica*

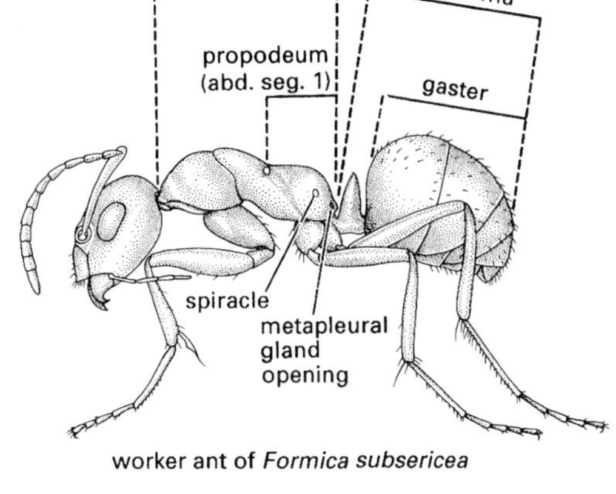

worker ant of *Formica subsericea*

The Hymenoptera is an order of about 100,000 described species of holometabolous neopterans, classified traditionally in two suborders, the "Symphyta" (wood wasps and sawflies) (which is a paraphyletic group) and Apocrita (wasps, bees, and ants). Within the Apocrita, the aculeate taxa (Chrysidoidea, Vespoidea, and Apoidea) form a monophyletic group (Fig. 12.2) characterized by the use of the ovipositor for stinging prey or enemies rather than for egg-laying. Adult hymenopterans range in size from minute (e.g. Trichogrammatidae, Fig. 16.3) to large (i.e. 0.15–120 mm long), and from slender (e.g. many Ichneumonidae) to robust (e.g. the bumble bee, Fig. 12.3). The head is hypognathous or prognathous, and the mouthparts range from generalized mandibulate to sucking and chewing, with mandibles in Apocrita often used for killing and handling prey, defense, and nest building. The compound eyes are often large; ocelli may be present, reduced, or absent. The antennae are long, multisegmented, and often prominently held forwardly or recurved dorsally. In Symphyta there are three conventional segments in the thorax, but in Apocrita the first abdominal segment (propodeum) is included in the thoracic tagma, which is then called a mesosoma (or in ants, alitrunk) (as illustrated for workers of the wasp *Vespula germanica* and a *Formica* ant). The wings have reduced venation, and the hind wings have rows of hooks (hamuli) along the leading edge that couple with the hind margin of the fore wing in flight. Abdominal segment 2 (and sometimes also 3) of Apocrita forms a constriction, or petiole, followed by the remainder of the abdomen, or gaster. The female genitalia include an ovipositor, comprising three valves and two major basal sclerites, which may be long and highly mobile allowing the valves to be directed vertically between the legs (Fig. 5.11). The ovipositor of aculeate Hymenoptera is modified as a sting associated with venom apparatus (Fig. 14.11).

The eggs of endoparasitic species are often deficient in yolk, and sometimes each may give rise to more than one individual (polyembryony; section 5.10.3). Symphytan larvae are eruciform (caterpillar-like) (Fig. 6.6c) with three pairs of thoracic legs with apical claws and some abdominal legs; most are phytophagous. Apocritan larvae are apodous (Fig. 6.6i), with the head capsule frequently reduced but with prominent strong mandibles; the larvae may vary greatly in morphology during development (heteromorphosis). Apocritan larvae have diverse feeding habits and may be parasitic (section 13.3), gall forming, or be fed with prey or nectar and pollen by their parent (or, if a social species, by other colony members). Adult hymenopterans mostly feed on nectar or honeydew, and sometimes drink hemolymph of other insects; only a few consume other insects.

Haplodiploidy allows a reproductive female to control the sex of offspring according to whether the egg is fertilized or not. Possible high relatedness amongst aggregated individuals facilitates well-developed social behaviors in many aculeate Hymenoptera.

For phylogenetic relationships of the Hymenoptera see section 7.4.2 and Figs. 7.2 and 12.2.

Box 12.3 Isoptera (termites)

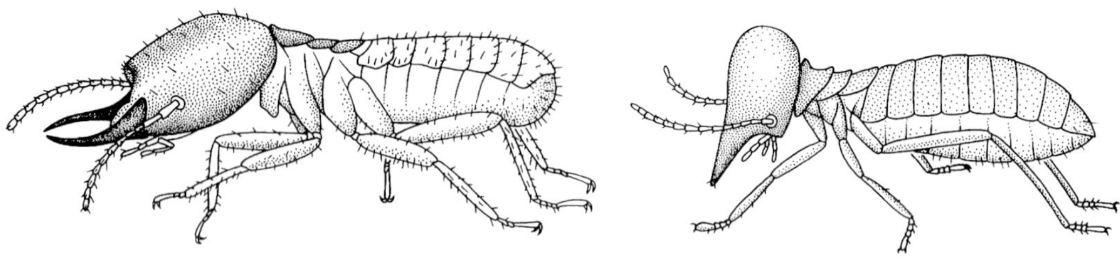

soldier of *Coptotermes*

soldier of *Nasutitermes*

The Isoptera is a small order of some 2600 described species of hemimetabolous neopterans, living socially with polymorphic caste systems of reproductives, workers, and soldiers (section 12.2.4; Fig. 12.8). All stages are small to moderately sized (even winged reproductives are usually <20 mm long). The head is hypognathous or prognathous, and the mouthparts typically blattoid and mandibulate, but varying between castes: soldiers often have bizarre development of their mandibles or possess a nasus (as illustrated on the left for the mandibulate *Coptotermes* and on the right for the nasute *Nasutitermes*, after Harris 1971). The compound eyes are frequently reduced, and the antennae are long and multisegmented with a variable number of segments. The wings are membranous with restricted venation; fore and hind wings are similar, except in *Mastotermes*, which has complex venation and an expanded hind-wing anal lobe. All castes have a pair of one- to five-segmented cerci terminally. External genitalia are absent, except in *Mastotermes* in which the female has a reduced blattoid ovipositor and the male a membranous copulatory organ. Gonads are poorly developed in adult soldiers and worker castes.

Internal anatomy is dominated by a convoluted alimentary tract including an elaborated hind gut, containing symbiotic bacteria and, in all species except the Termitidae, also protists (section 9.5.3 discusses fungal culture by Macrotermitinae). Food exchange between individuals (trophallaxis) is the sole means of replenishment of symbionts to young and newly molted individuals and is one explanation of the universal eusociality of termites.

Nests may be galleries or more complex structures within wood, such as rotting timber or even a sound tree, or above-ground nests (termitaria) such as prominent earth mounds (Figs. 12.9 & 12.10). Termites feed predominantly on cellulose-rich material; many harvest grasses and return the food to their subterranean or above-ground nest mounds.

Isoptera belong to a clade called Dictyoptera. Generally seven families and 14 subfamilies of termites are recognized. For phylogenetic relationships of the Dictyoptera see section 7.4.2 and Figs. 7.2 and 7.4.

Reproduction (eds. G.A. Kerkut & L.I. Gilbert), pp. 231–93. Pergamon Press, Oxford.

Robinson, G.E. (1992) Regulation of division of labor in insect societies. *Annual Review of Entomology* **37**, 637–65.

Sammataro, D., Gerson, U. & Needham, G. (2000) Parasitic mites of honey bees: life history, implications and impact. *Annual Review of Entomology* **45**, 519–48.

Tallamy, D.W. & Wood, T.K. (1986) Convergence patterns in subsocial insects. *Annual Review of Entomology* **31**, 369–90.

Thorne, B.L. (1997) Evolution of eusociality in termites. *Annual Review of Ecology and Systematics* **28**, 27–54.

Wilson, E.O. (1971) *The Insect Societies.* The Belknap Press of Harvard University Press, Cambridge, MA.

Chapter 13

INSECT PREDATION AND PARASITISM

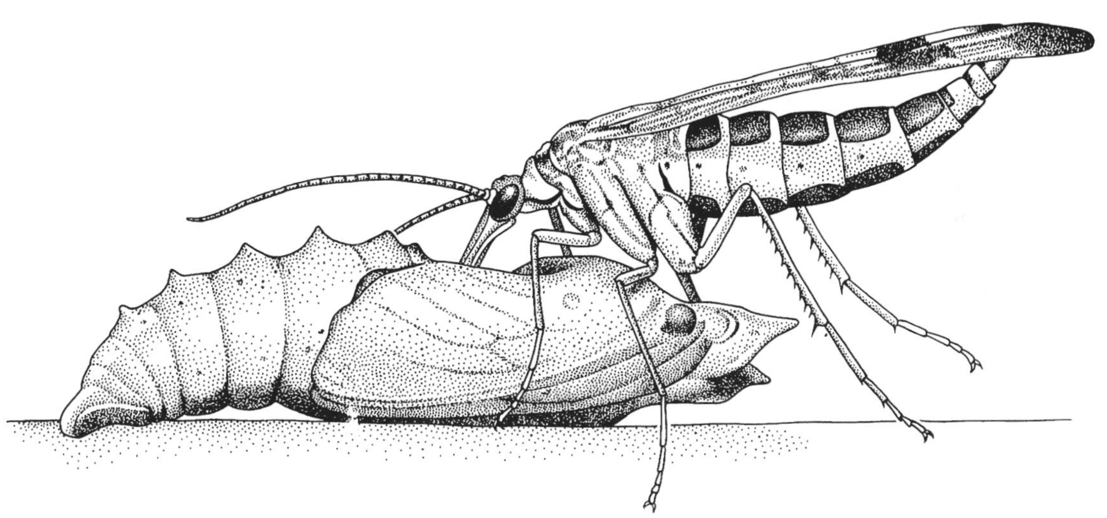

Scorpionfly feeding on a butterfly pupa. (After a photograph by P.H. Ward & S.L. Ward.)

We saw in Chapter 11 that many insects are phytophagous, feeding directly on primary producers, the algae and higher plants. These phytophages comprise a substantial food resource, which is fed upon by a range of other organisms. Individuals within this broad carnivorous group may be categorized as follows. A **predator** kills and consumes a number of **prey** animals during its life. **Predation** involves the interactions in space and time between predator foraging and prey availability, although often it is treated in a one-sided manner as if predation is what the predator does. Animals that live at the expense of another animal (a **host**) that eventually dies as a result are called **parasitoids**; they may live externally (**ectoparasitoids**) or internally (**endoparasitoids**). Those that live at the expense of another animal (also a host) that they do not kill are **parasites**, which likewise can be internal (**endoparasites**) or external (**ectoparasites**). A host attacked by a parasitoid or parasite is **parasitized**, and **parasitization** is the condition of being parasitized. **Parasitism** describes the relationship between parasitoid or parasite and the host. Predators, parasitoids, and parasites, although defined above as if distinct, may not be so clear-cut, as parasitoids may be viewed as specialized predators.

By some estimates, about 25% of insect species are predatory or parasitic in feeding habit in some life-history stage. Representatives from amongst nearly every order of insects are predatory, with adults and immature stages of the Odonata, Mantodea, Mantophasmatodea and the neuropteroid orders (Neuroptera, Megaloptera, and Raphidioptera), and adults of the Mecoptera being almost exclusively predatory. These orders are considered in Boxes 10.2, and 13.2–13.5, and the vignette for this chapter depicts a female mecopteran, *Panorpa communis* (Panorpidae), feeding on a dead pupa of a small tortoiseshell butterfly, *Aglais urticae*. The Hymenoptera (Box 12.2) are speciose, with a preponderance of parasitoid taxa using almost exclusively invertebrate hosts. The uncommon Strepsiptera are unusual in being endoparasites in other insects (Box 13.6). Other parasites that are of medical or veterinary importance, such as lice, adult fleas, and many Diptera, are considered in Chapter 15.

Insects are amenable to field and laboratory studies of predator–prey interactions as they are unresponsive to human attention, easy to manipulate, may have several generations a year, and show a range of predatory and defensive strategies and life histories. Furthermore, studies of predator–prey and parasitoid–host interactions are fundamental to understanding and effecting biological control strategies for pest insects. Attempts to model predator–prey interactions mathematically often emphasize parasitoids, as some simplifications can be made. These include the ability to simplify search strategies, as only the adult female parasitoid seeks hosts, and the number of offspring per unit host remains relatively constant from generation to generation.

In this chapter we show how predators, parasitoids, and parasites **forage**, i.e. locate and select their prey or hosts. We look at morphological modifications of predators for handling prey, and how some of the prey defenses covered in Chapter 14 are overcome. The means by which parasitoids overcome host defenses and develop within their hosts is examined, and different strategies of host use by parasitoids are explained. The host use and specificity of ectoparasites is discussed from a phylogenetic perspective. Finally, we conclude with a consideration of the relationships between predator/parasitoid/parasite and prey/host abundances and evolutionary histories. In the taxonomic boxes at the end of the chapter, the Mantodea, Mantophasmatodea, neuropteroid orders, Mecoptera, and Strepsiptera are described.

13.1 PREY/HOST LOCATION

The foraging behaviors of insects, like all other behaviors, comprise a stereotyped sequence of components. These lead a predatory or host-seeking insect towards the resource, and on contact, enable the insect to recognize and use it. Various stimuli along the route elicit an appropriate ensuing response, involving either action or inhibition. The foraging strategies of predators, parasitoids, and parasites involve trade-offs between profits or benefits (the quality and quantity of resource obtained) and cost (in the form of time expenditure, exposure to suboptimal or adverse environments, and the risks of being eaten). Recognition of the time component is important, as all time spent in activities other than reproduction can be viewed, in an evolutionary sense, as time wasted.

In an optimal foraging strategy, the difference between benefits and costs is maximized, either through increasing nutrient gain from prey capture, or reducing effort expended to catch prey, or both. Choices available are:

- where and how to search;
- how much time to expend in fruitless search in one area before moving;

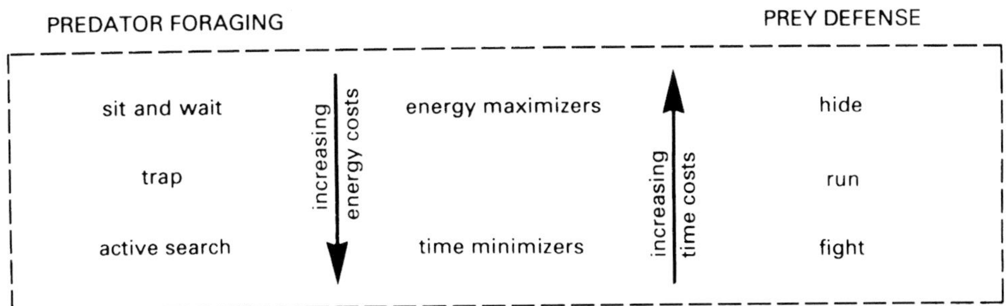

Fig. 13.1 The basic spectrum of predator foraging and prey defense strategies, varying according to costs and benefits in both time and energy. (After Malcolm 1990.)

• how much (if any) energy to expend in capture of suboptimal food, once located.

A primary requirement is that the insect be in the appropriate habitat for the resource sought. For many insects this may seem trivial, especially if development takes place in the area which contained the resources used by the parental generation. However, circumstances such as seasonality, climatic vagaries, ephemerality, or major resource depletion, may necessitate local **dispersal** or perhaps major movement (**migration**) in order to reach an appropriate location.

Even in a suitable habitat, resources rarely are evenly distributed but occur in more or less discrete microhabitat clumps, termed **patches**. Insects show a gradient of responses to these patches. At one extreme, the insect waits in a suitable patch for prey or host organisms to appear. The insect may be camouflaged or apparent, and a trap may be constructed. At the other extreme, the prey or host is actively sought within a patch. As seen in Fig. 13.1, the waiting strategy is economically effective, but time-consuming; the active strategy is energy intensive, but time-efficient; and trapping lies intermediate between these two. Patch selection is vital to successful foraging.

13.1.1 Sitting and waiting

Sit-and-wait predators find a suitable patch and wait for mobile prey to come within striking range. As the vision of many insects limits them to recognition of movement rather than precise shape, a sit-and-wait predator may need only to remain motionless in order to be unobserved by its prey. Nonetheless, amongst those that wait, many have some form of camouflage (crypsis). This may be defensive, being directed against highly visual predators such as birds, rather than evolved to mislead invertebrate prey. Cryptic predators modeled on a feature that is of no interest to the prey (such as tree bark, lichen, a twig, or even a stone) can be distinguished from those that model on a feature of some significance to prey, such as a flower that acts as an insect attractant.

In an example of the latter case, the Malaysian mantid *Hymenopus bicornis* closely resembles the red flowers of the orchid *Melastoma polyanthum* amongst which it rests. Flies are encouraged to land, assisted by the presence of marks resembling flies on the body of the mantid: larger flies that land are eaten by the mantid. In another related example of **aggressive foraging mimicry**, the African flower-mimicking mantid *Idolum* does not rest hidden in a flower, but actually resembles one due to petal-shaped, colored outgrowths of the prothorax and the coxae of the anterior legs. Butterflies and flies that are attracted to this hanging "flower" are snatched and eaten.

Ambushers include cryptic, sedentary insects such as mantids, which prey fail to distinguish from the inert, non-floral plant background. Although these predators rely on the general traffic of invertebrates associated with vegetation, often they locate close to flowers, to take advantage of the increased visiting rate of flower feeders and pollinators.

Odonate nymphs, which are major predators in many aquatic systems, are classic ambushers. They rest concealed in submerged vegetation or in the substrate, waiting for prey to pass. These predators may show dual strategies: if waiting fails to provide food, the hungry insect may change to a more active searching mode after a fixed period. This energy expenditure may

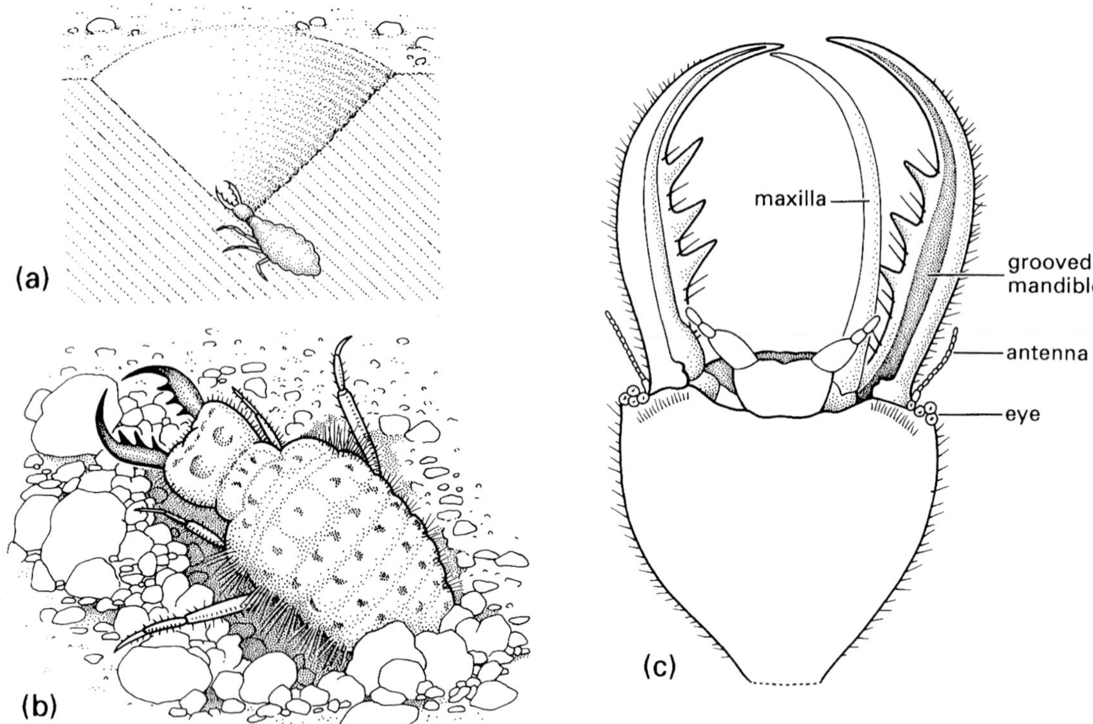

Fig. 13.2 An antlion of *Myrmeleon* (Neuroptera: Myrmeleontidae): (a) larva in its pit in sand; (b) detail of dorsum of larva; (c) detail of ventral view of larval head showing how the maxilla fits against the grooved mandible to form a sucking tube. (After Wigglesworth 1964.)

bring the predator into an area of higher prey density. In running waters, a disproportionately high number of organisms found drifting passively with the current are predators: this drift constitutes a low-energy means for sit-and-wait predators to relocate, induced by local prey shortage.

Sitting-and-waiting strategies are not restricted to cryptic and slow-moving predators. Fast-flying, diurnal, visual, rapacious predators such as many robber flies (Diptera: Asilidae) and adult odonates spend much time perched prominently on vegetation. From these conspicuous locations their excellent sight allows them to detect passing flying insects. With rapid and accurately controlled flight, the predator makes only a short foray to capture appropriately sized prey. This strategy combines energy saving, through not needing to fly incessantly in search of prey, with time efficiency, as prey is taken from outside the immediate area of reach of the predator.

Another sit-and-wait technique involving greater energy expenditure is the use of traps to ambush prey. Although spiders are the prime exponents of this method, in the warmer parts of the world the pits of certain larval antlions (Neuroptera: Myrmeleontidae) (Fig. 13.2a,b) are familiar. The larvae either dig pits directly or form them by spiraling backwards into soft soil or sand. Trapping effectiveness depends upon the steepness of the sides, the diameter, and the depth of the pit, which vary with species and instar. The larva waits, buried at the base of the conical pit, for passing prey to fall in. Escape is prevented physically by the slipperiness of the slope, and the larva may also flick sand at prey before dragging it underground to restrict its defensive movements. The location, construction, and maintenance of the pit are vitally important to capture efficiency but construction and repair is energetically very expensive. Experimentally it has been shown that even starved Japanese antlions (*Myrmeleon bore*) would not relocate their pits to an area where prey was provided artificially. Instead, larvae of this species of antlion reduce their metabolic rate to tolerate famine, even if death by starvation is the result.

In holometabolous ectoparasites, such as fleas and parasitic flies, immature development takes place away from their vertebrate hosts. Following pupation, the adult must locate the appropriate host. Since in many of these ectoparasites the eyes are reduced or absent, vision cannot be used. Furthermore, as many of these insects are flightless, mobility is restricted. In fleas and some Diptera, in which larval development often takes place in the nest of a host vertebrate, the adult insect waits quiescent in the pupal cocoon until the presence of a host is detected. The duration of this quiescent period may be a year or longer, as in the cat flea (*Ctenocephalides felis*) – a familiar phenomenon to humans that enter an empty dwelling that previously housed flea-infested cats. The stimuli to cease dormancy include some or all of: vibration, rise in temperature, increased carbon dioxide, or another stimulus generated by the host.

In contrast, the hemimetabolous lice spend their lives entirely on a host, with all developmental stages ectoparasitic. Any transfer between hosts is either through phoresy (see below) or when host individuals make direct contact, as from mother to young within a nest.

13.1.2 Active foraging

More energetic foraging involves active searching for suitable patches, and once there, for prey or for hosts. Movements associated with foraging and with other locomotory activities, such as seeking a mate, are so similar that the "motivation" may be recognized only in retrospect, by resultant prey capture or host finding. The locomotory search patterns used to locate prey or hosts are those described for general orientation in section 4.5, and comprise non-directional (random) and directional (non-random) locomotion.

Random, or non-directional foraging

The foraging of aphidophagous larval coccinellid beetles and syrphid flies amongst their clumped prey illustrates several features of random food searching. The larvae advance, stop periodically, and "cast" about by swinging their raised anterior bodies from side to side. Subsequent behavior depends upon whether or not an aphid is encountered. If no prey is encountered, motion continues, interspersed with casting and turning at a fundamental frequency. However, if contact is made and feeding has taken place or if the prey is encountered and lost, searching intensifies with an enhanced frequency of casting, and, if the larva is in motion, increased turning or direction-changing. Actual feeding is unnecessary to stimulate this more concentrated search: an unsuccessful encounter is adequate. For early-instar larvae that are very active but have limited ability to handle prey, this stimulus to search intensively near a lost feeding opportunity is important to survival.

Most laboratory-based experimental evidence, and models of foraging based thereon, are derived from single species of walking predators, frequently assumed to encounter a single species of prey randomly distributed within selected patches. Such premises may be justified in modeling grossly simplified ecosystems, such as an agricultural monoculture with a single pest controlled by one predator. Despite the limitations of such laboratory-based models, certain findings appear to have general biological relevance.

An important consideration is that the time allocated to different patches by a foraging predator depends upon the criteria for leaving a patch. Four mechanisms have been recognized to trigger departure from a patch:

1 a certain number of food items have been encountered (fixed number);
2 a certain time has elapsed (fixed time);
3 a certain searching time has elapsed (fixed searching time);
4 the prey capture rate falls below a certain threshold (fixed rate).

The fixed-rate mechanism has been favored by modelers of optimal foraging, but even this is likely to be a simplification if the forager's responsiveness to prey is non-linear (e.g. declines with exposure time) and/or derives from more than simple prey encounter rate, or prey density. Differences between predator–prey interactions in simplified laboratory conditions and the actuality of the field cause many problems, including failure to recognize variation in prey behavior that results from exposure to predation (perhaps multiple predators). Furthermore, there are difficulties in interpreting the actions of polyphagous predators, including the causes of predator/parasitoid/parasite behavioral switching between different prey animals or hosts.

Non-random, or directional foraging

Several more specific directional means of host finding can be recognized, including the use of chemicals, sound, and light. Experimentally these are rather difficult to establish, and to separate, and it may be that the use of these cues is very widespread, if little understood. Of the variety of cues available, many insects probably use more than one, depending upon distance or proximity to the resource sought. Thus, the European crabronid wasp *Philanthus*, which eats only bees, relies initially on vision to locate moving insects of appropriate size. Only bees, or other insects to which bee odors have been applied experimentally, are captured, indicating a role for odor when near the prey. However, the sting is applied only to actual bees, and not to bee-smelling alternatives, demonstrating a final tactile recognition.

Not only may a stepwise sequence of stimuli be necessary, as seen above, but also appropriate stimuli may have to be present simultaneously in order to elicit appropriate behavior. Thus, *Telenomus heliothidis* (Hymenoptera: Scelionidae), an egg parasitoid of *Heliothis virescens* (Lepidoptera: Noctuidae), will investigate and probe at appropriate-sized round glass beads that emulate *Heliothis* eggs, if they are coated with female moth proteins. However, the scelionid makes no response to glass beads alone, or to female moth proteins applied to improperly shaped beads.

Chemicals

The world of insect communication is dominated by chemicals, or pheromones (section 4.3). Ability to detect the chemical odors and messages produced by prey or hosts (kairomones) allows specialist predators and parasitoids to locate these resources. Certain parasitic tachinid flies and braconid wasps can locate their respective stink bug or coccoid host by tuning to their hosts' long-distance sex attractant pheromones. Several unrelated parasitoid hymenopterans use the aggregation pheromones of their bark and timber beetle hosts. Chemicals emitted by stressed plants, such as terpenes produced by pines when attacked by an insect, act as **synomones** (communication chemicals that benefit both producer and receiver); for example, certain pteromalid (Hymenoptera) parasitoids locate their hosts, the damage-causing scolytid timber beetles, in this way. Some species of tiny wasps (Trichogrammatidae) that are egg endoparasitoids (Fig. 16.3) are able to locate the eggs laid by their preferred host moth by the sex attractant pheromones released by the moth. Furthermore, there are several examples of parasitoids that locate their specific insect larval hosts by "frass" odors – the smells of their feces. Chemical location is particularly valuable when hosts are concealed from visual inspection, for example when encased in plant or other tissues.

Chemical detection need not be restricted to tracking volatile compounds produced by the prospective host. Thus, many parasitoids searching for phytophagous insect hosts are attracted initially, and at a distance, to host-plant chemicals, in the same manner that the phytophage located the resource. At close range, chemicals produced by the feeding damage and/or frass of phytophages may allow precise targeting of the host. Once located, the acceptance of a host as suitable is likely to involve similar or other chemicals, judging by the increased use of rapidly vibrating antennae in sensing the prospective host.

Blood-feeding adult insects locate their hosts using cues that include chemicals emitted by the host. Many female biting flies can detect increased carbon dioxide levels associated with animal respiration and fly upwind towards the source. Highly host-specific biters probably also are able to detect subtle odors: thus, human-biting black flies (Diptera: Simuliidae) respond to components of human exocrine sweat glands. Both sexes of tsetse flies (Diptera: Glossinidae) track the odor of exhaled breath, notably carbon dioxide, octanols, acetone, and ketones emitted by their preferred cattle hosts.

Sound

The sound signals produced by animals, including those made by insects to attract mates, have been utilized by some parasites to locate their preferred hosts acoustically. Thus, the blood-sucking females of *Corethrella* (Diptera: Corethrellidae) locate their favored host, hylid treefrogs, by following the frogs' calls. The details of the host-finding behavior of ormiine tachinid flies are considered in detail in Box 4.1. Flies of two other dipteran species are known to be attracted by the songs of their hosts: females of the larviparous tachinid *Euphasiopteryx ochracea* locate the male crickets of *Gryllus integer*, and the sarcophagid *Colcondamyia auditrix* finds its male cicada host, *Okanagana rimosa*, in this manner. This allows precise deposition of the parasitic immature stages in, or close to, the hosts in which they are to develop.

Predatory biting midges (Ceratopogonidae) that prey upon swarm-forming flies, such as midges (Chironomidae), appear to use cues similar to those used by their prey to locate the swarm; cues may include the sounds produced by wing-beat frequency of the members of the swarm. Vibrations produced by their hosts can be detected by ectoparasites, notably amongst the fleas. There is also evidence that certain parasitoids can detect at close range the substrate vibration produced by the feeding activity of their hosts. Thus, *Biosteres longicaudatus*, a braconid hymenopteran endoparasitoid of a larval tephritid fruit fly (Diptera: *Anastrepha suspensa*), detects vibrations made by the larvae moving and feeding within fruit. These sounds act as a behavioral releaser, stimulating host-finding behavior as well as acting as a directional cue for their concealed hosts.

Light

The larvae of the Australian cave-dwelling mycetophilid fly *Arachnocampa* and its New Zealand counterpart, *Bolitophila luminosa*, use bioluminescent lures to catch small flies in sticky threads that they suspend from the cave ceiling. Luminescence (section 4.4.5), as with all communication systems, provides scope for abuse; in this case, the luminescent courtship signaling between beetles is misappropriated. Carnivorous female lampyrids of some *Photurus* species, in an example of aggressive foraging mimicry, can imitate the flashing signals of females of up to five other firefly species. The males of these different species flash their responses and are deluded into landing close by the mimetic female, whereupon she devours them. The mimicking *Photurus* female will eat the males of her own species, but cannibalism is avoided or reduced as the *Photurus* female is most piratical only after mating, at which time she becomes relatively unresponsive to the signals of males of her own species.

13.1.3 Phoresy

Phoresy is a phenomenon in which an individual is transported by a larger individual of another species. This relationship benefits the carried and does not directly affect the carrier, although in some cases its progeny may be disadvantaged (as we shall see below). Phoresy provides a means of finding a new host or food source. An often observed example involves ischnoceran lice (Phthiraptera) transported by the winged adults of *Ornithomyia* (Diptera: Hippoboscidae). Hippoboscidae are blood-sucking ectoparasitic flies and *Ornithomyia* occurs on many avian hosts. When a host bird dies, lice can reach a new host by attaching themselves by their mandibles to a hippoboscid, which may fly to a new host. However, lice are highly host-specific but hippoboscids are much less so, and the chances of any hitchhiking louse arriving at an appropriate host may not be great. In some other associations, such as a biting midge (*Forcipomyia*) found on the thorax of various adult dragonflies in Borneo, it is difficult to determine whether the hitchhiker is actually parasitic or merely phoretic.

Amongst the egg-parasitizing hymenopterans (notably the Scelionidae, Trichogrammatidae, and Torymidae), some attach themselves to adult females of the host species, thereby gaining immediate access to the eggs at oviposition. *Matibaria manticida* (Scelionidae), an egg parasitoid of the European praying mantid (*Mantis religiosa*), is phoretic, predominantly on female hosts. The adult wasp sheds its wings and may feed on the mantid, and therefore can be an ectoparasite. It moves to the wing bases and amputates the female mantid's wings and then oviposits into the mantid's egg mass whilst it is frothy, before the ootheca hardens. Individuals of *M. manticida* that are phoretic on male mantids may transfer to the female during mating. Certain chalcid hymenopterans (including species of Eucharitidae) have mobile planidium larvae that actively seek worker ants, on which they attach, thereby gaining transport to the ant nest. Here the remainder of the immature life cycle comprises typical sedentary grubs that develop within ant larvae or pupae.

The human bot fly, *Dermatobia hominis* (Diptera: Cuterebridae) of the neotropical region (Central and South America), which causes myiasis (section 15.3) of humans and cattle, shows an extreme example of phoresy. The female fly does not find the vertebrate host herself, but uses the services of blood-sucking flies, particularly mosquitoes and muscoid flies. The female bot fly, which produces up to 1000 eggs in her lifetime, captures a phoretic intermediary and glues around 30 eggs to its body in such a way that flight is not impaired. When the intermediary finds a vertebrate host on which it feeds, an elevation of temperature induces the eggs to hatch rapidly and the larvae transfer to the host where they penetrate the skin via hair follicles and develop within the resultant pus-filled boil.

13.2 PREY/HOST ACCEPTANCE AND MANIPULATION

During foraging, there are some similarities in location of prey by a predator and of the host by a parasitoid or parasite. When contact is made with the potential prey or host, its acceptability must be established, by checking the identity, size, and age of the prey/host. For example, many parasitoids reject old larvae, which are close to pupation. Chemical and tactile stimuli are involved in specific identification and in subsequent behaviors including biting, ingestion, and continuance of feeding. Chemoreceptors on the antennae and ovipositor of parasitoids are vital in chemically detecting host suitability and exact location.

Different manipulations follow acceptance: the predator attempts to eat suitable prey, whereas parasitoids and parasites exhibit a range of behaviors regarding their hosts. A parasitoid either oviposits (or larviposits) directly or subdues and may carry the host elsewhere, for instance to a nest, prior to the offspring developing within or on it. An ectoparasite needs to gain a hold and obtain a meal. The different behavioral and morphological modifications associated with prey and host manipulation are covered in separate sections below, from the perspectives of predator, parasitoid, and parasite.

13.2.1 Prey manipulation by predators

When a predator detects and locates suitable prey, it must capture and restrain it before feeding. As predation has arisen many times, and in nearly every order, the morphological modifications associated with this lifestyle are highly convergent. Nevertheless, in most predatory insects the principal organs used in capture and manipulation of prey are the legs and mouthparts. Typically, **raptorial** legs of adult insects are elongate and bear spines on the inner surface of at least one of the segments (Fig. 13.3). Prey is captured by closing the spinose segment against another segment, which may itself be spinose, i.e. the femur against the tibia, or the tibia against the tarsus. As well as spines, there may be elongate spurs on the apex of the tibia, and the apical claws may be strongly developed on the raptorial legs. In predators with leg modifications, usually it is the anterior legs that are raptorial, but some hemipterans also employ the mid legs, and scorpionflies (Box 5.1) grasp prey with their hind legs.

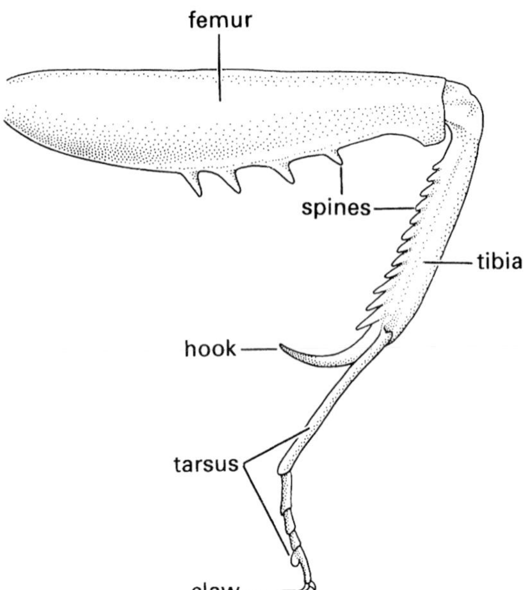

Fig. 13.3 Distal part of the leg of a mantid showing the opposing rows of spines that interlock when the tibia is drawn upwards against the femur. (After Preston-Mafham 1990.)

Mouthpart modifications associated with predation are of two principal kinds: (i) incorporation of a variable number of elements into a tubular rostrum to allow piercing and sucking of fluids; or (ii) development of strengthened and elongate mandibles. Mouthparts modified as a rostrum (Box 11.8) are seen in bugs (Hemiptera) and function in sucking fluids from plants or from dead arthropods (as in many gerrid bugs) or in predation on living prey, as in many other aquatic insects, including species of Nepidae, Belostomatidae, and Notonectidae. Amongst the terrestrial bugs, assassin bugs (Reduviidae), which use raptorial fore legs to capture other terrestrial arthropods, are major predators. They inject toxins and proteolytic saliva into captured prey, and suck the body fluids through the rostrum. Similar hemipteran mouthparts are used in blood sucking, as demonstrated by *Rhodnius*, a reduviid that has attained fame for its role in experimental insect physiology, and the family Cimicidae, including the bed bug, *Cimex lectularius*.

In the Diptera, mandibles are vital for wound production by the blood-sucking Nematocera (mosquitoes, midges, and black flies) but have been lost in the higher flies, some of which have regained the blood-sucking habit. Thus, in the stable flies (*Stomoxys*) and tsetse flies (*Glossina*), for example, alternative mouthpart structures have evolved; some specialized mouthparts of blood-sucking Diptera are described and illustrated in Box 15.5.

Many predatory larvae and some adults have hardened, elongate, and apically pointed mandibles capable of piercing durable cuticle. Larval neuropterans (lacewings and antlions) have the slender maxilla and sharply pointed and grooved mandible, which are pressed together to form a composite sucking tube (Fig. 13.2c). The composite structure may be straight, as in active pursuers of prey, or curved, as in the sit-and-wait ambushers such as antlions. Liquid may be sucked (or pumped) from the prey, using a range of mandibular modifications after enzymatic predigestion has liquefied the contents (**extra-oral digestion**).

An unusual morphological modification for predation is seen in the larvae of Chaoboridae (Diptera) that use modified antennae to grasp their planktonic cladoceran prey. Odonate nymphs capture passing prey by striking with a highly modified labium (Fig. 13.4), which is projected rapidly outwards by release of hydrostatic pressure, rather than by muscular means.

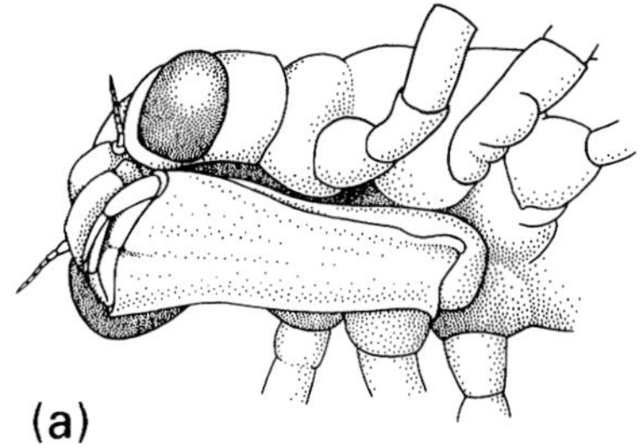

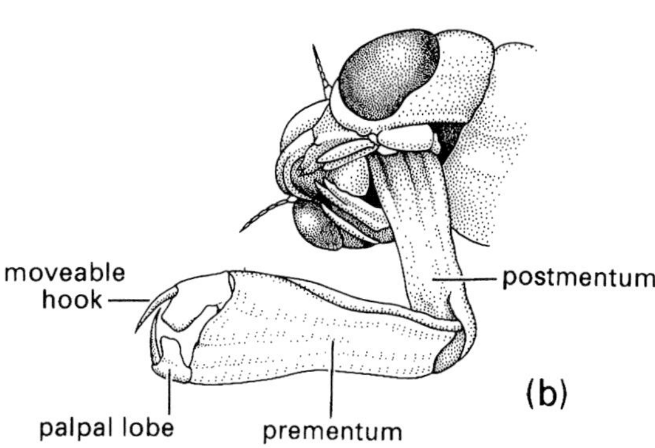

Fig. 13.4 Ventrolateral view of the head of a dragonfly nymph (Odonata: Aeshnidae: *Aeshna*) showing the labial "mask": (a) in folded position, and (b) extended during prey capture with opposing hooks of the palpal lobes forming claw-like pincers. (After Wigglesworth 1964.)

13.2.2 Host acceptance and manipulation by parasitoids

The two orders with greatest numbers and diversity of larval parasitoids are the Diptera and Hymenoptera. Two basic approaches are displayed once a potential host is located, though there are exceptions. Firstly, as seen in many hymenopterans, it is the adult that seeks out the actual larval development site. In contrast, in many Diptera it is often the first-instar planidium larva that makes the close-up host contact. Parasitic hymenopterans use sensory information from the elongate and constantly mobile antennae to precisely locate even a hidden host. The antennae and specialized ovipositor (Fig. 5.11) bear sensilla that allow host acceptance and accurate oviposition, respectively. Modification of the ovipositor as a sting in the aculeate Hymenoptera permits behavioral modifications (section 14.6), including provisioning of the immature stages with a food source captured by the adult and maintained alive in a paralyzed state.

Endoparasitoid dipterans, including the Tachinidae, may oviposit (or in larviparous taxa, deposit a larva) onto the cuticle or directly into the host. In several distantly related families, a convergently evolved "substitutional" ovipositor (sections 2.5.1 & 5.8) is used. Frequently, however, the parasitoid's egg or larva is deposited onto a suitable substrate and the mobile planidium larva is responsible for finding its host. Thus, *Euphasiopteryx ochracea*, a tachinid that responds phonotactically to the call of a male cricket, actually deposits larvae around the calling site, and these larvae locate and parasitize not only the vocalist, but other crickets attracted by the call. Hypermetamorphosis, in which the first-instar larva is morphologically and behaviorally different from subsequent larval instars (which are sedentary parasitic maggots), is common amongst parasitoids.

Certain parasitic and parasitoid dipterans and some hymenopterans use their aerial flying skills to gain access to a potential host. Some are able to intercept their hosts in flight, others can make rapid lunges at an alert and defended target. Some of the inquilines of social insects (section 12.3) can enter the nest via an egg laid upon a worker whilst it is active outside the nest. For example, certain phorid flies, lured by ant odors, may be seen darting at ants in an attempt to oviposit on them. A West Indian leaf-cutter ant (*Atta* sp.) cannot defend itself from such attacks whilst bearing leaf fragments in its mandibles. This problem frequently is addressed (but is unlikely to be completely overcome) by stationing a guard on the leaf during transport; the guard is a small (minima) worker (Fig. 9.6) that uses its jaws to threaten any approaching phorid fly.

The success of attacks of such insects against active and well-defended hosts demonstrates great rapidity in host acceptance, probing, and oviposition. This may contrast with the sometimes leisurely manner of many parasitoids of sessile hosts, such as scale insects, pupae, or immature stages that are restrained within confined spaces, such as plant tissue, and unguarded eggs.

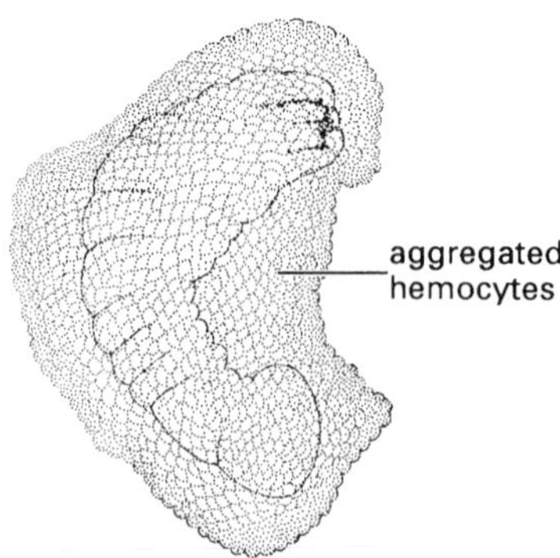

Fig. 13.5 Encapsulation of a living larva of *Apanteles* (Hymenoptera: Braconidae) by the hemocytes of a caterpillar of *Ephestia* (Lepidoptera: Pyralidae). (After Salt 1968.)

13.2.3 Overcoming host immune responses

Insects that develop within the body of other insects must cope with the active immune responses of the host. An adapted or compatible parasitoid is not eliminated by the cellular immune defenses of the host. These defenses protect the host by acting against incompatible parasitoids, pathogens, and biotic matter that may invade the host's body cavity. Host immune responses entail mechanisms for (i) recognizing introduced material as non-self, and (ii) inactivating, suppressing, or removing the foreign material. The usual host reaction to an incompatible parasitoid is **encapsulation**, i.e. surrounding the invading egg or larva by an aggregation of hemocytes (Fig. 13.5). The hemocytes become flattened onto the surface of the parasitoid and phagocytosis commences as the hemocytes build up, eventually forming a capsule that surrounds and kills the intruder. This type of reaction rarely occurs when parasitoids infect their normal hosts, presumably because the parasitoid or some factor(s) associated with it alters the host's ability to recognize the parasitoid as foreign and/or to respond to it. Parasitoids that cope successfully with the host immune system do so in one or more of the following ways:

• Avoidance – for example, ectoparasitoids feed externally on the host (in the manner of predators), egg parasitoids lay into host eggs that are incapable of immune response, and many other parasitoids at least temporarily occupy host organs (such as the brain, a ganglion, a salivary gland, or the gut) and thus escape the immune reaction of the host hemolymph.

• Evasion – this includes molecular mimicry (the parasitoid is coated with a substance similar to host proteins and is not recognized as non-self by the host), cloaking (e.g. the parasitoid may insulate itself in a membrane or capsule, derived from either embryonic membranes or host tissues; see also "subversion" below), and/or rapid development in the host.

• Destruction – the host immune system may be blocked by attrition of the host such as by gross feeding that weakens host defense reactions, and/or by destruction of responding cells (the host hemocytes).

• Suppression – host cellular immune responses may be suppressed by viruses associated with the parasitoids (Box 13.1); often suppression is accompanied by reduction in host hemocyte counts and other changes in host physiology.

• Subversion – in many cases parasitoid development occurs despite host response; for example, physical resistance to encapsulation is known for wasp parasitoids, and in dipteran parasitoids the host's hemocytic capsule is subverted for use as a sheath that the fly larva keeps open at one end by vigorous feeding. In many parasitic Hymenoptera, the serosa or trophamnion associated with the parasitoid egg fragments into individual cells that float free in the host hemolymph and grow to form giant cells, or teratocytes, that may assist in overwhelming the host defenses.

Obviously, the various ways of coping with host immune reactions are not discrete and most adapted parasitoids probably use a combination of methods to

Box 13.1 Viruses, wasp parasitoids, and host immunity

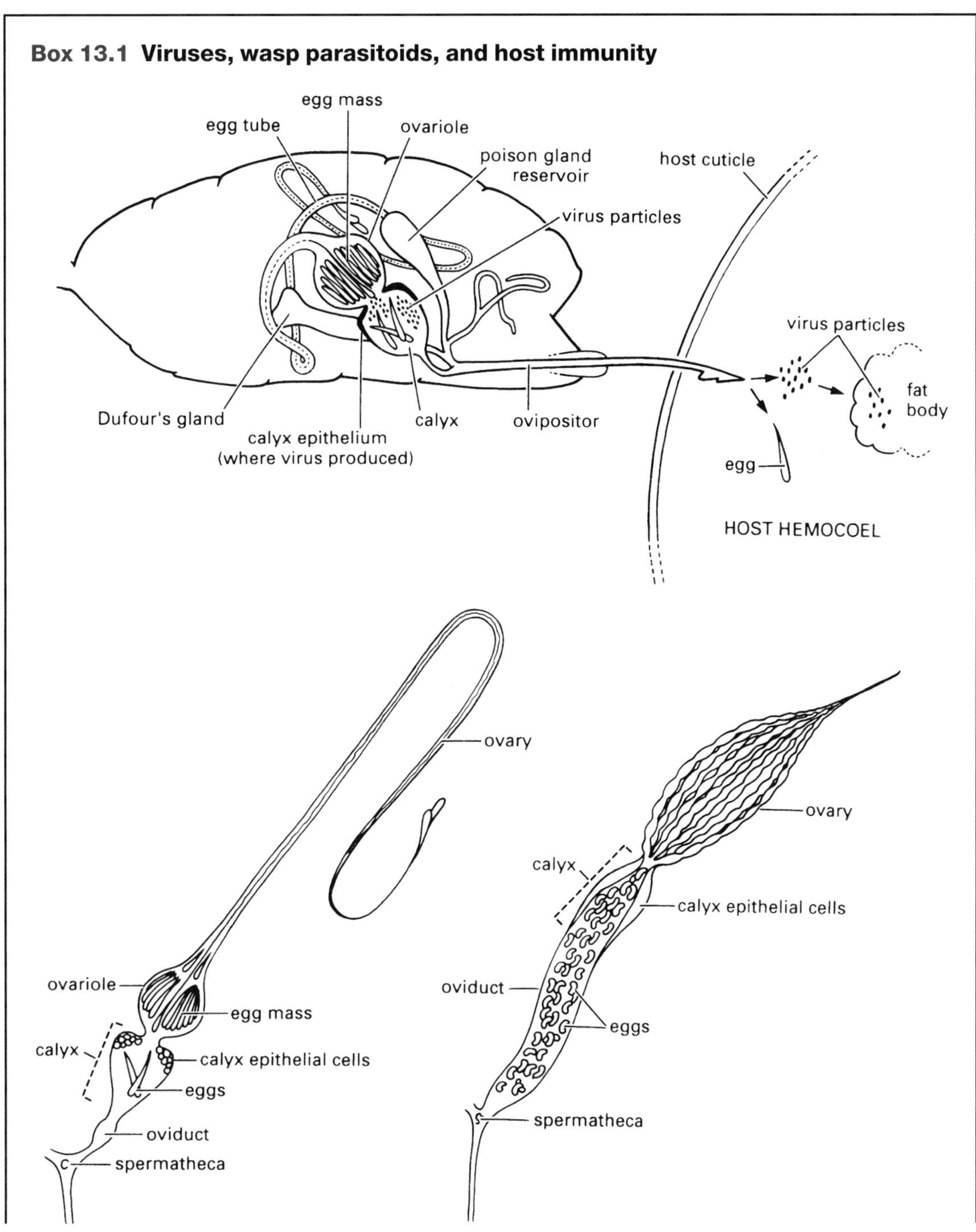

In certain endoparasitoid wasps in the families Ichneumonidae and Braconidae, the ovipositing female wasp injects the larval host not only with her egg(s), but also with accessory gland secretions and substantial numbers of viruses (as depicted in the upper drawing for the braconid *Toxoneuron* (formerly *Cardiochiles*) *nigriceps*, after Greany et al. 1984) or virus-like particles (VLPs). The viruses belong to a distinct group, the **polydnaviruses** (PDVs), which are characterized by the possession of multipartite double-stranded circular DNA. PDVs are transmitted between wasp generations through the germline. The PDVs of braconids (called bracoviruses) differ from the PDVs of ichneumonids (ichnoviruses) in morphology, morphogenesis, and in relation to their interaction with other wasp-derived factors in the parasitized host. The PDVs of different wasp species generally are considered to be distinct viral species. Furthermore, the evolutionary association of ichnoviruses with ichneumonids is known to be unrelated to the evolution of the braconid–bracovirus association and, within the braconids, PDVs occur only in the monophyletic microgastroid group of subfamilies and appear to have coevolved with their wasp hosts.

VLPs are known only in some ichneumonid wasps. It is not clear whether all VLPs are viruses, as the morphology of some VLPs is different from that of typical PDVs and some VLPs lack DNA. However, all PDVs and VLPs appear to be involved in overcoming the host's immune reaction and often are responsible for other symptoms in infected hosts. For example, the PDVs of some wasps apparently can induce most of the changes in growth, development, behavior, and hemocytic activity that are observed in infected host larvae. The PDVs of other parasitoids (usually braconids) seem to require the presence of accessory factors, particularly venoms, to completely prevent encapsulation of the wasp egg or to fully induce symptoms in the host.

The calyx epithelium of the female reproductive tract is the primary site of replication of PDVs (as depicted for the braconid *Toxoneuron nigriceps* in the lower left drawing, and for the ichneumonid *Campoletis sonorensis* on the lower right, after Stoltz & Vinson 1979) and is the only site of VLP assembly (as in the ichneumonid *Venturia canescens*). The lumen of the wasp oviduct becomes filled with PDVs or VLPs, which thus surround the wasp eggs. If VLPs or PDVs are removed artificially from wasp eggs, encapsulation occurs if the unprotected eggs are then injected into the host. If appropriate PDVs or VLPs are injected into the host with the washed eggs, encapsulation is prevented. The physiological mechanism for this protection is not clearly understood, although in the wasp *Venturia*, which coats its eggs in VLPs, it appears that molecular mimicry of a host protein by a VLP protein interferes with the immune recognition process of the lepidopteran host. The VLP protein is similar to a host hemocyte protein involved in recognition of foreign particles. In the case of PDVs, the process is more active and involves the expression of PDV-encoded gene products that directly interfere with the mode of action of hemocytes.

allow development within their respective hosts. Parasitoid–host interactions at the level of cellular and humoral immunity are complex and vary greatly among different taxa. Our understanding of these systems is still relatively limited but this field of research is producing exciting findings concerning parasitoid genomes and coevolved associations between insects and viruses.

13.3 PREY/HOST SELECTION AND SPECIFICITY

As we have seen in Chapters 9–11, insects vary in the breadth of food sources they use. Thus, some predatory insects are monophagous, utilizing a single species of prey; others are oligophagous, using few species; and many are polyphagous, feeding on a variety of prey species. As a broad generalization, predators are mostly polyphagous, as a single prey species rarely will provide adequate resources. However, sit-and-wait (ambush) predators, by virtue of their chosen location, may have a restricted diet – for example, antlions may predominantly trap small ants in their pits. Furthermore, some predators select gregarious prey, such as certain eusocial insects, because the predictable behavior and abundance of this prey allows monophagy. Although these prey insects may be aggregated, often they are aposematic and chemically defended. Nonetheless, if the defenses can be countered, these predictable and often abundant food sources permit predator specialization.

Predator–prey interactions are not discussed further; the remainder of this section concerns the more complicated host relations of parasitoids and parasites. In referring to parasitoids and their range of hosts, the terminology of monophagous, oligophagous, and polyphagous is applied, as for phytophages and predators. However, a different, parallel terminology exists for parasites: **monoxenous** parasites are restricted to a single host, **oligoxenous** to few, and **polyxenous** ones avail themselves of many hosts. In the following sections, we discuss first the variety of strategies for host selection by parasitoids, followed by the ways in which a parasitized host may be manipulated by the developing parasitoid. In the final section, patterns of host use by parasites are discussed, with particular reference to coevolution.

13.3.1 Host use by parasitoids

Parasitoids require only a single individual in which to complete development, they always kill their immature host, and rarely are parasitic in the adult stage. Insect-eating (**entomophagous**) parasitoids show a range of strategies for development on their selected insect hosts. The larva may be ectoparasitic, developing externally, or endoparasitic, developing within the host. Eggs (or larvae) of ectoparasitoids are laid close to or upon the body of the host, as are sometimes those of endoparasitoids. However, in the latter group, more often the eggs are laid within the body of the host, using a piercing ovipositor (in hymenopterans) or a substitutional ovipositor (in parasitoid dipterans). Certain parasitoids that feed within host pupal cases or under the covers and protective cases of scale insects and the like actually are ectophages (external feeding), living internal to the protection but external to the insect host body. These different feeding modes give different exposures to the host immune system, with endoparasitoids encountering and ectoparasitoids avoiding the host defenses (section 13.2.3). Ectoparasitoids are often less host specific than endoparasitoids, as they have less intimate association with the host than do endoparasitoids, which must counter the species-specific variations of the host immune system.

Parasitoids may be solitary on or in their host, or gregarious. The number of parasitoids that can develop on a host relates to the size of the host, its postinfected longevity, and the size (and biomass) of the parasitoid. Development of several parasitoids in one individual host arises commonly through the female ovipositing several eggs on a single host, or, less often, by polyembryony, in which a single egg laid by the mother divides and can give rise to numerous offspring (section 5.10.3). Gregarious parasitoids appear able to regulate the clutch size in relation to the quality and size of the host.

Most parasitoids **host discriminate**; i.e. they can recognize, and generally reject, hosts that are parasitized already, either by themselves, their conspecifics, or another species. Distinguishing unparasitized from parasitized hosts generally involves a marking pheromone placed internally or externally on the host at the time of oviposition.

However, not all parasitoids avoid already parasitized hosts. In **superparasitism**, a host receives multiple eggs either from a single individual or from several individuals of the same parasitoid species; although the host cannot sustain the total parasitoid burden to maturity. The outcome of multiple oviposition is discussed in section 13.3.2. Theoretical models, some of which have been substantiated experimentally, imply that superparasitism will increase:

- as unparasitized hosts are depleted;
- as parasitoid numbers searching any patch increase;
- in species with high fecundity and small eggs.

Although historically all such instances were deemed to have been "mistakes", there is some evidence of adaptive benefits deriving from the strategy. Superparasitism is adaptive for individual parasitoids when there is competition for scarce hosts, but avoidance is adaptive when hosts are abundant. Very direct benefits accrue in the case of a solitary parasitoid that uses a host that is able to encapsulate a parasitoid egg (section 13.2.3). Here, a first-laid egg may use all the host hemocytes, and a subsequent egg may thereby escape encapsulation.

In **multiparasitism**, a host receives eggs of more than one species of parasitoid. Multiparasitism occurs more often than superparasitism, perhaps because parasitoid species are less able to recognize the marking pheromones placed by species other than their own. Closely related parasitoids may recognize each others' marks, whereas more distantly related species may be unable to do so. However, secondary parasitoids, called **hyperparasitoids**, appear able to detect the odors left by a primary parasitoid, allowing accurate location of the site for the development of the hyperparasite.

Hyperparasitic development involves a secondary parasitoid developing at the expense of the primary parasitoid. Some insects are **obligate** hyperparasitoids, developing only within primary parasitoids, whereas others are **facultative** and may develop also as primary parasitoids. Development may be external or internal to the primary parasitoid host, with oviposition into the primary host in the former, or into the primary parasitoid in the latter (Fig. 13.6). External feeding is frequent, and hyperparasitoids are predominantly restricted to the host larval stage, sometimes the pupa;

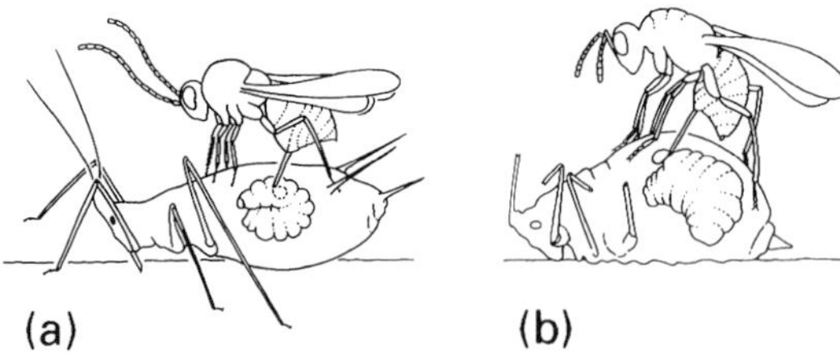

Fig. 13.6 Two examples of the ovipositional behavior of hymenopteran hyperparasitoids of aphids: (a) endophagous *Alloxysta victrix* (Hymenoptera: Figitidae) ovipositing into a primary parasitoid inside a live aphid; (b) ectophagous *Asaphes lucens* (Hymenoptera: Pteromalidae) ovipositing onto a primary parasitoid in a mummified aphid. (After Sullivan 1988.)

hyperparasitoids of eggs and adults of primary parasitoid hosts are very rare.

Hyperparasitoids belong to two families of Diptera (certain Bombyliidae and Conopidae), two families of Coleoptera (a few Rhipiphoridae and Cleridae), and notably the Hymenoptera, principally amongst 11 families of the superfamily Chalcidoidea, in four subfamilies of Ichneumonidae, and in Figitidae (Cynipoidea). Hyperparasitoids are absent among the Tachinidae and surprisingly do not seem to have evolved in certain parasitic wasp families such as Braconidae, Trichogrammatidae, and Mymaridae. Within the Hymenoptera, hyperparasitism has evolved several times, each originating in some manner from primary parasitism, with facultative hyperparasitism demonstrating the ease of the transition. Hymenopteran hyperparasitoids attack a wide range of hymenopteran-parasitized insects, predominantly amongst the hemipterans (especially Sternorrhyncha), Lepidoptera, and symphytans. Hyperparasitoids often have a broader host range than the frequently oligophagous or monophagous primary parasitoids. However, as with primary parasitoids, endophagous hyperparasitoids seem to be more host specific than those that feed externally, relating to the greater physiological problems experienced when developing within another living organism. Additionally, foraging and assessment of host suitability of a complexity comparable with that of primary parasitoids is known, at least for cynipoid hyperparasitoids of aphidophagous parasitoids (Fig. 13.7). As explained in section 16.5.1, hyperparasitism and the degree of host-specificity is fundamental information in biological control programs.

13.3.2 Host manipulation and development of parasitoids

Parasitization may kill or paralyze the host, and the developing parasitoid, called an **idiobiont**, develops rapidly, in a situation that differs only slightly from predation. Of greater interest and much more complexity is the **koinobiont** parasitoid that lays its egg(s) in a young host, which continues to grow, thereby providing an increasing food resource. Parasitoid development can be delayed until the host has attained a sufficient size to sustain it. **Host regulation** is a feature of koinobionts, with certain parasitoids able to manipulate host physiology, including suppression of its pupation to produce a "super host".

Many koinobionts respond to hormones of the host, as demonstrated by (i) the frequent molting or emergence of parasitoids in synchrony with the host's molting or metamorphosis, and/or (ii) synchronization of diapause of host and parasitoid. It is uncertain whether, for example, host ecdysteroids act directly on the parasitoid's epidermis to cause molting, or act indirectly on the parasitoid's own endocrine system to elicit synchronous molting. Although the specific mechanisms remain unclear, some parasitoids undoubtedly disrupt the host endocrine system, causing developmental arrest, accelerated or retarded metamorphosis, or inhibition of reproduction in an adult host. This may arise through production of hormones (including mimetic ones) by the parasitoid, or through regulation of the host's endocrine system, or both. In cases of **delayed parasitism**, such as is seen in certain platygastrine and braconid hymenopterans, development of an egg laid in the host egg is delayed for up to a year, until the host is a late-stage larva. Host hormonal changes approaching metamorphosis are implicated in the stimulation of parasitoid development. Specific interactions between the endocrine systems of endoparasitoids and their hosts can limit the range of hosts utilized. Parasitoid-introduced viruses or virus-like particles (Box 13.1) may also modify host physiology and determine host range.

The host is not a passive vessel for parasitoids – as we have seen, the immune system can attack all but the adapted parasitoids. Furthermore, host quality (size and age) can induce variation in size, fecundity, and even the sex ratio of emergent solitary parasitoids. Generally, more females are produced from high-quality (larger) hosts, whereas males are produced from poorer quality ones, including smaller and superparasitized

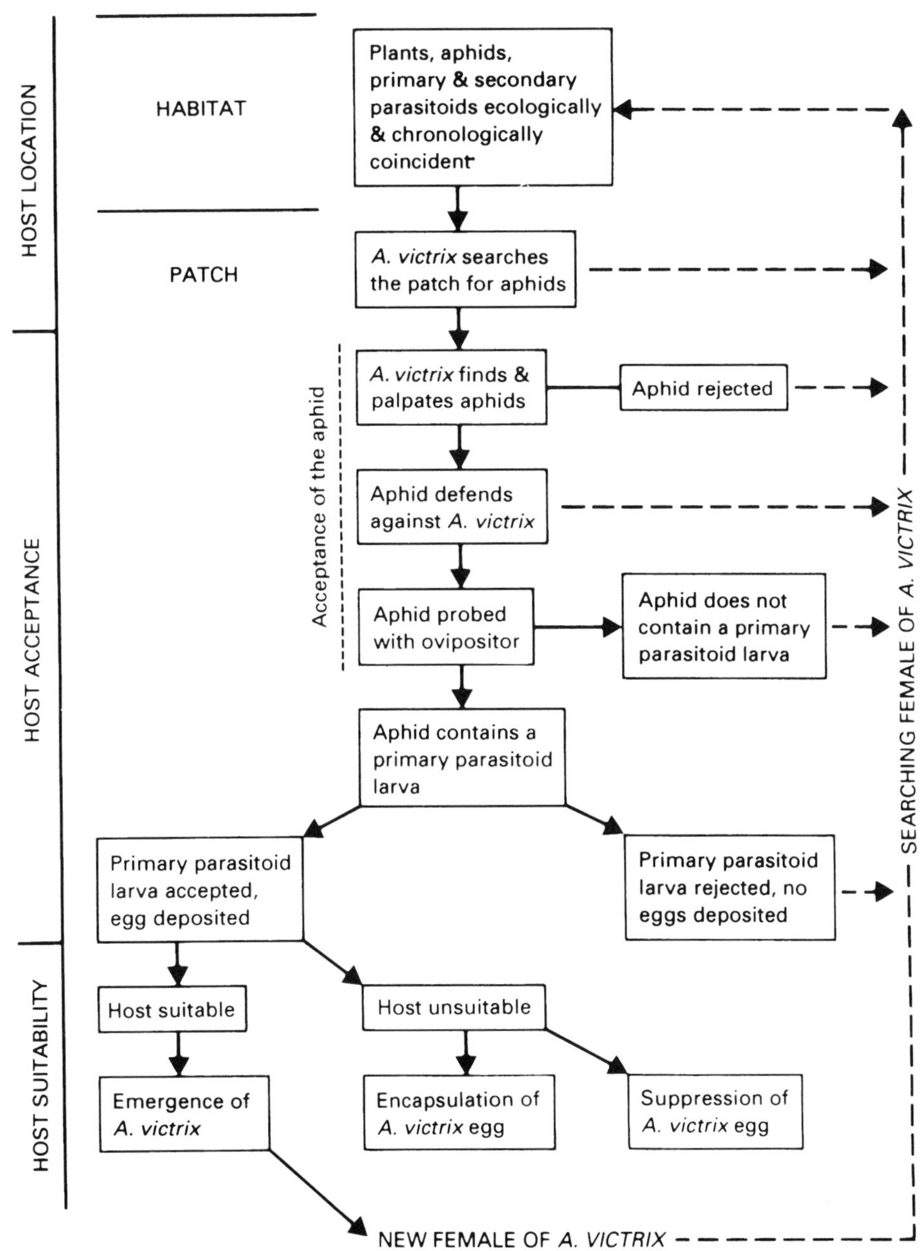

Fig. 13.7 Steps in host selection by the hyperparasitoid *Alloxysta victrix* (Hymenoptera: Figitidae). (After Gutierrez 1970.)

hosts. Host aphids reared experimentally on deficient diets (lacking sucrose or iron) produced *Aphelinus* (Hymenoptera: Aphelinidae) parasitoids that developed more slowly, produced more males, and showed lowered fecundity and longevity. The young stages of an endophagous koinobiont parasitoid compete with the host tissues for nutrients from the hemolymph. Under laboratory conditions, if a parasitoid can be induced to oviposit into an "incorrect" host (by the use of appropriate kairomones), complete larval development often occurs, showing that hemolymph is adequate nutritionally for development of more than just the adapted parasitoid. Accessory gland secretions (which may include paralyzing venoms) are injected by the ovipositing female parasitoid with the eggs, and appear to play a role in regulation of the host's hemolymph nutrient supply to the larva. The specificity of these substances may relate to the creation of a suitable host.

In superparasitism and multiparasitism, if the host cannot support all parasitoid larvae to maturity, larval competition often takes place. Depending on the nature of the multiple ovipositions, competition may involve aggression between siblings, other conspecifics, or interspecific individuals. Fighting between larvae, especially in mandibulate larval hymenopterans, can result in death and encapsulation of excess individuals. Physiological suppression with venoms, anoxia, or food deprivation also may occur. Unresolved larval overcrowding in the host can result in a few weak and small individuals emerging, or no parasitoids at all if the host dies prematurely or resources are depleted before pupation. Gregariousness may have evolved from solitary parasitism in circumstances in which multiple larval development is permitted by greater host size. Evolution of gregariousness may be facilitated when the potential competitors for resources within a single host are relatives. This is particularly so in polyembryony, which produces clonal, genetically identical larvae (section 5.10.3).

13.3.3 Patterns of host use and specificity in parasites

The wide array of insects that are ectoparasitic upon vertebrate hosts are of such significance to the health of humans and their domestic animals that we devote a complete chapter to them (Chapter 15) and medical issues will not be considered further here. In contrast to the radiation of ectoparasitic insects using vertebrate hosts and the immense numbers of species of insect parasitoids seen above, there are remarkably few insect parasites of other insects, or indeed, of other arthropods.

The largest group of endoparasitic insects using other insects as hosts belongs to the Strepsiptera, an order comprising a few hundred exclusively parasitic species (Box 13.6). The characteristically aberrant bodies of their predominantly hemipteran and hymenopteran hosts are termed "stylopized", so-called for a common strepsipteran genus, *Stylops*. Within the host's body cavity, growth of larvae and pupae of both sexes, and the adult female strepsipteran, causes malformations including displacement of the internal organs. The host's sexual organs degenerate, or fail to develop appropriately.

Although larval Dryinidae (Hymenoptera) develop parasitically part-externally and part-internally in hemipterans, virtually all other insect–insect parasitic interactions involve ectoparasitism. The Braulidae is a family of Diptera comprising some aberrant, mite-like flies belonging to a single genus, *Braula*, intimately associated with *Apis* (honey bees). Larval braulids scavenge on pollen and wax in the hive, and the adults usurp nectar and saliva from the proboscis of the bee. This association certainly involves phoresy, with adult braulids always found on their hosts' bodies, but whether the relationship is ectoparasitic is open to debate. Likewise, the relationship of several genera of aquatic chironomid larvae with nymphal hosts, such as mayflies, stoneflies, and dragonflies, ranges from phoresy to suggested ectoparasitism. Generally, there is little evidence that any of these ecto- and endoparasites using insects show a high degree of specificity at the species level. However, this is not necessarily the case for insect parasites with vertebrate hosts.

The patterns of host-specificity and preferences of parasites raise some of the most fascinating questions in parasitology. For example, most orders of mammals bear lice (Phthiraptera), many of which are monoxenic or found amongst a limited range of hosts. Even some marine mammals, namely certain seals, have lice, although whales do not. No Chiroptera (bats) harbor lice, despite their apparent suitability, although they host many other ectoparasitic insects, including the Streblidae and Nycteribiidae – two families of ectoparasitic Diptera that are restricted to bats.

Some terrestrial hosts are free of all ectoparasites, others have very specific associations with one or a few guests, and in Panama the opossum *Didelphis marsupialis* has been found to harbor 41 species of ectoparasitic insects and mites. Although four or five of these are commonly present, none are restricted to the opossum and the remainder are found on a variety of hosts, ranging from distantly related mammals to reptiles, birds, and bats.

We can examine some principles concerning the different patterns of distribution of parasites and their

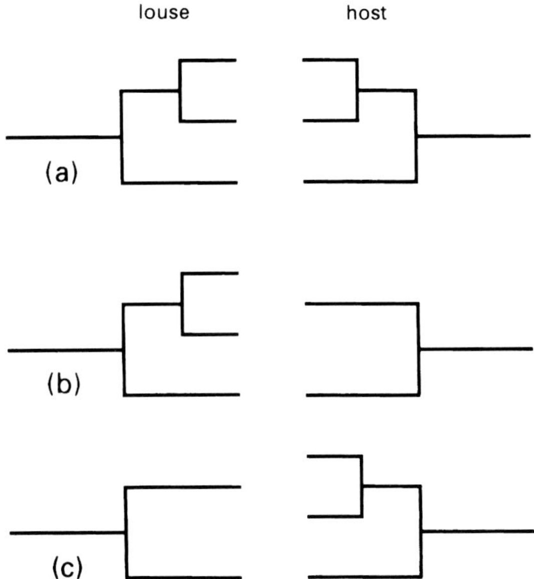

Fig. 13.8 Comparisons of louse and host phylogenetic trees: (a) adherence to Fahrenholz's rule; (b) independent speciation of the lice; (c) independent speciation of the hosts. (After Lyal 1986.)

hosts by looking in some detail at cases where close associations of parasites and hosts are expected. The findings can then be related to ectoparasite–host relations in general.

The Phthiraptera are obligate permanent ectoparasites, spending all their lives on their hosts, and lacking any free-living stage. Extensive surveys, such as one which showed that neotropical birds averaged 1.1 lice species per host across 127 species and 26 families of birds, indicate that lice are highly monoxenous (restricted to one host species). A high level of coevolution between louse and host might be expected, and in general, related animals have related lice. The widely quoted Fahrenholz's rule formally states that the phylogenies of hosts and parasites are identical, with every speciation event affecting hosts being matched by a synchronous speciation of the parasites, as shown in Fig. 13.8a.

It follows that:

- phylogenetic trees of hosts can be derived from the trees of their ectoparasites;
- ectoparasite phylogenetic trees are derivable from the trees of their hosts (the potential for circularity of reasoning is evident);
- the number of parasite species in the group under consideration is identical to the number of host species considered;
- no species of host has more than one species of parasite in the taxon under consideration;
- no species of parasite parasitizes more than one species of host.

Fahrenholz's rule has been tested for mammal lice selected from amongst the family Trichodectidae, for which robust phylogenetic trees, derived independently of any host mammal phylogeny, are available. Amongst a sample of these trichodectids, 337 lice species parasitize 244 host species, with 34% of host species parasitized by more than one trichodectid. Several possible explanations exist for these mismatches. Firstly, speciation may have occurred independently amongst certain lice on a single host (Fig. 13.8b). This is substantiated, with at least 7% of all speciation events in the sampled Trichodectidae showing this pattern of independent speciation. A second explanation involves secondary transfer of lice species to phylogenetically unrelated host taxa. Amongst extant species, when cases arising from human-induced unnatural host proximity are excluded (accounting for 6% of cases), unmistakable and presumed natural transfers (i.e. between marsupial and eutherian mammal, or bird and mammal) occur in about 2% of speciation events. However, hidden within the phylogenies of host and parasite are speciation events that involve lateral transfer between rather more closely-related host taxa, but these transfers fail to match precisely the phylogeny. Examination of the detailed phylogeny of the sampled Trichodectidae shows that a minimum of 20% of all speciation events are associated with distant and lateral secondary transfer, including historical transfers (lying deeper in the phylogenetic trees).

In detailed examinations of relationships between a smaller subset of trichodectids and eight of their pocket gopher (Rodentia: Geomyidae) hosts, substantial concordance was claimed between trees derived from biochemical data for hosts and parasites, and some evidence of co-speciation was found. However, many of the hosts were shown to have two lice species, and unconsidered data show most species of gopher to have a substantial suite of associated lice. Furthermore, a minimum of three instances of lateral transfer (host switching) appeared to have occurred, in all cases between hosts with geographically contiguous ranges. Although many speciation events in these lice "track" speciation in the host and some estimates even indicate similar ages of host and parasite species, it is evident from the Trichodectidae that strict co-speciation of host and parasite is not the sole explanation of the associations observed.

The reasons why apparently monoxenic lice sometimes do deviate from strict coevolution and co-speciation apply equally to other ectoparasites, many of which show similar variation in complexity of host relationships. Deviations from strict co-speciation arise if host speciation occurs without commensurate parasite speciation (Fig. 13.8c). This resulting pattern of relationships is identical to that seen if one of two parasite sister taxa generated by co-speciation in concert with the host subsequently became extinct. Frequently, a parasite is not present throughout the complete range of its host, resulting perhaps from the parasite being restricted in range by environmental factors independent of those controlling the range of the host. Hemimetabolous ectoparasites, such as lice, which spend their entire lives on the host, might be expected to closely follow the ranges of their hosts, but there are exceptions in which the ectoparasite distribution is restricted by external environmental factors. For holometabolous ectoparasites, which spend some of their lives away from their hosts, such external factors will be even more influential in governing parasite range. For example, a homeothermic vertebrate may tolerate environmental conditions that cannot be sustained by the free-living stage of a poikilothermic ectoparasite, such as a larval flea. As speciation may occur in any part of the distribution of a host, host speciation may be expected to occur without necessarily involving the parasite. Furthermore, a parasite may show geographical variation within all or part of the host range that is incongruent with the variation of the host. If either or both variations lead to eventual species formation, there will be incongruence between parasite and host phylogeny.

Furthermore, poor knowledge of host and parasite interactions may result in misleading conclusions. A true host may be defined as one that provides the conditions for parasite reproduction to continue indefinitely. When there is more than one true host, there may be a principal (preferred) or exceptional host, depending on the proportional frequencies of ectoparasite occurrence. An intermediate category may be recognized – the sporadic or secondary host – on which parasite development cannot normally take place, but an association arises frequently, perhaps through predator–prey interactions or environmental encounters (such as a shared nest). Small sample sizes and limited biological information can allow an accidental or secondary host to be mistaken for a true host, giving rise to a possible erroneous "refutation" of co-speciation. Extinctions of certain parasites and true hosts (leaving the parasite extant on a secondary host) will refute Fahrenholz's rule.

Even assuming perfect recognition of true host-specificity and knowledge of the historical existence of all parasites and hosts, it is evident that successful parasite transfers between hosts have taken place throughout the history of host–parasite interactions. Co-speciation is fundamental to host–parasite relations, but the factors encouraging deviations must be considered. Predominantly, these concern (i) geographical and social proximity of different hosts, allowing opportunities for parasite colonization of the new host, together with (ii) ecological similarity of different hosts, allowing establishment, survival, and reproduction of the ectoparasite on the novel host. The results of these factors have been termed **resource tracking**, to contrast with the **phyletic tracking** implied by Fahrenholz's rule. As with all matters biological, most situations lie somewhere along a continuum between these two extremes, and rather than forcing patterns into one category or the other, interesting questions arise from recognizing and interpreting the different patterns observed.

If all host–parasite relationships are examined, some of the factors that govern host-specificity can be identified:

- the stronger the life-history integration with that of the host, the greater the likelihood of monoxeny;
- the greater the vagility (mobility) of the parasite, the more likely it is to be polyxenous;
- the number of accidental and secondary parasite species increases with decreasing ecological specialization and with increase in geographical range of the host, as we saw earlier in this section for the opossum, which is widespread and unspecialized.

If a single host shares a number of ectoparasites, there may be some ecological or temporal segregation on the host. For example, in hematophagous (blood-sucking) black flies (Simuliidae) that attack cattle, the belly is more attractive to certain species, whereas others feed only on the ears. *Pediculus humanus capitis* and *P. humanus corporis* (Phthiraptera), human head and body lice respectively, are ecologically separated examples of sibling taxa in which strong reproductive isolation is reflected by only slight morphological differences.

13.4 POPULATION BIOLOGY – PREDATOR/PARASITOID AND PREY/HOST ABUNDANCE

Ecological interactions between an individual, its conspecifics, its predators and parasitoids (and other causes of mortality), and its abiotic habitat are fundamentally important aspects of population dynamics. Accurate estimation of population density and its regulation is at the heart of population ecology, biodiversity studies, conservation biology, and monitoring and management of pests. A range of tools are available to entomologists to understand the effects of the many factors that influence population growth and survivorship, including sampling methods, experimental designs, and manipulations and modeling programs.

Insects usually are distributed on a wider scale than investigators can survey in detail, and thus sampling must be used to allow extrapolation to the wider population. Sampling may be absolute, in which case all organisms in a given area or volume might be assessed, such as mosquito larvae per liter of water, or ants per cubic meter of leaf litter. Alternatively, relative measures, such as number of Collembola in pitfall trap samples, or micro-wasps per yellow pan trap, may be obtained from an array of such trapping devices (section 17.1). Relative measures may or may not reflect actual abundances, with variables such as trap size, habitat structure, and insect behavior and activity levels affecting **"trappability"** – the likelihood of capture. Measures may be integrated over time, for example a series of sticky, pheromone, or continuous running light traps, or instantaneous snap-shots such as the inhabitants of a submerged freshwater rock, the contents of a timed sweep netting, or the knock-down from an insecticidal fogging of a tree's canopy. Instantaneous samples may be unrepresentative, whereas longer duration sampling can overcome some environmental variability.

Sampling design is the most important component in any population study, with stratified random designs providing power to interpret data statistically. Such a design involves dividing the study site into regular blocks (subunits) and, within each of these blocks, sampling sites are allocated randomly. Pilot studies can allow understanding of the variation expected, and the appropriate matching of environmental variables between treatments and controls for an experimental study. Although more widely used for vertebrate studies, mark-and-recapture methods have been effective for adult odonates, larger beetles, moths and, with fluorescent chemical dyes, smaller pest insects.

A universal outcome of population studies is that the expectation that the number and density of individuals grows at an ever-increasing rate is met very rarely, perhaps only during short-lived pest outbreaks. Exponential growth is predicted because the rate of reproduction of insects potentially is high (hundreds of eggs per mother) and generation times are short – even with mortality as high as 90%, numbers increase dramatically. The equation for such geometric or exponential growth is:

$$dN/dt = rN$$

where N is population size or density, dN/dt is the growth rate, and r is the instantaneous per capita rate of increase. At $r = 0$ rates of birth and death are equal and the population is static; if $r < 0$ the population declines; when $r > 0$ the population increases.

Growth continues only until a point at which some resource(s) become limiting, called the carrying capacity. As the population nears the carrying capacity, the rate of growth slows in a process represented by:

$$dN/dt = rN - rN^2/K$$

in which K, representing the carrying capacity, contributes to the second term, called environmental resistance. Although this basic equation of population dynamics underpins a substantial body of theoretical work, evidently natural populations persist in more narrowly fluctuating densities, well below the carrying capacity. Observed persistence over evolutionary time (section 8.2) allows the inference that, averaged over time, birth rate equals death rate.

Parasitism and predation are major influences on population dynamics as they affect death rate in a manner that varies with host density. Thus, an increase in mortality with density (positive density dependence) contrasts with a decrease in death rate with density (negative density dependence). A substantial body of experimental and theoretical evidence demonstrates that predators and parasitoids impose density-dependent effects on components of their food webs, in a trophic cascade (see below). Experimental removal of the most important ("top") predator can induce a major shift in community structure, demonstrating that predators control the abundance of subdominant predators and certain prey species. Models of complex relationships between predators and prey frequently are motivated by a desire to understand interactions of

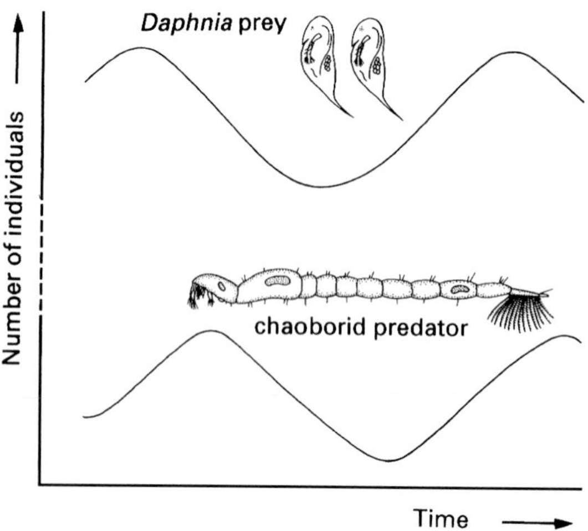

Fig. 13.9 An example of the regular cycling of numbers of predators and their prey: the aquatic planktonic predator *Chaoborus* (Diptera: Chaoboridae) and its cladoceran prey *Daphnia* (Crustacea).

native predators or biological control agents and target pest species.

Mathematical models may commence from simple interactions between a single monophagous predator and its prey. Experiments and simulations concerning the long-term trend in densities of each show regular cycles of predators and prey: when prey are abundant, predator survival is high; as more predators become available, prey abundance is reduced; predator numbers decrease as do those of prey; reduction in predation allows the prey to escape and rebuild numbers. The sinusoidal, time-lagged cycles of predator and prey abundances may exist in some simple natural systems, such as the aquatic planktonic predator *Chaoborus* (Diptera: Chaoboridae) and its cladoceran prey *Daphnia* (Fig. 13.9).

Examination of shorter-term feeding responses using laboratory studies of simple systems shows that predators vary in their responses to prey density. Early ecologists' assumptions of a linear relationship (increased prey density leading to increased predator feeding) have been superseded. A common functional response of a predator to prey density involves a gradual slowing of the rate of predation relative to increased prey density, until an asymptote is reached. This upper limit beyond which no increased rate of prey capture occurs is due to the time constraints of foraging and handling prey in which there is a finite limit to the time spent in feeding activities, including a recovery period. The rate of prey capture does not depend upon prey density alone: individuals of different instars have different feeding rate profiles, and in poikilothermic insects there is an important effect of ambient temperature on activity rates.

Assumptions of predator monophagy often may be biologically unrealistic, and more complex models include multiple prey items. Predator behavior is based upon optimal foraging strategies involving simulated prey selection varying with changes in proportional availability of different prey items. However, predators may not switch between prey items based upon simple relative numerical abundance; other factors include differences in prey profitability (nutritional content, ease of handling, etc.), the hunger-level of the predator, and perhaps predator learning and development of a search-image for particular prey, irrespective of abundance.

Models of prey foraging and handling by predators, including more realistic choice between profitable and less profitable prey items, indicate that:

- prey specialization ought to occur when the most profitable prey is abundant;
- predators should switch rapidly from complete dependence on one prey to the other, with partial preference (mixed feeding) being rare;
- the actual abundance of a less-abundant prey should be irrelevant to the decision of a predator to specialize on the most abundant prey.

Improvements can be made concerning parasitoid searching behavior which simplistically is taken to resemble a random-searching predator, independent of host abundance, the proportion of hosts already parasitized, or the distribution of the hosts. As we have seen above, parasitoids can identify and respond behaviorally to already-parasitized hosts. Furthermore, prey (and hosts) are not distributed at random, but occur in patches, and within patches the density is likely to vary. As predators and parasitoids aggregate in areas of high resource density, interactions between predators/parasitoids (**interference**) become significant, perhaps rendering a profitable area less profitable. For a number of reasons, there may be **refuges** from predators and parasitoids within a patch. Thus, amongst California red scale insects (Hemiptera: Diaspididae: *Aonidiella aurantii*) on citrus trees, those on the periphery of the tree may be up to 27 times more vulnerable to two species of parasitoids compared with individual scales in the center of the tree, which thus may be termed a refuge. Furthermore, the effectiveness of a refuge varies between taxonomic or ecological groups: external leaf-feeding insects support more parasitoid species than leaf-mining insects, which in turn support more than highly concealed insects such as root feeders or those living in structural refuges. These observations have important implications for the success of biological control programs.

The direct effects of a predator (or parasitoid) on its prey (or host) translate into changes in the prey's or host's energy supply (i.e. plants if the prey or host is a herbivore) in an interaction chain. The effects of resource consumption are predicted to cascade from the top consumers (predators or parasitoids) to the base of the energy pyramid via feeding links between inversely related trophic levels. The results of field experiments on such **trophic cascades** involving predator manipulation (removal or addition) in terrestrial arthropod-dominated food webs have been synthesized using meta-analysis. This involves the statistical analysis of a large collection of analysis results from individual studies for the purpose of integrating the findings. Meta-analysis found extensive support for the existence of trophic cascades, with predator removal mostly leading to increased densities of herbivorous insects and higher levels of plant damage. Furthermore, the amount of herbivory following relaxation of predation pressure was significantly higher in crop than in non-crop systems such as grasslands and woodlands. It is likely that "top-down" control (from predators) is more frequently observed in managed than in natural systems due to simplification of habitat and food-web structure in managed environments. These results suggest that natural enemies can be very effective in controlling plant pests in agroecosystems and thus conservation of natural enemies (section 16.5.1) should be an important aspect of pest control.

13.5 THE EVOLUTIONARY SUCCESS OF INSECT PREDATION AND PARASITISM

In Chapter 11 we saw how the development of angiosperms and their colonization by specific plant-eating insects explained a substantial diversification of phytophagous insects relative to their non-phytophagous sister taxa. Analogous diversification of Hymenoptera in relation to adoption of a parasitic lifestyle exists, because numerous small groups form a "chain" on the phylogenetic tree outside the primarily parasitic sister group, the suborder Apocrita. It is likely that Orussoidea (with only one family, Orussidae) is the sister group to Apocrita, and probably all are parasitic on wood-boring insect larvae. However, the next prospective sister group lying in the (paraphyletic) "Symphyta" is a small group of wood wasps. This sister group is non-parasitic (as are the remaining symphytans) and species-poor with respect to the speciose combined Apocrita plus Orussoidea. This phylogeny implies that, in this case, adoption of a parasitic lifestyle was associated with a major evolutionary radiation. An explanation may lie in the degree of host restriction: if each species of phytophagous insect were host to a more or less monophagous parasitoid, then we would expect to see a diversification (radiation) of insect parasitoids that corresponded to that of phytophagous insects. Two assumptions need examination in this context – the degree of host-specificity and the number of parasitoids harbored by each host.

The question of the degree of monophagy amongst parasites and parasitoids is not answered conclusively. For example, many parasitic hymenopterans are extremely small, and the basic taxonomy and host associations are yet to be fully worked out. However, there is no doubt that the parasitic hymenopterans are extremely speciose, and show a varying pattern of host-specificity from strict monophagy to oligophagy. Amongst parasitoids within the Diptera, the species-rich Tachinidae are relatively general feeders, specializing only in hosts belonging to families or even ordinal groups. Amongst the ectoparasites, lice are predominantly monoxenic, as are many fleas and flies. However, even if several species of ectoparasitic insects were borne by each host species, as the vertebrates are not numerous, ectoparasites contribute relatively little to biological diversification in comparison with the parasitoids of insect (and other diverse arthropod) hosts.

There is substantial evidence that many hosts support multiple parasitoids (much of this evidence is acquired by the diligence of amateur entomologists). This phenomenon is well known to lepidopterists that endeavor to rear adult butterflies or moths from wild-caught larvae – the frequency and diversity of parasitization is very high. Suites of parasitoid and hyperparasitoid species may attack the same species of host at different seasons, in different locations, and in different life-history stages. There are many records of more than 10 parasitoid species throughout the range of some widespread lepidopterans, and although this is true also for certain well-studied coleopterans, the situation is less clear for other orders of insects.

Finally, some evolutionary interactions between parasites and parasitoids and their hosts may be considered. Patchiness of potential host abundance throughout the host range seems to provide opportunity for increased specialization, perhaps leading to species formation within the guild of parasites/parasitoids. This can be seen as a form of niche differentiation, where the total range of a host provides a niche that is ecologically partitioned. Hosts may escape from parasitization within refuges within the range, or by modification of the life cycle, with the introduction of a phase that the parasitoid cannot track. Host diapause may be a mechanism for evading a parasite that is restricted to continuous generations, with an extreme example of escape perhaps seen in the periodic cicada. These species of *Magicicada* grow concealed for many years as nymphs beneath the ground, with the very visible adults appearing only every 13 or 17 years. This cycle of a prime number of years may allow avoidance of predators or parasitoids that are able only to adapt to a predictable cyclical life history. Life-cycle shifts as attempts to evade predators may be important in species formation.

Strategies of prey/hosts and predators/parasitoids have been envisaged as evolutionary arms races, with a stepwise sequence of prey/host escape by evolution of successful defenses, followed by radiation before the predator/parasitoid "catches-up", in a form of prey/host tracking. An alternative evolutionary model envisages both prey/host and predator/parasitoid evolving defenses and circumventing them in virtual synchrony, in an evolutionarily stable strategy termed the "Red Queen" hypothesis (after the description in *Alice in Wonderland* of Alice and the Red Queen running faster and faster to stand still). Tests of each can be devised and models for either can be justified, and it is unlikely that conclusive evidence will be found in the short term. What is clear is that parasitoids and predators do exert great selective pressure on their hosts or prey, and remarkable defenses have arisen, as we shall see in the next chapter.

Box 13.2 Mantodea (mantids)

The Mantodea is an order of about 2000 species of moderate to large (1–15 cm long) hemimetabolous predators classified in eight families. Males are generally smaller than females. The head is small, triangular and mobile, with slender antennae, large, widely separated eyes, and mandibulate mouthparts. The thorax comprises an elongate, narrow prothorax and shorter (almost subquadrate) meso- and metathorax. The fore wings form leathery tegmina, with the anal area reduced; the hind wings are broad and membranous, with long veins unbranched and many cross-veins. Aptery and subaptery are frequent. The fore legs are raptorial (Fig. 13.3 and as illustrated here for a mantid of a *Tithrone* species holding and eating a fly, after Preston-Mafham 1990), whereas the mid and hind legs are elongate for walking. On the abdomen, the 10th visible segment bears variably segmented cerci. The ovipositor is predominantly internal; the external male genitalia are asymmetrical.

Eggs are laid in an ootheca (see Plate 3.3, facing p. 14) produced from accessory gland frothy secretions that harden on contact with the air. Some females guard their ootheca. First-instar nymphs do not feed, but molt immediately. As few as three or as many as 12 instars follow; the nymphs resemble adults except for lack of wings and genitalia. Adult mantids are sit-and-wait predators (see section 13.1.1) which use their fully mobile head and excellent sight to detect prey. Female mantids sometimes consume the male during or after copulation (Box 5.2); males often display elaborate courtship.

Mantodea are undoubtedly the sister group to the Blattodea (cockroaches) and Isoptera, forming the Dictyoptera grouping (Figs. 7.2 & 7.4).

Box 13.3 Mantophasmatodea (heel walkers)

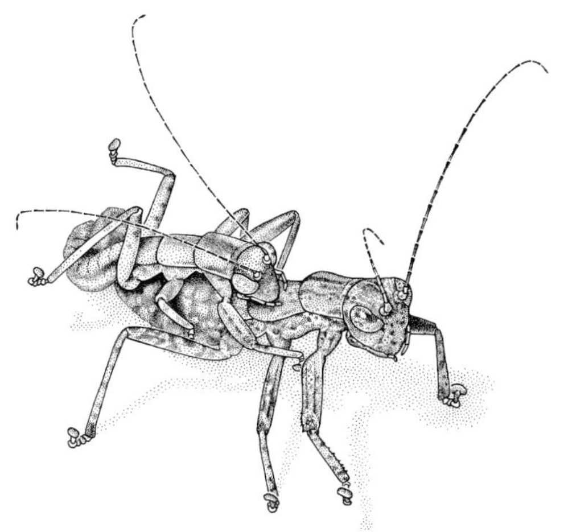

The discovery of a previously unrecognized order of insects is an unusual event. In the 20th century only two orders were newly described: Zoraptera in 1913 and Grylloblattodea in 1932. The opening of the 21st century saw a flurry of scientific and popular media interest concerning the unusual discovery and subsequent recognition of a new order, the Mantophasmatodea.

The first formal recognition of this new taxon was from a specimen preserved in 45 million year old Baltic amber, which bore a superficial resemblance to a stick-insect or a mantid, but evidently belonged to neither. Shortly thereafter a museum specimen from Tanzania was discovered, and comparison with more fossil specimens including adults showed that the fossil and recent insects were related. Further museum searches and appeals to curators uncovered specimens from rocky outcrops in Namibia (south-west Africa). An expedition found the living insects in several Namibian localities, and subsequently many specimens were identified in historic and recent collections from succulent karoo and fynbos vegetation of South Africa.

The Mantophasmatodea was named for its superficial resemblance to two other orders. In an assessment of the morphology it was difficult to place the new order, but relationships with the phasmids (Phasmatodea) and/or rock crawlers (Grylloblattodea) were suggested. Nucleotide sequencing data have justified the rank of order, and confirmed that it forms the sister group to Grylloblattodea. Currently there are three families, with two extinct and 10 extant genera, and 13 described extant species (mostly undescribed), now restricted to south-western and South Africa, and Tanzania in eastern Africa.

Mantophasmatids are moderate-sized (1.1–2.5 cm long in extant species, 1.5 cm in fossil species) hemimetabolous insects, with a hypognathous head with generalized mouthparts (mandibles with three small teeth) and long slender antennae with 26–32 segments and a sharply elbowed distal region. The prothoracic pleuron is large and exposed, not covered by pronotal lobes. Each tergum of the thorax narrowly overlaps and is smaller than the previous. All species are apterous, without any rudiments of wings. The coxae are elongate, the fore and mid femora are somewhat broadened and with bristles or spines ventrally. The tarsi are five-segmented with euplanulae on the basal four, the ariolum is very large and, characteristically, the distal tarsomere is held off the substrate (hence the name "heel walkers"). The hind legs are elongate and can be used in making small jumps. Male cerci are prominent (as on the male shown in the Appendix, after a photograph by M.D. Picker), clasping, and do not form a differentiated articulation with the 10th tergite. Female cerci are one-segmented and short. The ovipositor projects beyond the short subgenital lobe and there is no protective operculum (plate below ovipositor) as occurs in phasmids.

Copulation may be prolonged (up to three days uninterrupted) and, at least in captivity, the male is eaten after mating. The male mounts the female with his genitalia engaged from her right-hand side, as shown here for a copulating pair of South African mantophasmatids (after a photograph by S.I. Morita). A South African species was observed to lay 12 very large eggs in an egg pod made up of a foam mixed with sand and laid superficially in the soil. An ethanol-preserved specimen from Namibia had 40 eggs within its abdomen. The life cycle is not well known, although studies are in progress and it is known that the molted cuticle is eaten after ecdysis. At least one Namibian species seems to be diurnal, whereas South African species are nocturnal. Mantophasmatids are either ground dwelling or live on shrubs or in grass clumps. They are predatory, feeding for example on small flies, bugs, and moths. For this reason, the common name "gladiators" has been proposed for the order, although we prefer the more descriptive "heel walkers". Raptorial femora are grooved to receive the tibia during prey capture; at rest the raptorial limbs are not folded. Most species exhibit considerable color variation from light green to dark brown. Males generally are smaller and of a different color to females.

Based on molecular evidence, their sister group is the Grylloblattodea, one of the suggested relationships based on morphology (section 7.5 and Fig. 7.2). Winged insects otherwise resembling Mantophasmatodea are known from the Upper Carboniferous (300 mya). These belong to the extinct order Cnemidolestodea, and may reflect an ancient radiation of which Mantophasmatodea and perhaps Grylloblattodea are wingless relicts.

Box 13.4 Neuropterida, or neuropteroid orders

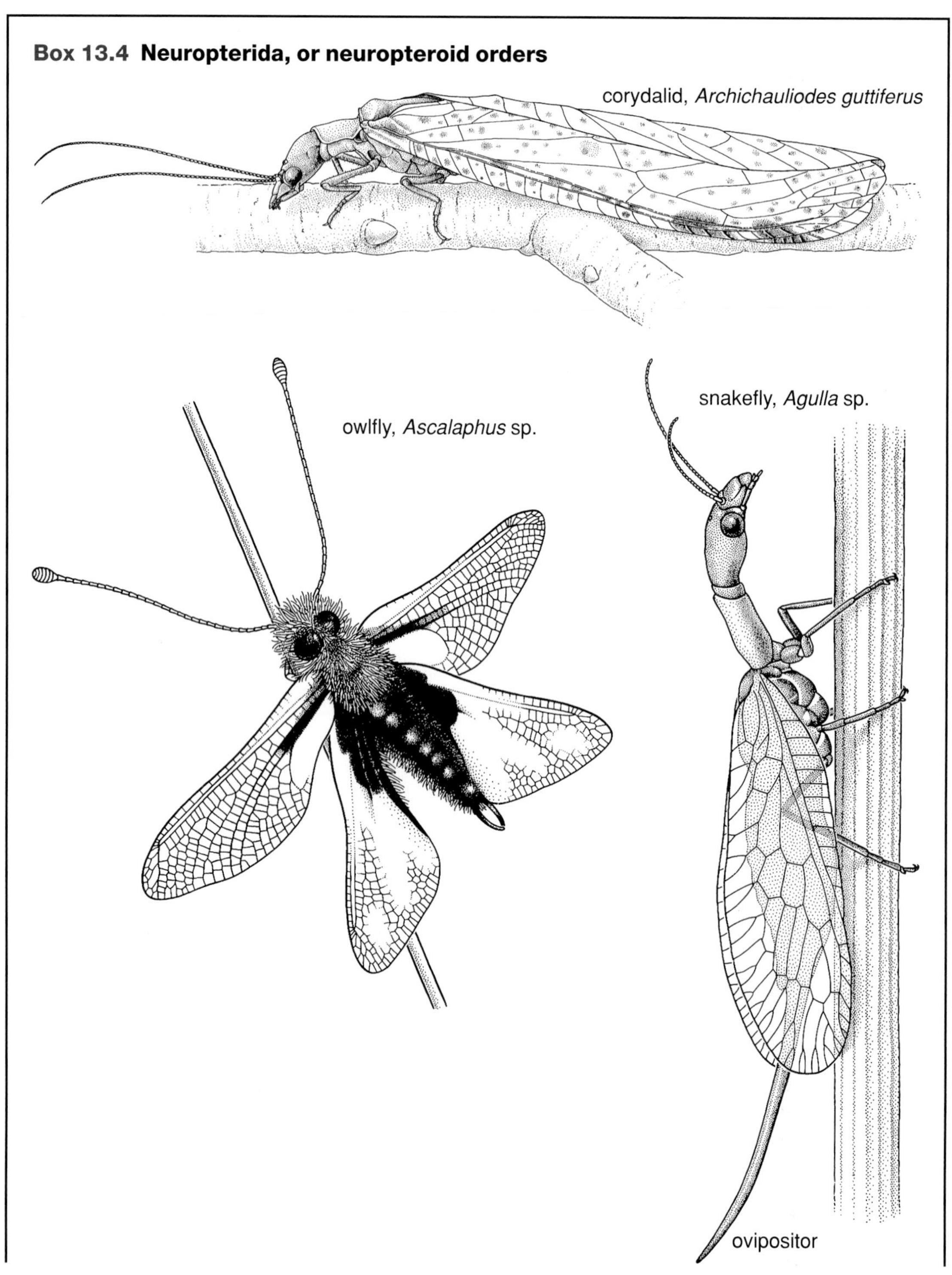

Members of these three small neuropteroid orders have holometabolous development, and are mostly predators. Approximate numbers of described species are: 5000–6000 for Neuroptera (lacewings, owlflies, antlions) in about 20 families; 300 in Megaloptera (alderflies and dobsonflies) in two widely recognized families; and 200 in Raphidioptera (snakeflies) in two families.

Adults have multisegmented antennae, large separated eyes, and mandibulate mouthparts. The prothorax may be larger than the meso- and metathorax, which are about equal in size. The legs may be modified for predation. Fore and hind wings are similar in shape and venation, with folded wings often extending beyond the abdomen. The abdomen lacks cerci.

Megaloptera (see Appendix) are predatory only in the aquatic larval stage (Box 10.6) – although the adults have strong mandibles, these are not used in feeding. Adults (such as the corydalid, *Archichauliodes guttiferus*, illustrated here) closely resemble neuropterans, except for the presence of an anal fold in the hind wing. The pupa (Fig. 6.7a) is mobile.

Raphidioptera are terrestrial predators both as adults and larvae. The adult is mantid-like, with an elongate prothorax – as shown here by the female snakefly of an *Agulla* sp. (Raphidiidae) (after a photograph by D.C.F. Rentz) – and mobile head used to strike, snake-like, at prey. The larva (illustrated in the Appendix) has a large prognathous head, and a sclerotized prothorax that is slightly longer than the membranous meso- and metathorax. The pupa is mobile.

Adult Neuroptera (illustrated in Fig. 6.12 and the Appendix, and exemplied here by an owlfly, *Ascalaphus* sp. (Ascalaphidae), after a photograph by C.A.M. Reid) possess wings typically with numerous cross-veins and "twigging" at ends of veins; many are predators, but nectar, honeydew, and pollen are consumed by some species. Neuropteran larvae (Fig. 6.6d) are usually specialized, active predators, with prognathous heads and slender, elongate mandibles and maxillae combined to form piercing and sucking mouthparts (Fig. 13.2c); all have a blind-ending hind gut. Larval dietary specializations include spider egg masses (for Mantispidae), freshwater sponges (for Sisyridae; Box 10.6), or soft-bodied hemipterans such as aphids and scale insects (for Chrysopidae, Hemerobiidae, and Coniopterygidae). Pupation is terrestrial, within shelters spun with silk from Malpighian tubules. The pupal mandibles are used to open a toughened cocoon.

The Megaloptera, Raphidioptera, and Neuroptera are treated here as separate orders; however, some authorities include the Raphidioptera in the Megaloptera, or all three may be united in the Neuroptera. Phylogenetic relationships are considered in section 7.4.2 and depicted in Fig. 7.2.

Box 13.5 Mecoptera (scorpionflies, hangingflies)

The Mecoptera is an order of about 550 known species in nine families, with common names associated with the two largest families – Bittacidae (hangingflies, see Box 5.1) and Panorpidae (scorpionflies, illustrated in the Appendix; see also Plate 5.3, facing p. 14, and the vignette of this chapter). Development is holometabolous. Adults have an elongate hypognathous rostrum; their mandibles and maxillae are elongate, slender, and serrate; the labium is elongate. They have large, separated eyes, and filiform, multisegmented antennae. The prothorax may be smaller than the equally developed meso- and metathorax, each with a scutum, scutellum, and postscutellum visible. The fore and hind wings are narrow and of similar size, shape, and venation; they are often reduced or absent. The legs may be modified for predation. The abdomen is 11-segmented, with the first tergite fused to the metathorax. The cerci have one or two segments. Larvae possess a heavily sclerotized head capsule, are mandibulate, and have compound eyes. Their thoracic segments are about equal; the short thoracic legs have a fused tibia and tarsus and a single claw. Prolegs usually occur on abdominal segments 1–8, and the (10th) terminal segment bears either paired hooks or a suction disc. The pupa (Fig. 6.7b) is immobile, exarate, and mandibulate.

The dietary habits of mecopterans vary among families, and often between adults and larvae within a family. The Bittacidae are predatory as adults but saprophagous as larvae; Panorpidae are scavengers, probably feeding mostly on dead arthropods, as both larvae and adults. Less is known of the diets of the other families but saprophagy and phytophagy, including moss-feeding, have been reported.

Copulation in certain mecopterans is preceded by elaborate courtship procedures that may involve nuptial feeding (Box 5.1). Oviposition sites vary, but known larval development is predominantly in moist litter, or aquatic in Gondwanan Nannochoristidae.

Phylogenetic relationships are considered in section 7.4.2 and depicted in Figs. 7.2 and 7.6.

Box 13.6 Strepsiptera

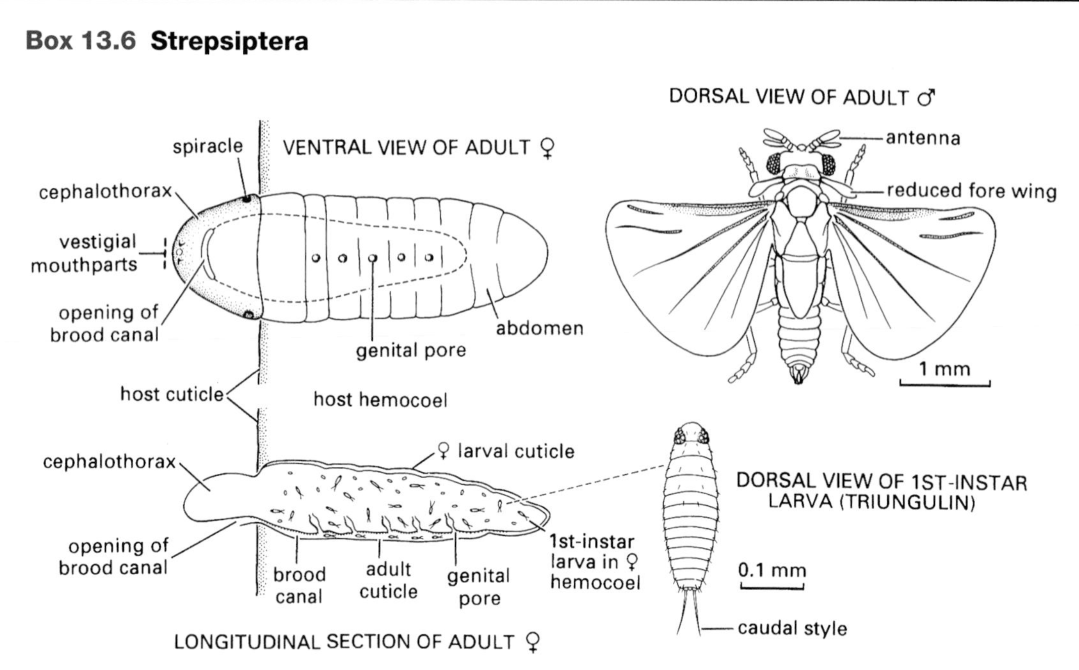

The Strepsiptera is an order of 500–550 species of highly modified endoparasites with extreme sexual dimorphism. The male (top right figure, after CSIRO 1970) has a large head and bulging eyes with few large facets, and no ocelli. Antennae of the male are flabellate or branched, with four to seven segments. The pro- and mesothorax are small; the fore wings stubby and without veins, and the hind wings broad, fan-shaped, and with few radiating veins. The legs lack trochanters and often also claws. An elongate metanotum overlies the anterior part of the tapering abdomen. The female is either coccoid-like or larviform, wingless, and usually retained in a pharate (cloaked) state, protruding from the host (as illustrated in ventral view and longitudinal section in the left figures, after Askew 1971). The triungulin (first-instar larva; bottom right figure) has three pairs of thoracic legs, but lacks antennae and mandibles; subsequent instars are maggot-like, lacking mouthparts or appendages. The pupa is exarate and adecticous, within a puparium formed from the final larval instar.

Strepsipterans are endoparasites of other insects, most commonly of Hemiptera and Hymenoptera. The host insects suffer morphological and physiological abnormalities and, although not killed prematurely, they rarely can reproduce. Strepsipteran eggs hatch within the mother, and active triungulins emerge via a brood canal (as shown here on the bottom left) and seek out a host, usually in its immature stage. In Stylopidae that parasitize hymenopterans, triungulins leave their host whilst on flowers, and from here seek a suitable adult bee or wasp to gain a ride to the nest, where they enter a host egg or larva.

Entry to the host is via the enzymatically softened cuticle, followed by an immediate molt to a maggot-like instar that develops as an endoparasite. The pupa protrudes from the host's body; the male emerges by pushing off a cephalothoracic cap, but the female remains within the cuticle. The virgin female releases pheromones to lure free-flying males, one of which copulates, inseminating through the brood canal on the female cephalothorax.

Eight families of Strepsiptera are recognized. Phylogenetic relationships of this order are controversial and are considered in section 7.4.2 and depicted in Fig. 7.2.

FURTHER READING

Askew, R.R. (1971) *Parasitic Insects.* Heinemann, London.

Beckage, N.E. (1998) Parasitoids and polydnaviruses. *Bio-Science* **48**, 305–11.

Byers, G.W. & Thornhill, R. (1983) Biology of Mecoptera. *Annual Review of Entomology* **28**, 303–28.

Eggleton, P. & Belshaw, R. (1993) Comparisons of dipteran, hymenopteran and coleopteran parasitoids: provisional phylogenetic explanations. *Biological Journal of the Linnean Society* **48**, 213–26.

Feener, D.H. Jr & Brown, B.V. (1997) Diptera as parasitoids. *Annual Review of Entomology* **42**, 73–97.

Fleming, J.G.W. (1992) Polydnaviruses: mutualists and pathogens. *Annual Review of Entomology* **37**, 401–25.

Gauld, I. & Bolton, B. (eds.) (1988) *The Hymenoptera.* British Museum (Natural History)/Oxford University Press, London.

Godfray, H.C.J. (1994) *Parasitoids: Behavioural and Evolutionary Ecology.* Princeton University Press, Princeton, NJ.

Halaj, J. & Wise, D.H. (2001) Terrestrial trophic cascades: how much do they trickle? *The American Naturalist* **157**, 262–81.

Hassell, M.P. & Southwood, T.R.E. (1978) Foraging strategies of insects. *Annual Review of Ecology and Systematics* **9**, 75–98.

Lyal, C.H.C. (1986) Coevolutionary relationships of lice and their hosts: a test of Fahrenholz's Rule. In: *Coevolution and Systematics* (eds. A.R. Stone & D.L. Hawksworth), pp. 77–91. Systematics Association, Oxford.

Marshall, A.G. (1981) *The Ecology of Ectoparasitic Insects.* Academic Press, London.

New, T.R. (1991) *Insects as Predators.* The New South Wales University Press, Kensington.

Quicke, D.L.J. (1997) *Parasitic Wasps.* Chapman & Hall, London.

Resh, V.H. & Cardé, R.T. (eds.) (2003) *Encyclopedia of Insects.* Academic Press, Amsterdam. [Particularly see articles on host seeking by parasitoids; Hymenoptera; hyperparasitism; parasitoids; predation and predatory insects.]

Schoenly, K., Cohen, J.E., Heong, K.L. et al. (1996) Food web dynamics of irrigated rice fields at five elevations in Luzon, Philippines. *Bulletin of Entomological Research* **86**, 451–66.

Stoltz, D. & Whitfield, J.B. (1992) Viruses and virus-like entities in the parasitic Hymenoptera. *Journal of Hymenoptera Research* **1**, 125–39.

Sullivan, D.J. (1987) Insect hyperparasitism. *Annual Review of Entomology* **32**, 49–70.

Symondson, W.O.C., Sunderland, K.D. & Greenstone, M.H. (2002) Can generalist predators be effective biocontrol agents? *Annual Review of Entomology* **47**, 561–94.

Vinson, S.B. (1984) How parasitoids locate their hosts: a case of insect espionage. In: *Insect Communication* (ed. T. Williams), pp. 325–48. Academic Press, London.

Whitfield, J.B. (1992) Phylogeny of the nonaculeate Apocrita and the evolution of parasitism in the Hymenoptera. *Journal of Hymenopteran Research* **1**, 3–14.

Whitfield, J.B. (1998) Phylogeny and evolution of host–parasitoid interactions in Hymenoptera. *Annual Review of Entomology* **43**, 129–51.

Chapter 14

INSECT DEFENSE

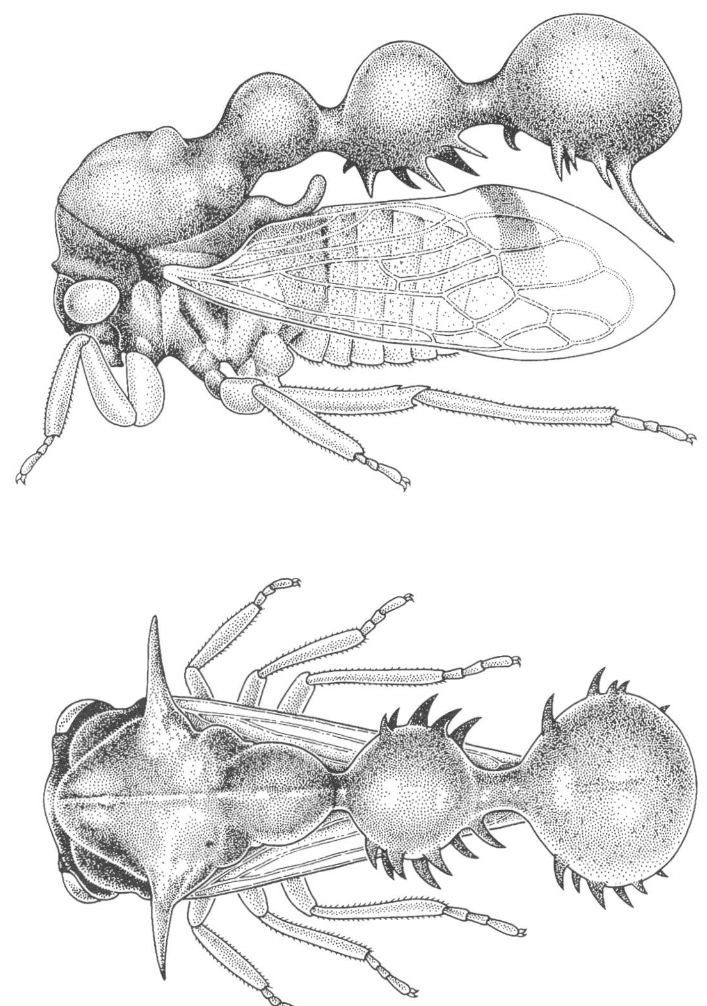

An African ant-mimicking membracid bug. (After Boulard 1968.)

Although some humans eat insects (section 1.6), many "western" cultures are reluctant to use them as food; this aversion extends no further than humans. For very many organisms, insects provide a substantial food source, because they are nutritious, abundant, diverse, and found everywhere. Some animals, termed **insectivores**, rely almost exclusively on a diet of insects; omnivores may eat them opportunistically; and many herbivores unavoidably consume insects. Insectivores may be vertebrates or invertebrates, including arthropods – insects certainly eat other insects. Even plants lure, trap, and digest insects; for example, pitcher plants (both New World Sarraceniaceae and Old World Nepenthaceae) digest arthropods, predominantly ants, in their fluid-filled pitchers (section 11.4.2), and the flypaper and Venus flytraps (Droseraceae) capture many flies. Insects, however, actively or passively resist being eaten, by means of a variety of protective devices – the insect defenses – which are the subject of this chapter.

A review of the terms discussed in Chapter 13 is appropriate. A predator is an animal that kills and consumes a number of prey animals during its life. Animals that live at the expense of another animal but do not kill it are parasites, which may live internally (endoparasites) or externally (ectoparasites). Parasitoids are those that live at the expense of one animal that dies prematurely as a result. The animal attacked by parasites or parasitoids is a host. All insects are potential prey or hosts to many kinds of predators (either vertebrate or invertebrate), parasitoids or, less often, parasites.

Many defensive strategies exist, including use of specialized morphology (as shown for the extraordinary, ant-mimicking membracid bug *Hamma rectum* from tropical Africa in the vignette of this chapter), behavior, noxious chemicals, and responses of the immune system. This chapter deals with aspects of defense that include death feigning, autotomy, crypsis (camouflage), chemical defenses, aposematism (warning signals), mimicry, and collective defensive strategies. These are directed against a wide range of vertebrates and invertebrates but, because much study has involved insects defending themselves against insectivorous birds, the role of these particular predators will be emphasized. Immunological defense against microorganisms is discussed in Chapter 3, and defenses used against parasitoids are considered in Chapter 13.

A useful framework for discussion of defense and predation can be based upon the time and energy inputs to the respective behaviors. Thus, hiding, escape by running or flight, and defense by staying and fighting involve increasing energy expenditure but diminishing costs in time expended (Fig. 14.1). Many insects will change to another strategy if the previous defense fails: the scheme is not clear-cut and it has elements of a continuum.

14.1 DEFENSE BY HIDING

Visual deception may reduce the probability of being found by a natural enemy. A well-concealed **cryptic** insect that either resembles its general background or an inedible (neutral) object may be said to "mimic" its surroundings. In this book, **mimicry** (in which an animal resembles another animal that is recognizable by natural enemies) is treated separately (section 14.5). However, crypsis and mimicry can be seen as similar in that both arise when an organism gains in fitness through developing a resemblance (to a neutral or animate object) evolved under selection. In all cases, it is

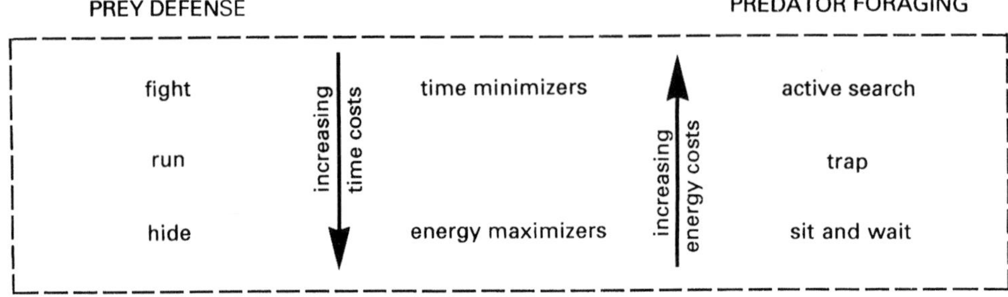

Fig. 14.1 The basic spectrum of prey defense strategies and predator foraging, varying according to costs and benefits in both time and energy. (After Malcolm 1990.)

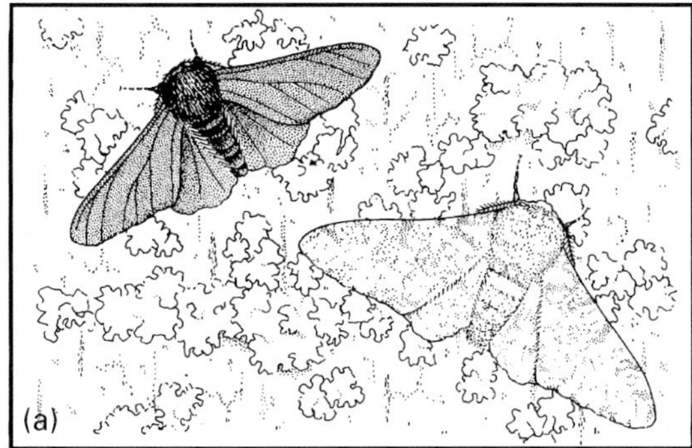

Fig. 14.2 Pale and melanic (*carbonaria*) morphs of the peppered moth *Biston betularia* resting on: (a) pale, lichen-covered; and (b) dark trunks.

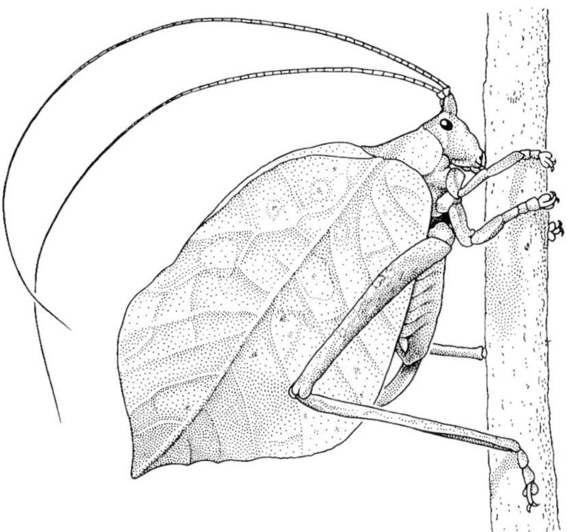

Fig. 14.3 A leaf-mimicking katydid, *Mimetica mortuifolia* (Orthoptera: Tettigoniidae), in which the fore wing resembles a leaf even to the extent of leaf-like venation and spots resembling fungal mottling. (After Belwood 1990.)

assumed that such defensive adaptive resemblance is under selection by predators or parasitoids, but, although maintenance of selection for accuracy of resemblance has been demonstrated for some insects, the origin can only be surmised.

Insect crypsis can take many forms. The insect may adopt **camouflage**, making it difficult to distinguish from the general background in which it lives, by:

- resembling a uniform colored background, such as a green geometrid moth on a leaf;
- resembling a patterned background, such as a mottled moth on tree bark (Fig. 14.2; see also Plate 6.1, facing p. 14);
- being countershaded – light below and dark above – as in some caterpillars and aquatic insects;
- having a pattern to disrupt the outline, as is seen in many moths that settle on leaf litter;
- having a bizarre shape to disrupt the silhouette, as demonstrated by some membracid leafhoppers.

In another form of crypsis, termed **masquerade** or **mimesis** to contrast with the camouflage described above, the organism deludes a predator by resembling an object that is a particular specific feature of its environment, but is of no inherent interest to a predator. This feature may be an inanimate object, such as the bird dropping resembled by young larvae of some butterflies such as *Papilio aegeus* (Papilionidae), or an animate but neutral object – for example, "looper" caterpillars (the larvae of geometrid moths) resemble twigs, some membracid bugs imitate thorns arising from a stem, and many stick-insects look very much like sticks and may even move like a twig in the wind. Many insects, notably amongst the lepidopterans and orthopteroids, resemble leaves, even to the similarity in venation (Fig. 14.3), and appearing to be dead or alive, mottled with fungus, or even partially eaten as if by a herbivore. However, interpretation of apparent resemblance to inanimate objects as simple crypsis may be revealed as more complex when subject to experimental manipulation (Box 14.2).

Crypsis is a very common form of insect concealment, particularly in the tropics and amongst nocturnally active insects. It has low energetic costs but relies on the insect being able to select the appropriate background. Experiments with two differently colored

Box 14.1 Avian predators as selective agents for insects

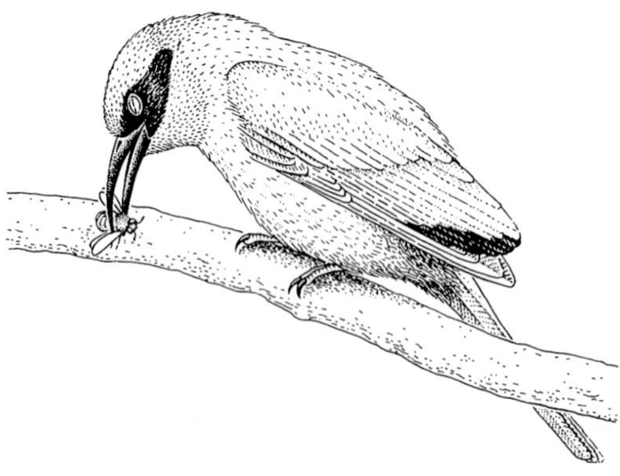

Henry Bates, who was first to propose a theory for mimicry, suggested that natural enemies such as birds selected among different prey such as butterflies, based upon an association between mimetic patterns and unpalatability. A century later, Henry Kettlewell argued that selective predation by birds on the peppered moth (Geometridae: *Biston betularia*) altered the proportions of dark- and light-colored morphs (Fig. 14.2) according to their concealment (crypsis) on natural and industrially darkened trees upon which the moths rested by day. Amateur lepidopterists recorded that the proportion of the dark ("melanic") *carbonaria* form dramatically increased as industrial pollution increased in northern England from the mid-19th century. Elimination of pale lichen on tree trunk resting areas was suggested to have made normal pale morphs more visible against the sooty, lichen-denuded trunks (as shown in Fig. 14.2b), and hence they were more susceptible to visual recognition by bird predators. This phenomenon, termed **industrial melanism**, often has been cited as a classic example of evolution through natural selection.

The peppered moth/avian predation story has been challenged for its experimental design and procedures, and biased interpretation. The case depended upon:

- birds being the major predators rather than night-flying, pattern-insensitive bats;
- moths resting "exposed" on trunks rather than under branches or in the canopy;
- dark and pale morphs favoring the cryptic background appropriate to their patterning;
- crypsis to the human eye being quantifiable and equating to that for moth-feeding birds;
- selection being concentrated in the adult stage of the moth's life cycle;
- genes responsible for origination of melanism acting in a particular way, and with very high levels of selection.

None of these components have been confirmed. Evolution undoubtedly has taken place. The proportions of dark morphs (alleles for melanism) have changed through time, increasing with industrialization, and reducing as "post-industrial" air quality improves. However, the centrality of avian predation acting as a force for natural selection in *B. betularia* is no longer so evident.

More convincing is the demonstration of directly observed predation, and inference from beak pecks on the wings of butterflies and from experiments with color-manipulated daytime-flying moths. Thus, winter-roosting monarch butterflies (*Danaus plexippus*) are fed upon by black-backed orioles (Icteridae), which browse selectively on poorly defended individuals, and by black-headed grosbeaks (Fringillidae), which appear to be completely insensitive to the toxins. Specialized predators such as Old World bee-eaters (Meropidae) and neotropical jacamars (Galbulidae) can deal with the stings of hymenopterans (the red-throated bee-eater, *Merops bullocki*, is shown here de-stinging a bee on a branch, after Fry et al. 1992) and the toxins of butterflies, respectively. A similar suite of birds selectively feeds on noxious ants. The ability of these specialist predators to distinguish between varying pattern and edibility may make them selective agents in the evolution and maintenance of defensive mimicry.

Birds are observable insectivores for laboratory studies: their readily recognizable behavioral responses to unpalatable foods include head-shaking, disgorging of food, tongue-extending, bill-wiping, gagging, squawking, and ultimately vomiting. For many birds, a single learning trial with noxious (Class I) chemicals appears to lead to long-term aversion to the particular insect, even with a substantial delay between feeding and illness. However, manipulative studies of bird diets are complicated by their fear of novelty (neophobia), which, for example, can lead to rejection of prey with startling and frightening displays (section 13.2). Conversely, birds rapidly learn preferred items, as in Kettlewell's experiments in which birds quickly recognized both *Biston betularia* morphs on tree trunks in his artificial set-up.

Perhaps no insect has completely escaped the attentions of predators and some birds can overcome even severe insect defenses. For example, the lubber grasshopper (Acrididae: *Romalea guttata*) is large, gregarious, and aposematic, and if attacked it squirts volatile, pungent chemicals accompanied by a hissing noise. The lubber is extremely toxic and is avoided by all lizards and birds except one, the loggerhead shrike (Laniidae: *Lanius ludovicanus*), which snatches its prey, including lubbers, and impales them "decoratively" upon spikes with minimal handling time. These impaled items serve both as food stores and in sexual or territorial displays. *Romalea*, which are emetic to shrikes when fresh, become edible after two days of lardering, presumably by denaturation of the toxins. The impaling behavior shown by most species of shrikes thus is preadaptive in permitting the loggerhead to feed upon an extremely well-defended insect. No matter how good the protection, there is no such thing as total defense in the arms race between prey and predator.

morphs of *Mantis religiosa* (Mantidae), the European praying mantid, have shown that brown and green morphs placed against appropriate and inappropriate colored backgrounds were fed upon in a highly selective manner by birds: they removed all "mismatched" morphs and found no camouflaged ones. Even if the correct background is chosen, it may be necessary to orientate correctly: moths with disruptive outlines or with striped patterns resembling the bark of a tree may be concealed only if orientated in a particular direction on the trunk.

The Indomalayan orchid mantid, *Hymenopus coronatus* (Hymenopodidae), blends beautifully with the pink flower spike of an orchid, where it sits awaiting prey. The crypsis is enhanced by the close resemblance of the femora of the mantid's legs to the flower's petals. Crypsis enables the mantid to avoid detection by its potential prey (flower visitors) (section 13.1.1) as well as conceal itself from predators.

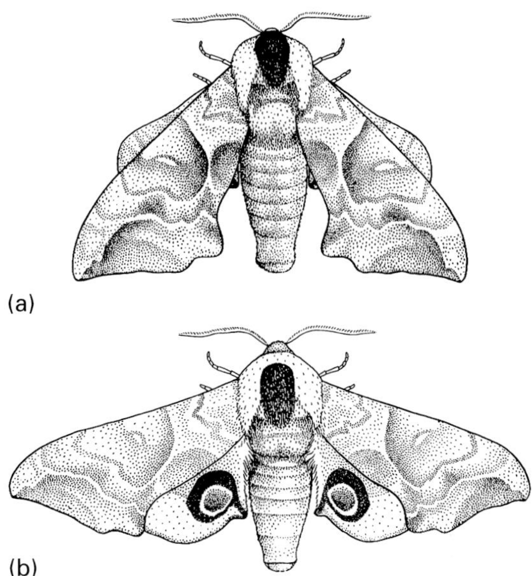

Fig. 14.4 The eyed hawkmoth, *Smerinthus ocellatus* (Lepidoptera: Sphingidae). (a) The brownish fore wings cover the hind wings of a resting moth. (b) When the moth is disturbed, the black and blue eyespots on the hind wings are revealed. (After Stanek 1977.)

14.2 SECONDARY LINES OF DEFENSE

Little is known of the learning processes of inexperienced vertebrate predators, such as insectivorous birds. However, studies of the gut contents of birds show that cryptic insects are not immune from predation (Box 14.1). Once found for the first time (perhaps accidentally), birds subsequently seem able to detect cryptic prey via a "search image" for some element(s) of the pattern. Thus, having discovered that some twigs were caterpillars, American blue jays were observed to continue to peck at sticks in a search for food. Primates can identify stick-insects by one pair of unfolded legs alone, and will attack actual sticks to which phasmatid legs have been affixed experimentally. Clearly, subtle cues allow specialized predators to detect and eat cryptic insects.

Once the deception is discovered, the insect prey may have further defenses available in reserve. In the energetically least demanding response, the initial crypsis may be exaggerated, as when a threatened masquerader falls to the ground and lies motionless. This behavior is not restricted to cryptic insects: even visually obvious prey insects may feign death (**thanatosis**). This behavior, used by many beetles (particularly weevils), can be successful, as predators lose interest in apparently dead prey or may be unable to locate a motionless insect on the ground. Another secondary line of defense is to take flight and suddenly reveal a flash of conspicuous color from the hind wings. Immediately on landing the wings are folded, the color vanishes, and the insect is cryptic once more. This behavior is common amongst certain orthopterans and underwing moths; the color of the flash may be yellow, red, purple, or, rarely, blue.

A third type of behavior of cryptic insects upon discovery by a predator is the production of a **startle display**. One of the commonest is to open the fore wings and reveal brightly colored "eyes" that are usually concealed on the hind wings (Fig. 14.4). Experiments using birds as predators have shown that the more perfect the eye (with increased contrasting rings to resemble true eyes), the better the deterrence. Not all eyes serve to startle: perhaps a rather poor imitation of an eye on a wing may direct pecks from a predatory bird to a nonvital part of the insect's anatomy.

An extraordinary type of insect defense is the convergent appearance of part of the body to a feature of a vertebrate, albeit on a much smaller scale. Thus, the head of a species of fulgorid bug, commonly called the alligator bug, bears an uncanny resemblance to that of a caiman. The pupa of a particular lycaenid butterfly looks like a monkey head. Some tropical sphingid larvae assume a threat posture which, together with false eyespots that actually lie on the abdomen, gives a snake-like impression. Similarly, the caterpillars of certain swallowtail butterflies bear a likeness to a snake's head (see Plate 5.7). These resemblances may deter predators (such as birds that search by "peering about") by their startle effect, with the incorrect scale of the mimic being overlooked by the predator.

14.3 MECHANICAL DEFENSES

Morphological structures of predatory function, such as the modified mouthparts and spiny legs described in Chapter 13, also may be defensive, especially if a fight ensues. Cuticular horns and spines may be used in deterrence of a predator or in combating rivals for mating, territory, or resources, as in *Onthophagus* dung beetles (section 5.3). For ectoparasitic insects, which are vulnerable to the actions of the host, body shape and sclerotization provide one line of defense. Fleas are laterally compressed, making these insects difficult to dislodge from host hairs. Biting lice are flattened dorsoventrally, and are narrow and elongate, allowing them to fit between the veins of feathers, secure from preening by the host bird. Furthermore, many ectoparasites have resistant bodies, and the heavily sclerotized cuticle of certain beetles must act as a mechanical antipredator device.

Many insects construct retreats that can deter a predator that fails to recognize the structure as containing anything edible or that is unwilling to eat inorganic material. The cases of caddisfly larvae (Trichoptera), constructed of sand grains, stones, or organic fragments (Fig. 10.6), may have originated in response to the physical environment of flowing water, but certainly have a defensive role. Similarly, a portable case of vegetable material bound with silk is constructed by the terrestrial larvae of bagworms (Lepidoptera: Psychidae). In both caddisflies and psychids, the case serves to protect during pupation. Certain insects construct artificial shields; for example, the larvae of certain chrysomelid beetles decorate themselves with their feces. The larvae of certain lacewings and reduviid bugs cover themselves with lichens and detritus and/or the sucked-out carcasses of their insect prey, which can act as barriers to a predator, and also may disguise themselves from prey (Box 14.2).

The waxes and powders secreted by many hemipterans (such as scale insects, woolly aphids, whiteflies, and fulgorids) may function to entangle the mouthparts of a potential arthropod predator, but also may have a waterproofing role. The larvae of many ladybird beetles (Coccinellidae) are coated with white wax, thus resembling their mealybug prey. This may be a disguise to protect them from ants that tend the mealybugs.

Body structures themselves, such as the scales of moths, caddisflies, and thrips, can protect as they detach readily to allow the escape of a slightly denuded insect from the jaws of a predator, or from the sticky threads of spiders' webs or the glandular leaves of insectivorous plants such as the sundews. A mechanical defense that seems at first to be maladaptive is **autotomy**, the shedding of limbs, as demonstrated by stick-insects (Phasmatodea) and perhaps crane flies (Diptera: Tipulidae). The upper part of the phasmatid leg has the trochanter and femur fused, with no muscles running across the joint. A special muscle breaks the leg at a weakened zone in response to a predator grasping the leg. Immature stick-insects and mantids can regenerate lost limbs at molting, and even certain autotomized adults can induce an adult molt at which the limb can regenerate.

Secretions of insects can have a mechanical defensive role, acting as a glue or slime that ensnares predators or parasitoids. Certain cockroaches have a permanent slimy coat on the abdomen that confers protection. Lipid secretions from the **cornicles** (also called **siphunculi**) of aphids may gum-up predator mouthparts or small parasitic wasps. Termite soldiers have a variety of secretions available to them in the form of cephalic glandular products, including terpenes that dry on exposure to air to form a resin. In *Nasutitermes* (Termitidae) the secretion is ejected via the nozzle-like nasus (a pointed snout or rostrum) as a quick-drying fine thread that impairs the movements of a predator such as an ant. This defense counters arthropod predators but is unlikely to deter vertebrates. Mechanical-acting chemicals are only a small selection of the total insect armory that can be mobilized for chemical warfare.

14.4 CHEMICAL DEFENSES

Chemicals play vital roles in many aspects of insect behavior. In Chapter 4 we considered the use of pheromones in many forms of communication, including alarm pheromones elicited by the presence of a

Box 14.2 Backpack bugs – dressed to kill?

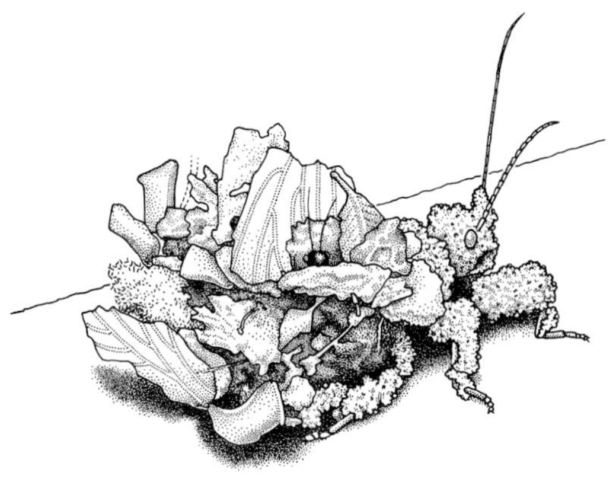

Certain West African predatory assassin bugs (Hemiptera: Reduviidae) decorate themselves with a coat of dust which they adhere to their bodies with sticky secretions from abdominal setae. To this undercoat, the nymphal instars (of several species) add vegetation and cast skins of prey items, mainly ants and termites. The resultant "backpack" of trash can be much larger than the animal itself (as in this illustration derived from a photograph by M. Brandt). It had been assumed that the bugs are mistaken by their predators or prey for an innocuous pile of debris; but rather few examples of such deceptive camouflage have been tested critically.

In the first behavioral experiment, investigators Brandt and Mahsberg (2002) exposed bugs to predators typical of their surroundings, namely spiders, geckos, and centipedes. Three groups of bugs were tested experimentally: naturally occurring ones with dustcoat and backpack, individuals only with a dustcoat, and naked ones lacking both dustcoat and backpack. Bug behavior was unaffected, but the predators' reactions varied: spiders were slower to capture the individuals with backpacks than individuals of the other two groups; geckos also were slower to attack backpack wearers; and centipedes never attacked backpackers although they ate most of the nymphs without backpacks. The implied anti-predatory protection certainly includes some visual disguise, but only the gecko is a visual predator: spiders are tactile predators, and centipedes hunt using chemical and tactile cues. Backpacks are conspicuous more than cryptic, but they confuse visual, tactile, and chemical-orientating predators by looking, feeling, and smelling wrong for a prey item.

Next, differently dressed bugs and their main prey, ants, were manipulated. Studied ants responded to individual naked bugs much more aggressively than they did to dustcoated or backpack-bearing nymphs. The backpack did not diminish the risk of hostile response (taken as equating to "detection") beyond that to the dustcoat alone, rejecting any idea that ants may be lured by the odor of dead conspecifics included in the backpack. One trialed prey item, an army ant, is highly aggressive but blind and although unable to detect the predator visually, it responded as did other prey ants – with aggression directed preferentially towards naked bugs. Evidently, any covering confers "concealment", but not by the visual protective mechanism assumed previously.

Thus, what appeared to be simple visual camouflage proved more a case of disguise to fool chemical- and touch-sensitive predators and prey. Additional protection is provided by the bugs' abilities to shed their backpacks – while collecting research specimens, Brandt and Mahsberg observed that bugs readily vacated their backpacks in an inexpensive autotomy strategy resembling the metabolically expensive lizard tail-shedding. Such experimental research undoubtedly will shed more light on other cases of visual camouflage/predator deception.

predator. Similar chemicals, called allomones, that benefit the producer and harm the receiver, play important roles in the defenses of many insects, notably amongst many Heteroptera and Coleoptera. The relationship between defensive chemicals and those used in communication may be very close, sometimes with the same chemical fulfilling both roles. Thus, a noxious chemical that repels a predator can alert conspecific insects to the predator's presence and may act as a stimulus to action. In the energy–time dimensions shown in Fig. 14.1, chemical defense lies towards the energetically expensive but time-efficient end of the spectrum. Chemically defended insects tend to have high **apparency** to predators, i.e. they are usually non-cryptic, active, often relatively large, long-lived, and frequently aggregated or social in behavior. Often they signal their distastefulness by **aposematism** – warning signaling that often involves bold coloring (see

Plates 5.6 & 6.2) but may include odor, or even sound or light production.

14.4.1 Classification by function of defensive chemicals

Amongst the diverse range of defensive chemicals produced by insects, two classes of compounds can be distinguished by their effects on a predator. Class I defensive chemicals are noxious because they irritate, hurt, poison, or drug individual predators. Class II chemicals are innocuous, being essentially anti-feedant chemicals that merely stimulate the olfactory and gustatory receptors, or aposematic indicator odors. Many insects use mixtures of the two classes of chemicals and, furthermore, Class I chemicals in low concentrations may give Class II effects. Contact by a predator with Class I compounds results in repulsion through, for example, emetic (sickening) properties or induction of pain, and if this unpleasant experience is accompanied by odorous Class II compounds, predators learn to associate the odor with the encounter. This conditioning results in the predator learning to avoid the defended insect at a distance, without the dangers (to both predator and prey) of having to feel or taste it.

Class I chemicals include both **immediate**-acting substances, which the predator experiences through handling the prey insect (which may survive the attack), and chemicals with **delayed**, often systemic, effects including vomiting or blistering. In contrast to immediate-effect chemicals sited in particular organs and applied topically (externally), delayed-effect chemicals are distributed more generally within the insect's tissues and hemolymph, and are tolerated systemically. Whereas a predator evidently learns rapidly to associate immediate distastefulness with particular prey (especially if it is aposematic), it is unclear how a predator identifies the cause of nausea some time after the predator has killed and eaten the toxic culprit, and what benefits this action brings to the victim. Experimental evidence from birds shows that at least these predators are able to associate a particular food item with a delayed effect, perhaps through taste when regurgitating the item. Too little is known of feeding in insects to understand if this applies similarly to predatory insects. Perhaps a delayed poison that fails to protect an individual from being eaten evolved through the education of a predator by a sacrifice, thereby allowing differential survival of relatives (section 14.6).

14.4.2 The chemical nature of defensive compounds

Class I compounds are much more specific and effective against vertebrate than arthropod predators. For example, birds are more sensitive than arthropods to toxins such as cyanides, cardenolides, and alkaloids. Cyanogenic glycosides are produced by zygaenid moths (Zygaenidae), *Leptocoris* bugs (Rhopalidae), and *Acraea* and *Heliconius* butterflies (Nymphalidae). Cardenolides are very prevalent, occurring notably in monarch or wanderer butterflies (Nymphalidae), certain cerambycid and chrysomelid beetles, lygaeid bugs, pyrgomorphid grasshoppers, and even an aphid. A variety of alkaloids similarly are acquired convergently in many coleopterans and lepidopterans.

Possession of Class I emetic or toxic chemicals is very often accompanied by aposematism, particularly coloration directed against visual-hunting diurnal predators. However, visible aposematism is of limited use at night, and the sounds emitted by nocturnal moths, such as certain Arctiidae when challenged by bats, may be aposematic, warning the predator of a distasteful meal. Furthermore, it seems likely that the bioluminescence emitted by certain larval beetles (Phengodidae, and Lampyridae and their relatives; section 4.4.5) is an aposematic warning of distastefulness.

Class II chemicals tend to be volatile and reactive organic compounds with low molecular weight, such as aromatic ketones, aldehydes, acids, and terpenes. Examples include the stink-gland products of Heteroptera and the many low molecular weight substances, such as formic acid, emitted by ants. Bitter-tasting but non-toxic compounds such as quinones are common Class II chemicals. Many defensive secretions are complex mixtures that can involve synergistic effects. Thus, the carabid beetle *Heluomorphodes* emits a Class II compound, formic acid, that is mixed with *n*-nonyl acetate, which enhances skin penetration of the acid giving a Class I painful effect.

The role of these Class II chemicals in aposematism, warning of the presence of Class I compounds, was considered above. In another role, these Class II chemicals may be used to deter predators such as ants that rely on chemical communication. For example, prey such as certain termites, when threatened by predatory ants, release mimetic ant alarm pheromones, thereby inducing inappropriate ant behaviors of panic and nest defense. In another case, ant-nest inquilines, which might provide prey to their host ants, are unrecognized

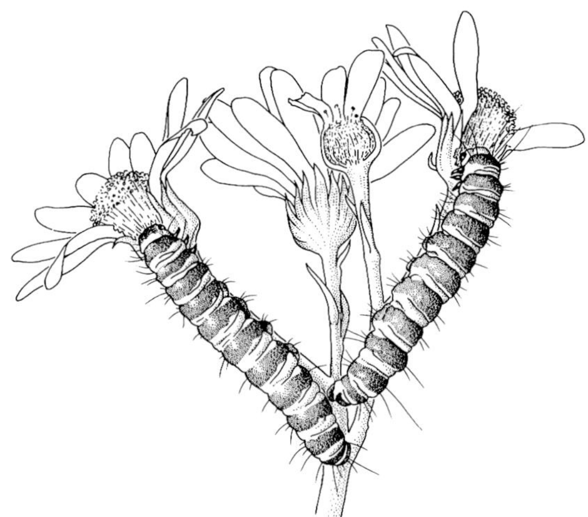

Fig. 14.5 The distasteful and warningly colored caterpillars of the cinnabar moth, *Tyria jacobaeae* (Lepidoptera: Arctiidae), on ragwort, *Senecio jacobaeae*. (After Blaney 1976.)

as potential food because they produce chemicals that appease ants.

Class II compounds alone appear unable to deter many insectivorous birds. For example, blackbirds (Turdidae) will eat notodontid (Lepidoptera) caterpillars that secrete a 30% formic acid solution; many birds actually encourage ants to secrete formic acid into their plumage in an apparent attempt to remove ectoparasites (so-called "anting").

14.4.3 Sources of defensive chemicals

Many defensive chemicals, notably those of phytophagous insects, are derived from the host plant upon which the larvae (Fig. 14.5; Box 14.3) and, less commonly, the adults feed. Frequently, a close association is observed between restricted host-plant use (monophagy or oligophagy) and the possession of a chemical defense. An explanation may lie in a coevolutionary "arms race" in which a plant develops toxins to deter phytophagous insects. A few phytophages overcome the defenses and thereby become specialists able to detoxify or sequester the plant toxins. These specialist herbivores can recognize their preferred host plants, develop on them, and use the plant toxins (or metabolize them to closely related compounds) for their own defense.

Although some aposematic insects are closely associated with toxic food plants, certain insects can produce their own toxins. For example, amongst the Coleoptera, blister beetles (Meloidae) synthesize cantharidin, jewel beetles (Buprestidae) make buprestin, and some leaf beetles (Chrysomelidae) can produce cardiac glycosides. The very toxic staphylinid *Paederus* synthesizes its own blistering agent, paederin. Many of these chemically defended beetles are aposematic (e.g. Coccinellidae, Meloidae) and will **reflex-bleed** their hemolymph from the femoro-tibial leg joints if handled (see Plate 6.3). Experimentally, it has been shown that certain insects that sequester cyanogenic compounds from plants can still synthesize similar compounds if transferred to toxin-free host plants. If this ability preceded the evolutionary transfer to the toxic host plant, the possession of appropriate biochemical pathways may have preadapted the insect to using them subsequently in defense.

A bizarre means of obtaining a defensive chemical is used by *Photurus* fireflies (Lampyridae). Many fireflies synthesize deterrent lucibufagins, but *Photurus* females cannot do so. Instead they mimic the flashing sexual signal of *Photinus* females, thus luring male *Photinus* fireflies, which they eat to acquire their defensive chemicals.

Defensive chemicals, either manufactured by the insect or obtained by ingestion, may be transmitted between conspecific individuals of the same or a different life stage. Eggs may be especially vulnerable to natural enemies because of their immobility and it is not surprising that some insects endow their eggs with chemical deterrents (Box 14.3). This phenomenon may be more widespread among insects than is recognized currently.

14.4.4 Organs of chemical defense

Endogenous defensive chemicals (those synthesized within the insect) generally are produced in specific glands and stored in a reservoir (Box 14.4). Release is through muscular pressure or by evaginating the organ, rather like turning the fingers of a glove inside-out. The Coleoptera have developed a wide range of glands, many eversible, that produce and deliver defensive chemicals. Many Lepidoptera use urticating (itching) hairs and spines to inject venomous chemicals into a predator. Venom injection by social insects is covered in section 14.6.

Box 14.3 Chemically protected eggs

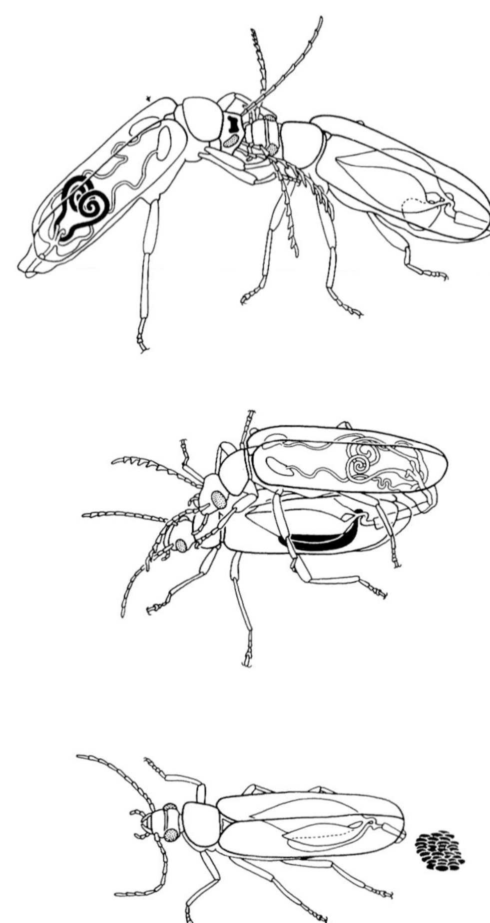

Some insect eggs can be protected by parental provisioning of defensive chemicals, as seen in certain arctiid moths and some butterflies. Pyrrolizidine alkaloids from the larval food plants are passed by the adult males to the females via seminal secretions, and the females transmit them to the eggs, which become distasteful to predators. Males advertise their possession of the defensive chemicals via a courtship pheromone derived from, but different to, the acquired alkaloids. In at least two of these lepidopteran species, it has been shown that males are less successful in courtship if deprived of their alkaloid.

Amongst the Coleoptera, certain species of Meloidae and Oedemeridae can synthesize cantharidin and others, particularly species of Anthicidae and Pyrochroidae, can sequester it from their food. Cantharidin ("Spanish fly") is a sesquiterpene with very high toxicity due to its inhibition of protein phosphatase, an important enzyme in glycogen metabolism. The chemical is used for egg-protective purposes, and certain males transmit this chemical to the female during copulation. In *Neopyrochroa flabellata* (Pyrochroidae) males ingest exogenous cantharidin and use it both as a precopulatory "enticing" agent and as a nuptial gift. During courtship, the female samples cantharidin-laden secretions from the male's cephalic gland (as in the top illustration, after Eisner et al. 1996a,b) and will mate with cantharidin-fed males but reject males devoid of cantharidin. The male's glandular offering represents only a fraction of his systemic cantharidin; much of the remainder is stored in his large accessory gland and passed, presumably with the spermatophore, to the female during copulation (as shown in the middle illustration). Eggs are impregnated with cantharidin (probably in the ovary) and, after oviposition, egg batches (bottom illustration) are protected from coccinellids and probably also other predators such as ants and carabid beetles.

An unsolved question is where do the males of *N. flabellata* acquire their cantharidin from under natural conditions? They may feed on adults or eggs of the few insects that can manufacture cantharidin and, if so, might *N. flabellata* and other cantharidiphilic insects (including certain bugs, flies, and hymenopterans, as well as beetles) be selective predators on each other?

In contrast to these endogenous chemicals, exogenous toxins, derived from external sources such as foods, are usually incorporated in the tissues or the hemolymph. This makes the complete prey unpalatable, but requires the predator to test at close range in order to learn, in contrast to the distant effects of many endogenous compounds. However, the larvae of some swallowtail butterflies (Papilionidae) that feed upon distasteful food plants concentrate the toxins and secrete them into a thoracic pouch called an **osmeterium**, which is everted if the larvae are touched. The color of the osmeterium often is aposematic and rein-

Box 14.4 Insect binary chemical weapons

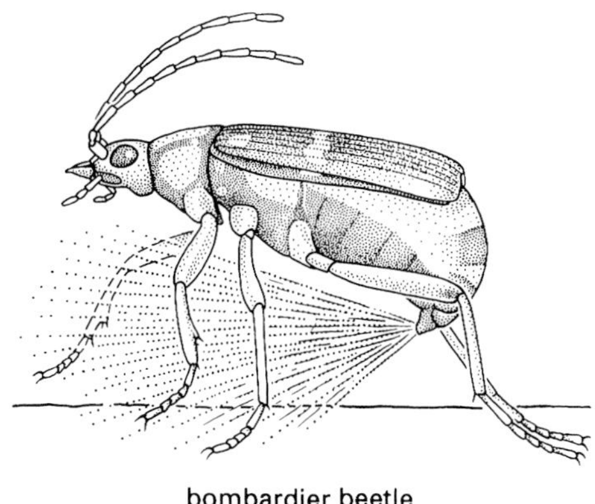

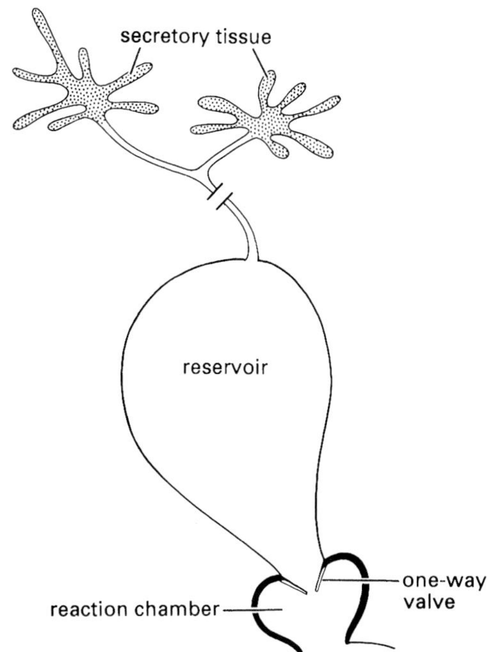

The common name of bombardier beetles (Carabidae: including genus *Brachinus*) derives from observations of early naturalists that the beetles released volatile defensive chemicals that appeared like a puff of smoke, accompanied by a "popping" noise resembling gunfire. The spray, released from the anus and able to be directed by the mobile tip of the abdomen, contains *p*-benzoquinone, a deterrent of vertebrate and invertebrate predators. This chemical is not stored, but when required is produced explosively from components held in paired glands. Each gland is double, comprising a muscular-walled compressible inner chamber containing a reservoir of hydroquinones and hydrogen peroxide, and a thick-walled outer chamber containing oxidative enzymes. When threatened, the beetle contracts the reservoir, and releases the contents through the newly opened inlet valve into the reaction chamber. Here an exothermic reaction takes place, resulting in the liberation of *p*-benzoquinone at a temperature of 100°C.

Studies on a Kenyan bombardier beetle, *Stenaptinus insignis* (illustrated here, after Dean et al. 1990) showed that the discharge is pulsed: the explosive chemical oxidation produces a build-up of pressure in the reaction chamber, which closes the one-way valve from the reservoir, thereby forcing discharge of the contents through the anus (as shown by the beetle directing its spray at an antagonist in front of it). This relieves the pressure, allowing the valve to open, permitting refilling of the reaction chamber from the reservoir (which remains under muscle pressure). Thus, the explosive cycle continues. By this mechanism a high-intensity pulsed jet is produced by the chemical reaction, rather than requiring extreme muscle pressure. Humans discovered the principles independently and applied them to engineering (as pulse jet propulsion) some millions of years after the bombardier beetles developed the technique!

forces the deterrent effect on a predator (Fig. 14.6). Larval sawflies (Hymenoptera: Pergidae), colloquially called "spitfires", store eucalypt oils, derived from the leaves that they eat, within a diverticulum of their fore gut and ooze this strong-smelling, distasteful fluid from their mouths when disturbed (Fig. 14.7).

14.5 DEFENSE BY MIMICRY

The theory of **mimicry**, an interpretation of the close resemblances of unrelated species, was an early application of the theory of Darwinian evolution. Henry Bates, a naturalist studying in the Amazon in

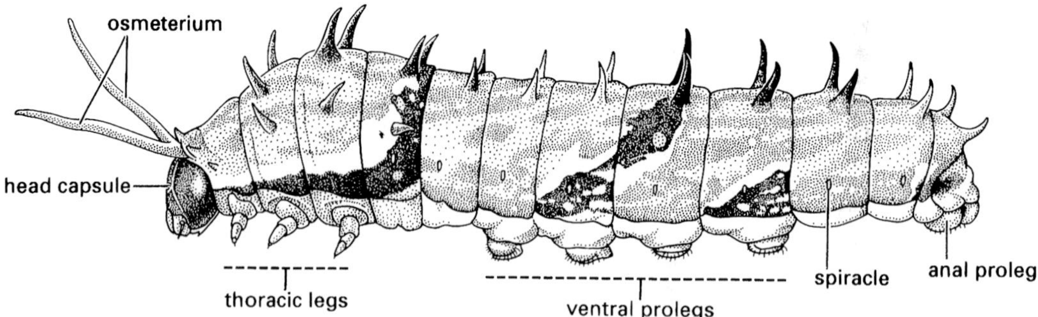

Fig. 14.6 A caterpillar of the orchard butterfly, *Papilio aegeus* (Lepidoptera: Papilionidae), with the osmeterium everted behind its head. Eversion of this glistening, bifid organ occurs when the larva is disturbed and is accompanied by a pungent smell.

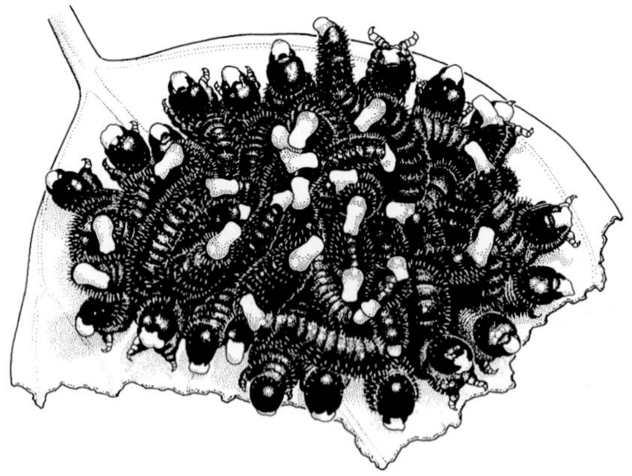

Fig. 14.7 An aggregation of sawfly larvae (Hymenoptera: Pergidae: *Perga*) on a eucalypt leaf. When disturbed, the larvae bend their abdomens in the air and exude droplets of sequestered eucalypt oil from their mouths.

the mid-19th century, observed that many similar butterflies, all slow-flying and brightly marked, seemed to be immune from predators. Although many species were common and related to each other, some were rare, and belonged to fairly distantly related families (see Plate 6.4). Bates believed that the common species were chemically protected from attack, and this was advertised by their aposematism – high apparency (behavioral conspicuousness) through bright color and slow flight. The rarer species, he thought, probably were not distasteful, but gained protection by their superficial resemblance to the protected ones. On reading the views that Charles Darwin had proposed newly in 1859, Bates realized that his own theory of mimicry involved evolution through natural selection. Poorly protected species gain increased protection from predation by differential survival of subtle variants that more resembled protected species in appearance, smell, taste, feel, or sound. The **selective agent** is the predator, which preferentially eats the inexact mimic. Since that time, mimicry has been interpreted in the light of evolutionary theory, and studies of insects, particularly butterflies, have remained central to mimicry theory and manipulation.

An understanding of the defensive systems of mimicry (and crypsis; section 14.1) can be gained by recognizing three basic components: the **model**, the **mimic**, and an **observer** that acts as a selective agent. These components are related to each other through signal generating and receiving systems, of which the basic association is the warning signal given by the model (e.g. aposematic color that warns of a sting or bad taste) and perceived by the observer (e.g. a hungry predator). The naïve predator must associate aposematism and consequent pain or distaste. When learnt, the predator subsequently will avoid the model. The model clearly benefits from this coevolved system, in which the predator can be seen to gain by not wasting time and energy chasing inedible prey.

Once such a mutually beneficial system has evolved, it is open to manipulation by others. The third component is the mimic: an organism that parasitizes the signaling system through deluding the observer, for example by false warning coloration. If this provokes a reaction from the observer similar to true aposematic coloration, the mimic is dismissed as unacceptable food. It is important to realize that the mimic need not be perfect, but only must elicit the appropriate avoidance response from the observer. Thus, only a limited subset of the signals given by the model may be required. For example, the black and yellow banding of venomous wasps is an aposematic color pattern that is displayed by countless species from amongst many orders of insects. The exactness of the match, at least to our eyes, varies considerably. This may be due to subtle differences between several different venomous models, or it may reflect the inability of the observer to discriminate: if only yellow and black banding is required to deter a predator there may be little or no selection to refine the mimicry more fully.

14.5.1 Batesian mimicry

In these mimicry triangles, each component has a positive or negative effect on each of the others. In **Batesian mimicry** an aposematic inedible model has an edible mimic. The model suffers by the mimic's presence because the aposematic signal aimed at the observer is diluted as the chances increase that the observer will taste an edible individual and fail to learn the association between aposematism and distastefulness. The mimic gains both from the presence of the protected model and the deception of the observer. As the mimic's presence disadvantages the model, interaction with the model is negative. The observer benefits by avoiding the noxious model, but misses a meal through failing to recognize the mimic as edible.

These Batesian mimicry relationships hold up only if the mimic remains relatively rare. However, should the model decline or the mimic become abundant, then the protection given to the mimic by the model will wane because the naïve observer increasingly encounters and tastes edible mimics. Evidently, some palatable butterfly mimics adopt different models throughout their range. For example, the mocker swallowtail, *Papilio dardanus*, is highly polymorphic with up to five mimetic morphs in Uganda (central Africa) and several more throughout its wide range. This polymorphism allows a larger total population of *P. dardanus* without prejudicing (by dilution) the successful mimetic system, as each morph can remain rare relative to its Batesian model. In this case, and for many other mimetic polymorphisms, males retain the basic color pattern of the species and only amongst females in some populations does mimicry of such a variety of models occur. The conservative male pattern may result from sexual selection to ensure recognition of the male by conspecific females of all morphs for mating, or by other conspecific males in territorial contests. An additional consideration concerns the effects of differential predation pressure on females (by virtue of their slower flight and conspicuousness at host plants) – meaning females may gain more by mimicry relative to the differently behaving males.

Larvae of the Old World tropical butterfly *Danaus chrysippus* (Nymphalidae: Danainae) feed predominantly on milkweeds (Asclepiadaceae) from which they can sequester cardenolides, which are retained to the aposematic, chemically protected adult stage. A variable but often high proportion of *D. chrysippus* develop on milkweeds lacking bitter and emetic chemicals, and the resulting adult is unprotected. These are intraspecific Batesian **automimics** of their protected relatives. Where there is an unexpectedly high proportion of unprotected individuals, this situation may be maintained by parasitoids that preferentially parasitize noxious individuals, perhaps using their cardenolides as kairomones in host finding. The situation is complicated further, because unprotected adults, as in many *Danaus* species, actively seek out sources of pyrrolizidine alkaloids from plants to use in production of sex pheromones; these alkaloids also may render the adult less palatable.

14.5.2 Müllerian mimicry

In a contrasting set of relationships, called **Müllerian mimicry**, the model(s) and mimic(s) are all distasteful and warningly colored and all benefit from coexistence, as observers learn from tasting any individual. Unlike Batesian mimicry, in which the system is predicted to fail as the mimic becomes relatively more abundant, Müllerian mimicry systems gain through enhanced predator learning when the density of component distasteful species increases. "Mimicry rings" of species may develop, in which organisms from distant families, and even orders, acquire similar aposematic patterns, although the source of protection varies greatly. In the species involved, the warning signal of the co-models differs markedly from that of their close relatives, which are non-mimetic.

Interpretation of mimicry may be difficult, particularly in distinguishing protected from unprotected mimetic components. For example, a century after discovery of one of the seemingly strongest examples of Batesian mimicry, the classical interpretation seems

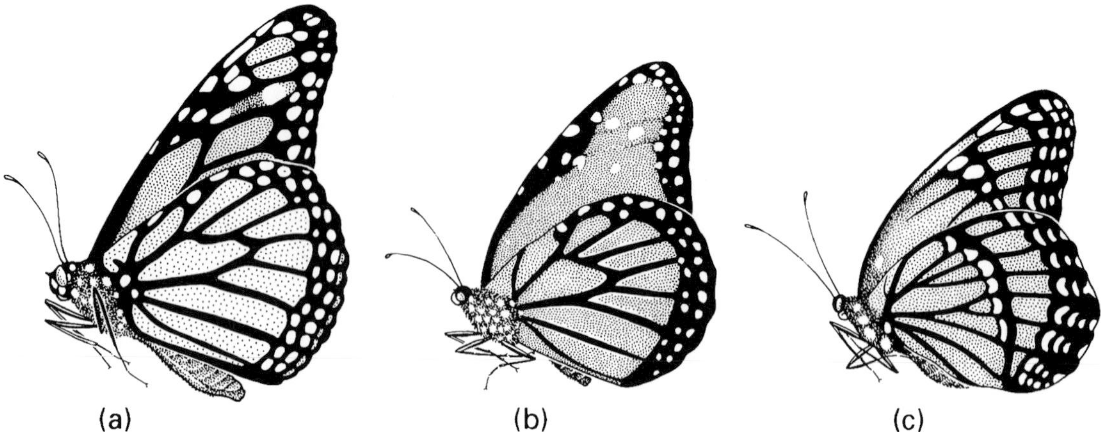

Fig. 14.8 Three nymphalid butterflies that are Müllerian co-mimics in Florida: (a) the monarch or wanderer (*Danaus plexippus*); (b) the queen (*D. gilippus*); (c) the viceroy (*Limenitis archippus*). (After Brower 1958.)

flawed. The system involves two North American danaine butterflies, *Danaus plexippus*, the monarch or wanderer, and *D. gilippus*, the queen, which are chemically defended models each of which is mimicked by a morph of the nymphaline viceroy butterfly (*Limenitis archippus*) (Fig. 14.8). Historically, larval food plants and taxonomic affiliation suggested that viceroys were palatable, and therefore Batesian mimics. This interpretation was overturned by experiments in which isolated butterfly abdomens were fed to natural predators (wild-caught redwing blackbirds). The possibility that feeding by birds might be affected by previous exposure to aposematism was excluded by removal of the aposematically patterned butterfly wings. Viceroys were found to be at least as unpalatable as monarchs, with queens least unpalatable. At least in the Florida populations and with this particular predator, the system is interpreted now as Müllerian, either with the viceroy as model, or with the viceroy and monarch acting as co-models, and the queen being a less well chemically protected member that benefits through the asymmetry of its palatability relative to the others. Few such appropriate experiments to assess palatability, using natural predators and avoiding problems of previous learning by the predator, have been reported and clearly more are required.

If all members of a Müllerian mimicry complex are aposematic and distasteful, then it can be argued that an observer (predator) is not deceived by any member – and this can be seen more as shared aposematism than mimicry. More likely, as seen above, distastefulness is unevenly distributed, in which case some specialist observers may find the least well-defended part of the complex to be edible. Such ideas suggest that true Müllerian mimicry may be rare and/or dynamic and represents one end of a spectrum of interactions.

14.5.3 Mimicry as a continuum

The strict differentiation of defensive mimicry into two forms – Müllerian and Batesian – can be questioned, although each gives a different interpretation of the ecology and evolution of the components, and makes dissimilar predictions concerning life histories of the participants. For example, mimicry theory predicts that in aposematic species there should be:

• limited numbers of co-modeled aposematic patterns, reducing the number that a predator has to learn;

• behavioral modifications to "expose" the pattern to potential predators, such as conspicuous display rather than crypsis, and diurnal rather than nocturnal activity;

• long post-reproductive life, with prominent exposure to encourage the naïve predator to learn of the distastefulness on a post-reproductive individual.

All of these predictions appear to be true in some or most systems studied. Furthermore, theoretically there should be variation in polymorphism with selection enforcing aposematic uniformity (monomorphism) in Müllerian cases, but encouraging divergence (mimetic polymorphism) in Batesian cases (section 14.5.1). Sex-limited (female-only) mimicry and divergence of the model's pattern away from that of the mimic (evolutionary escape) are also predicted in Batesian mimicry. Although these predictions are met in some mimetic species, there are exceptions to all of them. Polymorphism certainly occurs in Batesian mimetic swallowtails (Papilionidae), but is much rarer elsewhere, even within other butterflies; furthermore, there are polymorphic Müllerian mimics such as the viceroy. It is suggested now that some relatively undefended mimics may be fairly abundant relative to the distasteful model and need not have attained abundance via polymorphism. It is argued that this can arise and be maintained if the major predator is a generalist that requires only to be deterred relative to other more palatable species.

A complex range of mimetic relationships are based on mimicry of lycid beetles (see Plate 6.5), which are often aposematically odoriferous and warningly colored. The black and orange Australian lycid *Metriorrhynchus rhipidius* is protected chemically by odorous methoxyalkylpyrazine, and by bitter-tasting compounds and acetylenic antifeedants. Species of *Metriorrhynchus* provide models for mimetic beetles from at least six distantly related families (Buprestidae, Pythidae, Meloidae, Oedemeridae, Cerambycidae, and Belidae) and at least one moth. All these mimics are convergent in color; some have nearly identical alkylpyrazines and distasteful chemicals; others share the alkylpyrazines but have different distasteful chemicals; and some have the odorous chemical but appear to lack any distasteful chemicals. These aposematically colored insects form a mimetic series. The oedemerids clearly are Müllerian mimics, modeled precisely on the local *Metriorrhynchus* species and differing only in using cantharidin as an antifeedant. The cerambycid mimics use different repellent odors, whereas the buprestids lack warning odor but are chemically protected by buprestins. Finally, pythids and belids are Batesian mimics, apparently lacking any chemical defenses. After careful chemical examination, what appears to be a model with many Batesian mimics, or perhaps a Müllerian ring, is revealed to demonstrate a complete range between the extremes of Müllerian and Batesian mimicry.

Although the extremes of the two prominent mimicry systems are well studied, and in some texts appear to be the only systems described, they are but two of the possible permutations involving the interactions of model, mimic, and observer. Further complexity ensues if model and mimic are the same species, as in automimicry, or in cases where sexual dimorphism and polymorphism exist. All mimicry systems are complex, interactive, and never static, because population sizes change and relative abundances of mimetic species fluctuate so that density-dependent factors play an important role. Furthermore, the defense offered by shared aposematic coloring, and even shared distastefulness, can be circumvented by specialized predators able to learn and locate the warning, overcome the defenses and eat selected species in the mimicry complex. Evidently, consideration of mimicry theory demands recognition of the role of predators as flexible, learning, discriminatory, coevolving, and coexisting agents in the system (Box 14.1).

14.6 COLLECTIVE DEFENSES IN GREGARIOUS AND SOCIAL INSECTS

Chemically defended, aposematic insects are often clustered rather than uniformly distributed through a suitable habitat. Thus, unpalatable butterflies may live in conspicuous aggregations as larvae and as adults; the winter congregation of migratory adult monarch butterflies in California (see Plate 3.5) and Mexico is an example. Many chemically defended hemipterans aggregate on individual host plants, and some vespid wasps congregate conspicuously on the outside of their nests (as shown in the vignette of Chapter 12). Orderly clusters occur in the phytophagous larvae of sawflies (Hymenoptera: Pergidae; Fig. 14.7) and some chrysomelid beetles that form defended circles (**cycloalexy**). Some larvae lie within the circle and others form an outer ring with either their heads or abdomens directed outwards, depending upon which end secretes the noxious compounds. These groups often make synchronized displays of head and/or abdomen bobbing, which increase the apparency of the group.

Formation of such clusters is sometimes encouraged by the production of aggregation pheromones by early arriving individuals (section 4.3.2), or may result from the young failing to disperse after hatching from one or several egg batches. Benefits to the individual from the clustering of chemically defended insects may relate to the dynamics of predator training. However, these also may involve kin selection in subsocial insects, in which aggregations comprise relatives that benefit at the expense of an individual "sacrificed" to educate a predator.

This latter scenario for the origin and maintenance of group defense certainly seems to apply to the eusocial Hymenoptera (ants, bees, and wasps), as seen in Chapter 12. In these insects, and in the termites (Isoptera), defensive tasks are undertaken usually by morphologically modified individuals called soldiers. In all social insects, excepting the army ants, the focus for defensive action is the nest, and the major role of the soldier caste is to protect the nest and its inhabitants. Nest architecture and location is often a first line of defense, with many nests buried underground, or hidden within trees, with a few, easily defendable entrances. Exposed nests, such as those of savanna-zone termites, often have hard, impregnable walls.

Termite soldiers can be male or female, have weak sight or be blind, and have enlarged heads (sometimes exceeding the rest of the body length). Soldiers may have well-developed jaws, or be **nasute**, with small jaws but an elongate "nasus" or rostrum. They may protect the colony by biting, by chemical means, or, as in *Cryptotermes*, by **phragmosis** – the blocking of access to the nest with their modified heads. Amongst the most serious adversaries of termites are ants, and complex warfare takes place between the two. Termite soldiers have developed an enormous battery of chemicals, many produced in highly elaborated frontal and salivary glands. For example, in *Pseudacanthotermes spiniger* the salivary glands fill nine-tenths of the abdomen, and *Globitermes sulphureus* soldiers are filled to bursting with sticky yellow fluid used to entangle the predator – and the termite – usually fatally. This suicidal phenomenon is seen also in some *Camponotus* ants, which use hydrostatic pressure in the gaster to burst the abdomen and release sticky fluid from the huge salivary glands.

Some of the specialized defensive activities used by termites have developed convergently amongst ants. Thus, the soldiers of some formicines, notably the subgenus *Colobopsis*, and several myrmecines show phragmosis, with modifications of the head to allow the blocking of nest entrances (Fig. 14.9). Nest entrances are made by minor workers and are of such a size that the head of a single major worker (soldier) can seal it; in others such as the myrmecine *Zacryptocerus*, the entrances are larger, and a formation of guarding blockers may be required to act as "gatekeepers". A further defensive strategy of these myrmecines is for the head to be covered with a crust of secreted filamentous material, such that the head is camouflaged when it blocks a nest entrance on a lichen-covered twig.

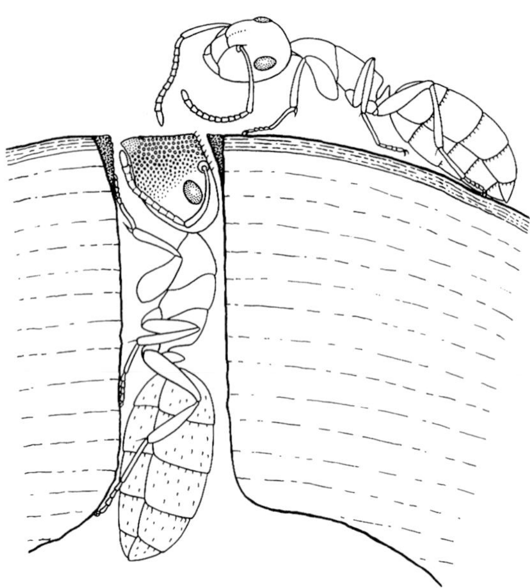

Fig. 14.9 Nest guarding by the European ant *Camponotus* (*Colobopsis*) *truncatus*: a minor worker approaching a soldier that is blocking a nest entrance with her plug-shaped head. (After Hölldobler & Wilson 1990; from Szabó-Patay 1928.)

Most soldiers use their strongly developed mandibles in colony defense as a means of injuring an attacker. A novel defense in termites involves elongate mandibles that snap against one another, as we might snap our fingers. A violent movement is produced as the pent-up elastic energy is released from the tightly appressed mandibles (Fig. 14.10a). In *Capritermes* and *Homallotermes*, the mandibles are asymmetric (Fig. 14.10b) and the released pressure results in the violent movement of only the right mandible; the bent left one, which provides the elastic tension, remains immobile. These soldiers can only strike to their left! The advantage of this defense is that a powerful blow can be delivered in a confined tunnel, in which there is inadequate space to open the mandibles wide enough to obtain conventional leverage on an opponent.

Major differences between termite defenses and those of social hymenopterans are the restriction of the defensive caste to females in Hymenoptera, and the frequent use of venom injected through an ovipositor modified as a sting (Fig. 14.11). Whereas parasitic hymenopterans use this weapon to immobilize prey, in social aculeate

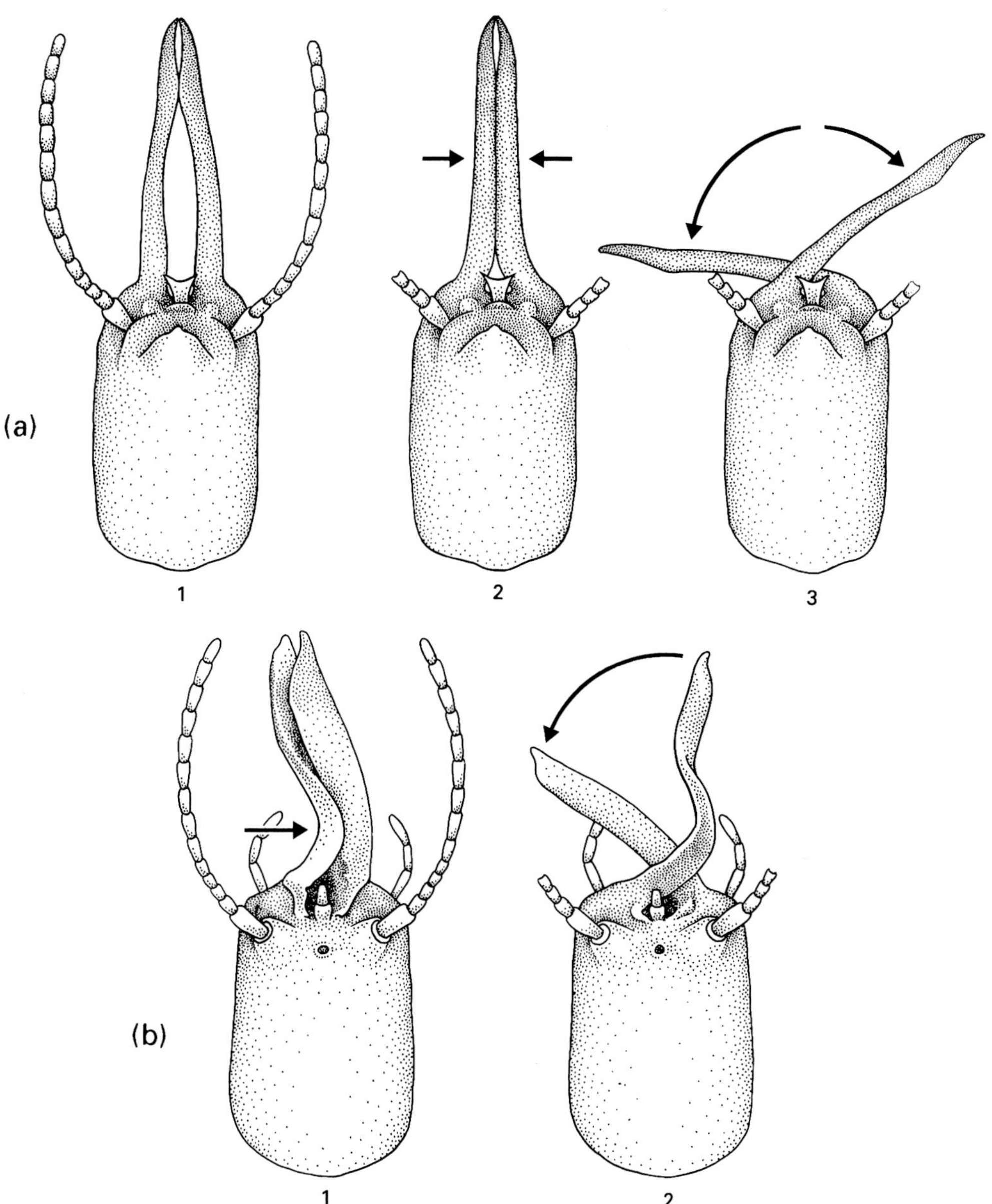

Fig. 14.10 Defense by mandible snapping in termite soldiers. (a) Head of a symmetric snapping soldier of *Termes* in which the long thin mandibles are pressed hard together (1) and thus bent inwards (2) before they slide violently across one another (3). (b) Head of an asymmetric snapping soldier of *Homallotermes* in which force is generated in the flexible left mandible by being pushed against the right one (1) until the right mandible slips under the left one to strike a violent blow (2). (After Deligne et al. 1981.)

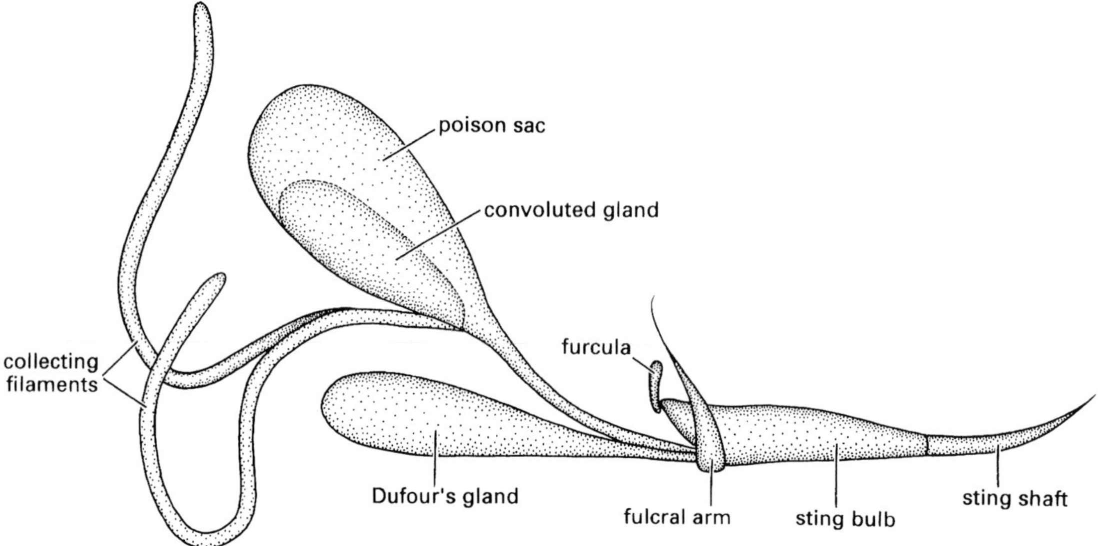

Fig. 14.11 Diagram of the major components of the venom apparatus of a social aculeate wasp. (After Hermann & Blum 1981.)

hymenopterans it is a vital weapon in defense against predators. Many subsocial and all social hymenopterans co-operate to sting an intruder *en masse*, thereby escalating the effects of an individual attack and deterring even large vertebrates. The sting is injected into a predator through a lever (the furcula) acting on a fulcral arm, though fusion of the furcula to the sting base in some ants leads to a less maneuverable sting.

Venoms include a wide variety of products, many of which are polypeptides. Biogenic amines such as any or all of histamine, dopamine, adrenaline (epinephrine), and noradrenaline (norepinephrine) (and serotonin in wasps) may be accompanied by acetylcholine, and several important enzymes including phospholipases and hyaluronidases (which are highly allergenic). Wasp venoms have a number of vasopeptides – pharmacologically active kinins that induce vasodilation and relax smooth muscle in vertebrates. Non-formicine ant venoms comprise either similar materials of proteinaceous origin or a pharmacopoeia of alkaloids, or complex mixtures of both types of component. In contrast, formicine venoms are dominated by formic acid.

Venoms are produced in special glands sited on the bases of the inner valves of the ninth segment, comprising free filaments and a reservoir store, which may be simple or contain a convoluted gland (Fig. 14.11). The outlet of Dufour's gland enters the sting base ventral to the venom duct. The products of this gland in eusocial bees and wasps are poorly known, but in ants Dufour's gland is the site of synthesis of an astonishing array of hydrocarbons (over 40 in one species of *Camponotus*). These exocrine products include esters, ketones, and alcohols, and many other compounds used in communication and defense.

The sting is reduced and lost in some social hymenopterans, notably the stingless bees and formicine ants. Alternative defensive strategies have arisen in these groups; thus many stingless bees mimic stinging bees and wasps, and use their mandibles and defensive chemicals if attacked. Formicine ants retain their venom glands, which disperse formic acid through an acidophore, often directed as a spray into a wound created by the mandibles.

Other glands in social hymenopterans produce additional defensive compounds, often with communication roles, and including many volatile compounds that serve as alarm pheromones. These stimulate one or more defensive actions: they may summon more individuals to a threat, marking a predator so that the attack is targeted, or, as a last resort, they may encourage the colony to flee from the danger. Mandibular glands produce alarm pheromones in many insects and also substances that cause pain when they enter wounds caused by the mandibles. The metapleural

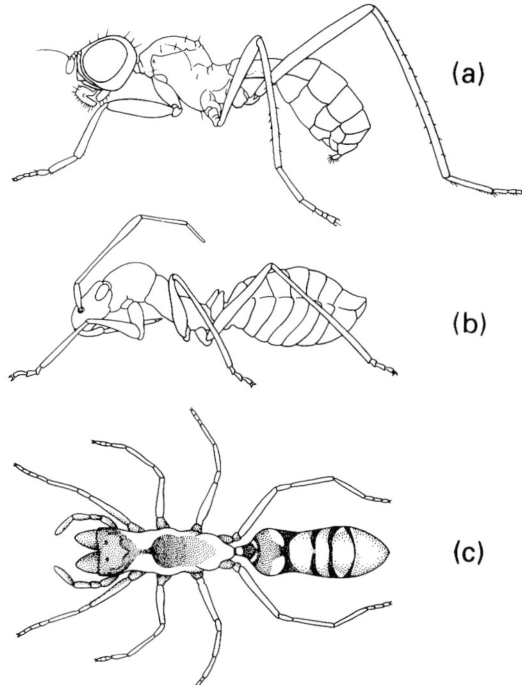

Fig. 14.12 Three ant mimics: (a) a fly (Diptera: Micropezidae: *Badisis*); (b) a bug (Hemiptera: Miridae: Phylinae); (c) a spider (Araneae: Clubionidae: *Sphecotypus*). ((a) After McAlpine 1990; (b) after Atkins 1980; (c) after Oliveira 1988.)

glands in some species of ants produce compounds that defend against microorganisms in the nest through antibiotic action. Both sets of glands may produce sticky defensive substances, and a wide range of pharmacological compounds is currently under study to determine possible human benefit.

Even the best-defended insects can be parasitized by mimics, and the best of chemical defenses can be breached by a predator (Box 14.1). Although the social insects have some of the most elaborate defenses seen in the Insecta, they remain vulnerable. For example, many insects model themselves on social insects, with representatives of many orders converging morphologically on ants (Fig. 14.12), particularly with regard to the waist constriction and wing loss, and even kinked antennae. Some of the most extraordinary ant-mimicking insects are tropical African bugs of the genus *Hamma* (Membracidae), as exemplified by *H. rectum* depicted in both side and dorsal view in the vignette for this chapter.

The aposematic yellow-and-black patterns of vespid wasps and apid bees provide models for hundreds of mimics throughout the world. Not only are these communication systems of social insects parasitized, but so are their nests, which provide many parasites and inquilines with a hospitable place for their development (section 12.3).

Defense must be seen as a continuing coevolutionary process, analogous to an "arms race", in which new defenses originate or are modified and then are selectively breached, stimulating improved defenses.

FURTHER READING

Blum, M.S. (1981) *Chemical Defenses of Arthropods*. Academic Press, New York.

Cook, L.M. (2000) Changing views on melanic moths. *Biological Journal of the Linnean Society* **69**, 431–41.

Dyer, L.A. (1995) Tasty generalists and nasty specialists? Antipredator mechanisms in tropical lepidopteran larvae. *Ecology* **76**, 1483–96.

Eisner, T. & Aneshansley, D.J. (1999) Spray aiming in the bombardier beetle: photographic evidence. *Proceedings of the National Academy of Sciences of the USA* **96**, 9705–9.

Evans, D.L. & Schmidt, J.O. (eds.) (1990) *Insect Defenses. Adaptive Mechanisms and Strategies of Prey and Predators.* State University of New York Press, Albany, NY.

Grant, B.S., Owen, D.F. & Clarke, C.A. (1996) Parallel rise and fall of melanic peppered moths in America and Britain. *Journal of Heredity* **87**, 351–7.

Gross, P. (1993) Insect behavioural and morphological defenses against parasitoids. *Annual Review of Entomology* **38**, 251–73.

Hölldobler, B. & Wilson, E.O. (1990) *The Ants.* Springer-Verlag, Berlin.

Hooper, J. (2002) *Of Moths and Men; An Evolutionary Tale: The Untold Story of Science and the Peppered Moth.* W.W. Norton & Co., New York.

Joron, M. & Mallet, J.L.B. (1998) Diversity in mimicry: paradox or paradigm? *Trends in Ecology and Evolution* **13**, 461–6.

McIver, J.D. & Stonedahl, G. (1993) Myrmecomorphy: morphological and behavioural mimicry of ants. *Annual Review of Entomology* **38**, 351–79.

Moore, B.P. & Brown, W.V. (1989) Graded levels of chemical defense in mimics of lycid beetles of the genus *Metriorrhynchus* (Coleoptera). *Journal of the Australian Entomological Society* **28**, 229–33.

Pasteels, J.M., Grégoire, J.-C. & Rowell-Rahier, M. (1983) The chemical ecology of defense in arthropods. *Annual Review of Entomology* **28**, 263–89.

Resh, V.H. & Cardé, R.T. (eds.) (2003) *Encyclopedia of Insects.* Academic Press, Amsterdam. [Particularly see articles on aposematic coloration; chemical defense; defensive behavior; industrial melanism; mimicry; venom.]

Ritland, D.B. (1991) Unpalatability of viceroy butterflies (*Limenitis archippus*) and their purported mimicry models, Florida queens (*Danaus gilippus*). *Oecologia* **88**, 102–8.

Starrett, A. (1993) Adaptive resemblance: a unifying concept for mimicry and crypsis. *Biological Journal of the Linnean Society* **48**, 299–317.

Turner, J.R.G. (1987) The evolutionary dynamics of Batesian and Muellerian mimicry: similarities and differences. *Ecological Entomology* **12**, 81–95.

Vane-Wright, R.I. (1976) A unified classification of mimetic resemblances. *Biological Journal of the Linnean Society* **8**, 25–56.

Wickler, W. (1968) *Mimicry in Plants and Animals.* Weidenfeld & Nicolson, London.

Yosef, R. & Whitman, D.W. (1992) Predator exaptations and defensive adaptations in evolutionary balance: no defense is perfect. *Evolutionary Ecology* **6**, 527–36.

Papers in *Biological Journal of the Linnean Society* (1981) **16**, 1–54 (includes a shortened version of H.W. Bates's classic 1862 paper).

Chapter 15

MEDICAL AND VETERINARY ENTOMOLOGY

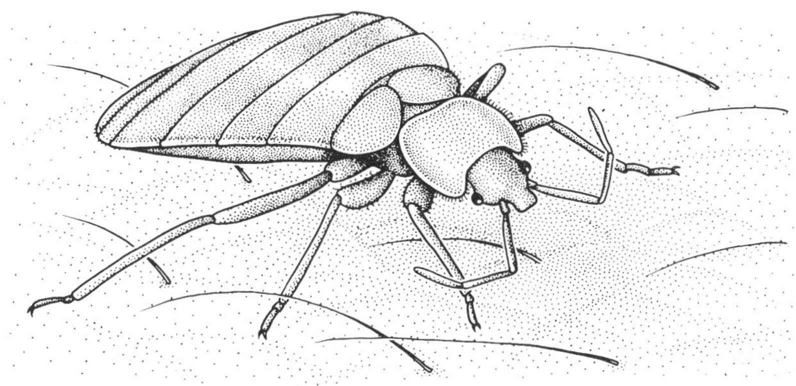

Bed bug on the skin of its host. (After Anon. 1991.)

Aside from their impact on agricultural and horticultural crops, insects impinge on us mainly through the diseases they can transmit to humans and our domestic animals. The number of insect species involved is not large, but those that transmit disease (**vectors**), cause wounds, inject venom, or create nuisance have serious social and economic consequences. Thus, the study of the veterinary and medical impact of insects is a major scientific discipline.

Medical and veterinary entomology differs from, and often is much broader in scope than, other areas of entomological pursuit. Firstly, the frequent motivation (and funding) for study is rarely the insect itself, but the insect-borne human or animal disease(s). Secondly, the scientist studying medical and veterinary aspects of entomology must have a wide understanding not only of the insect vector of disease, but of the biology of host and parasite. Thirdly, most practitioners do not restrict themselves to insects, but have to consider other arthropods, notably ticks, mites, and perhaps spiders and scorpions.

For brevity in this chapter, we refer to medical entomologists as those who study all arthropod-borne diseases, including diseases of livestock. The insect, though a vital cog in the chain of disease, need not be the central focus of medical research. Medical entomologists rarely work in isolation but usually function in multidisciplinary teams that may include medical practitioners and researchers, epidemiologists, virologists, and immunologists, and ought to include those with skills in insect control.

In this chapter, we deal with entomophobia, followed by allergic reactions, venoms, and urtication caused by insects. This is followed by details of transmission of a specific disease, malaria, an exemplar of insect-borne disease. This is followed by a review of additional diseases in which insects play an important role. We finish with a section on forensic entomology. At the end of the chapter are taxonomic boxes dealing with the Phthiraptera (lice), Siphonaptera (fleas), and Diptera (flies), especially medically significant ones.

15.1 INSECT NUISANCE AND PHOBIA

Our perceptions of nuisance may be little related to the role of insects in disease transmission. Insect nuisance is often perceived as a product of high densities of a particular species, such as bush flies (*Musca vetustissima*) in rural Australia, or ants and silverfish around the house. Most people have a more justifiable avoidance of filth-frequenting insects such as blow flies and cockroaches, biters such as some ants, and venomous stingers such as bees and wasps. Many serious disease vectors are rather uncommon and have inconspicuous behaviors, aside from their biting habits, such that the lay public may not perceive them as particular nuisances.

Harmless insects and arachnids sometimes arouse reactions such as unwarranted phobic responses (**arachnophobia** or **entomophobia** or **delusory parasitosis**). These cases may cause time-consuming and fruitless inquiry by medical entomologists, when the more appropriate investigations ought to be psychological. Nonetheless, there certainly are cases in which sufferers of persistent "insect bites" and persistent skin rashes, in which no physical cause can be established, actually suffer from undiagnosed local or widespread infestation with microscopic mites. In these circumstances, diagnosis of delusory parasitosis, through medical failure to identify the true cause, and referral to psychological counseling is unhelpful to say the least.

There are, however, some insects that transmit no disease, but feed on blood and whose attentions almost universally cause distress – bed bugs. Our vignette for this chapter shows *Cimex lectularius* (Hemiptera: Cimicidae), the cosmopolitan common bed bug, whose presence between the sheets often indicates poor hygiene conditions.

15.2 VENOMS AND ALLERGENS

15.2.1 Insect venoms

Some people's earliest experiences with insects are memorable for their pain. Although the sting of the females of many social hymenopterans (bees, wasps, and ants) can seem unprovoked, it is an aggressive defense of the nest. The delivery of venom is through the sting, a modified female ovipositor (Fig. 14.11). The honey-bee sting has backwardly directed barbs that allow only one use, as the bee is fatally damaged when it leaves the sting and accompanying venom sac in the wound as it struggles to retract the sting. In contrast, wasp and ant stings are smooth, can be retracted, and are capable of repeated use. In some ants, the ovipositor sting is greatly reduced and venom is either sprayed around liberally, or it can be directed with great accuracy into a wound made by the jaws. The venoms of social insects are discussed in more detail in section 14.6.

15.2.2 Blister and urtica (itch)-inducing insects

Some toxins produced by insects can cause injury to humans, even though they are not inoculated through a sting. Blister beetles (Meloidae) contain toxic chemicals, cantharidins, which are released if the beetle is crushed or handled (see Plate 6.3, facing p. 14). Cantharidins cause blistering of the skin and, if taken orally, inflammation of the urinary and genital tracts, which gave rise to its notoriety (as "Spanish fly") as a supposed aphrodisiac. Staphylinid beetles of the genus *Paederus* produce potent contact poisons including paederin, that cause delayed onset of severe blistering and long-lasting ulceration.

Lepidopteran caterpillars, notably moths, are a frequent cause of skin irritation, or urtication (a description derived from a similarity to the reaction to nettles, genus *Urtica*). Some species have hollow spines containing the products of a subcutaneous venom gland, which are released when the spine is broken. Other species have setae (bristles and hairs) containing toxins, which cause intense irritation when the setae contact human skin. Urticating caterpillars include the processionary caterpillars (Notodontidae) and some cup moths (Limacodidae). Processionary caterpillars combine frass (dry insect feces), cast larval skins, and shed hairs into bags suspended in trees and bushes, in which pupation occurs. If the bag is damaged by contact or by high wind, urticating hairs are widely dispersed.

The pain caused by hymenopteran stings may last a few hours, urtication may last a few days, and the most ulcerated beetle-induced blisters may last some weeks. However, increased medical significance of these injurious insects comes when repeated exposure leads to allergic disease in some humans.

15.2.3 Insect allergenicity

Insects and other arthropods are often implicated in **allergic disease**, which occurs when exposure to some arthropod allergen (a moderate-sized molecular weight chemical component, usually a protein) triggers excessive immunological reaction in some exposed people or animals. Those who handle insects in their occupations, such as in entomological rearing facilities, tropical fish food production, or research laboratories, frequently develop allergic reactions to one or more of a range of insects. Mealworms (beetle larvae of *Tenebrio* spp.), bloodworms (larvae of *Chironomus* spp.), locusts, and blow flies have all been implicated. Stored products infested with astigmatic mites give rise to allergic diseases such as baker's and grocer's itch. The most significant arthropod-mediated allergy arises through the fecal material of house-dust mites (*Dermatophagoides pteronyssinus* and *D. farinae*), which are ubiquitous and abundant in houses throughout many regions of the world. Exposure to naturally occurring allergenic arthropods and their products may be underestimated, although the role of house-dust mites in allergy is now well recognized.

The venomous and urticating insects discussed above can cause greater danger when some **sensitized** (previously exposed and allergy-susceptible) individuals are exposed again, as anaphylactic shock is possible, with death occurring if untreated. Individuals showing indications of allergic reaction to hymenopteran stings must take appropriate precautions, including allergen avoidance and carrying adrenaline (epinephrine).

15.3 INSECTS AS CAUSES AND VECTORS OF DISEASE

In tropical and subtropical regions, scientific, if not public, attention is drawn to the role of insects in transmitting protists, viruses, bacteria, and nematodes. Such pathogens are the causative agents of many important and widespread human diseases, including malaria, dengue, yellow fever, onchocerciasis (river blindness), leishmaniasis (oriental sore, kala-azar), filariasis (elephantiasis), and trypanosomiasis (sleeping sickness).

The causative agent of diseases may be the insect itself, such as the human body or head louse (*Pediculus humanus corporis* and *P. humanus capitis*, respectively), which cause pediculosis, or the mite *Sarcoptes scabiei*, whose skin-burrowing activities cause the skin disease scabies. In **myiasis** (from *myia*, the Greek for fly) the maggots or larvae of blow flies, house flies, and their relatives (Diptera: Calliphoridae, Sarcophagidae, and Muscidae) can develop in living flesh, either as primary agents or subsequently following wounding or damage by other insects, such as ticks and biting flies. If untreated, the animal victim may die. As death approaches and the flesh putrefies through bacterial activity, there may be a third wave of specialist fly larvae, and these colonizers are present at death. One particular form of myiasis affecting livestock is known as "strike" and is caused in the Old World by *Chrysomya bezziana* and in the Americas by the New World screw-worm fly, *Cochliomyia hominivorax* (Fig. 6.6h; section 16.10). The name "screw-worm" derives from the distinct rings of setae on the maggot resembling a screw. Virtually all myiases, including screw-worm, can affect humans, particularly under conditions of poor hygiene. Further groups of "higher" Diptera develop in mammals as endoparasitic larvae in the dermis, intestine, or, as in the sheep nostril fly, *Oestrus ovis*, in the nasal and head sinuses. In many parts of the world, losses caused by fly-induced damage to hides and meat, and death as a result of myiases, may amount to many millions of dollars.

Even more frequent than direct injury by insects is their action as vectors, transmitting disease-inducing pathogens from one animal or human **host** to another. This transfer may be by mechanical or biological means. **Mechanical transfer** occurs, for example, when a mosquito transfers myxomatosis from rabbit to rabbit in the blood on its proboscis. Likewise, when a cockroach or house fly acquires bacteria when feeding on feces it may physically transfer some bacteria from its mouthparts, legs, or body to human food, thereby transferring enteric diseases. The causative agent of the disease is passively transported from host to host, and does not increase in the vector. Usually in mechanical transfer, the arthropod is only one of several means of pathogen transfer, with poor public and personal hygiene often providing additional pathways.

In contrast, **biological transfer** is a much more specific association between insect vector, pathogen, and host, and transfer never occurs naturally without all three components. The disease agent **replicates** (increases) within the vector insect, and there is often close specificity between vector and disease agent. The insect is thus a vital link in biological transfer, and efforts to curb disease nearly always involve attempts to reduce vector numbers. In addition, biologically transferred disease may be controlled by seeking to interrupt contact between vector and host, and by direct attack on the pathogen, usually whilst in the host. Disease control comprises a combination of these approaches, each of which requires detailed knowledge of the biology of all three components – vector, pathogen, and host.

15.4 GENERALIZED DISEASE CYCLES

In all biologically transferred diseases, a biting (blood-feeding or sucking) adult arthropod, often an insect, particularly a true fly (Diptera), transmits a parasite from animal to animal, human to human, or from animal to human, or, very rarely, from human to animal. Some human pathogens (causative agents of human disease such as malaria parasites) can complete their parasitic life cycles solely within the insect vector and the human host. Human malaria exemplifies a disease with a **single cycle** involving *Anopheles* mosquitoes, malaria parasites, and humans. Although related malaria parasites occur in animals, notably other primates and birds, these hosts and parasites are not involved in the human malarial cycle. Only a few human insect-borne diseases have single cycles, as in malaria, because these diseases require coevolution of pathogen and vector and *Homo sapiens*. As *H. sapiens* is of relatively recent evolutionary origin, there has been only a short time for the development of unique insect-borne diseases that require specifically a human rather than any alternative vertebrate for completion of the disease-causing organism's life cycle.

In contrast to single-cycle diseases, many other insect-borne diseases that affect humans include a (non-human) vertebrate host, as for instance in yellow fever in monkeys, plague in rats, and leishmaniasis in desert rodents. Clearly, the non-human cycle is primary in these cases and the sporadic inclusion of humans in a **secondary** cycle is not essential to maintain the disease. However, when outbreaks do occur, these diseases can spread in human populations and may involve many cases.

Outbreaks in humans often stem from human actions, such as the spread of people into the natural ranges of the vector and animal hosts, which act as disease **reservoirs**. For example, yellow fever in native forested Uganda (central Africa) has a "sylvan" (woodland) cycle, remaining within canopy-dwelling primates with the exclusively primate-feeding mosquito *Aedes africanus* as the vector. It is only when monkeys and humans coincide at banana plantations close to or within the forest that *Aedes bromeliae* (formerly *Ae. simpsoni*), a second mosquito vector that feeds on both humans and monkeys, can transfer jungle yellow fever to humans. In a second example, in Arabia, *Phlebotomus* sand flies (Psychodidae) depend upon arid-zone burrowing rodents and, in feeding, transmit *Leishmania* parasites between rodent hosts. Leishmaniasis, a disfiguring ailment showing a dramatic increase in the

Neotropics, is transmitted to humans when suburban expansion places humans within this rodent reservoir, but unlike yellow fever, there appears to be no change in vector when humans enter the cycle.

In epidemiological terms, the natural cycle is maintained in animal reservoirs: sylvan primates for yellow fever and desert rodents for leishmaniasis. Disease control clearly is complicated by the presence of these reservoirs in addition to a human cycle.

15.5 PATHOGENS

The disease-causing organisms transferred by the insect may be viruses (termed "arboviruses", an abbreviation of *ar*thropod-*bo*rne *viruses*), bacteria (including rickettsias), protists, or filarial nematode worms. Replication of these parasites in both vectors and hosts is required and some complex life cycles have developed, notably amongst the protists and filarial nematodes. The presence of a parasite in the vector insect (which can be determined by dissection and microscopy and/or biochemical means) generally appears not to harm the host insect. When the parasite is at an appropriate developmental stage, and following multiplication or replication (amplification and/or concentration in the vector), transmission can occur. Transfer of parasites from vector to host or vice versa takes place when the blood-feeding insect takes a meal from a vertebrate host. The transfer from host to previously uninfected vector is through parasite-infected blood. Transmission to a host by an infected insect usually is by injection along with anticoagulant salivary gland products that keep the wound open during feeding. However, transmission may also be through deposition of infected feces close to the wound site.

In the following survey of major arthropod-borne disease, malaria will be dealt with in some detail. Malaria is the most devastating and debilitating disease in the world, and it illustrates a number of general points concerning medical entomology. This is followed by briefer sections reviewing the range of pathogenic diseases involving insects, arranged by phylogenetic sequence of parasite, from virus to filarial worm.

15.5.1 Malaria

The disease

Malaria affects more people, more persistently, throughout more of the world than any other insect-borne disease. Some 120 million new cases arise each year. The World Health Organization calculated that malaria control during the period 1950–72 reduced the proportion of the world's (excluding China's) population exposed to malaria from 64% to 38%. Since then, however, exposure rates to malaria in many countries have risen towards the rates of half a century ago, as a result of concern over the unwanted side-effects of dichlorodiphenyl-trichloroethane (DDT), resistance of insects to modern pesticides and of malaria parasites to antimalarial drugs, and civil unrest and poverty in a number of countries. Even in countries such as Australia, in which there is no transmission of malaria, the disease is on the increase among travelers, as demonstrated by the number of cases having risen from 199 in 1970, to 629 in 1980, and 700–900 in the 1990s with 1–5 deaths per annum.

The parasitic protists that cause malaria are sporozoans, belonging to the genus *Plasmodium*. Four species are responsible for the human malarias, with others described from, but not necessarily causing diseases in, primates, some other mammals, birds, and lizards. There is developing molecular evidence that at least some of these species of *Plasmodium* may not be restricted to humans, but are shared (under different names) with other primates. The vectors of mammalian malaria are always *Anopheles* mosquitoes, with other genera involved in bird plasmodial transmission.

The disease follows a course of a **prepatent period** between infective bite and **patenty**, the first appearance of parasites (sporozoites; see Box 15.1) in the erythrocytes (red blood cells). The first clinical symptoms define the end of an **incubation period**, some nine (*P. falciparum*) to 18–40 (*P. malariae*) days after infection. Periods of fever followed by severe sweating recur cyclically and follow several hours after synchronous rupture of infected erythrocytes (see below). The spleen is characteristically enlarged. The four malaria parasites each induce rather different symptoms:

1 *Plasmodium falciparum*, or malignant tertian malaria, kills many untreated sufferers through, for example, cerebral malaria or renal failure. Fever recurrence is at 48 h intervals (tertian is Latin for third day, the name for the disease being derived from the sufferer having a fever on day one, normal on day two, with fever recurrent on the third day). *P. falciparum* is limited by a minimum 20°C isotherm and is thus most common in the warmest areas of the world.

2 *Plasmodium vivax*, or benign tertian malaria, is a less serious disease that rarely kills. However, it is more widespread than *P. falciparum*, and has a wider

Box 15.1 Life cycle of *Plasmodium*

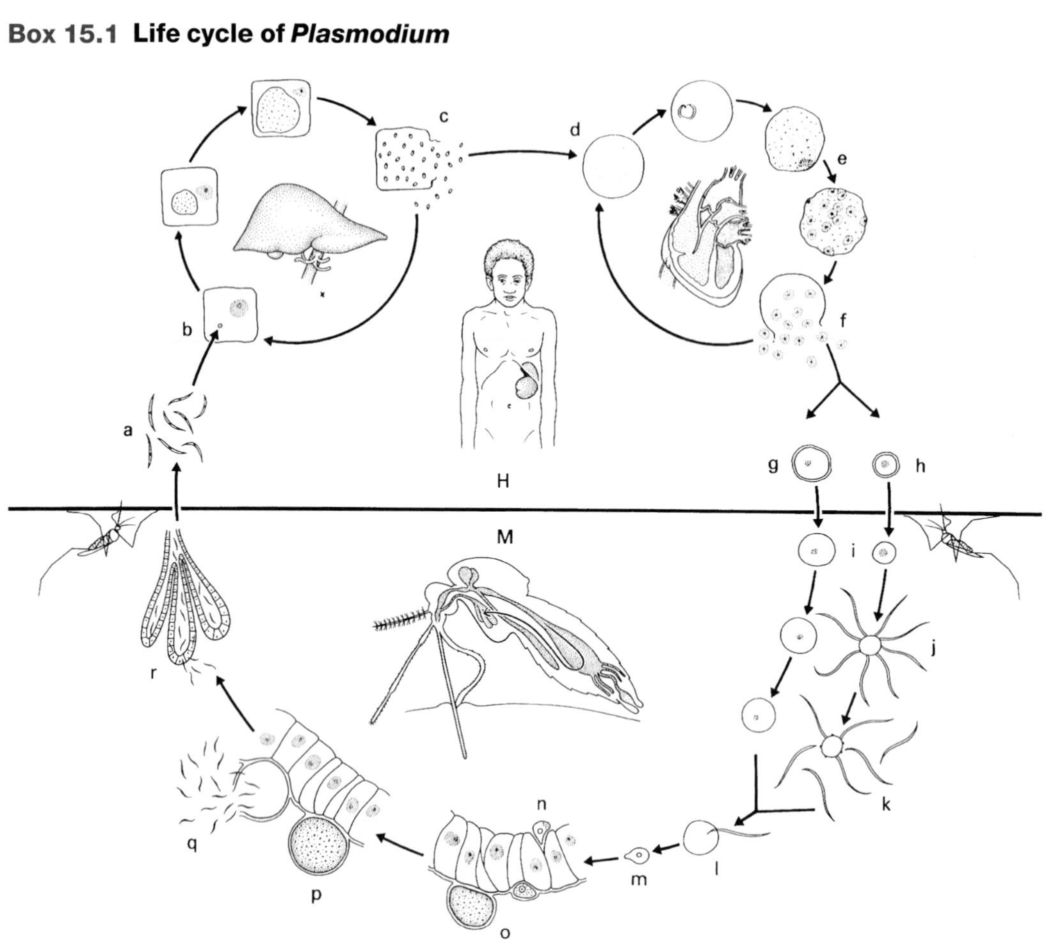

The malarial cycle, shown here modified after Kettle (1984) and Katz et al. (1989), commences with an infected female *Anopheles* mosquito taking a blood meal from a human host (H). Saliva contaminated with the **sporozoite** stage of the *Plasmodium* is injected (a). The sporozoite circulates in the blood until reaching the liver, where a **pre**- (or **exo**-) **erythrocytic schizogonous** cycle (b,c) takes place in the parenchyma cells of the liver. This leads to the formation of a large **schizont**, containing from 2000 to 40,000 **merozoites**, according to *Plasmodium* species. The prepatent period of infection, which started with an infective bite, ends when the merozoites are released (c) to either infect more liver cells or enter the bloodstream and invade the erythrocytes. Invasion occurs by the erythrocyte invaginating to engulf the merozoite, which subsequently feeds as a **trophozoite** (e) within a vacuole. The first and several subsequent **erythrocyte schizogonous** (d–f) cycles produce a trophozoite that becomes a schizont, which releases from 6 to 16 merozoites (f), which commence the repetition of the erythrocytic cycle. Synchronous release of merozoites from the erythrocytes liberates parasite products that stimulate the host's cells to release cytokines (a class of immunological mediators) and these provoke the fever and illness of a malaria attack. Thus, the duration of the erythrocyte schizogonous cycle is the duration of the interval between attacks (i.e. 48 h for tertian, 72 h for quartan).

After several erythrocyte cycles, some trophozoites mature to gametocytes (g,h), a process that takes eight days for *P. falciparum* but only four days for *P. vivax*. If a female *Anopheles* (M) feeds on an infected human host at this stage in the cycle, she ingests blood containing erythrocytes, some of which contain both types of gametocytes. Within a susceptible mosquito the erythrocyte is disposed of and both types of gametocytes (i) develop further: half are female gametocytes, which remain large and are termed **macrogametes**; the other half are males, which divide into eight flagellate **microgametes** (j), which rapidly deflagellate (k), and seek and fuse with a macrogamete to form a **zygote** (l). All this sexual activity has taken place in a matter of 15 min or so while within the female mosquito the blood meal passes towards the midgut. Here the initially inactive zygote becomes an active **ookinete** (m) which burrows into the epithelial lining of the midgut to form a mature **oocyst** (n–p).

Asexual reproduction (**sporogony**) now takes place within the expanding oocyst. In a temperature-dependent process, numerous nuclear divisions give rise to sporozoites. Sporogony does not occur below 16°C or above 33°C, thus explaining the temperature limitations for *Plasmodium* development noted in section 15.5.1. The mature oocyst may contain 10,000 sporozoites, which are shed into the hemocoel (q), from whence they migrate into the mosquito's salivary glands (r). This sporogonic cycle takes a minimum of 8–9 days and produces sporozoites that are active for up to 12 weeks, which is several times the complete life expectancy of the mosquito. At each subsequent feeding, the infective female *Anopheles* injects sporozoites into the next host along with the saliva containing an anticoagulant, and the cycle recommences.

temperature tolerance, extending as far as the 16°C summer isotherm. Recurrence of fever is every 48 h, and the disease may persist for up to eight years with relapses some months apart.

3 *Plasmodium malariae* is known as quartan malaria, and is a more widespread, but rarer parasite than *P. falciparum* or *P. vivax*. If allowed to persist for an extended period, death occurs through chronic renal failure. Recurrence of fever is at 72 h, hence the name quartan (fever on day one, recurrence on the fourth day). It is persistent, with relapses occurring up to half a century after the initial attack.

4 *Plasmodium ovale* is a rare tertian malaria with limited pathogenicity and a very long incubation period, with relapses at three-monthly intervals.

Malaria epidemiology

Malaria exists in many parts of the world but the incidence varies from place to place. As with other diseases, malaria is said to be **endemic** in an area when it occurs at a relatively constant incidence by natural transmission over successive years. Categories of endemicity have been recognized based on the incidence and severity of symptoms (spleen enlargement) in both adults and children. An **epidemic** occurs when the incidence in an endemic area rises or a number of cases of the disease occur in a new area. Malaria is said to be in a stable state when there is little seasonal or annual variation in the disease incidence, and it is predominantly transmitted by a strongly **anthropophilic** (human-loving) *Anopheles* vector species. Stable malaria is found in the warmer areas of the world where conditions encourage rapid sporogeny and usually is associated with the *P. falciparum* pathogen. In contrast, unstable malaria is associated with sporadic epidemics, often with a short-lived and more **zoophilic** (preferring other animals to humans) vector that may occur in massive numbers. Often ambient temperatures are lower than for areas with stable malaria, sporogeny is slower, and the pathogen is more often *P. vivax*.

Disease transmission can be understood only in relation to the potential of each vector to transmit the particular disease. This involves the variously complex relationship between:

- vector distribution;
- vector abundance;
- life expectancy (survivorship) of the vector;
- predilection of the vector to feed on humans (anthropophily);
- feeding rate of the vector;
- vector competence.

With reference to *Anopheles* and malaria, these factors can be detailed as follows.

Vector distribution

Anopheles mosquitoes occur almost worldwide, with the exception of cold temperate areas, and there are over 400 known species. However, the four species of human pathogenic *Plasmodium* are transmitted significantly in nature only by some 30 species of *Anopheles*. Some species have very local significance, others can be infected experimentally but have no natural role, and perhaps 75% of *Anopheles* species are rather refractory (intolerant) to malaria. Of the vectorial species, a handful are important in stable malaria, whereas others

Box 15.2 *Anopheles gambiae* complex

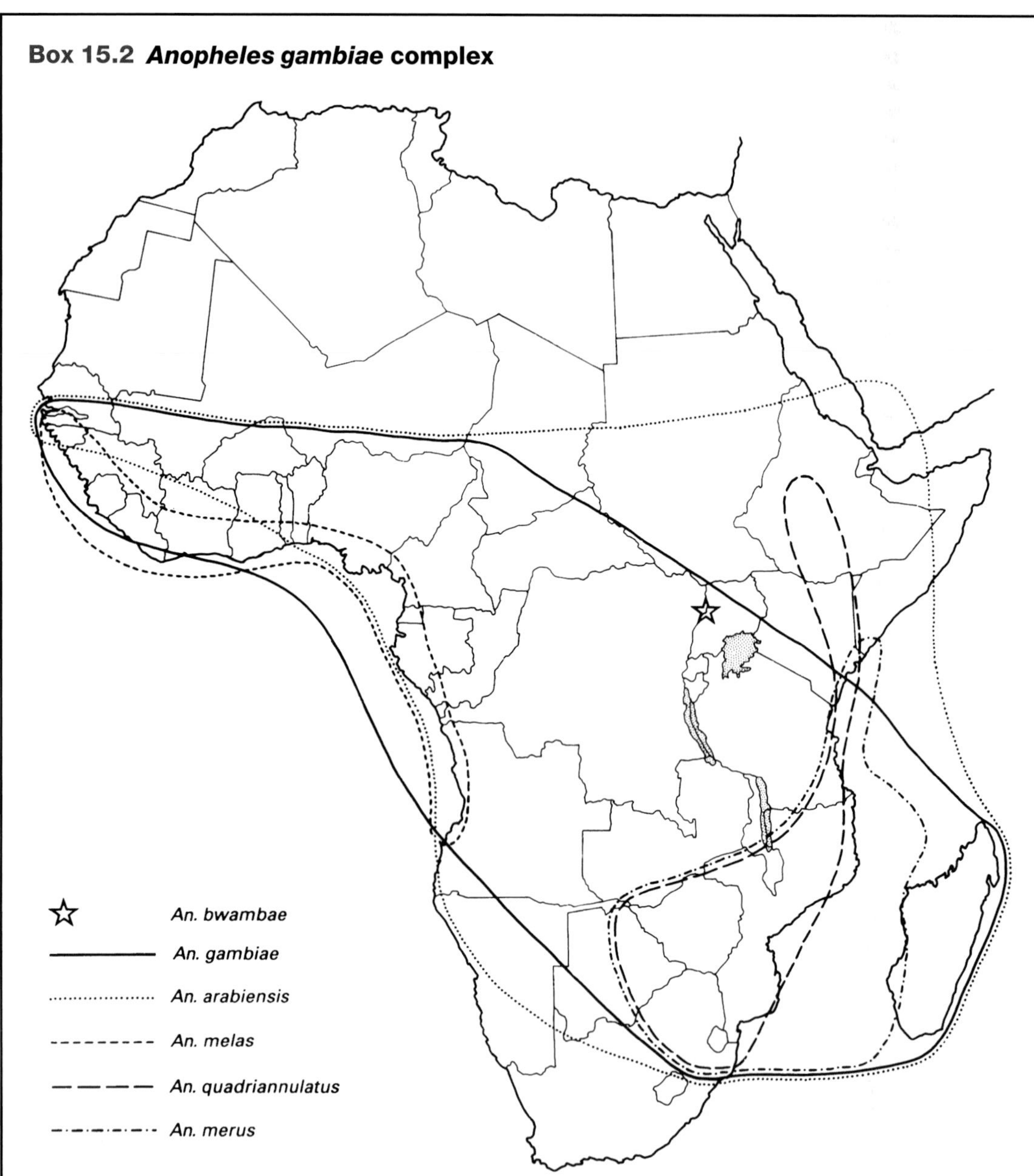

In the early days of African malariology, the common, predominantly pool-breeding *Anopheles gambiae* was found to be a highly anthropophilic, very efficient vector of malaria virtually throughout the continent. Subtle variation in morphology and biology suggested, however, that more than one species might be involved. Initial investigations allowed morphological segregation of West African *An. melas* and East African *An. merus*; both breed in saline waters, unlike the freshwater-breeding *An. gambiae*. Reservations remained as to whether the latter belonged to a single species, and studies involving meticulous rearing from single egg masses, cross-fertilization, and examination of fertility of thousands of hybrid offspring indeed revealed discontinuities in the *An. gambiae* gene pool. These were interpreted as supporting four species, a view that was substantiated by banding patterns of the larval salivary gland and ovarian nurse-cell giant chromosomes and by protein electrophoresis. Even with reliable cytologically determined specimens, morphological features do not allow segregation of the component species of the freshwater members of the *An. gambiae* complex of sibling (or cryptic) species.

An. gambiae is restricted now to one widespread African taxon; *An. arabiensis* was recognized for a second sibling taxon that in many areas is sympatric with *An. gambiae*; *An. quadriannulatus* is an East and southern African sibling; and *An. bwambae* is a rare and localized taxon from hot mineralized pools in Uganda. The maximum distributional limit of each sibling species is shown here on the map of Africa (data from White 1985). The siblings differ markedly in their vectorial status: *An. gambiae* and *An. arabiensis* are both endophilic (feeding indoors) and highly anthropophilic vectors of malaria and bancroftian filariasis. However, when cattle are present, *An. arabiensis* shows increased zoophily, much reduced anthropophily, and an increased tendency to exophily (feeding outdoors) compared with *An. gambiae*. In contrast to these two sibling species, *An. quadriannulatus* is entirely zoophilic and does not transmit disease of medical significance to humans. *An. bwambae* is a very localized vector of malaria that is endophilic if native huts are available.

become involved only in epidemic spread of unstable malaria. Vectorial status can vary across the range of a taxon, an observation that may be due to the hidden presence of sibling species that lack morphological differentiation, but differ slightly in biology and may have substantially different epidemiological significance, as in the *An. gambiae* complex (Box 15.2).

Vector abundance

Anopheles development is temperature dependent, as in *Aedes aegypti* (Box 6.2), with one or two generations per year in areas where winter temperatures force hibernation of adult females, but with generation times of perhaps six weeks at 16°C and as short as 10 days in tropical conditions. Under optimal conditions, with batches of over 100 eggs laid every two to three days, and a development time of 10 days, 100-fold increases in adult *Anopheles* can take place within 14 days.

As *Anopheles* larvae develop in water, rainfall significantly governs numbers. The dominant African malaria vector, *An. gambiae* (in the restricted sense; Box 15.2), breeds in short-lived pools that require replenishment; increased rainfall obviously increases the number of *Anopheles* breeding sites. On the other hand, rivers where other *Anopheles* species develop in lateral pools or streambed pools during a low- or no-flow period will be scoured out by excessive wet season rainfall. Adult survivorship clearly is related to elevated humidity and, for the female, availability of blood meals and a source of carbohydrate.

Vector survival rate

The duration of the adult life of the female infective *Anopheles* mosquito is of great significance in its effectiveness as a disease transmitter. If a mosquito dies within eight or nine days of an initial infected blood meal, no sporozoites will have become available and no malaria is transmitted. The age of a mosquito can be calculated by finding the physiological age based on the ovarian "relicts" left by each ovarian cycle (section 6.9.2). With knowledge of this physiological age and the duration of the sporogonic cycle (Box 15.1), the proportion of each *Anopheles* vector population of sufficient age to be infective can be calculated. In African *An. gambiae* (in the restricted sense; Box 15.2), three ovarian cycles are completed before infectivity is detected. Maximum transmission of *P. falciparum* to humans occurs in *An. gambiae* that has completed four to six ovarian cycles. Despite these old individuals forming only 16% of the population, they constitute 73% of infective individuals. Clearly, adult life expectancy (demography) is important in epidemiological calculations. Raised humidity prolongs adult life and the most important cause of mortality is desiccation.

Anthropophily of the vector

To act as a vector, a female *Anopheles* mosquito must feed at least twice; once to gain the pathogenic *Plasmodium* and a second time to transmit the disease. **Host preference** is the term for the propensity of a vector mosquito to feed on a particular host species. In malaria, the host preference for humans (anthropophily) rather than alternative hosts (zoophily) is crucial to human malaria epidemiology. Stable malaria is associated with strongly anthropophilic vectors that may never feed on other hosts. In these circumstances the probability of two consecutive meals being taken from a human is very high, and disease transmission can take place even when mosquito densities are low. In contrast, if the vector has a low rate of anthropophily (a low probability of human feeding) the probability of consecutive blood meals being taken from humans is slight and human malarial transmission by this particular vector is correspondingly low. Transmission will take place only when the vector is very numerous, as in epidemics of unstable malaria.

Feeding interval

The frequency of feeding of the female *Anopheles* vector is important in disease transmission. This frequency can be estimated from mark–release–recapture data or from survey of the ovarian-age classes of indoor resting mosquitoes. Although it is assumed that one blood meal is needed to mature each batch of eggs, some mosquitoes may mature a first egg batch without a meal, and some anophelines require two meals. Already-infected vectors may experience difficulty in feeding to satiation at one meal, because of blockage of the feeding apparatus by parasites, and may probe many times. This, as well as disturbance during feeding by an irritated host, may lead to feeding on more than one host.

Vector competence

Even if an uninfected *Anopheles* feeds on an infectious host, either the mosquito may not acquire a viable infection, or the *Plasmodium* parasite may fail to replicate within the vector. Furthermore, the mosquito may not transmit the infection onwards at a subsequent meal. Thus, there is scope for substantial variation, both within and between species, in the competence to act as a disease vector. Allowance must also be made for the density, infective condition, and age profiles of the human population, as human immunity to malaria increases with age.

Vectorial capacity

The **vectorial capacity** of a given *Anopheles* vector to transmit malaria in a circumscribed human population can be modeled. This involves a relationship between the:

- number of female mosquitoes per person;
- daily biting rate on humans;
- daily mosquito survival rate;
- time between mosquito infection and sporozoite production in the salivary glands;
- vectoral competence;
- some factor expressing the human recovery rate from infection.

This vectorial capacity must be related to some estimate concerning the biology and prevalence of the parasite when modeling disease transmission, and in monitoring disease control programs. In malarial studies, the **infantile conversion rate** (ICR), the rate at which young children develop antibodies to malaria, may be used. In Nigeria (West Africa), the Garki Malaria Project found that over 60% of the variation in the ICR derived from the human-biting rate of the two dominant *Anopheles* species. Only 2.2% of the remaining variation is explained by all other components of vectorial capacity, casting some doubt on the value of any measurements other than human-biting rate. This was particularly reinforced by the difficulties and biases involved in obtaining reasonably accurate estimates of many of the vectorial factors listed above.

15.5.2 Arboviruses

Viruses which multiply in an invertebrate vector and a vertebrate host are termed arboviruses. This definition excludes the mechanically transmitted viruses, such as the myxoma virus that causes myxomatosis in rabbits. There is no viral amplification in myxomatosis vectors such as the rabbit flea, *Spilopsyllus cuniculi*, and, in Australia, *Anopheles* and *Aedes* mosquitoes. Arboviruses are united by their ecologies, notably their ability to replicate in an arthropod. It is an unnatural grouping rather than one based upon virus phylogeny, as arboviruses belong to several virus families. These include some Bunyaviridae, Reoviridae, and Rhabdoviridae, and notably many Flaviviridae and Togaviridae. *Alphavirus* (Togaviridae) includes exclusively mosquito-transmitted viruses, notably the agents of equine encephalitides. Members of *Flavivirus* (Flaviviridae), which includes yellow fever, dengue, Japanese encephalitis, West Nile, and other encephalitis viruses, are borne by mosquitoes or ticks.

Yellow fever exemplifies a flavivirus life cycle. A similar cycle to the African sylvan (forest) one seen in section 15.4 involves a primate host in Central and South America, although with different mosquito vectors from those in Africa. Sylvan transmission to humans does occur, as in Ugandan banana plantations, but the disease makes its greatest fatal impact in urban epidemics. The urban and peri-domestic insect vector in Africa and the Americas is the female of the yellow-fever mosquito, *Aedes (Stegomyia) aegypti*. This mosquito acquires the virus by feeding on a human yellow-fever sufferer in the early stages of disease, from 6 h preclinical to four days later. The viral cycle in the mosquito is 12 days long, after which the yellow-fever virus reaches the mosquito saliva and remains there for the rest of the mosquito's life. With every subsequent blood meal the female mosquito transmits virus-contaminated saliva. Infection results, and yellow-fever symptoms develop in the host within a week. An urban disease cycle must originate from individuals infected with yellow fever from the sylvan (rural) cycle moving to an urban environment. Here, disease outbreaks may persist, such as those in which hundreds or thousands of people have died, including in New Orleans as recently as 1905. In South America, monkeys may die of yellow fever, but African ones are asymptomatic: perhaps neotropical monkeys have yet to develop tolerance to the disease. The common urban mosquito vector, *Ae. aegypti*, may have been transported relatively recently from West Africa to South America, perhaps aboard slave ships, together with yellow fever. The range of *Ae. aegypti* is greater than that of the disease, being present in southern USA, where it is spreading, and in Australia, and much of Asia. However, only in India are there susceptible but, as yet, uninfected monkey hosts of the disease.

Other Flaviviridae affecting humans and transmitted by mosquitoes cause dengue, dengue hemorrhagic fever, and a number of diseases called encephalitis (or encephalitides), because in clinical cases inflammation of the brain occurs. Each encephalitis has a preferred mosquito host, frequently an *Aedes (Stegomyia)* species such as *Ae. aegypti* for dengue, and often a *Culex* species for encephalitis. The reservoir hosts for these diseases vary, and, at least for encephalitis, include wild birds, with amplification cycles in domestic mammals, for example pigs for Japanese encephalitis. Horses can be carriers of togaviruses, giving rise to the name for a subgroup of diseases termed "equine encephalitides".

West Nile virus belongs to the Japanese encephalitis virus complex, preferentially transmitted by *Culex* species, with wild birds as reservoirs able to amplify the virus during outbreaks. The virus is distributed widely from the western Mediterranean eastward through the Middle East, Africa to India and Indonesia. Human symptoms and mortality vary with age, health, and virus strain, but encephalitis is uncommon. The disease entered New York from an unknown source in 1999, causing seven human deaths from 61 confirmed cases (and many more asymptomatic infections), and the mortality of many wild birds, especially *Corvus* species (crows). In subsequent years the geographic distribution has spread rapidly south and westward and reached the Pacific coast states of the USA in 2003. Many wild birds and more humans have died, and the range of potential vector mosquitoes has expanded beyond the species of *Culex* identified in New York. As in some European outbreaks, horses have proved susceptible and a vaccine has become available.

Several flaviviruses are transmitted by ixodid ticks, including more viruses that cause encephalitis and hemorrhagic fevers of humans, but more significantly of domestic animals. Bunyaviruses may be tick-borne, notably hemorrhagic diseases of cattle and sheep, particularly when conditions encourage an explosion of tick numbers and disease alters from normal hosts (**enzootic**) to epidemic (**epizootic**) conditions. Mosquito-borne bunyaviruses include African Rift Valley fever, which can produce high mortality amongst African sheep and cattle during mass outbreaks.

Amongst the Reoviridae, bluetongue virus is the best known, most debilitating, and most significant economically. The disease, which is virtually worldwide and has many different serotypes, causes tongue ulceration (hence "bluetongue") and an often terminal fever in sheep. Bluetongue is one of the few diseases in which biting midges of *Culicoides* (Ceratopogonidae) have been clearly established as the sole vectors of an arbovirus of major significance, although many arboviruses have been isolated from these biting flies.

Studies of the epidemiology of arboviruses have been complicated by the discovery that some viruses may persist between generations of vector. Thus, La Crosse virus, a bunyavirus that causes encephalitis in the USA, can pass from the adult mosquito through the egg (**transovarial transmission**) to the larva, which overwinters in a near-frozen tree-hole. The first emerging female of the spring generation is capable of transmitting La Crosse virus to chipmunk, squirrel, or human with her first meal of the year. Transovarial transmission is suspected in other diseases and is substantiated in increasing numbers of cases, including Japanese encephalitis in *Culex tritaenorhynchus* mosquitoes.

15.5.3 Rickettsias and plague

Rickettsias are bacteria (Proteobacteria: Rickettsiales) associated with arthropods. The genus *Rickettsia* includes virulent pathogens of humans. *R. prowazekii*, which causes endemic typhus, has influenced world affairs as much as any politician, causing the deaths of millions of refugees and soldiers in times of social upheaval, such as the years of Napoleonic invasion of Russia and those following World War I. Typhus symptoms are headache, high fever, spreading rash, delirium, and aching muscles, and in epidemic typhus from 10% to 60% of untreated patients die. The vectors of typhus are lice (Box 15.3), notably the body louse, *Pediculus humanus corporis*. Infestation of lice indicates unsanitary conditions and in western nations, after years of decline, is resurgent in homeless people. Although the head louse (*P. humanus capitis*), pubic louse (*Pthirus pubis*), and some fleas experimentally can transmit *R. prowazekii*, they are of little or no epidemiological significance. After the rickettsias of *R. prowazekii* have multiplied in the louse epithelium, they rupture the cells and are voided in the feces. Because the louse dies, the rickettsias are demonstrated to be rather poorly adapted to the louse host. Human hosts are infected by scratching infected louse feces (which remain infective for up to two months after deposition) into the itchy site where the louse has fed. There is evidence of low level persistence of rickettsias in those who recover from typhus. These act as endemic reservoirs for resurgence of the disease, and domestic and a few wild animals may be disease reservoirs. Lice are also vectors of relapsing fever, a spirochete disease that historically occurred together with epidemic typhus.

Other rickettsial diseases include murine typhus, transmitted by flea vectors, scrub typhus through trombiculid mite vectors, and a series of spotted fevers, termed tick-borne typhus. Many of these diseases have a wide range of natural hosts, with antibodies to the widespread American Rocky Mountain spotted fever (*Rickettsia rickettsii*) reported from numerous bird and mammal species. Throughout the range of the disease from Virginia to Brazil, several species of ticks with broad host ranges are involved, with transmission through feeding activity alone. Bartonellosis (Oroya fever) is a rickettsial infection transmitted by South American phlebotomine sand flies, with symptoms of exhaustion, anemia, and high fever, followed by wart-like eruptions on the skin.

Plague is a rodent–flea–rodent disease caused by the bacterium *Yersinia pestis*, also known as *Y. pseudotuberculosis* var. *pestis*. Plague-bearing fleas are principally *Xenopsylla cheopis*, which is ubiquitous between 35°N and 35°S, but also include *X. brasiliensis* in India, Africa, and South America, and *X. astia* in south-east Asia. Although other species including *Ctenocephalides felis* and *C. canis* (cat and dog fleas) can transmit plague, they play a minor role at most. The major vector fleas occur especially on peri-domestic (house-dwelling) species of *Rattus*, such as the black rat (*R. rattus*) and brown rat (*R. norvegicus*). Reservoirs for plague in specific localities include the bandicoot (*Bandicota bengalensis*) in India, rock squirrels (*Spermophilus* spp.) in western USA and related ground squirrels (*Citellus* sp.) in south-east Europe, gerbils (*Meriones* spp.) in the Middle East, and *Tatera* spp. in India and South Africa. Between plague outbreaks, the bacterium circulates within some or all of these rodents without evident mortality, thus providing silent, long-term reservoirs of infection.

When humans become involved in plague outbreaks (such as the pandemic called the "Black Death" that ravaged the northern hemisphere during the 14th century) mortality may approach 90% in undernourished people and around 25% in previously well-fed, healthy people. The plague epidemiological cycle commences amongst rats, with fleas naturally transmitting *Y. pestis* between peri-domestic rats. In an outbreak of plague, when the preferred-host brown rats die, some infected fleas move on to and eventually kill the secondary preference, black rats. As *X. cheopis* readily bites humans, infected fleas switch host again in the absence of the rats. Plague is a particular problem where rat (and flea) populations are high, as occurs in overcrowded, unsanitary urban conditions. Outbreak conditions require appropriate preceding conditions of mild temperatures and high humidity that encourage build-up of flea populations by increased larval survival and adult longevity. Thus, natural variations in the intensity of plague epidemics relate to the previous year's climate. Even during prolonged plague outbreaks, periods of fewer cases used to occur when hot, dry conditions prevented recruitment, because flea larvae are very susceptible to desiccation, and low humidity reduced adult survival in the subsequent year.

During its infective lifetime the flea varies in its ability to transmit plague, according to internal physiological changes induced by *Y. pestis*. If the flea takes an infected blood meal, *Y. pestis* increases in the proventriculus and midgut and may form an impassable plug. Further feeding involves a fruitless attempt by the pharyngeal pump to force more blood into the gut, with the result that a contaminated mixture of blood and bacteria is regurgitated. However, the survival time of *Y. pestis* outside the flea (of no more than a few hours) suggests that mechanical transmission is unlikely. More likely,

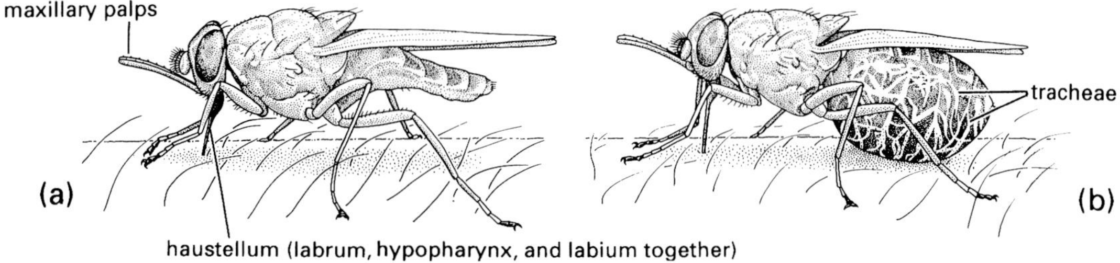

Fig. 15.1 A tsetse fly, *Glossina morsitans* (Diptera: Glossinidae), at the commencement of feeding (a) and fully engorged with blood (b). Note that the tracheae are visible through the abdominal cuticle in (b). (After Burton & Burton 1975.)

even if the proventricular blockage is alleviated, it fails to function properly as a one-way valve, and at every subsequent attempt at feeding, the flea regurgitates a contaminated mixture of blood and pathogen into the feeding wound of each successive host.

15.5.4 Protists other than malaria

Some of the most important insect-borne pathogens are protists (protozoans), which affect a substantial proportion of the world's population, particularly in subtropical and tropical areas. Malaria has been covered in detail above (section 15.5.1) and two important flagellate protists of medical significance are described below.

Trypanosoma

Trypanosoma is a large genus of parasites of vertebrate blood that are transmitted usually by blood-feeding "higher" flies. However, throughout South America blood-feeding triatomine reduviid bugs ("kissing bugs"), notably *Rhodnius prolixus* and *Triatoma infestans*, transmit trypanosomes that cause Chagas' disease. Symptoms of the disease, also called American trypanosomiasis, are predominantly fatigue, with cardiac and intestinal problems if untreated. The disease affects 16–18 million people in the Neotropics, perhaps 350,000 in Brazil, and causes 45–50,000 deaths each year. From a public health perspective in the USA, some percentage of the millions of Latino migrants into the USA inevitably must have the disease, and localized transmission can occur. Other such diseases, termed **trypanosomiasis**, include sleeping sicknesses transmitted to African humans and their cattle by tsetse flies (species of *Glossina*) (Fig. 15.1). In this and other diseases, the development cycle of the *Trypanosoma* species is complex. Morphological change occurs in the protist as it migrates from the tsetse-fly gut, around the posterior free end of the peritrophic membrane, then anteriorly to the salivary gland. Transmission to human or cattle host is through injection of saliva. Within the vertebrate, symptoms depend upon the species of trypanosome: in humans, a vascular and lymphatic infection is followed by an invasion of the central nervous system that gives rise to "sleeping" symptoms, followed by death.

Leishmania

A second group of flagellates belong to the genus *Leishmania*, which includes parasites that cause internal visceral or disfiguring external ulcerating diseases of humans and dogs. The vectors are exclusively phlebotomines (Psychodidae) – small to minute sand flies that can evade mosquito netting and, in view of their usual very low biting rates, have impressive abilities to transmit disease. Most cycles cause infections in wild animals such as desert and forest rodents, canines, and hyraxes, with humans becoming involved as their homes expand into areas naturally home to these animal reservoirs. Some two million new cases are diagnosed each year, with approximately 12 million people infected at any given time. Visceral leishmaniasis (also known as kala-azar) inevitably kills if untreated; cutaneous leishmaniasis disfigures and leaves scars; mucocutaneous leishmaniasis destroys the mucous membranes of the mouth, nose, and throat.

15.5.5 Filariases

Two of the five main debilitating diseases transmitted by insects are caused by nematodes, namely filarial worms. The diseases are bancroftian and brugian filariases, commonly termed elephantiasis and onchocerciasis (or river blindness). Other filariases cause minor ailments in humans, and *Dirofilaria immitis* (canine heartworm) is one of the few significant veterinary diseases caused by this type of parasite. These filarial nematodes are dependent on *Wolbachia* bacteria for embryo development and thus infection can be reduced or eliminated with antibiotics (see also section 5.10.4).

Bancroftian and brugian filariasis

Two worms, *Wuchereria bancrofti* and *Brugia malayi*, are responsible for over a hundred million active cases of filariasis worldwide. The worms live in the lymphatic system, causing debilitation, and edema, culminating in extreme swellings of the lower limbs or genitals called elephantiasis. Although the disease is less often seen in the extreme form, the number of sufferers is increasing as one major vector, the worldwide peridomestic mosquito, *Culex quinquefasciatus*, increases.

The cycle starts with uptake of small microfilariae with blood taken up by the vector mosquito. The microfilariae move from the mosquito gut through the hemocoel into the flight muscles, where they mature into an infective larva. The 1.5 mm long larvae migrate through the hemocoel into the mosquito head where, when the mosquito next feeds, they rupture the labella and invade the host through the puncture of the mosquito bite. In the human host the larvae mature slowly over many months. The sexes are separate, and pairing of mature worms must take place before further microfilariae are produced. These microfilariae cannot mature without the mosquito phase. Cyclical (nocturnal periodic) movement of microfilariae into the peripheral circulatory system may make them more available to feeding mosquitoes.

Onchocerciasis

Onchocerciasis actually kills no-one directly but debilitates millions of people by scarring their eyes, which leads to blindness. The common name of "river blindness" refers to the impact of the disease on people living alongside rivers in West Africa and South America, where the insect vectors, *Simulium* black flies (Diptera: Simuliidae), live in flowing waters. The pathogen is a filarial worm, *Onchocerca volvulus*, in which the female is up to 50 mm long and the male smaller at 20–30 mm. The adult filariae live in subcutaneous nodules and are relatively harmless. It is the microfilariae that cause the damage to the eye when they invade the tissues and die there. The major black-fly vector has been shown to be one of the most extensive complexes of sibling species: "*Simulium damnosum*" has more than 40 cytologically determined species known from West and East Africa; in South America similar sibling species diversity in *Simulium* vectors is apparent. The larvae, which are common filter-feeders in flowing waters, are fairly readily controlled, but adults are strongly migratory and re-invasion of previously controlled rivers allows the disease to recur.

15.6 FORENSIC ENTOMOLOGY

As seen in section 15.3, some flies develop in living flesh, with two waves discernible: primary colonizers that cause initial myiases, with secondary myiases developing in pre-existing wounds. A third wave may follow before death. This ecological **succession** results from changes in the attractiveness of the substrate to different insects. An analogous succession of insects occurs in a corpse following death (section 9.4), with a somewhat similar course taken whether the corpse is a pig, rabbit, or human. This rather predictable succession in corpses has been used for medico-legal purposes by **forensic entomologists** as a faunistic method to assess the elapsed time (and even prevailing environmental conditions) since death for human corpses.

The generalized sequence of colonization is as follows. A fresh corpse is rapidly visited by a first wave of *Calliphora* (blow flies) and *Musca* (house flies), which oviposit or drop live larvae onto the cadaver. Their subsequent development to mature larvae (which depart the corpse to pupariate away from the larval development site) is temperature-dependent. Given knowledge of the particular species, the larval development times at different temperatures, and the ambient temperature at the corpse, an estimate of the age of a corpse may be made, perhaps accurate to within half a day if fresh, but with diminishing accuracy with increasing exposure.

As the corpse ages, larvae and adults of *Dermestes* (Coleoptera: Dermestidae) appear, followed by cheese-skipper larvae (Diptera: Piophilidae). As the body becomes drier, it is colonized by a sequence of other dipteran larvae, including those of Drosophilidae (fruit flies) and *Eristalis* (Diptera: Syrphidae: the rat-tailed maggot, a hover fly). After some months, when the

Text continues on p. 393.

Box 15.3 Phthiraptera (lice)

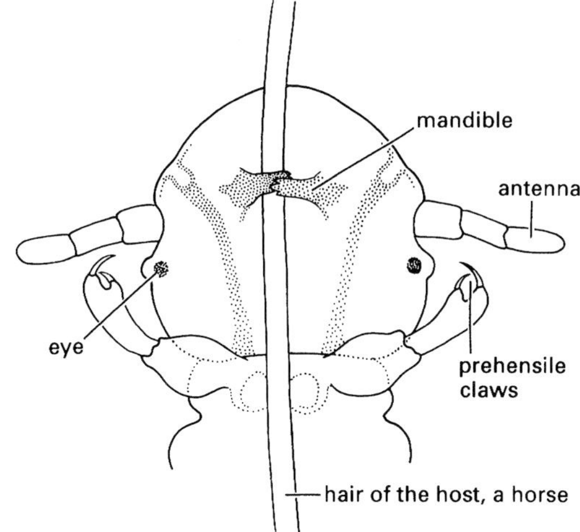

VENTRAL VIEW OF HEAD

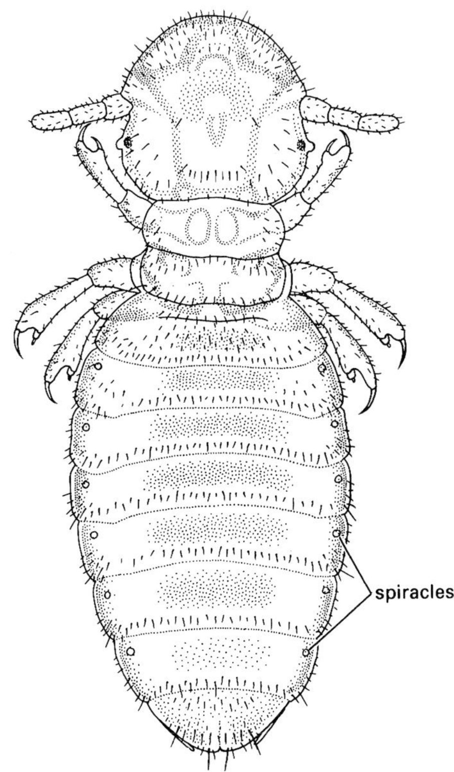

DORSAL VIEW OF LOUSE

The Phthiraptera is an order of some 5000 species of highly modified, apterous, dorsoventrally flattened ectoparasites, as typified by *Werneckiella equi*, the horse louse (Ischnocera: Trichodectidae) illustrated here. Lice are classified in four suborders: Rhynchophthirina (a small group found only on elephants and wart hogs), Amblycera and Ischnocera (the chewing or biting lice, formerly called Mallophaga), and Anoplura (sucking lice). Development is hemimetabolous. Mouthparts are mandibulate in Amblycera and Ischnocera, and beak-like for piercing and sucking in Anoplura (Fig. 2.14), which also lack maxillary palps. The eyes are either absent or reduced; the antennae are either held in grooves (Amblycera) or extended, filiform (and sometimes modified as claspers) in Ischnocera and Anoplura. The thoracic segments are variably fused, and completely fused in Anoplura. The legs are well developed and stout with strong claws used in grasping host hair or fur. Eggs are laid on the hair or feathers of the host. The nymphs resemble smaller, less pigmented adults, and all stages live on the host.

Lice are obligate ectoparasites lacking any free-living stage and occurring on all orders of birds and most orders of mammals (with the notable exception of bats). Ischnocera and Amblycera feed on bird feathers and mammal skin, with a few amblycerans feeding on blood. Anoplura feed solely on mammal blood.

The degree of host-specificity amongst lice is high and many monophyletic groups of lice occur on monophyletic groups of hosts. However, host speciation and parasite speciation do not match precisely, and historically many transfers have taken place between ecologically proximate but unrelated taxa (section 13.3.3). Furthermore, even when louse and host phylogenies match, a lag in timing between host speciation and lice differentiation may be evident, although gene transfer must have been interrupted simultaneously.

As with most parasitic insects, some Phthiraptera are involved in disease transmission. *Pediculus humanus corporis*, the human body louse, is one vector of typhus (section 15.5.3). It is notable that the subspecies *P. humanus capitis*, the human head louse (and *Pthirus pubis*, the pubic louse, illustrated on the right in the louse diagnosis in the Appendix), are insignificant typhus vectors, although often co-occurring with the body louse.

Phthiraptera are derived from within Psocoptera, all members of which are free living. Phylogenetic relationships are considered in section 7.4.2 and depicted in Fig. 7.2.

Box 15.4 Siphonaptera (fleas)

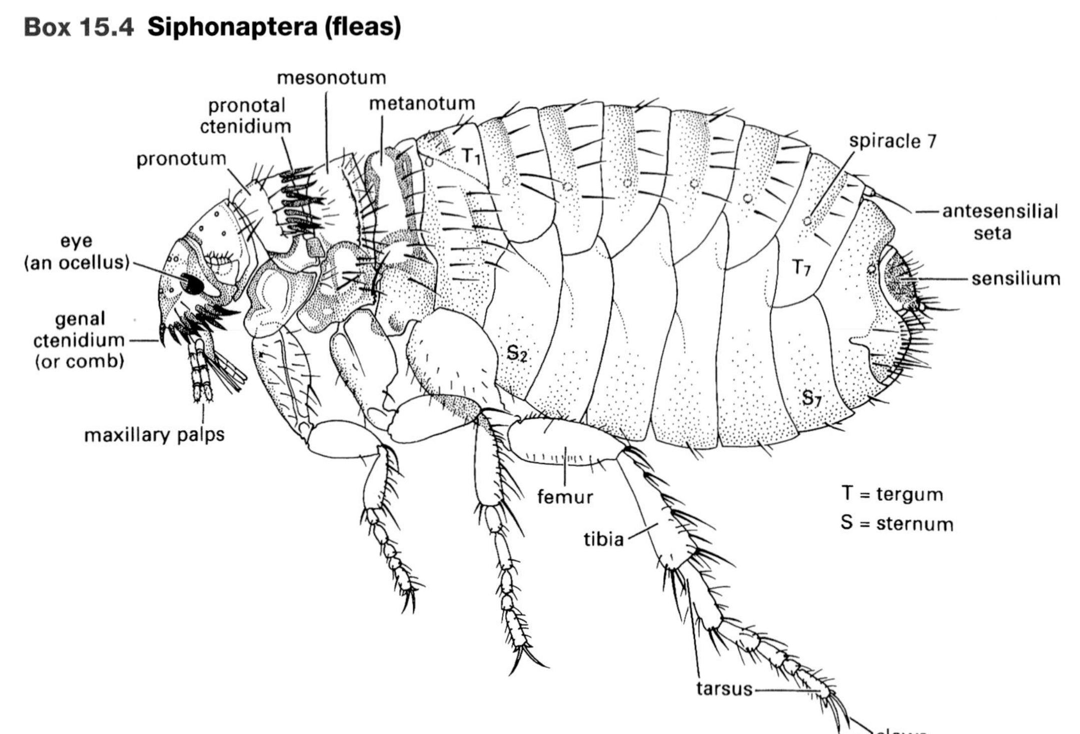

The Siphonaptera is an order of some 2500 species, all of which are highly modified, apterous, and laterally compressed ectoparasites. Development is holometabolous. The mouthparts (Fig. 2.15) are modified for piercing and sucking, without mandibles but with a stylet derived from the epipharynx and two elongate, serrate lacinial blades within a sheath formed from the labial palps. The gut has a salivary pump to inject saliva into the wound, and cibarial and pharyngeal pumps to suck up blood. Compound eyes are absent, and ocelli range from absent to well developed. Each antenna lies in a deep lateral groove. The body has many backwardly directed setae and spines; some may be grouped into combs (ctenidia) on the gena (part of the head) and thorax (especially the prothorax). The large metathorax houses the hind-leg muscles. The legs are long and strong, terminating in strong claws for grasping host hairs.

The large eggs are laid predominantly into the host's nest, where free-living worm-like larvae (illustrated in the Appendix) develop on material such as shed skin debris from the host. High temperatures and humidity are required for development by many fleas, including those on domestic cats (*Ctenocephalides felis*) (illustrated here), dogs (*C. canis*), and humans (*Pulex irritans*). The pupa is exarate and adecticous in a loose cocoon. Both sexes of adult take blood from a host, some species being **monoxenous** (restricted to one host), but many others being **polyxenous** (occurring on several to many hosts). The plague flea *Xenopsylla chiopis* belongs to the latter group, with polyxeny facilitating transfer of plague from rat to human host (section 15.5.3). Fleas transmit some other diseases of minor significance from other mammals to humans, including murine typhus and tularemia, but apart from plague, the most common human health threat is from allergic reaction to frequent bites from the fleas of our pets, *C. felis* and *C. canis*.

Fleas predominantly use mammalian hosts, with relatively few birds having fleas, these being derived from many lineages of mammal flea. Some hosts (e.g. *Rattus fuscipes*) have been reported to harbor more than 20 different species of flea, and conversely, some fleas have over 30 recorded hosts, so host-specificity is clearly much less than for lice.

Phylogenetic relationships are considered in section 7.4.2 and depicted in Fig. 7.6.

Box 15.5 Diptera (flies)

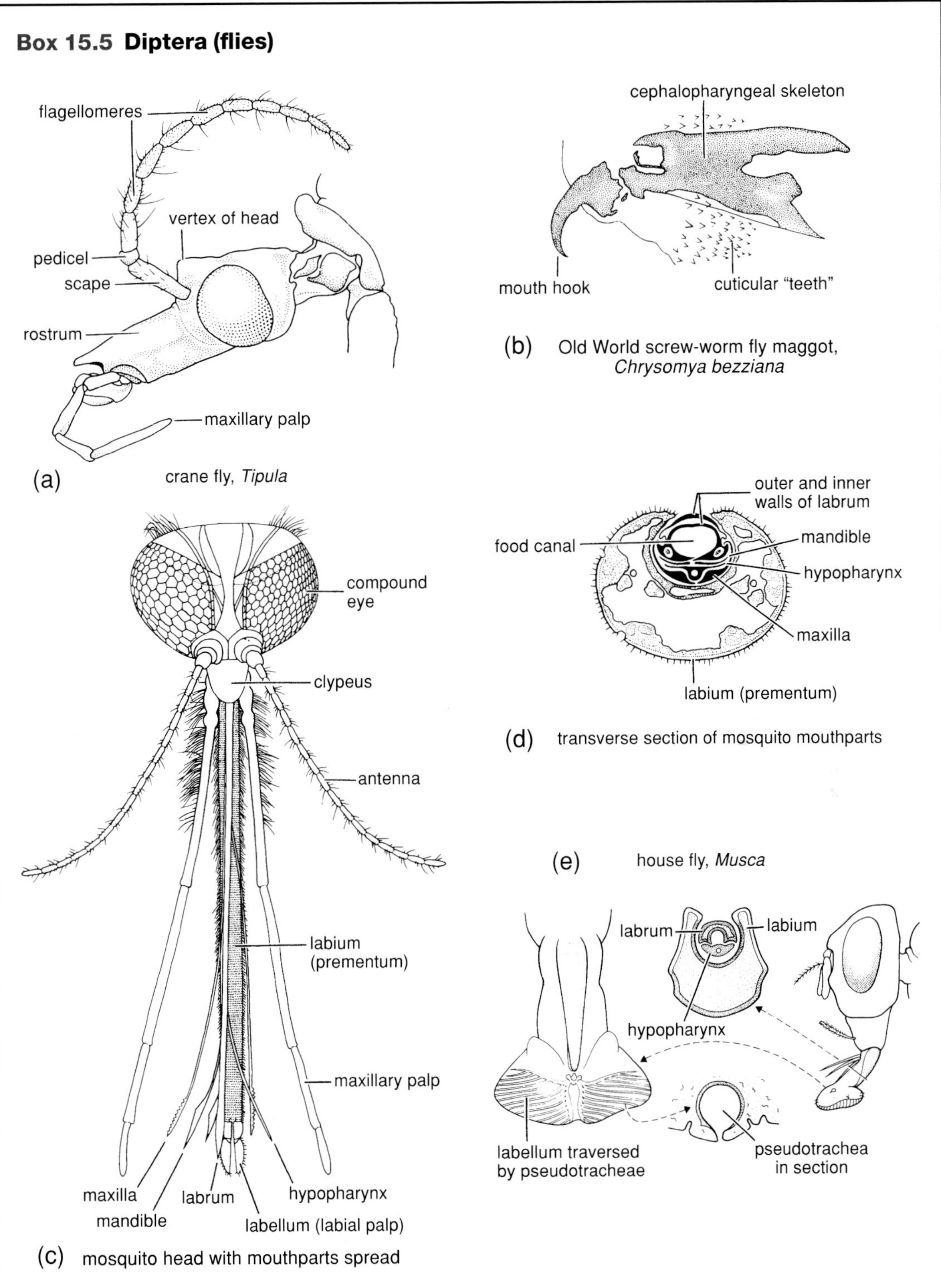

(a) crane fly, *Tipula*

(b) Old World screw-worm fly maggot, *Chrysomya bezziana*

(c) mosquito head with mouthparts spread

(d) transverse section of mosquito mouthparts

(e) house fly, *Musca*

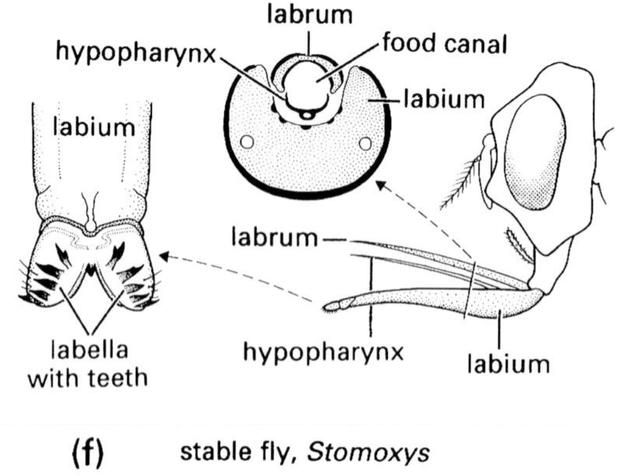

(f) stable fly, *Stomoxys*

The Diptera is an order containing perhaps some 125,000 described species, in roughly 130 families, with several thousands of species of medical or veterinary importance. Development is holometabolous, and adults have variously modified mouthparts, mesothoracic wings, and metathoracic halteres (balancers) (Fig. 2.22f). The larvae lack true legs (Fig. 6.6h), and their head structure ranges from a complete sclerotized capsule to acephaly with no external capsule and only an internal skeleton. The pupae are adecticous and obtect, or exarate in a puparium (Fig. 6.7e,f).

The paraphyletic Nematocera comprises crane flies, mosquitoes, midges, gnats, and relatives; these have slender antennae with upwards of six flagellomeres and a three- to five-segmented maxillary palp (illustrated for a crane fly (Tipulidae: *Tipula*) in (a), after McAlpine 1981). Brachycera contains heavier-built flies including hover flies, bee flies, dung flies, and blow flies. They have more solid, often shorter antennae with fewer (less than seven) flagellomeres, often with terminal arista (Fig. 2.17i); the maxillary palps have only one or two segments (illustrated for Muscidae in (e) and (f)). Within the Brachycera, schizophoran Cyclorrhapha use a ptilinum to aid emergence from the puparium.

Fly larvae have a wide variety of habits. Many nematoceran larvae are aquatic (Box 10.5), and brachyceran larvae show a phylogenetic radiation into drier and more specialized larval habits, including phytophagy, predation and parasitization of other arthropods, and myiasis-induction in vertebrates (section 15.3). Myiasis-inducing maggots have a much reduced head but with sclerotized mouthparts known as mouth hooks (illustrated for a third-instar larva of the Old World screwworm fly, *Chrysomya bezziana* (Calliphoridae) in (b), after Ferrar 1987), which scrape the living flesh of the host.

Adult dipteran mouthparts are illustrated in frontal view (c) (after Freeman & Bracegirdle 1971) and transverse section (d) (after Jobling 1976) for a female mosquito. All dipterans typically have a tubular sucking organ, the proboscis, comprising elongate mouthparts (usually including the labrum). A biting-and-sucking type of proboscis appears to be a primitive dipteran feature. Although biting functions have been lost and regained with modifications more than once, blood-feeding is frequent, and leads to the importance of the Diptera as vectors of disease. The blood-feeding female nematocerans – Culicidae (mosquitoes); Ceratopogonidae (biting midges); Psychodidae: Phlebotominae (sand flies); and Simuliidae (black flies) – have generally similar mouthparts, but differ in proboscis length, allowing penetration of the host to different depths. Mosquitoes can probe deep in search of capillaries, but other blood-feeding nematocerans operate more superficially where a pool of blood is induced in the wound. The labium ends in two sensory labella (singular: labellum), forming a protective sheath for the functional mouthparts. Enclosed are serrate-edged, cutting mandibles and maxillary lacinia, the curled labrum–epipharynx, and the hypopharynx, all of which are often termed stylets. When feeding, the labrum, mandibles, and laciniae act as a single unit driven through the skin of the host. The flexible labium remains bowed outside the wound. Saliva, which may contain anticoagulant, is injected through a salivary duct that runs the length of the sharply pointed and often toothed hypopharynx. Blood is transported up a food canal formed from the curled labrum sealed by either the paired mandibles or the hypopharynx. Capillary blood can flow unaided, but blood must be sucked or pumped from a pool with pumping action from two muscular pumps: the cibarial located at the base of the food canal, and the pharyngeal in the pharynx between the cibarium and midgut.

Many mouthparts are lost in the "higher" flies, and the remaining mouthparts are modified for lapping food using pseudotracheae of the labella as "sponges" (illustrated for a house fly (Muscidae: *Musca*) in (e), after Wigglesworth 1964). With neither mandibles nor maxillary lacinia to make a wound, blood-feeding cyclorrhaphans often use modified labella, in which the inner surfaces are adorned with sharp teeth (illustrated for a stable fly (Muscidae: *Stomoxys*) in (f), after Wigglesworth 1964). Through muscular contraction and relaxation, the labellar lobes dilate and contract repeatedly, creating an often painful rasping of the labellar teeth to give a pool of blood. The hypopharynx applies saliva which is dissipated via the labellar pseudotracheae. Uptake of blood is via capillary action through "food furrows" lying dorsal to the pseudotracheae, with the aid of three pumps operating synchronously to produce continuous suction from labella to pharynx. A prelabral pump produces the contractions in the labella, with a more proximal labral pump linked via a feeding tube to the cibarial pump.

The mouthparts and their use in feeding have implications for disease transmission. Shallow-feeding species such as black flies are more involved in transmission of microfilariae, such as those of *Onchocerca*, which aggregate just beneath the skin, whilst deeper feeders such as mosquitoes transmit pathogens that circulate in the blood. The transmission from fly to host is aided by the introduction of saliva into the wound, and many parasites aggregate in the salivary glands or ducts. Filariae, in contrast, are too large to enter the wound through this route, and leave the insect host by rupturing the labium or labella during feeding.

Phylogenetic relationships are considered in section 7.4.2 and depicted in Fig. 7.6.

corpse is completely dry, more species of Dermestidae appear and several species of clothes moth (Lepidoptera: Tineidae) scavenge the desiccated remnants.

This simple outline is confounded by a number of factors including:

1 geography, with different insect species (though perhaps relatives) present in different regions, especially if considered on a continental scale;

2 difficulty in identifying the early stages, of especially blow fly larvae, to species;

3 variation in ambient temperatures, with direct sunlight and high temperatures speeding the succession (even leading to rapid mummification), and shelter and cold conditions retarding the process;

4 variation in exposure of the corpse, with burial, even partial, slowing the process considerably, and with a very different entomological succession;

5 variation in cause and site of death, with death by drowning and subsequent degree of exposure on the shore giving rise to a different necrophagous fauna from those infesting a terrestrial corpse, with differences between freshwater and marine stranding.

Problems with identification of larvae using morphology are being alleviated using DNA-based approaches. Entomological forensic evidence has proved crucial to post-mortem investigations. Forensic entomological evidence has been particularly successful in establishing disparities between the location of a crime scene and the site of discovery of the corpse, and between the time of death (perhaps homicide) and subsequent availability of the corpse for insect colonization.

FURTHER READING

Dye, C. (1992) The analysis of parasite transmission by blood-sucking insects. *Annual Review of Entomology* **37**, 1–19.

Hinkle, N.C. (2000) Delusory parasitosis. *American Entomologist* **46**, 17–25.

Kettle, D.S. (1995) *Medical and Veterinary Entomology*, 2nd edn. CAB International, Wallingford.

Lane, R.P. & Crosskey, R.W. (eds.) (1993) *Medical Insects and Arachnids*. Chapman & Hall, London.

Lehane, M.J. (1991) *Biology of Blood-sucking Insects*. Harper Collins Academic, London.

Mullen, G. & Durden, L. (eds.) (2002) *Medical and Veterinary Entomology*. Academic Press, San Diego, CA.

Smith, K.G.V. (1986) *A Manual of Forensic Entomology*. The Trustees of the British Museum (Natural History), London.

Chapter 16

PEST MANAGEMENT

Biological control of aphids by coccinellid beetles. (After Burton & Burton 1975.)

Insects become pests when they conflict with our welfare, aesthetics, or profits. For example, otherwise innocuous insects can provoke severe allergic reactions in sensitized people, and reduction or loss of food-plant yield is a universal result of insect-feeding activities and pathogen transmission. Pests thus have no particular ecological significance but are defined from a purely anthropocentric point of view. Insects may be pests of people either directly through disease transmission (Chapter 15), or indirectly by affecting our domestic animals, cultivated plants, or timber reserves. From a conservation perspective, introduced insects become pests when they displace native species, often with ensuing effects on other non-insect species in the community. Some introduced and behaviorally dominant ants, such as the big-headed ant, *Pheidole megacephala*, and the Argentine ant, *Linepithema humile*, impact negatively on native biodiversity in many islands including those of the tropical Pacific (Box 1.2). Honey bees (*Apis mellifera*) outside their native range form feral nests and, although they are generalists, may out-compete local insects. Native insects usually are efficient pollinators of a smaller range of native plants than are honey bees, and their loss may lead to reduced seed set. Research on insect pests relevant to conservation biology is increasing, but remains modest compared to a vast literature on pests of our crops, garden plants, and forest trees.

In this chapter we deal predominantly with the occurrence and control of insect pests of agriculture, including horticulture or silviculture, and with the management of insects of medical and veterinary importance. We commence with a discussion of what constitutes a pest, how damage levels are assessed, and why insects become pests. Next, the effects of insecticides and problems of insecticide resistance are considered prior to an overview of integrated pest management (IPM). The remainder of the chapter discusses the principles and methods of management applied in IPM, namely: chemical control, including insect growth regulators and neuropeptides; biological control using natural enemies (such as the coccinellid beetles shown eating aphids in the vignette of this chapter) and microorganisms; host-plant resistance; mechanical, physical, and cultural control; the use of attractants such as pheromones; and finally genetic control of insect pests. A more comprehensive list than for other chapters is provided as further reading because of the importance and breadth of topics covered in this chapter.

16.1 INSECTS AS PESTS

16.1.1 Assessment of pest status

The pest status of an insect population depends on the abundance of individuals as well as the type of nuisance or injury that the insects inflict. **Injury** is the usually deleterious effect of insect activities (mostly feeding) on host physiology, whereas **damage** is the measurable loss of host usefulness, such as yield quality or quantity or aesthetics. Host injury (or insect number used as an injury estimate) does not necessarily inflict detectable damage and even if damage occurs it may not result in appreciable economic loss. Sometimes, however, the damage caused by even a few individual insects is unacceptable, as in fruit infested by codling moth or fruit fly. Other insects must reach high or plague densities before becoming pests, as in locusts feeding on pastures. Most plants tolerate considerable leaf or root injury without significant loss of vigor. Unless these plant parts are harvested (e.g. leaf or root vegetables) or are the reason for sale (e.g. indoor plants), certain levels of insect feeding on these parts should be more tolerable than for fruit, which "sophisticated" consumers wish to be blemish-free. Often the effects of insect feeding may be merely cosmetic (such as small marks on the fruit surface) and consumer education is more desirable than expensive controls. As market competition demands high standards of appearance for food and other commodities, assessments of pest status often require socio-economic as much as biological judgments.

Pre-emptive measures to counter the threat of arrival of particular novel insect pests are sometimes taken. Generally, however, control becomes economic only when insect density or abundance cause (or are expected to cause if uncontrolled) financial loss of productivity or marketability greater than the costs of control. Quantitative measures of insect density (section 13.4) allow assessment of the pest status of different insect species associated with particular agricultural crops. In each case, an **economic injury level (EIL)** is determined as the pest density at which the loss caused by the pest equals in value the cost of available control measures or, in other words, the lowest population density that will cause economic damage. The formula for calculating the EIL includes four factors:

1 costs of control;
2 market value of the crop;
3 yield loss attributable to a unit number of insects;
4 effectiveness of the control;

and is as follows:

$$EIL = C/VDK$$

in which EIL is pest number per production unit (e.g. insects ha^{-1}), C is cost of control measure(s) per production unit (e.g. $ ha^{-1}), V is market value per unit of product (e.g. $ kg^{-1}), D is yield loss per unit number of insects (e.g. kg reduction of crop per n insects), and K is proportionate reduction of insect population caused by control measures.

The calculated EIL will not be the same for different pest species on the same crop or for a particular insect pest on different crops. The EIL also may vary depending on environmental conditions, such as soil type or rainfall, as these can affect plant vigor and compensatory growth. Control measures normally are instigated before the pest density reaches the EIL, as there may be a time lag before the measures become effective. The density at which control measures should be applied to prevent an increasing pest population from attaining the EIL is referred to as the **economic threshold (ET)** (or an "action threshold"). Although the ET is defined in terms of population density, it actually represents the time for instigation of control measures. It is set explicitly at a different level from the EIL and thus is predictive, with pest numbers being used as an index of the time when economic damage will occur.

Insect pests may be described as being one of the following:

- Non-economic, if their populations are never above the EIL (Fig. 16.1a).
- Occasional pests, if their population densities exceed the EIL only under special circumstances (Fig. 16.1b), such as atypical weather or inappropriate use of insecticides.
- Perennial pests, if the general equilibrium population of the pest is close to the ET so that pest population density reaches the EIL frequently (Fig. 16.1c).
- Severe or key pests, if their numbers (in the absence of controls) always are higher than the EIL (Fig. 16.1d). Severe pests must be controlled if the crop is to be grown profitably.

The EIL fails to consider the influence of variable external factors, including the role of natural enemies, resistance to insecticides, and the effects of control measures in adjoining fields or plots. Nevertheless, the virtue of the EIL is its simplicity, with management depending on the availability of decision rules that can be comprehended and implemented with relative ease. The concept of the EIL was developed primarily as a means for more sensible use of insecticides, and its application is confined largely to situations in which control measures are discrete and curative, i.e. chemical or microbial insecticides. Often EILs and ETs are difficult or impossible to apply due to the complexity of many agroecosystems and the geographic variability of pest problems. More complex models and dynamic thresholds are needed but these require years of field research.

The discussion above applies principally to insects that directly damage an agricultural crop. For forest pests, estimation of almost all of the components of the EIL is difficult or impossible, and EILs are relevant only to short-term forest products such as Christmas trees. Furthermore, if insects are pests because they can transmit (vector) disease of plants or animals, then the ET may be their first appearance. The threat of a virus affecting crops or livestock and spreading via an insect vector requires constant vigilance for the appearance of the vector and the presence of the virus. With the first occurrence of either vector or disease symptoms, precautions may need to be taken. For economically very serious disease, and often in human health, precautions are taken before any ET is reached, and insect vector and virus population monitoring and modeling is used to estimate when pre-emptive control is required. Calculations such as the vectorial capacity, referred to in Chapter 15, are important in allowing decisions concerning the need and appropriate timing for pre-emptive control measures. However, in human insect-borne disease, such rationales often are replaced by socio-economic ones, in which levels of vector insects that are tolerated in less developed countries or rural areas are perceived as requiring action in developed countries or in urban communities.

A limitation of the EIL is its unsuitability for multiple pests, as calculations become complicated. However, if injuries from different pests produce the same type of damage, or if effects of different injuries are additive rather than interactive, then the EIL and ET may still apply. The ability to make management decisions for a pest complex (many pests in one crop) is an important part of integrated pest management (section 16.3).

16.1.2 Why insects become pests

Insects may become pests for one or more reasons. First, some previously harmless insects become pests

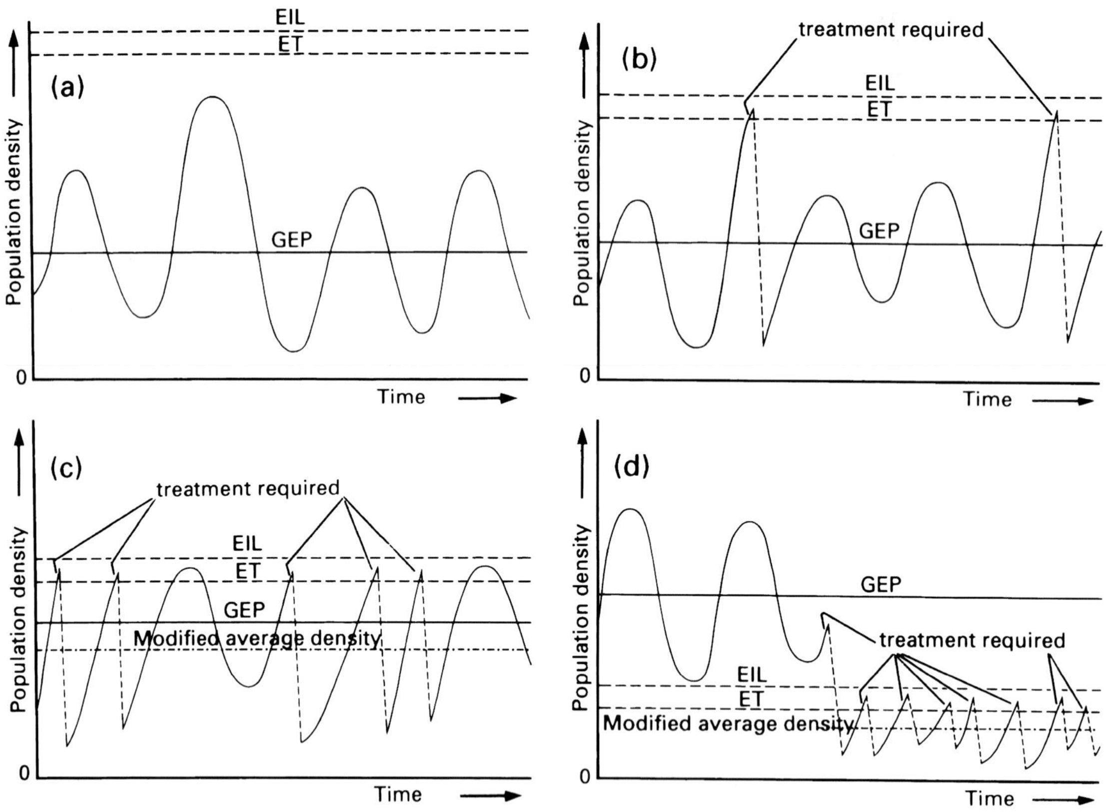

Fig. 16.1 Schematic graphs of the fluctuations of theoretical insect populations in relation to their general equilibrium population (GEP), economic threshold (ET), and economic injury level (EIL). From comparison of the general equilibrium density with the ET and EIL, insect populations can be classified as: (a) non-economic pests if population densities never exceed the ET or EIL; (b) occasional pests if population densities exceed the ET and EIL only under special circumstances; (c) perennial pests if the general equilibrium population is close to the ET so that the ET and EIL are exceeded frequently; or (d) severe or key pests if population densities always are higher than the ET and EIL. In practice, as indicated here, control measures are instigated before the EIL is reached. (After Stern et al. 1959.)

after their accidental (or intentional) introduction to areas outside their native range, where they escape from the controlling influence of their natural enemies. Such range extensions have allowed many previously innocuous phytophagous insects to flourish as pests, usually following the deliberate spread of their host plants through human cultivation. Second, an insect may be harmless until it becomes a vector of a plant or animal (including human) pathogen. For example, mosquito vectors of malaria and filariasis occur in the USA, England, and Australia but the diseases are absent currently. Third, native insects may become pests if they move from native plants onto introduced ones; such host switching is common for polyphagous and oligophagous insects. For example, the oligophagous Colorado potato beetle switched from other solanaceous host plants to potato, *Solanum tuberosum*, during the 19th century (Box 16.5), and some polyphagous larvae of *Helicoverpa* and *Heliothis* (Lepidoptera: Noctuidae) have become serious pests of cultivated cotton and other crops within the native range of the moths.

A fourth, related, problem is that the simplified, virtually monocultural, ecosystems in which our food crops and forest trees are grown and our livestock are raised create dense aggregations of predictably available resources that encourage the proliferation of specialist and some generalist insects. Certainly, the pest status of many native noctuid caterpillars is elevated by the provision of abundant food resources. Moreover, natural enemies of pest insects generally require more diverse habitat or food resources and are discouraged from agro-monocultures. Fifth, in addition to large-scale monocultures, other farming or cultivating methods can lead to previously benign species or minor pests becoming major pests. Cultural practices such as continuous cultivation without a fallow period allow build-up of insect pest numbers. The inappropriate or prolonged use of insecticides can eliminate natural enemies of phytophagous insects while inadvertently selecting for insecticide resistance in the latter. Released from natural enemies, other previously non-pest species sometimes increase in numbers until they reach ETs. These problems of insecticide use are discussed in more detail below.

Sometimes the primary reason why a minor nuisance insect becomes a serious pest is unclear. Such a change in status may occur suddenly and none of the conventional explanations given above may be totally satisfactory either alone or in combination. An example is the rise to notoriety of the silverleaf whitefly, which is variously known as *Bemisia tabaci* biotype B or *B. argentifolii*, depending on whether this insect is regarded as a distinct species or a form of *B. tabaci* (Box 16.1).

Box 16.1 *Bemisia tabaci* biotype B: a new pest or an old one transformed?

Bemisia tabaci, often called the tobacco or sweetpotato whitefly, is a polyphagous and predominantly tropical–subtropical whitefly (Hemiptera: Aleyrodidae) that feeds on numerous fiber (particularly cotton), food, and ornamental plants. Nymphs suck phloem sap from minor veins (as illustrated diagrammatically on the left of the figure, after Cohen et al. 1998). Their thread-like mouthparts (section 11.2.3; Fig. 11.4) must contact a suitable vascular bundle in order for the insects to feed successfully. The whiteflies cause plant damage by inducing physiological changes in some hosts, such as irregular ripening in tomato and silverleafing in squash and zucchini (courgettes), by fouling with excreted honeydew and subsequent sooty mold growth, and by the transmission of more than 70 viruses, particularly geminiviruses (Geminiviridae).

Infestations of *B. tabaci* have increased in severity since the early 1980s owing to intensive continuous cropping with heavy reliance on insecticides and the possibly related spread of what is either a virulent form of the insect or a morphologically indistinguishable sibling species. The likely area of origin of this pest, often called *B. tabaci* biotype B, is the Middle East, perhaps Israel. Certain entomologists (especially in the USA) recognize the severe pest as a separate species, *B. argentifolii*, the silverleaf whitefly (the fourth-instar nymph or "puparium" is depicted on the right, after Bellows et al. 1994), so-named because of the leaf symptoms it causes in squash and zucchini. *B. argentifolii* exhibits minor and labile cuticular differences from the true *B. tabaci* (often called biotype A) but comparisons extended to morphologies of eight biotypes of

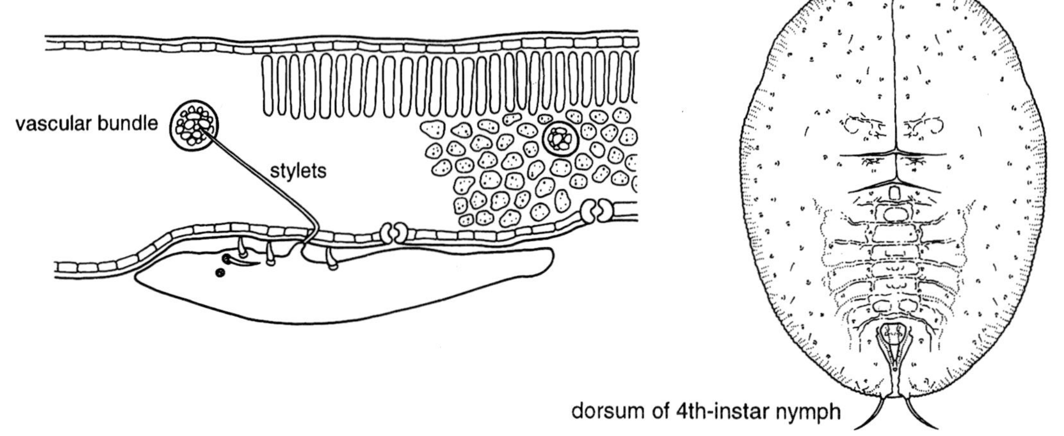

B. tabaci found no reliable features to separate them. However, clear allozyme, nuclear, and mitochondrial genetic information allows separation of the non-B biotypes of *B. tabaci*. Nucleotide sequences of the 18S rDNAs of biotypes A and B and the 16S rDNAs of their bacterial endosymbionts are essentially identical, suggesting that these two whiteflies are either the same or very recently evolved species. Some biotypes show variable reproductive incompatibility, as shown by crossing experiments, which may be due to the presence of strain- or sex-specific bacteria, resembling the *Wolbachia* and similar endosymbiont activities observed in other insects (section 5.10.4). Populations of *B. tabaci* biotype A are eliminated wherever biotype B is introduced, suggesting that incompatibility might be mediated by microorganisms. Indeed, the bacterial faunas of *B. tabaci* biotypes A and B show some differences in composition, consistent with the hypothesis that symbiont variation may be associated with biotype formation. For example, recently it was shown that biotype A, but not biotype B, is infected by a chlamydia species (Simkaniaceae: *Fritschea bemisiae*) and it is possible that the presence of this bacterium influences the fitness of its host whitefly. Furthermore, endosymbionts in some other Hemiptera have been associated with enhanced virus transmission (section 3.6.5), and it is possible that endosymbionts mediate the transmission of geminiviruses by *B. tabaci* biotypes.

The sudden appearance and spread of this apparently new pest, *B. tabaci* biotype B, highlights the importance of recognizing fine taxonomic and biological differences among economically significant insect taxa. This requires an experimental approach, including hybridization studies with and without bacterial associates. It is probable that *B. tabaci* is a sibling species complex, in which most of the species currently are called biotypes, but some forms (e.g. biotypes A and B) may be conspecific although biologically differentiated by endosymbiont manipulation. In addition, it is feasible that strong selection, resulting from heavy insecticide use, may select for particular strains of whitefly or bacterial symbionts that are more resistant to the chemicals.

Effective biological control of *Bemisia* whiteflies is possible using host-specific parasitoid wasps, such as *Encarsia* and *Eretmocerus* species (Aphelinidae). However, the intensive and frequent application of broad-spectrum insecticides adversely affects biological control. Even *B. tabaci* biotype B can be controlled if insecticide use is reduced.

16.2 THE EFFECTS OF INSECTICIDES

The chemical insecticides developed during and after World War II initially were effective and cheap. Farmers came to rely on the new chemical methods of pest control, which rapidly replaced traditional forms of chemical, cultural, and biological control. The 1950s and 1960s were times of an insecticide boom, but use continued to rise and insecticide application is still the single main pest control tactic employed today. Although pest populations are suppressed by insecticide use, undesirable effects include the following:

1 Selection for insects that are genetically resistant to the chemicals (section 16.2.1).

2 Destruction of non-target organisms, including pollinators, the natural enemies of the pests, and soil arthropods.

3 **Pest resurgence** – as a consequence of effects **1** and **2**, a dramatic increase in numbers of the targeted pest(s) can occur (e.g. severe outbreaks of cottony-cushion scale as a result of dichlorodiphenyltrichloroethane (DDT) use in California in the 1940s (Box 16.2; see also Plate 6.6, facing p. 14)) and if the natural enemies recover much more slowly than the pest population, the latter can exceed levels found prior to insecticide treatment.

4 **Secondary pest outbreak** – a combination of suppression of the original target pest and effects **1** and **2** can lead to insects previously not considered pests being released from control and becoming major pests.

5 Adverse environmental effects, resulting in contamination of soils, water systems, and the produce itself with chemicals that accumulate biologically (especially in vertebrates) as the result of biomagnification through food chains.

6 Dangers to human health either directly from the handling and consumption of insecticides or indirectly via exposure to environmental sources.

Despite increased insecticide use, damage by insect pests has increased; for example, insecticide use in the USA increased 10-fold from about 1950 to 1985, whilst the proportion of crops lost to insects roughly doubled (from 7% to 13%) during the same period. Such figures do not mean that insecticides have not controlled insects, because non-resistant insects clearly are killed by chemical poisons. Rather, an array of factors accounts for this imbalance between pest problems and control measures. Human trade has

Box 16.2 The cottony-cushion scale

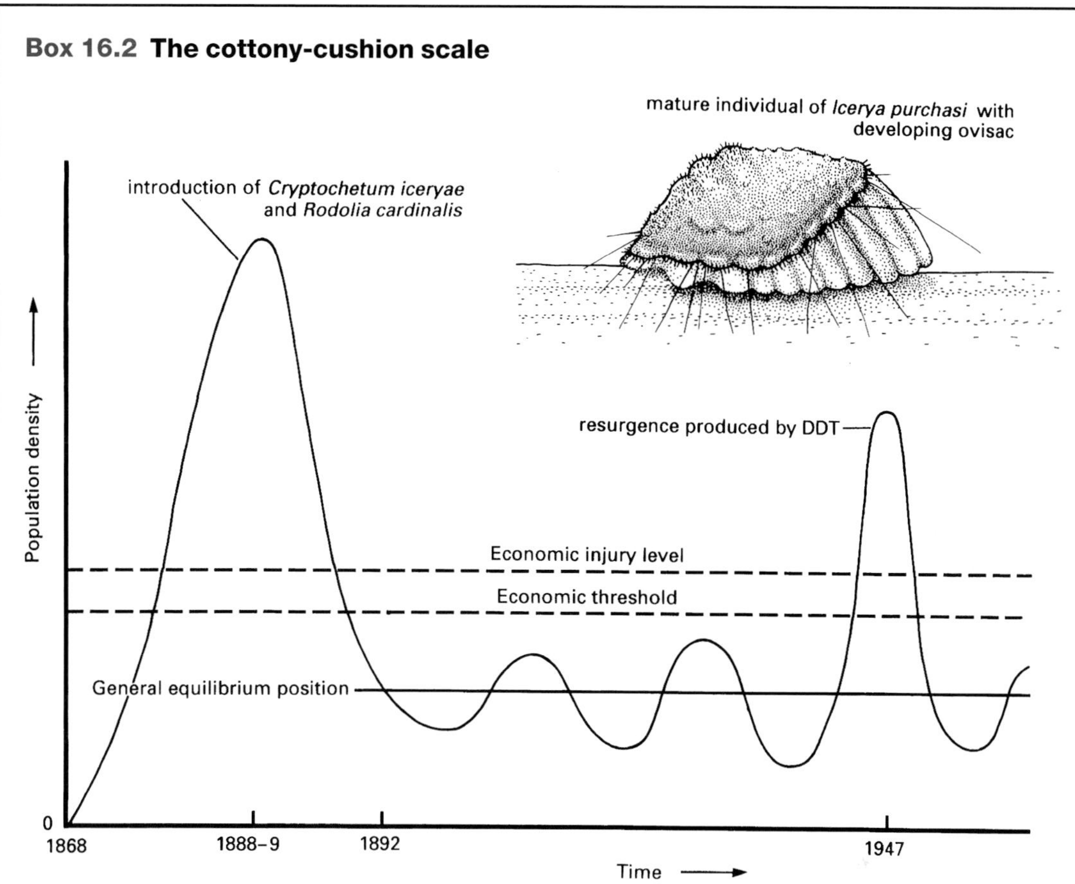

An example of a spectacularly successful classical biological control system is the control of infestations of the cottony-cushion scale, *Icerya purchasi* (Hemiptera: Margarodidae), in Californian citrus orchards from 1889 onwards, as illustrated in the accompanying graph (after Stern et al. 1959). Control has been interrupted only by DDT use, which killed natural enemies and allowed resurgence of cottony-cushion scale.

The hermaphroditic, self-fertilizing adult of this scale insect produces a very characteristic fluted white ovisac (see inset on graph; see also Plate 6.6, facing p. 14), under which several hundred eggs are laid. This mode of reproduction, in which a single immature individual can establish a new infestation, combined with polyphagy and capacity for multivoltinism in warm climates, makes the cottony-cushion scale a potentially serious pest. In Australia, the country of origin of the cottony-cushion scale, populations are kept in check by natural enemies, especially ladybird beetles (Coleoptera: Coccinellidae) and parasitic flies (Diptera: Cryptochetidae).

Cottony-cushion scale was first noticed in the USA in about 1868 on a wattle (*Acacia*) growing in a park in northern California. By 1886, it was devastating the new and expanding citrus industry in southern California. Initially, the native home of this pest was unknown but correspondence between entomologists in the USA, Australia, and New Zealand identified Australia as the source. The impetus for the introduction of exotic natural enemies came from C.V. Riley, Chief of the Division of Entomology of the US Department of Agriculture. He arranged for A. Koebele to collect natural enemies in Australia and New Zealand from 1888 to 1889 and ship them to D.W. Coquillett for rearing and release in Californian orchards. Koebele obtained many cottony-cushion scales infected with flies of *Cryptochetum iceryae* and also coccinellids of *Rodolia cardinalis*, the vedalia ladybird. Mortality during several shipments was high and only about 500 vedalia beetles arrived alive in the USA; these were bred and distributed to all Californian citrus growers with outstanding results. The vedalia beetles ate their way through infestations of cottony-cushion scale, the citrus industry was saved and biological control became popular. The parasitic fly was largely forgotten in these early days of enthusiasm for coccinellid predators. Thousands of flies were imported as a result of Koebele's collections but establishment from this source is doubtful. Perhaps the major or only source of the present populations of *C. iceryae* in California was a batch sent in late 1887 by F. Crawford of Adelaide, Australia, to W.G. Klee, the California State Inspector of Fruit Pests, who made releases near San Francisco in early 1888, before Koebele ever visited Australia.

Today, both *R. cardinalis* and *C. iceryae* control populations of *I. purchasi* in California, with the beetle dominant in the hot, dry inland citrus areas and the fly most important in the cooler coastal region; interspecific competition can occur if conditions are suitable for both species. Furthermore, the vedalia beetle, and to a lesser extent the fly, have been introduced successfully into many countries worldwide wherever *I. purchasi* has become a pest. Both predator and parasitoid have proved to be effective regulators of cottony-cushion scale numbers, presumably owing to their specificity and efficient searching ability, aided by the limited dispersal and aggregative behavior of their target scale insect. Unfortunately, few subsequent biological control systems involving coccinellids have enjoyed the same success.

accelerated the spread of pests to areas outside the ranges of their natural enemies. Selection for high-yield crops often inadvertently has resulted in susceptibility to insect pests. Extensive monocultures are commonplace, with reduction in sanitation and other cultural practices such as crop rotation. Finally, aggressive commercial marketing of chemical insecticides has led to their inappropriate use, perhaps especially in developing countries.

16.2.1 Insecticide resistance

Insecticide **resistance** is the result of selection of individuals that are predisposed genetically to survive an insecticide. **Tolerance**, the ability of an individual to survive an insecticide, implies nothing about the basis of survival. Over the past few decades more than 500 species of arthropod pests have developed resistance to one or more insecticides (Fig. 16.2).

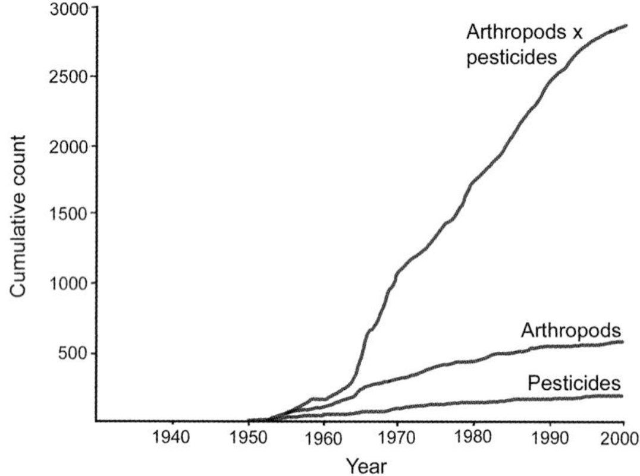

Fig. 16.2 Cumulative increase in the number of arthropod species (mostly insects and mites) known to be resistant to one or more insecticides. (After Bills et al. 2000.)

The tobacco or silverleaf whitefly (Box 16.1), the Colorado potato beetle (Box 16.5), and the diamondback moth (see discussion of Bt in section 16.5.2) are resistant to virtually all chemicals available for control. Chemically based pest control of these and many other pests may soon become virtually ineffectual because many show cross- or multiple resistance. **Cross-resistance** is the phenomenon of a resistance mechanism for one insecticide giving tolerance to another. **Multiple resistance** is the occurrence in a single insect population of more than one defense mechanism against a given compound. The difficulty of distinguishing cross-resistance from multiple resistance presents a major challenge to research on insecticide resistance. Mechanisms of insecticide resistance include:

- increased behavioral avoidance, as some insecticides, such as neem and pyrethroids, can repel insects;
- physiological changes, such as sequestration (deposition of toxic chemicals in specialized tissues), reduced cuticular permeability (penetration), or accelerated excretion;
- biochemical detoxification (called **metabolic resistance**) mediated by specialized enzymes;
- increased tolerance as a result of decreased sensitivity

to the presence of the insecticide at its target site (called **target-site resistance**).

The tobacco budworm, *Heliothis virescens* (Lepidoptera: Noctuidae), a major pest of cotton in the USA, exhibits behavioral, penetration, metabolic, and target-site resistance. Phytophagous insects, especially polyphagous ones, frequently develop resistance more rapidly than their natural enemies. Polyphagous herbivores may be preadapted to evolve insecticide resistance because they have general detoxifying mechanisms for secondary compounds encountered among their host plants. Certainly, detoxification of insecticidal chemicals is the most common form of insecticide resistance. Furthermore, insects that chew plants or consume non-vascular cell contents appear to have a greater ability to evolve pesticide resistance compared with phloem- and xylem-feeding species. Resistance has developed also under field conditions in some arthropod natural enemies (e.g. some lacewings, parasitic wasps, and predatory mites), although few have been tested. Intraspecific variability in insecticide tolerances has been found among certain populations subjected to differing insecticide doses.

Insecticide resistance in the field is based on relatively few or single genes (monogenic resistance), i.e. owing to allelic variants at just one or two loci. Field applications of chemicals designed to kill all individuals lead to rapid evolution of resistance, because strong selection favors novel variants such as a very rare allele for resistance present at a single locus. In contrast, laboratory selection often is weaker, producing polygenic resistance. Single-gene insecticide resistance could be due also to the very specific modes of action of certain insecticides, which allow small changes at the target site to confer resistance.

Management of insecticide resistance requires a program of controlled use of chemicals with the primary goals of: (i) avoiding or (ii) slowing the development of resistance in pest populations; (iii) causing resistant populations to revert to more susceptible levels; and/or (iv) fostering resistance in selected natural enemies. The tactics for resistance management can involve maintaining reservoirs of susceptible pest insects (either in refuges or by immigration from untreated areas) to promote dilution of any resistant genes, varying the dose or frequency of insecticide applications, using less-persistent chemicals, and/or applying insecticides as a rotation or sequence of different chemicals or as a mixture. The optimal strategy for retarding the evolution of resistance is to use insecticides only when control by natural enemies fails to curtail economic damage. Furthermore, resistance monitoring should be an integral component of management, as it allows the anticipation of problems and assessment of the effectiveness of operational management tactics.

Recognition of the problems discussed above, cost of insecticides, and also a strong consumer reaction to environmentally damaging agronomic practices and chemical contamination of produce have led to the current development of alternative pest control methods. In some countries and for certain crops, chemical controls increasingly are being integrated with, and sometimes replaced by, other methods.

16.3 INTEGRATED PEST MANAGEMENT

Historically, **integrated pest management (IPM)** was promoted first during the 1960s as a result of the failure of chemical insecticides, notably in cotton production, which in some regions required at least 12 sprayings per crop. IPM philosophy is to limit economic damage to the crop and simultaneously minimize adverse effects on non-target organisms in the crop and surrounding environment and on consumers of the produce. Successful IPM requires a thorough knowledge of the biology of the pest insects, their natural enemies, and the crop to allow rational use of a variety of cultivation and control techniques under differing circumstances. The key concept is integration of (or compatibility among) pest management tactics. The factors that regulate populations of insects (and other organisms) are varied and interrelated in complex ways. Thus, successful IPM requires an understanding of both population processes (e.g. growth and reproductive capabilities, competition, and effects of predation and parasitism) and the effects of environmental factors (e.g. weather, soil conditions, disturbances such as fire, and availability of water, nutrients, and shelter), some of which are largely stochastic in nature and may have predictable or unpredictable effects on insect populations. The most advanced form of IPM also takes into consideration societal and environmental costs and benefits within an ecosystem context when making management decisions. Efforts are made to conserve the long-term health and productivity of the ecosystem, with a philosophy approaching that of organic farming. One of the rather few examples of this advanced IPM is insect pest management in tropical irrigated rice, in which there is co-ordinated training of farmers by other farmers and field research involving local communities in implementing successful IPM. Worldwide, other functional IPM systems include the field crops of cotton, alfalfa, and citrus in certain regions, and many greenhouse crops.

Despite the economic and environmental advantages of IPM, implementation of IPM systems has been slow. For example, in the USA, true IPM is probably being practiced on much less than 10% of total crop area, despite decades of Federal government commitments to increased IPM. Often what is called IPM is simply "integrated pesticide management" (sometimes called first-level IPM) with pest consultants monitoring crops to determine when to apply insecticides. Universal reasons for lack of adoption of advanced IPM include:

• lack of sufficient data on the ecology of many insect pests and their natural enemies;
• requirement for knowledge of EILs for each pest of each crop;
• requirement for interdisciplinary research in order to obtain the above information;
• risks of pest damage to crops associated with IPM strategies;
• apparent simplicity of total insecticidal control combined with the marketing pressures of pesticide companies;
• necessity of training farmers, agricultural extension officers, foresters, and others in new principles and methods.

Successful IPM often requires extensive biological research. Such applied research is unlikely to be financed by many industrial companies because IPM may reduce their insecticide market. However, IPM does incorporate the use of chemical insecticides, albeit at a reduced level, although its main focus is the establishment of a variety of other methods of controlling insect pests. These usually involve modifying the insect's physical or biological environment or, more rarely, entail changing the genetic properties of the insect. Thus, the control measures that can be used in IPM include: insecticides, biological control, cultural control, plant resistance improvement, and techniques that interfere with the pest's physiology or reproduction, namely genetic (e.g. sterile insect technique; section 16.10), semiochemical (e.g. pheromone), and insect growth-regulator control methods. The remainder of this chapter discusses the various principles and methods of insect pest control that could be employed in IPM systems.

16.4 CHEMICAL CONTROL

Despite the hazards of conventional insecticides, some use is unavoidable. However, careful chemical choice and application can reduce ecological damage. Carefully timed suppressant doses can be delivered at vulnerable stages of the pest's life cycle or when a pest population is about to explode in numbers. Appropriate and efficient use requires a thorough knowledge of the pest's field biology and an appreciation of the differences among available insecticides.

An array of chemicals has been developed for the purposes of killing insects. These enter the insect body either by penetrating the cuticle, called contact action or dermal entry, by inhalation into the tracheal system, or by oral ingestion into the digestive system. Most **contact poisons** also act as **stomach poisons** if ingested by the insect, and toxic chemicals that are ingested by the insect after translocation through a host are referred to as **systemic insecticides**. Fumigants used for controlling insects are **inhalation poisons**. Some chemicals may act simultaneously as inhalation, contact, and stomach poisons. Chemical insecticides generally have an acute effect and their mode of action (i.e. method of causing death) is via the nervous system, either by inhibiting acetylcholinesterase (an essential enzyme for transmission of nerve impulses at synapses) or by acting directly on the nerve cells. Most synthetic insecticides (including pyrethroids) are nerve poisons. Other insecticidal chemicals affect the developmental or metabolic processes of insects, either by mimicking or interfering with the action of hormones, or by affecting the biochemistry of cuticle production.

16.4.1 Insecticides (chemical poisons)

Chemical insecticides may be synthetic or natural products. Natural plant-derived products, usually called botanical insecticides, include:

• **alkaloids**, including nicotine from tobacco;
• **rotenone** and other **rotenoids** from roots of legumes;
• **pyrethrins**, derived from flowers of *Tanacetum cinerariifolium* (formerly in *Pyrethrum* and then *Chrysanthemum*);
• **neem**, i.e. extracts of the tree *Azadirachta indica*, have a long history of use as insecticides (Box 16.3).

Insecticidal alkaloids have been used since the 1600s

Box 16.3 Neem

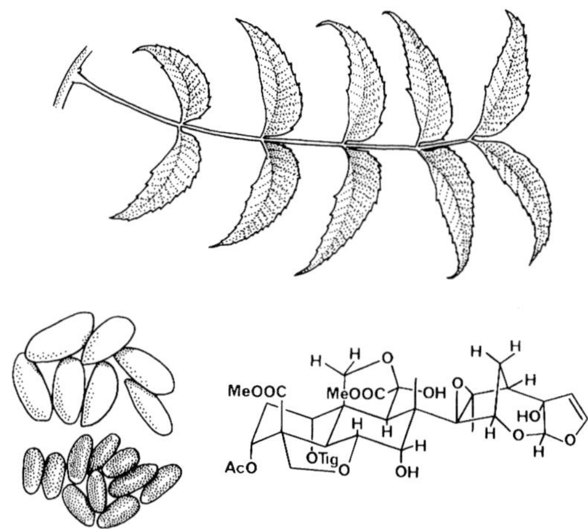

The neem tree, *Azadirachta indica* (family Meliaceae), is native to tropical Asia but has been planted widely in the warmer parts of Africa, Central and South America, and Australia. It is renowned, especially in India and some areas of Africa, for its anti-insect properties. For example, pressed leaves are put in books to keep insects away, and bags of dried leaves are placed in cupboards to deter moths and cockroaches. Extracts of neem seed kernels and leaves act as repellents, antifeedants, and/or growth disruptants. The kernels (brown colored and shown here below the entire seeds, after Schmutterer 1990) are the most important source of the active compounds that affect insects, although leaves (also illustrated here, after Corner 1952) are a secondary source. The main active compound in kernels is azadirachtin (AZ), a limonoid, but a range of other active compounds also are present. Various aqueous and alcoholic extracts of kernels, neem oil, and pure AZ have been tested for their effects on many insects. These neem derivatives can repel, prevent settling and/or inhibit oviposition, inhibit or reduce food intake, interfere with the regulation of growth (as discussed in section 16.4.2), as well as reduce the fecundity, longevity, and vigor of adults. In lepidopteran species, AZ seems to reduce the feeding activity of oligophagous species more than polyphagous ones. The antifeedant (phagodeterrent) action of neem apparently has a gustatory (regulated by sensilla on the mouthparts) as well as a non-gustatory component, as injected or topically applied neem derivatives can reduce feeding even though the mouthparts are not affected directly.

Neem-based products appear effective under field conditions against a broad spectrum of pests, including phytophagous insects of most orders (such as Hemiptera, Coleoptera, Diptera, Lepidoptera, and Hymenoptera), stored-product pests, certain pests of livestock, and even some mosquito vectors of human disease. Fortunately, honey bees and many predators of insect pests, such as spiders and coccinellid beetles, are less susceptible to neem, making it very suitable for IPM. Furthermore, neem derivatives are non-toxic to warm-blooded vertebrates. Unfortunately, the complex structures of limonoids such as AZ (illustrated here, after Schmutterer 1990) preclude their economical chemical synthesis, but they are readily available from plant sources. The abundance of neem trees in many developing countries means that resource-poor farmers can have access to non-toxic insecticides for controlling crop and stored-product pests.

and pyrethrum since at least the early 1800s. Although nicotine-based insecticides have been phased out for reasons including high mammalian toxicity and limited insecticidal activity, the new generation **nicotinoids** or **neonicotinoids**, which are modeled on natural nicotine, have a large market, in particular the systemic insecticide imidacloprid, which is used especially against sucking insects. Rotenoids are mitochondrial poisons that kill insects by respiratory failure, but they also poison fish, and must be kept out of waterways. Neem derivatives act as feeding poisons for most nymphs and larvae as well as altering behavior and disrupting normal development; they are dealt with in section 16.4.2 and in Box 16.3. Pyrethrins (and the structurally related synthetic **pyrethroids**) are especially effective against lepidopteran larvae, kill on contact even at low doses, and have low environmental persistence. An advantage of most pyrethrins and pyrethroids, and also neem derivatives, is their much lower mammalian and avian toxicity compared with synthetic insecticides, although pyrethroids are highly toxic to fish. A number of insect pests already have developed resistance to pyrethroids.

The other major classes of insecticides have no

natural analogs. These are the synthetic **carbamates** (e.g. aldicarb, carbaryl, carbofuran, methiocarb, methomyl, propoxur), **organophosphates** (e.g. chlorpyrifos, dichlorvos, dimethoate, malathion, parathion, phorate), and **organochlorines** (also called chlorinated hydrocarbons, e.g. aldrin, chlordane, DDT, dieldrin, endosulfan, gamma-benzene hexachloride (BHC) (lindane), heptachlor). Certain organochlorines (e.g. aldrin, chlordane, dieldrin, endosulfan, and heptachlor) are known as **cyclodienes** because of their chemical structure. A new class of insecticides, the **phenylpyrazoles** (or fiproles, e.g. fipronil), has similarities to DDT.

Most synthetic insecticides are **broad spectrum** in action, i.e. they have non-specific killing action, and most act on the insect (and incidentally on the mammalian) nervous system. Organochlorines are stable chemicals and persistent in the environment, have a low solubility in water but a moderate solubility in organic solvents, and accumulate in mammalian body fat. Their use is banned in many countries and they are unsuitable for use in IPM. Organophosphates may be highly toxic to mammals but are not stored in fat and, being less environmentally damaging and non-persistent, are suitable for IPM. They usually kill insects by contact or upon ingestion, although some are systemic in action, being absorbed into the vascular system of plants so that they kill most phloem-feeding insects. Non-persistence means that their application must be timed carefully to ensure efficient kill of pests. Carbamates usually act by contact or stomach action, more rarely by systemic action, and have short to medium persistence. Neonicotinoids such as imidacloprid are highly toxic to insects due to their blockage of nicotinic acetylcholine receptors, less toxic to mammals, and relatively non-persistent. Fipronil is a contact and stomach poison that acts as a potent inhibitor of gamma-aminobutyric acid (GABA) regulated chloride channels in neurons of insects, but is less potent in vertebrates. However, the poison and its degradates are moderately persistent and one photo-degradate appears to have an acute toxicity to mammals that is about 10 times that of fipronil itself. Although human and environmental health concerns are associated with its use, it is very effective in controlling many soil and foliar insects, for treating seed, and as a bait formulation to kill ants, vespid wasps, termites, and cockroaches.

In addition to the chemical and physical properties of insecticides, their toxicity, persistence in the field, and method of application are influenced by how they are formulated. **Formulation** refers to what and how other substances are mixed with the active ingredient, and largely constrains the mode of application. Insecticides may be formulated in various ways, including as solutions or emulsions, as unwettable powders that can be dispersed in water, as dusts or granules (i.e. mixed with an inert carrier), or as gaseous fumigants. Formulation may include abrasives that damage the cuticle and/or baits that attract the insects (e.g. fipronil often is mixed with fishmeal bait to attract and poison pest ants and wasps). The same insecticide can be formulated in different ways according to the application requirements, such as aerial spraying of a crop versus domestic use.

16.4.2 Insect growth regulators

Insect growth regulators (IGRs) are compounds that affect insect growth via interference with metabolism or development. They offer a high level of efficiency against specific stages of many insect pests, with a low level of mammalian toxicity. The two most commonly used groups of IGRs are distinguished by their mode of action. Chemicals that interfere with the normal maturation of insects by disturbing the hormonal control of metamorphosis are the **juvenile hormone mimics**, such as **juvenoids** (e.g. fenoxycarb, hydroprene, methoprene, pyriproxyfen). These halt development so that the insect either fails to reach the adult stage or the resulting adult is sterile and malformed. As juvenoids deleteriously affect adults rather than immature insects, their use is most appropriate to species in which the adult rather than the larva is the pest, such as fleas, mosquitoes, and ants. The **chitin synthesis inhibitors** (e.g. diflubenzuron, triflumuron) prevent the formation of chitin, which is a vital component of insect cuticle. Many conventional insecticides cause a weak inhibition of chitin synthesis, but the benzoylureas (also known as benzoylphenylureas or acylureas, of which diflubenzuron and triflumuron are examples) strongly inhibit formation of cuticle. Insects exposed to chitin synthesis inhibitors usually die at or immediately after ecdysis. Typically, the affected insects shed the old cuticle partially or not all and, if they do succeed in escaping from their exuviae, their body is limp and easily damaged as a result of the weakness of the new cuticle.

IGRs, which are fairly persistent indoors, usefully control insect pests in storage silos and domestic premises. Typically, juvenoids are used in urban pest control and inhibitors of chitin synthesis have greatest application in controlling beetle pests of stored grain. However, IGRs (e.g. pyriproxyfen) have been used also in field crops, for example in citrus in southern Africa. This use has led to severe secondary pest outbreaks because of their adverse effects on natural enemies, especially coccinellids but also wasp parasitoids. Spray drift from IGRs applied in African orchards also has affected the development of non-target beneficial insects, such as silkworms. In the USA, in the citrus-growing areas of California, many growers are interested in using IGRs, such as pyriproxyfen and buprofezin, to control California red scale (Diaspididae: *Aonidiella aurantii*); however, trials have shown that such chemicals have high toxicity to the predatory coccinellids that control several scale pests. The experimental application of methoprene (often used as a mosquito larvicide) to wetlands in the USA resulted in benthic communities that were impoverished in non-target insects, as a result of both direct toxic and indirect food-web effects, although there was a 1–2 year lag-time in the response of the insect taxa to application of this IGR.

Neem derivatives are another group of growth-regulatory compounds with significance in insect control (Box 16.3). Their ingestion, injection, or topical application disrupts molting and metamorphosis, with the effect depending on the insect and the concentration of chemical applied. Treated larvae or nymphs fail to molt, or the molt results in abnormal individuals in the subsequent instar; treated late-instar larvae or nymphs generally produce deformed and non-viable pupae or adults. These physiological effects of neem derivatives are not fully understood but are believed to result from interference with endocrine function; in particular, the main active principle of neem, azadirachtin (AZ), may act as an anti-ecdysteroid by blocking binding sites for ecdysteroid on the protein receptors. AZ may inhibit molting in insects by preventing the usual molt-initiating rise in ecdysteroid titer. Cuticle structures known to be particularly sensitive to ecdysteroids develop abnormally at low doses of AZ.

The newest group of IGRs developed for commercial use comprises the molting hormone mimics (e.g. tebufenozide), which are ecdysone agonists that appear to disrupt molting by binding to the ecdysone receptor protein. They have been used successfully against immature insect pests, especially lepidopterans. There are a few other types of IGRs, such as the anti-juvenile hormone analogs (e.g. precocenes), but these currently have little potential in pest control. Anti-juvenile hormones disrupt development by accelerating termination of the immature stages.

16.4.3 Neuropeptides and insect control

Insect neuropeptides are small peptides that regulate most aspects of development, metabolism, homeostasis, and reproduction. Their diverse functions have been summarized in Table 3.1. Although neuropeptides are unlikely to be used as insecticides *per se*, knowledge of their chemistry and biological actions can be applied in novel approaches to insect control. Neuroendocrine manipulation involves disrupting one or more of the steps of the general hormone process of synthesis–secretion–transport–action–degradation. For example, developing an agent to block or over-stimulate at the release site could alter the secretion of a neuropeptide. Alternatively, the peptide-mediated response at the target tissue could be blocked or over-stimulated by a peptide mimic. Furthermore, the protein nature of neuropeptides makes them amenable to control using recombinant DNA technology and genetic engineering. However, neuropeptides produced by transgenic crop plants or bacteria that express neuropeptide genes must be able to penetrate either the insect gut or cuticle. Manipulation of insect viruses appears more promising for control. Neuropeptide or "anti-neuropeptide" genes could be incorporated into the genome of insect-specific viruses, which then would act as expression vectors of the genes to produce and release the insect hormone(s) within infected insect cells. Baculoviruses have the potential to be used in this way, especially in Lepidoptera. Normally, such viruses cause slow or limited mortality in their host insect (section 16.5.2), but their efficacy might be improved by creating an endocrine imbalance that kills infected insects more quickly or increases viral-mediated mortality among infected insects. An advantage of neuroendocrine manipulation is that some neuropeptides may be insect- or arthropod-specific – a property that would reduce deleterious effects on many non-target organisms.

16.5 BIOLOGICAL CONTROL

Regulation of the abundance and distributions of species is influenced strongly by the activities of naturally occurring enemies, namely predators, parasites/parasitoids, pathogens, and/or competitors. In most managed ecosystems these biological interactions are severely restricted or disrupted in comparison with natural ecosystems, and certain species escape from their natural regulation and become pests. In **biological control**, deliberate human intervention attempts to restore some balance, by introducing or enhancing the natural enemies of target organisms such as insect pests or weedy plants. One advantage of natural enemies is their host-specificity, but a drawback (shared with other control methods) is that they do not eradicate pests. Thus, biological control may not necessarily alleviate all economic consequences of pests, but control systems are expected to reduce the abundance of a target pest to below ET levels. In the case of weeds, natural enemies include phytophagous insects; biological control of weeds is discussed in section 11.2.6. Several approaches to biological control are recognized but these categories are not discrete and published definitions vary widely, leading to some confusion. Such overlap is recognized in the following summary of the basic strategies of biological control.

Classical biological control involves the importation and establishment of natural enemies of exotic pests and is intended to achieve control of the target pest with little further assistance. This form of biological control is appropriate when insects that spread or are introduced (usually accidentally) to areas outside their natural range become pests mainly because of the absence of natural enemies. Two examples of successful classical biological control are outlined in Boxes 16.2 and 16.4. Despite the many beneficial aspects of this control strategy, negative environmental impacts can arise through ill-considered introductions of exotic natural enemies. Many introduced agents have failed to control pests; for example, over 60 predators and parasitoids have been introduced into north-eastern North America with little effect thus far on the target gypsy moth, *Lymantria dispar* (Lymantriidae) (see Plate 6.7). Some introductions have exacerbated pest problems, whereas others have become pests themselves. Exotic introductions generally are irreversible and non-target species can suffer worse consequences from efficient natural enemies than from chemical insecticides, which are unlikely to cause total extinctions of native insect species.

There are documented cases of introduced biological control agents annihilating native invertebrates. A number of endemic Hawai'ian insects (target and non-target) have become extinct apparently largely as a result of biological control introductions. The endemic snail fauna of Polynesia has been almost completely replaced by accidentally and deliberately introduced

Box 16.4 Taxonomy and biological control of the cassava mealybug

Cassava (manioc, or tapioca – *Manihot esculenta*) is a staple food crop for 200 million Africans. In 1973 a new mealybug (Hemiptera: Pseudococcidae) was found attacking cassava in central Africa. Named in 1977 as *Phenacoccus manihoti*, this pest spread rapidly until by the early 1980s it was causing production losses of over 80% throughout tropical Africa. The origin of the mealybug was considered to be the same as the original source of cassava – the Americas. In 1977, the apparent same insect was located in Central America and northern South America and parasitic wasps attacking it were found. However, as biological control agents they failed to reproduce on the African mealybugs.

Working from existing collections and fresh samples, taxonomists quickly recognized that two closely related mealybug species were involved. The one infesting African cassava proved to be from central South America, and not from further north. When the search for natural enemies was switched to central South America, the true *P. manihoti* was eventually found in the Paraguay basin, together with an encyrtid wasp, *Apoanagyrus* (formerly known as *Epidinocarsis*) *lopezi* (J.S. Noyes, pers. comm.). This wasp gave spectacular biological control when released in Nigeria, and by 1990 had been established successfully in 26 African countries and had spread to more than 2.7 million km^2. The mealybug is now considered to be under almost complete control throughout its range in Africa.

When the mealybug outbreak first occurred in 1973, although it was clear that this was an introduction of neotropical origin, the detailed species-level taxonomy was insufficiently refined, and the search for the mealybug and its natural enemies was misdirected for three years. The search was redirected thanks to taxonomic research. The savings were enormous: by 1988, the total expenditure on attempts to control the pest was estimated at US$14.6 million. In contrast, accurate species identification has led to an annual benefit of an estimated US$200 million, and this financial saving may continue indefinitely.

alien species. The introduction of the fly *Bessa remota* (Tachinidae) from Malaysia to Fiji, which led to extinction of the target coconut moth, *Levuana iridescens* (Zygaenidae), has been argued to be a case of biological control induced extinction of a native species. However, this seems to be an oversimplified interpretation, and it remains unclear as to whether the pest moth was indeed native to Fiji or an adventitious insect of no economic significance elsewhere in its native range. Moth species most closely related to *L. iridescens* predominantly occur from Malaysia to New Guinea, but their systematics are poorly understood. Even if *L. iridescens* had been native to Fiji, habitat destruction, especially replacement of native palms with coconut palms, also may have affected moth populations that probably underwent natural fluctuations in abundance.

At least 84 parasitoids of lepidopteran pests have been released in Hawai'i, with 32 becoming established mostly on pests at low elevation in agricultural areas. Suspicions that native moths were being impacted in natural habitats at higher elevation have been confirmed in part. In a massive rearing exercise, over 2000 lepidopteran larvae were reared from the remote, high elevation Alaka'i Swamp on Kauai, producing either adult moths or emerged parasitoids, each of which was identified and categorized as native or introduced. Parasitization, based on the emergence of adult parasitoids, was approximately 10% each year, higher based on dissections of larvae, and rose to 28% for biological control agents in certain native moth species. Some 83% of parasitoids belonged to one of three biological control species (two braconids and an ichneumonid), and there was some evidence that these competed with native parasitoids. These substantial non-target effects appear to have developed over many decades, but the progression of the incursion into native habitat and hosts was not documented.

A controversial form of biological control, sometimes referred to as **neoclassical biological control**, involves the importation of non-native species to control native ones. Such new associations have been suggested to be very effective at controlling pests because the pest has not coevolved with the introduced enemies. Unfortunately, the species that are most likely to be effective neoclassical biological control agents because of their ability to utilize new hosts are also those most likely to be a threat to non-target species. An example of the possible dangers of neoclassical control is provided by the work of Jeffrey Lockwood, who campaigned against the introduction of a parasitic wasp and an entomophagous fungus from Australia as control agents of native rangeland grasshoppers in the western USA. Potential adverse environmental effects of such introductions include the suppression or extinction of many non-target grasshopper species, with probable concomitant losses of biological diversity and existing weed control, and disruptions to food chains and plant community structure. The inability to predict the ecological outcomes of neoclassical introductions means that they are high risk, especially in systems where the exotic agent is free to expand its range over large geographical areas.

Polyphagous agents have the greatest potential to harm non-target organisms, and native species in tropical and subtropical environments may be especially vulnerable to exotic introductions because, in comparison with temperate areas, biotic interactions can be more important than abiotic factors in regulating their populations. Sadly, the countries and states that may have most to lose from inappropriate introductions are exactly those with the most lax quarantine restrictions and few or no protocols for the release of alien organisms.

Biological control agents that are present already or are non-persistent may be preferred for release. **Augmentation** is the supplementation of existing natural enemies, including **periodic release** of those that do not establish permanently but nevertheless are effective for a while after release. Periodic releases may be made regularly during a season so that the natural enemy population is gradually increased (augmented) to a level at which pest control is very effective. Augmentation or periodic release may be achieved in one of two ways, although in some systems a distinction between the following methods may be inapplicable. **Inoculation** is the periodic release of a natural enemy unable either to survive indefinitely or to track an expanding pest range. Control depends on the progeny of the natural enemies, rather than the original release. **Inundation** resembles insecticide use as control is achieved by the individuals released or applied, rather than by their progeny; control is relatively rapid but short-term. Examples of inundation include entomopathogens used as microbial insecticides (section 16.5.2) and *Trichogramma* wasps, which are mass reared and released into glasshouses. For cases in which short-term control is mediated by the original release and pest suppression is maintained for a period by the activities of the progeny of the original natural enemies, then the control process is neither strictly inoculative nor inundative. Augmentative releases are particularly appropriate for pests that combine good dispersal abilities with high reproductive rates – features that make them unsuitable candidates for classical biological control.

Conservation is another broad strategy of biological control that aims to protect and/or enhance the activities of natural enemies. In some ecosystems this may involve **preservation** of existing natural enemies through practices that minimize disruption to natural ecological processes. For example, the IPM systems for rice in south-east Asia encourage management practices, such as reduction or cessation of insecticide use, that interfere minimally with the predators and parasitoids that control rice pests such as brown planthopper (*Nilaparvata lugens*). The potential of biological control is much higher in tropical than in temperate countries because of high arthropod diversity and year-round activity of natural enemies. Complex arthropod food webs and high levels of natural biological control have been demonstrated in tropical irrigated rice fields. Furthermore, for many crop systems, **environmental manipulation** can greatly enhance the impact of natural enemies in reducing pest populations. Typically, this involves altering the habitat available to insect predators and parasitoids to improve conditions for their growth and reproduction by the provision or maintenance of shelter (including overwintering sites), alternative foods, and/or oviposition sites. Similarly, the effectiveness of entomopathogens of insect pests sometimes can be improved by altering environmental conditions at the time of application, such as by spraying a crop with water to elevate the humidity during release of fungal pathogens.

All biological control systems should be underpinned by sound taxonomic research on both pest and natural enemy species. Failure to invest adequate resources in systematic studies can result in incorrect identifications of the species involved, and ultimately may cost more in time and resources than any other step in the biological control system. The value of taxonomy in biological control is exemplified by the cassava mealybug in Africa (Box 16.4) and in management of *Salvinia* (Box 11.3).

The next two subsections cover more specific aspects of biological control by natural enemies. Natural enemies are divided somewhat arbitrarily into arthropods (section 16.5.1) and smaller, non-arthropod organisms (section 16.5.2) that are used to control various insect pests. In addition, many vertebrates, especially birds, mammals, and fish, are insect predators and their significance as regulators of insect populations should not be underestimated. However, as biological control agents the use of vertebrates is limited because most are dietary generalists and their times and places of activity are difficult to manipulate. An exception may be the mosquito fish, *Gambusia*, which has been released in many subtropical and tropical waterways worldwide in an effort to control the immature stages of biting flies, particularly mosquitoes. Although some control has been claimed, competitive interactions have been severely detrimental to small native fishes. Birds, as visually hunting predators that influence insect defenses, are discussed in Box 14.1.

16.5.1 Arthropod natural enemies

Entomophagous arthropods may be predatory or parasitic. Most predators are either other insects or arachnids, particularly spiders (order Araneae) and mites (Acarina, also called Acari). Predatory mites are important in regulating populations of phytophagous mites, including the pestiferous spider mites (Tetranychidae). Some mites that parasitize immature and adult insects or feed on insect eggs are potentially useful control agents for certain scale insects, grasshoppers, and stored-product pests. Spiders are diverse and efficient predators with a much greater impact on insect populations than mites, particularly in tropical ecosystems. The role of spiders may be enhanced in IPM by preservation of existing populations or habitat manipulation for their benefit, but their lack of feeding specificity is restrictive. Predatory beetles (Coleoptera: notably Coccinellidae and Carabidae) and lacewings (Neuroptera: Chrysopidae and Hemerobiidae) have been used successfully in biological control of agricultural pests, but many predatory species are polyphagous and inappropriate for targeting particular pest insects. Entomophagous insect predators may feed on several or all stages (from egg to adult) of their prey and each predator usually consumes several individual prey organisms during its life, with the predaceous habit often characterizing both immature and adult instars. The biology of predatory insects is discussed in Chapter 13 from the perspective of the predator.

The other major type of entomophagous insect is parasitic as a larva and free-living as an adult. The larva develops either as an endoparasite within its insect host or externally as an ectoparasite. In both cases the host

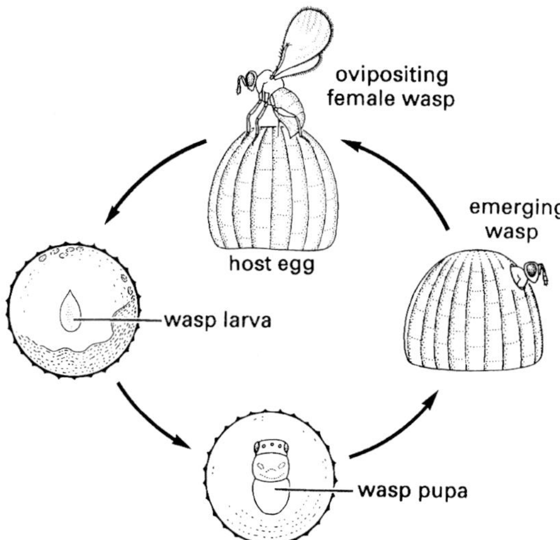

Fig. 16.3 Generalized life cycle of an egg parasitoid. A tiny female wasp of a *Trichogramma* species (Hymenoptera: Trichogrammatidae) oviposits into a lepidopteran egg; the wasp larva develops within the host egg, pupates, and emerges as an adult, often with the full life cycle taking only one week. (After van den Bosch & Hagen 1966.)

is consumed and killed by the time that the fully fed larva pupates in or near the remains of the host. Such insects, called parasitoids, all are holometabolous insects and most are wasps (Hymenoptera: especially superfamilies Chalcidoidea, Ichneumonoidea, and Platygasteroidea) or flies (Diptera: especially the Tachinidae). The Chalcidoidea contains 20 families and perhaps 100,000–500,000 species (mostly undescribed), of which most are parasitoids, including egg parasitoids such as the Mymaridae and Trichogrammatidae (Fig. 16.3), and the speciose ecto- and endoparasitic Aphelinidae and Encyrtidae, which are biological control agents of aphids, mealybugs (Box 16.4), other scale insects, and whiteflies. The Ichneumonoidea includes two speciose families, the Braconidae and Ichneumonidae, which contain numerous parasitoids mostly feeding on insects and often exhibiting quite narrow host-specificity. The Platygasteroidea contains the Platygasteridae, which are parasitic on insect eggs and larvae, and the Scelionidae, which parasitize the eggs of insects and spiders. Parasitoids from many of these wasp groups have been utilized for biological control, whereas within the Diptera only the tachinids are commonly used as biological control agents.

Parasitoids often are parasitized themselves by secondary parasitoids, called hyperparasitoids (section 13.3.1), which may reduce the effectiveness of the primary parasitoid in controlling the primary host – the pest insect. In classical biological control, usually great care is taken specifically to exclude the natural hyperparasitoids of primary parasitoids, and also the parasitoids and specialized predators of other introduced exotic natural enemies. However, some highly efficient natural enemies, especially certain predatory coccinellids, sometimes eliminate their food organisms so effectively that their own populations die out, with subsequent uncontrolled resurgence of the pest. In such cases, limited biological control of the pest's natural enemies may be warranted. More commonly, exotic parasitoids that are imported free of their natural hyperparasitoids are utilized by indigenous hyperparasitoids in the new habitat, with varying detrimental effects on the biological control system. Little can be done to solve this latter problem, except to test the host-switching abilities of some indigenous hyperparasitoids prior to introductions of the natural enemies. Of course, the same problem applies to introduced predators, which may become subject to parasitization and predation by indigenous insects in the new area. Such hazards of classical biological control systems result from the complexities of food webs, which can be unpredictable and difficult to test in advance of introductions.

Some positive management steps can facilitate long-term biological control. For example, there is clear evidence that providing a stable, structurally and floristically diverse habitat near or within a crop can foster the numbers and effectiveness of predators and parasitoids. Habitat stability is naturally higher in perennial systems (e.g. forests, orchards, and ornamental gardens) than in annual or seasonal crops (especially monocultures), because of differences in the duration of the crop. In unstable systems, the permanent provision or maintenance of ground cover, hedgerows, or strips or patches of cultivated or remnant native vegetation enable natural enemies to survive unfavorable periods, such as winter or harvest time, and then reinvade the next crop. Shelter from climatic extremes, particularly during winter in temperate areas, and alternative food resources (when the pest insects are unavailable) are essential to the continuity of predator and parasitoid populations. In particular, the free-living adults of parasitoids generally require different food sources from their larvae, such as nectar and/or pollen from flowering plants. Thus, appropriate cultivation practices can contribute significant benefits to biological control. Diversification of agroecosystems also can provide refuges for pests, but densities are likely to be low, with damage only significant for crops with low EILs. For these crops, biological control must be integrated with other methods of IPM.

Pest insects must contend with predators and parasitoids, but also with competitors. Competitive interactions appear to have little regulative influence on most phytophagous insects, but may be important for species that utilize spatially or temporally restricted resources, such as rare or dispersed prey/host organisms, dung, or animal carcasses. Interspecific competition can occur within a guild of parasitoids or predators, particularly for generalist feeders and facultative hyperparasitoids, and may inhibit biological control agents.

Biological control using natural enemies is particularly successful within the confines of greenhouses (glasshouses) or within certain crops. The commercial use of inundative and seasonal inoculative releases of natural enemies is common in many greenhouses, orchards, and fields in Europe and the USA. In Europe, more than 80 species of natural enemies are available commercially, with the most commonly sold arthropods being various species of parasitoid wasps (including *Aphidius*, *Encarsia*, *Leptomastix*, and *Trichogramma* spp.), predatory insects (especially coccinellid beetles such as *Cryptolaemus montrouzieri* and *Hippodamia convergens*, and mirid (*Macrolophus*) and anthocorid (*Orius* spp.) bugs), and predatory mites (*Amblyseius* and *Hypoaspis* spp.).

16.5.2 Microbial control

Microorganisms include bacteria, viruses, and small eukaryotes (e.g. protists, fungi, and nematodes). Some are pathogenic, usually killing insects, and of these many are host-specific to a particular insect genus or family. Infection is from spores, viral particles, or organisms that persist in the insect's environment, often in the soil. These pathogens enter insects by several routes. Entry via the mouth (*per os*) is common for viruses, bacteria, nematodes, and protists. Cuticular and/or wound entry occurs in fungi and nematodes; the spiracles and anus are other sites of entry. Viruses and protists also can infect insects via the female ovipositor or during the egg stage. The microorganisms then multiply within the living insect but have to kill it to release more infectious spores, particles or, in the case of nematodes, juveniles. Disease is common in dense insect populations (pest or non-pest) and under environmental conditions suitable to the microorganisms. At low host density, however, disease incidence is often low as a result of lack of contact between the pathogens and their insect hosts.

Microorganisms that cause diseases in natural or cultured insect populations can be used as biological control agents in the same way as other natural enemies (section 16.5.1). The usual strategies of control are appropriate, namely:

- classical biological control (i.e. an introduction of an exotic pathogen such as the bacterium *Paenibacillus* (formerly *Bacillus*) *popilliae* established in the USA for the control of the Japanese beetle *Popillia japonica* (Scarabaeidae));
- augmentation via either:
  (i) inoculation (e.g. a single treatment that provides season-long control, as in the fungus *Verticillium lecanii* used against *Myzus persicae* aphids in glasshouses), or
  (ii) inundation (i.e. entomopathogens such as *Bacillus thuringiensis* used as microbial insecticides; see pp. 414–15);
- conservation of entomopathogens through manipulation of the environment (e.g. raising the humidity to enhance the germination and spore viability of fungi).

Some disease organisms are fairly host-specific (e.g. viruses) whereas others, such as fungal and nematode species, often have wide host ranges but possess different strains that vary in their host adaptation. Thus, when formulated as a stable microbial insecticide, different species or strains can be used to kill pest species with little or no harm to non-target insects. In addition to virulence for the target species, other advantages of microbial insecticides include their compatibility with other control methods and the safety of their use (non-toxic and non-polluting). For some **entomopathogens** (insect pathogens) further advantages include the rapid onset of feeding inhibition in the host insect, stability and thus long shelf-life, and often the ability to self-replicate and thus persist in target populations. Obviously, not all of these advantages apply to every pathogen; many have a slow action on host insects, with efficacy dependent on suitable environmental conditions (e.g. high humidity or protection from sunlight) and appropriate host age and/or density. The very selectivity of microbial agents also can have practical drawbacks as when a single crop has two or more unrelated pest species, each requiring separate microbial control. All entomopathogens are more expensive to produce than chemicals and the cost is even higher if several agents must be used. However, bacteria, fungi, and nematodes that can be mass-produced in liquid fermenters (*in vitro* culture) are much cheaper to produce than those microorganisms (most viruses and protists) requiring living hosts (*in vivo* techniques). Some of the problems with the use of microbial agents are being overcome by research on formulations and mass-production methods.

Insects can become resistant to microbial pathogens as evidenced by the early success in selecting honey bees and silkworms resistant to viral, bacterial, and protist pathogens. Furthermore, many pest species exhibit significant intraspecific genetic variability in their responses to all major groups of pathogens. The current rarity of significant field resistance to microbial agents probably results from the limited exposure of insects to pathogens rather than any inability of most pest insects to evolve resistance. Of course, unlike chemicals, pathogens do have the capacity to coevolve with their hosts and over time there is likely to be a constant trade-off between host resistance, pathogen virulence, and other factors such as persistence.

Each of the five major groups of microorganisms (viruses, bacteria, protists, fungi, and nematodes) has different applications in insect pest control. Insecticides based on the bacterium *Bacillus thuringiensis* have been used most widely, but entomopathogenic fungi, nematodes, and viruses have specific and often highly successful applications. Although protists, especially microsporidia such as *Nosema*, are responsible for natural disease outbreaks in many insect populations and can be appropriate for classical biological control, they have less potential commercially than other microorganisms because of their typical low pathogenicity (infections are chronic rather than acute) and the present difficulty of large-scale production for most species.

Nematodes

Nematodes from four families, the Mermithidae, Heterorhabditidae, Steinernematidae, and Neotylenchidae, include useful or potentially useful control agents for insects. The infective stages of entomopathogenic nematodes are usually applied inundatively, although establishment and continuing control is feasible under particular conditions. Genetic engineering of nematodes is expected to improve their biological control efficacy (e.g. increased virulence), production efficiency, and storage capacity. However, entomopathogenic nematodes are susceptible to desiccation, which restricts their use to moist environments.

Mermithid nematodes are large and infect their host singly, eventually killing it as they break through the cuticle. They kill a wide range of insects, but aquatic larvae of black flies and mosquitoes are prime targets for biological control by mermithids. A major obstacle to their use is the requirement for *in vivo* production, and their environmental sensitivity (e.g. to temperature, pollution, and salinity).

Heterorhabditids and steinernematids are small, soil-dwelling nematodes, associated with symbiotic gut bacteria (of the genera *Photorhabdus* and *Xenorhabdus*) that are pathogenic to host insects, killing them by septicemia. In conjunction with their respective bacteria, nematodes of *Heterorhabditis* and *Steinernema* can kill their hosts within two days of infection. They can be mass-produced easily and cheaply and applied with conventional equipment, and have the advantage of being able to search for their hosts. The infective stage is the third-stage juvenile (or dauer stage) – the only stage found outside the host. Host location is an active response to chemical and physical stimuli. Although these nematodes are best at controlling soil pests, some plant-boring beetle and moth pests can be controlled as well. Mole crickets (Gryllotalpidae: *Scapteriscus* spp.) are soil pests that can be infected with nematodes by being attracted to acoustic traps containing infective-phase *Steinernema scapterisci*, and then being released to inoculate the rest of the cricket population.

The Neotylenchidae contains the parasitic *Deladenus siricidicola*, which is one of the biological control agents of the sirex wood wasp, *Sirex noctilio* – a serious pest of forestry plantations of *Pinus radiata* in Australia. The juvenile nematodes infect larvae of *S. noctilio*, leading to sterilization of the resulting adult female wasp. This nematode has two completely different forms – one with a parasitic life cycle completely within the sirex wood wasp and the other with a number of cycles feeding within the pine tree on the fungus introduced by the ovipositing wasp. The fungal feeding cycle of *D. siricidicola* is used to mass culture the nematode and thus obtain infective juvenile nematodes for classical biological control purposes.

Fungi

Fungi are the commonest disease organisms in insects, with approximately 750 species known to infect arthropods, although only a few dozen naturally infect agriculturally and medically important insects. Fungal spores that contact and adhere to an insect germinate and send out hyphae. These penetrate the cuticle, invade the hemocoel and cause death either rapidly owing to release of toxins, or more slowly owing to massive hyphal proliferation that disrupts insect body functions. The fungus then sporulates, releasing spores that can establish infections in other insects; and thus the fungal disease may spread through the insect population.

Sporulation and subsequent spore germination and infection of entomopathogenic fungi often require moist conditions. Although formulation of fungi in oil improves their infectivity at low humidity, water requirements may restrict the use of some species to particular environments, such as soil, glasshouses, or tropical crops. Despite this limitation, the main advantage of fungi as control agents is their ability to infect insects by penetrating the cuticle at any developmental stage. This property means that insects of all ages and feeding habits, even sap-suckers, are susceptible to fungal disease. However, fungi can be difficult to mass-produce, and the storage life of some fungal products can be limited unless kept at low temperature. A novel application method uses felt bands containing living fungal cultures applied to the tree trunks or branches, as is done in Japan using a strain of *Beauveria brongniartii* against longhorn beetle borers in citrus and mulberry. Useful species of entomopathogenic fungi belong to genera such as *Beauveria*, *Entomophthora*, *Hirsutella*, *Metarhizium*, *Nomuraea*, and *Verticillium*. Many of these fungi overcome their hosts after very little growth in the insect hemocoel, in which case toxins are believed to cause death.

Verticillium lecanii is used commercially to control aphids and scale insects in European glasshouses. *Entomophthora* species also are useful for aphid control in glasshouses. Species of *Beauveria* and *Metarhizium*, known as white and green muscardines, respectively (depending on the color of the spores), are pathogens of soil pests, such as termites and beetle larvae, and can affect other insects, such as spittle bugs of sugarcane and certain moths that live in moist microhabitats. One *Metarhizium* species, *M. anisopliae* (= *flavoviride*) var. *acridum*, has been developed as a successful myco-insecticide for locusts and other grasshoppers in Africa.

Bacteria

Bacteria rarely cause disease in insects, although saprophytic bacteria, which mask the real cause of death, frequently invade dead insects. Relatively few bacteria are used for pest control, but several have proved to be useful entomopathogens against particular pests. *Paenibacillus popilliae* is an obligate pathogen of scarab beetles (Scarabaeidae) and causes milky disease (named for the white appearance of the body of infected larvae). Ingested spores germinate in the larval gut and lead to septicemia. Infected larvae and adults are slow to die, which means that *P. popilliae* is unsuitable as a microbial insecticide, but the disease can be transmitted to other beetles by spores that persist in the soil. Thus, *P. popilliae* is useful in biological control by introduction or inoculation, although it is expensive to produce. Two species of *Serratia* are responsible for amber disease in the scarab *Costelytra zealandica*, a pest of pastures in New Zealand, and have been developed for scarab control. *Bacillus sphaericus* has a toxin that kills mosquito larvae. The strains of *Bacillus thuringiensis* have a broad spectrum of activity against larvae of many species of Lepidoptera, Coleoptera, and aquatic Diptera, but can be used only as inundative insecticides because of lack of persistence in the field.

Bacillus thuringiensis, usually called Bt, was isolated first from diseased silkworms (*Bombyx mori*) by a Japanese bacteriologist, S. Ishiwata, about a century ago. He deduced that a toxin was involved in the pathogenicity of Bt and, shortly afterwards, other Japanese researchers demonstrated that the toxin was a protein present only in sporulated cultures, was absent from culture filtrates, and thus was not an exotoxin. Of the many isolates of Bt, several have been commercialized for insect control. Bt is produced in large liquid fermenters and formulated in various ways, including as dusts and granules that can be applied to plants as aqueous sprays. Currently, the largest market for Bt-based products (other than in transgenic plants) is the North American forestry industry.

Bt forms spores, each containing a proteinaceous inclusion called a crystal, which is the source of the toxins that cause most larval deaths. The mode of action of Bt varies among different susceptible insects. In some species insecticidal action is associated with the toxic effects of the crystal proteins alone (as for some moths and black flies). However, in many others (including a number of lepidopterans) the presence of the spore enhances toxicity substantially, and in a few insects death results from septicemia following spore germination in the insect midgut rather than from the toxins. For insects affected by the toxins, paralysis occurs in mouthparts, the gut, and often the body, so that feeding is inhibited. Upon ingestion by a larval insect, the crystal is dissolved in the midgut, releasing proteins called delta-endotoxins. These proteins are protoxins that must be activated by midgut proteases before they can interact with gut epithelium and disrupt its integrity, after which the insect ultimately dies. Early-instar larvae generally are more susceptible to Bt than older larvae or adult insects.

Effective control of insect pests by Bt depends on the following factors:

- the insect population being uniformly young to be susceptible;
- active feeding of insects so that they consume a lethal dose;
- evenness of spraying of Bt;
- persistence of Bt, especially lack of denaturation by ultraviolet light;
- suitability of the strain and formulation of Bt for the insect target.

Different Bt isolates vary greatly in their insecticidal activity against a given insect species, and a single Bt isolate usually displays very different activity in different insects. At present there are about 80 recognized Bt subspecies (or serovars) based on serotype and certain biochemical and host-range data. There is disagreement, however, concerning the basis of the Bt classification scheme, as it may be more appropriate to use a system based on the crystal toxin genes, which directly determine the level and range of Bt activity. The nomenclature and classification scheme for crystal genes (*cry*) is based on their phenotype, types of crystal proteins produced, and the protein's host range as insecticidal toxins. Toxins are encoded by the *cry*I, *cry*II, *cry*III, *cry*IV and *cyt*, and *cry*V gene classes: *cry*I genes are associated with bipyramidal crystals that are toxic to lepidopteran larvae; *cry*II with cuboidal crystals active against both lepidopteran and dipteran larvae; *cry*III with flat, square crystals toxic to coleopteran larvae; *cry*IV and *cyt* with various-shaped crystals that kill dipteran larvae; and *cry*V, which is toxic to lepidopteran and some coleopteran larvae. *B. t. israelensis*, for example, has *cry*IV and *cyt* genes, whereas *B. t. tenebrionis* has *cry*III genes, and *B. t. kurstaki* has *cry*I and *cry*II genes. In addition, some cultures of Bt produce exotoxins, which are effective against various insects including larvae of the Colorado potato beetle. Thus, the nature and insecticidal effects of the various isolates of Bt are far from simple and further research on the modes of action of the toxins is desirable, especially for understanding the basis of potential and actual resistance to Bt.

Bt products have been used increasingly for control of various Lepidoptera (such as caterpillars on crucifers and in forests) since 1970. For the first two decades of use, resistance was rare or unknown, except in a stored-grain moth (Pyralidae: *Plodia interpunctella*). The first insect to show resistance in the field was a major plant pest, the diamondback moth (Plutellidae: *Plutella xylostella*), which is believed to be native to South Africa. Watercress growers in Japan and Hawai'i complained that Bt had reduced ability to kill this pest, and by 1989 further reports of resistant moths in Hawai'i were confirmed in areas where frequent high doses of Bt had been used. Similarly in Japan, by 1988 an extremely high level of Bt resistance was found in moths in greenhouses where watercress had been grown year-round with a total of 40–50 applications of Bt over three to four years. Moths resistant to Bt also were reported in Thailand, the Philippines, and mainland USA. Furthermore, laboratory studies and field reports have indicated that more than a dozen other insect species have naturally evolved or could be bred to show differing levels of resistance. Bt resistance mechanisms of the diamondback moth have been shown to derive from a single gene that confers resistance to four different Bt toxins.

Problems with chemical insecticides have stimulated interest in the use of Bt products as an alternative method of pest control. In addition to conventional applications of Bt, genetic engineering with Bt genes has produced transgenic plants ("Bt plants") that manufacture their own protective toxins (section 16.6.1), such as INGARD cotton, which carries the *cry*IA(c) Bt gene, and transgenic varieties of corn and soybean that are grown widely in the USA. Current optimism has led to the belief that insects are unlikely to develop extremely high levels of Bt resistance in the field, as a result of both instability of resistance and dilution by immigrants from susceptible populations. Strategies to prevent or slow down the evolution of resistance to Bt are the same as those used to retard resistance to synthetic insecticides. Obviously, the continued success of Bt products and the benefits of technological advances will depend on appropriate use as well as understanding and limiting resistance to the Bt crystal proteins.

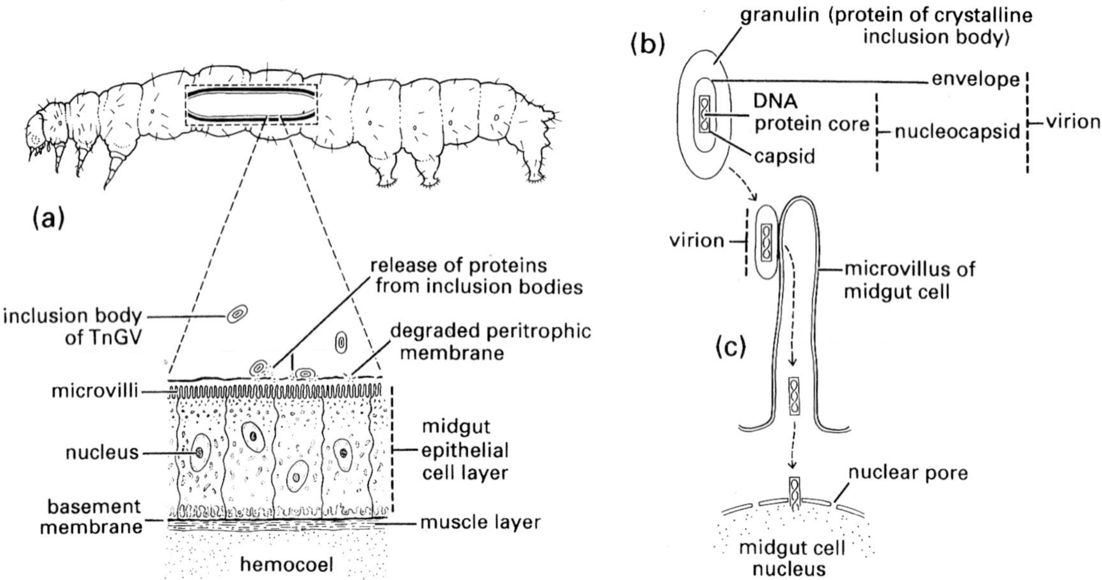

Fig. 16.4 The mode of infection of insect larvae by baculoviruses. (a) A caterpillar of the cabbage looper, *Trichoplusia ni* (Lepidoptera: Noctuidae), ingests the viral inclusion bodies of a granulosis virus (called TnGV) with its food and the inclusion bodies dissolve in the alkaline midgut releasing proteins that destroy the insect's peritrophic membrane, allowing the virions access to the midgut epithelial cells. (b) A granulosis virus inclusion body with virion in longitudinal section. (c) A virion attaches to a microvillus of a midgut cell, where the nucleocapsid discards its envelope, enters the cell and moves to the nucleus in which the viral DNA replicates. The newly synthesized virions then invade the hemocoel of the caterpillar where viral inclusion bodies are formed in other tissues (not shown). (After Entwistle & Evans 1985; Beard 1989.)

Viruses

Many viruses infect and kill insects, but those with potential for insect control are from just three viral groups, all with proteinaceous inclusion bodies, which enclose the virions (virus particles). These "occluded" viral species are considered safe because they have been found only in arthropods and appear unable to replicate in vertebrates or vertebrate cell cultures, although distant relatives of two of these groups have wider host ranges. Many "non-occluded" viruses that infect insects are considered unsafe for pest control because of their lack of specificity and possible adverse side-effects (such as infection of vertebrates and/or beneficial insects).

The useful entomopathogenic groups are the nuclear polyhedrosis viruses (NPVs), granulosis viruses (GVs) (both belonging to Baculoviridae – the baculoviruses or BVs), the cytoplasmic polyhedrosis viruses (CPVs) (Reoviridae: *Cypovirus*), and the entomopoxviruses (EPVs) (Poxviridae: Entomopoxvirinae). Baculoviruses replicate within the nuclei of the host cells, whereas the CPVs and EPVs replicate in the host cell cytoplasm. Baculoviruses have DNA genomes and are found mostly in endopterygotes, such as moth and beetle larvae, which become infected when they ingest the inclusion bodies with their food. Inclusion bodies dissolve in the high pH of the insect midgut and release the virion(s) (Fig. 16.4). These infect the gut epithelial cells and usually spread to other tissues, particularly the fat body. The inclusion bodies of NPVs are usually very stable and may persist in the environment for years (if protected from ultraviolet light, as in the soil), increasing their utility as biological control agents or microbial insecticides. The host-specificity of different viruses also influences their potential usefulness as pest control agents; some baculoviruses (such as the *Helicoverpa* NPV) are specific to an insect genus. CPVs have RNA genomes and have been found in more than 200 insect species, mainly of Lepidoptera and Diptera. Their inclusion bodies are less stable than those of NPVs. EPVs have large DNA genomes and infect a wide range of hosts in the Orthoptera, Lepidoptera,

Coleoptera, and Diptera, but individual viral isolates generally have a narrow host range. Infection of insect cells follows a similar path to that of baculoviruses.

For certain pests, viral insecticides provide feasible alternatives to chemical controls but several factors may restrict the usefulness of different viruses. Ideally, viral insecticides should be host-specific, virulent, kill quickly, persist for a reasonable time in the environment after application, and be easy to provide in large amounts. CPVs fulfill these requirements poorly, whereas the other viruses score better on these criteria, although they are inactivated by ultraviolet light within hours or days, often they kill larvae slowly and/or have a low virulence, and production costs can be high. At present, viral pesticides are produced mostly by *in vivo* or small-scale *in vitro* methods, which are expensive because of the costs of rearing the host larvae; although an *in vivo* technology called HeRD (high efficiency rearing device) greatly improves the cost/benefit ratio for producing baculovirus pesticide. Also, the use of new tissue culture technology has significantly reduced the very high cost of *in vitro* production methods. Potency problems may be overcome by genetic engineering to increase either the speed of action or the virulence of naturally occurring viruses, such as the baculoviruses that infect the heliothine pests (Lepidoptera: Noctuidae: *Helicoverpa* and *Heliothis* spp.) of cotton. The presence of particular proteins appears to enhance the action of baculoviruses; viruses can be altered to produce much more protein or the gene controlling protein production can be added to viruses that lack it. There is considerable commercial interest in the manufacture of toxin-producing viral insecticides by inserting genes encoding insecticidal products, such as insect-specific neurotoxins, into baculoviruses. However, the environmental safety of such genetically engineered viruses must be evaluated carefully prior to their wide-scale application.

Insect pests that damage valuable crops, such as bollworms of cotton and sawflies of coniferous forest trees, are suitable for viral control because substantial economic returns offset the large costs of development (including genetic engineering) and production. The other way in which insect viruses could be manipulated for use against pests is to transform the host plants so that they produce the viral proteins that damage the gut lining of phytophagous insects. This is analogous to the engineering of host-plant resistance by incorporating foreign genes into plant genomes using the crown-gall bacterium as a vector (section 16.6.1).

16.6 HOST-PLANT RESISTANCE TO INSECTS

Plant resistance to insects consists of inherited genetic qualities that result in a plant being less damaged than another (susceptible one) that is subject to the same conditions but lacks these qualities. Plant resistance is a relative concept, as spatial and temporal variations in the environment influence its expression and/or effectiveness. Generally, the production of plants resistant to particular insect pests is accomplished by selective breeding for resistance traits. The three functional categories of plant resistance to insects are:

1 **antibiosis**, in which the plant is consumed and adversely affects the biology of the phytophagous insect;
2 **antixenosis**, in which the plant is a poor host, deterring any insect feeding;
3 **tolerance**, in which the plant is able to withstand or recover from insect damage.

Antibiotic effects on insects range from mild to lethal, and antibiotic factors include toxins, growth inhibitors, reduced levels of nutrients, sticky exudates from glandular trichomes (hairs), and high concentrations of indigestible plant components such as silica and lignin. Antixenosis factors include plant chemical repellents and deterrents, pubescence (a covering of simple or glandular trichomes), surface waxes, and foliage thickness or toughness – all of which may deter insect colonization. Tolerance involves only plant features and not insect–plant interactions, as it depends only on a plant's ability to outgrow or recover from defoliation or other damage caused by insect feeding. These categories of resistance are not necessarily discrete – any combination may occur in one plant. Furthermore, selection for resistance to one type of insect may render a plant susceptible to another or to a disease.

Selecting and breeding for host-plant resistance can be an extremely effective means of controlling pest insects. The grafting of susceptible *Vitis vinifera* cultivars onto naturally resistant American vine rootstocks confers substantial resistance to grape phylloxera (Box 11.2). At the International Rice Research Institute (IRRI), numerous rice cultivars have been developed with resistance to all of the major insect pests of rice in southern and south-east Asia. Some cotton cultivars are tolerant of the feeding damage of certain insects, whereas other cultivars have been developed for their chemicals (such as gossypol) that inhibit insect growth.

Box 16.5 The Colorado potato beetle

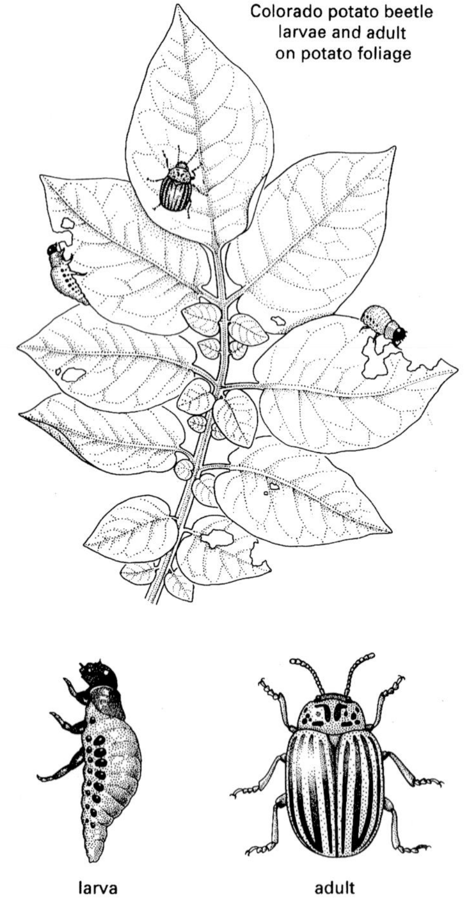

Leptinotarsa decemlineata (Coleoptera: Chrysomelidae), commonly known as the Colorado potato beetle, is a striking beetle (illustrated here, after Stanek 1969) that has become a major pest of cultivated potatoes in the northern hemisphere. Originally probably native to Mexico, it expanded its host range about 150 years ago and then spread into Europe from North America in the 1920s, and is still expanding its range. Its present hosts are about 20 species in the family Solanaceae, especially *Solanum* spp. and in particular *S. tuberosum*, the cultivated potato. Other occasional hosts include *Lycopersicon esculentum*, the cultivated tomato, and *Solanum melongena*, eggplant. The adult beetles are attracted by volatile chemicals released by the leaves of *Solanum* species, on which they feed and lay eggs. Female beetles live for about two months, in which time they can lay a few thousand eggs each. Larvae defoliate potato plants (as illustrated here) resulting in yield losses of up to 100% if damage occurs prior to tuber formation. The Colorado potato beetle is the most important defoliator of potatoes and, where it is present, control measures are necessary if crops are to be grown successfully.

Insecticides effectively controlled the Colorado potato beetle until it developed resistance to DDT in the 1950s. Since then the beetle has developed resistance to each new insecticide (including synthetic pyrethroids) at progressively faster rates. Currently, many beetle populations are resistant to all traditional insecticides, although new, narrow-spectrum insecticides became available in the late 1990s to control resistant populations. Feeding can be inhibited by application to leaf surfaces of antifeedants, including neem products (Box 16.3) and certain fungicides; however, deleterious effects on the plants and/or slow suppression of beetle populations has made antifeedants unpopular. Cultural control, via rotation of crops, delays infestation of potatoes and can reduce the build-up of early-season beetle populations. Diapausing adults mostly overwinter in the soil of fields where potatoes were grown the previous year and are slow to colonize new fields because much post-diapause dispersal is by walking. However, populations of second-generation beetles may or may not be reduced in size compared with those in non-rotated crops. Attempts to produce potato varieties resistant to the Colorado potato beetle have failed to combine useful levels of resistance (either from chemicals or glandular hairs) with a commercially suitable product. Even biological control has been unsuccessful because known natural enemies generally do not reproduce rapidly enough nor individually consume sufficient prey to regulate populations of the Colorado potato beetle effectively, and most natural enemies cannot survive the cold winters of temperate potato-growing areas. However, mass rearing and augmentative releases of certain predators (e.g. two species of pentatomid bugs) and an egg parasitoid (a eulophid wasp) may provide substantial control. Sprays of bacterial insecticides can produce effective microbial control if applications are timed to target the vulnerable early-instar larvae. Two strains of the bacterium *Bacillus thuringiensis* produce toxins that kill the larvae of Colorado potato beetle. The bacterial genes responsible for producing the toxin of *B. thuringiensis* ssp. *tenebrionis* (= *B. t.* var. *san diego*) have been genetically engineered into potato plants by inserting the genes into another bacterium, *Agrobacterium tumefaciens*, which is capable of inserting its DNA into that of the host plant. Remarkably, these transgenic potato plants are resistant to both adult and larval stages of the Colorado potato beetle, and also produce high-quality potatoes. However, their use has been restricted by concerns that consumers will reject transgenic potatoes and because the Bt plants do not deter certain other pests that still must be controlled with insecticides. Of course, even if Bt potatoes become popular, the Colorado potato beetle may rapidly develop resistance to the "new" toxins.

In general, there are more cultivars of insect-resistant cereal and grain crops than insect-resistant vegetable or fruit crops. The former often have a higher value per hectare and the latter have a low consumer tolerance of any damage but, perhaps more importantly, resistance factors can be deleterious to food quality.

Conventional methods of obtaining host-plant resistance to pests are not always successful. Despite more than 50 years of intermittent effort, no commercially suitable potato varieties resistant to the Colorado potato beetle (Chrysomelidae: *Leptinotarsa decemlineata*) have been developed. Attempts to produce potatoes with high levels of toxic glycoalkaloids mostly have stopped, partly because potato plants with high foliage levels of glycoalkaloids often have tubers rich in these toxins, resulting in risks to human health. Breeding potato plants with glandular trichomes also may have limited utility, because of the ability of the beetle to adapt to different hosts. The most promising resistance mechanism for control of the Colorado potato beetle on potato is the production of genetically modified potato plants that express a foreign gene for a bacterial toxin that kills many insect larvae (Box 16.5). Attempts to produce resistance in other vegetables often have failed because the resistance factor is incompatible with product quality, resulting in poor taste or toxicity introduced with the resistance.

16.6.1 Genetic engineering of host resistance and the potential problems

Molecular biologists have used genetic engineering techniques to produce insect-resistant varieties of a number of crop plants, including corn, cotton, tobacco, tomato, and potato, that can manufacture foreign antifeedant or insecticidal proteins under field conditions. The genes encoding these proteins are obtained from bacteria or other plants and are inserted into the recipient plant mostly via two common methods: (i) using an electric pulse or a metal fiber or particle to pierce the cell wall and transport the gene into the nucleus, or (ii) via a plasmid of the crown-gall bacterium, *Agrobacterium tumefaciens*. This bacterium can move part of its own DNA into a plant cell during infection because it possesses a tumor-inducing (Ti) plasmid containing a piece of DNA that can integrate into the chromosomes of the infected plant. Ti plasmids can be modified by removal of their tumor-forming capacity, and useful foreign genes, such as insecticidal toxins, can be inserted. These plasmid vectors are introduced into plant cell cultures, from which the transformed cells are selected and regenerated as whole plants.

Insect control via resistant genetically modified (transgenic) plants has several advantages over insecticide-based control methods, including continuous protection (even of plant parts inaccessible to insecticide sprays), elimination of the financial and environmental costs of unwise insecticide use, and cheaper modification of a new crop variety compared to development of a new chemical insecticide. Whether such genetically modified (GM) plants lead to increased or reduced environmental and human safety is currently a highly controversial issue. Problems with GM plants that produce foreign toxins include complications concerning registration and patent applications for these new biological entities, and the potential for the development of resistance in the target insect populations. For example, insect resistance to the toxins of *Bacillus thuringiensis* (Bt) (section 16.5.2) is to be expected after continuous exposure to these proteins in transgenic plant tissue. This problem might be overcome by restricting expression of the toxins to certain plant parts (e.g. the bolls of cotton rather than the whole cotton plant) or to tissues damaged by insects. A specific limitation of plants modified to produce Bt toxins is that the spore, and not just the toxin, must be present for maximum Bt activity with some pest insects.

It is possible that plant resistance based on toxins (allelochemicals) from genes transferred to plants might result in exacerbation rather than alleviation of pest problems. At low concentrations, many toxins are more active against natural enemies of phytophagous insects than against their pest hosts, adversely affecting biological control. Alkaloids and other allelochemicals ingested by phytophagous insects affect development of or are toxic to parasitoids that develop within hosts containing them, and can kill or sterilize predators. In some insects, allelochemicals sequestered whilst feeding pass into the eggs with deleterious consequences for egg parasitoids. Furthermore, allelochemicals can increase the tolerance of pests to insecticides by selecting for detoxifying enzymes that lead to cross-reactions to other chemicals. Most other plant resistance mechanisms decrease pest tolerance to insecticides and thus improve the possibilities of using pesticides selectively to facilitate biological control.

In addition to the hazards of inadvertent selection of insecticide resistance, there are several other environmental risks resulting from the use of transgenic plants.

First, there is the concern that genes from the modified plants may transfer to other plant varieties or species leading to increased weediness in the recipient of the transgene, or the extinction of native species by hybridization with transgenic plants. Second, the transgenic plant itself may become weedy if genetic modification improves its fitness in certain environments. Third, non-target organisms, such as beneficial insects (pollinators and natural enemies) and other non-pest insects, may be affected by accidental ingestion of genetically modified plants, including their pollen. A potential hazard to monarch butterfly populations from larvae eating milkweed foliage dusted with pollen from Bt corn attained some notoriety. Milkweeds, the host plants of the monarch larvae, and commercial cornfields commonly grow in close proximity in the USA. Following detailed assessment of the distance and Bt content of pollen drift, the exposure of caterpillars to corn pollen was quantified. A comprehensive risk assessment concluded that the threat to the butterfly populations was low.

Crop plants engineered genetically for resistance to herbicides may impact deleteriously on non-target insects. For example, the widespread use of weed control chemicals in fields of herbicide-resistant corn in the mid-western USA is leading to the loss of milkweeds used by the larvae and flowering annuals used as nectar sources by the adults of the monarch butterfly. The monarch has received much attention because it is a charismatic, flagship species (section 1.7), and similar effects on populations of numerous other insects are unlikely to be noticed so readily.

16.7 PHYSICAL CONTROL

Physical control refers to non-chemical, non-biological methods that destroy pests or make the environment unsuitable for the entry or survival of pests. Most of these control methods may be classified as passive (e.g. fences, trenches, traps, inert dusts, and oils) or active (e.g. mechanical, impact, and thermal treatments). Physical control measures generally are limited to confined environments such as glasshouses, food storage structures (e.g. silos), and domestic premises, although certain methods, such as exclusion barriers or trenches, can be employed in fields of crops. The best known mechanical method of pest control is the "fly swatter", but the sifting and separating procedure used in flour mills to remove insects is another example. An obvious method is physical exclusion such as packaging of food products, semi-hermetic sealing of grain silos, or provision of mesh screens on glasshouses. In addition, products may be treated or stored under controlled conditions of temperature (low or high), atmospheric gas composition (e.g. low oxygen or high carbon dioxide), or low relative humidity, which can kill or reduce reproduction of insect pests. Ionizing radiation can be used as a quarantine treatment for insects inside exported fruit, and hot-water immersion of mangoes has been used to kill immature tephritid fruit flies. The use of certain physical control methods should be increased in order to replace methyl bromide, which is used as a fumigant for many stored and exported products but will be phased out by 2005 because it depletes ozone in the atmosphere.

Traps that use long-wave ultraviolet light (e.g. "insect-o-cutors" or "zappers" that lure flying insects towards an electrified metal grid) or adhesive surfaces can be effective in domestic or food retail buildings or in glasshouses, but should not be used outdoors because of the likelihood of catching native or introduced beneficial insects. One study of the insect catches from electric traps in suburban yards in the USA showed that insects from more than a hundred non-target families were killed; about half of the insects caught were non-biting aquatic insects, over 13% were predators and parasitoids, and only about 0.2% was nuisance biting flies.

16.8 CULTURAL CONTROL

Subsistence farmers have utilized cultural methods of pest control for centuries, and many of their techniques are applicable to large-scale as well as small-scale, intensive agriculture. Typically, cultural practices involve reducing insect populations in crops by one or a combination of the following techniques: crop rotation, tillage or burning of crop stubble to disrupt pest life cycles, careful timing or placement of plantings to avoid synchrony with pests, destruction of wild plants that harbor pests and/or cultivation of non-crop plants to conserve natural enemies, and use of pest-free rootstocks and seeds. Intermixed plantings of several crops (called **intercropping** or **polyculture**) may reduce crop apparency (plant apparency hypothesis) or resource concentration for the pests (resource concentration hypothesis), increase protection for susceptible plants growing near resistant plants (associational resistance hypothesis), and/or promote natural enemies (the natural enemies hypothesis). Recent agroecology research has compared densities of insect pests and their natural enemies in monocultures and polycultures (including di- and tricultures) to test whether the success of intercropping can be explained better by a particular hypothesis; however, the hypotheses are not mutually exclusive and there is some support for each one.

In medical entomology, cultural control methods consist of habitat manipulations, such as draining marshes and removal or covering of water-holding containers to limit larval breeding sites of disease-transmitting mosquitoes, and covering rubbish dumps to prevent access and breeding by disease-disseminating flies. Examples of cultural control of livestock pests include removal of dung that harbors pestiferous flies and simple walk-through traps that remove and kill flies resting on cattle. These exclusion and trapping methods also could be classified as physical methods of control.

16.9 PHEROMONES AND OTHER INSECT ATTRACTANTS

Insects use a variety of chemical odors called semiochemicals to communicate within and between species (Section 4.3.2). Pheromones are particularly important chemicals used for signaling between members of the same species – these are often mixtures of two, three, or more components, which, when released by one individual, elicit a specific response in another individual. Other members of the species, for example prospective mates, arrive at the source. Naturally derived or synthetic pheromones, especially sex pheromones, can be used in pest management to misdirect the behavior and prevent reproduction of pest insects. The pheromone is released from point-source dispensers, often in association with traps that are placed in the crop. The strength of the insect response depends upon dispenser design, placement, and density. The rate and duration of pheromone emission from each dispenser depends upon the method of release (e.g. from impregnated rubber, microcapsules, capillaries, or wicks), strength of formulation, original volume, surface area from which it is volatilized, and longevity and/or stability of the formulation. Male lures, such as **cuelure**, **trimedlure**, and **methyl eugenol** (sometimes called **parapheromones**), which are strongly attractive to many male tephritid fruit flies, can be dispensed in a manner similar to pheromones. Methyl eugenol is thought to attract males of the oriental fruit fly *Bactrocera dorsalis* because of the benefit its consumption confers on their mating success (see p. 102). Sometimes other attractants, such as food baits or oviposition site lures, can be incorporated into a pest management scheme to function in a manner analogous to pheromones (and parapheromones), as discussed below.

There are three main uses for insect pheromones (and sometimes other attractants) in horticultural, agricultural, and forest management. The first use is in **monitoring**, initially to detect the presence of a particular pest and then to give some measure of its abundance. A trap containing the appropriate pheromone (or other lure) is placed in the susceptible crop and checked at regular intervals for the presence of any individuals of the pest lured to the trap. In most pest species, females emit sex pheromone to which males respond and thus the presence of males of the pest (and by inference, females) can be detected even at very low population densities, allowing early recognition of an impending outbreak. Knowledge of the relationship between trap-catch size and actual pest density allows a decision about when the ET for the crop will be reached and thus facilitates the efficient use of control measures, such as insecticide application. Monitoring is an essential part of IPM.

Pheromone mass trapping is another method of using pheromones in pest management and has been used primarily against forest pests. It is one form of **attraction–annihilation** – a more general method in which individuals of the targeted pest species are lured and killed. Lures may be light (e.g. ultraviolet), color (e.g. yellow is a common attractant), or semiochemicals such as pheromones or odors produced by the mating or oviposition site (e.g. dung), host plant, host animal, or empirical attractants (e.g. fruit fly chemical lures). Sometimes the lure, as with methyl eugenol for tephritid fruit flies, is more attractive than any other substance used by the insect. The insects may be attracted into container or sticky traps, onto an electrocutor grid, or onto surfaces treated with toxic chemicals or pathogens. The effectiveness of the attraction–annihilation technique appears to be inversely related to the population density of the pest and the size of the infested area. Thus, this method is likely to be most effective for control of non-resident insect pests that become abundant through annual or seasonal immigration, or pests that are geographically restricted or always present at low density. Pheromone mass trapping systems have been undertaken mostly for certain moths, such as the gypsy moth (Lymantriidae: *Lymantria dispar*) (see Plate 6.7), using their female sex pheromones, and for bark and ambrosia beetles (Curculionidae: Scolytinae) using their aggregation pheromones (section 4.3.2). An advantage of this technique for scolytines is that both sexes are caught. Success has been difficult to demonstrate because of the difficulties of designing controlled, large-scale experiments. Nevertheless, mass trapping appears effective in isolated gypsy moth populations and at low scolytine beetle densities. If beetle populations are high, even removal of part of the pest population may be beneficial, because in tree-killing beetles there is a positive feedback between population density and damage.

The third method of practical pheromone use involves sex pheromones and is called **mating disruption** (previously sometimes called "male confusion", which as we shall see is an inappropriate term). It has been applied very successfully in the field to a number of moth species, such as the pink bollworm (Gelechiidae: *Pectinophora gossypiella*) in cotton, the oriental fruit moth (Tortricidae: *Grapholita molesta*) in stone-fruit orchards, and the tomato pinworm (Gelechiidae: *Keiferia lycopersicella*) in tomato fields. Basically, numerous synthetic pheromone dispensers are placed within the crop so that the level of female sex pheromone in the orchard or field becomes higher than the background level. A reduction in the number of males locating female moths means fewer matings and a lowered population in subsequent generations. The exact behavioral or physiological mechanism(s) responsible for mating disruption are far from resolved but relate to altered behavior in males and/or females. Disruption of male behavior may be through **habituation** – temporary modifications within the central nervous system – rather than adaptation of the receptors on the antennae or confusion resulting in the following of false plumes. The high background levels promoted by use of synthetic pheromones also may mask the natural pheromone plumes of the females so that males can no longer differentiate them. Understanding the mechanism(s) of disruption is important for production of the appropriate type of formulation and quantities of synthetic pheromone needed to cause disruption, and thus control.

All of the above three pheromone methods have been used most successfully for certain moth, beetle, and fruit fly pests. Pest control using pheromones appears most effective for species that: (i) are highly dependent on chemical (rather than visual) cues for locating dispersed mates or food sources; (ii) have a limited host range; and (iii) are resident and relatively sedentary so that locally controlled populations are not constantly supplemented by immigration. Advantages of using pheromone mass trapping or mating disruption include:

- non-toxicity, leaving fruit and other products free of toxic chemicals (insecticides);
- application may be required only once or a few times per season;
- confinement of suppression to the target pest, unless predators or parasitoids use the pest's own pheromone for host location;
- enhancement of biological control (except for the circumstance mentioned in the previous point).

The limitations of pheromone use include the following:

- high selectivity and therefore no effect on other primary or secondary pests;
- cost-effective only if the target pest is the main pest for which insecticide schedules are designed;
- requirement that the treated area be isolated or large to avoid mated females flying in from untreated crops;
- requirement for detailed knowledge of pest biology in the field (especially of flight and mating activity), as timing of application is critical to successful control if continuous costly use is to be avoided;
- the possibility that artificial use will select for a shift in natural pheromone preference and production, as has been demonstrated for some moth species.

The latter three limitations apply also to pest management using chemical or microbial insecticides; for example, appropriate timing of insecticide applications is particularly important to target vulnerable stages of the pest, to reduce unnecessary and costly spraying, and to minimize detrimental environmental effects.

16.10 GENETIC MANIPULATION OF INSECT PESTS

Cochliomyia hominivorax (Calliphoridae), the New World screw-worm fly, is a devastating pest of livestock in tropical America, laying eggs into wounds, where the larvae cause myiasis (section 15.3) by feeding in the growing suppurating wounds of the living animals, including some humans. The fly perhaps was present historically in the USA, but seasonally spread into the southern and south-western states, where substantial economic losses of stock hides and carcasses required a continuing control campaign. As the female of *C. hominivorax* mates only once, control can be achieved by swamping the population with infertile males, so that the first male to arrive and mate with each female is likely to be sterile and the resultant eggs inviable. The **sterile male technique** (also called the **sterile insect technique, SIT**, or the **sterile insect release method, SIRM**) in the Americas depends upon mass-rearing facilities, sited in Mexico, where billions of screw-worm flies are reared in artificial media of blood and casein. The larvae (Fig. 6.6h) drop to the floor of the rearing chambers, where they form a puparium. At a crucial time, after gametogenesis, sterility of the developing adult is induced by gamma-irradiation of the five-day-old puparia. This treatment sterilizes the males, and although the females cannot be separated in the pupal stage and are also released, irradiation prevents their ovipositing. The released sterile males mix with the wild population, and with each mating the fertile proportion diminishes, with eradication a theoretical possibility.

The technique has eradicated the screw-worm fly, first from Florida, then Texas and the western USA, and more recently from Mexico, from whence reinvasions of the USA once originated. The goal to create a fly-free buffer zone from Panama northwards has been attained, with progressive elimination from Central American countries and releases continuing in a permanent "sterile fly barrier" in eastern Panama. In 1990, when *C. hominivorax* was introduced accidentally to Libya (North Africa), the Mexican facility was able to produce enough sterile flies to prevent the establishment of this potentially devastating pest. The impressive cost/benefit ratio of screw-worm control and eradication using the sterile insect technique has induced the expenditure of substantial sums in attempts to control similar economic pests. Other examples of successful pest insect eradications involving sterile insect releases are the Mediterranean fruit fly or "medfly", *Ceratitis capitata* (Tephritidae), from Mexico and northern Guatemala, the melon fly, *Bactrocera cucurbitae* (Tephritidae), from the Ryukyu Archipelago of Japan, and the Queensland fruit fly, *Bactrocera tryoni*, from Western Australia. The frequent lack of success of other ventures can be attributed to difficulties with one or more of the following:

- inability to mass culture the pest;
- lack of competitiveness of sterile males, including discrimination against captive-reared sterile males by wild females;
- genetic and phenotypic divergence of the captive population so that the sterile insects mate preferentially with each other (assortative mating);
- release of an inadequate number of males to swamp the females;
- failure of irradiated insects to mix with the wild population;
- poor dispersal of the sterile males from the release site, and rapid reinvasion of wild types.

Attempts have been made to introduce deleterious genes into pest species that can be mass cultured and released, with the intention that the detrimental genes spread through the wild population. The reasons for the failure of these attempts are likely to include those cited above for many sterile insect releases, particularly their lack of competitiveness, together with genetic drift and recombination that reduces the genetic effects.

FURTHER READING

Altieri, M.A. (1991) Classical biological control and social equity. *Bulletin of Entomological Research* **81**, 365–9.

Barbosa, P. (ed.) (1998) *Conservation Biological Control*. Academic Press, San Diego, CA.

Beegle, C.C. & Yamamoto, T. (1992) History of *Bacillus thuringiensis* Berliner research and development. *Canadian Entomologist* **124**, 587–616.

Boake, C.R.B., Shelly, T.E. & Kaneshiro, K.Y. (1996) Sexual selection in relation to pest-management strategies. *Annual Review of Entomology* **41**, 211–29.

Bonning, B.C. & Hammock, B.D. (1996) Development of recombinant baculoviruses for insect control. *Annual Review of Entomology* **41**, 191–210.

Brown, J.K., Frohlich, D.R. & Rosell, R.C. (1995) The sweet-potato or silverleaf whiteflies: biotypes of *Bemisia tabaci* or a species complex? *Annual Review of Entomology* **40**, 511–34.

Caltagirone, L.E. (1981) Landmark examples in classical biological control. *Annual Review of Entomology* **26**, 213–32.

Caltagirone, L.E. & Doutt, R.L. (1989) The history of the vedalia beetle importation to California and its impact on the development of biological control. *Annual Review of Entomology* **34**, 1–16.

Cannon, R.J.C. (1996) *Bacillus thuringiensis* use in agriculture: a molecular perspective. *Biological Reviews* **71**, 561–636.

Cardé, R.T. & Minks, A.K. (1995) Control of moth pests by mating disruption: successes and constraints. *Annual Review of Entomology* **40**, 559–85.

Collins, W.W. & Qualset, C.O. (eds.) (1999) *Biodiversity in Agroecosystems*. CRC Press, Boca Raton, FL.

DeBach, P. & Rosen, D. (1991) *Biological Control by Natural Enemies*, 2nd edn. Cambridge University Press, New York.

Denholm, I. & Rowland, M.W. (1992) Tactics for managing pesticide resistance in arthropods: theory and practice. *Annual Review of Entomology* **37**, 91–112.

Denholm, I., Pickett J.A. & Devonshire, A.L. (eds.) (1999) *Insecticide Resistance: From Mechanisms to Management*. CABI Publishing, CAB International, Wallingford.

Dent, D. (2000) *Insect Pest Management*, 2nd edn. CABI Publishing, CAB International, Wallingford.

Dent, D.R. & Walton, M.P. (1997) *Methods in Ecological and Agricultural Entomology*. CAB International, Wallingford.

Ehler, L.E. & Bottrell, D.G. (2000) The illusion of integrated pest management. *Issues in Science and Technology* **16**(3), 61–4.

Flint, M.L. & Dreistadt, S.H. (1998) *Natural Enemies Handbook. The Illustrated Guide to Biological Pest Control*. University of California Press, Berkeley, CA.

Gerson, U. & Smiley, R.L. (1990) *Acarine Biocontrol Agents*. Chapman & Hall, London.

Gill, S.S., Cowles, E.A. & Pietrantonio, P.V. (1992) The mode of action of *Bacillus thuringiensis* endotoxins. *Annual Review of Entomology* **37**, 615–36.

Hare, J.D. (1990) Ecology and management of the Colorado potato beetle. *Annual Review of Entomology* **35**, 81–100.

Higley, L.G. & Pedigo, L.P. (1993) Economic injury level concepts and their use in sustaining environmental quality. *Agriculture, Ecosystems and Environment* **46**, 233–43.

Hill, D.S. (1997) *The Economic Importance of Insects*. Chapman & Hall, London.

Howarth, F.G. (1991) Environmental impacts of classical biological control. *Annual Review of Entomology* **36**, 485–509.

Howse, P.E., Stevens, I.D.R. & Jones, O.T. (1998) *Insect Pheromones and their Use in Pest Management*. Chapman & Hall, London.

Jervis, M. & Kidd, N. (eds.) (1996) *Insect Natural Enemies: Practical Approaches to their Study and Evaluation*. Chapman & Hall, London.

Kennedy, G. & Sutton, T.B. (eds.) (2000) *Emerging Technologies for Integrated Pest Management: Concepts, Research, and Implementation*. APS Press, St Paul, MN.

Kogan, M. (1998) Integrated pest management: historical perspectives and contemporary developments. *Annual Review of Entomology* **43**, 243–70.

Lacey, L.A., Frutos, R., Kaya, H.K. & Vail, P. (2001) Insect pathogens as biological control agents: do they have a future? *Biological Control* **21**, 230–48.

Landis, D.A., Wratten, S.D. & Gurr, G.M. (2000) Habitat management to conserve natural enemies of arthropod pests in agriculture. *Annual Review of Entomology* **45**, 175–201.

Lockwood, J.A. (1993) Environmental issues involved in biological control of rangeland grasshoppers (Orthoptera: Acrididae) with exotic agents. *Environmental Entomology* **22**, 503–18.

Louda, S.M., Pemberton, R.W., Johnson, M.T. & Follett, P.A. (2003) Nontarget effects – the Achille's heel of biological control? Retrospective analyses to reduce risk associated with biocontrol introductions. *Annual Review of Entomology* **48**, 365–96.

Matteson, P.C. (2000) Insect pest management in tropical Asian irrigated rice. *Annual Review of Entomology* **45**, 549–74.

Metcalf, R.L. & Metcalf, R.A. (1992) *Destructive and Useful Insects: Their Habits and Control*, 5th edn. McGraw-Hill, New York.

Metz, M. (ed.) (2003) *Bacillus thuringiensis: A Cornerstone of Modern Agriculture*. Haworth Press, Binghamton, NY.

Neuenschwander, P. (2001) Biological control of the cassava mealybug in Africa: a review. *Biological Control* **21**, 214–29.

Obrycki, J.J. & Kring, T.J. (1998) Predaceous Coccinellidae in biological control. *Annual Review of Entomology* **43**, 295–321.

Parrella, M.P., Heinz, K.M. & Nunney, L. (1992) Biological control through augmentative releases of natural enemies: a strategy whose time has come. *American Entomologist* **38**, 172–9.

Pedigo, L.P. (2001) *Entomology and Pest Management*, 4th edn. Prentice-Hall, Upper Saddle, NJ.

Resh, V.H. & Cardé, R.T. (eds.) (2003) *Encyclopedia of Insects*. Academic Press, Amsterdam. [See articles on biological control of insect pests; genetically modified plants; insecticide and acaricide resistance; integrated pest management; physical control of insect pests; sterile insect technique.]

Robinson, W. (2004) *Urban Entomology*. Cambridge University Press, Cambridge.

Sands, D.P.A. (1997) The "safety" of biological control agents: assessing their impact on beneficial and other nontarget hosts. *Memoirs of the Museum of Victoria* **56**, 611–16.

Schmutterer, H. (1990) Properties and potential of natural pesticides from the neem tree, *Azadirachta indica*. *Annual Review of Entomology* **35**, 271–97.

Sears, M.K., Hellmich, R.L., Stanley-Horn, D.E. et al. (2001) Impact of Bt corn pollen on monarch butterfly populations: a risk assessment. *Proceedings of the National Academy of Science USA* **98**, 11937–42.

Settle, W.H., Ariawan, H., Astuti, E.T. et al. (1996) Managing tropical rice pests through conservation of generalist natural enemies and alternative prey. *Ecology* **77**, 1975–88.

Smith, S.M. (1996) Biological control with *Trichogramma*: advances, successes, and potential of their use. *Annual Review of Entomology* **41**, 375–406.

Tabashnik, B.E. (1994) Evolution of resistance to *Bacillus thuringiensis*. *Annual Review of Entomology* **39**, 47–79.

Thacker, J.R.M. (2002) *An Introduction to Arthropod Pest Control*. Cambridge University Press, Cambridge.

van Emden, H.F. & Dabrowski, Z.T. (1994) Biodiversity and habitat modification in pest management. *Insect Science and its Applications* **15**, 605–20.

van Lenteren, J.C., Roskam, M.M. & Timmer, R. (1997) Commercial mass production and pricing of organisms for biological control of pests in Europe. *Biological Control* **10**, 143–9.

Vincent, C., Hallman, G., Panneton, B. & Fleurat-Lessard, F. (2003) Management of agricultural insects with physical control methods. *Annual Review of Entomology* **48**, 261–81.

Walter, G.H. (2003) *Insect Pest Management and Ecological Research*. Cambridge University Press, Cambridge.

Waterhouse, D.F. (1998) *Biological Control of Insect Pests: South-east Asian Prospects*. Australian Centre for International Agricultural Research, Canberra.

Weeden, C.R., Shelton, A.M., Li, Y. & Hoffman, M.P. (continuing) Biological Control: A Guide to Natural Enemies in North America. http://www.nysaes.cornell.edu/ent/biocontrol/

Williams, D.F. (ed.) (1994) *Exotic Ants: Biology, Impact, and Control of Introduced Species*. Westview Press, Boulder, CO.

Chapter 17

METHODS IN ENTOMOLOGY: COLLECTING, PRESERVATION, CURATION, AND IDENTIFICATION

Alfred Russel Wallace collecting butterflies. (After various sources, especially van Oosterzee 1997; Gardiner 1998.)

For many entomologists, questions of how and what to collect and preserve are determined by the research project (see also section 13.4). Choice of methods may depend upon the target taxa, life-history stage, geographical scope, kind of host plant or animal, disease vector status, and most importantly, sampling design and cost-effectiveness. One factor common to all such studies is the need to communicate the information unambiguously to others, not least concerning the identity of the study organism(s). Undoubtedly, this will involve identification of specimens to provide names (section 1.4), which are necessary not only to tell others about the work, but also to provide access to previously published studies on the same, or related, insects. Identification requires material to be appropriately preserved so as to allow recognition of morphological features which vary among taxa and life-history stages. After identifications have been made, the specimens remain important, and even have added value, and it is important to preserve some material (**vouchers**) for future reference. As information grows, it may be necessary to revisit the specimens to confirm identity, or to compare with later-collected material.

In this chapter we review a range of collecting methods, mounting and preservation techniques, and specimen curation, and discuss methods and principles of identification.

17.1 COLLECTION

Those who study many aspects of vertebrate and plant biology can observe and manipulate their study organisms in the field, identify them and, for larger animals, also capture, mark, and release them with few or no harmful effects. Amongst the insects, these techniques are available perhaps only for butterflies and dragonflies, and the larger beetles and bugs. Most insects can be identified reliably only after collection and preservation. Naturally, this raises ethical considerations, and it is important to:

- collect responsibly;
- obtain the appropriate **permit(s)**;
- ensure that **voucher specimens** are deposited in a well-maintained museum collection.

Responsible collecting means collecting only what is needed, avoidance or minimization of habitat destruction, and making the specimens as useful as possible to all researchers by providing labels with detailed collection data. In many countries or in designated reserve areas, permission is needed to collect insects. It is the collector's responsibility to apply for permits and fulfill the demands of any permit-issuing agency. Furthermore, if specimens are worth collecting in the first place, they should be preserved as a record of what has been studied. Collectors should ensure that all specimens (in the case of taxonomic work) or at least representative voucher specimens (in the case of ecological, genetic, or behavioral research) are deposited in a recognized museum. Voucher specimens from surveys or experimental studies may be vital to later research.

Depending upon the project, collection methods may be active or passive. Active collecting involves searching the environment for insects, and may be preceded by periods of observation before obtaining specimens for identification purposes. Active collecting tends to be quite specific, allowing targeting of the insects to be collected. Passive collecting involves erection or installation of traps, lures, or extraction devices, and entrapment depends upon the activity of the insects themselves. This is a much more general type of collecting, being relatively unselective in what is captured.

17.1.1 Active collecting

Active collecting may involve physically picking individuals from the habitat, using a wet finger, fine-hair brush, forceps, or an **aspirator** (also known in Britain as a pooter). Such techniques are useful for relatively slow-moving insects, such as immature stages and sedentary adults that may be incapable of flying or reluctant to fly. Insects revealed by searching particular habitats, as in turning over stones, removing tree bark, or observed at rest by night, are all amenable to direct picking in this manner. Night-flying insects can be selectively picked from a **light sheet** – a piece of white cloth with an ultraviolet light suspended above it (but be careful to protect eyes and skin from exposure to ultraviolet light).

Netting has long been a popular technique for capturing active insects. The vignette for this chapter depicts the naturalist and biogeographer Alfred Russel Wallace attempting to net the rare butterfly, *Graphium androcles*, in Ternate in 1858. Most insect nets have a handle about 50 cm long and a bag held open by a hoop of 35 cm diameter. For fast-flying, mobile insects such as butterflies and flies, a net with a longer handle and a wider mouth is appropriate, whereas a net with a narrower mouth and a shorter handle is sufficient for small and/or less agile insects. The net bag should always be deeper than the diameter so that the insects caught may be trapped in the bag when the net is twisted over. Nets can be used to capture insects whilst on the wing, or by using sweeping movements over the substrate to capture insects as they take wing on being disturbed, as for example from flower heads or other vegetation. Techniques of **beating** (**sweeping**) the vegetation require a stouter net than those used to intercept flight. Some insects when disturbed drop to the ground: this is especially true of beetles. The technique of beating vegetation whilst a net or tray is held beneath allows the capture of insects with this defensive behavior. Indeed, it is recommended that even when seeking to pick individuals from exposed positions, that a net or tray be placed beneath for the inevitable specimen that will evade capture by dropping to the ground (where it may be impossible to locate). Nets should be emptied frequently to prevent damage to the more fragile contents by more massive objects. Emptying depends upon the methods to be used for preservation. Selected individuals can be removed by picking or aspiration, or the complete contents can be emptied into a container, or onto a white tray from which targeted taxa can be removed (but beware of fast fliers departing).

The above netting techniques can be used in aquatic habitats, though specialist nets tend to be of different materials from those used for terrestrial insects, and of smaller size (resistance to dragging a net through water is much greater than through air). Choice of mesh size is an important consideration – the finer mesh net required to capture a small aquatic larva compared with an adult beetle provides more resistance to being dragged through the water. Aquatic nets are usually shallower and triangular in shape, rather than the circular shape used for trapping active aerial insects. This allows for more effective use in aquatic environments.

17.1.2 Passive collecting

Many insects live in microhabitats from which they are difficult to extract – notably in leaf litter and similar soil debris or in deep tussocks of vegetation. Physical inspection of the habitat may be difficult and in such cases the behavior of the insects can be used to separate them from the vegetation, detritus, or soil. Particularly useful are negative phototaxic and thermotaxic and positive hygrotaxic responses in which the target insects move away from a source of strong heat and/or light along a gradient of increasing moisture, at the end of which they are concentrated and trapped. The **Tullgren funnel** (sometimes called a **Berlese funnel**) comprises a large (e.g. 60 cm diameter) metal funnel tapering to a replaceable collecting jar. Inside the funnel a metal mesh supports the sample of leaf litter or vegetation. A well-fitting lid containing illuminating lights is placed just above the sample and sets up a heat and humidity gradient that drives the live animals downwards in the funnel until they drop into the collecting jar, which contains ethanol or other preservative.

The **Winkler bag** operates on similar principles, with drying of organic matter (litter, soil, leaves) forcing mobile animals downwards into a collecting chamber. The device consists of a wire frame enclosed with cloth that is tied at the top to ensure that specimens do not escape and to prevent invasion by scavengers, such as ants. Pre-sieved organic matter is placed into one or more mesh sleeves, which are hung from the metal frame within the bag. The bottom of the bag tapers into a screw-on plastic collecting jar containing either preserving fluid or moist tissue paper for live material. Winckler bags are hung from a branch or from rope tied between two objects, and operate via the drying effects of the sun and wind. However, even mild windy conditions cause much detritus to fall into the residue, thus defeating the major purpose of the trap. They are extremely light, require no electric power and are very useful for collecting in remote areas, although when housed inside buildings or in areas subject to rain or high humidity, they can take many days to dry completely and thus extraction of the fauna may be slow.

Separating bags rely on the positive phototaxic (light) response of many flying insects. The bags are made from thick calico with the upper end fastened to a supporting internal ring on top of which is a clear Perspex lid; they are either suspended on strings or supported on a tripod. Collections made by sweeping or specialized collections of habitat are introduced by quickly tipping the net contents into the separator and closing the lid. Those mobile (flying) insects that are attracted to light will fly to the upper, clear surface, from which they can be collected with a long-tubed aspirator inserted through a slit in the side of the bag.

Insect flight activity is seldom random, and it is possible for the observer to recognize more frequently used routes and to place barrier traps to intercept the flight path. Margins of habitats (ecotones), stream lines, and

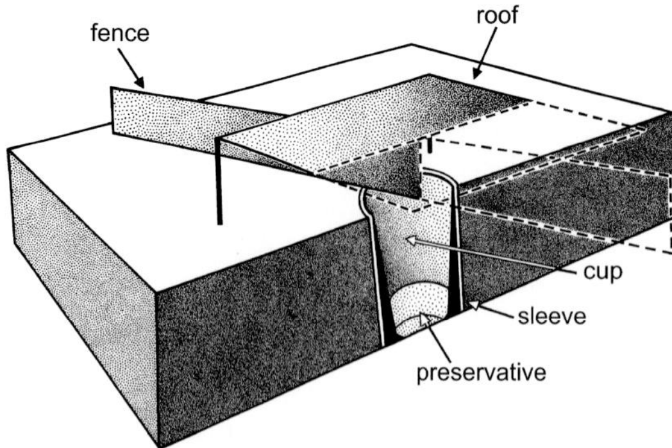

Fig. 17.1 A diagrammatic pitfall trap cut away to show the inground cup filled with preserving fluid. (After an unpublished drawing by A. Hastings.)

gaps in vegetation are evidently more utilized routes. Traps that rely on the interception of flight activity and the subsequent predictable response of certain insects include Malaise traps and window traps. The **Malaise trap** is a kind of modified tent in which insects are intercepted by a transverse barrier of net material. Those that seek to fly or climb over the vertical face of the trap are directed by this innate response into an uppermost corner and from there into a collection jar, usually containing liquid preservative. A modified Malaise trap, with a fluid-filled gutter added below, can be used to trap and preserve all those insects whose natural reaction is to drop when contact is made with a barrier. Based on similar principles, the **window trap** consists of a window-like vertical surface of glass, Perspex, or black fabric mesh, with a gutter of preserving fluid lying beneath. Only insects that drop on contact with the window are collected when they fall into the preserving fluid. Both traps are conventionally placed with the base to the ground, but either trap can be raised above the ground, for example into a forest canopy, and still function appropriately.

On the ground, interception of crawling insects can be achieved by sinking containers into the ground to rim-level such that active insects fall in and cannot climb out. These **pitfall traps** vary in size, and may feature a roof to restrict dilution with rain and preclude access by inquisitive vertebrates (Fig. 17.1). Trapping can be enhanced by construction of a fence-line to guide insects to the pitfall, and by baiting the trap. Specimens can be collected dry if the container contains insecticide and crumpled paper, but more usually they are collected into a low-volatile liquid, such as propylene glycol or ethylene glycol, and water, of varying composition depending on the frequency of visitation to empty the contents. Adding a few drops of detergent to the pitfall trap fluid reduces the surface tension and prevents the insects from floating on the surface of the liquid. Pitfall traps are used routinely to estimate species richness and relative abundances of ground active insects. However, it is too rarely understood that strong biases in trapping success may arise between compared sites of differing habitat structure (density of vegetation). This is because the probability of capture of an individual insect (trappability) is affected by the complexity of the vegetation and/or substrate that surrounds each trap. Habitat structure should be measured and controlled for in such comparative studies. Trappability is affected also by the activity levels of insects (due to their physiological state, weather, etc.), their behavior (e.g. some species avoid traps or escape from them), and by trap size (e.g. small traps may exclude larger species). Thus, the capture rate (C) for pitfall traps varies with the population density (N) and trappability (T) of the insect according to the equation $C = TN$. Usually, researchers are interested in estimating the population density of captured insects or in determining the presence or absence of species, but such studies will be biased if trappability changes between study sites or over the time interval of the study. Similarly, comparisons of the abundances of different species will be biased if one species is more trappable than another.

Many insects are attracted by **baits** or **lures**, placed in or around traps; these can be designed as "generic" to lure many insects, or "specific", designed for a single target. Pitfall traps, which trap a broad spectrum of mobile ground insects, can have their effectiveness increased by baiting with meat (for carrion attraction), dung (for coprophagous insects such as dung beetles), fresh or rotting fruit (for certain Lepidoptera, Coleoptera, and Diptera), or pheromones (for specific target insects such as fruit flies). A sweet, fermenting mixture of alcohol plus brown sugar or molasses can be daubed on surfaces to lure night-flying insects, a method termed "**sugaring**". Carbon dioxide and volatiles such as butanol can be used to lure vertebrate-host-seeking insects such as mosquitoes and horseflies.

Colors differentially attract insects: yellow is a strong lure for many hymenopterans and dipterans. This behavior is exploited in **yellow pan traps** which are simple yellow dishes filled with water and a surface-tension reducing detergent and placed on the ground to lure flying insects to death by drowning. Outdoor swimming pools act as giant pan traps.

Light trapping (see section 17.1.1 for light sheets) exploits the attraction to light of many nocturnal flying insects, particularly to the ultraviolet light emitted by fluorescent and mercury vapor lamps. After attraction to the light, insects may be picked or aspirated individually from a white sheet hung behind the light, or they may be funneled into a container such as a tank filled with egg carton packaging. There is rarely a need to kill all insects arriving at a light trap, and live insects may be sorted and inspected for retention or release.

In flowing water, strategic placement of a stationary net to intercept the flow will trap many organisms, including live immature stages of insects that may otherwise be difficult to obtain. Generally, a fine mesh net is used, secured to a stable structure such as bank, tree, or bridge, to intercept the flow in such a way that drifting insects (either deliberately or by dislodgement) enter the net. Other passive trapping techniques in water include **emergence traps**, which are generally large inverted cones, into which adult insects fly on emergence. Such traps also can be used in terrestrial situations, such as over detritus or dung, etc.

17.2 PRESERVATION AND CURATION

Most adult insects are pinned or mounted and stored dry, although the adults of some orders and all soft-bodied immature insects (eggs, larvae, nymphs, pupae or puparia) are preserved in vials of 70–80% **ethanol** (ethyl alcohol) or mounted onto microscope slides. Pupal cases, cocoons, waxy coverings, and exuviae may be kept dry and either pinned, mounted on cards or points, or, if delicate, stored in gelatin capsules or in preserving fluid.

17.2.1 Dry preservation

Killing and handling prior to dry mounting

Insects that are intended to be pinned and stored dry are best killed either in a **killing bottle** or **tube** containing a volatile poison, or in a freezer. Freezing avoids the use of chemical killing agents but it is important to place the insects into a small, airtight container to prevent drying out and to freeze them for at least 12–24 h. Frozen insects must be handled carefully and properly thawed before being pinned, otherwise the brittle appendages may break off. The safest and most readily available liquid killing agent is **ethyl acetate**, which although flammable, is not especially dangerous unless directly inhaled. It should not be used in an enclosed room. More poisonous substances, such as cyanide and chloroform, should be avoided by all except the most experienced entomologists. Ethyl acetate killing containers are made by pouring a thick mixture of plaster of Paris and water into the bottom of a tube or wide-mouthed bottle or jar to a depth of 15–20 mm; the plaster must be completely dried before use. To "charge" a killing bottle, a small amount of ethyl acetate is poured onto and absorbed by the plaster, which can then be covered with tissue or cellulose wadding. With frequent use, particularly in hot weather, the container will need to be recharged regularly by adding more ethyl acetate. Crumpled tissue placed in the container will prevent insects from contacting and damaging each other. Killing bottles should be kept clean and dry, and insects should be removed as soon as they die to avoid color loss. Moths and butterflies should be killed separately to avoid them contaminating other insects with their scales. For details of the use of other killing agents, refer to either Martin (1977) or Upton (1991) under Further reading.

Dead insects exhibit *rigor mortis* (stiffening of the muscles), which makes their appendages difficult to handle, and it is usually better to keep them in the killing bottle or in a hydrated atmosphere for 8–24 h (depending on size and species) until they have relaxed (see below), rather than pin them immediately after death. It should be noted that some large insects, especially weevils, may take many hours to die in ethyl acetate vapors and a few insects do not freeze easily and thus may not be killed quickly in a normal household freezer.

It is important to eviscerate (remove the gut and other internal organs of) large insects or gravid females (especially cockroaches, grasshoppers, katydids, mantids, stick-insects, and very large moths), otherwise the abdomens may rot and the surface of the specimens go greasy. Evisceration, also called **gutting**, is best carried out by making a slit along the side of the abdomen (in the membrane between the terga and sterna) using fine, sharp scissors and removing the body contents with a pair of fine forceps. A mixture of 3 parts talcum powder and 1 part boracic acid can be dusted into the body cavity, which in larger insects may be stuffed carefully with cotton wool.

The best preparations are made by mounting insects while they are fresh, and insects that have dried out must be relaxed before they can be mounted. **Relaxing** involves placing the dry specimens in a water-saturated atmosphere, preferably with a mold deterrent, for one to several days depending on the size of the insects. A suitable relaxing box can be made by placing a wet sponge or damp sand in the bottom of a plastic container or a wide jar and closing the lid firmly. Most smaller insects will be relaxed within 24 h, but larger specimens will take longer, during which time they should be checked regularly to ensure they do not become too wet.

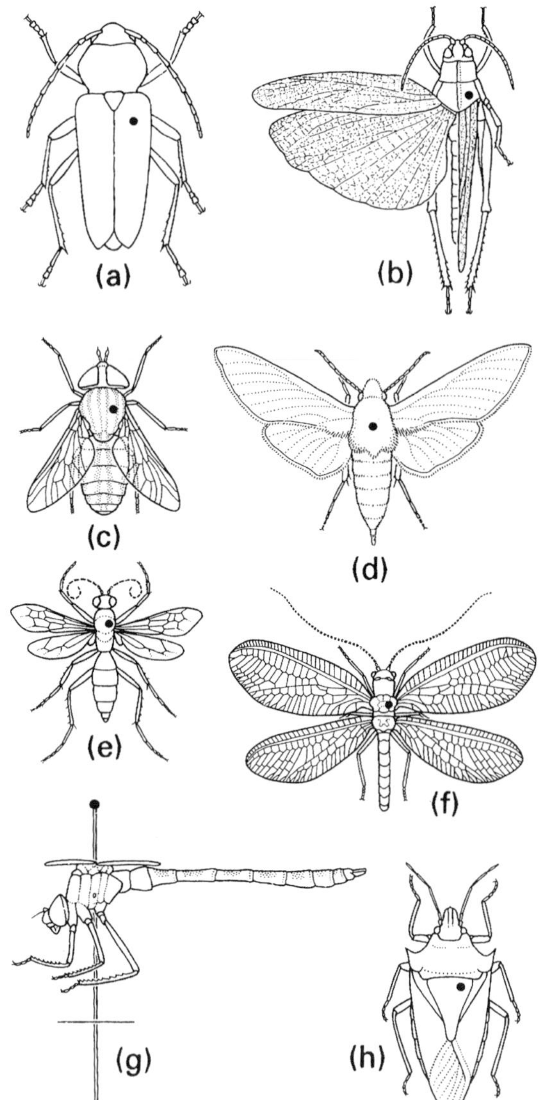

Fig. 17.2 Pin positions for representative insects: (a) larger beetles (Coleoptera); (b) grasshoppers, katydids, and crickets (Orthoptera); (c) larger flies (Diptera); (d) moths and butterflies (Lepidoptera); (e) wasps and sawflies (Hymenoptera); (f) lacewings (Neuroptera); (g) dragonflies and damselflies (Odonata), lateral view; (h) bugs, cicadas, and leaf- and planthoppers (Hemiptera: Heteroptera, Cicadomorpha, and Fulgoromorpha).

Pinning, staging, pointing, carding, spreading, and setting

Specimens should be mounted only when they are fully relaxed, i.e. when their legs and wings are freely movable, rather than stiff or dry and brittle. All dry-mounting methods use entomological **macropins** – these are stainless steel pins, mostly 32–40 mm long, and come in a range of thicknesses and with either a solid or a nylon head. *Never use dressmakers' pins for mounting insects*; they are too short and too thick. There are three widely used methods for mounting insects and the choice of the appropriate method depends on the kind of insect and its size, as well as the purpose of mounting. For scientific and professional collections, insects are either pinned directly with a macropin, micropinned, or pointed, as follows.

Direct pinning

This involves inserting a macropin, of appropriate thickness for the insect's size, directly through the insect's body; the correct position for the pin varies among insect orders (Fig. 17.2; section 17.2.4) and it is important to place the pin in the suggested place to avoid damaging structures that may be useful in identification. Specimens should be positioned about three-quarters of the way up the pin with at least 7 mm protruding above the insect to allow the mount to be gripped below the pin head using entomological forceps (which have a broad, truncate end) (Fig. 17.3). Specimens then are held in the desired positions on a piece of polyethylene foam or a cork board until they dry, which may take up to three weeks for large specimens. A desiccator or other artificial drying methods are recommended in humid climates, but oven temperature should not rise above 35°C.

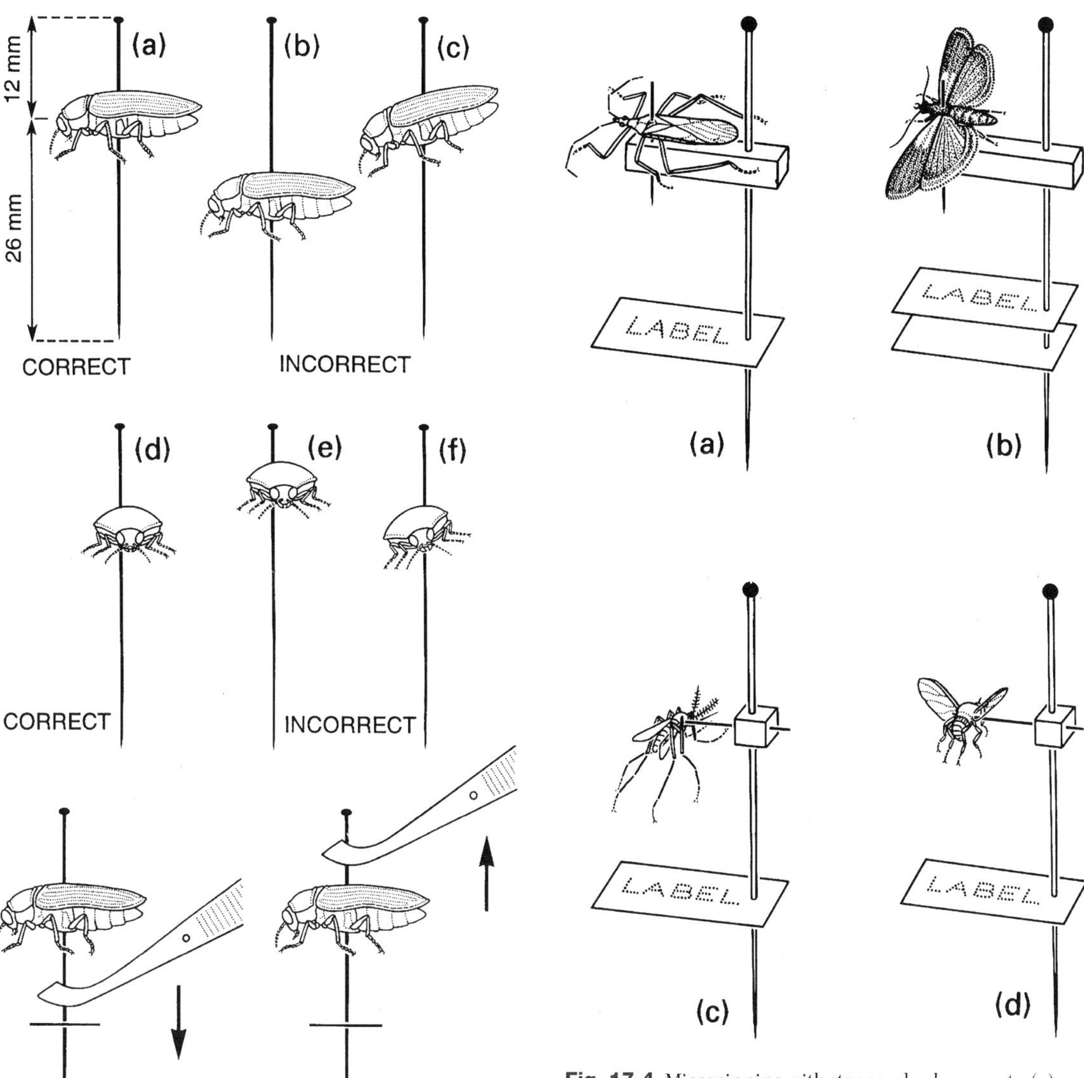

Fig. 17.3 Correct and incorrect pinning: (a) insect in lateral view, correctly positioned; (b) too low on pin; (c) tilted on long axis, instead of horizontal; (d) insect in front view, correctly positioned; (e) too high on pin; (f) body tilted laterally and pin position incorrect. Handling insect specimens with entomological forceps: (g) placing specimen mount into foam or cork; (h) removing mount from foam or cork. ((g,h) After Upton 1991.)

Fig. 17.4 Micropinning with stage and cube mounts: (a) a small bug (Hemiptera) on a stage mount, with position of pin in thorax as shown in Fig. 17.2h; (b) moth (Lepidoptera) on a stage mount, with position of pin in thorax as shown in Fig. 17.2d; (c) mosquito (Diptera: Culicidae) on a cube mount, with thorax impaled laterally; (d) black fly (Diptera: Simuliidae) on a cube mount, with thorax impaled laterally. (After Upton 1991.)

Micropinning (staging or double mounting)

This is used for many small insects and involves pinning the insect with a micropin to a stage that is mounted on a macropin (Fig. 17.4a,b); **micropins** are very fine, headless, stainless steel pins, from 10 to 15 mm long,

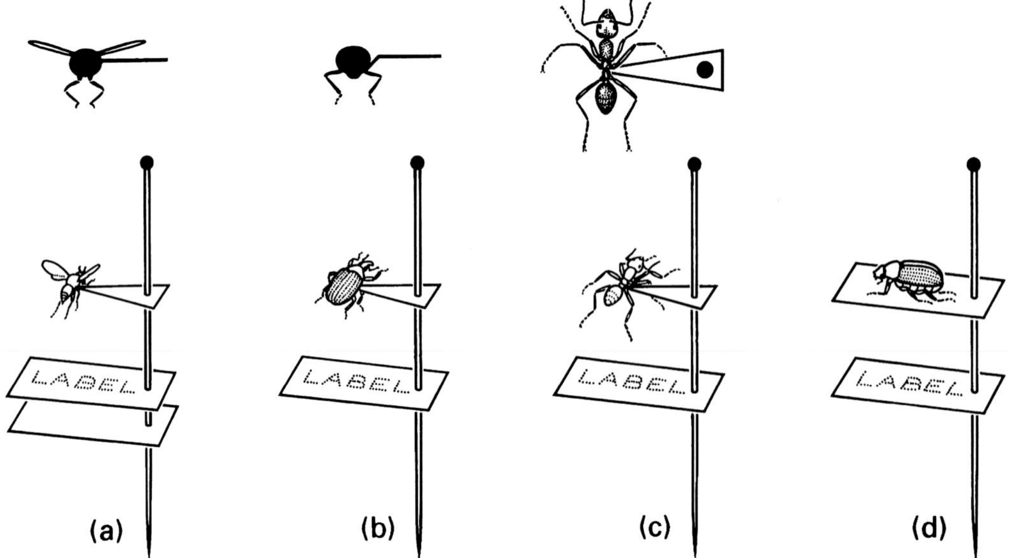

Fig. 17.5 Point mounts: (a) a small wasp; (b) a weevil; (c) an ant. Carding: (d) a beetle glued to a card mount. (After Upton 1991.)

and **stages** are small square or rectangular strips of white polyporus pith or synthetic equivalent. The micropins are inserted through the insect's body in the same positions as used in macropinning. Small wasps and moths are mounted with their bodies parallel to the stage with the head facing away from the macropin, whereas small beetles, bugs, and flies are pinned with their bodies at right angles to the stage and to the left of the macropin. Some very small and delicate insects that are difficult to pin, such as mosquitoes and other small flies, are pinned to **cube mounts**; a cube of pith is mounted on a macropin and a micropin is inserted horizontally through the pith so that most of its length protrudes, and the insect then is impaled ventrally or laterally (Fig. 17.4c,d).

Pointing

This is used for small insects that would be damaged by pinning (Fig. 17.5a) (but *not* for small moths because the glue does not adhere well to scales, nor flies because important structures are obscured), for very sclerotized, small to medium-sized insects (especially weevils and ants) (Fig. 17.5b,c) whose cuticle is too hard to pierce with a micropin, or for mounting small specimens that are already dried. Points are made from small triangular pieces of white cardboard which either can be cut out with scissors or punched out using a special point punch. Each point is mounted on a stout macropin that is inserted centrally near the base of the triangle and the insect is then glued to the tip of the point using a minute quantity of water-soluble glue, for example based on gum arabic. The head of the insect should be to the right when the apex of the point is directed away from the person mounting. For most very small insects, the tip of the point should contact the insect on the vertical side of the thorax below the wings. Ants are glued to the upper apex of the point, and two or three points, each with an ant from the same nest, can be placed on one macropin. For small insects with a sloping lateral thorax, such as beetles and bugs, the tip of the point can be bent downwards slightly before applying the glue to the upper apex of the point.

Carding

For hobby collections or display purposes, insects (especially beetles) are sometimes **carded**, which involves gluing each specimen, usually by its venter, to a rectangular piece of card through which a macropin passes (Fig. 17.5d). Carding is not recommended for adult insects because structures on the underside are

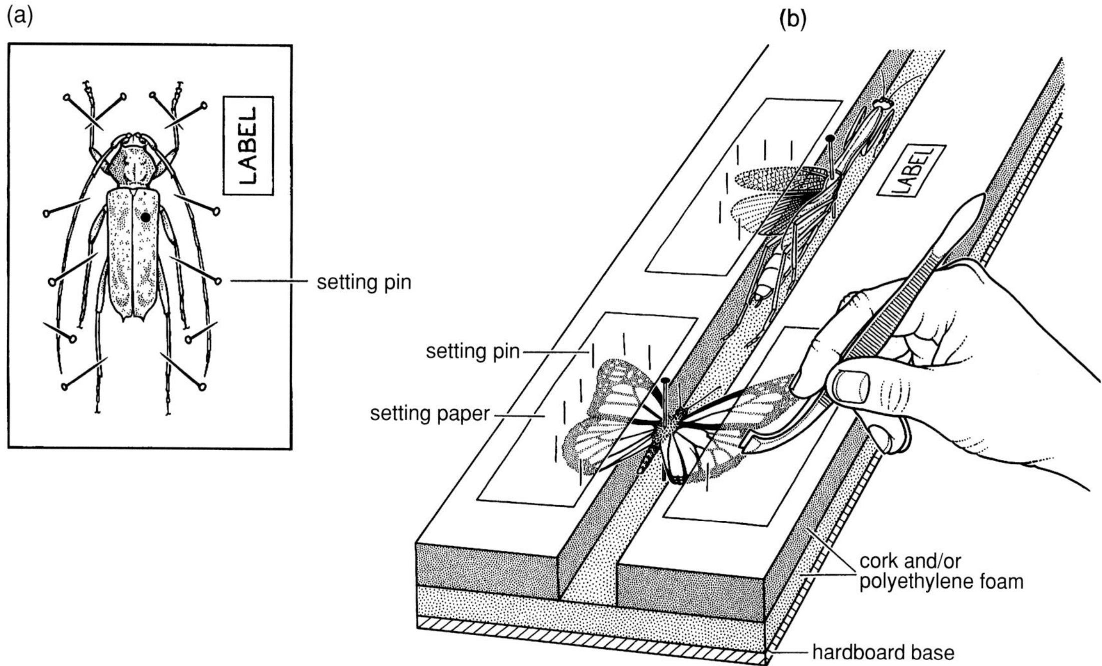

Fig. 17.6 Spreading of appendages prior to drying of specimens: (a) a beetle pinned to a foam sheet showing the spread antennae and legs held with pins; (b) setting board with mantid and butterfly showing spread wings held in place by pinned setting paper. ((b) After Upton 1991.)

obscured by being glued to the card; however, carding may be suitable for mounting exuviae, pupal cases, puparia, or scale covers.

Spreading and setting

It is important to display the wings, legs, and antennae of many insects during mounting because features used for identification are often on the appendages. Specimens with open wings and neatly arranged legs and antennae also are more attractive in a collection. **Spreading** involves holding the appendages away from the body while the specimens are drying. Legs and antennae can be held in semi-natural positions with pins (Fig. 17.6a) and the wings can be opened and held out horizontally on a setting board using pieces of tracing paper, cellophane, greaseproof paper, etc. (Fig. 17.6b). **Setting boards** can be constructed from pieces of polyethylene foam or soft cork glued to sheets of plywood or masonite; several boards with a range of groove and board widths are needed to hold insects of different body sizes and wingspans. Insects must be left to dry thoroughly before removing the pins and/or setting paper, but it is essential to keep the collection data associated correctly with each specimen during drying. A permanent data label must be placed on each macropin below the mounted insect (or its point or stage) after the specimen is removed from the drying or setting board. Sometimes two labels are used – an upper one for the collection data and a second, lower label for the taxonomic identification. See section 17.2.5 for information on the data that should be recorded.

17.2.2 Fixing and wet preservation

Most eggs, nymphs, larvae, pupae, puparia, and soft-bodied adults are preserved in liquid because drying usually causes them to shrivel and rot. The most commonly used preservative for the long-term storage of insects is **ethanol** (ethyl alcohol) mixed in various concentrations (but usually 75–80%) with water. However, aphids and scale insects are often preserved in **lactic-alcohol**, which is a mixture of 2 parts 95%

ethanol and 1 part 75% lactic acid, because this liquid prevents them from becoming brittle and facilitates subsequent maceration of body tissue prior to slide mounting. Most immature insects will shrink, and pigmented ones will discolor if placed directly into ethanol. Immature and soft-bodied insects, as well as specimens intended for study of internal structures, must first be dropped alive into a **fixative** solution prior to liquid preservation. All fixatives contain ethanol and glacial acetic acid, in various concentrations, combined with other liquids. Fixatives containing **formalin** (40% formaldehyde in water) should never be used for specimens intended for slide mounting (as internal tissues harden and will not macerate), but are ideal for specimens intended for histological study. Recipes for some commonly employed fixatives are:

KAA – 2 parts glacial acetic acid, 10 parts 95% ethanol, and 1 part kerosene (dye free).

Carnoy's fluid – 1 part glacial acetic acid, 6 parts 95% ethanol, and 3 parts chloroform.

FAA – 1 part glacial acetic acid, 25 parts 95% ethanol, 20 parts water, and 5 parts formalin.

Pampel's fluid – 2–4 parts glacial acetic acid, 15 parts 95% ethanol, 30 parts water, and 6 parts formalin.

AGA – 1 part glacial acetic acid, 6 parts 95% ethanol, 4 parts water, and 1 part glycerol.

Each specimen or collection should be stored in a separate glass vial or bottle that is sealed to prevent evaporation. The data label (section 17.2.5) should be inside the vial to prevent its separation from the specimen. Vials can be stored in racks or, to provide greater protection against evaporation, they can be placed inside a larger jar containing ethanol.

17.2.3 Microscope slide mounting

The features that need to be seen for the identification of many of the smaller insects (and their immature stages) often can be viewed satisfactorily only under the higher magnification of a compound microscope. Specimens must therefore be mounted either whole on glass microscope slides or dissected before mounting. Furthermore, the discrimination of minute structures may require the staining of the cuticle to differentiate the various parts or the use of special microscope optics such as phase- or interference-contrast microscopy. There is a wide choice of **stains** and mounting media, and the preparation methods largely depend on which type of mountant is employed. **Mountants** are either aqueous gum-chloral-based (e.g. **Hoyer's mountant**, **Berlese fluid**) or resin-based (e.g. **Canada balsam**, **Euparal**). The former are more convenient for preparing temporary mounts for some identification purposes but deteriorate (often irretrievably) over time, whereas the latter are more time-consuming to prepare but are permanent and thus are recommended for taxonomic specimens intended for long-term storage.

Prior to slide mounting, the specimens generally are "**cleared**" by soaking in either alkaline solutions (e.g. 10% potassium hydroxide (KOH) or 10% sodium hydroxide (NaOH)) or acidic solutions (e.g. lactic acid or lactophenol) to macerate and remove the body contents. Hydroxide solutions are used where complete maceration of soft tissues is required and are most appropriate for specimens that are to be mounted in resin-based mountants. In contrast, most gum-chloral mountants continue to clear specimens after mounting and thus gentler macerating agents can be used or, in some cases, very small insects can be mounted directly into the mountant without any prior clearing. After hydroxide treatment, specimens must be washed in a weak acidic solution to halt the maceration. Cleared specimens are mounted directly into gum-chloral mountants, but must be stained (if required) and dehydrated thoroughly prior to placing in resin-based mountants. Dehydration involves successive washes in a graded alcohol (usually ethanol) series with several changes in absolute alcohol. A final wash in **propan-2-ol** (isopropyl alcohol) is recommended because this alcohol is hydrophilic and will remove all trace of water from the specimen. If a specimen is to be stained (e.g. in **acid fuchsin** or **chlorazol black E**), then it is placed, prior to dehydration, in a small dish of stain for the length of time required to produce the desired depth of color.

The last stage of mounting is to put a drop of the mountant centrally on a glass slide, place the specimen in the liquid, and carefully lower a cover slip onto the preparation. A small amount of mountant on the underside of the cover slip will help to reduce the likelihood of bubbles in the preparation. The slides should be maintained in the flat (horizontal) position during drying, which can be hastened in an oven at 40–45°C; slides prepared using aqueous mountants should be oven dried for only a few days but resin-based mountants may be left for several weeks (Canada balsam mounts may take many months to harden unless oven dried). If longer-term storage of gum-chloral slides is required, then they must be "**ringed**" with an insulating varnish to give an airtight seal around the edge of the cover slip. Finally, it is essential to label each dried slide mount with the collection data and, if available, the identification (section 17.2.5). For more detailed explanations of slide-mounting methods, refer to Upton (1991, 1993) or Brown (1997) under Further reading.

17.2.4 Habitats, mounting, and preservation of individual orders

The following list is alphabetical (by order) and gives a summary of the usual habitats or collection methods, and recommendations for mounting and preserving each kind of insect or other hexapod. Insects that are to be pinned and stored dry can be killed either in a freezer or in a killing bottle (section 17.2.1); the list also specifies those insects that should be preserved in ethanol or fixed in another fluid prior to preservation (section 17.2.2). Generally, 75–80% ethanol is suggested for liquid storage, but the preferred strength often differs between collectors and depends on the kind of insect. For detailed instructions on how to collect and preserve different insects, refer to the publications in the further reading list at the end of this chapter.

Archaeognatha (bristletails)
These occur in leaf litter, under bark, or similar situations. Collect into and preserve in 80% ethanol.

Blattodea (roaches, cockroaches)
These are ubiquitous, found in sites ranging from peri-domestic to native vegetation, including caves and burrows; they are predominantly nocturnal. Eviscerate large specimens, and pin through the center of the metanotum – wings of the left side may be spread. They may also be preserved in 80% ethanol.

Coleoptera (beetles)
Beetles are found in all habitats. Pin adults and store dry; pin through the right elytron near its front so that the pin emerges between the mid and hind legs (Figs. 17.2a, 17.3, & 17.6a). Mount smaller specimens on card points with the apex of the point bent down slightly (Fig. 17.5b) and contacting the posterior lateral thorax between the mid and hind pair of legs. Immature stages are preserved in fluid (stored in 85–90% ethanol, preferably after fixation in KAA or Carnoy's fluid).

Collembola (springtails)
These are found in soil, litter, and at water surfaces (fresh and intertidal). Collect into 95–100% ethanol and preserve on microscope slides.

Dermaptera (earwigs)
Favored locations include litter, under bark or logs, dead vegetation (including along the shoreline), and in caves; exceptionally they are ectoparasitic on bats. Pin through the right elytron and with the left wings spread. Collect a representative sample of immature stages into Pampel's fluid and then 75% ethanol.

Diplura (diplurans)
These occur in damp soil under rocks or logs. Collect into 75% ethanol; preserve in 75% ethanol or slide mount.

Diptera (flies)
Flies are found in all habitats. Pin adult specimens and store dry, or preserve in 75% ethanol; pin most adults to right of center of the mesothorax (Fig. 17.2c); stage or cube mount smaller specimens (Fig. 17.4c,d) (card pointing not recommended). Collect immature stages into Pampel's fluid (larger) or 75% ethanol (smaller). Slide mount smaller adults and the larvae of some families.

Embiidina (or Embioptera; embiids or webspinners)
Typical locations for the silken galleries of webspinners are in or on bark, lichens, rocks, or wood. Preserve and store in 75% ethanol or slide mount; winged adults can be pinned through the center of the thorax with wings spread.

Ephemeroptera (mayflies)
Adults occur beside water. Preserve in 75% ethanol (preferably after fixing in Carnoy's fluid or FAA) or pin through the center of the thorax with the wings spread. Immature stages are aquatic. Collect these into and preserve in 75% ethanol or first fix in Carnoy's fluid or FAA, or store dissected on slides or in microvials.

Grylloblattodea (rock crawlers or ice crawlers)
These can be collected on or under rocks, or on snow or ice. Preserve specimens in 75% ethanol (preferably after fixing in Pampel's fluid), or slide mount.

Hemiptera
The Cicadomorpha (cicadas, leafhoppers, spittle bugs),

Fulgoromorpha (planthoppers), and Heteroptera (true bugs) are associated with their host plants or are predaceous and free-living; aquatic forms also have these habits. Preserve the adults dry; pin through the scutellum or thorax to the right of center (Fig. 17.2h); spread the wings of cicadas and fulgoroids, point or stage smaller specimens (Fig. 17.4a). Preserve nymphs in 80% ethanol.

The Aphidoidea (aphids) and Coccoidea (scale insects, mealybugs) are found associated with their host plants, including leaves, stems, roots, and galls. Store nymphs and adults in lactic-alcohol or 80% ethanol, or dry on a plant part; slide mount to identify.

Aleyrodoidea (whiteflies) are associated with their host plants. The sessile final-instar nymph ("puparium") or its exuviae ("pupal case") are of taxonomic importance. Preserve all stages in 80–95% ethanol; slide mount puparia or pupal cases.

The Psylloidea (psyllids, lerps) are associated with host plants; rear nymphs to obtain adults. Preserve nymphs in 80% ethanol, dry mount galls or lerps (if present). Preserve adults in 80% ethanol or dry mount on points; slide mount dissected parts.

Hymenoptera (ants, bees, sawflies, and wasps)

Hymenoptera are ubiquitous, and many are parasitic, in which case the host association should be retained. Collect bees, sawflies, and wasps into 80% ethanol or pin and store dry – pin larger adults to the right of center of the mesothorax (Fig. 17.2e) (sometimes with the pin angled to miss the base of the fore legs); point mount smaller adults (Fig. 17.5a); slide mount if very small. Immature stages should be preserved in 80% ethanol, often after fixing in Carnoy's fluid or KAA. Ants require stronger ethanol (90–95%); point a series of ants from each nest, with each ant glued on to the upper apex of the point between the mid and hind pairs of legs (Fig. 17.5c); two or three ants from a single nest can be mounted on separate points on a single macropin.

Isoptera (termites or "white ants")

Collect termites from colonies in mounds, on live or dead trees, or below ground. Preserve all castes available in 80% ethanol.

Lepidoptera (butterflies and moths)

Lepidoptera are ubiquitous. Collect by netting and (especially moths) at a light. Pin vertically through the thorax and spread the wings so that the hind margins of the fore wings are at right angles to the body (Figs. 17.2d & 17.6b). Microlepidopterans are best micropinned (Fig. 17.4b) immediately after death. Immature stages are killed in KAA or boiling water, and transferred to 85–95% ethanol.

Mantodea (mantids)

These are generally found on vegetation, sometimes attracted to light at night. Rear the nymphs to obtain adults. Eviscerate larger specimens. Pin between the wing bases and set the wings on the left side (Fig. 17.6b).

Mantophasmatodea (heel walkers)

These are found on mountains in Namibia by day, and also at lower elevations in South Africa at night. Pin mid-thorax, or transfer into 80–90% ethanol.

Mecoptera (scorpionflies or hangingflies)

Mecoptera often occur in damp habitats, near streams or wet meadows. Pin adults to the right of center of the thorax with the wings spread. Alternatively, all stages may be fixed in KAA, FAA, or 80% ethanol and preserved in 80% ethanol.

Megaloptera (alderflies or dobsonflies)

These are usually found in damp habitats, often near streams and lakes. Pin adults to the right of center of the thorax with the wings spread. Alternatively, all stages can be fixed in FAA or 80% ethanol and preserved in ethanol.

Neuroptera (lacewings and antlions)

Neuroptera are ubiquitous, associated with vegetation, sometimes in damp places. Pin adults to the right of center of the thorax with the wings spread (Fig. 17.2f) and the body supported. Alternatively, preserve in 80% ethanol. Immature stages are fixed in KAA, Carnoy's fluid, or 80% ethanol, and preserved in ethanol.

Odonata (damselflies and dragonflies)

Although generally found near water, adult odonates may disperse and migrate; the nymphs are aquatic. If possible keep the adult alive and starve for 1–2 days before killing (this helps to preserve body colors after death). Pin through the mid-line of the thorax between the wings, with the pin emerging between the first and second pair of legs (Fig. 17.2g); set the wings with the front margins of the hind wings at right angles to the body (a good setting method is to place the newly pinned odonate upside down with the head of the pin pushed into a foam drying board). Preserve immature stages in 80% ethanol; the exuviae should be placed on a card associated with adult.

Orthoptera (grasshoppers, locusts, crickets, katydids)
Orthoptera are found in most terrestrial habitats. Remove the gut from all but the smallest specimens, and pin vertically through the right posterior quarter of the prothorax, spreading the left wings (Fig. 17.2b). Nymphs and soft-bodied adults should be fixed in Pampel's fluid then preserved in 75% ethanol.

Phasmatodea (phasmatids, phasmids, stick-insects or walking sticks)
These are found on vegetation, usually nocturnally (sometimes attracted to light). Rear the nymphs to obtain adults, and remove the gut from all but the smallest specimens. Pin through the base of the mesothorax with the pin emerging between bases of the mesothoracic legs, spread the left wings, and fold the antennae back along the body.

Phthiraptera (lice)
Lice can be seen on their live hosts by inspecting the plumage or pelt, and can be removed using an ethanol-soaked paintbrush. Lice depart recently dead hosts as the temperature drops – and can be picked from a dark cloth background. Ectoparasites also can be removed from a live host by keeping the host's head free from a bag enclosing the rest of the body and containing chloroform to kill the parasites, which can be shaken free, and leaving the host unharmed. Legislation concerning the handling of hosts and of chloroform render this a specialized technique. Lice are preserved in 80% ethanol and slide mounted.

Plecoptera (stoneflies)
Adult plecopterans are restricted to the proximity of aquatic habitats. Net or pick from the substrate, infrequently attracted to light. Nymphs are aquatic, being found especially under stones. Pin adults through the center of the thorax with the wings spread, or preserve in 80% ethanol. Immature stages are preserved in 80% ethanol, or dissected on slides or in microvials.

Protura (proturans)
Proturans are most easily collected by extracting from litter using a Tullgren funnel. Collect into and preserve in 80% ethanol, or slide mount.

Psocoptera (psocids; barklice and booklice)
Psocids occur on foliage, bark, and damp wooden surfaces, sometimes in stored products. Collect with an aspirator or ethanol-laden paintbrush into 80% ethanol; slide mount small specimens.

Raphidioptera (snakeflies)
These are typically found in damp habitats, often near streams and lakes. Pin adults or fix in FAA or 80% ethanol; immature stages are preserved in 80% ethanol.

Siphonaptera (fleas)
Fleas can be removed from a host bird or mammal by methods similar to those outlined above for Phthiraptera. If free-living in a nest, use fine forceps or an alcohol-laden brush. Collect adults and larvae into 75–80% ethanol; preserve in ethanol or by slide mounting.

Strepsiptera (strepsipterans)
Adult males are winged, whereas females and immature stages are endoparasitic, especially in leafhoppers and planthoppers (Hemiptera) and Hymenoptera. Preserve in 80% ethanol or by slide mounting.

Thysanoptera (thrips)
Thrips are common in flowers, fungi, leaf litter, and some galls. Collect adults and nymphs into AGA or 60–90% ethanol and preserve by slide mounting.

Trichoptera (caddisflies)
Adult caddisflies are found beside water and attracted to light, and immature stages are aquatic in all waters. Pin adults through the right of center of the mesonotum with the wings spread, or preserve in 80% ethanol. Immature stages are fixed in FAA or 75% ethanol, and preserved into 80% ethanol. Micro-caddisflies and dissected nymphs are preserved by slide mounting.

Zoraptera (zorapterans)
These occur in rotten wood and under bark, with some found in termite nests. Preserve in 75% ethanol or slide mount.

Zygentoma (or Thysanura; silverfish)
Silverfish are peri-domestic, and also occur in leaf litter, under bark, in caves, and with termites and ants. They

are often nocturnal, and elusive to normal handling. Collect by stunning with ethanol, or using Tullgren funnels; preserve with 80% ethanol.

17.2.5 Curation

Labeling

Even the best-preserved and displayed specimens are of little or no scientific value without associated data such as location, date of capture, and habitat. Such information should be uniquely associated with the specimen. Although this can be achieved by a unique numbering or lettering system associated with a notebook or computer file, it is essential that it appears also on a permanently printed label associated with the specimen. The following is the minimal information that should be recorded, preferably into a field notebook at the time of capture rather than from memory later.

- Location – usually in descending order from country and state (your material may be of more than local interest), township, or distance from map-named location. Include map-derived names for habitats such as lakes, ponds, marshes, streams, rivers, forests, etc.
- Co-ordinates – preferably using a Geographic Position System (GPS) and citing latitude and longitude rather than non-universal metrics. Increasingly, these locations are used in Geographic Information Systems (GIS) and climate-derived models that depend upon accurate ground positioning.
- Elevation – derived from map or GPS as elevational accuracy has increased.
- Date – usually in sequence of day in Arabic numerals, month preferably in abbreviated letters or in Roman numerals (to avoid the ambiguity of, say, 9.11.2001 – which is 9th November in many countries but 11th September in others), and year, from which the century might be omitted. Thus, 2.iv.1999, 2.iv.99, and 2 Apr. 99 are all acceptable.
- Collector's identity, brief project identification, and any codes that refer to notebook.
- Collection method, any host association or rearing record, and any microhabitat information.

On another label, record details of the identity of the specimen including the name of the person who made the identification and the date on which it was made. It is important that subsequent examiners of the specimen know the history and timing of previous study, notably in relation to changes in taxonomic concepts in the intervening period. If the specimen is used in taxonomic description, such information should also be appended to pre-existing labels or additional label(s). It is important never to discard previous labels – transcription may lose useful evidence from handwriting and, at most, vital information on status, location, etc. Assume that all specimens valuable enough to conserve and label have potential scientific significance into the future, and thus print labels on high-quality acid-free paper using permanent ink – which can be provided now by high-quality laser printers.

Care of collections

Collections start to deteriorate rapidly unless precautions are taken against pests, mold, and vagaries of temperature and humidity. Rapid alteration in temperature and humidity should be avoided, and collections should be kept in as dark a place as possible because light causes fading. Application of some insecticides may be necessary to kill pests such as *Anthrenus* ("museum beetles") but use of all dangerous chemicals should conform to local regulations. Deep freezing (below −20°C for 48 h) also can be used to kill any pest infestation. Vials of ethanol should be securely capped, with a triple-ring nylon stopper if available, and preferably stored in larger containers of ethanol. Larger ethanol collections must be maintained in separate, ventilated, fireproof areas. Collections of glass slides preferably are stored horizontally, but with major taxonomic collections of groups preserved on slides, some vertical storage of *well-dried* slides may be required on grounds of costs and space-saving.

Other than small personal ("hobby") collections of insects, it is good scientific practice to arrange for the eventual deposition of collections into major local or national institutions such as museums. This guarantees the security of valuable specimens, and enters them into the broader scientific arena by facilitating the sharing of data, and the provision of loans to colleagues and fellow scientists.

17.3 IDENTIFICATION

Identification of insects is at the heart of almost every entomological study, but this is not always recognized. Rather too often a survey is made for one of a variety of reasons (e.g. ranking diversity of particular sites or detecting pest insects), but with scant regard to the eventual need, or even core requirement, to identify the organisms accurately. There are several possible routes to attaining accurate identification, of which the most satisfying may be to find an interested taxonomic expert in the insect group(s) under study. This person must have time available and be willing to undertake the exercise solely out of interest in the project and the insects collected. If this possibility was ever commonplace, it is no longer so because the pool of expertise has diminished and pressures upon remaining taxonomic experts have increased. A more satisfactory solution is to incorporate the identification requirements into each research proposal at the outset of the investigation, including producing a realistic budget for the identification component. Even with such planning, there may be some further problems. There may be:

- logistical constraints that prevent timely identification of mass (speciose) samples (e.g. canopy fogging samples from rainforest, vacuum sampling from grassland), even if the taxonomic skills are available;
- no entomologists who are both available and have the skills required to identify all, or even selected groups, of the insects that are encountered;
- no specialist with knowledge of the insects from the area in which your study takes place – as seen in Chapter 1, entomologists are distributed in an inverse manner to the diversity of insects;
- no specialists able or prepared to study the insects collected because the condition or life-history stage of the specimens prevents ready identification.

There is no single answer to such problems, but certain difficulties can be minimized by early consultation with local experts or with relevant published information, by collecting the appropriate life-history stage, by preserving material correctly, and by making use of vouchered material. It should be possible to advance the identification of specimens using taxonomic publications, such as field guides and keys, which are designed for this purpose.

17.3.1 Identification keys

The output of taxonomic studies usually includes keys for determining the names (i.e. for identification) of organisms. Traditionally, keys involve a series of questions, concerning the presence, shape, or color of a structure, which are presented in the form of choices. For example, one might have to determine whether the specimen has wings or not – in the case that the specimen of interest has wings then all possibilities without wings are eliminated. The next question might concern whether there is one or two pairs of wings, and if there are two pairs, whether one pair of wings is modified in some way relative to the other pair. This means of proceeding by a choice of one out of two (couplets), thereby eliminating one option at each step, is termed a "**dichotomous key**" because at each consecutive step there is a dichotomy, or branch. One works down the key until eventually the choice is between two alternatives that lead no further: these are the terminals in the key, which may be of any rank (section 1.4) – families, genera, or species. This final choice gives a name and although it is satisfying to believe that this is the "answer", it is necessary to check the identification against some form of description. An error in interpretation early on in a key (by either the user or the compiler) can lead to correct answers to all subsequent questions but a wrong final determination. However, an erroneous conclusion can be recognized only following comparison of the specimen with some "diagnostic" statements for the taxon name that was obtained from the key.

Sometimes a key may provide several choices at one point, and as long as each possibility is mutually exclusive (i.e. all taxa fall clearly into one of the multiple choices), this can provide a shorter route through the available choices. Other factors that can assist in helping the user through such keys is to provide clear illustrations of what is expected to be observed at each point. Of necessity, as we discuss in the introduction to the Glossary, there is a language associated with the morphological structures that are used in keys. This nomenclature can be rather off-putting, especially if different names are used for structures that appear to be the same, or very similar, between different taxonomic groups.

A good illustration can be worth a thousand words – but nonetheless there are also lurking problems with illustrated keys. It is difficult to relate a drawing of a structure to what is seen in the hand or under the microscope. Photography, which seems to be an obvious aid, actually can hinder because it is always tempting to look at the complete organism or structure (and in doing so to recognize or deny overall similarity to the study organism) and fail to see that the key requires only a particular detail. Another major difficulty with any branching key, even if well illustrated, is that the compiler enforces the route through the key – and even if the feature required to be observed is elusive, the structure must be recognized and a choice made between alternatives in order to proceed. There is little or no room for error by compiler or user. Even the best constructed keys may require information on a structure that the best intentioned user cannot see – for example, a choice in a key may require assessment of a feature of one sex, and the user has only the alternative sex, or an immature specimen.

The answer to identification undoubtedly requires a different structure to the questioning, using the power of computers to allow multiple access to the data needed for identification. Instead of a dichotomous structure, the compiler builds a matrix of all features that in any way can help in identification, and allows the user to select (with some guidance available for those that want it) which features to examine. Thus, it may not matter if a specimen lacks a head (through damage), whereas a conventional key may require assessment of the antennal features at an early stage. Using a computer-based, so-called **interactive key**, it may be possible to proceed using options that do not involve "missing" anatomy, and yet still make an identification. Possibilities of linking illustrations and photographs, with choices of looking "like this, or this, or this", rather than dichotomous choice, can allow efficient movement through less-constrained options than paper keys. Computer keys proceed by elimination of possible answers until one (or a few) possibilities remain – at which stage detailed descriptions may be called up to allow optimal comparisons. The ability to attach compendious information concerning the included taxa allows confirmation of identifications against illustrations and summarized diagnostic features. Furthermore, the compiler can attach all manner of biological data about the organisms, plus references. Advances such as these, as implemented in proprietal software such as Lucid (www.lucidcentral.com), suggest that interactive keys inevitably will be the preferred method by which taxonomists present their work to those who need to identify insects.

17.3.2 Unofficial taxonomies

As explained elsewhere in this book, the sheer diversity of the insects means that even some fairly commonly encountered species are not described formally yet. Only in Britain can it really be said that the total fauna is described and recognizable using identification keys. Elsewhere, the undescribed and unidentifiable proportion of the fauna can be substantial. This is an impediment to understanding how to separate species and communicate information about them. In response to the lack of formal names and keys, some "informal" taxonomies have arisen, which bypass the time-consuming formal distinguishing and naming of species. Although these taxonomies are not intended to be permanent, they do fulfill a need and can be effective. One practical system is the use of voucher numbers or codes as unique identifiers of species or morphospecies, following comparative morphological analysis across the complete geographical range of the taxa but prior to the formal act of publishing names as Latin binomens (section 1.4). If the informal name is in the form of a species name, these are referred to as **manuscript names** – and sometimes they never do become published. However, in this system, taxa can be compared across their distributional and ecological range in identical manner to taxa provided with formal names.

In narrower treatments, informal codes refer only to the biota of a limited region, typically in association with an inventory (survey) of a restricted area. The codes allocated in these studies typically represent morphospecies (estimates of species based on morphological criteria), which may not have been compared with specimens from other areas. Furthermore, the informal coded units may include taxa that may have been described formally from elsewhere. This system suffers lack of comparability of units with those from other areas – it is impossible to assess beta diversity (species turnover with distance). Furthermore, vouchers (**morphospecies**) may or may not correspond to real biological units – although strictly this criticism applies to a greater or lesser extent to all forms of taxonomic arrangements. For simple number-counting exercises at sites, with no further questions being asked of the data, a morphospecies voucher system can approximate reality, unless confused by, for example, polymorphism, cryptic species, or unassociated life-history stages.

Essential to all informal taxonomies is the need to retain voucher specimens for each segregate. This allows contemporary and future researchers to integrate informal taxa into the standardized system, and retain the association of biological information with the names, be they formal or informal. In many cases where informality is advocated, ignorance of the taxonomic process is at the heart – but in others, the sheer numbers of readily segregated morphospecies that lack formal identification requires such an approach.

17.3.3 DNA-based identifications and voucher specimens

Insect DNA is acquired for population studies, to assist with species delimitation or for phylogenetic purposes and, as recently publicized, may be used for DNA-based identification in which the sequence of base pairs of one or more genes is used as the main criterion for recognizing species (called DNA barcoding). The optimal preservation of insects for subsequent DNA extraction, amplification, and sequencing usually requires fresh specimens preserved and stored in a freezer at −80°C, or in absolute ethanol and refrigerated. It is essential that appropriate voucher specimens are retained and, if possible, most or part of the actual specimens from which the DNA is extracted. For example, DNA can be extracted from a single leg of larger insects or, for smaller insects, such as thrips and scale insects, there are methods for obtaining DNA from the whole specimen while retaining the relatively intact cuticle as the voucher.

FURTHER READING

Regional texts for identifying insects

Africa

Picker, M., Griffiths, C. & Weaving, A. (2002) *Field Guide to Insects of South Africa.* Struik Publishers, Cape Town.

Scholtz, C.H. & Holm, E. (eds.) (1985) *Insects of Southern Africa.* University of Pretoria, Pretoria.

Australia

CSIRO (1991) *The Insects of Australia*, 2nd edn. Vols. I and II. Melbourne University Press, Carlton.

Europe

Richards, O.W. & Davies, R.G. (1977) *Imms' General Textbook of Entomology*, 10th edn. Vol. 1: *Structure, Physiology and Development*; Vol. 2: *Classification and Biology*. Chapman & Hall, London.

The Americas

Arnett, R.H. (1993) *American Insects – A Handbook of the Insects of America North of Mexico.* Sandhill Crane Press, Gainesville, FL.

Arnett, R.H. & Thomas, M.C. (2001) *American Beetles*, Vol. I: *Archostemata, Myxophaga, Adephaga, Polyphaga: Staphyliniformia.* CRC Press, Boca Raton, FL.

Arnett, R.H., Thomas, M.C., Skelley, P.E. & Frank, J.J. (2002) *American Beetles*, Vol. II: *Polyphaga: Scarabaeoidea through Curculionoidea.* CRC Press, Boca Raton, FL.

Hogue, C.L. (1993) *Latin American Insects and Entomology.* University of California Press, Berkeley, CA.

Johnson, N.F. & Triplehorn, C.A. (2005) *Borror and DeLong's Introduction to the Study of Insects*, 7th edn. Brooks/Cole, Belmont, CA.

Merritt, R.W. & Cummins, K.W. (eds.) (1996) *An Introduction to the Aquatic Insects of North America*, 3rd edn. Kendall/Hunt Publishing, Dubuque, IA.

Identification of immature insects

Chu, H.F. & Cutkomp, L.K. (1992) *How to Know the Immature Insects.* William C. Brown Communications, Dubuque, IA.

Stehr, F.W. (ed.) (1987) *Immature Insects*, Vol. 1. Kendall/Hunt Publishing, Dubuque, IA. [Deals with non-insect hexapods, apterygotes, Trichoptera, Lepidoptera, Hymenoptera, plus many small orders.]

Stehr, F.W. (ed.) (1991) *Immature Insects*, Vol. 2. Kendall/Hunt Publishing, Dubuque, IA. [Deals with Thysanoptera, Hemiptera, Megaloptera, Raphidioptera, Neuroptera, Coleoptera, Strepsiptera, Siphonaptera, and Diptera.]

Collecting and preserving methods

Brown, P.A. (1997) A review of techniques used in the preparation, curation and conservation of microscope slides at the Natural History Museum, London. *The Biology Curator*, Issue 10, special supplement, 34 pp.

Martin, J.E.H. (1977) Collecting, preparing, and preserving insects, mites, and spiders. In: *The Insects and Arachnids of Canada*, Part 1. Canada Department of Agriculture, Biosystematics Research Institute, Ottawa.

McGavin, G.C. (1997) *Expedition Field Techniques. Insects and Other Terrestrial Arthropods.* Expedition Advisory Centre, Royal Geographical Society, London.

New, T.R. (1998) *Invertebrate Surveys for Conservation.* Oxford University Press, Oxford.

Upton, M.S. (1991) *Methods for Collecting, Preserving, and Studying Insects and Allied Forms*, 4th edn. Australian Entomological Society, Brisbane.

Upton, M.S. (1993) Aqueous gum-chloral slide mounting media: an historical review. *Bulletin of Entomological Research* **83**, 267–74.

Museum collections

Arnett, R.H. Jr, Samuelson, G.A. & Nishida, G.M. (1993) *The Insect and Spider Collections of the World*, 2nd edn. Flora & Fauna Handbook No. 11. Sandhill Crane Press, Gainesville, FL. (http://www.bishop.hawaii.org/bishop/ento/codens-r-us.html)

GLOSSARY

Each scientific and technical field has a particular vocabulary: entomology is no exception. This is not an attempt by entomologists to restrict access to their science. It results from the need for precision in communication, whilst avoiding, for example, misplaced anthropocentric terms derived from human anatomy. Many terms are derived from Latin or Greek: when competence in these languages was a prerequisite for scholarship (including in the natural sciences), these terms were comprehensible to the educated, whatever their mother tongue. The utility of these terms remains, although fluency in the languages from which they are derived does not. In this glossary we have tried to define the terms we use in a straightforward manner to complement the definitions used on first mention in the main body of the book. Comprehensive glossaries of entomological terms are provided by Torre-Bueno (1989) and Gordh & Headrick (2001).

Bold within the text of an entry indicates a relevant cross-reference to another headword. The following abbreviations are used:

adj. adjective
dim. diminutive
n. noun
pl. plural
sing. singular
Am. American spelling.

abdomen The third (posterior) major division (**tagma**) of an insect body.

acanthae Fine, unicellular, cuticular extensions (Fig. 2.6c).

accessory gland(s) A gland subsidiary to a major one; more specifically, a gland opening into the **genital chamber** (Figs. 3.1 & 3.20a,b).

accessory pulsatile organs Valved pumps aiding the circulation of **hemolymph**, usually lying close to the **dorsal vessel**.

acclimation Physiological changes to a changed environment (especially temperature) that allow tolerance of more extreme conditions than prior to acclimation.

acrotergite The anterior part of a secondary segment, sometimes large (then called **postnotum**), often reduced (Fig. 2.7).

acrotrophic ovariole *See* **telotrophic ovariole**.

activation (in embryology) The commencement of embryonic development within the egg.

aculeate Belonging to the aculeate Hymenoptera (Fig. 12.2) – wasps in which the **ovipositor** is modified as a sting.

adecticous Describing a **pupa** with immovable **mandibles** (Fig. 6.7); *see also* **decticous**.

adenotrophic viviparity Viviparity (producing living offspring) in which there is no free-living larval stage; eggs develop within the female uterus, nourished by special **milk glands** until the larvae mature, at which stage they are laid and immediately pupate; occurring only in some Diptera (Hippoboscidae and *Glossina*).

aedeagus (*Am.* edeagus) The male copulatory organ, variably constructed (sometimes refers to the **penis** alone) (Figs. 2.24b & 5.4).

aeriferous trachea Trachea with surface bearing a system of evaginated spiral tubules with permeable cuticle allowing aeration of surrounding tissues, especially in the ovary.

aestivate To undergo **quiescence** or **diapause** during seasonal hot or dry conditions.

age-grading Determination of the physiological age of an insect.

air sac Any of the thin-walled, dilated sections of the **tracheae** (Fig. 3.11b).

akinesis A state of immobility, caused by lack of any stimuli.

alary muscles Paired muscles that support the heart.

alate Possessing wings.

alinotum The wing-bearing plate on the dorsum of the **mesothorax** or **metathorax** (Fig. 2.18).

alitrunk The fused thorax and first abdominal segment (**propodeum**) of adult ants (*see* **mesosoma**) (see Box 12.2).

alkaloids Chemicals found in plants, many with important pharmacological actions.

allelochemical A chemical functioning in interspecific communication; *see also* **allomone**, **kairomone**, **pheromone**, **synomone**.

allochthonous Originating from elsewhere, as of nutrients entering an aquatic ecosystem; *see also* **autochthonous**.

allomone A communication chemical that benefits the producer by the effect it invokes in the receiver.

allopatric Non-overlapping geographical distributions of organisms or taxa; *see also* **sympatric**.

altruism Behavior costly to the individual but beneficial to others.

ametabolous Lacking **metamorphosis**, i.e. with no change in body form during development to the adult, with the immature stages lacking only genitalia.

amnion (in embryology) The layer covering the **germ band** (Fig. 6.5).

amphimixis True sexual reproduction, each female inherits a **haploid** genome from both her mother and father.

amphitoky (amphitokous parthenogenesis; deuterotoky) A form of **parthenogenesis** in which the female produces offspring of both sexes.

amplexiform A form of wing-coupling in which there is extensive overlap between the fore and hind wing, but without any specific coupling mechanism.

anal In the direction or position of the **anus**, near the anus or on the last abdominal segment.

anal area The posterior part of the wing, supported by the anal vein(s) (Fig. 2.20).

anal fold (vannal fold) A distinctive fold in the **anal area** of the wing (Fig. 2.20).

anamorphic Describing development in which the immature stages have fewer abdominal segments than the adult; *see also* **epimorphic**.

anautogenous Requiring a protein meal to develop eggs.

anemophily Wind pollination.

anholocyclic (*adj.*) Of aphids, describing a life cycle in which the only host plant used is a summer annual, and in which reproduction is solely by **parthenogenesis**.

anlage A cell cluster in an immature individual that will give rise to a specific organ in the adult; *see also* **imaginal disc**.

annulate Comprised of ring(s).

anoxic Lacking oxygen.

antecostal suture (intersegmental groove) A groove marking the position of the intersegmental fold between the primary segments (Figs. 2.7 & 2.18).

antenna (*pl.* antennae) Paired, segmented, sensory appendages, lying usually anterodorsally, on the head (Figs. 2.9, 2.10, & 2.17); derived from the second head segment.

antennomere A subdivision of the **antenna**.

anterior At or towards the front (Fig. 2.8).

anthophilous Flower-loving.

anthropogenic Caused by humans.

anthropophilic Associated with humans.

antibiosis A property of an organism that adversely affects the wellbeing of another organism that consumes or contacts it.

antixenosis In plant resistance, the unsuitability of a plant to a feeding insect.

anus The posterior opening of the digestive tract (Fig. 2.23).

apical At or towards the apex (Fig. 2.8).

apneustic A respiratory system without functional **spiracles**; *see also* **oligopneustic**, **polypneustic**.

apocritan Belonging to the suborder of Hymenoptera (Apocrita) in which the first abdominal segment is fused to the thorax; *see also* **propodeum**.

apod A **larva** lacking true legs (Fig. 6.6); *see also* **oligopod**, **polypod**.

apode (*adj.* apodous) An organism without legs.

apodeme An ingrowth of the **exoskeleton**, tendon-like, to which muscles are attached (Figs. 2.23 & 3.2c).

apolysial space The space between the old and the new **cuticle** that forms during **apolysis**, prior to **ecdysis**.

apolysis The separation of the old from the new **cuticle** during **molting**.

apomixis Parthenogenesis (q.v.) in which eggs are produced mitotically (no meiosis); *see also* **automixis**.

apomorphy (synapomorphy) A derived feature (shared by two or more groups).

apophysis (*pl.* apophyses) An elongate **apodeme**, an internal projection of the **exoskeleton**.

aposematic Warning of unpalatability, particularly using color.

aposematism A communication system based on warning signals.

apparency Obviousness (e.g. of a plant to an insect herbivore).

appendicular ovipositor The true **ovipositor** formed from appendages of segments 8 and 9; *see also* **substitutional ovipositor**.

appendix dorsalis The medial caudal appendage arising from the **epiproct**, lying above the **anus**; present in apterygotes, most mayflies, and some fossil insects.

apterous Wingless.

arachnophobia Fear of arachnids (spiders and relatives).

aroliar pad Pretarsal pad-like structure (Fig. 2.19).

arolium (*pl.* arolia) Pretarsal pad-like or sac-like structure(s) lying between the **claws** (Fig. 2.19).

arrhenotoky (arrhenotokous parthenogenesis)

Production of **haploid** male offspring from unfertilized eggs; *see also* **haplodiploidy**.

arthrodial membrane Soft, stretchable **cuticle**, e.g. between segments (Fig. 2.4).

articular sclerites Separate, small, movable plates that lie between the body and a wing.

asynchronous muscle A muscle that contracts many times per nerve impulse, as in many flight muscles and those controlling the cicada tymbal.

atrium (*pl.* atria) A chamber, especially inside a tubular conducting system, such as the **tracheal system** (Fig. 3.10a).

augmentation The supplementation of existing natural enemies by the release of additional individuals.

autapomorphy A feature unique to a taxonomic group; *see also* **apomorphy**, **plesiomorphy**.

autochthonous Originating from within, as of nutrients generated in an aquatic ecosystem, e.g. primary production; native; *see also* **allochthonous**.

automimic A condition of **Batesian mimicry** in which palatable members of a species are defended by their resemblance to members of the same species that are chemically unpalatable.

automixis Parthenogenesis (q.v.) in which eggs are produced after meiosis, but ploidy is restored.

autotomy The shedding of appendage(s), notably for defense.

axillary area An area at the wing base bearing the wing articulation (Fig. 2.20).

axillary plates Two (anterior and posterior) articulating plates that are fused with the veins in an odonate wing; the anterior supports the **costal** vein, the posterior supports the remaining veins; in Ephemeroptera there is only a posterior plate.

axillary sclerites Three or four **sclerites**, which together with the **humeral plate** and **tegula** comprise the **articular sclerites** of the neopteran wing base (Fig. 2.21).

axon A nerve cell fiber that transmits a nerve impulse away from the cell body (Fig. 3.5); *see also* **dendrite**.

azadirachtin (AZ) The chemically active principal of the **neem** tree.

basal At or towards either the base or the main body, or closer to the point of attachment (Fig. 2.8).

basalare (*pl.* basalaria) A small **sclerite**, one of the **epipleurites** that lies anterior to the pleural wing process; an attachment for the **direct flight** muscles (Fig. 2.18).

basisternum The main **sclerite** of the **eusternum**, lying between the anterior **presternum** and posterior **sternellum** (Fig. 2.18).

Batesian mimicry A mimetic system in which a palatable species obtains protection from predation by resembling an unpalatable species; *see also* **Müllerian mimicry**.

benthos The bottom sediments of aquatic habitats and/or the organisms that live there.

benzoylureas A class of chemicals that inhibit **chitin** synthesis.

biogeography The study of biotic distribution in space and time.

biological control (biocontrol) The human use of selected living organisms to control populations of plant or animal pest species.

biological monitoring Using plants or animals to detect changes in the environment.

biological transfer The movement of a disease organism from one **host** to another by one or more **vectors** in which there is a biological cycle of disease.

bioluminescence The production by an organism of cold light, commonly involving the action of an enzyme (luciferase) on a substrate (luciferin).

biotype A biologically differentiated form of a purported single species.

bivoltine Having two generations in one year; *see also* **univoltine**, **multivoltine**, **semivoltine**.

bivouac An army ant camp during the mobile phase.

borer (*adj.* boring) A maker of burrows in dead or living tissue.

brachypterous Having shortened wings.

brain In insects, the supraoesophageal ganglion of the nervous system (Fig. 3.6), comprising **protocerebrum**, **deutocerebrum**, and **tritocerebrum**.

broad-spectrum insecticide An insecticide with a broad range of targets, usually acting generally on the insect nervous system.

brood A clutch of individuals that hatch at the same time from eggs produced by one set of parents.

bud *See* **imaginal disc**.

bursa copulatrix The female **genital chamber** if functioning as a copulatory pouch; in Lepidoptera, the primary receptacle for sperm (Fig. 5.6).

bursicon A neuropeptide **hormone** that controls hardening and darkening of the **cuticle** after **ecdysis**.

caecum (*pl.* caeca) (*Am.* cecum) A blind-ending tube or sac (Fig. 3.1).

calliphorin A protein produced in the **fat body** and stored in the **hemolymph** of larval Calliphoridae (Diptera).

calyx (*pl.* calyces) A cup-like expansion, especially of the oviduct into which the **ovaries** open (Fig. 3.20a).

camouflage crypsis The state in which an organism is indistinguishable from its background.

campaniform sensillum A mechanoreceptor that detects stress on the **cuticle**, comprising a dome of thin cuticle overlying one **neuron** per **sensillum**, located especially on joints (Fig. 4.2).

cantharophily Plant pollination by beetles.

cap cell The outermost cell of a sense organ such as a **chordotonal organ** (Fig. 4.3).

carbamate A synthetic insecticide.

cardiopeptide A neuropeptide **hormone** that stimulates the **dorsal vessel** ("heart") causing **hemolymph** movement.

cardo The proximal part of the maxillary base (Fig. 2.10).

castes (*sing.* caste) Morphologically distinctive groups of individuals within a single species of social insect, usually differing in behavior.

caudal At or towards the **anal** (tail) end.

caudal filament One of two or three terminal filaments (Fig. 8.4, see also Box 10.1).

caudal lamellae One of two or three terminal gills (see Box 10.2).

cavernicolous (troglodytic; troglobiont) Living in caves.

cecidology The study of plant **galls**.

cecidozoa Gall-inducing animals; *see also* **gallicola**.

cell An area of the wing membrane partially or completely surrounded by veins; *see* **closed cell**, **open cell**.

cement layer The outermost layer of the **cuticle** (Fig. 2.1), often absent.

central nervous system In insects, the central series of **ganglia** extending for the length of the body (Fig. 3.6); *see also* **brain**.

cephalic Pertaining to the head.

cercus (*pl.* cerci) One of a pair of appendages originating from abdominal segment 11 but usually visible as if on segment 10 (Fig. 2.23).

cervical sclerite(s) Small **sclerite(s)** on the membrane between the head and thorax (actually the first thoracic segment) (Fig. 2.9).

chitin The major component of arthropod **cuticle**, a polysaccharide composed of subunits of acetylglucosamine and glucosamine (Fig. 2.2).

chitin synthesis inhibitor An insecticide that prevents **chitin** formation.

chloride cells Osmoregulatory cells found in the epithelium of the abdominal gills of aquatic insects.

chordotonal organs Sense organs (mechanoreceptors) that perceive vibrations, comprising one to several elongate cells called **scolopidia** (Figs. 4.3 & 4.4). Examples include the **tympanum**, **subgenual organ**, and **Johnston's organ**.

chorion The outermost shell of an insect egg, which may be multilayered, including the exochorion, endochorion, and wax layer (Fig. 5.10).

cibarium The dorsal food pouch, lying between the **hypopharynx** and the inner wall of the **clypeus**, often with a muscular pump (Figs. 2.14 & 3.14).

circadian rhythms Repeated periodic behavior with an interval of about 24 h.

clade A group of organisms proposed to be **monophyletic**, i.e. all descendants of one common ancestor.

cladistics A classification system in which **clades** are the only permissible groupings.

cladogram A diagram illustrating the branching sequence of purported relationships of organisms, based on the distribution of shared derived features (synapomorphies) (Fig. 7.2).

classical biological control Control of an exotic pest by natural enemies from the pest's area of origin.

claval furrow A **flexion-line** on the wing that separates the **clavus** from the **remigium** (Fig. 2.20).

clavus An area of the wing delimited by the **claval furrow** and the posterior margin (Figs. 2.20 & 2.22e).

claw (pretarsal claw; unguis) A hooked structure on the distal end of the **pretarsus**, usually paired (Fig. 2.19); more generally, any hooked structure.

closed cell An area of the wing membrane bounded entirely by veins; *see also* **open cell**.

closed tracheal system A gas-exchange system comprising **tracheae** and **tracheoles** but lacking **spiracles** and therefore closed to direct contact with the atmosphere (Fig. 3.11d–f); *see also* **open tracheal system**.

clypeus The part of the insect head to which the **labrum** is attached anteriorly (Figs. 2.9 & 2.10); it lies below the **frons**, with which it may be fused in a **frontoclypeus** or separated by a suture.

coevolution Evolutionary interactions between two organisms, such as plants and pollinators, hosts and parasites; the degree of specificity and reciprocity varies; *see also* **guild coevolution**, **phyletic tracking**, **specific coevolution**.

colleterial glands Accessory glands of the female internal genitalia that produce secretions used to cement eggs to the substrate.

collophore The **ventral tube** of Collembola.

colon The **hindgut** between the **ileum** and **rectum** (Figs. 3.1 & 3.13).

comb In a social hymenopteran nest, a layer of regularly arranged cells (Fig. 12.6 & Box 12.1).

common (median) oviduct In female insects, the tubes leading from the fused **lateral oviducts** to the **vagina** (Fig. 3.20a).

competency The potential of a termite of one **caste** to become another, e.g. a worker to become a soldier.

compound eye An aggregation of **ommatidia**, each acting as a single facet of the eye (Figs. 2.9 & 4.11).

conjunctiva (conjunctival membrane) *See* **intersegmental membrane**.

connective Anything that connects; more specifically, the paired longitudinal nerve cords that connect the **ganglia** of the **central nervous system**.

conservation (in biological control) Measures that protect and/or enhance the activities of natural enemies.

constitutive defense Part of the normal chemical composition; *see also* **induced defense**.

contact poisons Insecticides that poison by direct contact.

coprophage (*adj.* coprophagous) A feeder on dung or excrement (Fig. 9.5).

corbicula The pollen basket of bees (Fig. 12.4).

coremata (*sing.* corema) Eversible, thin-walled abdominal organs of male moths used for dissemination of sex **pheromone**.

corium A section of the heteropteran **hemelytron** (fore wing), differentiated from the **clavus** and membrane, usually leathery (Fig. 2.22e).

cornea The cuticle covering the eye or **ocellus** (Figs. 4.10 & 4.11).

corneagenous cell One of the translucent cells beneath the **cornea** which secretes and supports the corneal lens (Fig. 4.9).

cornicle (siphunculus) Paired tubular structures on the abdomen of aphids that discharge defensive lipids and alarm **pheromones**.

corpora allata (*sing.* corpus allatum) Paired endocrine glands associated with the stomodeal ganglia behind the **brain** (Fig. 3.8), the source of **juvenile hormone**.

corpora cardiaca (*sing.* corpus cardiacum) Paired glands lying close to the aorta and behind the **brain** (Fig. 3.8), acting as stores and producers of **neurohormones**.

cosmopolitan Distributed worldwide (or nearly so).

costa (*adj.* costal) The most anterior longitudinal wing vein, running along the costal margin of the wing and ending near the apex (Fig. 2.21).

costal fracture A break or weakness in the costal margin in Heteroptera that divides the **corium**, separating the **cuneus** from the embolium (Fig. 2.22e).

coxa (*pl.* coxae) The proximal (basal) leg segment (Fig. 2.19).

crepuscular Active at low light intensities, dusk or dawn; *see also* **diurnal**, **nocturnal**.

crista acustica (crista acoustica) The main **chordotonal organ** of the tibial tympanal organ of katydids (Orthoptera: Tettigoniidae) (Fig. 4.4).

crochets Curved hooks, spines, or spinules on prolegs.

crop The food storage area of the digestive system, posterior to the oesophagus (Figs. 3.1 & 3.13).

cross-resistance Resistance of an insect to one insecticide providing resistance to a different insecticide.

cross-veins Transverse wing veins that link the longitudinal veins.

cryoprotection Mechanisms that allow organisms to survive periods of, often extreme, cold.

crypsis Camouflage by resemblance to environmental features.

cryptic Hidden, camouflaged, concealed.

cryptobiosis The state of a living organism during which there are no signs of life and metabolism virtually ceases.

cryptonephric system A condition of the excretory system in which the **Malpighian tubules** form an intricate contact with the **rectum**, allowing production of dry excreta (see Box 3.4).

crystalline cone A hard crystalline body lying beneath the **cornea** in an **ommatidium** (Fig. 4.10).

crystalline lens A lens lying beneath the cuticle of the **stemma** of some insects (Fig. 4.11).

ctenidium (*pl.* ctenidia) A comb (see Box 15.4).

cubitus (Cu) The sixth longitudinal vein, lying posterior to the **media**, often divided into an anterior two-branched CuA_1 and CuA_2 and a posterior unbranched CuP_1 (Fig. 2.21).

cuneus The distal section of the **corium** in the heteropteran wing (Fig. 2.22e).

cursorial Running or adapted for running.

cuticle The external skeletal structure, secreted by the **epidermis**, composed of **chitin** and protein comprising several differentiated layers (Fig. 2.1).

cycloalexy Forming aggregations in defensive circles (Fig. 14.7).

cyclodienes A class of organochlorine insecticides.

cytoplasmic incompatibility Reproductive incompatibility arising from cytoplasmic-inherited microorganisms that causes embryological failure; can be unidirectional or bidirectional.

day-degrees A measure of physiological time, the product of time and temperature above a **threshold**.

deciduous Falling off, detaching (e.g. at maturity).

decticous Describing an **exarate** pupa in which the **mandibles** are articulated (Fig. 6.7); *see also* **adecticous**.

delayed effect (of a defensive chemical) An effect that appears after a time lapse from first encounter; *see also* **immediate effect**.

delayed parasitism Parasitism in which hatching of the **parasite** (or **parasitoid**) egg is delayed until the **host** is mature.

delusory parasitosis A psychotic illness in which parasitic infection is imagined.

dendrite A fine branch of a nerve cell (Fig. 3.5); *see also* **axon**.

dermal gland A unicellular epidermal gland that may secrete molting fluid, cements, wax, etc., and probably **pheromones** (Fig. 2.1).

determinate Describing growth or development in which there is a distinctive final, adult, instar; *see also* **indeterminate**.

detritivore (*adj.* detritivorous) An eater of organic detritus of plant or animal origin.

deuterotoky *See* **amphitoky**.

deutocerebrum The middle part of the insect **brain**; the **ganglion** of the second segment, comprising antennal and olfactory lobes.

developmental threshold (growth threshold) The temperature below which no development takes place.

diapause Delayed development that is not the direct result of prevailing environmental conditions.

diapause hormone A **hormone** produced by neurosecretory cells in the **suboesophageal ganglion** that affects the timing of future development of eggs.

dicondylar Describing an articulation (as of a mandible) with two points of articulation (condyles).

diplodiploidy (diploid males) The genetic system found in most insects in which each male receives a **haploid** genome from both his mother and his father, and these two genomes have equal probability of being transmitted through his sperm; *see also* **haplodiploidy, paternal genome elimination**.

diploid With two sets of chromosomes; *see also* **haploid**.

direct flight muscles Flight muscles that are attached directly to the wing (Fig. 3.4); *see also* **indirect flight muscles**.

disjunct Widely separated ranges, as in populations or species geographically separated so as to prevent gene flow.

dispersal Movement of an individual or population away from its birth site.

distal At or near the furthest end from the attachment of an appendage (opposite to **proximal**) (Fig. 2.8).

diurnal Day active; *see also* **crepuscular, nocturnal**.

domatia Plant chambers produced specifically to house certain arthropods, especially ants.

dorsal On the upper surface (Fig. 2.8).

dorsal closure The embryological process in which the dorsal wall of an embryo is formed by growth of the **germ band** to surround the yolk.

dorsal diaphragm The main fibromuscular septum that divides the **hemocoel** into the **pericardial** and **perivisceral sinuses** (compartments) (Fig. 3.9).

dorsal vessel The "aorta" and "heart", the main pump for **hemolymph**; a longitudinal tube lying in the dorsal pericardial sinus (Fig. 3.9).

dorsum The upper surface.

drift Passive movement caused by water or air currents.

drone The male bee, especially of honey bees and bumble bees, derived from an unfertilized egg.

Dufour's gland In **aculeate** hymenopterans, a sac opening into the poison duct near the sting (Fig. 14.11); the site of production of **pheromones** and/or poison components.

dulosis A slave-like relationship between parasitic ant species and the captured brood from another species.

Dyar's rule An observational "rule" governing the size increment found between subsequent **instars** of the same species (Fig. 6.11).

ecdysis (*adj.* ecdysial) The final stage of **molting**, the process of casting off the **cuticle** (Fig. 6.8).

ecdysone The steroid **hormone** secreted by the **prothoracic** gland that is converted to 20-hydroxyecdysone, which stimulates molting fluid excretion.

ecdysteroid The general term for steroids that induce **molting** (Figs. 5.13, 6.9, & 6.10).

ecdysterone An old term for 20-hydroxyecdysone, the major molt-inducing steroid.

eclosion The release of the adult insect from the cuticle of the previous instar; sometimes used of hatching from the egg.

eclosion hormone A **neurohormone** with several functions associated with adult **eclosion**, including increasing cuticle extensibility.

economic injury level (EIL) The level at which economic pest damage equals the costs of their control.

economic threshold (ET) The pest density at which control must be applied to prevent the **economic injury level** being reached.

ectognathy Having exposed mouthparts.

ectoparasite A **parasite** that lives externally on and at the expense of another organism, which it does not kill; *see also* **ectoparasitoid, endoparasite**.

ectoparasitoid A **parasite** that lives externally on and at the expense of another organism, which it kills; *see also* **ectoparasite, endoparasitoid**.

ectoperitrophic space The space between the **peritrophic membrane** and the midgut wall (Fig. 3.16).

ectothermy (*adj.* ectothermic) The inability to regulate the body temperature relative to the surrounding environment.

edeagus (*Am.*) *See* **aedeagus**.

ejaculatory duct The duct that leads from the fused **vas deferens** to the **gonopore** (Fig. 3.20b), through which semen or sperm is transported.

elaiosome A food body forming an appendage on a plant seed (Fig. 11.9).

elytron (*pl.* elytra) The modified, hardened, fore wing of a beetle that protects the hind wing (Fig. 2.22d).

empodium (*pl.* empodia) A central spine or pad on the **pretarsus** of Diptera.
encapsulation A reaction of the **host** to an **endoparasitoid** in which the invader is surrounded by **hemocytes** that eventually form a capsule.
endemic Describing a taxon or disease that is restricted to a particular geographical area.
endite An inwardly directed (mesal) appendage or lobe of a limb segment (Fig. 8.4).
endocuticle The flexible, unsclerotized inner layer of the **procuticle** (Fig. 2.1); *see also* **exocuticle**.
endogenous rhythms Clock-like or calendar-like activity patterns, commonly with **circadian rhythms**, unaffected by external conditions.
endoparasite A **parasite** that lives internally at the expense of another organism, which it does not kill; *see also* **endoparasitoid**, **ectoparasite**.
endoparasitoid A **parasite** that lives internally at the expense of another organism, which it kills; *see also* **endoparasite**, **ectoparasitoid**.
endoperitrophic space In the gut, the space enclosed within the **peritrophic membrane** (Fig. 3.16).
endophallus (vesica) The inner, eversible tube of the **penis** (Fig. 5.4).
endophilic Indoor loving, as of an insect that feeds inside a dwelling; *see also* **exophilic**.
endopterygote Describing development in which the wings form within pockets of the integument, with eversion taking place only at the larval–pupal molt (as in the **monophyletic** grouping Endopterygota).
endosymbiont Intracellular symbionts, typically bacteria, that usually have a mutualistic association with their insect hosts.
endothermy (*adj.* endothermic) The ability to regulate the body temperature higher than the surrounding environment.
energids In an embryo, the daughter nuclei cleavage products and their surrounding cytoplasm.
entomopathogen A pathogen (disease-causing organism) that attacks insects particularly.
entomophage (*adj.* entomophagous) An eater of insects.
entomophily Pollination by insects.
entomophobia Fear of insects.
environmental manipulation Alteration of the environment, particularly to enhance natural populations of insect predators and parasitoids.
enzootic A disease present in a natural **host** within its natural range.
epicoxa A basal leg segment (Fig. 8.4), forming the **articular sclerites** in all extant insects, believed to have borne the **exites** and **endites** that fused to form the evolutionary precursors of wings.
epicranial suture A Y-shaped line of weakness on the **vertex** of the head where the split at **molting** occurs.
epicuticle The inextensible and unsupportive outermost layer of **cuticle**, lying outside the **procuticle** (Fig. 2.1).
epidemic The spread of a disease from its **endemic** area and/or from its normal host(s).
epidermis The unicellular layer of ectodermally derived **integument** that secretes the **cuticle** (Fig. 2.1).
epimeron (*pl.* epimera) The posterior division of the **pleuron** of a thorax, separated from the **episternum** by the **pleural suture** (Fig. 2.18).
epimorphic Describing development in which the segment number is fixed in the embryo before hatching; *see also* **anamorphic**.
epipharynx The ventral surface of the **labrum**, a membranous roof to the mouth (Fig. 2.15).
epipleurite (1) The more dorsal of the **sclerites** formed when the **pleuron** is divided longitudinally. (2) One of two small sclerites of a wing-bearing segment: the anterior **basalare** and posterior **subalare** (Fig. 2.18).
epiproct The dorsal relic of segment 11 (Fig. 2.23).
episternum (*pl.* episterna) The anterior division of the **pleuron**, separated from the **epimeron** by the **pleural suture** (Fig. 2.18).
epizootic Of a disease, when **epidemic** (there is an unusually high number of cases and/or deaths).
ergatoid (apterous neotenic) In termites, a supplementary reproductive derived from a worker, held in a state of arrested development and lacking wings, able to replace reproductives if they die; *see also* **neotenic**.
erythrocyte A red blood cell.
erythrocyte schizogonous cycle The stage in the asexual development of a malaria parasite (*Plasmodium* spp.) in which **trophozoites** within the vertebrate host's red blood cells divide to form **merozoites** (Box 15.1).
esophagus (*Am.*) *See* **oesophagus**.
euplantula (*pl.* euplantulae) A pad-like structure on the ventral surface of some **tarsomeres** of the leg.
eusocial Exhibiting co-operation in reproduction and division of labor, with overlap of generations.
eusternum (*pl.* eusterna) The dominant ventral plate of the thorax that frequently extends into the pleural region (Fig. 2.18).
eutrophication Nutrient enrichment, especially of water bodies.
evolutionary systematics A classification system in which **clades** (**monophyletic** groups) and **grades** (**paraphyletic** groups) are recognized.

exaptation A morphological–physiological predisposition or preadaptation to evolve into a new function.

exarate Describing a **pupa** in which the appendages are free from the body (Fig. 6.7), as opposed to being cemented; *see also* **obtect**.

excretion The elimination of metabolic wastes from the body, or their internal storage in an insoluble form.

exite An outer appendage or lobe of a limb segment (Fig. 8.4).

exocuticle The rigid, sclerotized outer layer of the **procuticle** (Fig. 2.1); *see also* **endocuticle**.

exogenous rhythms Activity patterns governed by variations in the external environment (e.g. light, temperature, etc.).

exophilic (*n.* exophily) Out-of-door loving, used of biting insects that do not enter buildings; *see also* **endophilic**.

exopterygote Describing development in which the wings form progressively in sheaths that lie externally on the dorsal or dorsolateral surface of the body (as in the **paraphyletic** grouping Exopterygota).

exoskeleton The external, hardened, cuticular skeleton to which muscles are attached internally.

external On the outside.

extra-oral digestion Digestion that takes place outside the organism, by secretion of salivary enzymes onto or into the food, with soluble digestive products being sucked up.

facultative Not compulsory, optional behavior, such as facultative parasitism, in which a free-living organism may adopt a parasitic mode of life.

fat body A loose or compact aggregation of cells, mostly **trophocytes** suspended in the **hemocoel**, responsible for storage and excretion.

femur (*pl.* femora) The third segment of an insect leg, following the **coxa** and **trochanter**; often the stoutest leg segment (Fig. 2.19).

fermentation Breakdown of complex molecules by microbes, as of carbohydrates by yeast.

file A toothed or ridged structure used in sound production by **stridulation** through contact with a **scraper**.

filter chamber Part of the alimentary canal of many hemipterans, in which the anterior and posterior parts of the **midgut** are in intimate contact, forming a system in which most fluid bypasses the absorptive midgut (see Box 3.3).

fitness (1) **Darwinian fitness** The contribution of an individual to the gene pool through its offspring. (2) **Inclusive (extended) fitness** The contribution of an individual to the gene pool by enhanced success of its kin.

flabellum In bees, the lobe at the tip of the **glossae** ("tongue") (Fig. 2.11).

flagellomere One of the subdivisions of a "multisegmented" (actually multi-annulate) antennal **flagellum**.

flagellum The third part of an antenna, distal to the **scape** and **pedicel**; more generally, any whip or whip-like structure.

flexion-line A line along which a wing flexes (bends) when in flight (Fig. 2.20).

fluctuating asymmetry The level of deviation from absolute symmetry in a bilaterally symmetrical organism, argued to be due to variable stress during development.

fold-line A line along which a wing is folded when at rest (Fig. 2.20).

follicle The **oocyte** and **follicular** epithelium; more generally, any sac or tube.

follicular Relic morphological evidence left in the **ovary** showing that an egg has been laid (or resorbed), which may include dilation of the lumen and/or pigmentation.

food canal A canal anterior to the **cibarium** through which fluid food is ingested (Figs. 2.11 & 2.12).

forage To seek and gather food.

fore Anterior, towards the head.

fore wings The anterior pair of wings, usually on the **mesothorax**.

foregut (stomodeum) The part of the gut that lies between the mouth and the **midgut** (Fig. 3.13), derived from the ectoderm.

forensic entomologist A scientist that studies the role of insects in criminal matters.

formulation The components and proportions of additional substances that accompany an insecticide when prepared for application.

fossorial Digging, or adapted for digging (Fig. 9.2).

frass Solid excreta of an insect, particularly a larva.

frenate coupling A form of wing-coupling in which one or more hind-wing structures (**frenulum**) attach to a retaining structure (**retinaculum**) on the fore wing.

frenulum Spine or group of bristles on the **costa** of the hind wing of Lepidoptera that locks with the fore-wing **retinaculum** in flight.

frons The single medio-anterior **sclerite** of the insect head, usually lying between the epicranium and the **clypeus** (Fig. 2.9).

frontoclypeal suture (epistomal suture) A groove that runs across the insect's face, often separating the **frons** from the **clypeus** (Fig. 2.9).

frontoclypeus The combined **frons** and **clypeus**.

fructivore (*adj.* fructivorous) A fruit-eater.

fundatrix (*pl.* fundatrices) An apterous viviparous

parthenogenetic female aphid that develops from the overwintering egg.

fungivore (*adj.* fungivorous) A fungus-eater.

furca (furcula) The abdominal springing organ of Collembola (see Box 9.2); with the fulcral arm, the lever of the hymenopteran sting (Fig. 14.11).

galea The lateral lobe of the maxillary stipes (Figs. 2.10, 2.11, & 2.12).

gall An aberrant plant growth produced in response to the activities of another organism, often an insect (Fig. 11.5).

gallicola (*pl.* gallicolae) A **gall**-dweller, more particularly a stage in certain aphids (including Phylloxeridae) that induces aerial galls on the host plant; *see also* **radicicola**.

gametocyte A cell from which a gamete (egg or spermatozoon) is produced.

ganglion (*pl.* ganglia) A nerve center; in insects, forming fused pairs of white, ovoid bodies lying in a row ventrally in the body cavity, linked by a double nerve cord (Fig. 3.1).

gas exchange The system of oxygen uptake and carbon dioxide elimination.

gas gills Specific respiratory (gas-exchange) surfaces on aquatic insects, often as abdominal lamellae, but may be present almost anywhere on the body.

gaster The swollen part of the abdomen of **aculeate** Hymenoptera, lying posterior to the **petiole** (waist) (see Box 12.2).

gena (*pl.* genae) Literally, a cheek; on each side of the head, the part lying beneath the **compound eye**.

genital chamber A cavity of the female body wall that contains the **gonopore** (Fig. 3.20a), also known as the **bursa copulatrix** if functioning as a copulatory pouch.

genitalia All ectodermally derived structures of both sexes associated with reproduction (copulation, fertilization, and oviposition).

genus (*pl.* genera, *adj.* generic) The name of the taxonomic category ranked between species and family; an assemblage of one or more species united by one or more derived features and therefore believed to be of a single evolutionary origin (i.e. **monophyletic**).

germ anlage (in embryology) The germ disc that denotes the first indication of a developing embryo (Fig. 6.5).

germ band (in embryology) The postgastrulation band of thickened cells on the ventral gastroderm, destined to form the ventral part of the developing embryo (Fig. 6.5).

germarium The structure within an **ovariole** in which the oogonia give rise to **oocytes** (Fig. 3.20a).

giant axon A nerve fiber that conducts impulses rapidly from the sense organ(s) to the muscles.

gill A gas-exchange organ, found in various forms in aquatic insects.

glossa (*pl.* glossae) The "tongue", one of a pair of lobes on the inner apex of the **prementum** (Fig. 2.11).

gonapophysis (*pl.* gonapophyses) A valve (part of the shaft) of the **ovipositor** (Fig. 2.23); also in the genitalia of many male insects (Fig. 2.24).

gonochorism Sexual reproduction in which males and females are separate individuals.

gonocoxite The base of an appendage, formed of **coxa + trochanter**, of a genital segment (8 or 9) (also called a **valvifer** in females) (Figs. 2.23 & 2.24).

gonopore The opening of the genital duct; in the unmodified female the opening of the common oviduct (Fig. 3.20a), in the male the opening of the ejaculatory duct.

gonostyle The style (rudimentary appendage) of the ninth segment (Fig. 2.23), often functioning as a male clasper (Fig. 2.24).

grade A **paraphyletic** group, one which does not include all descendants of a common ancestor, united by shared primitive features.

gregarious Forming aggregations.

gressorial Walking, or adapted for walking.

guild coevolution (diffuse coevolution) Concerted evolutionary change that takes place between groups of organisms, as opposed to between two species; *see also* **specific coevolution**.

gula A ventromedian sclerotized plate on the head of **prognathous** insects (Fig. 2.10).

gyne A reproductive female hymenopteran, a **queen**.

habituation Reduction in the response to a stimulus with repeated exposure, through modification of the central nervous system.

haemo- *See* **hemo-** (*Am.*).

hair A cuticular extension, also called a **macrotrichium** or **seta**.

hair plate A group of sensory hairs that act as a **proprioceptor** for movement of articulating parts of the body (Fig. 4.2a).

haltere The modified hind wing in Diptera, acting as a balancer (Fig. 2.22f).

hamuli (*adj.* hamulate) Hooks along the anterior (costal) margin of the hind wing of Hymenoptera which couple the wings in flight by catching on a fold of the fore wing.

haplodiploidy (*adj.* haplodiploid) A genetic system in which the male is **haploid** and transmits only his mother's genome; *see also* **arrhenotoky**, **diplodiploidy**, **paternal genome elimination**.

haploid With one set of chromosomes; *see also* **diploid**.

haustellate Sucking, as of mouthparts.

haustellum Sucking mouthparts (Fig. 15.1).

head The anterior of the three major divisions (**tagmata**) of an insect body.

hellgrammite The larva of Megaloptera.

hematophage (*or* haematophage) (*adj.* hematophagous) An eater of blood (or like fluid).

hemelytron (*pl.* hemelytra) The fore wing of Heteroptera, with a thickened basal section and membranous apical section (Fig. 2.22e).

hemimetaboly (*adj.* hemimetabolous) Development in which the body form gradually changes at each molt, with wing buds growing larger at each molt; incomplete **metamorphosis**; *see also* **holometaboly**.

hemocoel (*or* haemocoel) The main body cavity of many invertebrates including insects, formed from an expanded "blood" system.

hemocoelous (*or* haemocoelous) **viviparity** Viviparity (producing live offspring) in which the immature stages develop within the **hemocoel** of the parent female, e.g. as in Strepsiptera.

hemocyte (*or* haemocyte) An insect blood cell.

hemolymph (*or* haemolymph) The fluid filling the **hemocoel**.

hermaphroditism Having individuals that possess both testes and ovaries.

heterochrony Alteration in the relative timing of activation of different developmental pathways.

heteromorphosis (hypermetamorphosis) Undergoing a major change in morphology between larval instars, as from **triungulin** to grub.

hibernate To undergo **quiescence** or **diapause** during seasonal cold conditions.

hind At or towards the posterior.

hind wings The wings on the metathoracic segment.

hindgut (proctodeum) The posterior section of the gut, extending from the end of the **midgut** to the **anus** (Fig. 3.13).

holoblastic (in embryology) A complete cleavage of the egg.

holocyclic In aphids, describing a life cycle in which reproduction is solely sexual, and in which winter survival is on a (often scarce) host plant, with no migration to a summer host plant.

holometaboly (*adj.* holometabolous) Development in which the body form abruptly changes at the pupal molt; complete **metamorphosis**, as in the group Endopterygota; *see also* **hemimetaboly**.

homeosis (*adj.* homeotic) The genetic or developmental modification of a structure (e.g. an appendage) on one segment to resemble a morphologically similar or different structure on another segment (**serial homology**).

homeostasis Maintenance of a prevailing condition (physiological or social) by internal feedback.

homeothermy The maintenance of an even body temperature despite variation in the ambient temperature.

homology (*adj.* homologous) Morphological identity or similarity of a structure in two (or more) organisms as a result of common evolutionary origin.

homoplasy Characters acquired convergently or in parallel, rather than by direct inheritance from an ancestor.

honeydew A watery fluid containing sugars eliminated from the anus of some Hemiptera.

hormone A chemical messenger that regulates some activity at a distance from the endocrine organ that produced it.

host An organism that harbors another, especially a **parasite** or **parasitoid**, either internally or externally.

host discriminate To choose between different **hosts**.

host preference To prefer one **host** over another.

host regulation The ability of a **parasitoid** to manipulate the **host's** physiology.

humeral plate One of the articular **sclerites** of the neopteran wing base (Fig. 2.21); *see also* **axillary sclerites**, **tegula**.

humus Organic soil.

hydrostatic skeleton Turgid structural support provided by fluid pressure maintained by muscle contractions on a fixed volume of liquid, especially within larval insects.

hypermetamorphosis *See* **heteromorphosis**.

hyperparasite (*adj.* hyperparasitic) A **parasite** that lives upon another parasite.

hyperparasitoid A secondary **parasitoid** that develops upon another **parasite** or parasitoid.

hypognathous With the head directed vertically and mouthparts directed ventrally; *see also* **opisthognathous**, **prognathous**.

hypopharynx A median lobe of the preoral cavity of the mouthparts (Fig. 2.10).

hyporheic Living in the substrate beneath the bed of a waterbody.

idiobiont A **parasitoid** that prevents its **host** from developing any further, by paralysis or death; *see also* **konobiont**.

ileum The second section of the **hindgut**, preceding the **colon** (Figs. 3.1 & 3.13).

imaginal disc (imaginal bud) Latent adult structure in an immature insect, visible as group of undifferentiated cells (Fig. 6.4).

imago (*pl.* imagines *or* imagos) The adult insect.

immediate effect (of a defensive chemical) An effect that appears immediately on first encounter; *see also* **delayed effect**.

indeterminate Describing growth or development in which there is no distinctive final, adult instar, with no definitive terminal molt; *see also* **determinate**.

indirect flight muscles Muscles that power flight by deforming the thorax rather than directly moving the wings (Fig. 3.4); *see also* **direct flight muscles**.

indirect system (of flight) With power for flight coming from regular deformation of the thorax by **indirect flight muscles**, in contrast to predominantly from muscular connection to the wings.

induced defense A chemical change, deleterious to herbivores, induced in foliage as a result of feeding damage.

industrial melanism The phenomenon of dark **morphs** occurring in a higher than usual frequency in areas in which industrial pollution darkens tree trunks and other surfaces upon which insects may rest.

inhalation poison An insecticide with a fumigant action.

innate Describing behavior requiring no choice or learning.

inner epicuticle The innermost of the three layers of **epicuticle**, with the **procuticle** beneath it (Fig. 2.1); *see also* **outer epicuticle**, **superficial layer**.

inoculation To infect with a disease by introducing it into the blood; in biological control to periodically release a natural enemy unable to survive naturally in an area or unable to track the spread of a pest.

inquiline An organism that lives in the home of another, sharing food; in entomology, used particularly of residents in the nests of social insects (*see also* **integrated inquiline**, **non-integrated inquiline**) or in plant galls induced by another organism.

insectivore (*adj.* insectivorous) An insect eater; *see also* **entomophage**.

instar The growth stage between two successive molts.

integrated inquiline An **inquiline** that is incorporated into the social life of the host by behavioral modification of both inquiline and host; *see also* **non-integrated inquiline**.

integrated pest management (IPM) Integration of chemical means of insect control with other methods, notably **biological control** and habitat manipulation.

integument The epidermis plus **cuticle**; the outer covering of the living tissues of an insect.

intercropping Mixed planting of agricultural crops.

interference (1) (in colors) Iridescent colors produced by variable reflection of light by narrowly separated surfaces (as in the scales of lepidopterans). (2) (in population dynamics) A reduction in the profitability of an otherwise high resource density caused by intra- and interspecific interactions between predators and parasitoids.

intermediate organ A **chordotonal organ** in the **subgenual organ** of the fore leg of some orthopterans (Fig. 4.4), associated with the **tympanum** and believed to respond to sound frequencies of 2–14 kHz.

intermolt period *See* **stadium**.

intersegmental groove *See* **antecostal suture**.

intersegmental membrane (conjunctiva; conjunctival membrane) A membrane between segments, particularly of the abdomen (Fig. 2.7).

intersternite An intersegmental sternal plate posterior to the **eusternum**, known as the **spinasternum** except on the metasternum (Fig. 2.7).

inundative release (in biological control) Swamping a pest with large numbers of control agents, with control deriving from the released organisms rather than from any of their progeny.

IPM system The operating system used by farmers in order to manage the control of crop pests; *see also* **integrated pest management**.

Johnston's organ A **chordotonal** (sensory) organ within the antennal **pedicel**.

jugal area (**jugum**; *pl.* juga) The posterobasal area of the wing, delimited by the **jugal fold** and the wing margin (Fig. 2.20).

jugal fold A **fold-line** of the wing, dividing the **jugal area** from the **clavus** (Fig. 2.20).

jugate coupling A mechanism for coupling the fore and hind wings in flight by a large **jugal area** of the fore wing overlapping the hind wing.

juvenile hormone (JH) A hormone, based on a chain of 16, 17, or 18 carbon atoms, that is released by the **corpora allata** into the **hemolymph**, and involved in many aspects of insect physiology, including modification of the expression of a molt.

juvenile hormone mimics (juvenoids) Synthetic chemicals that mimic the effect of **juvenile hormone** on development.

juvenoid *See* **juvenile hormone mimics**.

kairomone A communication chemical that benefits the receiver and is disadvantageous to the producer; *see also* **allomone**, **synomone**.

katatrepsis Adoption by the embryo of the final position in the egg, involving moving from dorsal on ovum to ventral aspect.

kinesis (*pl.* kineses) Movement of an organism in response to a stimulus, usually restricted to response to stimulus intensity only.

king The male primary reproductive in termites (Fig. 12.8).

kinship component An indirect contribution to an individual's inclusive fitness (*see* **fitness**) derived from increased reproductive success of the individual's kin (relatives) through the altruistic assistance of the individual.

klinokinesis Movement resembling a "random walk", in which changes in direction are made when unfavorable stimuli are encountered, with a frequency of turning dependent upon the stimulus intensity.

klinotaxis A movement in a definite direction relative to a stimulus, either directly towards or away from the source.

koinobiont A **parasitoid** that allows its **host** to continue to develop; *see also* **idiobiont**.

labella (*sing.* labellum) In certain flies, paired lobes at the apex of the **proboscis**, derived from **labial palps** (see Box 15.5).

labial palp One- to five-segmented appendage of the **labium** (Figs. 2.9 & 2.10).

labium (*adj.* labial) The "lower lip", forming the floor of the mouth, often with a pair of palps and two pairs of median lobes (Figs. 2.9 & 2.10); derived from the sixth head segment.

labrum (*adj.* labral) The "upper lip", forming the roof of the preoral cavity and mouth (Figs. 2.9 & 2.10); derived from the first head segment.

lacinia The mesal lobe of the maxillary stipes (Fig. 2.10).

larva (*pl.* larvae) An immature insect after emerging from the egg, often restricted to insects in which there is complete metamorphosis (**holometaboly**), but sometimes used for any immature insect that differs strongly from its adult; *see also* **nymph**.

lateral At, or close to, the side (Fig. 2.8).

lateral oviducts In female insects, the paired tubes leading from the ovaries to the **common oviduct** (Fig. 3.20a).

laterosternite The result of a fusion of the **eusternum** and a pleural sclerite.

leaf miner A feeder on the mesophyll layer between the upper and lower epidermis of a leaf (Fig. 11.2).

lek A male mating aggregation associated with a defended territory that contains no resources other than available courting males.

lentic Of standing water.

ligation An experimental technique which isolates one part of the body of a living insect from another, usually by tightening a ligature.

ligula The **glossae** plus **paraglossae** of the **prementum** of the **labium**, whether fused or separate.

litter A layer of dead vegetative matter overlying the soil.

longitudinal In the direction of the long axis of the body.

longitudinal muscles Muscles running along the long axis of the body (Fig. 2.7a).

lotic Of flowing water.

macrogamete The female **gametocyte** of a malaria parasite, *Plasmodium* spp. (Box 15.1); *see also* **microgamete**.

macrophage An eater of large particles; *see also* **microphage**.

macrotrichium (*pl.* macrotrichia) A **trichoid sensillum**, also called a **seta** or **hair**.

maggot A legless larval insect, usually with a reduced head, frequently a fly.

major worker An individual of the largest-sized worker **caste** of termites and ants, specialized for defense; *see also* **media worker**, **minor worker**.

Malpighian tubules Thin, blind-ending tubules, originating near the junction of the **midgut** and **hindgut** (Figs. 3.1 & 3.13), predominantly involved in regulation of salt, water, and nitrogenous waste excretion.

mandible (*adj.* mandibular) The jaws, either jaw-like in shape in biting and chewing (mandibulate) insects (Figs. 2.9 & 2.10), or modified as narrow **stylets** in piercing and sucking insects (Fig. 2.13); the first pair of jaws; derived from the fourth head segment.

mandibulate Possessing **mandibles**.

masquerade (mimesis) A form of **crypsis** in which an organism resembles a feature of its environment that is of no interest to a predator.

mating disruption A form of insect control in which synthetic sex **pheromones** (usually of the female) are maintained artificially at a higher level than the background, interfering with mate location.

matrifilial Describing **eusocial** hymenopterans whose colonies consist of mothers and their daughters.

maxilla (*pl.* maxillae) The second pair of jaws, jaw-like in chewing insects (Figs. 2.9 & 2.10), variously modified in others (Fig. 2.13); derived from the fifth head segment.

maxillary palp A one- to seven-segmented sensory appendage borne on the **stipes** of the **maxilla** (Figs. 2.9 & 2.10).

mechanical transfer The movement of a disease organism from one **host** to another by passive transfer, with no biological cycle in the **vector**.

meconium The first excreta of a newly emerged adult following the pupal stage.

media In wing venation, the fifth longitudinal vein, lying between the **radius** and the **cubitus**, with a maximum of eight branches (Fig. 2.21).

media worker An individual of the medium-sized worker **caste** of termites and ants; *see also* **major worker**, **minor worker**.

medial Towards the middle.

median At or towards the middle (Fig. 2.8).

median flexion-line A **fold-line** that runs longitudinally through the approximate middle of the wing (Fig. 2.20).
melanic Darkened.
melanism Darkening caused by increased pigmentation.
melittophily Pollination by bees.
menotaxis Movement (including the light compass response) that maintains a constant orientation relative to a light source.
mentum The ventral fused plate derived from the **labium** (Fig. 2.16).
merozoite The third stage in the asexual cycle of a malaria parasite (*Plasmodium* spp.) (Box 15.1), derived from division of the **schizont**.
mesal (medial) Nearer to the midline of the body.
mesenteron *See* **midgut**.
mesosoma The middle of the three major divisions (**tagmata**) of the insect body, equivalent to the **thorax** but in **apocritan** Hymenoptera including the **propodeum**; called the **alitrunk** in adult ants (see Box 12.2).
mesothorax The second (and middle) segment of the **thorax** (Fig. 2.18).
metabolic resistance The ability to avoid harm by biochemical detoxification of an insecticide.
metamorphosis The relatively abrupt change in body form between the end of immature development and the onset of the imaginal (adult) phase.
metasoma In **apocritan** Hymenoptera, the **petiole** plus **gaster** (see Box 12.2).
metathorax The third (and last) segment of the **thorax** (Fig. 2.18).
microgamete The male **gametocyte** of a malaria parasite, *Plasmodium* spp., flagellate initially, after which the flagellum is lost (Box 15.1); *see also* **macrogamete**.
microlecithal Describing an egg lacking large yolk reserves.
microphage A feeder on small particles, such as spores; *see also* **macrophage**.
micropyle A minute opening in the **chorion** of an insect egg (Fig. 5.10), through which sperm enter.
microtrichium (*pl.* microtrichia) A subcellular cuticular extension, usually several to very many per cell (Fig. 2.6d).
midgut (mesenteron) The middle section of the gut, extending from the end of the **proventriculus** to the start of the **ileum** (Figs. 3.1 & 3.13).
migration Directional movement to more appropriate conditions.
milk glands Specialized accessory glands in certain adenotrophically viviparous (*see* **adenotrophic viviparity**) flies (Hippoboscidae and *Glossina*) that produce secretions fed upon by larvae.
mimesis Resemblance to an inedible object in the environment; *see also* **masquerade**.
mimic (*adj.* mimetic) One of the three components of a mimicry system, the emitter of false signal(s) received by an **observer**; an individual, population, or species that resembles a **model**, usually another species or part thereof; *see also* **automimic**, **Batesian mimicry**, **Müllerian mimicry**.
mimicry The resemblance of a **mimic** to a **model**, by which the mimic derives protection from predation provided to the model (e.g. by unpalatability).
minor worker An individual of the smaller-sized worker **caste** of termites and ants; *see also* **media worker**, **major worker**.
model One of the three components of a **mimicry** system, the emitter of signal(s) received by the **observer**; the organism resembled by a **mimic**, protected from predation, for instance, by distastefulness.
molar area The grinding surface of the **mandible** (Fig. 2.10).
molt increment The increase in size between successive instars (Fig. 6.11).
molting The formation of new **cuticle** followed by **ecdysis** (Figs. 6.9 & 6.10).
monocondylar Describing an articulation (as of a **mandible**) with one point of articulation (condyle).
monogynous Describing a colony of eusocial insects dominated by one **queen**.
monophage (*adj.* monophagous) An eater of only one kind of food, used particularly of specialized **phytophages**; *see also* **oligophage**.
monophyletic Evolutionarily derived from a single ancestor, recognized by the joint possession of shared derived feature(s).
monoxene (*adj.* monoxenous) A **parasite** restricted to one **host**.
morph A genetic form or variant.
motor neuron A nerve cell with an **axon** that transmits stimuli from an interneuron to muscles (Fig. 3.5).
mouth hooks The head skeleton of the maggot larva of higher flies (see Box 15.5).
Müllerian mimicry A **mimicry** system in which two or more unpalatable species obtain protection from predation by resembling each other; *see also* **Batesian mimicry**.
multiparasitism Parasitization of a **host** by two or more **parasites** or **parasitoids**.
multiple resistance The concurrent existence in a single insect population of two or more defense mechanisms against insecticide.
multiporous Having several small openings.
multivoltine Having several generations in one year; *see also* **bivoltine**, **univoltine**, **semivoltine**.

mushroom body A cluster of **seminal vesicles** and **accessory gland** tubules forming a single mushroom-shaped structure, found in certain orthopteroid or blattoid insects.
mycetocyte A cell containing symbiotic microorganisms, scattered throughout the body, particularly within the **fat body**, or aggregated in organs called **mycetomes**.
mycetome An organ containing aggregations of **mycetocytes**, usually located in the **fat body** or gonads.
mycophage (*adj.* mycophagous) An eater of fungi; *see also* **fungivore**.
myiasis Disease or injury caused by feeding of larval flies on live flesh, of humans or other animals.
myofibrils Contractile fibers that run the length of a muscle fiber, comprising actin sandwiched between myosin fibers.
myophily Plant pollination by flies.
myrmecochory The collection and dispersal of seeds by ants.
myrmecophily Plant pollination by ants.
myrmecophyte ("ant plant") A plant that contains **domatia** to house ants.
myrmecotrophy The feeding of plants by ants, notably through the waste products of an ant colony.
naiad An alternative name for the immature stages of aquatic hemimetabolous insects; *see also* **larva**, **nymph**.
nasus A nose, the snout of certain termite soldiers (**nasutes**).
nasute A **soldier** termite possessing a snout.
natatorial Swimming.
necrophage (*adj.* necrophagous) An eater of dead and/or decaying animals.
neem *Azadirachta indica* (Meliaceae), the neem tree.
neoclassical biological control The use of exotic natural enemies to control native pests.
neonicotinamides A class of insecticides modeled on, and similar to, natural nicotine, e.g. imidacloprid.
neotenic In termites, a supplementary reproductive, arrested in its development, that has the potential to take on the reproductive role should the primary reproductives be lost; *see also* **ergatoid**.
neoteny (*adj.* neotenous) The retention of juvenile features into the adult stage.
nephrocyte (pericardial cell) Cell that sieves the **hemolymph** for metabolic products.
neuroendocrine cells *See* **neurosecretory cells**.
neurohormone (neuropeptide) Any of the largest class of insect **hormones**, being small proteins secreted within different parts of the nervous system (Fig. 5.13).
neuron A nerve cell, comprising a cell body, **dendrite**, and **axon** (Figs. 3.5 & 4.3).
neuropeptide *See* **neurohormone**.
neurosecretory cells (neuroendocrine cells) Modified **neurons** found throughout the nervous system (Fig. 3.8), producing insect **hormones** excepting **ecdysteroids** and **juvenile hormones**.
neuston (*adj.* neustic) The water surface.
nocturnal Night active; *see also* **crepuscular**, **diurnal**.
nomenclature The science of naming (living organisms).
non-integrated inquiline An **inquiline** that is adapted ecologically to the nest of the host, but does not interact socially with the host; *see also* **integrated inquiline**.
notum (*pl.* nota) A thoracic **tergum**.
nulliparous Describing a female that has laid no eggs.
nymph An immature insect after emerging from the egg, usually restricted to insects in which there is incomplete metamorphosis (**hemimetaboly**); *see also* **larva**.
obligatory Compulsory or exclusive; e.g. obligatory diapause is a resting stage that occurs in every individual of each generation of a univoltine insect.
observer One of the three components of **mimicry** systems, the receiver of the signal(s) emitted by the **model** and **mimic**.
obtect Describing a **pupa** with body appendages fused (cemented) to the body; not free (Fig. 6.7); *see also* **exarate**.
occipital foramen The opening of the back of the head.
occiput The dorsal part of the posterior cranium (Fig. 2.9).
ocellus (*pl.* ocelli) The "simple" eye (Fig. 4.10b) of adult and nymphal insects, typically three in a triangle on the **vertex**, with one median and two lateral ocelli (Figs. 2.9 & 2.11); the **stemma** of some holometabolous larvae.
oenocyte A cell of the **hemocoel**, **epidermis**, or especially the **fat body**, probably with many functions, most of which are unclear, but including synthesis of cuticle paraffins and/or hemoglobin.
oesophagus (*Am.* esophagus) The **foregut** that lies posterior to the **pharynx**, anterior to the **crop** (Figs. 2.14, 3.1, & 3.13).
oligophage (*adj.* oligophagous) An eater of few kinds of food, e.g. several plant species within one genus or one family; used particularly of **phytophages**; *see also* **monophage**.
oligopneustic Describing a respiratory system with one to two functional **spiracles** on each side of the body; *see also* **apneustic**, **polypneustic**.
oligopod A **larva** with legs on the thorax and not on the abdomen (Fig. 6.6); *see also* **apod**, **polypod**.

oligoxene (*adj.* oligoxenous) A **parasite** or **parasitoid** restricted to a few **hosts**.
ommatidium (*pl.* ommatidia) A single element of the **compound eye** (Fig. 4.11).
ontogeny The process of development from egg to adult.
oocyst A stage in the sexual cycle of a malaria parasite (*Plasmodium* spp.) within the mosquito midgut lining, formed from the **ookinete** and which undergoes sporogony (asexual reproduction) (Box 15.1).
oocyte An immature egg cell formed from the **oogonium** within the **ovariole**.
oogonium The first stage in the development in the **germarium** of an egg from a female germ cell.
ookinete An active sexual stage in the life cycle of a malaria parasite (*Plasmodium* spp.) formed from the zygote and which penetrates the body cavity within the mosquito (Box 15.1).
oostatic hormone A peptide **hormone**, produced either by the ovary or associated neurosecretory tissue, with inhibitory activities on one or more ovarian functions.
ootheca (*pl.* oothecae) A protective surrounding for eggs (see Box 9.8).
open cell An area of the wing membrane partially bounded by veins but including part of the wing margin; *see also* **closed cell**.
open tracheal system A gas-exchange system comprising **tracheae** and **tracheoles** and with spiracular contact with the atmosphere (Fig. 3.11a–c); *see also* **closed tracheal system**.
opisthognathous With the head deflexed such that the mouthparts are directed posteriorly, as in many Hemiptera; *see also* **hypognathous**, **prognathous**.
organochlorine Any of a group of organic chemicals that contain chlorine, including several insecticides.
organophosphate Any of a group of organic chemicals that contain phosphorus, including several insecticides.
orthokinesis A response to a stimulus in which the rate of response (such as speed of movement) is positively proportional to the intensity of the stimulus.
osmeterium (*pl.* osmeteria) An eversible tubular pouch on the **prothorax** of some larval swallowtail butterflies (Lepidoptera: Papilionidae) (Fig. 14.6), used to disseminate volatile toxic, defensive compounds.
osmoregulation The regulation of water balance, maintaining the **homeostasis** of osmotic and ionic content of the body fluids.
ostium (*pl.* ostia) A slit-like opening in the **dorsal vessel** ("heart") present usually on each thoracic and the first nine abdominal segments, each ostium having a one-way valve that permits flow of **hemolymph** from the **pericardial sinus** into the **dorsal vessel** (Fig. 3.9).
outer epicuticle The middle of the three layers of epicuticle, with the **inner epicuticle** beneath it (Fig. 2.1); *see also* **superficial layer**.
ovarian cycle The length of time between successive ovipositions.
ovariole One of several ovarian tubes that form the **ovary** (Fig. 3.1a), each consisting of a **germarium**, a **vitellarium**, and a stalk or pedicel (Fig. 3.20a).
ovary (*pl.* ovaries) One of the paired gonads of female insects, each comprising several **ovarioles**.
oviparity Reproduction in which eggs are laid; *see also* **ovoviviparity**, **viviparity**.
ovipositor The organ used for laying eggs; *see also* **appendicular ovipositor**, **substitutional ovipositor**.
ovoviviparity Retention of the developing fertilized egg within the mother, considered to be a form of **viviparity** (producing live offspring) but in which there is no nutrition of the hatched young; *see also* **oviparity**.
paedogenesis (*adj.* paedogenetic, *Am.* pedogenetic) Reproduction in an immature stage.
pair-wise coevolution *See* **specific coevolution**.
palp (palpus; *pl.* palpi) A finger-like, usually segmented appendage of the maxilla (**maxillary palp**) and labium (**labial palp**) (Figs. 2.9 & 2.10).
panoistic ovariole An **ovariole** that lacks nurse cells; *see also* **polytrophic ovariole**, **telotrophic ovariole**.
paraglossa (*pl.* paraglossae) One of a pair of lobes distolateral on the **prementum** of the **labium**, lying outside the **glossae**, but mesal to the **labial palp** (Fig. 2.10).
paramere One of a pair of lobes lying lateral to the **penis**, forming part of the **aedeagus** (Fig. 2.24).
paranota (*sing.* paranotum, *adj.* paranotal) Postulated lobes of the thoracic terga from which, it has been argued, the wings derive.
parapheromone A chemical that functions as a male lure, e.g. methyl eugenol which attracts male tephritid fruit flies.
paraphyletic Describing a group (**grade**) that is evolutionarily derived from a single ancestor, but which does not contain all descendants, recognized by the joint possession of shared primitive character(s); rejected in **cladistics** but often accepted in **evolutionary systematics**; *see also* **monophyletic**, **polyphyletic**.
paraproct Ventral relic of segment 11 (Fig. 2.23).
parasite An organism that lives at the expense of another (**host**), which it does not usually kill; *see also* **endoparasite**, **ectoparasite**, **parasitoid**.
parasitism The relationship between a **parasitoid** or **parasite** and its **host**.

parasitization The condition of being parasitized, by either a **parasitoid** or **parasite**.

parasitized Describing the state of a **host** that supports a **parasitoid** or **parasite**.

parasitoid A **parasite** that kills its **host**; *see also* **ectoparasitoid**, **endoparasitoid**.

parous Describing a female that has laid at least one egg.

parthenogenesis Development from an unfertilized egg; *see also* **amphitoky**, **arrhenotoky**, **paedogenesis**, **thelytoky**.

patch A discrete area of microhabitat.

patenty In the course of the malaria disease, the first appearance of parasites in the red blood cells.

paternal genome elimination Loss of the paternal genome during the development of an initially **diploid** male, so that his sperm carries only his mother's genes.

pedicel (1) The stem or stalk of an organ. (2) The stalk of an **ovariole** (Fig. 3.20a). (3) The second antennal segment (Fig. 2.10). (4) The "waist" of an ant.

pedogenesis *See* **paedogenesis**.

penis (*pl.* penes) **(phallus)** The median intromittent organ (Figs. 2.24 & 3.20b), variously derived in different insect orders; *see also* **aedeagus**.

pericardial sinus The body compartment that contains the **dorsal vessel** ("heart") (Fig. 3.9).

perineural sinus The ventral body compartment that contains the nerve cord, separated from the **perivisceral sinus** by the **ventral diaphragm** (Fig. 3.9).

periodic release The regular release of biological control agents that are effective in control but unable to establish permanently.

peripheral nervous system The network of nerve fibers and cells associated with the muscles.

peritreme A sclerotized plate surrounding an orifice, notably around a **spiracle**.

peritrophic membrane A thin sheath lining the midgut epithelium of many insects (Fig. 3.16).

perivisceral sinus The central body compartment, delimited by the **ventral** and **dorsal diaphragms**.

pest resurgence The rapid increase in numbers of a pest following cessation of control measures or resulting from development of **resistance** and/or elimination of natural enemies.

petiole A stalk; in **apocritan** Hymenoptera, the narrow second (and sometimes third) abdominal segments that precede the **gaster**, forming the "waist" (see Box 12.2).

phalaenophily Plant pollination by moths.

phallobase In male genitalia, the support for the **aedeagus** (Figs. 2.24b & 5.4).

phallomere A lobe lateral to the **penis**.

pharate Within the cuticle of the previous stadium; "cloaked".

pharynx The anterior part of the **foregut**, anterior to the **oesophagus** (Figs. 2.14 & 3.13).

phenetic Describing a classification system in which overall resemblance between organisms is the criterion for grouping; *see also* **cladistics**, **evolutionary systematics**.

phenylpyrazoles A class of insecticides with some similarities to DDT (dichlorodiphenyl-trichloroethane), e.g. fipronil.

pheromone A chemical used in communication between individuals of the same species, that releases a specific behavior or development in the receiver. Pheromones have roles in aggregation, alarm, courtship, queen recognition, sex, sex attraction, spacing (epideictic or dispersion), and trail-marking.

pheromone mass trapping The use of **pheromones** to lure pest insects, which are then killed.

phoresy (*adj.* phoretic) The phenomenon of one individual being carried on the body of a larger individual of another species.

photoperiod The duration of the light (and therefore also dark) part of the 24 h daily cycle.

photoreceptor A sense organ that responds to light.

phragma (*pl.* phragmata) A plate-like **apodeme**, notably those of the **antecostal suture** of the thoracic segments that support the longitudinal flight muscles (Figs. 2.7d & 2.18).

phragmosis The closing of a nest opening with part of the body.

phyletic tracking Strict **coevolution** in which the phylogenies of each taxon (e.g. host and parasite, plant and pollinator) match precisely.

phylogenetic Relating to **phylogeny**.

phylogeny The evolutionary history (of a taxon).

physiological time A measure of development time based upon the amount of heat required rather than calendar time elapsed.

physogastry Having a swollen abdomen, as in mature **queen** termites (Fig. 12.8), ants, and bees.

phytophage (*adj.* phytophagous) An eater of plants.

phytophagy The eating of plants.

plant resistance A range of mechanisms by which plants resist insect attack; *see also* **antibiosis**, **antixenosis**, **tolerance**.

plasma The aqueous component of **hemolymph**.

plastron The air–water interface (or the air film itself) on an external surface of an aquatic insect, the site of gaseous exchange.

pleiotropic Describing a single gene that has multiple effects on morphology and physiology.

pleometrosis The foundation of a colony of social insects by more than one **queen**.

plesiomorphy (symplesiomorphy) An ancestral feature (shared by two or more groups).

pleural coxal process The anterior end of the **pleural ridge** providing reinforcement for the coxal articulation (Fig. 2.18).

pleural ridge The internal ridge dividing the **pterothorax** into the anterior **episternum** and posterior **epimeron**.

pleural suture The externally visible indication of the **pleural ridge**, running from the leg base to the **tergum** (Fig. 2.18).

pleural wing process The posterior end of the **pleural ridge** providing reinforcement for the wing articulation (Fig. 2.18).

pleuron (*pl.* pleura, *adj.* pleural, *dim.* pleurite) The lateral region of the body, bearing the limb bases.

poikilothermy The inability to maintain an invariate body temperature independent of the ambient temperature.

poison glands A class of **accessory glands** that produce poison, as in the stings of Hymenoptera (Fig. 14.11).

pollination The transfer of pollen from male to female flower parts.

polyculture The cultivation of several crops intermingled.

polydnaviruses (PDVs) A group of viruses found in the ovaries of some parasitic wasps, involved in overcoming host immune responses when injected with the wasp eggs.

polyembryony The production of more than one (often many) embryos from a single egg, notably in parasitic insects.

polyethism Within a social insect **caste**, the division of labor either by specialization throughout the life of an individual or by different ages performing different tasks.

polygyny Social insects that have several **queens**, either at the same time or sequentially (serial polygyny).

polymorphic Describing a species with two or more variants (morphs).

polyphage (*adj.* polyphagous) An eater of many kinds of food, e.g. many plant species from a range of families; used particularly of **phytophages**.

polyphenism Environmentally induced differences between successive generations or different **castes** of social insects, lacking a genetic basis.

polyphyletic Describing a group that is evolutionarily derived from more than one ancestor, recognized by the possession of one or more features evolved convergently; rejected in **cladistics** and **evolutionary systematics**.

polypneustic Describing a respiratory system with at least eight functional **spiracles** on each side of the body; *see also* **apneustic**, **oligopneustic**.

polypod A type of **larva** with jointed legs on the thorax and prolegs on the abdomen (Fig. 6.6); *see also* **apod**, **oligopod**.

polytrophic ovariole An **ovariole** in which several nurse cells remain closely attached to each **oocyte** as it moves down the ovariole; *see also* **panoistic ovariole**, **telotrophic ovariole**.

polyxene (*adj.* polyxenous) A **parasite** or **parasitoid** with a wide range of **hosts**.

pore canals Fine tubules that run through the **cuticle** and carry epidermally derived compounds to the **wax canals** and thus to the epicuticular surface.

pore kettle The chamber within a chemoreceptor **sensillum** that has many pores (slits) in the wall (Box 4.3).

postcoxal bridge The pleural area behind the **coxa**, often fused with the **sternum** (Fig. 2.18).

posterior At or towards the rear (Fig. 2.8).

posterior cranium The posterior, often horseshoe-shaped area of the head.

postgena The lateral part of the occipital arch posterior to the **postoccipital suture** (Fig. 2.9).

postmentum The proximal part of the **labium** (Figs. 2.10 & 13.4).

postnotum The posterior part of a pterothoracic notum, bearing the **phragmata** that support longitudinal muscles (Figs. 2.7d & 2.18).

postoccipital suture A groove on the head that indicates the original head segmentation, separating the **postocciput** from the remainder of the head (Fig. 2.9).

postocciput The posterior rim of the head behind the **postoccipital suture** (Fig. 2.9).

post-tarsus *See* **pretarsus**.

precosta The most anterior wing vein (Fig. 2.21).

precoxal bridge The **pleural** area anterior to the **coxa**, often fused with the **sternum** (Fig. 2.18).

predation (1) Preying on other organisms. (2) Interactions between predator foraging and prey availability.

predator An organism that eats more than one other organism during its life; *see also* **parasitoid**.

pre-erythrocytic schizogonous cycle (exo-erythrocytic schizogonous cycle) In the asexual cycle of a malaria parasite (*Plasmodium* spp.), the cycle within the parenchyma cells of the liver, producing a large **schizont** (Box 15.1).

pregenital segments The first seven abdominal segments.

prementum The free distal end of the **labium**, usually bearing **labial palps**, **glossae**, and **paraglossae** (Figs. 2.10 & 13.4).

prepatent period In the course of a disease, the time between infection and first symptoms.

prescutum The anterior third of the **alinotum** (either meso- or metanotum), in front of the **scutum** (Fig. 2.18).

presoldier In termites, an intervening stage between **worker** and **soldier**.

press The process on the proximal apex of the tarsus of a bee that pushes pollen into the **corbicula** (basket) (Fig. 12.4).

presternum A smaller **sclerite** of the **eusternum**, lying anterior to the **basisternum** (Fig. 2.18).

pretarsus (*pl.* pretarsi) **(post-tarsus)** The distal segment of the insect leg (Fig. 2.19).

prey A food item for a **predator**.

primary cycle In a disease, the cycle that involves the typical host(s); *see also* **secondary cycle**.

primary reproductives In termites, the king and queen founders of a colony (Fig. 12.8).

proboscis A general term for elongate mouthparts (Fig. 2.12); *see also* **rostrum**.

procephalon (in embryology) The anterior head formed by fusion of the primitive anterior three segments (Fig. 6.5).

proctodeum (*adj.* proctodeal) *See* **hindgut**.

procuticle The thicker layer of **cuticle**, which in sclerotized cuticle comprises an outer **exocuticle** and inner **endocuticle**; lying beneath the thinner **epicuticle** (Fig. 2.1).

prognathous With the head horizontal and the mouthparts directed anteriorly; *see also* **hypognathous**, **opisthognathous**.

proleg An unsegmented leg of a **larva**.

pronotum The upper (dorsal) plate of the **prothorax**.

propodeum In **apocritan** Hymenoptera, the first abdominal segment if fused with the thorax to form a **mesosoma** (or **alitrunk** in ants) (see Box 12.2).

proprioceptors Sense organs that respond to the position of body organs.

prothoracic gland The thoracic or cephalic glands (Fig. 3.8) that secrete **ecdysteroids** (Fig. 5.13).

prothoracicotropic hormone A neuropeptide **hormone** secreted by the brain that controls aspects of **molting** and **metamorphosis** via action on the **corpora cardiaca**.

prothorax The first segment of the **thorax** (Fig. 2.18).

protocerebrum The anterior part of the insect **brain**, the **ganglia** of the first segment, comprising the ocular and associative centers.

protraction Withdrawal, the converse of extension.

protrusible vesicle (exsertile vesicle) A small sac or bladder, capable of being extended or protruded.

proventriculus (gizzard) The grinding organ of the **foregut** (Figs. 3.1 & 3.13).

proximal Describing the part of an appendage closer to or at the body (opposite to **distal**) (Fig. 2.8).

pseudergate In "lower" termites, the equivalent to the worker caste, comprising immature nymphs or undifferentiated larvae.

pseudocopulation The attempted copulation of an insect with a flower.

pseudoplacental viviparity Viviparity (producing live offspring) in which a **microlecithal** egg develops via nourishment from a presumed placenta.

pseudotrachea A ridged groove on the ventral surface of the **labellum** of some higher Diptera (see Box 15.5), used to uptake liquid food.

psychophily Plant pollination by butterflies.

pterostigma A pigmented (and denser) spot near the anterior margin of the fore and sometimes hind wings (Figs. 2.20, 2.21, & 2.22b).

pterothorax The enlarged second and third segments of the thorax bearing the wings in pterygotes.

ptilinum A sac everted from a fissure between the antennae of schizophoran flies (Diptera) that aids **puparium** fracture at emergence.

pubescent (*adj.*) Clothed in fine short **setae**.

puddling The action of drinking from pools, especially evident in butterflies, to obtain scarce salts.

pulvillus (*pl.* pulvilli) A bladder-like pretarsal appendage (Fig. 2.19).

pupa (*adj.* pupal) The inactive stage between the **larva** and adult in holometabolous insects; also termed a chrysalis in butterflies.

puparium The hardened skin of the final-instar larva ("higher" Diptera, Strepsiptera), in which the **pupa** forms, or last nymphal instar (Aleyrodidae).

pupation Becoming a **pupa**.

pylorus The anterior **hindgut** where the **Malpighian tubules** enter, sometimes indicated by a muscular valve.

pyrethrin One of the insecticidal chemicals present in the plant pyrethrum (*Tanacetum cinerariifolium*).

pyrethroids Synthetic chemicals with similarity in structure to **pyrethrins**.

quasisocial Social behavior in which individuals of the same generation co-operate and nest-share without division of labor.

queen A female belonging to the reproductive caste in eusocial or semisocial insects (Fig. 12.8), called a **gyne** in social Hymenoptera.

quiescence A slowing down of metabolism and development in response to adverse environmental conditions; *see also* **diapause**.

radicicola (*pl.* radicicolae) A gall-dweller, more particularly a stage in certain aphids (including Phylloxeridae) that induces root tuberosities on the host plant; *see also* **gallicola**.

radius In wing venation, the fourth longitudinal

vein, posterior to the **subcosta**; with a maximum of five branches R_{1-5} (Fig. 2.21).

rake The process on the distal apex of the **tibia** of a bee that gathers pollen into the **press** (Fig. 12.4).

rank The classificatory level in a taxonomic hierarchy, e.g. species, genus, family, order.

raptorial Adapted for capturing prey by grasping.

rectal pad Thickened sections of the epithelium of the rectum involved in water uptake from the feces (Figs. 3.17 & 3.18).

rectum (*adj.* rectal) The posterior part of the **hindgut** (Figs. 3.1 & 3.13).

reflex A simple response to a simple stimulus.

refractory period (1) The time interval during which a nerve will not initiate another impulse. (2) In reproduction, the period during which a mated female will not re-mate.

refuge A safe place.

release To stimulate a particular behavior.

releaser A particular stimulus whose signal stimulates a specific behavior.

remigium The anterior part of the wing, usually more rigid than the posterior **clavus** and with more veins (Fig. 2.20).

reniform Kidney-shaped.

replete An individual ant that is distended by liquid food.

replicate (of disease organisms) To increase in numbers.

reservoir (of diseases) The natural host and geographical range.

resilin A rubber-like or elastic protein in some insect **cuticles**.

resistance The ability to withstand (e.g. temperature extremes, insecticides, insect attack).

resource tracking A relationship, e.g. between parasite and host or plant and pollinator, in which the evolution of the association is based on ecology rather than phylogeny; *see also* **coevolution**, **phyletic tracking**.

respiration (1) A metabolic process in which substrates (food) are oxidized using molecular oxygen. (2) Used inappropriately to mean breathing, as through spiracles or gas exchange across thin cuticle.

retinaculum (1) The specialized hooks or scales on the base of the fore wing that lock with the **frenulum** of lepidopteran hind wings in flight. (2) The retaining hook of the springtail **furca** (spring) (see Box 9.2).

retinula cell A nerve cell of the light receptor organs (**ommatidia**, **stemma**, or **ocellus**) comprising a **rhabdom** of several **rhabdomeres** and connected by nerve axons to the optic lobe (Fig. 4.9).

rhabdom The central zone of the retinula consisting of microvilli filled with visual pigment; comprising **rhabdomeres** belonging to several different **retinula cells** (Fig. 4.10).

rhabdomere One of the seven or eight units comprising a **rhabdom**, the inner part of a **retinula cell** (Fig. 4.10).

rheophilic Liking running water.

rhizosphere A zone surrounding the roots of plants, usually richer in fungi and bacteria than elsewhere in the soil.

riparian Associated with or relating to the waterside.

river continuum concept A formulation of the idea that energy inputs into a river are **allochthonous** in the upper parts and increasingly **autochthonous** in the lower reaches.

rostrum A facial extension that bears the mouthparts at the end (see Box 15.5); *see also* **proboscis**.

rotenone A particular legume-derived chemical with insecticidal and other toxic effects.

round dance A communication dance of honey bees (see Box 12.1).

salivarium (salivary reservoir) The cavity into which the **salivary gland** opens, between the **hypopharynx** and the **labium** (Figs. 3.1 & 3.14).

salivary gland The gland that produces saliva (Fig. 3.1).

saltatorial Adapted for jumping or springing.

saprophage (*adj.* saprophagous) An eater of decaying organisms.

sarcolemma The outer sheath of a striated muscle fiber.

scale A flattened **seta**; a unicellular outgrowth of the **cuticle**.

scape The first segment of the **antenna** (Fig. 2.10).

schizont The second stage in the asexual cycle of a malaria parasite (*Plasmodium* spp.), produced by the **pre-erythrocytic schizogonous cycle** or by division of a **trophozoite** (Box 15.1).

Schwann cell A cell surrounding the axon of a **scolopidium** (Fig. 4.3).

sclerite A plate on the body wall surrounded by membrane or sutures.

sclerophyllous (*n.* sclerophylly) (of plants) Bearing tough leaves, strengthened with sclerenchyma.

sclerotization Stiffening of the **cuticle** by cross-linkage of protein chains.

scolopale cell In a **chordotonal organ**, the **sheath cell** that envelops the **dendrite** (Fig. 4.3).

scolopidia In a **chordotonal organ**, the combination of three cells, the **cap cell**, **scolopale cell**, and **dendrite** (Fig. 4.3).

scopa A brush of thick hair on the hind **tibia** of adult bees.

scraper The ridged surface drawn over a **file** to produce stridulatory sounds.

scutellum The posterior third of the **alinotum**

(either meso- or metanotum), lying behind the **scutum** (Fig. 2.18).

scutum The middle third of the **alinotum** (either meso- or metanotum), in front of the **scutellum** (Fig. 2.18).

secondary cycle In a disease, the cycle that involves an atypical host; *see also* **primary cycle**.

secondary pest outbreak Previously harmless insects becoming pests following insecticide treatment for a primary pest.

secondary plant compounds Plant chemicals assumed to be produced for defensive purposes.

secondary segmentation Any segmentation that fails to match the embryonic segmentation; more specifically, the insect external skeleton in which each apparent segment includes the posterior (intersegmental) parts of the primary segment preceding it (Fig. 2.7).

sector A major wing vein branch and all of its subdivisions.

semi-aquatic Living in saturated soils, but not immersed in free water.

seminal vesicle Male sperm storage organ (Fig. 3.20b).

semiochemical Any chemical used in intra- and interspecific communication.

semisocial Describing social behavior in which individuals of the same generation co-operate and nest-share with some division of reproductive labor.

semivoltine Having a life cycle of greater than one year; *see also* **bivoltine**, **multivoltine**, **univoltine**.

sensillum (*pl.* sensilla) A sense organ, either simple and isolated, or part of a more complex organ.

sensory neuron A nerve cell that receives and transmits stimuli from the environment (Fig. 3.5).

serial homology The occurrence of identically derived features on different segments (e.g. legs).

serosa The membrane covering the embryo (Fig. 6.5).

seta (*pl.* setae) A cuticular extension, a **trichoid sensillum**; also called a **hair** or **macrotrichium**.

sexuales Sexually reproductive aphids of either sex.

sexupara (*pl.* sexuparae) An **alate** parthenogenetic female aphid that produces both sexes of offspring.

sheath cell Any cell enveloping another (Fig. 4.3).

sibling Full brother or sister.

single cycle disease A disease involving one species of **host**, one **parasite**, and one insect **vector**.

siphunculus (*pl.* siphunculi) *See* **cornicle**.

sister group The closest related group of the same taxonomic rank as the group under study.

smell The olfactory sense, the detection of airborne chemicals.

sociality The condition of living in an organized community.

soldier In social insects, an individual worker belonging to a subcaste involved in colony defense.

solitary Non-colonial, occurring singly or in pairs.

species (*adj.* specific) A group of all individuals that can interbreed, mating within the group (sharing a gene pool) and producing fertile offspring, usually similar in appearance and behavior (but see **polymorphic**) and sharing a common evolutionary history.

specific coevolution (pair-wise coevolution) Concerted evolutionary change that takes place between two species, in which the evolution of a trait in one leads to reciprocal development of a trait in a second organism in a feature that evolved initially in response to a trait of the first species; *see also* **guild coevolution**.

sperm competition In multi-mated females, the syndrome by which sperm from one mating compete with other sperm to fertilize the eggs.

sperm precedence The preferential use by the female of the sperm of one mating over others.

spermatheca The female receptacle for sperm deposited during mating (Fig. 3.20a).

spermathecal gland A tubular gland off the **spermatheca**, producing nourishment for sperm stored in the spermatheca (Fig. 3.20a).

spermatophore An encapsulated package of spermatozoa (Fig. 5.6).

spermatophylax In katydids, a proteinaceous part of the **spermatophore** eaten by the female after mating (Box 5.2).

sphecophily Plant pollination by wasps.

spina An internal **apodeme** of the intersegmental sternal plate called the **spinasternum**.

spinasternum An **intersternite** bearing a **spina** (Fig. 2.18), sometimes fused with the **eusterna** of the **prothorax** and **mesothorax**, but never the **metathorax**.

spine A multicellular unjointed cuticular extension, often thorn-like (Fig. 2.6a).

spiracle An external opening of the **tracheal system** (Fig. 3.10a).

sporogony The asexual cycle of a malaria parasite (*Plasmodium* spp.) in the mosquito, in which an **oocyst** undergoes nuclear division to produce **sporozoites** (Box 15.1).

sporozoite In the asexual cycle of a malaria parasite (*Plasmodium* spp.) in the mosquito, a product of **sporogony**; stored in the salivary gland and passed to the vertebrate host during mosquito feeding (Box 15.1).

spur An articulated **spine**.

stadium (*pl.* stadia) The period between molts, the **instar** duration or intermolt period.

startle display A display made by some **cryptic**

insects upon discovery, involving exposure of a startling color or pattern, such as eyespots.

statary The sedentary, stationary phase of army ants.

stemma (*pl.* stemmata) The "simple" eye of many larval insects, sometimes aggregated into a more complex visual organ.

stenogastrous Having a shortened or narrow abdomen.

sterile male technique A means of controlling insects by swamping populations with large numbers of artificially sterilized males.

sternellum The small **sclerite** of the **eusternum**, lying posterior to the **basisternum** (Fig. 2.18).

sternite The diminutive of **sternum**; a subdivision of a sternum.

sternum (*pl.* sterna, *adj.* sternal, *dim.* sternite) The ventral surface of a segment (Fig. 2.7).

stipes The distal part of the **maxilla**, bearing a **galea**, a **lacinia**, and a **maxillary palp** (Fig. 2.10).

stomach poison An insecticidal poison that acts after ingestion into the insect gut.

stomatogastric nervous system The nerves associated with the **foregut** and **midgut**.

stomodeum (*adj.* stomodeal) *See* **foregut**.

striated muscle Muscles in which myosin and actin filaments overlap to give a striated effect.

stridulation The production of sound by rubbing two rough or ridged surfaces together.

style In apterygote insects, small appendages on abdominal segments, homologous to abdominal legs.

stylet One of the elongate parts of piercing–sucking mouthparts (Figs. 2.13, 2.14, & 11.4), a needle-like structure.

subalare (*pl.* subalaria) A small **sclerite**, one of the **epipleurites** that lies posterior to the **pleural wing process**, forming an attachment for the **direct flight muscles** (Fig. 2.18).

subcosta In wing venation, the third longitudinal vein, posterior to the **costa** (Fig. 2.21).

subgenual organ A **chordotonal organ** on the proximal **tibia** that detects substrate vibration (Fig. 4.4).

subimaginal instar *See* **subimago**.

subimago (subimaginal instar) In Ephemeroptera, the winged penultimate instar; subadult.

suboesophageal ganglion The fused **ganglia** of the mandibular, maxillary, and labial segments, forming a ganglionic center beneath the **oesophagus** (Figs. 3.6 & 3.14).

subsocial Describing a social system in which adults look after immature stages for a certain period.

substitutional ovipositor An **ovipositor** formed from extensible posterior abdominal segments; *see also* **appendicular ovipositor**.

succession An ecological sequence.

superficial layer The outermost of the three layers of **epicuticle**, often bearing a lipid, wax, and/or cement layer, with **outer** and **inner epicuticle** beneath it (Fig. 2.1).

superlingua (*pl.* superlinguae) A lateral lobe of the **hypopharynx** (Fig. 2.10), the remnant of a leg appendage of the third head segment.

superparasitism The occurrence of more **parasitoids** within a **host** than can complete their development within the host.

supplementary reproductive In termites, a potential replacement reproductive within its natal nest, which does not become **alate**; also called a **neotenic** or **ergatoid**.

supraoesophageal ganglion *See* **brain**.

suture An external groove that may show the fusion of two plates (Fig. 2.10).

swarm An aerial aggregation of insects, for the purposes of mating.

symbiont An organism that lives in **symbiosis** with another.

symbiosis A long-lasting, close, and dependent relationship between organisms of two different species.

sympatric Describing overlapping geographical distributions of organisms or taxa; *see also* **allopatric**.

synanthropic Associated with humans or their dwellings.

synapse The site of approximation of two nerve cells at which they may communicate.

synchronous muscle A muscle that contracts once per nerve impulse.

syncytium (*adj.* syncytial) A multinucleate tissue without cell division.

synergism The enhancement of the effects of two substances that is greater than the sum of their individual effects.

synomone A communication chemical that benefits both the receiver and the producer; *see also* **allomone**, **kairomone**.

systematics The science of biological classification.

systemic insecticide An insecticide taken into the body of a host (plant or animal) that kills insects feeding on the host.

taenidium (*pl.* taenidia) The spiral thickening of the tracheal wall that prevents collapse.

tagma (*pl.* tagmata) A group of segments that forms a major body unit (**head**, **thorax**, **abdomen**).

tagmosis The organization of the body into major units (**head**, **thorax**, **abdomen**).

tapetum A reflective layer at the back of the eye formed from small **tracheae**.

target-site resistance Increased tolerance by an

insect to an insecticide through reduced sensitivity at the target site.

tarsomere A subdivision of the **tarsus** (Fig. 2.19).

tarsus (*pl.* tarsi) The leg segment distal to the **tibia**, comprising one to five **tarsomeres** and apically bearing the **pretarsus** (Fig. 2.19).

taste Chemoreception of chemicals in a liquid dissolved form.

taxis (*pl.* taxes) An orientated movement of an organism.

taxon (*pl.* taxa) A taxonomic unit (species, genus, family, phylum, etc.).

taxonomy (*adj.* taxonomic) The theory and practice of naming and classifying organisms.

tegmen (*pl.* tegmina) A leathery, hardened fore wing (Fig. 2.22c).

tegula One of the articular **sclerites** of the neopteran wing, lying at the base of the **costa** (Fig. 2.21); *see also* **axillary sclerites, humeral plate**.

telotaxis Orientation and movement directly towards a visual stimulus.

telotrophic ovariole (acrotrophic ovariole) An **ovariole** in which the nurse cells are only within the germarium; the nurse cells remain connected to the **oocytes** by long filaments as the oocytes move down the ovariole; *see also* **panoistic ovariole, polytrophic ovariole**.

teneral The condition of a newly eclosed adult insect, which is unsclerotized and unpigmented.

tentorium The endoskeletal cuticular invaginations of the head, including anterior and posterior tentorial arms.

tergite The diminutive of **tergum**; a subdivision of the tergum.

tergum (*pl.* terga, *adj.* tergal, *dim.* tergite) The dorsal surface of a segment (Fig. 2.7).

terminalia The terminal abdominal segments involved in the formation of the genitalia.

testis (*pl.* testes) One of (usually) a pair of male gonads (Figs. 3.1b & 3.20b).

thanatosis Feigning death.

thelytoky (thelytokous parthenogenesis) A form of **parthenogenesis** producing only female offspring.

thorax The middle of the three major divisions (**tagma**) of the body, comprising **prothorax**, **mesothorax**, and **metathorax** (Fig. 2.18).

threshold The minimum level of stimulus required to initiate (release) a response.

tibia (*pl.* tibiae) The fourth leg segment, following the **femur** (Fig. 2.19).

tolerance The ability of a plant to withstand insect attack and recover.

tonofibrillae Fibrils of **cuticle** that connect a muscle to the **epidermis** (Fig. 3.2).

tormogen cell The socket-forming epidermal cell associated with a **seta** (Figs. 2.6 & 4.1).

trachea (*pl.* tracheae) A tubular element of the insect gas-exchange system, within which air moves (Figs. 3.10 & 3.11).

tracheal system The insect gas-exchange system, comprising **tracheae**, **tracheoles**, and **spiracles** (Figs. 3.10 & 3.11); *see also* **closed tracheal system, open tracheal system**.

tracheole The fine tubules of the insect gas-exchange system (Fig. 3.10b).

transgenic plants Plants containing genes introduced from another organism by genetic engineering.

transovarial transmission (vertical transmission) The transmission of microorganisms between generations via the eggs.

transverse At right angles to the longitudinal axis.

traumatic insemination In Cimicidae and some Nabidae (Hemiptera), unorthodox mating behavior in which the male punctures the female's cuticle with the phallus to deposit sperm instead of utilizing the female reproductive tract.

triad (*adj.* triadic) A triplet of long wing veins (paired main veins and an intercalated longitudinal vein).

trichogen cell A hair-forming epidermal cell associated with a **seta** (Figs. 2.6 & 4.1).

trichoid sensillum A hair-like cuticular projection; a **seta**, **hair**, or **macrotrichium** (Figs. 2.6b, 3.5, & 4.1).

tritocerebrum The posterior (or posteroventral) paired lobes of the insect **brain**, the **ganglia** of the third segment, functioning in handling the signals from the body.

triungulin An active, dispersive first-instar larva of insects including many that undergo **heterometamorphosis**.

trochanter The second leg segment, following the **coxa** (Fig. 2.19).

trochantin A small **sclerite** anterior to the **coxa** (Figs. 2.18 & 2.19).

troglobite (troglobiont) An obligate cave-dweller.

trophallaxis (oral = stomodeal, anal = proctodeal) In social and subsocial insects, the transfer of alimentary fluid from one individual to another; may be mutual or unidirectional.

trophamnion In **parasitoids**, the enveloping membrane surrounding the polyembryonically derived multiple individuals that arise from a single egg, derived from the host's **hemolymph**.

trophic (1) Relating to food. (2) Describing an egg of a social insect that is degenerate and used in feeding other members of the colony.

trophic cascade The ecosystem-wide effects of the removal or introduction of predators on primary production via herbivores.

trophocyte The dominant metabolic and storage cell of the **fat body**.

trophogenesis (*adj.* trophogenic) In social insects, the determination of **caste** type by differential feeding of the immature stages (in contrast to genetic determination of caste).

trophozoite The first stage in the asexual cycle of a malaria parasite (*Plasmodium* spp.), derived from a **merozoite** (either from a liver cell or an **erythrocyte**) (Box 15.1).

trypanosomiasis A disease caused by *Trypanosoma* protozoans, transmitted to humans predominantly by reduviid bugs (Chagas' disease) or tsetse flies (sleeping sickness).

tymbal (tymbal organ) A stretched elastic membrane capable of sound production when flexed.

tympanum (*pl.* tympana) **(tympanal organ)** Any organ sensitive to vibration, comprising a tympanic membrane (thin cuticle), an air sac, and a sensory **chordotonal organ** attached to the tympanic membrane (Fig. 4.4).

unguis (*pl.* ungues) A **claw** (Fig. 2.19).

unguitractor plate The ventral **sclerite** of the **pretarsus** that articulates with the **claws** (Fig. 2.19).

uniporous Having a single opening.

univoltine Having one generation in one year; *see also* **bivoltine**, **multivoltine**, **semivoltine**.

urea A minor component of insect nitrogenous excretion, $CO(NH_2)_2$ (Fig. 3.19).

uric acid The main nitrogenous excretion product, $C_5H_4N_4O_3$ (Fig. 3.19).

uricotelism An excretory system based on **uric acid** excretion.

urocyte (urate cell) A cell that acts as a temporary store for urate excretion products.

vagility The propensity to move or disperse.

vagina A pouch-like or tubular genital chamber of the female genitalia.

valve (1) Generally, any unidirectional opening flap or lid. (2) In female genitalia, the blade-like structures comprising the ovipositor shaft (also called **gonapophysis**) (Fig. 2.23b).

valvifer In female insect genitalia, derivations of **gonocoxites** 8 and 9, supporting the valves of the ovipositor (Fig. 2.23b).

vannus The **anal area** of the wing anterior to the **jugal area** (Fig. 2.20).

vas deferens (*pl.* vasa deferentia) One of the ducts that carry sperm from the testes (Fig. 3.20b).

vector Literally "a bearer", specifically a **host** of a disease that transmits the pathogen to another species of organism.

vectorial capacity A mathematical expression of the probability of disease transmission by a particular **vector**.

venter The lower surface of the body.

ventilate To pass air or oxygenated water over a gas-exchange surface.

ventral Towards or at the lower surface (Fig. 2.8).

ventral diaphragm A membrane lying horizontally above the nerve cord in the body cavity, separating the **perineural sinus** from the **perivisceral sinus** (Fig. 3.9).

ventral nerve cord The chain of ventral **ganglia**.

ventral tube (collophore) In Collembola, a ventral sucker (see Box 9.2).

ventriculus The tubular part of the **midgut**, the main digestive section of the gut (Fig. 3.13).

vertex The top of the head, posterior to the **frons** (Fig. 2.9).

vertical transmission *See* **transovarial transmission**.

vesica *See* **endophallus**.

vibration dance A communication dance of honey bees (see Box 12.1).

vicariance Division of the range of a species by an earth history event (e.g. ocean or mountain formation).

visceral (sympathetic) nervous system The nerve system that innervates the gut, reproductive organs, and tracheal system.

vitellarium The structure within the **ovariole** in which **oocytes** develop and yolk is provided to them (Fig. 3.20a).

vitelline membrane The outer layer of an **oocyte**, surrounding the yolk (Fig. 5.10).

vitellogenesis The process by which **oocytes** grow by yolk deposition.

viviparity The bearing of live young (i.e. post-egg hatching) by the female; *see also* **adenotrophic viviparity**, **hemocoelous viviparity**, **oviparity**, **ovoviviparity**, **pseudoplacental viviparity**.

voltinism The number of generations per year.

vulva The external opening of the copulatory pouch (**bursa copulatrix**) or vagina of the female genitalia (Fig. 3.20a).

waggle dance A communication dance of honey bees (see Box 12.1).

Wasmannian mimicry A form of **mimicry** that allows an insect of another species to be accepted into a social insect colony.

wax A complex lipid mixture giving waterproofing to the **cuticle** or providing covering or building material.

wax canals (wax filaments) Fine tubules that transport lipids from the **pore canals** to the surface of the **epicuticle** (Fig. 2.1).

wax layer The lipid or waxy layer outside the **epicuticle** (Fig. 2.1); it may be absent.

wax mirrors Overlapping plates on the venter of the fourth to seventh abdominal segments of social bees that serve to direct the wax flakes that are produced beneath each mirror.

weed Any organism "in the wrong place", particularly used of plants away from their natural range or invading human monocultural crops.

wood borer An insect that tunnels into live or dead wood.

worker In social insects, a member of the sterile caste that assists the reproductives.

xylophage (*adj*. xylophagous) An eater of wood.

zoocecidia (*sing*. zoocecidium) Plant **galls** induced by animals such as insects, mites, and nematodes, as opposed to those formed by the plant response to microorganisms.

zoophilic Preferring other animals to humans, especially used of feeding preference of blood-feeding insects.

zygote A fertilized egg; in malaria parasites (*Plasmodium* spp.) resulting from fusion of a microgamete and macrogamete (Box 15.1).

FURTHER READING

Gordh, G. & Headrick, D. (2001) *A Dictionary of Entomology*. CABI Publishing, Wallingford.

Torre-Bueno, J.R. de la (1989) *The Torre-Bueno Glossary of Entomology*, 2nd edn. The New York Entomological Society in cooperation with the American Museum of Natural History, New York.

REFERENCES

Alcock, J. (1979) Selective mate choice by females of *Harpobittacus australis* (Mecoptera: Bittacidae). *Psyche* **86**, 213–17.

Alexander, B.A. (1992) An exploratory analysis of cladistic relationships within the superfamily Apoidea, with special reference to sphecid wasps (Hymenoptera). *Journal of Hymenoptera Research* **1**, 25–61.

Alstein, M. (2003) Neuropeptides. In: *Encyclopedia of Insects* (eds. V.H. Resh & R.T. Cardé), pp. 782–5. Academic Press, Amsterdam.

Ando, H. (ed.) (1982) *Biology of the Notoptera.* Kashiyo-Insatsu Co. Ltd, Nagano, Japan.

Anon. (1991) *Ladybirds and Lobsters, Scorpions and Centipedes.* British Museum (Natural History), London.

Askew, R.R. (1971) *Parasitic Insects.* Heinemann, London.

Atkins, M.D. (1980) *Introduction to Insect Behaviour.* Macmillan, New York.

Austin, A.D. & Browning, T.O. (1981) A mechanism for movement of eggs along insect ovipositors. *International Journal of Insect Morphology and Embryology* **10**, 93–108.

Badonnel, A. (1951) Ordre des Psocoptères. In: *Traité de Zoologie: Anatomie, Systématique, Biologie.* Tome X. *Insectes Supérieurs et Hémiptéroïdes,* Fascicule II (ed. P.-P. Grassé), pp. 1301–40. Masson, Paris.

Bandsma, A.T. & Brandt, R.T. (1963) *The Amazing World of Insects.* George Allen & Unwin, London.

Bartell, R.J., Shorey, H.H. & Barton Browne, L. (1969) Pheromonal stimulation of the sexual activity of males of the sheep blowfly *Lucilia cuprina* (Calliphoridae) by the female. *Animal Behaviour* **17**, 576–85.

Barton Browne, L., Smith, P.H., van Gerwen, A.C.M. & Gillott, C. (1990) Quantitative aspects of the effect of mating on readiness to lay in the Australian sheep blowfly, *Lucilia cuprina. Journal of Insect Behavior* **3**, 637–46.

Bar-Zeev, M. (1958) The effect of temperature on the growth rate and survival of the immature stages of *Aedes aegypti* (L.). *Bulletin of Entomological Research* **49**, 157–63.

Beard, J. (1989) Viral protein knocks the guts out of caterpillars. *New Scientist* **124**(1696–7), 21.

Beccari, O. (1877) Piante nuove o rare dell'Arcipelago Malese e della Nuova Guinea, raccolte, descritte ed illustrate da O. Beccari. *Malesia (Genova)* **1**, 167–92.

Bellows, T.S. Jr, Perring, T.M., Gill, R.J. & Headrick, D.H. (1994) Description of a species of *Bemisia* (Homoptera: Aleyrodidae). *Annals of the Entomological Society of America* **87**, 195–206.

Belwood, J.J. (1990) Anti-predator defences and ecology of neotropical forest katydids, especially the Pseudophyllinae. In: *The Tettigoniidae: Biology, Systematics and Evolution* (eds. W.J. Bailey & D.C.F. Rentz), pp. 8–26. Crawford House Press, Bathurst.

Bennet-Clark, H.C. (1989) Songs and the physics of sound production. In: *Cricket Behavior and Neurobiology* (eds. F. Huber, T.E. Moore & W. Loher), pp. 227–61. Comstock Publishing Associates (Cornell University Press), Ithaca, NY.

Bills, P.S., Mota-Sanchez, D. & Whalon, M. (2000) The Database of Arthropods Resistant to Insecticides. http://www.cips.msu.edu/resistance/rmdb/

Binnington, K.C. (1993) Ultrastructure of the attachment of *Serratia entomophila* to scarab larval cuticle and a review of nomenclature for insect epicuticular layers. *International Journal of Insect Morphology and Embryology* **22**(2–4), 145–55.

Birch, M.C. & Haynes, K.F. (1982) *Insect Pheromones.* Studies in Biology no. 147. Edward Arnold, London.

Blaney, W.M. (1976) *How Insects Live.* Elsevier-Phaidon, Oxford.

Bonhag, P.F. & Wick, J.R. (1953) The functional anatomy of the male and female reproductive systems of the milkweed bug, *Oncopeltus fasciatus* (Dallas) (Heteroptera: Lygaeidae). *Journal of Morphology* **93**, 177–283.

Borror, D.J., Triplehorn, C.A. & Johnson, N.F. (1989) *An Introduction to the Study of Insects,* 6th edn. Saunders College Publishing, Philadelphia, PA.

Boulard, M. (1968) Description de cinq Membracides nouveaux du genre *Hamma* accompagnée de précisions sur *H. rectum. Annales de la Societé Entomologique de France (N.S.)* **4**(4), 937–50.

Bourgoin, T. & Campbell, B.C. (2002) Inferring a phylogeny for Hemiptera: Falling into the "Autapomorphy Trap". *Denisia* **176**, 7–82.

Brackenbury, J. (1990) Origami in the insect world. *Australian Natural History* **23**(7), 562–9.

Bradley, T.J. (1985) The excretory system: structure and physiology. In: *Comprehensive Insect Physiology, Biochemistry, and Pharmacology*, Vol. 4: *Regulation. Digestion, Nutrition, Excretion* (eds. G.A. Kerkut & L.I. Gilbert), pp. 421–65. Pergamon Press, Oxford.

Brandt, M. & Mahsberg, D. (2002) Bugs with a backpack: the function of nymphal camouflage in the West African assassin bugs: *Paredocla* and *Acanthiaspis* spp. *Animal Behaviour* **63**, 277–84.

Brothers, D.J. (1999) Phylogeny and evolution of wasps, ants and bees (Hymenoptera, Chrysidoidea, Vespoidea and Apoidea). *Zoologica Scripta* **28**, 233–49.

Brower, J.V.Z. (1958) Experimental studies of mimicry in some North American butterflies. Part III. *Danaus gilippus berenice* and *Limenitis archippus floridensis*. *Evolution* **12**, 273–85.

Brower, L.P., Brower, J.V.Z. & Cranston, F.P. (1965) Courtship behavior of the queen butterfly, *Danaus gilippus berenice* (Cramer). *Zoologica* **50**, 1–39.

Burton, M. & Burton, R. (1975) *Encyclopedia of Insects and Arachnids*. Octopus Books, London.

Calder, A.A. & Sands, D.P.A. (1985) A new Brazilian *Cyrtobagous* Hustache (Coleoptera: Curculionidae) introduced into Australia to control salvinia. *Journal of the Entomological Society of Australia* **24**, 57–64.

Carroll, S.B. (1995) Homeotic genes and the evolution of arthropods and chordates. *Nature* **376**, 479–85.

Caudell, A.N. (1920) Zoraptera not an apterous order. *Proceedings of the Entomological Society of Washington* **22**, 84–97.

Chapman, R.F. (1982) *The Insects. Structure and Function*, 3rd edn. Hodder and Stoughton, London.

Chapman, R.F. (1991) General anatomy and function. In: *The Insects of Australia*, 2nd edn. (CSIRO), pp. 33–67. Melbourne University Press, Carlton.

Cherikoff, V. & Isaacs, J. (1989) *The Bush Food Handbook*. Ti Tree Press, Balmain.

Chu, H.F. (1949) *How to Know the Immature Insects*. William C. Brown, Dubuque, IA.

Clements, A.N. (1992) *The Biology of Mosquitoes*, Vol. 1: *Development, Nutrition and Reproduction*. Chapman & Hall, London.

Cohen, A.C., Chu, C.-C., Henneberry, T.J. et al. (1998) Feeding biology of the silverleaf whitefly (Homoptera: Aleyrodidae). *Chinese Journal of Entomology* **18**, 65–82.

Cohen, E. (1991) Chitin biochemistry. In: *Physiology of the Insect Epidermis* (eds. K. Binnington & A. Retnakaran), pp. 94–112. CSIRO Publications, Melbourne.

Common, I.B.F. (1990) *Moths of Australia*. Melbourne University Press, Carlton.

Common, I.F.B. & Waterhouse, D.F. (1972) *Butterflies of Australia*. Angus & Robertson, Sydney.

Corner, E.J.H. (1952) *Wayside Trees of Malaya*, Vol. II. Government Printing Office, Singapore.

Cornwell, P.B. (1968) *The Cockroach*. Hutchinson, London.

Cox, J.M. (1987) Pseudococcidae (Insecta: Hemiptera). *Fauna of New Zealand* **11**, 1–228.

Coyne, J.A. (1983) Genetic differences in genital morphology among three sibling species of *Drosophila*. *Evolution* **37**, 1101–17.

Cranston, P.S., Edward, D.H.D. & Cook, L.G. (2002) New status, distribution records and phylogeny for Australian mandibulate Chironomidae (Diptera). *Australian Journal of Entomology* **41**, 357–66.

CSIRO (1970) *The Insects of Australia*, 1st edn. Melbourne University Press, Carlton.

CSIRO (1991) *The Insects of Australia*, 2nd edn. Melbourne University Press, Carlton.

Currie, D.C. (1986) An annotated list of and keys to the immature black flies of Alberta (Diptera: Simuliidae). *Memoirs of the Entomological Society of Canada* **134**, 1–90.

Daly, H.V., Doyen, J.T. & Ehrlich, P.R. (1978) *Introduction to Insect Biology and Diversity*. McGraw-Hill, New York.

Darlington, A. (1975) *The Pocket Encyclopaedia of Plant Galls in Colour*, 2nd edn. Blandford Press, Dorset.

Dean, J., Aneshansley, D.J., Edgerton, H.E. & Eisner, T. (1990) Defensive spray of the bombardier beetle: a biological pulse jet. *Science* **248**, 1219–21.

DeFoliart, G.R. (1989) The human use of insects as food and as animal feed. *Bulletin of the Entomological Society of America* **35**, 22–35.

De Klerk, C.A., Ben-Dov, Y. & Giliomee, J.H. (1982) Redescriptions of four vine infesting species of *Margarodes* Guilding (Homoptera: Coccoidea: Margarodidae) from South Africa. *Phytophylactica* **14**, 61–76.

Deligne, J., Quennedey, A. & Blum, M.S. (1981) The enemies and defence mechanisms of termites. In: *Social Insects*, Vol. II (ed. H.R. Hermann), pp. 1–76. Academic Press, New York.

Devitt, J. (1989) Honeyants: a desert delicacy. *Australian Natural History* **22**(12), 588–95.

Deyrup, M. (1981) Deadwood decomposers. *Natural History* **90**(3), 84–91.

Dodson, G. (1989) The horny antics of antlered flies. *Australian Natural History* **22**(12), 604–11.

Dodson, G.N. (1997) Resource defence mating system in antlered flies, *Phytalmia* spp. (Diptera: Tephritidae). *Annals of the Entomological Society of America* **90**, 496–504.

Dolling, W.R. (1991) *The Hemiptera*. Natural History Museum Publications, Oxford University Press, Oxford.

Dow, J.A.T. (1986) Insect midgut function. *Advances in Insect Physiology* **19**, 187–328.

Downes, J.A. (1970) The feeding and mating behaviour of the specialized Empidinae (Diptera); observations on four species of *Rhamphomyia* in the high Arctic and a general discussion. *Canadian Entomologist* **102**, 769–91.

Duffy, E.A.J. (1963) *A Monograph of the Immature Stages of Australasian Timber Beetles (Cerambycidae)*. British Museum (Natural History), London.

Eastham, L.E.S. & Eassa, Y.E.E. (1955) The feeding mechanism of the butterfly *Pieris brassicae* L. *Philosophical Transactions of the Royal Society of London B* **239**, 1–43.

Eberhard, W.G. (1985) *Sexual Selection and Animal Genitalia.* Harvard University Press, Cambridge, MA.

Edwards, D.S. (1994) *Belalong: a Tropical Rainforest.* The Royal Geographical Society, London, and Sun Tree Publishing, Singapore.

Eibl-Eibesfeldt, I. & Eibl-Eibesfeldt, E. (1967) Das Parasitenabwehren der Minima-Arbeiterinnen der Blattschneider-Ameise (*Atta cephalotes*). *Zeitschrift für Tierpsychologie* **24**, 278–81.

Eidmann, H. (1929) Morphologische und physiologische Untersuchungen am weiblichen Genitalapparat der Lepidopteren. I. Morphologischer Teil. *Zeitschrift für Angewandte Entomologie* **15**, 1–66.

Eisenbeis, G. & Wichard, W. (1987) *Atlas on the Biology of Soil Arthropods*, 2nd edn. Springer-Verlag, Berlin.

Eisner, T., Smedley, S.R., Young, D.K., Eisner, M., Roach, B. & Meinwald, J. (1996a) Chemical basis of courtship in a beetle (*Neopyrochroa flabellata*): cantharidin as precopulatory "enticing agent". *Proceedings of the National Academy of Sciences of the USA* **93**, 6494–8.

Eisner, T., Smedley, S.R., Young, D.K., Eisner, M., Roach, B. & Meinwald, J. (1996b) Chemical basis of courtship in a beetle (*Neopyrochroa flabellata*): cantharidin as "nuptial gift". *Proceedings of the National Academy of Sciences of the USA* **93**, 6499–503.

Elliott, J.M. & Humpesch, U.H. (1983) A key to the adults of the British Ephemeroptera. *Freshwater Biological Association Scientific Publication* **47**, 1–101.

Entwistle, P.F. & Evans, H.F. (1985) Viral control. In: *Comprehensive Insect Physiology, Biochemistry and Pharmacology*, Vol. 12: *Insect Control* (eds. G.A. Kerkut & L.I. Gilbert), pp. 347–412. Pergamon Press, Oxford.

Evans, E.D. (1978) Megaloptera and aquatic Neuroptera. In: *An Introduction to the Aquatic Insects of North America* (eds. R.W. Merritt & K.W. Cummins), pp. 133–45. Kendall/Hunt, Dubuque, IA.

Ferrar, P. (1987) *A Guide to the Breeding Habits and Immature Stages of Diptera Cyclorrhapha.* Pt. 2, Entomonograph Vol. 8. E.J. Brill, Leiden, and Scandinavian Science Press, Copenhagen.

Filshie, B.K. (1982) Fine structure of the cuticle of insects and other arthropods. In: *Insect Ultrastructure*, Vol. 1 (eds. R.C. King & H. Akai), pp. 281–312. Plenum, New York.

Fjellberg, A. (1980) *Identification Keys to Norwegian Collembola.* Utgitt av Norsk Entomologisk Forening, Norway.

Foldi, I. (1983) Structure et fonctions des glandes tégumentaires de Cochenilles Pseudococcines et de leurs sécrétions. *Annales de la Societé Entomologique de France (N.S.)* **19**, 155–66.

Freeman, W.H. & Bracegirdle, B. (1971) *An Atlas of Invertebrate Structure.* Heinemann Educational Books, London.

Frisch, K. von (1967) *The Dance Language and Orientation of Bees.* The Belknap Press of Harvard University Press, Cambridge, MA.

Froggatt, W.W. (1907) *Australian Insects.* William Brooks Ltd, Sydney.

Frost, S.W. (1959) *Insect Life and Insect Natural History*, 2nd edn. Dover Publications, New York.

Fry, C.H., Fry, K. & Harris, A. (1992) *Kingfishers, Bee-Eaters and Rollers.* Christopher Helm, London.

Futuyma, D.J. (1986) *Evolutionary Biology*, 2nd edn. Sinauer Associates, Sunderland, MA.

Gäde, G., Hoffman, K.-H. & Spring, J.H. (1997) Hormonal regulation in insects: facts, gaps, and future directions. *Physiological Reviews* **77**, 963–1032.

Galil, J. & Eisikowitch, D. (1968) Pollination ecology of *Ficus sycomorus* in East Africa. *Ecology* **49**, 259–69.

Gardiner, B.G. (1998) Editorial. *The Linnean* **14**(3), 1–3.

Gauld, I. & Bolton, B. (eds.) (1988) *The Hymenoptera.* British Museum (Natural History), London, in association with Oxford University Press, Oxford.

Gibbons, B. (1986) *Dragonflies and Damselflies of Britain and Northern Europe.* Country Life Books, Twickenham.

Gray, E.G. (1960) The fine structure of the insect ear. *Philosophical Transactions of the Royal Society of London B* **243**, 75–94.

Greany, P.D., Vinson, S.B. & Lewis, W.J. (1984) Insect parasitoids: finding new opportunities for biological control. *BioScience* **34**, 690–6.

Grimstone, A.V., Mullinger, A.M. & Ramsay, J.A. (1968) Further studies on the rectal complex of the mealworm *Tenebrio molitor* L. (Coleoptera: Tenebrionidae). *Philosophical Transactions of the Royal Society B* **253**, 343–82.

Gutierrez, A.P. (1970) Studies on host selection and host specificity of the aphid hyperparasite *Charips victrix* (Hymenoptera: Cynipidae). 6. Description of sensory structures and a synopsis of host selection and host specificity. *Annals of the Entomological Society of America* **63**, 1705–9.

Gwynne, D.T. (1981) Sexual difference theory: Mormon crickets show role reversal in mate choice. *Science* **213**, 779–80.

Gwynne, D.T. (1990) The katydid spermatophore: evolution of a parental investment. In: *The Tettigoniidae: Biology, Systematics and Evolution* (eds. W.J. Bailey & D.C.F. Rentz), pp. 27–40. Crawford House Press, Bathurst.

Hadley, N.F. (1986) The arthropod cuticle. *Scientific American* **255**(1), 98–106.

Hadlington, P. (1987) *Australian Termites and Other Common Timber Pests.* New South Wales University Press, Kensington.

Harris, W.V. (1971) *Termites: Their Recognition and Control*, 2nd edn. Longman, London.

Haynes, K.F. & Birch, M.C. (1985) The role of other pheromones, allomones and kairomones in the behavioural responses of insects. In: *Comprehensive Insect Physiology, Biochemistry, and Pharmacology*, Vol. 9: *Behaviour* (eds.

G.A. Kerkut & L.I. Gilbert), pp. 225–55. Pergamon Press, Oxford.

Hely, P.C., Pasfield, G. & Gellatley, J.G. (1982) *Insect Pests of Fruit and Vegetables in NSW*. Inkata Press, Melbourne.

Hepburn, H.R. (1985) Structure of the integument. In: *Comprehensive Insect Physiology, Biochemistry and Pharmacology*, Vol. 3: *Integument, Respiration and Circulation* (eds. G.A. Kerkut & L.I. Gilbert), pp. 1–58. Pergamon Press, Oxford.

Hermann, H.R. & Blum, M.S. (1981) Defensive mechanisms in the social Hymenoptera. In: *Social Insects*, Vol. II (ed. H.R. Hermann), pp. 77–197. Academic Press, New York.

Herms, W.B. & James, M.T. (1961) *Medical Entomology*, 5th edn. Macmillan, New York.

Hölldobler, B. (1984) The wonderfully diverse ways of the ant. *National Geographic* **165**, 778–813.

Hölldobler, B. & Wilson, E.O. (1990) *The Ants*. Springer-Verlag, Berlin.

Holman, G.M., Nachman, R.J. & Wright, M.S. (1990) Insect neuropeptides. *Annual Review of Entomology* **35**, 201–17.

Horridge, G.A. (1965) Arthropoda: general anatomy. In: *Structure and Function in the Nervous Systems of Invertebrates*, Vol. II (eds. T.H. Bullock & G.A. Horridge), pp. 801–964. W.H. Freeman, San Francisco, CA.

Hoy, M. (2003) *Insect Molecular Genetics: An Introduction to Principles and Applications*, 2nd edn. Academic Press, San Diego, CA.

Hungerford, H.B. (1954) The genus *Rheumatobates* Bergroth (Hemiptera–Gerridae). *University of Kansas Science Bulletin* **36**, 529–88.

Huxley, J. & Kettlewell, H.B.D. (1965) *Charles Darwin and His World*. Thames & Hudson, London.

Imms, A.D. (1913) Contributions to a knowledge of the structure and biology of some Indian insects. II. On *Embia major*. *Sp. nov.*, from the Himalayas. *Transactions of the Linnean Society of London* **11**, 167–95.

Jobling, B. (1976) On the fascicle of blood-sucking Diptera. *Journal of Natural History* **10**, 457–61.

Johnson, W.T. & Lyon, H.H. (1991) *Insects that Feed on Trees and Shrubs*, 2nd edn. Comstock Publishing Associates of Cornell University Press, Ithaca, NY.

Julien, M.H. (ed.) (1992) *Biological Control of Weeds: A World Catalogue of Agents*. CAB International, Wallingford.

Katz, M., Despommier, D.D. & Gwadz, R.W. (1989) *Parasitic Diseases*, 2nd edn. Springer-Verlag, New York.

Keeley, L.L. & Hayes, T.K. (1987) Speculations on biotechnology applications for insect neuroendocrine research. *Insect Biochemistry* **17**, 639–61.

Kettle, D.S. (1984) *Medical and Veterinary Entomology*. Croom Helm, London.

Kristensen, N.P. & Skalski, A.W. (1999) Phylogeny and paleontology. In: *Lepidoptera: Moths and Butterflies 1. Handbuch der Zoologie/Handbook of Zoology*, Vol. IV, Part 35 (ed. N.P. Kristensen), pp. 7–25. Walter de Gruyter, Berlin.

Kukalová, J. (1970) Revisional study of the order Palaeodictyoptera in the Upper Carboniferous shales of Commentry, France. Part III. *Psyche* **77**, 1–44.

Kukalová-Peck, J. (1991) Fossil history and the evolution of hexapod structures. In: *The Insects of Australia*, 2nd edn. (CSIRO), pp. 141–79. Melbourne University Press, Carlton.

Labandeira, C.C. (1998) Plant–insect associations from the fossil record. *Geotimes*, September 1998.

Landsberg, J. & Ohmart, C. (1989) Levels of insect defoliation in forests: patterns and concepts. *Trends in Ecology and Evolution* **4**, 96–100.

Lane, R.P. & Crosskey, R.W. (eds.) (1993) *Medical Insects and Arachnids*. Chapman & Hall, London.

Lewis, T. (1973) *Thrips: Their Biology, Ecology and Economic Importance*. Academic Press, London.

Lindauer, M. (1960) Time-compensated sun orientation in bees. *Cold Spring Harbor Symposia on Quantitative Biology* **25**, 371–7.

Lloyd, J.E. (1966) Studies on the flash communication system in *Photinus* fireflies. *University of Michigan Museum of Zoology, Miscellaneous Publications* **130**, 1–95.

Loudon, C. (1989) Tracheal hypertrophy in mealworms: design and plasticity in oxygen supply systems. *Journal of Experimental Biology* **147**, 217–35.

Lubbock, J. (1873) *Monograph of the Collembola and Thysanura*. The Ray Society, London.

Lüscher, M. (1961) Air-conditioned termite nests. *Scientific American* **205**(1), 138–45.

Lyal, C.H.C. (1986) Coevolutionary relationships of lice and their hosts: a test of Fahrenholz's Rule. In: *Coevolution and Systematics* (eds. A.R. Stone & D.L. Hawksworth), pp. 77–91. Systematics Association, Oxford.

Majer, J. (1985) Recolonisation by ants of rehabilitated mineral sand mines on North Stradbroke Island, Queensland, with particular reference to seed removal. *Australian Journal of Ecology* **10**, 31–4.

Malcolm, S.B. (1990) Mimicry: status of a classical evolutionary paradigm. *Trends in Ecology and Evolution* **5**, 57–62.

Matsuda, R. (1965) Morphology and evolution of the insect head. *Memoirs of the American Entomological Institute* **4**, 1–334.

McAlpine, D.K. (1990) A new apterous micropezid fly (Diptera: Schizophora) from Western Australia. *Systematic Entomology* **15**, 81–6.

McAlpine, J.F. (ed.) (1981) *Manual of Nearctic Diptera*, Vol. 1. Monograph No. 27. Research Branch, Agriculture Canada, Ottawa.

McAlpine, J.F. (ed.) (1987) *Manual of Nearctic Diptera*, Vol. 2. Monograph No. 28. Research Branch, Agriculture Canada, Ottawa.

McIver, S.B. (1985) Mechanoreception. In: *Comprehensive Insect Physiology, Biochemistry, and Pharmacology*, Vol. 6: *Nervous System: Sensory* (eds. G.A. Kerkut & L.I. Gilbert), pp. 71–132. Pergamon Press, Oxford.

Mercer, W.F. (1900) The development of the wings in the Lepidoptera. *New York Entomological Society* **8**, 1–20.

Merritt, R.W., Craig, D.A., Walker, E.D., Vanderploeg, H.A. & Wotton, R.S. (1992) Interfacial feeding behavior and particle flow patterns of *Anopheles quadrimaculatus* larvae (Diptera: Culicidae). *Journal of Insect Behavior* **5**, 741–61.

Michelsen, A. & Larsen, O.N. (1985) Hearing and sound. In: *Comprehensive Insect Physiology, Biochemistry, and Pharmacology*, Vol. 6: *Nervous System: Sensory* (eds. G.A. Kerkut & L.I. Gilbert), pp. 495–556. Pergamon Press, Oxford.

Michener, C.D. (1974) *The Social Behavior of Bees*. The Belknap Press of Harvard University Press, Cambridge, MA.

Miyazaki, M. (1987a) Morphology of aphids. In: *Aphids: Their Biology, Natural Enemies and Control*, Vol. 2A (eds. A.K. Minks & P. Harrewijn), pp. 1–25. Elsevier, Amsterdam.

Miyazaki, M. (1987b) Forms and morphs of aphids. In: *Aphids: Their Biology, Natural Enemies and Control*, Vol. 2A (eds. A.K. Minks & P. Harrewijn), pp. 27–50. Elsevier, Amsterdam.

Moczek, A. & Emlen, D.J. (2000) Male horn dimorphism in the scarab beetle, *Onthophagus taurus*: do alternative reproductive tactics favour alternative phenotypes? *Animal Behaviour* **59**, 459–66.

Monteith, S. (1990) Life inside an ant-plant. *Wildlife Australia* **27**(4), 5.

Nagy, L. (1998) Changing patterns of gene regulation in the evolution of arthropod morphology. *American Zoologist* **38**, 818–28.

Nosek, J. (1973) *The European Protura*. Muséum D'Histoire Naturelle, Geneva.

Novak, V.J.A. (1975) *Insect Hormones*. Chapman & Hall, London.

Oliveira, P.S. (1988) Ant-mimicry in some Brazilian salticid and clubionid spiders (Araneae: Salticidae, Clubionidae). *Biological Journal of the Linnean Society* **33**, 1–15.

Palmer, M.A. (1914) Some notes on life history of lady-beetles. *Annals of the Entomological Society of America* **7**, 213–38.

Pivnick, K.A. & McNeil, J.N. (1987) Puddling in butterflies: sodium affects reproductive success in *Thymelicus lineola*. *Physiological Entomology* **12**, 461–72.

Poisson, R. (1951) Ordre des Hétéroptères. In: *Traité de Zoologie: Anatomie, Systématique, Biologie*, Tome X: *Insectes Supérieurs et Hémiptéroïdes*, Fascicule II (ed. P.-P. Grassé), pp. 1657–1803. Masson, Paris.

Preston-Mafham, K. (1990) *Grasshoppers and Mantids of the World*. Blandford, London.

Pritchard, G., McKee, M.H., Pike, E.M., Scrimgeour, G.J. & Zloty, J. (1993) Did the first insects live in water or in air? *Biological Journal of the Linnean Society* **49**, 31–44.

Purugganan, M.D. (1998) The molecular evolution of development. *Bioessays* **20**, 700–11.

Raabe, M. (1986) Insect reproduction: regulation of successive steps. *Advances in Insect Physiology* **19**, 29–154.

Richards, A.G. & Richards, P.A. (1979) The cuticular protuberances of insects. *International Journal of Insect Morphology and Embryology* **8**, 143–57.

Richards, G. (1981) Insect hormones in development. *Biological Reviews of the Cambridge Philosophical Society* **56**, 501–49.

Richards, O.W. & Davies, R.G. (1959) *Outlines of Entomology*. Methuen, London.

Richards, O.W. & Davies, R.G. (1977) *Imms' General Textbook of Entomology*, Vol. I: *Structure, Physiology and Development*, 10th edn. Chapman & Hall, London.

Riddiford, L.M. (1991) Hormonal control of sequential gene expression in insect epidermis. In: *Physiology of the Insect Epidermis* (eds. K. Binnington & A. Retnakaran), pp. 46–54. CSIRO Publications, Melbourne.

Robert, D., Read, M.P. & Hoy, R.R. (1994) The tympanal hearing organ of the parasitoid fly *Ormia ochracea* (Diptera, Tachinidae, Ormiini). *Cell and Tissue Research* **275**, 63–78.

Rossel, S. (1989) Polarization sensitivity in compound eyes. In: *Facets of Vision* (eds. D.G. Stavenga & R.C. Hardie), pp. 298–316. Springer-Verlag, Berlin.

Rumbo, E.R. (1989) What can electrophysiology do for you? In: *Application of Pheromones to Pest Control, Proceedings of a Joint CSIRO–DSIR Workshop, July 1988* (ed. T.E. Bellas), pp. 28–31. Division of Entomology, CSIRO, Canberra.

Sainty, G.R. & Jacobs, S.W.L. (1981) *Waterplants of New South Wales*. Water Resources Commission, New South Wales.

Salt, G. (1968) The resistance of insect parasitoids to the defence reactions of their hosts. *Biological Reviews* **43**, 200–32.

Samson, P.R. & Blood, P.R.B. (1979) Biology and temperature relationships of *Chrysopa* sp., *Micromus tasmaniae* and *Nabis capsiformis*. *Entomologia Experimentalis et Applicata* **25**, 253–9.

Santos Oliveira, J.F., Passos de Carvalho, J., Bruno de Sousa, R.F. & Madalena Simão, M. (1976) The nutritional value of four species of insects consumed in Angola. *Ecology of Food and Nutrition* **5**, 91–7.

Schmutterer, H. (1990) Properties and potential of natural pesticides from the neem tree, *Azadirachta indica*. *Annual Review of Entomology* **35**, 271–97.

Schwabe, J. (1906) Beiträge zur Morphologie und Histologie der tympanalen Sinnesapparate der Orthopteren. *Zoologica, Stuttgart* **50**, 1–154.

Sheppard, A.W. (1992) Predicting biological weed control. *Trends in Ecology and Evolution* **7**, 290–1.

Sivinski, J. (1978) Intrasexual aggression in the stick insects *Diapheromera veliei* and *D. covilleae* and sexual dimorphism in the Phasmatodea. *Psyche* **85**, 395–405.

Smedley, S.R. & Eisner, T. (1996) Sodium: a male moth's gift to its offspring. *Proceedings of the National Academy of Sciences of the USA* **93**, 809–13.

Smith, P.H., Gillott, C., Barton Browne, L. & van Gerwen, A.C.M. (1990) The mating-induced refractoriness of *Lucilia cuprina* females: manipulating the male contribution. *Physiological Entomology* **15**, 469–81.

Smith, R.L. (1997) Evolution of paternal care in the giant water bugs (Heteroptera: Belostomatidae). In: *The Evolution of Social Behavior in Insects and Arachnids* (eds. J.C. Choe & B.J. Crespi), pp. 116–49. Cambridge University Press, Cambridge.

Smithers, C.N. (1982) Psocoptera. In: *Synopsis and Classification of Living Organisms*, Vol. 2 (ed. S.P. Parker), pp. 394–406. McGraw-Hill, New York.

Snodgrass, R.E. (1935) *Principles of Insect Morphology*. McGraw-Hill, New York.

Snodgrass, R.E. (1946) The skeletal anatomy of fleas (Siphonaptera). *Smithsonian Miscellaneous Collections* **104**(18), 1–89.

Snodgrass, R.E. (1956) *Anatomy of the Honey Bee*. Comstock Publishing Associates, Ithaca, NY.

Snodgrass, R.E. (1957) A revised interpretation of the external reproductive organs of male insects. *Smithsonian Miscellaneous Collections* **135**(6), 1–60.

Snodgrass, R.E. (1967) *Insects: Their Ways and Means of Living*. Dover Publications, New York.

Spencer, K.A. (1990) *Host Specialization in the World Agromyzidae (Diptera)*. Kluwer Academic Publishers, Dordrecht.

Spradbery, J.P. (1973) *Wasps: an Account of the Biology and Natural History of Solitary and Social Wasps*. Sidgwick & Jackson, London.

Stanek, V.J. (1969) *The Pictorial Encyclopedia of Insects*. Hamlyn, London.

Stanek, V.J. (1977) *The Illustrated Encyclopedia of Butterflies and Moths*. Octopus Books, London.

Stern, V.M., Smith, R.F., van den Bosch, R. & Hagen, K.S. (1959) The integrated control concept. *Hilgardia* **29**, 81–101.

Stoltz, D.B. & Vinson, S.B. (1979) Viruses and parasitism in insects. *Advances in Virus Research* **24**, 125–71.

Struble, D.L. & Arn, H. (1984) Combined gas chromatography and electroantennogram recording of insect olfactory responses. In: *Techniques in Pheromone Research* (eds. H.E. Hummel & T.A. Miller), pp. 161–78. Springer-Verlag, New York.

Sullivan, D.J. (1988) Hyperparasites. In: *Aphids. Their Biology, Natural Enemies and Control*, Vol. B (eds. A.K. Minks & P. Harrewijn), pp. 189–203. Elsevier, Amsterdam.

Sutherst, R.W. & Maywald, G.F. (1985) A computerised system for matching climates in ecology. *Agriculture, Ecosystems and Environment* **13**, 281–99.

Sutherst, R.W. & Maywald, G.F. (1991) Climate modelling and pest establishment. *Plant Protection Quarterly* **6**, 3–7.

Suzuki, N. (1985) Embryonic development of the scorpionfly, *Panorpodes paradoxa* (Mecoptera, Panorpodidae) with special reference to the larval eye development. In: *Recent Advances in Insect Embryology in Japan* (eds. H. Ando & K. Miya), pp. 231–8. ISEBU, Tsukubo.

Szabó-Patay, J. (1928) A kapus-hangya. *Természettudományi Közlöny* **60**, 215–19.

Terra, W.R. & Ferreira, C. (1981) The physiological role of the peritrophic membrane and trehalase: digestive enzymes in the midgut and excreta of starved larvae of *Rhynchosciara*. *Journal of Insect Physiology* **27**, 325–31.

Thornhill, R. (1976) Sexual selection and nuptial feeding behavior in *Bittacus apicalis* (Insecta: Mecoptera). *American Naturalist* **110**, 529–48.

Torre-Bueno, J.R. de la (1989) *The Torre-Bueno Glossary of Entomology*, 2nd edn. The New York Entomological Society in cooperation with the American Museum of Natural History, New York.

Trueman, J.W.H. (1991) Egg chorionic structures in Corduliidae and Libellulidae (Anisoptera). *Odonatologica* **20**, 441–52.

Upton, M.S. (1991) *Methods for Collecting, Preserving, and Studying Insects and Allied Forms*, 4th edn. Australian Entomological Society, Brisbane.

Uvarov, B. (1966) *Grasshoppers and Locusts*. Cambridge University Press, Cambridge.

van den Bosch, R. & Hagen, K.S. (1966) Predaceous and parasitic arthropods in Californian cotton fields. *Californian Agricultural Experimental Station Bulletin* **820**, 1–32.

van Oosterzee, P. (1997) *Where Worlds Collide. The Wallace Line*. Reed, Kew, Victoria.

Waage, J.K. (1986) Evidence for widespread sperm displacement ability among Zygoptera (Odonata) and the means for predicting its presence. *Biological Journal of the Linnean Society* **28**, 285–300.

Wasserthal, L.T. (1997) The pollinators of the Malagasy star orchids *Angraecum sesquipedale*, *A. sororium* and *A. compactum* and the evolution of extremely long spurs by pollinator shift. *Botanica Acta* **110**, 343–59.

Waterhouse, D.F. (1974) The biological control of dung. *Scientific American* **230**(4), 100–9.

Watson, J.A.L. & Abbey, H.M. (1985) Seasonal cycles in *Nasutitermes exitiosus* (Hill) (Isoptera: Termitidae). *Sociobiology* **10**, 73–92.

Wheeler, W.C. (1990) Insect diversity and cladistic constraints. *Annals of the Entomological Society of America* **83**, 91–7.

Wheeler, W.M. (1910) *Ants: Their Structure, Development and Behavior*. Columbia University Press, New York.

White, D.S., Brigham, W.U. & Doyen, J.T. (1984) Aquatic Coleoptera. In: *An Introduction to the Aquatic Insects of North America*, 2nd edn. (eds. R.W. Merritt & K.W. Cummins), pp. 361–437. Kendall/Hunt, Dubuque, IA.

White, G.B. (1985) *Anopheles bwambae* sp. n., a malaria vector in the Semliki Valley, Uganda, and its relationships with other sibling species of the *An. gambiae* complex (Diptera: Culicidae). *Systematic Entomology* **10**, 501–22.

Whitfield, J. (2002) Social insects: The police state. *Nature* **416**, 782–4.

Whiting, M.F. (2002) Phylogeny of the holometabolous insect orders: molecular evidence. *Zoologica Scripta* **31**, 3–15.

Wiggins, G.B. (1978) Trichoptera. In: *An Introduction to the Aquatic Insects of North America* (eds. R.W. Merritt & K.W. Cummins), pp. 147–85. Kendall/Hunt, Dubuque, IA.

Wigglesworth, V.B. (1964) *The Life of Insects.* Weidenfeld & Nicolson, London.

Wigglesworth, V.B. (1972) *The Principles of Insect Physiology*, 7th edn. Chapman & Hall, London.

Wikars, L.-O. (1997) Effects of forest fire and the ecology of fire-adapted insects. PhD Thesis, Uppsala University, Sweden.

Williams, J.L. (1941) The relations of the spermatophore to the female reproductive ducts in Lepidoptera. *Entomological News* **52**, 61–5.

Wilson, M. (1978) The functional organisation of locust ocelli. *Journal of Comparative Physiology* **124**, 297–316.

Winston, M.L. (1987) *The Biology of the Honey Bee.* Harvard University Press, Cambridge, MA.

Womersley, H. (1939) *Primitive Insects of South Australia.* Government Printer, Adelaide.

Youdeowei, A. (1977) *A Laboratory Manual of Entomology.* Oxford University Press, Ibadan.

Zacharuk, R.Y. (1985) Antennae and sensilla. In: *Comprehensive Insect Physiology, Biochemistry, and Pharmacology*, Vol. 6: *Nervous System: Sensory* (eds. G.A. Kerkut & L.I. Gilbert), pp. 1–69. Pergamon Press, Oxford.

Zanetti, A. (1975) *The World of Insects.* Gallery Books, New York.

INDEX

Readers are referred also to the Glossary: glossary entries are not cross-referenced in this index. Parenthetical entries that follow non-italic entries refer to plural and/or diminutive endings for morphological terms. All *italicized* entries are the scientific names of insects, except where indicated by the names of other organisms, abbreviated where necessary as: Bact., bacterial name; Bot., botanical name; Mam., mammal name; Nem., nematode name; Prot., protist name; Vir., virus name. Page numbers in *italics* refer to figures; those in **bold** refer to tables.

APPENDIX: A REFERENCE GUIDE TO ORDERS

Summary of the diagnostic features of the three non-hexapod orders and the 30 orders of Insecta.

Protura (proturans)		Very small, wingless, eyeless, without antennae, entognathous (mouthparts within folds of head), fore legs held forward, thoracic segments like those of abdomen, legs five-segmented, adult abdomen 12-segmented without cerci; immature stages like small adult but with fewer abdominal segments.	Chapters 7, 9 Box 9.2
Collembola (springtails)	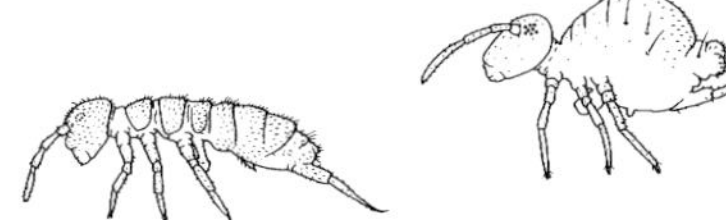	Small, wingless, mouthparts entognathous, antennae present, thoracic segments like those of abdomen, legs four-segmented, abdomen six-segmented with sucker-like ventral tube and forked jumping organ, without cerci; immature stages like small adult, with constant segment number.	Chapters 7, 9 Box 9.2
Diplura (diplurans)	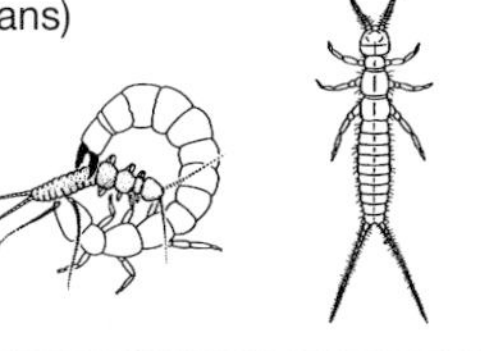	Small to medium, wingless, eyeless, entognathous, long antennae like string of beads, thoracic segments like those of abdomen, legs five-segmented, abdomen 10-segmented, some with small protrusions, terminal cerci filiform to forceps-like; immature stages like small adult.	Chapters 7, 9 Box 9.2
Archaeognatha (bristletails)		Medium, wingless, with humped thorax, hypognathous (mouthparts directed downwards), large compound eyes in near contact, some abdominal segments with paired styles and vesicles, with three "tails" – paired cerci shorter than single median caudal appendage; immature stages like small adult.	Chapters 7, 9 Box 9.3

Zygentoma or **Thysanura** (silverfish)	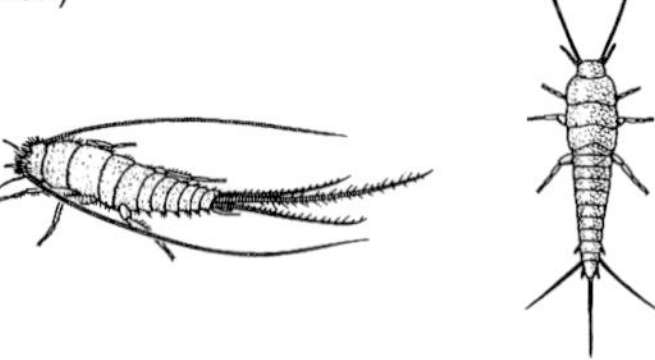	Medium, flattened, silvery-scaled, wingless, hypognathous to prognathous (mouthparts directed downwards to forwards), compound eyes small, widely separated or absent, some abdominal segments with ventral styles, with three "tails" – paired cerci nearly as long as median caudal appendage; immature stages like small adult.	Chapters 7, 9 Box 9.3
Ephemeroptera (mayflies)	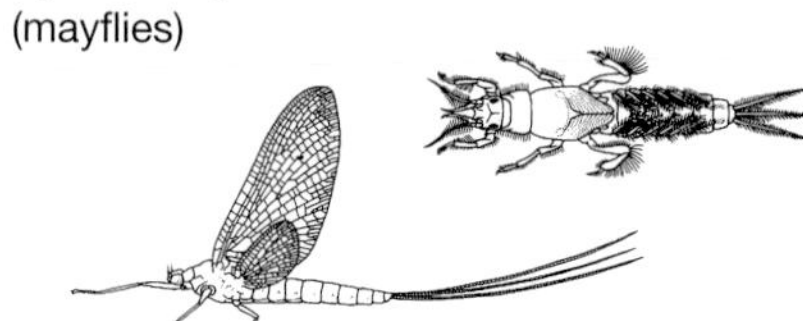	Small to large, winged with large triangular fore wings and smaller hind wings, mouthparts reduced, compound eyes large, short filiform antennae, abdomen slender compared to stout thorax, with three "tails" – paired cerci often as long as median caudal appendage; immature stages (nymphs) aquatic, with three "tails" and plate-like abdominal gills, penultimate instar a winged subimago.	Chapters 7, 10 Box 10.1
Odonata (dragonflies, damselflies)	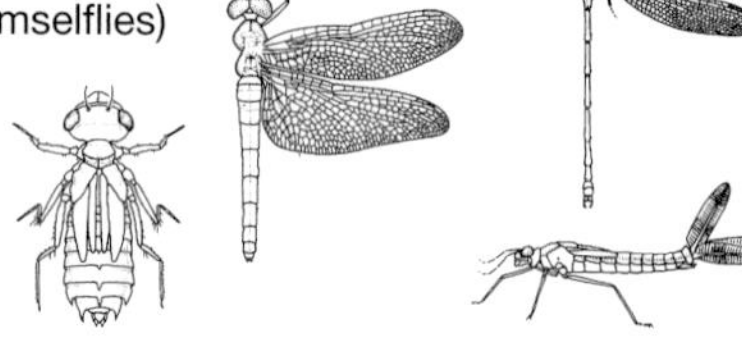	Medium to large, winged, with fore and hind wings equal (Zygoptera) or hind wings wider than fore wings (Anisoptera), head mobile, with large compound eyes separated (Zygoptera) or nearly in contact (Anisoptera), mouthparts mandibulate, antennae short, thorax stout, abdomen slender; immature stages (nymphs) aquatic, stout or narrow, with extensible labial "mask", terminal or rectal gills.	Chapters 7, 10 Boxes 5.3, 10.2
Plecoptera (stoneflies)	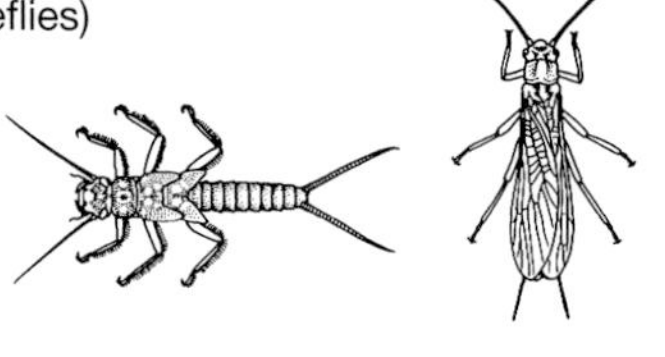	Medium, with fore and hind wings nearly equal (subequal) in size, at rest wings partly wrap abdomen and extend beyond abdominal apex but wing reduction frequent, legs weak, abdomen soft with filamentous cerci; immature stages (nymphs) aquatic resembling wingless adults, often with gills on abdomen.	Chapters 7, 10 Box 10.3
Isoptera (termites or "white ants")	 	Small to medium, mandibulate (with variable mouthpart development in different castes), antennae long, compound eyes often reduced, in winged forms fore and hind wings usually similar, often with reduced venation, body terminates in one- to five-segmented cerci; immature stages morphologically variable (polymorphic) according to caste.	Chapters 7, 12 Box 12.3

Order	Description	References
Blattodea (roaches, cockroaches) 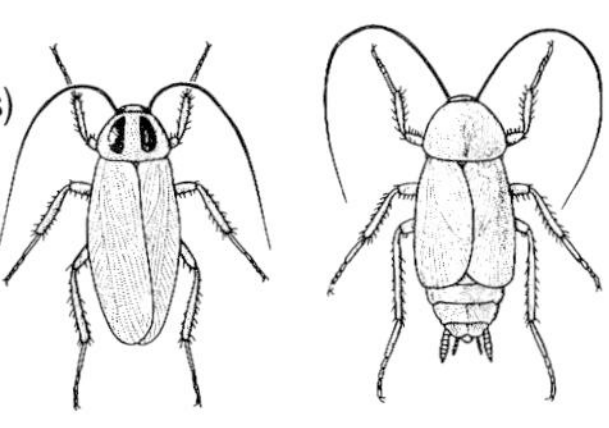	Small to large, dorsoventrally flattened, hypognathous, compound eyes well developed (except in cave dwellers), prothorax large and shield-like (may cover head), fore wings form leathery tegmina, protecting large hind wings, large anal lobe on hind wing, coxae large and abutting ventrally, cerci usually multisegmented; immature stages (nymphs) like small adults.	Chapters 7, 9 Box 9.8
Mantodea (mantids) 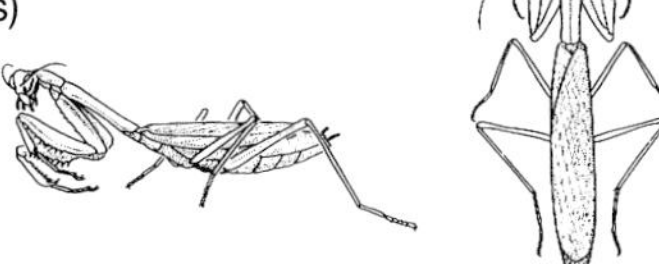	Moderate to large, head small, mobile, and triangular, compound eyes large and separated, thorax narrow, fore wings form tegmina, hind wings broad, fore legs predatory (raptorial), mid and hind legs elongate; immature stages (nymphs) resemble small adults.	Chapters 7, 13 Boxes 5.2, 13.2
Grylloblattodea (rock crawlers or ice crawlers)  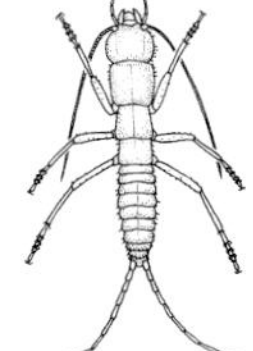	Medium, soft-bodied, elongate, pale, wingless and often eyeless, prognathous, with stout coxae on legs adapted for running, cerci five- to nine-segmented, female with short ovipositor; immature stages (nymphs) like small pale adults.	Chapters 7, 9 Box 9.4
Mantophasmatodea (heel walkers) 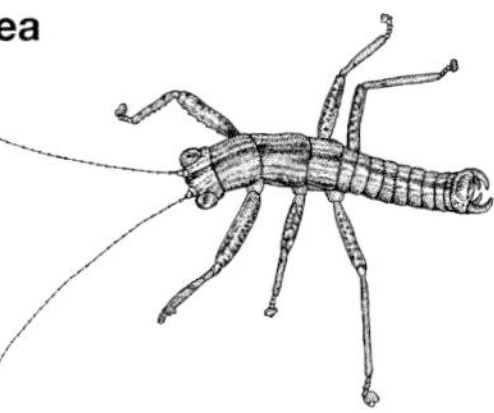	Small to medium, somewhat cylindrical, hypognathous, antennae long, multi-segmented, compound eyes large, fore and mid legs raptorial, wings absent, cerci small in female, prominent in male; immature stages (nymphs) resemble small adults.	Chapters 7, 13 Box 13.3
Phasmatodea (phasmatids, phasmids, stick-insects or walking sticks) 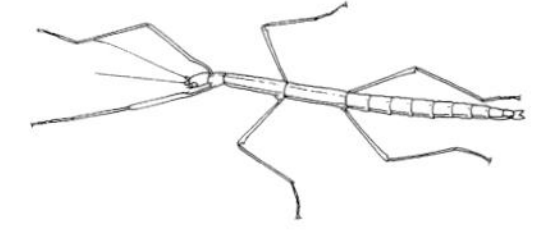 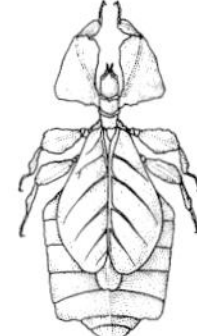	Medium to large, cylindrical stick-like or flattened leaf-like, prognathous, mandibulate, compound eyes small and laterally placed, fore wings form leathery tegmina, hind wings broad with toughened fore margin, legs elongate for walking, cerci one-segmented; immature stages (nymphs) like small adults.	Chapters 7, 11, 14 Box 11.6
Embiidina or **Embioptera** (embiids, webspinners)  	Small to medium, elongate, cylindrical, prognathous, compound eyes kidney-shaped, wingless in all females, some males with soft, flexible wings, legs short, basal fore tarsus swollen with silk gland, cerci two-segmented; immature stages like small adults.	Chapters 7, 9 Box 9.5

Orthoptera (grasshoppers, locusts, katydids, crickets) 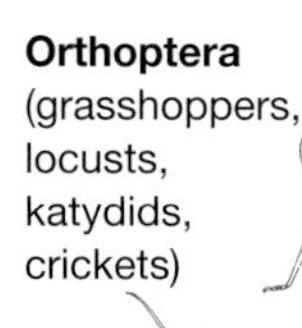 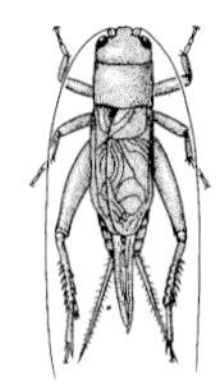	Medium to large, hypognathous, usually winged, fore wings forming leathery tegmina, hind wings broad, at rest pleated beneath tegmina, pronotum curved over pleura, hind legs often enlarged for jumping, cerci one-segmented; immature stages (nymphs) like small adults.	Chapters 5, 7, 11 Boxes 5.2, 11.5
Dermaptera (earwigs) 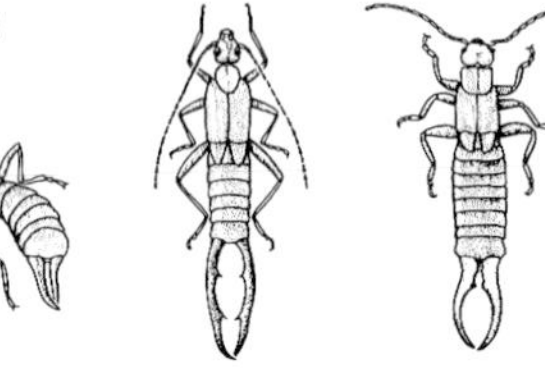	Small to medium, elongate and flattened, prognathous, antennae short to moderate, legs short, if wings present the fore wings are small leathery tegmina, hind wings semi-circular, abdomen with overlapping terga, cerci modified as forceps; immature stages (nymphs) resemble small adults.	Chapters 7, 9 Box 9.7
Zoraptera (zorapterans) 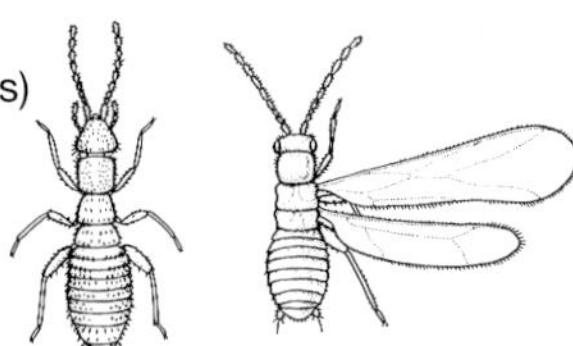	Small, termite-like, hypognathous, winged species with eyes and ocelli, wingless species lack either, if winged then wings with simple venation and readily shed, coxae well-developed, abdomen 11-segmented, short and swollen; immature stages (nymphs) resemble small adults.	Chapters 7, 9 Box 9.6
Psocoptera (psocids, or barklice, booklice) 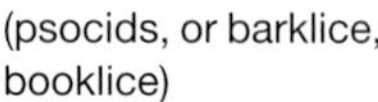 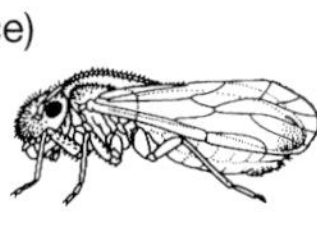 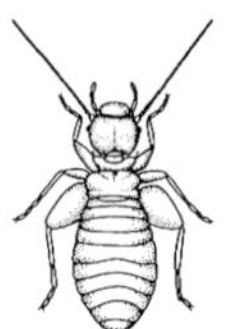	Small to medium, head large and mobile, chewing mouthparts asymmetrical, compound eyes large, antennae long and slender, wings often reduced or absent, if present venation simple, coupled in flight, held roof-like at rest, cerci absent; immature stages (nymphs) like small adults.	Chapters 7, 11 Box 11.9
Phthiraptera (lice) 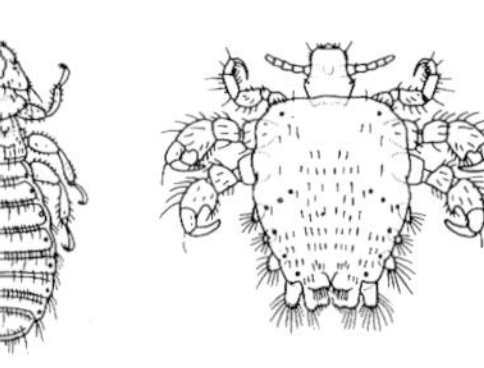	Small, dorsoventrally flattened, wingless ectoparasites, mouthparts mandibulate or beak-like, compound eyes small or absent, antennae either in grooves or extended, legs stout with strong claw(s) for grasping host hair or feathers; immature stages (nymphs) like small, pale adults.	Chapters 7, 13, 15 Box 15.3
Thysanoptera (thrips) 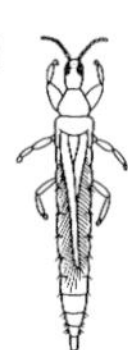 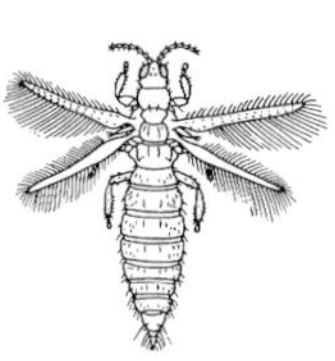	Small, slender, hypognathous with a feeding tube formed from three stylets – the maxillary laciniae plus the left mandible, with or without wings, if present wings subequal, strap-like, with long fringe; immature stages like small adults.	Chapters 7, 11 Box 11.7

Order	Illustrations	Description	References
Hemiptera (bugs, cicadas, plant- and leafhoppers, aphids, psyllids, scale insects, whiteflies, etc.)	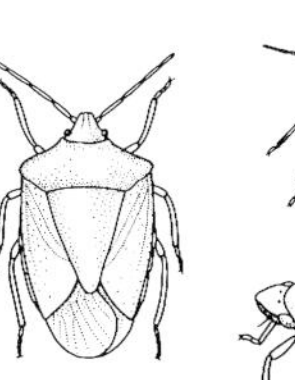 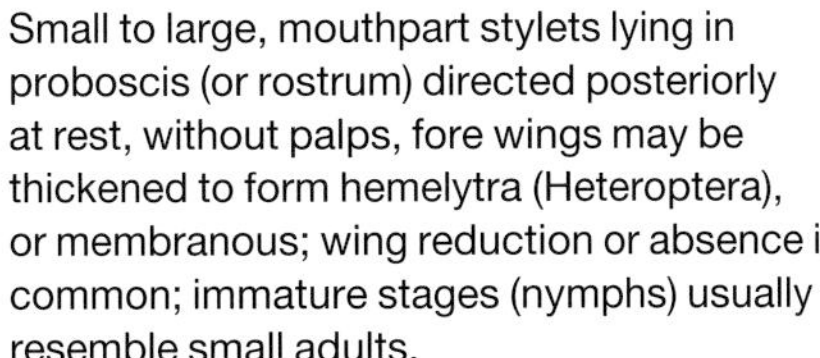	Small to large, mouthpart stylets lying in proboscis (or rostrum) directed posteriorly at rest, without palps, fore wings may be thickened to form hemelytra (Heteroptera), or membranous; wing reduction or absence is common; immature stages (nymphs) usually resemble small adults.	Chapters 7, 9, 10, 11 Boxes 3.3, 5.5, 9.1, 10.6, 11.2, 11.8, 14.2, 16.1, 16.2, 16.4
Neuroptera (lacewings, antlions)	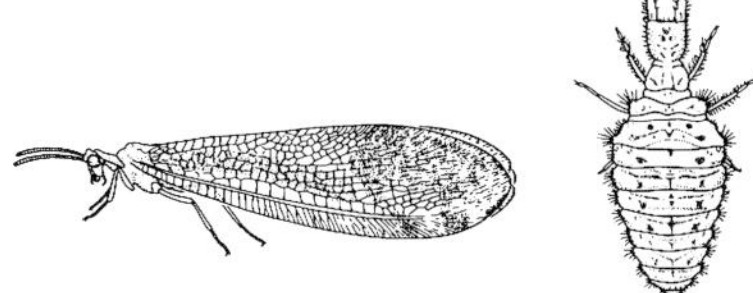	Medium, compound eyes large and separated, mandibulate, antennae multisegmented, prothorax often larger than meso- and metathorax, wings held roof-like over abdomen at rest, fore and hind wings subequal with numerous cross-veins and distal "twigging" of veins, without anal fold; immature stages (larvae) predominantly terrestrial, prognathous, with slender mandibles and maxillae usually forming piercing/sucking mouthparts, with jointed legs only on thorax, lacking abdominal gills.	Chapters 7, 13 Boxes 10.6, 13.4
Megaloptera (alderflies, dobsonflies, fishflies)	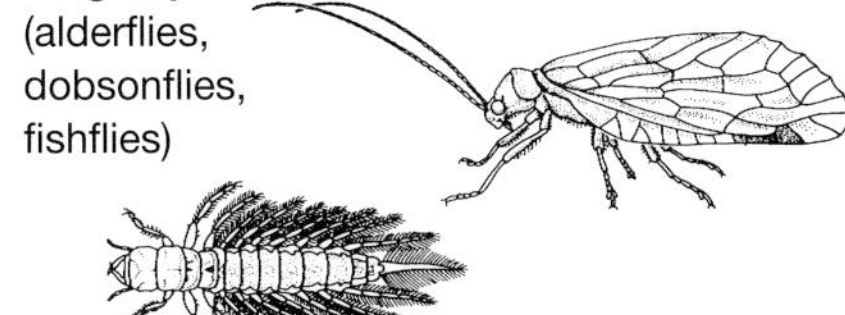	Medium to large, compound eyes large and separated, prognathous, mandibulate, antennae multisegmented, prothorax only slightly longer than meso- and metathorax, fore and hind wings subequal with anal fold in hind wing; immature stages (larvae) aquatic, prognathous, with stout mandibles, jointed legs only on thorax, with lateral abdominal gills.	Chapters 7, 13 Boxes 10.6, 13.4
Raphidioptera (snakeflies)	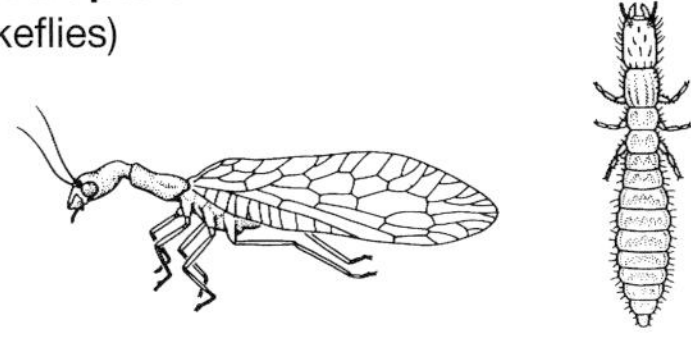	Medium, prognathous, mandibulate, antennae multisegmented, compound eyes large and separated, prothorax much longer than meso- and metathorax, fore wings rather longer than otherwise similar hind wings, without anal fold; immature stages (larvae) terrestrial, prognathous with jointed legs only on thorax, without abdominal gills.	Chapters 7, 13 Box 13.4
Coleoptera (beetles)	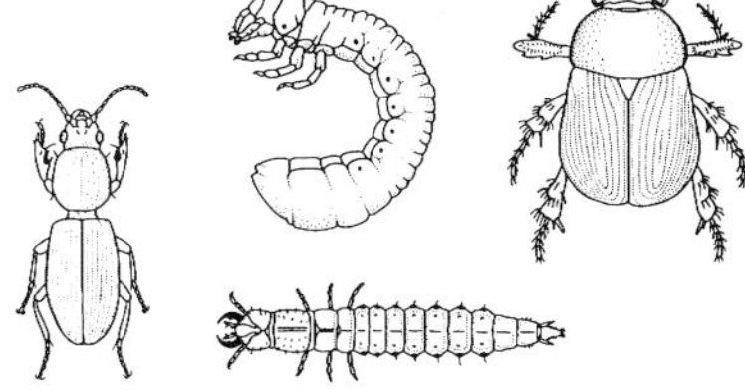	Small to large, often sturdy and compact, heavily sclerotized or armored, mandibulate, with fore wings modified as rigid elytra covering folded hind wings at rest, legs variously modified, often with claws and adhesive structures; immature stages (larvae) terrestrial or aquatic with sclerotized head capsule, opposable mandibles and usually five-segmented thoracic legs, without abdominal legs or labial silk glands.	Chapters 7, 10, 11 Boxes 10.6, 11.3, 11.10, 14.3, 14.4, 16.5

Strepsiptera (strepsipterans) 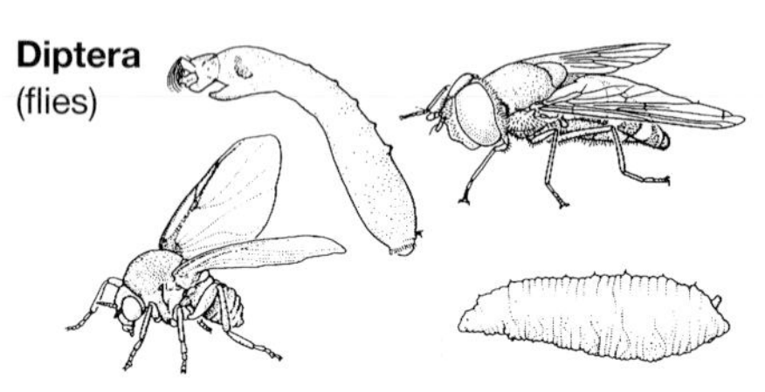	Small, aberrant endoparasites; male with large head, bulging eyes with few facets, antennae with fan-shaped branches, fore wings stubby, without veins, hind wings fan-shaped with few veins; female larviform, wingless, retained in host; immature stages (larvae) initially a triungulin with three pairs of thoracic legs, later maggot-like without mouthparts.	Chapters 7, 13 Box 13.6
Diptera (flies) 	Small to medium, wings restricted to mesothorax, metathorax with balancing organs (halteres), mouthparts vary from non-functional, to biting and sucking; immature stages (larvae, maggots) variable, without jointed legs, with sclerotized head capsule or variably reduced ultimately to remnant mouth hooks.	Chapters 7, 10, 15 Boxes 4.1, 5.4, 6.2, 6.3, 10.5, 15.2, 15.5
Mecoptera (scorpionflies, hangingflies) 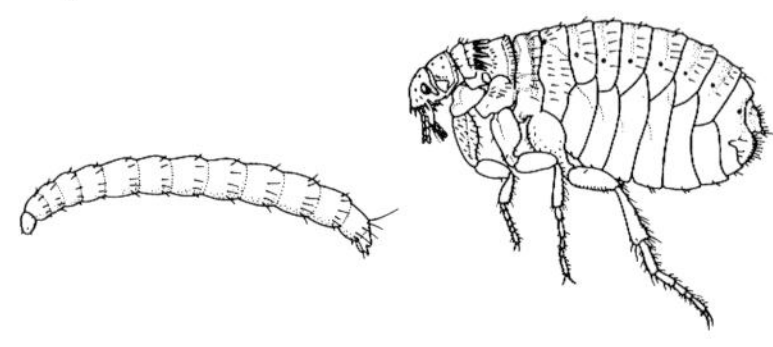	Medium, hypognathous with elongate rostrum formed from slender, serrate mandibles and maxillae and elongate labium, fore and hind wings narrow and subequal, legs raptorial; immature stages (larvae) mostly terrestrial, with heavily sclerotized head capsule, compound eyes, short, jointed thoracic legs, abdomen usually with prolegs.	Chapters 5, 7, 13 Boxes 5.1, 13.5
Siphonaptera (fleas) 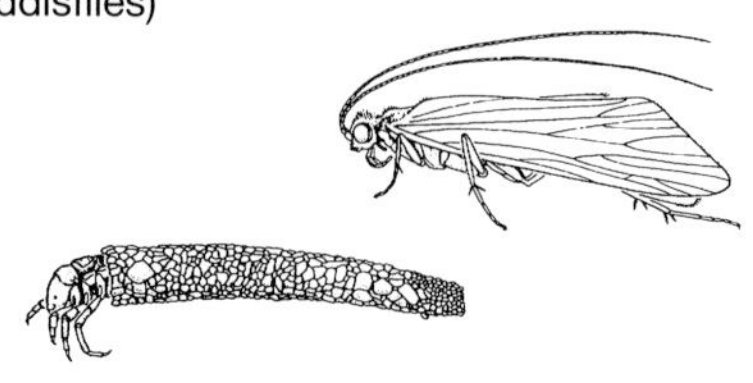	Small, highly modified, laterally compressed ectoparasites, mouthparts piercing and sucking, without mandibles, antennae lying in grooves, body with many backwardly directed setae and spines, some as combs, legs strong, terminating in strong claws for grasping host; immature stages (larvae) terrestrial, apodous (legless), with distinct head capsule.	Chapters 7, 15 Box 15.4
Trichoptera (caddisflies)	Small to large, with long, multisegmented antennae, reduced mouthparts (no proboscis) but well-developed maxillary and labial palps, hairy (or rarely scaly) wings, lacking discal cell and with fore wing anal veins looped (cf. Lepidoptera); immature stages (larvae) aquatic, often case-bearing, but many free-living, with three pairs of segmented thoracic legs and lacking abdominal prolegs.	Chapters 7, 10 Box 10.4

Lepidoptera (moths, butterflies) 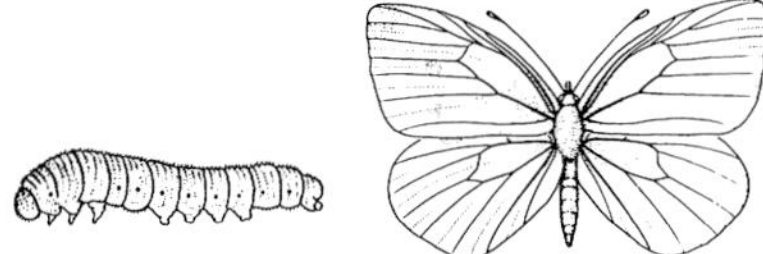	Small to large, hypognathous, nearly all with long coiled proboscis, antennae multisegmented and often comb-like (pectinate), clubbed in butterflies, wings with double layer of scales and large cells including the discal; immature stages (larvae, caterpillars) with sclerotized mandibulate head, labial spinnerets producing silk, jointed legs on thorax and some abdominal prolegs.	Chapters 7, 11, 14 Boxes 1.3, 11.11, 14.1
Hymenoptera (sawflies, wasps, ants, bees) 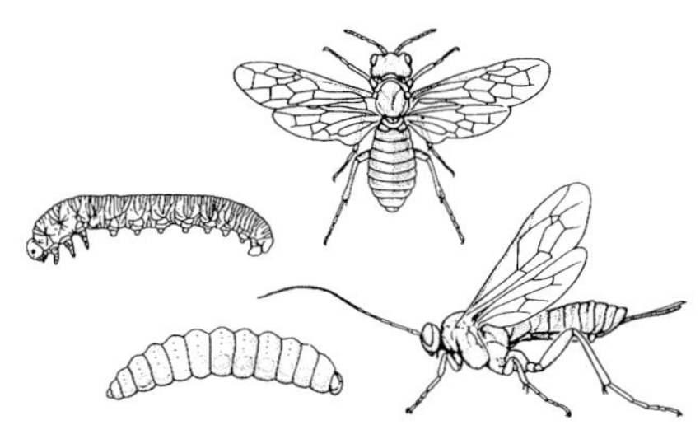	Minute to large, mouthparts mandibulate to sucking and chewing, antennae multisegmented often long and held forward, thorax either three-segmented or forms a mesosoma by incorporation of first abdominal segment in which case the abdomen is petiolate (waisted), wings with simple venation, fore and hind wings coupled together by hooks on hind wing; immature stages (larvae) very variable, many lack legs completely, all have distinct mandibles even if head is reduced.	Chapters 7, 12, 13, 14 Boxes 11.4, 12.1, 12.2, 13.1

RODIN

THE THINKER. DETAIL OF PLATE 16.

RODIN

Selected by Ludwig Goldscheider Photographed by Ilse Schneider-Lengyel Introduced by Sommerville Story

Φ

Phaidon Press Limited
Regent's Wharf
All Saints Street
London N1 9PA

First published 1939
Second edition 1949
Third edition 1951
Fourth edition 1953
Fifth edition 1956
Sixth edition 1961
Seventh edition 1964
Reprinted 1966, 1969
Eighth edition 1970
Ninth edition 1979
Reprinted 1985, 1988
Tenth edition 1996

ISBN 0 7148 3577 3

A CIP catalogue record for this book is available from the British Library.

Printed in Hong Kong

Cover illustration: THE PRODIGAL SON. DETAIL (Plate 57).

Auguste Rodin and his Work

EVERY GREAT THINKER is a mystery and a puzzle to his contemporaries. Auguste Rodin was not only the greatest sculptor since Michelangelo, but he was the greatest thinker in stone of modern times, perhaps ever since the prehistoric age when men first began to think it possible to shape in stone the presentments of living things. Who will venture to contemplate his wonderful collection of groups, figures and busts, and not require some guidance in the master's meaning? There is the great art and the technique, than which none are more remarkable, and behind these there is the soul in his work. How are we to read that soul?

The Burghers of Calais, for instance (Pl. 41). This is not merely a group of six men marching off to execution, but every man in the group is alive with character and with a meaning which those who will may read. The St. John has a significance different from the mystical figures of the Prophet with which centuries of religious art have familiarized us. So with his Venuses, his allegorical figures, his Naiads, his studies of men (and one is struck with how much more frequently, in comparison with some other sculptors, he depicts the force of man rather than the grace of woman). And his Balzac (Pls. 63-7)—was it without deep thought that such a man represented the author of the 'Comédie Humaine' as he did, in a way that set all the critics and dilettantes agog? Rodin himself realized that his meaning was sometimes hard to fathom. "I am," he said once, "like the Roman singer who replied to the yells of the populace, 'Equitibus cano'—I only sing for the knights" (that is, for the select few). All great thinkers and artists have felt this at times. Yet Rodin contradicted this, as he was right to do, when he left all his work to the nation with a view to helping others to attain education in art. He knew that nothing great and durable can really be for the few, either in art or in any other manifestation of human activity, and Rodin is for the world, as he wished to be.

RODIN'S LIFE

AUGUSTE RODIN was a Parisian, born in a working-class quarter, close to the Latin Quarter—No. 3, rue de l'Arbalète—on November 12, 1840. His father, Jean-Baptiste Rodin, a native of Normandy, born at Yvetot, a modest office employee at the time his son was born, later became an inspector at the Prefecture of Police. His mother, Marie Cheffer, was from Lorraine. The family seems to have been very united and highly religious. Rodin was the younger of two children. His sister, who was two years older than he, became a nun, but died in 1862 after a short illness. He had been greatly attached to her, and her death caused him much grief and suffering. In spite of the modesty of the family's position, the two children were brought up and educated with great care, the girl Marie at a convent and Auguste first at the friars' school in the Val-de-Grâce quarter and afterwards at Beauvais, at a school founded by his uncle, a man of considerable culture who had taken up teaching as a profession and succeeded in making an interesting position for himself. Auguste remained with his uncle until his fourteenth year.

The child's taste for drawing was already so pronounced that his father placed him then at the school of drawing and mathematics in the Rue de l'École de Médecine. This institution was known as the 'little school', to distinguish it from the official Fine Arts School; it was especially intended for the education of young artisans going in for artistic industries. No higher future was thought of for the boy at that time. The director of this school was a man named Belloc, but there was also a professor, Lecocq de Boisbaudran, who has left behind him a great name not as artist but as teacher, for he created a remarkable system of teaching drawing by the training of the memory, and thus helped to form a number of famous artists. Among his pupils were Fantin-Latour, Cazin, Legros, Lhermitte, the sculptor Dalou and the medallist Chaplain. There was another professor, a sculptor named Fort, who also had great influence over young Rodin, who in fact in later years declared that he owed his vocation to him. The youth also studied at the Gobelins school, where there was a professor named Lucas, for whom he expressed almost equal gratitude. These men all doubtless discovered in the studious and earnest young fellow the qualities that form an artist. As his bent became evident, it was decided to consult a sculptor named Maindron, who then enjoyed a considerable reputation. His verdict being favourable, young Rodin gave in his name at the École des Beaux-Arts. He was rejected three times.

Then for some years he earned money by exercising various crafts more or less allied to the sculptor's profession—moulder, ornamenter, goldsmith, and so on, all the time gathering valuable experience for his career of an artist. At the age of twenty-three Rodin, impelled by grief at the death of his beloved sister, took religious vows and entered the monastery of the Eudistes, in the Faubourg St. Jacques. This devotional impulse, however, did not last long, and five or six months afterwards he returned home. A little later he united his life to Rose Beuret, the faithful companion who was by his side for fifty years and only predeceased him by a few months.

In 1864 Rodin started to follow the lessons of Barye, the animal sculptor (1795–1875), who was an admirable and powerful artist, but as a teacher is said to have lacked enthusiasm. Then he entered the studio of Carrier-Belleuse (1824–87), another sculptor of great talent, a pupil of David d'Angers, and remained as assistant to him more or less for six years. But although Rodin called himself the pupil of both these artists, in order to conform to the rules of the Salon, he only followed the public lectures of the one and was the employee of the other. Up to that time, the known works by Rodin are very few, except several commercial orders, which were commissioned, such as the Caryatids of the Gobelins theatre, the chimney piece of the Gaîté Theatre, decorative reliefs in the Long Gallery of the Louvre Museum (Salle de Rubens), etc. There is a bust of his father (Pl. 1), which according to a legend in the family was made when he was seventeen, but which Rodin himself said dated from a year or two before his next work—the bust of Father Eymard (Pl. 2), the superior of the Societas Sanctissimi Sacramenti, to whom the young man became greatly attached during his short stay at the monastery. The date of this work was 1863. It is not surprising that this excellent priest, for whose memory Rodin always felt a great veneration, soon felt that the young man's true vocation was not the cloister. He arranged a studio for him, and after a while gently urged him to change his views and go home. This bust shows great powers of observation, fidelity to and comprehension of nature. The third work known is the famous 'Man with the broken nose' (Pl. 3), which he sent to the Salon and which was refused.

During the siege of Paris Rodin remained in the city and shouldered a gun as a National Guard. But in February 1871, having been judged unfit for military service, he went with Carrier-Belleuse to Belgium, with the idea of continuing to work beside him. But a dispute or misunderstanding arose, and Rodin left him and joined the Belgian sculptor Van Rasbourg, who continued the decoration of the Stock Exchange at Brussels begun by the French artist, and also worked on the Palais des Académies. He did work also on his own account, and in 1875 (a year in which he made a first timid appearance at the Salon with two busts), having put by a little money, he carried out a long nurtured ambition to go to Italy and study the works of Donatello and of Michelangelo. But his resources, naturally, were modest and he did not stay long. Most of his time was spent at Florence and Rome, but he also visited Venice and Naples. He came back enthusiastic and filled with the influence of the two great masters.

His next work shows the influence of Donatello, although it was sufficiently original to make the young sculptor immediately famous and to arouse a remarkable controversy. This was the statue first called 'Man awakening to Nature', now known as the 'Age of Brass'. Some of his confrères and critics professed to see in this work an imposture, and accused him of taking a mould from Nature. It led to the first scandal of Rodin's career (it was followed by numerous others), and the young sculptor was defended in a collective protest signed by a number of sculptors and painters. The State soon afterwards repaired the offence by acquiring the bronze of the incriminated statue, and a third-class medal was awarded to the sculptor. A little later, too, the State acquired his St. John the Baptist (Pl. 6), which was placed in the Luxembourg. The appearance of these two works brought him a great many new friends in the artistic and intellectual worlds. About this time Rodin was working at the decoration of the

Trocadéro Palace, Paris (now demolished), with the sculptor Legrain, and he modelled a number of terracotta busts in the style and manner of the eighteenth century. Then the Under-Secretary for Fine Arts, as an *amende honorable* for the aspersions that had been cast on him, offered him the choice of a State command. Rodin, still burning with enthusiasm from his Italian studies, asked to be allowed to execute a door for the future Museum of Decorative Arts, and, inspired by Dante, chose for his subject 'The Gate of Hell' (Pls. 8-16).

In 1877 Rodin returned to Paris, where he settled definitely after making a tour of France to visit the Cathedrals. It was the first of several such tours, for, like his great predecessor, Michelangelo, Rodin was intensely interested in cathedral and other architecture. He dreamed at one time indeed of being an architect, and during his various journeys was fond of studying architectural details, sketching and taking notes. (In the year 1910 his book 'Les Cathédrales de France' was published, illustrated with drawings by him.) Like some of his great predecessors, he touched various branches of art and shone in all—portraiture, landscape, etching, and even ceramics; for when he was employed for a time at the State Porcelain Manufactory at Sèvres (1879–82, with his old chief Carrier-Belleuse) he made several highly creditable vases. His journeys in France, therefore, like his trip to Italy, left a deep impression on him. On the occasion of a visit he made to Marseilles to work on the Fine Arts Palace, with an artist named Fourquet, he was greatly struck with the Greek type of beauty of the women in the Provençal towns.

The order for the 'Gate of Hell' was confirmed in 1880, and Rodin started work on it. This enormous task occupied and fascinated him for over twenty years, and on it he expended all the varied sides of his stupendous genius. As with all his other work, its chief characteristic is life in movement. It contains some 186 virile or graceful figures in all the phases of anguish, terror or voluptuousness. The famous statue of 'The Thinker' (which at first he called 'The Poet': Pls. 14-16) was to command and contemplate all these varied scenes. For a long time this was the work on which Rodin concentrated all his best energies; he made hundreds of drawings and models for it. Later on he seemed to tire of it or at any rate, he criticized his own conceptions; and for some years he used it as a sort of reservoir from which to evoke the varied forms of his work. It may be said indeed that nearly all his passionate figures came from the Gate. But in later years again he returned to this great work and set about the task of getting it definitely fixed in marble and bronze. It was never, however, completely finished.

About 1882 a studio was placed at Rodin's disposal free of charge by the Government in the State marble repository in the Rue de l'Université. During these years numbers of his works were shown at the Salon—for instance, St. John preaching (Pl. 6), the Creation of Man, Ugolino (Pl. 21), and the busts of J. P. Laurens (Pl. 18), Legros (Pl. 17), Dalou (Pl. 25), Victor Hugo and others. It was in these years that the sculptor came to know some of the leading literary men of the age, such as Alphonse Daudet and the Goncourts. In 1880 he competed for the monument of the National Defence to be erected at Courbevoie, at the gates of Paris, where a great stand of the French had taken place in 1871, when they were defeated. The work he sent in, 'The Genius of War', was not selected, as it was judged to be 'too ferocious'. This monument was chosen in 1916 by a Dutch committee who wished to offer a monument to the city of Verdun to commemorate the heroic defence of that place, which had gone far towards crushing the attempt made against the liberties of humanity. Rodin presented the work to the committee, and they accepted it with enthusiasm. In 1883 he received an order for an equestrian statue of General Lynch for South America. In 1884 the town of Calais opened a competition for a monument in memory of Eustache de St. Pierre, the historic burgher who had delivered the keys of the town to the English King. Rodin competed, and instead of a single figure he showed the hero accompanied by five of his companions in misfortune (Pls. 36-45). He obtained the order for the work, though not without difficulty, and devoted ten years of study and labour to it. The monument was inaugurated in 1895, and again it aroused criticism and discussion.

All or nearly all Rodin's greater works led to intense and often bitter controversies. Such were the Claude Lorrain, executed for Nancy in 1889, and erected in 1892; the Victor Hugo, which was refused by the commission because the sculptor had made a seated figure instead of an erect one, as was required for the architectural ensemble (See Pl. 65 and note). The seated statue was, however, placed in the Luxembourg and the Minister of the day ordered a Victor Hugo erect for the Panthéon. It was the same with the monument he made of President Sarmiento, of the Argentine Republic, for Buenos Aires. But the most bitter controversy of all was over the famous Balzac monument, which was ordered by the Société des Gens de Lettres and was exhibited at the Salon of 1898 (Pls. 63-4, 66-7). The society refused the work. The president, the poet Jean Aicard, in disgust at this decision, resigned; the statue was returned to the artist's studio at Meudon, and the order was passed on to Falguière. In London again the committee who had ordered from Rodin a monument to Whistler did not accept it.

Nevertheless there were those who recognized and acclaimed the genius of the artist, and these, whether French or foreign, were enthusiastic. Among other valuable testimonies to his worth about this time, a group of admirers contributed in 1886 for the casting of 'The Kiss' (Pl. 49). Critics like Gustave Geffroy and Octave Mirbeau supported him with their pens during this most difficult period. His portraits, as the late Léonce Bénédite, curator of the Rodin museum in Paris, remarked, would alone have brought fame to any other artist—such works as the busts of Madame Rodin (Pls. 59, 62), Carrier-Belleuse, Dalou, Puvis de Chavannes, Henri Rochefort.

Late Works · Rodin's Death

For England he did the busts of Henley, the poet, who was then editor of the 'Magazine of Art', George Wyndham, Bernard Shaw (Pls. 78-80), Lord Howard de Walden, Mrs. Hunter, Lady Sackville, Lady Warwick, and Miss Fairfax. For America his busts included those of Mr. Harriman, Mrs. Simpson, Mrs. Potter Palmer, Mr. Kay and Mr. Ryan. It was owing to the last-named gentleman's initiative and generosity that the Rodin collection at the Metropolitan Museum in New York was inaugurated.

The monument to Bastien Lepage was erected at Damvilliers (Meuse) in 1889. In this year during the Universal Exibition, a number of his works were shown at the Georges Petit gallery in Paris. These included the Perseus and Gorgon and the Danaïd (Pl. 34), among his most recently completed subjects (besides the Galatea, Walkyrie, the Fall of a Soul into Hades, Perseus, St. George, Bellona, St. John, the Burghers of Calais, the Thinker, Bastien Lepage, and others). The result was highly satisfactory for the sculptor. In the same year the foundation of the Société Nationale des Beaux-Arts, in which he occupied a prominent position, assured Rodin's status as the chief of a school. An exhibition held in a special pavilion in the Place de l'Alma in 1900 constituted a veritable triumph for him. Tardy fame had now begun to catch him up. His 'Thinker' was made the object of a subscription among his admirers, who offered it to the people of Paris, and it was placed, in 1906, in front of the Panthéon. Four reliefs made for the villa of Baron Vitta at Evian were shown at the Luxembourg museum in 1905. Altogether forty-two of his works were shown at the Luxembourg.

The one great passion which filled the career of this man of great genius was work, and his life was filled with it. It was indeed scarcely interrupted except for a few journeys to Belgium, England and Rome. In 1903 he went to Prague to attend an exhibition of his work. Temptations were held out to induce him to go to America, but he could never make up his mind. In later years he regretted that he had never been to Greece. Among his latest works were busts of the Duc de Rohan and Georges Clemenceau (Pl. 91), and of Pope Benedict XV, done at Rome in 1915. The last work of all was the bust of M. Etienne Clémentel, then Minister of Commerce (Pl. 90). In 1916 the sculptor was struck down by illness and had to take to his bed. He recovered slowly, but had to give up all work. He then began to set his artistic property in order and to occupy himself with the Rodin Museum. His beloved wife on February 13, 1917, died and, never recovering from this blow, Rodin himself succumbed on November 17, 1917. He was buried in the garden of his villa at Meudon, with a replica of the 'Thinker' watching over the grave.

All that last summer he scarcely left his villa at Meudon, where he lived surrounded by two lady cousins who looked after him, by M. Léonce Bénédite, and one or two other devoted friends. Occasionally, but with increasing rarity, he would go to the

AUGUSTE RODIN.
DETAIL FROM A PORTRAIT BY JACQUES-EMILE BLANCHE

SELF-PORTRAIT OF RODIN. DRAWING

Hôtel Biron (the Rodin Museum) or make up a little party with one or two friends. He showed himself affectionate and gay, and grateful for all attentions, his old combative spirit seeming to have left him.

Rodin was singularly indifferent to worldly honours, the only distinction he desired being the recognition of his work. Academical honours he never sought, but he keenly appreciated the approval of his foreign admirers, and the distinctions conferred on him by foreign bodies, such as the doctorate *honoris causa* given him by Oxford.

Rodin, as his friend Camille Mauclair described him, was of medium stature, with an enormous head on a massive trunk; prominent nose, flowing grey beard, and small bright eyes, in which there was the appearance of shortsightedness with a gentle irony. The impression of power in the man was emphasized by the rolling gait from his haunches, the 'rocky' aspect of the troubled brow under rough bushy hair, the bony aquiline nose and ample beard. This impression, however, was partly contradicted by the reticent fold of the mouth, the quick regard, which was at once penetrating, malicious and naive ("one of the most complex I have ever seen", adds M. Mauclair), and above all by the husky voice, modulated with difficulty and mixed with grave inflexions. He appeared simple, precise, reserved, courteous and cordial without excess. By degrees his timidity would give place to a tranquil and simple authority. He showed neither over-emphasis nor awkwardness, and his general manner was rather sad than inspired. An immense energy seemed to issue from his sober and measured gestures. The slowness and apparent embarrassment of his speech and the pauses in his conversation gave it particular significance.

In the presence of one of his works Rodin had a way of explaining it which was elliptical but clear. He said what was essential, finishing his words with a gesture. He used to contemplate his creations lovingly, and sometimes even seemed to be astonished and contemplative at the idea of having created them, speaking as if they existed apart from himself. Gradually under Rodin's simplicity were discovered qualities that had at first been hidden; he was ironical, sensual, nervous and proud. In himself were found *in posse* all the passions he expressed with such troubled magnificence, and one began to understand the secret bonds between this calm man and the art he revealed. He was the companion of these white mute creatures of his, he loved them and entered into their abstract lives, and had moral obligations towards them. They were his sole preoccupation. He was restrained in the expression of his ideas and opinions and rather summary in his dealings with people. His cordiality gave the impression that he wished to have done quickly with social duties. Very capable of friendship, he limited it to a tacit understanding on the essential subjects of thought; he had less faith in individuals than in general ideas. Devoted to his work as he was, he only tolerated other things with polite boredom, and had a horror of discussions or of being disturbed.

As to bad artists, he took no notice of them—he did not criticize them. He endured all the violent discussions and all the unjust criticism heaped on his own works without saying anything, for he was too proud to discuss the matter. At the time of the Balzac quarrel, his friends advised him to fight, since the terms of his contract were entirely to his advantage. But he thanked them for the advice and withdrew his statue without saying a word.

This was not weakness, for Rodin's life had been a hard one. It arose from the dignity of his inner being and indifference to the passing phases of life. He cared for few things, but those few he was much attached to. Reading but little, he was intense in his admiration of certain authors. One of these was Baudelaire. He was passionately fond of music, especially Gluck, but rarely talked of it.

He simplified all things in his life and disliked all that was unessential. When one knew him well it was impossible to separate him from his work. His statues were the states of his soul. Just as Rodin seemed to break the fragments around the statue away from the block in which it had been concealed, so he himself seemed to be a sort of rock hiding various forms and crystallized growths. For him work lovingly accomplished was the secret of happiness, and to love life and natural forms, and do nothing to outrage Nature's laws, was all his morality. He seemed to be isolated in his time by his genius as well as by the character of his artistic conception, but in reality this isolation was only apparent. After his years of hard struggle and neglect or contumely he had become a master to whom young artists looked up. As the artist Pierre Roche expressed it, he had 'opened a vast window in the pale house of modern statuary, and had made of sculpture, which had been a timid, compromised art, one that was audacious and full of hope'.

RODIN'S WORK

WORK WAS THE LAW of Rodin's life, and his career was one long-sustained effort. No man ever worked harder, and there are few who have any idea of his vast output over a long life. With the passion of genius and an insatiable desire to create, he had the patience of a monk of the Middle Ages. Towards the end of his life he declared that he was only just beginning to understand the laws of sculpture. He was often criticized or ridiculed on account of the length of time he took over his work, but this arose from excessive conscientiousness. "A work even when finished is never perfect," he would say. He spent five years in modelling the two female statues that represent the Muses in his Victor Hugo statue. Over the 'Balzac' he spent seven years, and as it was leaving his studio he regretted that he could not keep it with him still a while so as to add touches which he thought would help to perfect it. He warned youthful workers against believing in inspiration and told them they would achieve nothing except by hard work. In later years he was, in fact, apt to underrate the rôle of inspiration in artistic work, and would contemptuously call such talk 'a dream'. Work was to him the true religion. But inspiration represented something else to Rodin; and he liked to keep a work by him so that he could, as he contemplated it long after others might have considered it finished, make touches here and there suggested by sudden ideas of improvement. He has been compared to one of those nameless artificers of the Middle Ages who spent all their lives in chiselling the figures round a cathedral; but in the general breadth of his character and his sympathies he was a descendant of the Greeks, with the naturalism of the Greeks allied to deep religious sentiment.

No artist ever understood the conditions of matter or the laws of plastic like Rodin. For long he oscillated, as he himself said, between the influence of Michelangelo and that of Phidias, the Christian and the pagan—the expressive and the plastic. Finally his allegiance lay between the Greeks and the Gothic. Follow nature, was his watchword all through his life. He was intensely imbued with the mystery of Nature and the freshness of the art of the ancients.

He would not admit the existence of ugliness in nature. "That which one commonly calls ugliness in nature may in art become great beauty," he would say. "As it is simply the power of character which makes beauty in art," he explained on one occasion, "it often happens that the more a creature is ugly in nature, the more it is beautiful in art. In art only those things that are characterless are ugly—such as have no beauty without or within. For the artist worthy of the name all is beautiful in nature, because his eyes boldly accepting all truth shown outwardly, read the inner truth as in an open book. That which is ugly in art is that which is false and artificial—that which aims at being pretty or even beautiful instead of being expressive"—in other words, that which is mannered, all parade, without soul and without truth. To disguise or attenuate the facts of Nature in order to please the undiscerning was for him the work of an inferior artist. One sees in works like the 'St. John' (a simple peasant, though an illuminated one, as it were, with his mouth open) and the 'Burghers of Calais' (simple rough men, nude under their 'sack' shirts, and with nothing heroic about them either in face or posture) how implacably he insists upon the absolute interpretation of nature, refusing to sacrifice anything to conventional ideas of beauty. No sculptor understood better than he all the possibilities of the nude. He used to draw comparisons between the modelling of a body and that of a landscape.

Up to his thirtieth year Rodin was almost exclusively engaged in helping to carry out the work of others. But he was all the time learning, and those long years of apprenticeship were invaluable in giving him varied experience in the study of effects in grouping and dexterity in the use of the sculptor's tools, while they afforded him endless opportunities for the criticism of defects. In his Brussels experiences he had the advantage of doing work on a large scale out of doors, and of observing the effects of climate on sculpture, and how each figure appeared under the changing aspects of the day. He early realized that the spontaneous attitudes of the living model are the only ones that should be represented in statuary, and that any attempt to dictate gesture or pose must destroy the harmonious relation existing between the various parts of the body. In the observance of this law primarily resided the superiority of antique over modern sculpture. Another of his discoveries was that as, under the suggestion of successive impulses, the outlines of the body are continually changing—muscles swelling or relaxing in a sort of rhythmic flow or ripple—the sculptor has ample opportunity of choosing the reliefs and curves that most faithfully and most effectively interpret the pose they accompany.

Rodin's method of working, writes Paul Gsell, was singular. He had in his studio a number of nude models, men and women, in movement or reposing. Rodin paid them to supply him constantly with the picture of nudity in various attitudes and with all the liberty of ordinary life. He was constantly looking at them, and thus was always familiar with the spectacle of muscles in movement. Thus the nude, which to-day people rarely see, and which even sculptors only see during the short period of the pose, was for Rodin an ordinary spectacle. Thus he acquired that knowledge of the human body unclothed which was common to the Greeks through constant exercise in sports and games, and learned to read the expression of feeling in all parts of the body. The face is usually regarded as the only mirror of the soul, and mobility of features is supposed to be the only exteriorization of spiritual life. But in reality there is not a muscle of the body which does not reveal thoughts and feeling.

So when Rodin caught a pose that pleased him and that seemed expressive—the delicate grace of a woman raising her arms to arrange her hair or inclining the bust to lift something, or the nervous vigour of a man walking—he wanted the pose kept, then quickly he seized his implements and a model was made, and then just as swiftly he passed to another. Herein is shown the difference between Rodin's method in watching for the attitudes of nature and that of so many of his contemporaries, who pose their models on pedestals and arrange the positions they wish them to assume as if they were lay figures. It was never his habit, as we have seen, to undertake a work, complete it, and have done with it. He always had by him a number of ideas and thoughts on which he meditated patiently for years as they ripened in his mind.

It was after 1900, when his reputation was firmly established, that Rodin conceived the daring idea of exhibiting human figures (such as 'Meditation', 'The Walking Man', etc.) deprived of a head, legs or arms, which at first shock the beholder, but on examination are found to be so well balanced and so perfectly harmonized that one can only find beauty in them. And his reason for this is artistically profound, though it might be thought that it was pushing the 'religion of art' to extremes. In the development of a leading idea—of thought, of meditation, of the action of walking, his desire was to eliminate all that might counteract or draw attention from this central thought. "As to polishing nails or ringlets of hair, that has no interest for me," he said; "it detracts attention from the leading lines and the soul which I wish to interpret." Nevertheless he did not by any means underestimate the technique of his art. Though he regarded it as only a means to an end, he said the artist who neglects it will never attain his object, which is the interpretation of feelings and ideas. "One must have a consummate sense of technique," he said, "to hide what one knows."

On one occasion he told Paul Gsell of a valuable lesson in the science of modelling which he had received from a certain Constant, who worked in the studio where he started as a sculptor. He was one day watching him modelling a capital with foliage in clay, and said, "Rodin, you are going about that badly. All your leaves look flat—that's why they don't seem real. Make them so that some of them will be shooting their points out at you. Then on seeing them one will have the sense of depth. Never look at forms in extent but in depth. Never consider a surface except as the end of a volume, and the more or less broad point which it directs towards you." He followed this valuable advice and discovered later that the ancients had practised exactly this method of modelling; hence the vigour and suppleness of their works.

The influence of Michelangelo on Rodin was great, as we have seen. He made copies at one time of some of Rubens' pictures from memory, so deeply had he studied them. He also declared that from Rembrandt he had learned much in the way of chiaroscuro in sculpture. Later he was interested in Japanese art, which the Goncourt brothers did so much to popularize in France; the attitudes of some of his models are said to have been suggested by Japanese paintings. But all forms of art interested him—Egyptian, Aztec, Far Eastern, or French eighteenth century, and he was always learning. He used to say that a visit to the Louvre affected him like beautiful music or a deep emotion and incited him to renewed efforts. Rodin was particularly interested in the depiction of elemental natures. His mind harked back to the origins of things, and over and over again he represented beings that were blossoming into humanity from primitive or elemental conditions, creatures that are not very distant from the nature in which they have grown—part animal, part human, or such as have sprung from trees, fountains and the material of elemental life. For him, as for the early Greeks, all Nature was teeming with semi-human form: he saw dryads in the woods and nereids in the streams and their physical natures and aspect differed but little from the elements that gave them birth. Numbers of his creatures are issuing from matter, others are assimilated with trees, with rocks or the earth itself. Nature being the source of all life, it seems as if he could hardly conceive of life removed from this source of energy.

PLATES

All the sculptures reproduced in this volume are exhibited in the Rodin Museum in Paris, with the exception of Plate 49, which is in the Tate Gallery, London.

1. JEAN-BAPTISTE RODIN, THE SCULPTOR'S FATHER. BRONZE. 1860.

2. FATHER PIERRE-JULIEN EYMARD. BRONZE. 1863.

3. THE MAN WITH THE BROKEN NOSE. BRONZE MASK. 1864.

4. THE AGE OF BRASS. BRONZE. 1876.

5. THE AGE OF BRASS. DETAIL.

6. SAINT JOHN THE BAPTIST PREACHING. BRONZE. 1878.

7. SAINT JOHN THE BAPTIST. DETAIL.

8. THE GATE OF HELL. BRONZE. 1880–1917.

9. THE GATE OF HELL. DETAIL.

10. THE GATE OF HELL. DETAIL.

11. THE THREE SHADOWS. DETAIL.

12. THE SHADOW. FROM 'THE GATE OF HELL'. BRONZE. 1880.

13. THE THREE SHADOWS. FROM 'THE GATE OF HELL'. BRONZE. 1880.

14. THE THINKER. PLASTER MODEL. 1879–1880.

15. THE THINKER. FROM 'THE GATE OF HELL'. BRONZE. 1880–1900.

16. THE THINKER. FROM 'THE GATE OF HELL'. BRONZE. 1880–1900.

17. BUST OF THE PAINTER ALPHONSE LEGROS. BRONZE. 1881.

18. BUST OF THE PAINTER JEAN-PAUL LAURENS. BRONZE. 1881.

19. THE HEAD OF SORROW. BRONZE. 1882.

20. THE CARYATID. BRONZE. 1880–1.

21. UGOLINO. BRONZE. 1882.

22. EVE. BRONZE. 1881.

23. THE THREE FAUNS. PLASTER. 1882.

24. BUST OF HENRI ROCHEFORT. BRONZE. 1897.

25. BUST OF THE SCULPTOR DALOU. BRONZE. 1883.

26. THE CROUCHING WOMAN. BRONZE. 1882.

27. THE CROUCHING WOMAN. DETAIL.

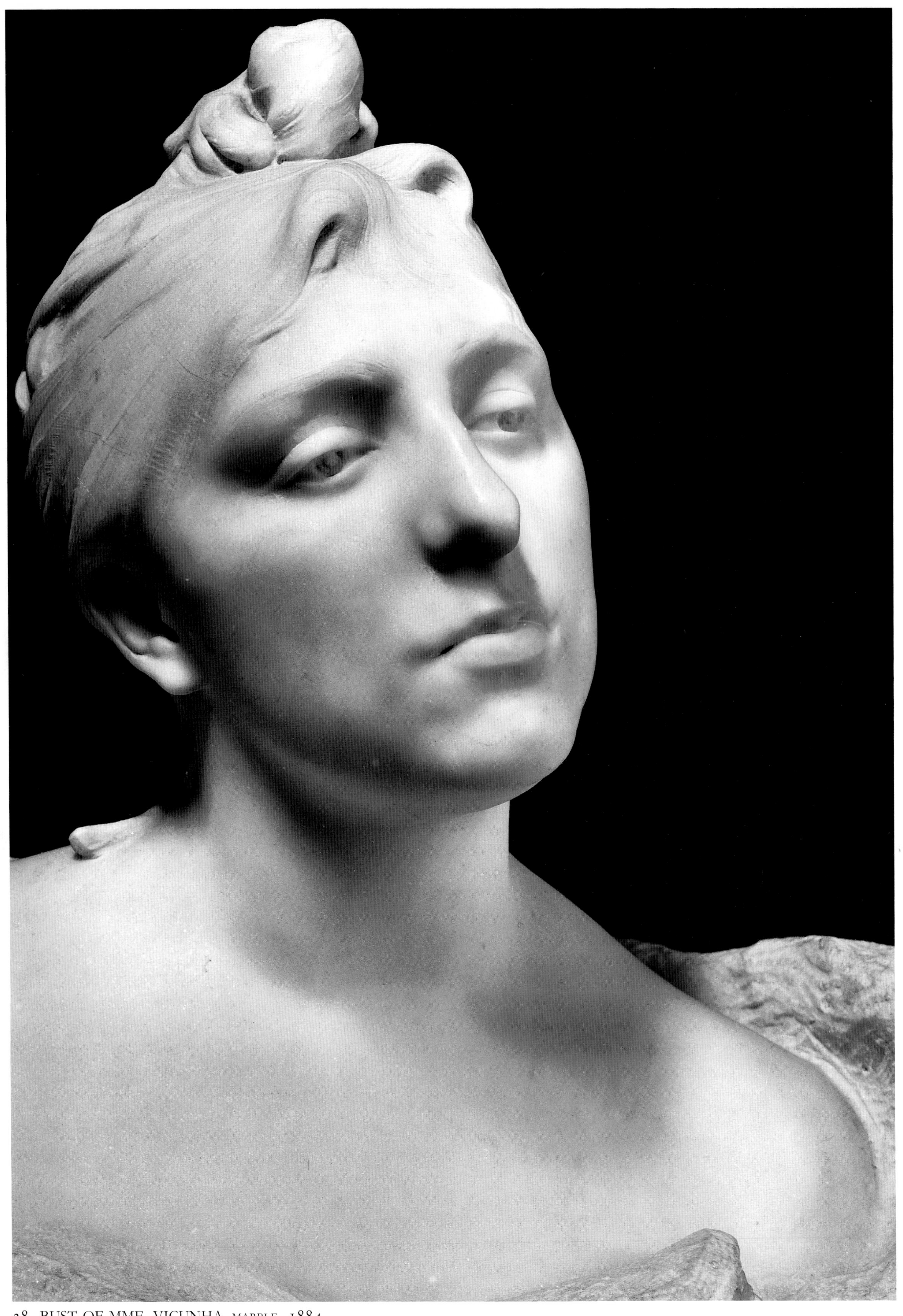

28. BUST OF MME. VICUNHA. MARBLE. 1884.

29. BUST OF MRS. RUSSELL. WAX. BEFORE 1888.

30. 'SHE WHO ONCE WAS THE HELMET-MAKER'S BEAUTIFUL WIFE'. BRONZE. BEFORE 1885.

31. 'THE HELMET-MAKER'S WIFE'. BACK VIEW.

32. KNEELING FAUN. PLASTER. 1884.

33. ERECT FAUN. PLASTER. 1884.

34. DANAÏD. MARBLE. 1885.

35. DANAÏD. DETAIL.

36. STUDY FOR A BURGHER OF CALAIS. BRONZE.

37. STUDY FOR A BURGHER OF CALAIS. BRONZE. 1884.

38. THE BURGHERS OF CALAIS.
PLASTER. 1884–1886.

39–49. STUDIES FOR THE BURGHERS OF CALAIS. BRONZE. 1884.

41. STUDY FOR THE BURGHERS OF CALAIS. BRONZE. 1884.

42. THE BURGHERS OF CALAIS. DETAIL.

43. THE BURGHERS OF CALAIS. DETAIL.

44. STUDY FOR THE BURGHERS OF CALAIS. BRONZE. 1884.

45. STUDY FOR THE BURGHERS OF CALAIS. BRONZE. 1884.

46. AURORA. MARBLE. 1885.

47. THOUGHT. MARBLE. 1886.

48. INVOCATION. PLASTER. 1886.

49. THE KISS. MARBLE. 1886.

50. FAUN AND NYMPH. (THE MINOTAUR.) PLASTER. BEFORE 1886.

51. POLYPHEMUS. BRONZE. 1888.

52. THE METAMORPHOSES OF OVID. BRONZE. BEFORE 1886.

53. FLYING FIGURE. BRONZE. 1889–90.

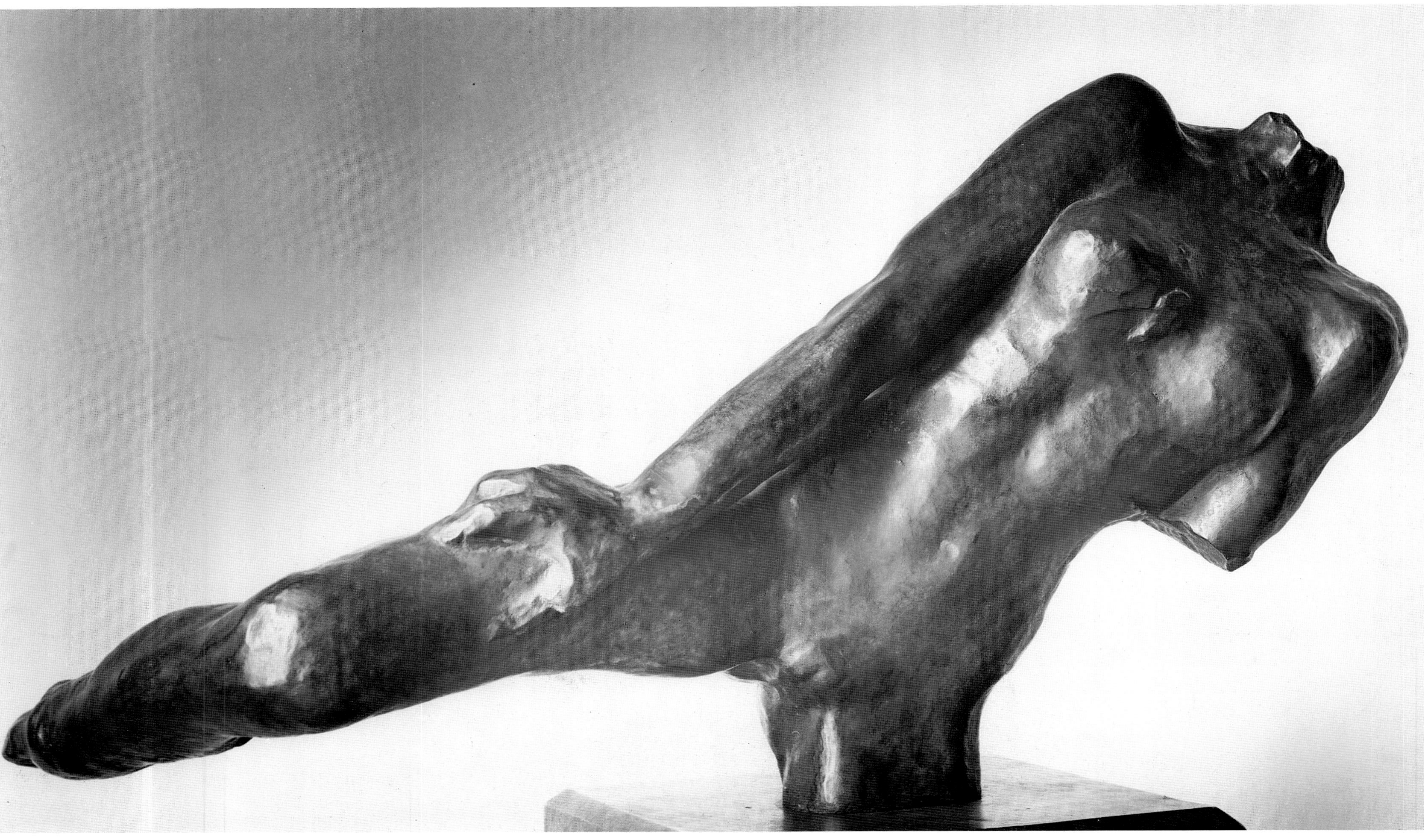

54. THE ETERNAL IDOL. PLASTER. 1889.

55. THE ETERNAL IDOL. DETAIL OF THE BRONZE. 1889.

56. THE PRODIGAL SON. BRONZE. BEFORE 1889.

57. THE PRODIGAL SON. DETAIL.

58. OCTAVE MIRBEAU. TERRACOTTA. 1889.

59. BUST OF ROSE BEURET (MME. AUGUSTE RODIN). MARBLE. 1890.

60. BROTHER AND SISTER. BRONZE. 1890.

. DESPAIR. BRONZE. 1890.

62. BUST OF ROSE BEURET (MME. AUGUSTE RODIN). BRONZE MASK. 1890.

63. STUDY FOR THE MONUMENT TO BALZAC. BRONZE. 1893.

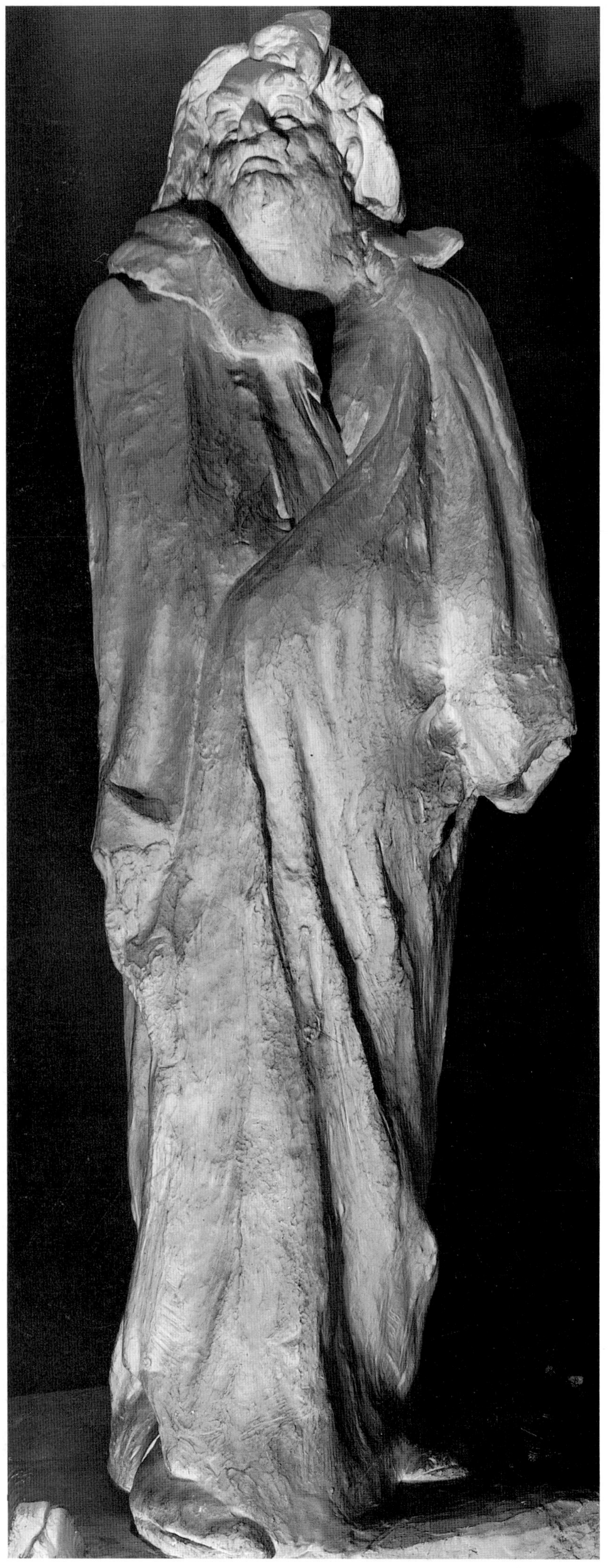

64. BALZAC. PLASTER. 1897.

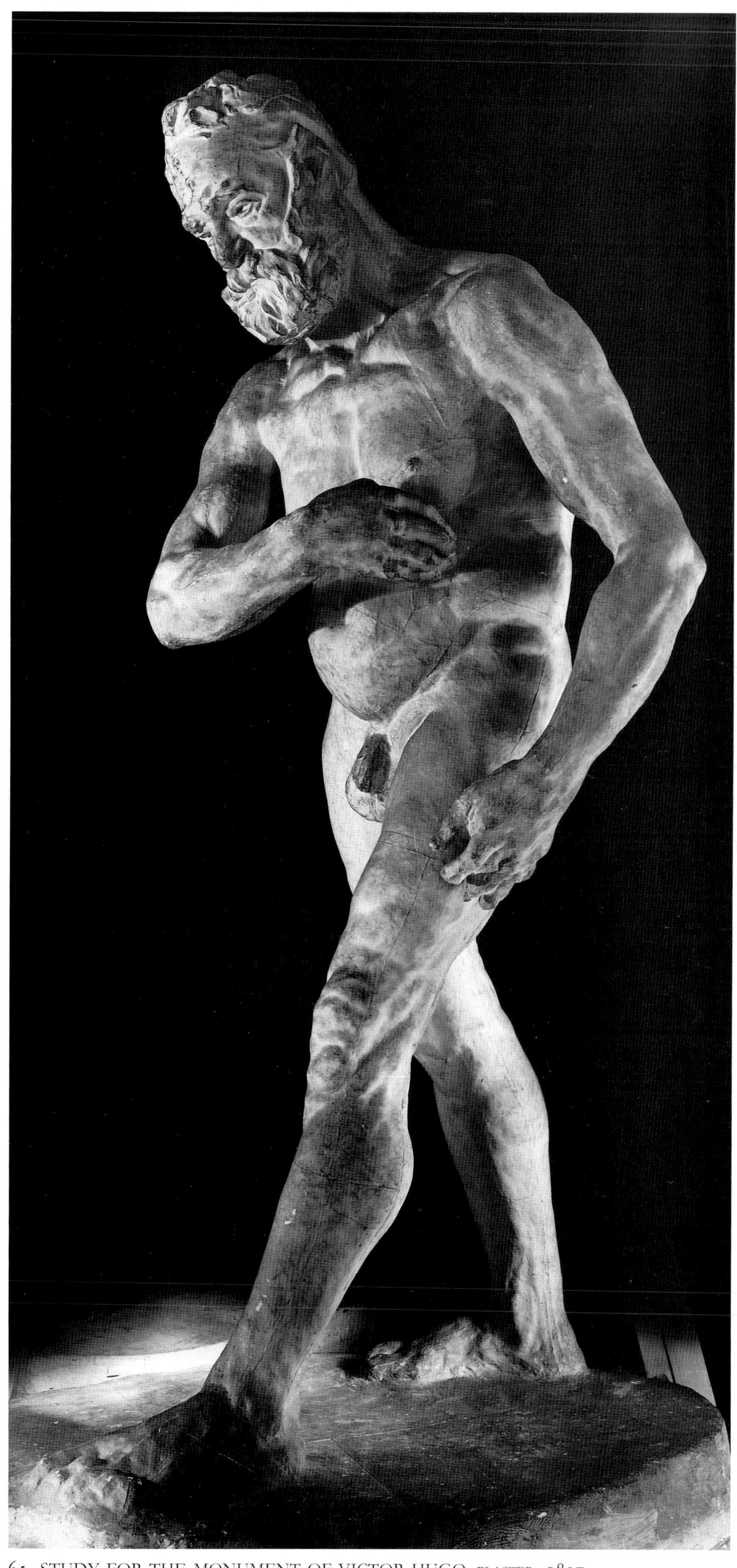

65. STUDY FOR THE MONUMENT OF VICTOR HUGO. PLASTER. 1897.

66. BALZAC. DETAIL.

67. BALZAC. DETAIL.

68. BUST OF THE SCULPTOR FALGUIÈRE. BRONZE. 1897.

69. BUST OF VICTOR HUGO. BRONZE. 1897.

70. THE HAND OF GOD. MARBLE. 1897–98.

71. PAN AND NYMPH. MARBLE. 1898.

72. BAUDELAIRE. PLASTER. 1898.

73. BUST OF MME. F... MARBLE. 1898.

74. A MUSE. STUDY FOR THE WHISTLER MONUMENT. MARBLE. 1902–3.

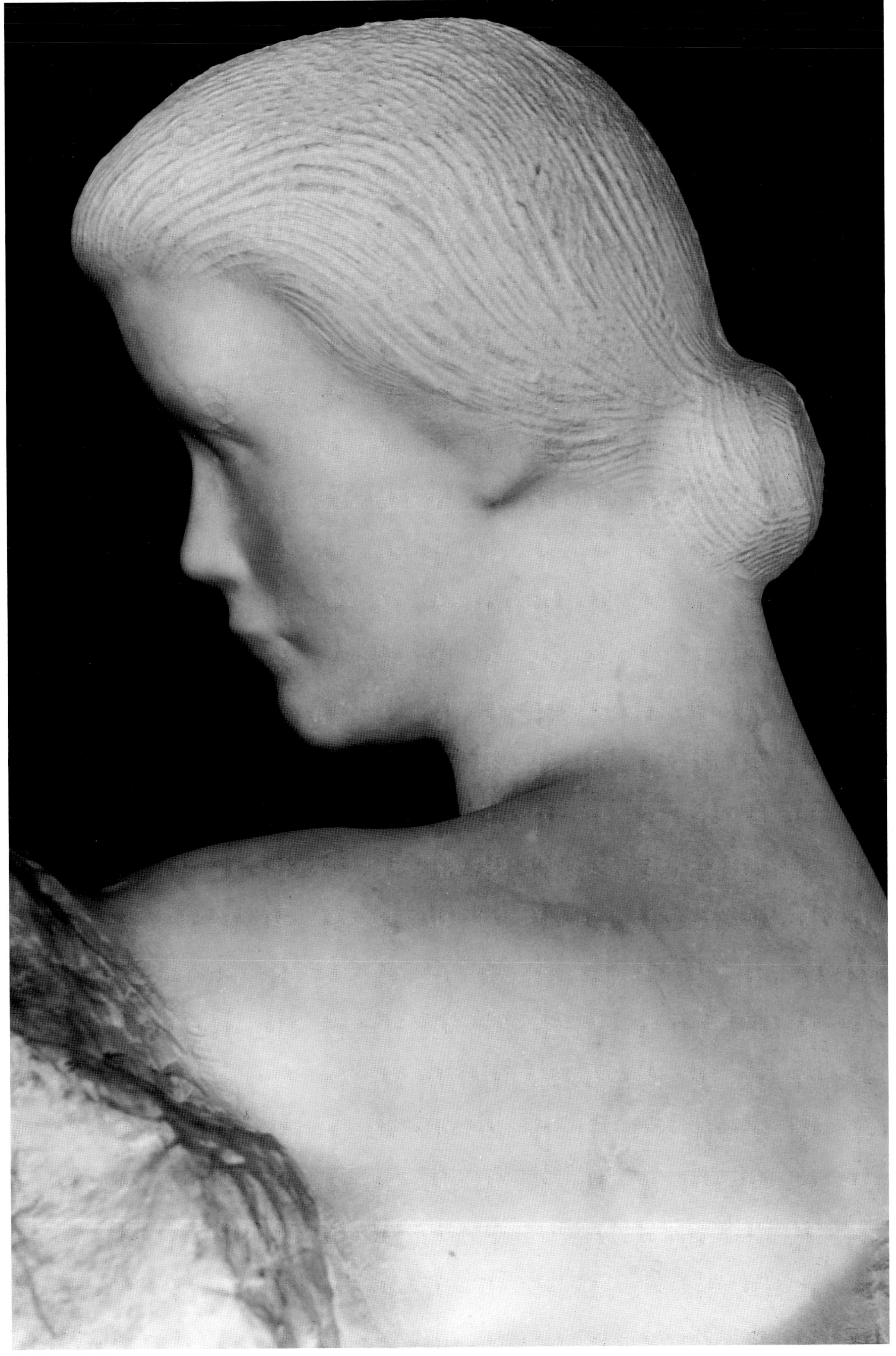

75. SLAV GIRL. MARBLE. 1905–6.

76. BUST OF THE SCULPTOR EUGÈNE GUILLAUME. BRONZE. 1903.

77. BUST OF MARCELIN BERTHELOT. BRONZE. 1906.

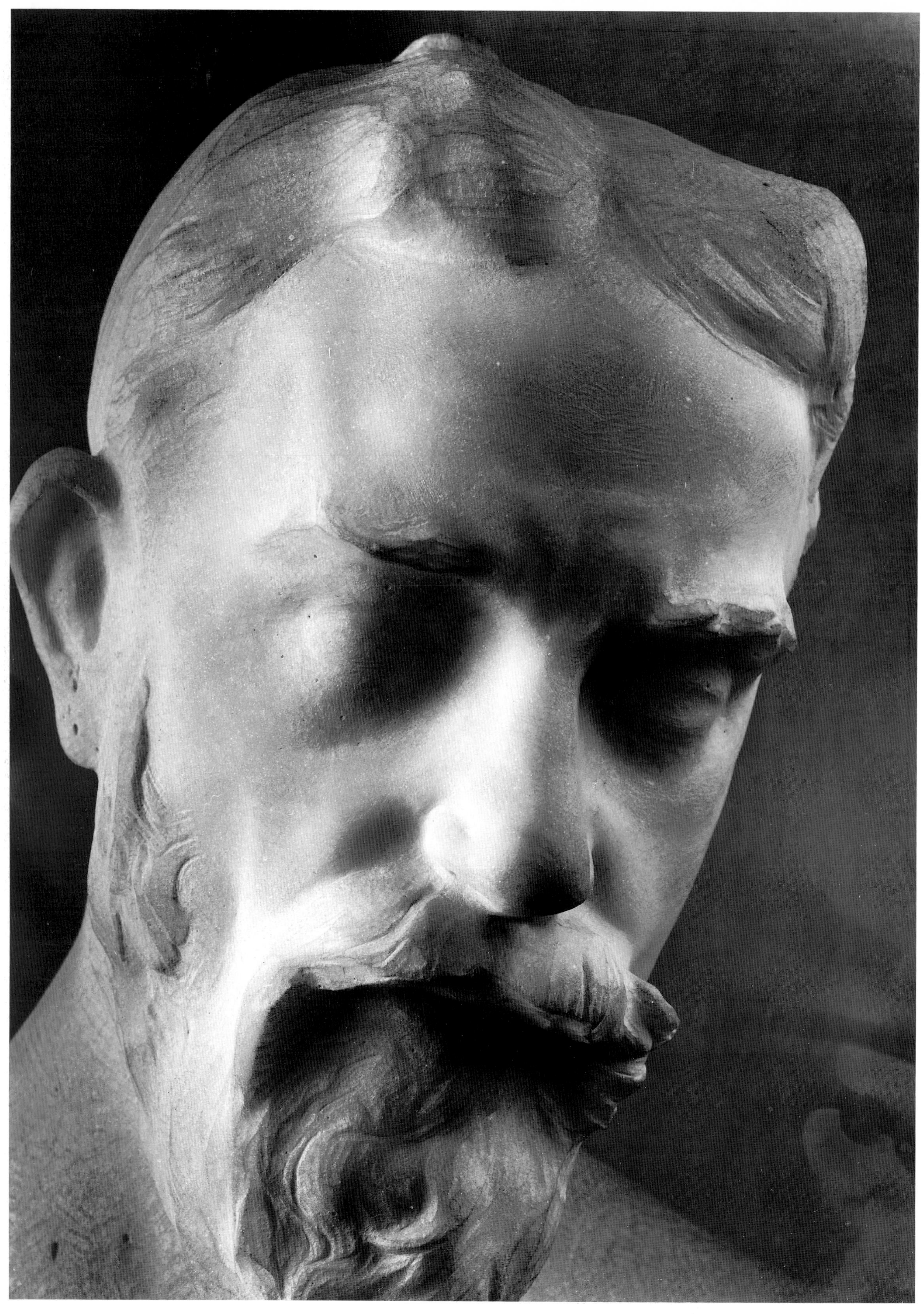

78. BUST OF BERNARD SHAW. MARBLE. 1906.

79. BUST OF BERNARD SHAW. MARBLE. 1906.

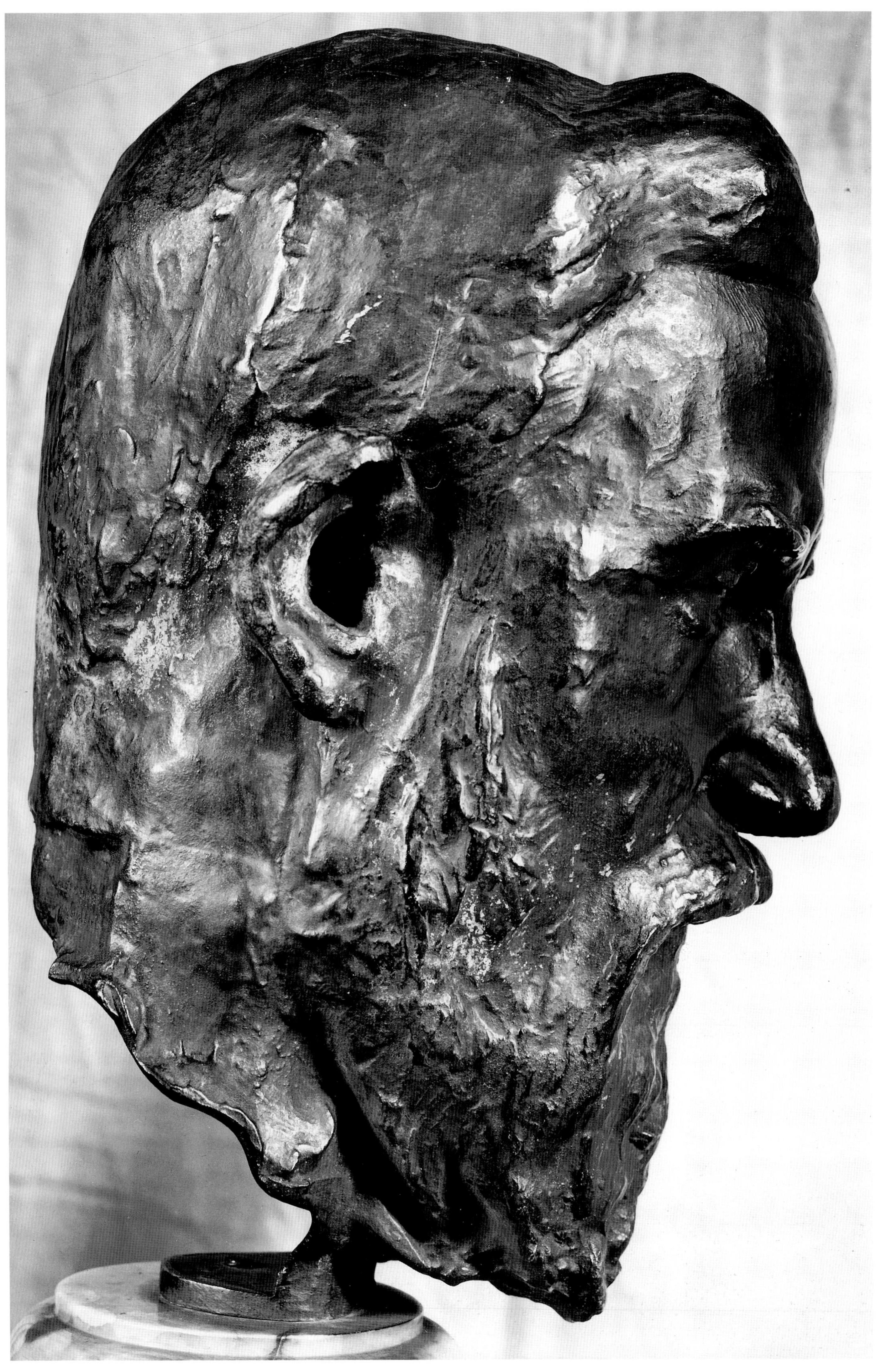

80. BUST OF BERNARD SHAW. BRONZE. 1906.

81. BUST OF MME. DE GOLOUBEFF. BRONZE. 1906.

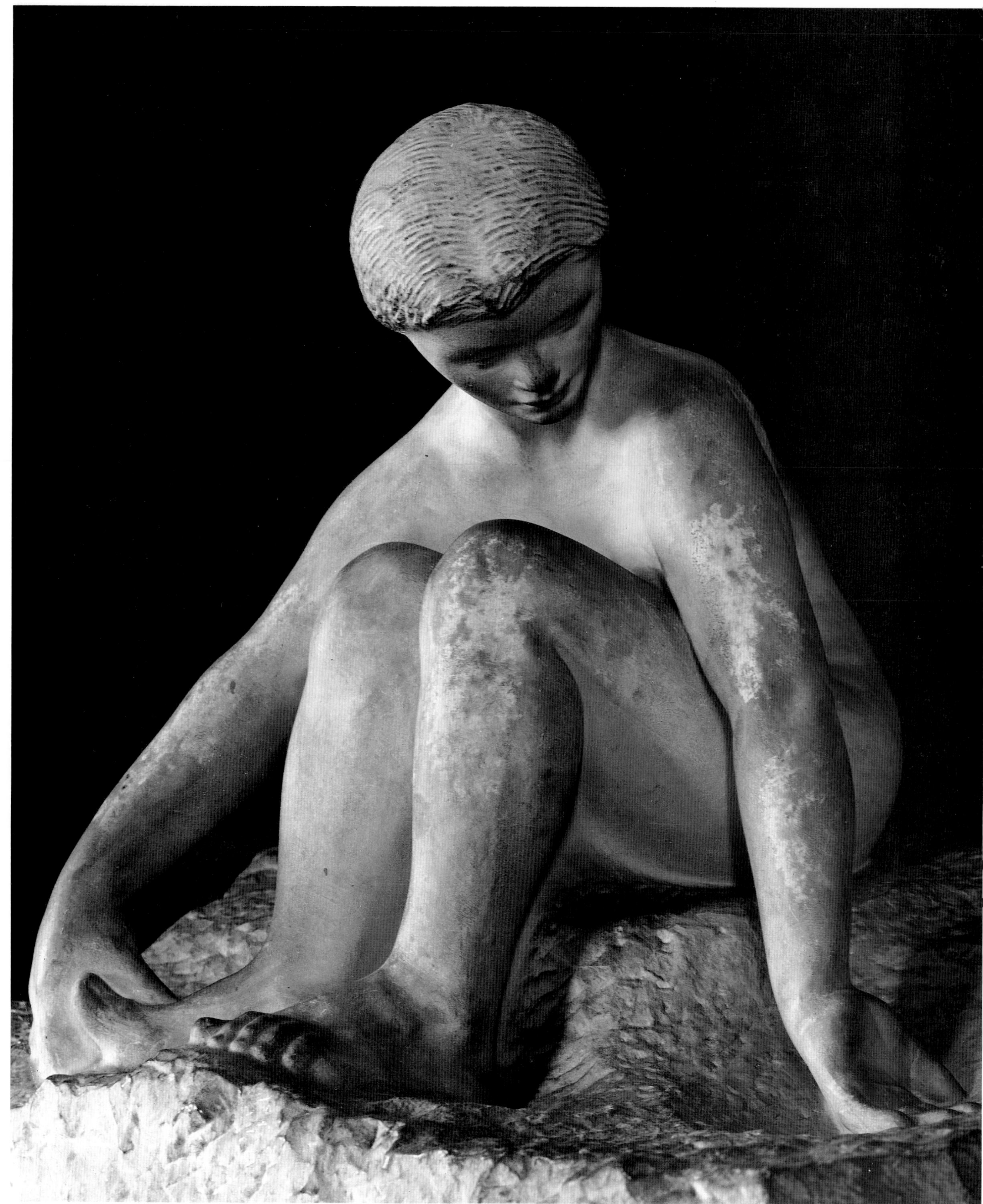

82. BY THE SEA. PLASTER. 1906–7.

83. MOTHER AND DYING CHILD. MARBLE. 1908.

84. PORTRAIT OF THE JAPANESE GIRL-DANCER HANAKO. BRONZE. 1908.

85. BUST OF GUSTAV MAHLER. BRONZE. 1909.

86. BUST OF THE PAINTER PUVIS DE CHAVANNES. MARBLE. 1910.

87. MOZART. MARBLE. 1910.

88. FEMALE TORSO. BRONZE. 1910.

89. THE DUCHESS OF CHOISEUL. BRONZE. 1908.

90. BUST OF ÉTIENNE CLÉMENTEL. BRONZE. 1916.

91. BUST OF GEORGES CLEMENCEAU. BRONZE. 1911.

92. THE CATHEDRAL. STONE. 1908.

NOTES ON THE PLATES

NOTES ON THE PLATES

Measurements are given in inches, first height, then width, then depth.

1. RODIN'S FATHER. 1860. Bronze, $16\frac{1}{2}\times11\frac{1}{8}\times9\frac{1}{2}$.
This bust is the artist's first known work. J.-B. Rodin, who was a Norman, came to Paris in the reign of Charles X, and was engaged as office boy in the Prefecture of Police. He retired with the rank of inspector in 1861 and died in 1883. This bust, modelled in Roman fashion, recalls an effigy of Cato or of Minucius Felix. Although inspired by antique statuary, it already manifests a very strong personality and remarkable technical skill. This youthful work was built up by successive profiles according to the method Rodin was to adopt each time he executed a portrait.

2. THE BLESSED FATHER EYMARD. 1863. Bronze, $23\frac{1}{4}\times11\frac{3}{8}\times11\frac{3}{8}$.
In 1863, Rodin's sister, who was a nun, died in the springtime of life. The young artist felt the loss so keenly that he sought admission to a religious order close to the family dwelling, rue St. Jacques, just founded by Father Eymard. On joining the Fathers of the Holy Sacrament, he took the name of 'Brother Augustine'. But Father Eymard, who was a rare psychologist, was not slow to realize that this vocation, prompted as it was by fraternal grief, was unlikely to prove lasting. He therefore requested his novice to sculpture his bust. While the monk was posing, he managed to persuade Rodin to return to the world, to fulfil his artistic destiny there. The Church has since beatified Father Eymard, and we are constrained to join in the homage paid to this holy personage, without whom, perhaps, we might not have had the magnificent work executed by Rodin and dating from this epoch.
The bust of Father Eymard, vital among all those the artist sculptured, is characterized by an intense and luminous expression which shows Rodin's growing mastery of statuary.

3. MASK OF THE MAN WITH THE BROKEN NOSE. 1864. Bronze mask, $9\frac{1}{2}\times8\frac{3}{4}\times10\frac{1}{4}$.
This work has some affinity with the bust of the artist's father (Plate 1). It, too, breathes classical inspiration and reminds one of the Greek marbles. Thus we are advised that Rodin, although refused admission thrice to the School of Fine Arts, was working in solitude to acquire that classical education he was deemed unworthy to receive.
This mask is the more interesting as being that of an old Bohemian whom the young artist encountered on the slopes of the Rue Mouffetard, as busy a thoroughfare as one could wish, and close to the Horse Fair, where Rodin was fond of observing life. The work, refused by the Salon for being offered, with audacious simplicity, under the title 'Mask of Bibi' (the Bohemian's name), remodelled in 1872 as bust and in marble and sent in under the title: 'Bust of Mr. B. . . . ', was received this time with favour.

4–5. THE AGE OF BRASS. 1876. Bronze, $68\frac{1}{2}\times23\frac{5}{8}\times23\frac{5}{8}$.
The history of the conception and carrying out of this work is particularly interesting. It came after a period when the artist had been taking long country walks in the beautiful regions round Brussels, either alone or with his wife. This communing with Nature had made a deep impression on him, which was emphasized by the reading of Jean-Jacques Rousseau, who at the time and for long afterwards was his favourite author. Thus it was that the idea came to him to depict a purely natural man in the infancy of comprehension. He chose for his model a Belgian soldier who in private life was a carpenter. He was a physically perfect man, and, though not possessing much education, was very intelligent and able to understand the higher thoughts and ambitions of the sculptor. Rodin made a full size nude in plaster —a representation, as one critic put it, of one of the men who sprang to life from the stones cast behind him by Deucalion (the impressions of his visit to Italy a year previous were still very strong in the sculptor).
Completed early in 1877, the work was first exhibited at the Cercle Artistique in Brussels. Already charges were made against him of having taken a mould from Nature, though they attracted little attention. He was completely unknown. The work was then exhibited at the Paris Salon of the same year. The jury were quite unaccustomed to such realistic work. He was accused of imposture—of having taken a mould from the living subject. Such tricks were not unknown among inferior artists for small portions of the figure, but, as far as was known, had never been done for the whole body—and, as Rodin ultimately showed, could not possibly be done. The Ministerial authorities sided with the sculptor's critics, and the controversy was long and bitter. Rodin had moulds taken of the model's torso, and these were photographed to show that the procedure of which he was accused would have distorted the figure. Auguste Neyt, the Belgian model, offered to come to Paris to confound the detractors; and the controversy was not ended until a group of Rodin's fellow artists and a few critics, in a collective protest addressed to the Fine Arts Ministry, declared their faith in his genius, from other work which they had seen him execute.

Notes on Plates 6-10

As a compensation to the artist, the State, in 1880, acquired the bronze which had been shown at the Salon, and it was placed in the gardens of the Luxembourg museum. In 1890 it was removed inside the museum.

In contemplating this statue the student must bear in mind that originally the left hand held a stick on which the man was leaning heavily, as if to push himself up from the ground, on which his feet were still heavy. This staff was removed, as it threw a shadow on the moulding of the body on that side; but the position of the hand is only explained by imagining this staff still in position.

6–7. SAINT JOHN THE BAPTIST PREACHING. 1878. Bronze, $79 \times 21\frac{5}{8} \times 38\frac{5}{8}$.

The model was an Italian peasant who had never before been a model and knew none of the 'tricks of the trade'. The sculptor made him move about the studio, and when he was in an attitude that pleased him, he stopped him, saying simply 'Keep that pose'. This method was afterwards frequently followed by Rodin, as stated above. He himself repeatedly said he tried to interpret internal sentiments by the mobility of the muscles.

The Italian peasant afterwards became a well-known model. This work, so daringly original, and so different from the age-long religio-artistic conception of the Baptist, like so many other of the works of Rodin, aroused a storm of controversy. But, like the others too, it 'made its way', and the leading critics of several countries saw its greatness.

This St. John, said Grant Allen, is 'no seer of mysteries filled with prophetic fire, but a plain worn man of the people, an itinerant preacher.' He is 'marching ahead in a clumsy sort of way, ignoring the obstacles and the length of the road,' adds Gustave Geffroy. Edmund Gosse, in his speech at the dinner to Rodin in London in 1903, described the St. John as a 'wasted and bitter anchorite'.

Rodin worked arduously on this statue, which he first exhibited at the Ghent Exhibition in 1879, where it earned for him the great gold medal. The same year he sent to the Salon, in Paris, the bust of the Forerunner. Then he exhibited this statue in the Salons of 1880, 1881 and 1883, thereby enlarging the circle of his admirers. Sent to the Vienna and Munich Exhibitions by the Beaux-Arts, it scored a notable success there and assisted to spread the artist's fame in German-speaking countries. To-day still, in its decayed grandeur, in its wonderful simplicity, 'Saint John the Baptist,' in spite of the works of a more acute modernism later executed by the artist, remains one of his most moving creations, one of the most powerful.

8–10. THE GATE OF HELL. 1880–1917. Bronze, $248 \times 157 \times 34$.

This monumental Gate, which was ordered by the State from Rodin in 1880, and was never finished, though he worked on it for over twenty years, was originally intended as a door for the Museum of Decorative Arts, which was then planned. Later it was proposed to place it in the disaffected chapel of the old Seminary of St. Sulpice, in the Place St. Sulpice, Paris, which was to become an annexe to the Luxembourg Museum.

Never since Michelangelo has a sculptor planned a work of such greatness and magnificence. His original idea was to make direct interpretations from the scenes in the 'Inferno' of Dante, while in the scheme of arrangement he adopted the plan of Ghiberti's door of the Baptistry at Florence, with symmetrical panels. By degrees Rodin departed from this plan, as he also drifted more and more from Dante's conception of 'Hell' and conceived symbolical figures of his own inspiration. The only actual groups taken from Dante's poem which are now left are 'Paolo and Francesca' and 'Ugolino', besides the representation of Charon's bark depositing the souls to be judged by Minos.

The other poet who greatly influenced Rodin as his ideas progressed for the 'Gate' was Baudelaire, the tragic and sensual author of 'Les Fleurs du Mal', but it may be said that most of the figures are due to his own keen sense of the tragic perversity of life and destiny.

As the Gate exists to-day it contains 186 figures, and, as has been said elsewhere, Rodin was continually adding, withdrawing or exchanging figures or subjects, so that it became a perfect repertory for his various works.

On the top of the Gate the three 'Shadows' look down into the abyss as if seeing or reading there the famous words of the poet:

'Lasciate ogni speranza, o voi ch'entrate!'

Below them, in the centre of the tympanum, the 'Thinker' also contemplates the tragedy. On the panels are represented all the passions and vices of humanity. Nearer to the ground the figures become more independent of the subject. Right at the bottom are the lost women, as Baudelaire conceived them, mingled with figures from pagan mythology—centaurs, faunesses, satyrs, and abstract vices, especially the remarkable 'Avarice and Luxury', representing a miser and a courtesan. On the cornice at the top are thirty heads showing varied characteristics, as if they were a summary or analysis of the whole. Rodin had said to the Fine Arts Minister Turquem, who gave him the commission, 'I will cover the door with a lot of small figures, so that nobody can accuse me of moulding from life'. Anatole France wrote of this wonderful collection of groups: 'Recall the damned who are placed on the left of God on the fine portal of Bourges cathedral. In those representations of the theological hell, the sinners are tormented by horned devils who have two faces, one of which is not on their shoulders. You will not find these monsters in the hell of Rodin. There are no demons there, or rather, the demons hide themselves in the damned. The evil demons through whom these men and women suffer are their own passions, their loves and hatreds; they are their own flesh, their own thought. These couples who "pass so lightly on the wind" cry to us: "Our eternal torments are in ourselves! We bear within us the fire that

burns us. Hell is earth, and human existence, and the flight of time; it is this life, in which one is incessantly dying." The hell of lovers is the desperate effort to put the infinite into an hour, to make life pause in one of those kisses which, on the contrary, proclaim its finality. The hell of the voluptuous is the decay of their flesh in the midst of the eternal joy and triumph of the race. The hell of Rodin is not a hell of vengeance, but one of tenderness and pity.'
The actual size of the Gate was never decided.

11, 13. THE THREE SHADOWS. 1880. Bronze, $38\frac{3}{4} \times 35\frac{1}{2} \times 17\frac{3}{4}$.
These three statues together crown 'The Gate of Hell'. It was the author's first idea that in front of them should be an unrolled phylactery bearing the famous inscription from Dante; 'Lasciate ogni speranza, o voi ch'entrate!' This intention, which explains to some extent the attitude of the three men, was not carried out.

12. THE SHADOW. 1880. Bronze, $75\frac{1}{2} \times 44 \times 19\frac{3}{4}$.
'The Gate of Hell' consists of three figures of half natural size, who with slow and weary gesture seem ready to enter into eternity. This work is, so to speak, the pendant of 'Adam', also in this trinity of brass, and which was conceived after Rodin's trip to Italy in 1875. The influence of Michelangelo is manifest, but the artist's personality was already too marked to be utterly overwhelmed by Buonarroti. 'The Shadow', modelled in 1880, was exhibited in the 1902 Salon and acquired by the State in 1910.

14–16. THE THINKER. 1880–1900. Bronze, $78 \times 51 \times 52\frac{3}{4}$. (Plaster model. 1880. $27 \times 15\frac{3}{4} \times 19\frac{5}{8}$.)
'The Thinker' also was part of the original conception of 'The Gate of Hell', of which it might be said that it constitutes the soul. Moreover, Rodin at first wanted his statue to be called 'The Poet'. In the artist's view, the poet was that Dante whose work he loved so passionately and whence he had drawn so many wonderful designs. But this Dante met the same fate as the 'Balzac' at a later date. Rodin went beyond his first conception and widened the theme he had thus first chosen until it became a universal symbol. 'The Thinker', executed in half life-size for setting up in 'The Gate' about 1879, was exhibited at a height of 6 feet in 1900 in the Alma Pavilion which Rodin had built in order to house his masterpieces. A subscription was raised in 1906 to present it to the people of Paris. It is under this bronze at Meudon that the great master rests to-day close to his wife.

17. THE PAINTER ALPHONSE LEGROS. 1881. Bronze, $11\frac{1}{2} \times 7 \times 9$.
In 1871 Rodin paid a visit to London, during which he ran against his friend Legros, whom he had known in 1854 at the School of Design in the Rue de l'École de Médecine, conducted by Professor Lecocq de Boisbaudran. Painter, engraver, sculptor, Legros had emigrated to England after the Commune, and was to remain there until his death in 1911. The two artists resumed their friendship, which took on a more intimate character from this moment and proved lasting. Ten years later, in 1881, Rodin executed the fine bust of Legros, and, in the same year, Legros painted a vivid and precious portrait of Rodin, at this decisive epoch of his career. It was Legros who taught the sculptor the art of engraving, in which he excelled, and to this fact we owe the well-known masterpieces of Rodin in this medium.

18. THE PAINTER JEAN-PAUL LAURENS. 1881. Bronze, $22 \times 13 \times 12\frac{1}{4}$.
Very soon after his return to France, Rodin became intimate with J. P. Laurens, upon whose bust he started work in 1881. The model also attracted him by the classical type of his expression, thus giving him an opportunity to pursue the study of antique sculpture. This large-shouldered bust recalls a sage of Greece. By way of return, J. P. Laurens put his friend in the frescoes of the Panthéon under the guise of a personage of the Merovingian Court. In 1900 Laurens wrote to the art critic Arsène Alexandre: 'You know my admiration for the great sculptor. He is of the race of those who walk alone, of those who are unceasingly attacked, but whom nothing can hurt. His procession of marble and bronze creations will always suffice to defend him—he may rely on them.'

19. THE HEAD OF SORROW. 1882. Bronze, $8\frac{3}{4} \times 9 \times 10\frac{3}{4}$.
Rodin was often prompted to reproduce this or that fragment of one of his prior works, either singly or as part of a new composition. Thus this head, of striking intensity of expression, is a retaking of that of one of the children in the groups of 'Ugolino' and 'The Prodigal Son' (cf. pl. 21 and 56). Moreover, in 1905, having had a chance to meet 'La Duse' he was filled with such enthusiasm for her wonderful genius that he thought of adapting the head of 'Sorrow' to one of her passionate impersonations. He also thought of using it for the face of a statue of 'Joan of Arc at the Stake', which it was proposed in 1913 to remove to the United States.

20. THE FALLEN CARYATID CARRYING ITS STONE. 1880–1. Bronze, $17\frac{1}{2} \times 12\frac{1}{2} \times 12\frac{1}{2}$.
This statue was originally part of 'The Gate of Hell', where it figured above, in the left angle uncovered by drapery. It was exhibited from 1883 in a gallery and reappeared in 1892 in the Salon, where it was praised by Rodin's admirers, who henceforth grew ever more numerous.

21. UGOLINO. 1882. Bronze, $15\frac{3}{4} \times 17 \times 13\frac{3}{4}$.
This group was one of the subjects for the 'Gate of Hell'. Rodin's imagination seems to have been greatly stimulated by this incident which is one of the most terrific in the whole of the 'Divine Comedy'.

Notes on Plates 22-28

'The Pisans,' wrote the old Italian chronicler Villani, the contemporary of Froissart, 'who had imprisoned the Count Ugolino, with his two sons and two of his grandsons, the offspring of his son Count Guelfo, in a tower on the Piazza of the Anziani, caused the tower to be locked, the key thrown into the Arno, and all food to be withheld from them. In a few days they died of hunger; but the Count first with loud cries declared his penitence, and yet neither priest nor friar was allowed to shrive him. From then on the tower was called the Tower of Famine, and so shall ever be.' The incident depicted is related by Dante as follows, the old Count speaking to the poet in Hell:

'When a faint beam
Had to our doleful prison made its way,
And in four countenances I descried
The image of my own, on either hand
Through agony I bit; and they, who thought
I did it through desire of feeding, rose
O' the sudden, and cried, "Father, we should grieve
Far less, if thou wouldst eat of us: thou gavest
These weeds of miserable flesh we wear:
And do thou strip them off from us again."
Then, not to make them sadder, I kept down
My spirit in stillness. That day and the next
We all were silent'. (Hell, canto XXXIII; Cary)

Ugolino seems in this powerful group almost to have been changed into a beast by his sufferings and the gnawings of his stomach. He drags himself on his knees over the lifeless or scarcely living bodies of his descendants. Perhaps he is thinking of how the children urged him to appease his hunger on their own flesh, and a contest is taking place in his mind between the beast who would fain eat and the higher being who revolts at such a monstrous idea, as is shown by the violence with which his head is thrown to one side. At the same time the sculptor lets us see the wolf-like look in the eyes of the emaciated creature who is scarcely human any longer:

'Whence I betook me, now grown blind, to grope
Over them all, and for three days aloud
Call'd on them who were dead. The fasting got
The mastery of grief.'

In the first rough model which Rodin did for this subject Ugolino was shown seated with one of his sons lying over his knees, and another standing by his side.

22. EVE. 1881. Bronze, $68\frac{1}{2} \times 25 \times 30\frac{1}{2}$.
This statue was executed during the period when the artist conceived his 'Gate of Hell'. It was intended to be placed by the side of Adam, but subsequently it was detached from the *ensemble* and offered separately to the 1899 Salon, in its natural size. For more than twenty years the work has been known in reduced copies, and has enjoyed great success. It is one of the most beautiful incarnations of the female form since the Greeks, full of robust life and in a natural attitude, suggesting her presentiment of coming motherhood as well as her natural anguish at thinking of the sorrow to which the coming generation is destined. The face is beautiful and tense with thought; the arms are folded over the breasts, one hand being raised as if to shield her face, the other grasping her left breast as if in pain or anxiety.

23. THE THREE FAUNS. 1882. Plaster, $6\frac{1}{2} \times 11 \times 7$.
This charming group is often called 'The Three Bretons'. Its model is the same figure reproduced three times. It has its place in 'The Gate of Hell' under the tympan on a level with the left folding door. It may certainly be dated 1882.

24. HENRI ROCHEFORT. 1897. Bronze, $29\frac{1}{2} \times 22\frac{1}{4} \times 13\frac{1}{2}$.
A plaster model of this bust was executed in 1884 and first appeared at an exhibition at the Georges Petit Gallery, where it was much admired, both because of the quality of the work and the personality of the model, who was a famous journalist. The bust, which is considerably bigger than the plaster model, was exhibited at the Salon of 1897.

25. THE SCULPTOR DALOU. 1883. Bronze, $20\frac{1}{2} \times 15 \times 8\frac{3}{4}$.
This bust, one of Rodin's most famous, of which there are casts in many of the great museums of the world, is dated 1883. Dalou was one of the sculptor's youthful companions: in their student years they both frequented the same 'Petite École', which turned out so many fine artists. Rodin met him again in London, in Legros' circle, and, on returning to Paris, they continued to see each other frequently. It was at this period that Rodin decided to stamp his friend's image in a work that, despite its modern air, reminds one of the finest busts of the Italian Renaissance.

26–27. THE CROUCHING WOMAN. 1882. Bronze, $33\frac{1}{2} \times 23\frac{3}{4} \times 19\frac{3}{4}$.
At this period the majority of the works which Rodin conceived, and which were not imposed upon him by a competition or definite order, were, so to speak, commissioned by 'The Gate of Hell'. In this monumental creation those visions disported themselves which he had carried about with him until the age of 40, when he cast forth impetuously all the people of his dreams. 'The Crouching Woman', who at first glance startled the critics, was one of the crowd of the damned who writhe in the tympan, as they used to do on the portals of cathedrals. This magnificent bronze, on the back of a prodigious example of anatomy, is without doubt one of the most amazing fragments of the sculptor, who here delighted in difficulties.

28. MADAME LYNCH DE MORLA VICUNHA. 1884. Marble, $22\frac{1}{2} \times 19\frac{1}{2} \times 14$.
Quite at the commencement of his career, towards 1882, Rodin chanced to meet Madame Lynch de Morla Vicunha,

wife of a South American diplomat; and he became a friend of the household. The young woman, who knew the artist's passion for classic music, often played Mozart or Beethoven to him, and, by way of recompense, the latter executed her bust. Done in 1884, it was exhibited in the 1888 Salon, where it aroused real enthusiasm. The State then acquired it for the Musée National.
Not one of Rodin's busts is more filled with the living and subtle charm of female beauty. The look in the eyes, the pose of the head, and the set of the mouth, which seems just to have been speaking, are all characteristic. The moulding of the marble is exquisitely delicate. The nosegay of flowers on the base of the bust strikes a note of homage paid to a charming person.

29. MRS. RUSSELL. Before 1888. Wax, $18\frac{1}{2} \times 11\frac{1}{2} \times 10\frac{5}{8}$.
This young woman, who was of Italian origin, married an Australian painter, a friend of Claude Monet, affiliated to the Impressionists, and living most of his time in Belle-Isle en Mer, where he had a property. Herself a genuine artist, she probably became acquainted with Rodin through Monet, and the sculptor, struck by her classical beauty, requested permission to make her bust. In 1888, he executed the work, which later was even cast in silver. The young woman gave the Master numerous sittings.

30–31. SHE WHO ONCE WAS THE HELMET-MAKER'S BEAUTIFUL WIFE. Before 1885. Bronze, $20 \times 10 \times 12$.
Here Rodin has taken a text from the old French poet François Villon, whose poem 'Les Regrets de la Belle Heaulmière jà parvenue à vieillesse' ('Lament of the Old Helmet-Maker's Wife on Reaching Old Age') is one of the most remarkable of his works. The former courtesan, once radiant with youth and grace, is now repellent with ugliness and decrepitude. Once proud of her charm, she is now equally ashamed of her decay. Bent double, she contemplates her withered breasts, her abdomen in folds, her arms and legs which are knotted like vine-stalks, while the skin falls over the hardly veiled skeleton. Beauty is only skin deep, says the sculptor, echoing the 'vanity of vanities' of the preacher. The sculptor's work is perhaps even more grimly expressive. Grotesque as it is, the spectacle is ineffably sad, for it is the distress of a pour soul who, still so tardily yearning after youth and beauty, and powerless before its decay, is the antithesis of anything spiritual. There is a pendant to this statue (notes M. Paul Gsell) in a strange statue by Donatello now at the Baptistery at Florence, which represents an old woman, nude, or rather simply draped, with her long hair falling thinly and sadly over her withered body. It represents Mary Magdalen, who has retired to the desert, burdened with years, and now macerates her poor body, in honour of God and to punish it for the care she formerly bestowed on it.
But the sentiment of the two works is widely different. The Magdalen, in her desire for renunciation, is filled with the mysticism of the middle ages. The old woman of Villon and Rodin still longs for her lost beauty, and is horrified to find herself resembling a corpse. The original model of this statue was a withered old Italian woman who had come to Paris to seek out her missing son. Falling into want, some of her fellow country people advised her to go and ask Rodin for help, and he, struck with her appearance, invited her to pose for him.
At first he gave the subject the form of a bas-relief on the left doorpost of the 'Gate', but about 1885 he returned to this subject and gave it the moving and admirable form under which we see it here.

32. KNEELING FAUN. 1884. Plaster, $22 \times 8 \times 12$.

33. ERECT FAUN. 1884. Plaster, $24\frac{1}{2} \times 11\frac{3}{4} \times 10$.
These two figures bring us back to 'The Gate of Hell'. They were among the first which were placed in the tympan, but most certainly they were not slow to leave the crowd of the damned to lead an individual life. We know of a bronze cast of the 'Kneeling Faun' belonging to a Romanian collector, which is dated 1884.

34–35. DANAÏD. 1885. Marble, $13\frac{3}{4} \times 28\frac{1}{2} \times 22\frac{1}{2}$.
This splendid marble ranks amongst the most justly renowned works of Rodin. It represents one of the daughters of Danaus, King of Argos, condemned in Hades to fill perpetually a vessel full of holes, as punishment for the murder of her husband. What exquisite rhythm and beauty of contour there are in this slim girlish figure, who has thrown herself down, face downwards, beside the stream in despairing abandonment, with the bottomless vessel under her arm. It was sculptured in 1885, in view of 'The Gate of Hell', but was suppressed in its final state. Exhibited for the first time in the Salon in 1890, it could be seen in the following years in Venice and Oslo under the title 'The Spring', and excited the same enthusiasm in these cities.

36–45. THE BURGHERS OF CALAIS. 1884–1886. Plaster, $82 \times 94 \times 75$.
(Plates 36, 37, 38, 39, 40, 44, 45. Studies for Burghers of Calais. 1884. Bronze).
The episode of the Burghers of Calais and the patriotic sacrifice of Eustache de Saint-Pierre and his comrades has been related by Froissart and Jean le Bel, his Belgian predecessor as chronicler and troubadour.
British military annals furnish few cases of more determined and noble resistance than that maintained for eleven months (1346–1347) by the burghers of Calais, under the command of Jean de Vienne, a 'commander worthy of the commanded'. Famine attacked them even more fiercely than the soldiers of King Edward, and still they resisted. It was only when, after almost incredible fortitude, they saw their last

hope dashed to the ground, at the very moment that they anticipated relief—it was only when Philip the Sixth came towards Calais, and then, not liking the aspect of the English defence, turned and went back again, that they allowed themselves to think of submission. Philip's cruel desertion was the death blow. They sent to Edward, who, however, would listen to no terms except unconditional surrender. The noble Sir Walter Mannay, however, spoke for them; and at last mercy was promised to all but six of the chief burghers, who were to come to him bare-headed, bare-footed, with ropes about their necks and the keys of the town and castle in their hands. The people of Calais were summoned by bell into the market place, and there the conditions of mercy were made known to them.

In 1884 the town of Calais opened a competition for a monument commemorating the heroic act of Eustache de Saint-Pierre. Rodin, struck by the narrative of the old chronicler, decided to make a composition which should commemorate the six hostages. The work occupied him some ten years, though the delays were not always his fault.

The Municipality, on inspection of the work, declared that the artist had not made 'the burghers sufficiently heroic'. It required all the persuasive powers of Dewavrin, the Mayor of Calais, who through seeing the sculptor at work had conceived admiration and friendship for him, the intervention of Alphonse Legros and of Charles Cazin—who had posed for the nude of Eustache de Saint Pierre—to quell the opposition. The work, inspired by the great masters of mediaeval sculpture and of Claus Sluter, was unveiled on June 8, 1895, in front of the Calais Town Hall and created a deep impression. Very fine replicas exist in London, Belgium and Copenhagen.

As with so many other of his works, Rodin was grieved at finding that many of his contemporaries did not understand his group, for a hot controversy again arose, as had been the case with the 'Bronze Age' and the 'Claude Lorrain' (to be repeated later with the 'Balzac'). 'They would have preferred,' Rodin said, in answer to some of his critics, 'gestures *à la Marseillaise*' (a reference to Rude's bas-relief on the Arc de Triomphe in Paris). 'I intended to show my burghers sacrificing themselves as people did in those days, without proclaiming their names.' Still there were more than sufficient admirers of the work to satisfy the artist.

The remarkable differences in character of these six hostages is evident to anyone who studies the group. The central figure is the aged Eustache, whose venerable head with its long hair is bowed, but not with fear or hesitation; he seems rather, in sorrowful resignation, to be in deep contemplation in the spirit of his own words, 'I have so good trust in the Lord God'. If his step is a little halting, it is from the privations of the long siege; his firmness is calculated to inspire his fellows. 'He was the one who said, "we must",' Rodin remarked.

The one next to him, who is probably Jehan d'Aire, holding the key which he is to hand to the King, also has no fear or hesitation, but his whole body is tense with the effort to get strength sufficient to go through the ordeal and humiliation. His face—a clean-shaven, lawyer-like face—is set in grim sorrow at the pass to which his city is reduced. Behind him is the figure known as the 'Weeping Burgess', whose face is covered by his two hands, as if he were indeed faltering or regretting his decision, and were thinking of wife and children. Just behind Eustache, one of the men looks back to the city, while he passes his hands before his eyes as if to drive away some terrible vision, for his resolution is evidently not so stern as that of the two leaders. Of the final two, who may be the two brothers, the one in advance, whose movement is more hasty and nervous than that of Eustache, as if he would fain have the ordeal over, may be encouraging by his gesture the one behind him, the youngest of the group, who hesitates at now leaving life and its sweetness behind.

The three figures in the second row are all of them less resolute and less brave than the three in front, as is shown by their attitude. Their act is no less heroic for all that. Their feet are heavy, but their wills urge them on.

In this great group the physical aspect is subordinated to the spiritual, and the composition is instinct with the shadow of the impending interview with the redoubtable King and the forfeit they are to pay. The greatness of the deed they are accomplishing pervades them with an atmosphere of august sadness, and eliminates all meaner sentiments. They move one by their simplicity and the absence of gesture.

Each one of the six is visible from any leading point from which one may look at them, and so one feels the crowd of sorrowful citizens, women and children gazing after them from the battlements of the stricken town.

Rodin had wished to have the group placed on a high pedestal, but as this would have entailed having the figures very much more than life size, the suggestion was not adopted. Failing that, he said he would have preferred the group to stand on the soil, so that they might seem to be part of the population, but even this plan did not seem practical and was not adopted.

46. AURORA. 1885. Marble, $22\frac{3}{8} \times 22\frac{3}{8} \times 13\frac{3}{4}$.

It was Mademoiselle Camille Claudel, Rodin's pupil and a sculptress herself, who posed for this radiant marble, worthy of the name given it. From this masterpiece emanates a light that once seen can never be forgotten.

47. THOUGHT. 1886. Marble, $29\frac{1}{4} \times 21\frac{3}{4} \times 20\frac{1}{2}$.

Again it was Mademoiselle Claudel who served as model for this head which, despite its unusual presentation, aroused deserved admiration from the start. Executed in 1886, 'Thought' was not exhibited until 1896 in the Salon.

'There is no disturbing strain,' says Grant Allen, 'but the calmness and the remote expression of one absorbed by the inward working of the mind'; and Paul Gsell says: 'Abstract thought blossoms out of the midst of inert matter and illumines it with the reflection of its splendour, though it tries in vain to escape from the shackles of reality.' Such are the ideas which Rodin sought—and with what success!—to express in this work, which is at once instinct with spirituality, and yet in the lucidity of the eyes, beauty of features and charm of pose, is unmistakably feminine. There is an intentional suppression of any sensuous element, even to the absence of the hair, which is covered by a cap. It has justly been described as the very symbol of Rodin's art.

48. INVOCATION. 1886. Plaster, 22×10×9¼.
This fine figure belongs to a period when Rodin was working with ardour at the 'Gate of Hell', for which he had just received the commission. A certain resemblance to the 'Old Man Suppliant' suggests that it was a feminine adaptation of the same subject. Madame Abruzzezzi, the model whom Rodin employed several times, seems to have posed for this work.

49. THE KISS. 1886. Marble, 75×47¼×45¼.
This piece is considered by far the most important of Rodin's classical works. It was shown for the first time in Paris in 1887, and was later exhibited at the Salon of 1898 together with the statue of Balzac.

50. FAUN AND NYMPH. (THE MINOTAUR.) Before 1886. Plaster, 13½×10×10.
Once more it is to the remotest ages of mankind that we are transported by this group, to resonances as deep and mysterious as some lines from Racine's *Phèdre*. Certain critics re-christened it 'Jupiter Taurus', but Rodin, whose imagination was haunted by the images evoked by reading about the Mycenaean civilization, did not accept this new title. Moreover, the work seems to have been in the first place a study for the group of 'Pygmalion and Galatea' excuted by the artist before 1886.

51. POLYPHEMUS. 1888. Bronze, 9⅞×5½×6¼.
Among the subjects which decorate the very fine Medici fountain in the Jardin du Luxembourg there is a statue of Polyphemus about to crush Acis and Galatea. This work of Ottin must have struck Rodin, who often had occasion to pass it during his youth, and when he conceived 'The Gate of Hell', there was an opportunity to introduce this mythological episode into his creation. But, after having modelled 'Polyphemus and Acis' in the right folding door, the sculptor realized that, plastically, the two characters, owing to their position, created gaps in the composition. The same year—1888—that he executed this group, he smothered the Acis in the plaster, and only preserved the Polyphemus.

52. THE METAMORPHOSES OF OVID. Before 1886. Bronze, 13×15¾×10½.
Inserted in the 'Gate of Hell', in the right angle of the attique, the work, almost as soon as born, like many of its sisters, was detached from the monument to take on a life of its own. Two other titles that it bore, 'Volupté' and 'Les Fleurs du Mal', testify that it was conceived under the influence of the poems of Baudelaire, which, together with the 'Divine Comedy', were his bedside books.

53. FLYING FIGURE. 1889–90. Bronze, 20⅛×27⅛×11¾.
This figure, of an amazing audacity, in its agitating nudity, might well have been a study for the characters intended to surround the 'Victor Hugo seated', 'Meditation' and 'The Tragic Muse', a hypothesis which is equally valid for 'Iris, Messenger of the Gods', this prodigious creation of the artist. This 'Flying Figure', which is probably of 1889, was, however, utilized by Rodin, this time completed, in his group 'Avarice and Luxury'.

54–55. THE ETERNAL IDOL. 1889. Plaster, 29¼×15¾×20½. (1889. Bronze, 6⅞×5⅞×3⅜.)
A young woman is half seated half kneeling, with her head bending forward and a dreamy look in her face, while a man, kneeling before her, restraining his desire—his arms are behind his back—softly bends his head and plants a kiss under the left breast, over the heart. He has a restrained fervour that is at once mystical and amorous. She has a reserved sphinx-like expression. Is she, too, awakening to the current of love, or is she wondering at man's passion, which is unfathomable, for the beauty which he as yet hardly knows? The originality of the pose is unique in sculpture, rendering it one of the loveliest commentaries on the relations between man and woman that exist in art. Rodin is said to have adopted the title from the remark of a visitor, to whom he was explaining his idea. 'I see,' said the visitor—'the eternal idol!' And Rodin understood at once how suitable it was.
It certainly originated prior to 1889, for then it was known under the name of 'The Host'. It was exhibited in the 1896 Salon.

56–57. THE PRODIGAL SON. Before 1889. Bronze, 54¾×41½×27⅝.
This work, which figures in the right hand folding door of 'The Gate', was probably conceived between 1885 and 1887, for the subject is utilized in the 'Fugit Amor,' which is prior to 1887. It was first called 'The Child of the Age', and was exhibited under this title in the 'Salon de la Plume', to re-appear in 1905 in the Salon d'Automne, with an appellation invented by the critic, 'The Dying Warrior'. This interpretation of the Biblical episode reminds one, in the largeness and humanity of the conception, of Rembrandt's finest etchings.

58. OCTAVE MIRBEAU. 1889. Terracotta, 11×7×6⅝.
Mirbeau was one of Rodin's oldest friends, and always defended him with his unsparing fire against the representatives of an obsolete academicism. He broke many lances in

the 'Balzac' controversy, and no one spoke better of the Master's water-colours than the author of 'Calvaire'. Moreover, Mirbeau, having prepared an édition de luxe of 'Le Jardin des Supplices', wanted it to be illustrated with his friend's works. For a copy of 'Sébastien Roch', belonging to Edmond de Goncourt, Rodin, as was his favourite custom, designed three profiles of the novelist on the cover. Mirbeau's bust, in terracotta, was done in 1889. There is also a medallion of the writer, which was exhibited in the 1895 Salon.

59. ROSE BEURET (MADAME AUGUSTE RODIN). 1890. Marble, $18\frac{5}{8} \times 15\frac{3}{4} \times 19\frac{3}{4}$.
Rose Beuret, born in 1845, of a humble family, became Rodin's companion at a very early age and shared his life in good times and bad, always keeping in the background. She died in Meudon shortly before Rodin, who had married her in his old age. Rodin made many busts of his wife, and was even inspired by her features for works of a more general character. This marble was placed by Madame Rodin, in 1916, at the disposal of her husband, for the Museum then being formed. Rodin had it cast in bronze.

60. BROTHER AND SISTER. 1890. Bronze, $15\frac{5}{8} \times 7 \times 7\frac{7}{8}$.
Perhaps under the influence of Carpeaux, which was stronger than is believed, Rodin always had a passion for sketching and sculpturing children. No doubt, 'Brother and Sister', the masterpiece of his maturity and dated 1890, is the most charming and affecting of the groups that inspired him with this taste from early youth.

61. DESPAIR. 1890. Bronze, $13\frac{1}{2} \times 10\frac{1}{4} \times 11\frac{3}{4}$.
Another figure originally intended for 'The Gate of Hell', where it is placed above the right folding door.
Retaken in the same part of the monument in three different settings, it shows how conscientiously Rodin worked at his subjects. This same year, 1890, Rodin detached the character to lend it a life of its own.

62. ROSE BEURET (MADAME AUGUSTE RODIN). 1890. Bronze mask, $10\frac{1}{8} \times 6\frac{3}{4} \times 6\frac{1}{4}$.
This portrait was done in 1890, in the full maturity of this fine woman's face, which had inspired the artist in the springtime of his union with more smiling images. The gravity of this face, combined with the sculptor's perfect command of his faculties, make this work an admirable and affecting creation.

63–64, 66–67. BALZAC. 1897. Plaster, $118 \times 47 \times 47$.
(Plate 63. Study for Balzac statue. 1893. Bronze, $10\frac{5}{8} \times 11\frac{3}{8} \times 7$.)
The most remarkable incident in Rodin's career was that which was connected with this statue of Balzac. The Société des Gens de Lettres, the leading literary society in France, after the Académie Française, had wanted a statue of the great romanticist, to be placed on a site in Paris. The work had, in 1888, been entrusted to Chapu, but he died in 1891, leaving only the commencement of his work, and having spent a portion of the 36,000 francs which was to be allocated for it.
Rodin then, at the suggestion of some of his literary friends, wrote to the Société (the president of which at the time was Emile Zola) offering to do the work, and the offer was accepted. Unfortunately he made the mistake of undertaking to deliver the statue in eighteen months. A great deal longer time than that went by, and no statue was delivered nor even a model submitted, so that the delay became the subject of popular jokes and quips. Relations between Rodin and the society became strained and some disagreeable correspondence passed. Zola among others grew inimical. There was a lawsuit, after which a new engagement was entered into, by which the commission was confirmed to Rodin, but without any time stipulation. He also returned the 10,000 francs, which had been advanced to him for the purchase of material, and agreed that he should receive nothing until the statue was delivered.
The material which Rodin had to work with was very meagre, but he set about a study of Balzac, his works, his habits and the country where he was born and lived, which was typical of the man but is probably very unusual in our days.
There is a bust of Balzac in existence by David d'Angers, and Rodin also had access to a daguerrotype of the novelist taken half-length standing, with his collar open showing his massive neck. But the best portrait of the novelist is the description by Lamartine:
'It was the face,' he said, 'of an element; a big head, hair dishevelled over his collar and cheeks, like a mane which the scissors never clipped; very obtuse, eye of flame, and colossal body. He was big, stout, square at the base and shoulders—much of the ampleness of Mirabeau, but no heaviness. There was so much soul that it carried it all lightly; the weight seemed to give him force, not to take it away from him; his short arms gesticulated with ease.' Théophile Gautier added that the novelist's usual expression was of intense mirth—a Rabelaisian hilarity ennobled by great power.
So thorough was Rodin's preparation for his task that at Tours, where he found an old tailor who had made clothes for Balzac forty years previously, he got him to make a suit from the old measurements.
Rodin started by making several nude models (he nearly always began by making nude models, whether he intended to clothe his statues or not) and others which did not satisfy him. He found that this was the most difficult sculptural interpretation he had ever undertaken. He was puzzled by the elements of the strange, abnormal lineaments of the man, his complex character and extraordinary personality, beside which there was his prodigious literary production to

be considered. Rodin set himself, as usual, to incarnate the great thinker in action, but also to create a Balzac with all his idiosyncracies. For this he had to continue the Gothic conception which he had partially and successfully tried with the Burghers of Calais, the subordination of form to the main idea and spiritual meaning of the character, and the simplification of the trunk of the statue into large surfaces adapted to the play of light and shade, in order to concentrate all the vigour of expression on the head. The result was this huge man ('a very living Balzac') with the powerful head and bull's neck, who throws upon the crowd that gazes up at him a look of deep but smiling irony tinged with sadness. His hands are crossed under the white gown, resembling a monk's garb, which he always wore when at work, the sleeves of which hang empty at his sides.

When the statue was at last exhibited at the Salon of 1898, side by side with 'Le Baiser', it is not too much to say that it was greeted with a chorus of scorn and ridicule. 'A cow,' said some of the people who went to see it, while critics with a choicer fund of expression talked of 'a snow man', 'formless lava', etc.

M. Roger Marx, that staunch friend, who wrote an interesting explanation of the work before the opening of the Salon, was unfortunately but little heeded. 'Rodin,' he said, 'has made it his business to seek for that which in this broad, frank, open face betokened will, power and genius.' He had got the man's personality—the height of the brow, the deep setting of the eyes and their keen brilliance, the bulky nose, and the sensuality of the thick lips. But besides all this, there was the complexity of the character—an indefinable smile made up of kindness and sarcasm, the defiance shown by the pose of the towering head, indifference to past insults, just satisfaction at the work he has accomplished, and faith in the judgement of posterity.

The Société des Gens de Lettres accepted the popular verdict and passed a resolution protesting against 'M. Rodin's rough model and refusing to recognize it as a statue of Balzac'! A split in the ranks of the society ensued, and Jean Aicard, the poet, who was then president, resigned with several others. The commission for a new statue was given to Alexandre Falguière. His seated statue is the one now to be found in the Avenue Friedland, Paris. Falguière's work should not be condemned so utterly as it has been by enthusiastic admirers of Rodin, and none who contemplate this statue can fail to note in it the influence of the great sculptor.

But the ridicule and contempt were by no means universal. Rodin was inclined at first to fight the society, but in face of the applause and admiration that came from other quarters he desisted. M. Auguste Pellerin, a wealthy art collector, asked him to sell the statue for 20,000 francs; and he received invitations from artistic bodies in London and Brussels to send the statue for exhibition to those cities. He replied to some of these friends in a charming letter acknowledging the encouraging testimonies of esteem and declaring that he had decided not to sell the statue—that it belonged to Paris and he would bide his time.

At the same time, it must not be forgotten that honest discerning criticism does not hurt, and that all those who criticized the 'Balzac' and thought that Rodin had carried his interpretation of the Gothic in sculpture to extremes were not mere scoffers and Philistines. Henri Rochefort, who could not be accused of being such, wittily said: 'To pretend to express the forty volumes which he (Balzac) left to the world in the contortions of the lips and the quivering of the nostrils, is carrying the *spirituel* a little far. It is the first time one has ever had the idea of extracting the brains of a man and putting them on his face! Let Rodin give us simply the nose of Balzac, his mouth, his forehead, with the structure of his powerful head, and in this transcription our own eyes will read the genius of Balzac. But for the love of art, let the sculptor spare us his commentaries.' Benjamin Constant, the artist, who passionately admired most of Rodin's work, was nonplussed at the 'Balzac', and it was the same with other distinguished critics and fellow artists. On the other hand, Octave Mirbeau prophesied that a day would come when the changing crowd, being more educated, would frantically applaud this work of genius.

An extraordinary thing is that there were not wanting some who accused Rodin of having voluntarily made his Balzac grotesque. His previous work should surely have supplied answer enough to such an absurd charge and proven the absolute and intense earnestness of the man. What! Rodin, whose eyes are like nothing else in modern sculpture, had made his Balzac with cavernous eyes as a joke! Rodin, who moulded hands as no others could do, had left the hands of Balzac (which are said to have been quite remarkable) out of sight, so that they should not draw attention from the spirit of contemplation in the face. Rodin replied to these charges in a few proud but scathing sentences: 'I shall fight for my sculpture no longer; it has for a long time been able to defend itself. To say that I patched up my Balzac as a practical joke is an insult that would formerly have made me writhe with indignation, but to-day I let it pass and go on with my work. My life has been a long course of study, and to accuse me of making fun of others is the same as to say I make fun of myself. If truth is to die, my Balzac will be torn to pieces by the next generation, but if truth is imperishable, I predict that my statue will make its way....

'This work, which has been laughed at, which is being scoffed at because it cannot be destroyed, is the result of my whole life's study—the very pivot of my aesthetic feelings. I was another man the very day I conceived it. My evolution was complete, and I had knitted a bond between the old lost traditions and my own time, which every day will help to strengthen.

'By force or by persuasion, it will make its way in men's minds. Young sculptors come to see it here in the studio, and think of it as they descend the stairs again, to go in the direction whither their ideals call them. There are men of

Notes on Plates 65, 68-72

the people who have understood—workers and those who, rare though they are in the crowd, continue the old traditions of the crafts in which each one did his work according to his conscience, and did not learn his art in the official catechisms.

'As to the public, they are not to blame. The fault lies with their educators. The sense of beauty and taste for reason are lost. There is no room for and no esteem among us for men who model their souls alone. And the huge majority are no longer interested in art, and see nothing more of art except through the eyes of a few elected judges. As for myself, knowing that life is short and my task is great, I shall continue my work far from polemics.'

Proud words—in the spirit of which he acted! How much of his own indomitable spirit Rodin put into his 'Balzac' it is not hard to see, for every artist puts into his work himself as well as his 'sitter'; and here Rodin had no sitter but an 'element', as Lamartine put it. Rodin's own words make one think of an English critic's judgement. 'No one in the intelligent world,' said Edmund Gosse, 'looks at sculpture to-day exactly as he did before Rodin put his mark on it.'

65. VICTOR HUGO STANDING. 1897. Plaster, $88\frac{3}{8}\times35\frac{3}{8}\times56\frac{1}{4}$.

As we have already had occasion to point out, Victor Hugo was one of the great subjects that preoccupied the genius of Rodin. Already in 1883 he had executed a first bust of the poet. When he received from the Beaux-Arts a commission for a statue for the Panthéon, the Director, M. Gustave Larroumet, was uncertain which position to assign to the hero; should he be seated or standing? It seems that there was some doubt upon this matter, since we have several studies of Victor Hugo standing. This one, the most important, shows how conscientiously Rodin grappled with his tasks. This nude, executed in 1897, is an enlargement of a plaster model done ten years earlier and shows that the sculptor thought of exhibiting the author of 'Les Châtiments' in front of the flood breaking on the shore.

68. THE SCULPTOR FALGUIÈRE. 1897. Bronze, $16\frac{7}{8}\times9\frac{1}{2}\times10\frac{1}{4}$.

When the Société des Gens de Lettres declined, in 1897, the statue of Balzac executed by Rodin, it commissioned Alexandre Falguière (1831–1900) to take the place of his illustrious colleague. But the latter bore no grudge against his friend, and to emphasize that he harboured no rancour for the substitution, he executed his bust, and had his own done by his happy and embarrassed rival. The two busts were simultaneously exhibited in the 1899 Salon.

69. BUST OF VICTOR HUGO. 1897. Bronze, $27\frac{1}{2}\times19\times19$.

Rodin was introduced to the poet by his friend Roger Marx, who held an official post in the Fine Arts Ministry, and there is a letter in existence addressed to the 'dear and illustrious Master' in which Rodin expresses in ardent terms his desire to perpetuate in marble or bronze the features of the greatest poet of his time. Victor Hugo, who was then living in the Avenue d'Eylau (near to the present Place Victor-Hugo), and was old and tired, was not at all enthusiastic at the idea, especially as he had just before been pestered by an inferior sculptor to whom he had given a number of sittings, the result of which had been unsatisfactory, or rather nil. However, Hugo consented to allow the sculptor to come to the house to lunch whenever he liked and make sketches of him, so long as he himself should not be disturbed in his habits. The Hugo family kept practically open house in those days, and there were always others to lunch. The arrangement lasted for several months, Rodin going to lunch not every day, but sometimes several days in succession, and sometimes leaving out a number of days. During the meal the sculptor would sketch the poet continually and feverishly, with the result that he himself rarely got any lunch, for when it was over he rushed off to his studio to transpose his rough sketches to the clay.

When Hugo saw the bust, neither he nor the family were pleased with it; they considered it not flattering enough.

In reality, the work is only a replica of the head and shoulders of the 'Victor Hugo Standing' (plate 65). It was exhibited in 1897, and in 1900 occupied the best position in the Alma Pavilion.

70. THE HAND OF GOD. 1897–8. Marble, $24\frac{3}{4}\times31\frac{1}{2}\times20\frac{5}{8}$.

This magnificent fragment of the great Master, almost contemporary with the 'Balzac', and which is perhaps a meditation in the margin of this famous statue, is like a symbol of the semi-divine creation of the prodigious sculptor. It seems that when he was struggling to express the supernatural gift of the writing genius, he was tempted to create the image of the genius of statuary.

71. PAN AND NYMPH. 1898. Marble, $53\frac{1}{4}\times30\frac{3}{4}\times27\frac{5}{8}$.

'Pan and Nymph' was done at about the time when Rodin, who was nearing his sixtieth year, returned to those themes of Greek mythology which had always captured his imagination. Perhaps he was influenced by a fresh reading of the 'Metamorphoses' of Ovid, one of the three most abundant sources, with Dante and Baudelaire, of his inspiration.

72. BAUDELAIRE. 1898. Plaster, $7\frac{7}{8}\times7\frac{1}{2}\times9$.

The admiration which Rodin entertained for Baudelaire was certain to prompt him one day to attempt to model the features of the great poet. Having made acquaintance with a young writer, now forgotten, M. Louis Malteste, an enthusiast for his sculpture, and who bore a curious resemblance to the author of 'Les Fleurs du Mal', Rodin sought to discover in his living countenance the physiognomy of the great master. The sculptor, however, never brought this attempt beyond the stage of a sketch, which was executed in 1898.

73. MADAME F . . . 1898. Marble, $25\frac{5}{8} \times 24\frac{3}{4} \times 20\frac{1}{2}$.
In the admirable gallery of feminine busts executed by Rodin, none is more delicately affecting than this. The model was the wife of a great amateur of the arts, who was one of the artist's most devoted friends, and who had supported his genius from the first. The work is dated 1898, and comprises several variations.

74. MUSE FOR THE WHISTLER MONUMENT. 1902–3. Marble, $19\frac{3}{4} \times 11\frac{1}{2} \times 11\frac{1}{2}$.
Rodin, who had the cult of friendship, projected several monuments to the glory of great artists with whom he had been associated. One of these was to Whistler, and about 1902 he carried out a certain number of studies from a model, Mary Jones. This marble is one of the finest fragments conceived for this design, which went no further.

75. STUDY OF A SLAV GIRL. 1905–6. Marble, $25\frac{1}{4} \times 28\frac{3}{4} \times 21\frac{5}{8}$.
Rodin made a marble bust about 1906 using as a model this young Slav girl, who has not been identified. The following year, still under the charm of her unusual face, he executed another graceful work in her likeness (plate 82).

76. EUGÈNE GUILLAUME. 1903. Bronze, $13\frac{3}{8} \times 12\frac{1}{4} \times 11\frac{3}{4}$.
The sculptor Eugène Guillaume was an 'official' artist, member of the Académie des Beaux-Arts, of the Académie Française, director of the École des Beaux-Arts and of the Académie de France in Rome. After having long resisted the aims of Rodin, he conceived in his old age a great admiration for the works of the great sculptor and became his intimate friend.

77 MARCELIN BERTHELOT. 1906. Bronze, $17\frac{3}{8} \times 9\frac{1}{2} \times 9\frac{1}{2}$.
Rodin was associated with the great chemist, as is proved by copies of the latter's works dedicated to the sculptor. But it was not until 1906 that he executed this bust, one of the most admirable that came from his hands. At this period of his career, the artist, having penetrated the secrets of his profession, like a Rembrandt, a Titian, a Goya or a Delacroix, thought only of expressing the essential in a work. He had neither desire nor need to multiply detail. He had reached the stage when detail appears useless and even harmful.

78–80. BERNARD SHAW. 1906. Marble, $23\frac{5}{8} \times 22\frac{7}{8} \times 15\frac{3}{4}$. 1906. Bronze, $11\frac{3}{8} \times 7\frac{7}{8} \times 4\frac{3}{4}$.
It was perhaps through Henley and the circle of English writers who supported him almost from the start that Rodin made the acquaintance of Bernard Shaw. The famous dramatist, hater of all poses, would allow none but the author of 'The Thinker' to make his bust. The work, executed in 1906, earned the artist the praise of the formidable satirist, and Mrs. Bernard Shaw, in her turn, wrote Rodin that the resemblance of the work to the model was so faithful that it frightened her.

81. MADAME DE GOLOUBEFF. 1906. Bronze, $19\frac{3}{8} \times 16\frac{1}{2} \times 9\frac{7}{8}$.
Madame de Goloubeff was one of the musicians who supplied Rodin with some of the purest joys of his life. In interpreting for him in a pure and expressive voice the finest songs of classical art, she evoked the sculptor's gratitude, which was admirably expressed by this bust done in 1906.

82. BY THE SEA. 1906–7. Plaster, $22\frac{7}{8} \times 32\frac{3}{4} \times 22\frac{7}{8}$.
The model for this charming study is the young, unidentified Slav girl who sat as a model for the marble bust made by Rodin the previous year (plate 75).

83. MOTHER AND DYING CHILD. 1908. Marble, $41 \times 39\frac{1}{4} \times 27\frac{1}{2}$.
Rodin's glory had become so great that he received commissions from beyond the seas not only for busts but also for funereal monuments. In 1908, Mrs. Thomas Merrill, desiring art to perpetuate the memory of her daughter who had died in tender years, applied to the great Master. It is one of the most moving works of the artist, himself gradually nearing eternity.

84. THE DANCER HANAKO. 1908. Bronze, $6\frac{1}{4} \times 4\frac{3}{4} \times 3\frac{1}{2}$.
This dance theme could hardly fail to arouse the enthusiasm of a sculptor who was so interested in movement. In his latter days he grew fond of noting in their spontaneity all the gestures of his models left to themselves in his studios and he liked to paint all their attitudes in water-colours. On the other hand, about 1900, there was in France a prolonged infatuation for dancing and the admirable interpreters of this art, Isadora Duncan, Loie Fuller, the Javanese dancers, Hanako, and the chorus of the Russian ballet. This mask of the Japanese Hanako is contemporary—1908—with these important studies.

85, 87. GUSTAV MAHLER. 1909. Bronze, $13\frac{3}{8} \times 9\frac{5}{8} \times 8\frac{5}{8}$.
MOZART. 1910. Marble, $12\frac{5}{8} \times 37\frac{1}{2} \times 24\frac{1}{2}$.
It seems that it was through Madame de Nostitz, a German friend of Rodin, that the latter made the acquaintance of the Austrian composer, famous for his symphonies and his great work 'The Song of the Earth'. Rodin found that he resembled Mozart, and when he had made his bust in 1909, he was inspired by his face to execute a bust of Mozart in marble (1910).

86. PUVIS DE CHAVANNES. 1910. Marble, $29\frac{5}{8} \times 49 \times 23\frac{5}{8}$.
The painter of the 'Life of St. Geneviève' at the Panthéon, of the 'Summer' and 'Winter' at the Hotel de Ville, Paris, of

the decoration of the Amphitheatre at the Sorbonne, and notable works at Rouen, Amiens and other French cities, and the decoration of the library at Boston, Mass., was an aristocrat and religious mystic and the 'utterer' of 'sweet but pallid harmonies' showing great elevation of thought.
There was no modern artist for whom Rodin had a greater admiration. 'To think he lived amongst us!' he exclaimed to M. Paul Gsell (reported in the latter's book, already quoted from). 'To think this genius worthy of the most radiant epochs of art has spoken to us, that I have seen him and shaken hands with him! It is as if I had shaken the hand of Nicolas Poussin. He always bore his head high,' continued the sculptor. 'His skull, which was solid and rounded, seemed made to don a helmet. His round thorax might have been accustomed to wearing a cuirass. One could easily have imagined him at Pavia fighting for honour beside Francis I.'
In 1891 Rodin exhibited in the Salon the bust of his friend, and, the following year, he presented to it the marble. In 1910 he intended to begin executing a new marble destined for the monument it was decided to raise to the great painter's memory. This latter bust, although no more than sketched (the war, then the sculptor's death, having suspended the work) is of impressive appearance and makes a strange impression, from the fact that the face emerges, like the sea, from the enormous block of marble.

88. FEMALE TORSO. 1910. Bronze, 29¼×13¾×23⅝.
This bronze, in its fragmentary state, has all the grace of a mutilated antique excavated from the earth. It radiates, through its youthful and supple limbs, so much life and beauty as to give the illusion of a complete and charming statue of this young woman, apparently carried away in the dizziness of this sacred dance. This figure dates from 1910.

89. THE DUCHESS OF CHOISEUL. 1908. Bronze, 11¾×9×6¼.
Rodin made three busts of the Duchess of Choiseul, one in bronze, one in marble, and one in terracotta; all of them were executed in 1908. He was on intimate terms with this American lady, who through her marriage became one of the prominent members of the French aristocracy.

90. ÉTIENNE CLÉMENTEL. 1916. Bronze, 21¾×14⅝×11.
This bust, executed in 1916, the year before the Master's death, is the last work which he completed. Étienne Clémentel was a prominent politician, who was several times a minister. He became one of the sculptor's three testamentary executors. This work, particularly moving because it is the artist's supreme labour, shows that, even on the threshold of death, the artist rough-hewed with great facility and retained a clear eye and hand of incomparable firmness.

91. GEORGES CLEMENCEAU. 1911. Bronze, 18⅝×11×11.
Rodin was from an early date on terms of intimate friendship with the great statesman, who, both as an influential politician and as a journalist, always stoutly defended and supported him. In 1911, Clemenceau had become one of the most prominent parliamentary figures of the day, yet desired that in his spare moments the sculptor should model his features. Before achieving the amazing resemblance of this bust, Rodin made several studies of it, but it is manifestly this unique bronze which will supply posterity with the truest image of this noble spirit and great Frenchman.

92. THE CATHEDRAL. 1908. Stone, 25¼×13⅝×12⅝.
Rodin always had a passion for modelling hands, so expressive in his view, so capable of displaying in themselves so many human emotions. Still in this period of the opening century pre-occupied with symbolism, more and more attached to the history of the religious architecture of the Middle Ages, the idea came to him one day in 1908 of representing the high pointed naves by two tapering hands joined in a gesture of prayer. Perhaps to accentuate the relationship uniting the symbol to the reality, he executed this work in stone.